Deep-Sea Mining and the Water Column

Rahul Sharma
Editor

Deep-Sea Mining and the Water Column

Advances, Monitoring and Related Issues

Springer

Editor
Rahul Sharma
Consultant, Deep-sea Mining and Former Chief Scientist
National Institute of Oceanography
Dona Paula, Goa, India

ISBN 978-3-031-59059-7 ISBN 978-3-031-59060-3 (eBook)
https://doi.org/10.1007/978-3-031-59060-3

Cover image: Schematic of Deep-sea mining system and associated impacts on the pelagic and benthic ecosystems (Modified from original diagram from NIWA-IUCN, figure provided by Malcolm Clark).

This Springer imprint is published by the registered company Springer Nature Switzerland AG
The registered company address is: Gewerbestrasse 11, 6330 Cham, Switzerland

Paper in this product is recyclable.

Foreword

I am delighted at receiving the invitation to contribute a foreword for the fourth book by Dr. Rahul Sharma on the topic of deep-sea mining. As with previous volumes, it is a great honour for me to accede to his request.

As the International Seabed Authority progresses its work on the regulatory framework for the inevitable transition from deep-sea mineral exploration to exploitation, the public debate on the topic has become increasingly polarized and frequently misinformed. A few sectors of society have even been calling for a moratorium on all deep-sea mineral activities, notwithstanding that these activities are fully regulated by international law and that deep-sea exploration has in fact provided considerable benefit to humankind through increased scientific knowledge and technological development.

Dr. Sharma's earlier volumes, on the resource potential and technical considerations for deep sea mining (2017), on environmental issues of deep-sea mining (2019), and on sustainability, technology, environmental policy and management (2022), contributed greatly to the scholarly literature on deep-sea mining and brought together the perspectives of experts in the field, including many of the world's leading experts in their subjects.

The present volume addresses the topic from the perspective of the water column, which has a unique significance being the large volume of water mass that will likely be impacted upon and in turn influence decisions related to offshore mining in future. Of particular importance is the fact that, rather than advancing political positions, most of the contributions to the book critically address the way in which the main expected impacts of deep-sea mining can be predicted, measured and mitigated, not least through the rapid maturation of technology that, for the most part, remains under development.

There is no doubt that, as with every other human endeavour, there are pros and cons to deep-sea mining. However, in the short to medium term, it seems likely that a mix of critical minerals from land and sea can more sustainably support the energy transition than relying on terrestrial sources alone. Deep-sea mining thus presents an important opportunity for development. But more technical, impartial and informed discussion is needed, especially on economic, geological and technological aspects, and Dr. Sharma's book makes an important contribution in this respect.

I congratulate Dr. Sharma and all the experts who have contributed to this volume.

Secretary-General
International Seabed Authority,
Kingston, Jamaica
September 2023

Michael W. Lodge

Preface

Deep-sea mining has been a topic of interest in recent years, as after the signing of exploration contracts between the International Seabed Authority (ISA) and several entities for different types of mineral deposits from the deep-sea regions in international areas, the focus is shifting to development and testing of various subsystems for mining and processing of these deep-sea minerals, as well as on the deliberations that are underway at ISA for developing a mining code. Similar efforts are underway for exploration and exploitation of deep-sea minerals within the Exclusive Economic Zones of several countries.

In this series on deep-sea mining, while the earlier three editions have focused on *Deep-sea mining: Resource Potential, Technical and Environmental Considerations*, *Environmental Issues of Deep-Sea Mining: Impacts, Consequences and Policy Perspectives* and *Perspectives on Deep-Sea Mining: Sustainability, Technology, Environmental Policy and Management*, this book deals with *Deep-Sea Mining and the Water Column: Advances, Monitoring and Related Issues*.

It is anticipated that the water column could be the largest volume of water mass that would come in contact with various subsystems as well as the effluents that may be discharged during deep-sea mining operations, leading to impacts on the marine ecosystem. This could lead to various environmental as well as legal issues, since the area impacted in the water column could be larger than the one impacted on the seafloor where mining would take place.

Hence, this book contains sections dealing with general issues of deep-sea mining; engineering concepts; approaches to sea surface and water column monitoring; regional assessment of chemical, physical and biological characteristics of water column; and the legal, policy and economic issues. This book has been possible due to the excellent contributions by researchers from around the world, who have shared their expertise for the benefit of deep-sea community in particular and the humankind in general.

As this book contains articles from authors from diverse backgrounds, readers are requested to note that the views expressed are those of individual authors and do not reflect the view of either the editor or that of the publisher.

The editor is extremely grateful to all the authors for sparing their time and contributing to this book. Mr. Michael Lodge, His Excellency, the Secretary General of ISA, deserves a special mention for his Foreword that sets the tone of the book. A special thanks goes to Dr. Virginie Tilot, Attaché honoraire, Museum National D'Histoire Naturelle, Paris, France, not only for suggesting the theme 'Deep-sea mining and the water column' but also for her constant support during the compilation of the book.

It is expected that this book will serve as an important source of information on different aspects of deep-sea mining with a special focus on the water column, for all stakeholders including the researchers, technologists, contractors, mining companies, regulators and the NGOs involved in deep-sea mining and marine environmental conservation.

Dona Paula, Goa, India Rahul Sharma

Contents

Part I
General Issues and Interdisciplinary Approach to Deep-Sea Mining

Chapter 1
Deep-Sea Mining and the Water Column: An Introduction

Rahul Sharma

Abstract Deep-sea mineral resources within and beyond the national jurisdictions offer several opportunities for exploration and possible exploitation owing to their potential as alternate sources of metals such as Cu, Ni, Co, rare earths and others. These are considered critical for meeting mankind's green energy requirements, among others, especially in view of depleting or low-grade terrestrial resources. This chapter gives an overview of the current status of deep-sea mining, its potential impacts, including on the water column, and the possible mitigation measures and regulations for the same.

Keywords Deep-sea mining · Environmental impacts · Water column · Mitigation · Regulations

1 Introduction

There is a growing need to find alternative sources for metals that are critical not only for meeting the day-to-day demands of mankind that include consumer and biomedical goods, alloys, batteries, pigments and catalysts, as well as heavy machinery and electric transmission, but also to transition from fossil fuel to green energy options including solar and wind farms, electric car batteries and high-capacity energy storage devices (Hein et al., 2020; Cronan, 2022). So far, the metal demand has been met from terrestrial sources, or to some extent from recycling, but the rising demand due to increasing population and economic growth around the world 'will require either expansion of existing mining projects or establishing new ones' (Koschinsky et al., 2018). According to Petersen et al. (2018), issues such as

R. Sharma (✉)
Consultant, Deep-sea Mining and Former Chief Scientist, National Institute of Oceanography, Dona Paula, Goa, India
e-mail: rsharmagoa@gmail.com

R. Sharma (ed.), *Deep-Sea Mining and the Water Column*,
https://doi.org/10.1007/978-3-031-59060-3_1

low-grade deposits, geopolitical concerns and security of metals supply could become the key drivers for deep-sea mining becoming a potential source for the supply of metals in the future.

It is anticipated that the demand for cobalt will be 423%, nickel 136% and lithium 280% of the currently known reserves by 2050 (www.mineralsindepth.org). The European Union has updated its list of critical raw materials (CRMs) to 34, which includes Cu and Ni as strategic metals (https://single-market-economy.ec.europa.eu/sectors/raw-materials/areas-specific-interest/critical-raw-materials_en), and the USGS has listed 50 CRMs (www.usgs.gov) that are essential for sustained growth of the industry. Whereas the major supply (60%) of raw Co, a critical component in lithium-ion batteries, comes from the Democratic Republic of Congo; 80% of refined Co comes from China (Al Barazi et al., 2018). Sourcing these metals from such conflict zones combined with the need to become self-reliant has led to the quest for alternative sources of metals.

Deep-sea minerals such as polymetallic nodules, ferromanganese crusts and hydrothermal sulphides, are emerging as potential sources of metals such as Cu, Ni, Co, Mn, Fe and rare earth elements that can contribute towards the world's growing demands for certain metals critical for sustaining the industrial growth as well as green energy alternatives (Hein et al., 2013; Miller et al., 2018). According to Weaver and Billet (2019), 'Each of these mineral deposits are known to form under specific environmental conditions, associated with discrete seafloor morphology and depth, having different concentrations of metals and found in different parts of the world oceans'. Prospecting of seabed mineral deposits has been underway for several decades, leading to the estimation of their resource potential in millions of tonnes worth billions of dollars, which could meet the global demands in the future (Van Nijen et al., 2018; Hein et al., 2020; Sharma, 2022).

This becomes more critical considering the increasing production rate of different metals (ranging from 1.1% for Au to 8.3% for Co) (Sharma & Smith, 2019) and that the available land reserves of some of these metals will last for just over two decades (e.g. Mn, Cu, Ni, Co), whereas a few others for just about a decade (e.g. Pb and Zn) (Sharma et al., 2019). With the objective of reducing dependency on a few countries that have the monopoly for supply of some of these metals, several entities (called contractors) have embarked upon exploration programmes for deep-sea minerals. Moreover, mining on the seabed area for any one of these mineral deposits can offer several metals as opposed to operating several mines for different minerals on land (Hein et al., 2020).

Estimates of metal resources in nodule deposits in Clarion-Clipperton Zone (CCZ) in the Pacific Ocean alone suggest that some of the key metals, such as Mn, Ni, Co and Y, are higher by 1.15–3.4 times than the entire global resources on land (Hein et al., 2020), underlining the potential of deep-sea minerals as an alternative source for some of these metals. In a typical contract area of 75,000 km^2 for polymetallic nodules with a cut-off abundance of 5 kg/m^2 (UNOET, 1987), the total resource available in the area could be 375 Mt. (wet) or 281.25 Mt. (dry), with a total metal equal to 67.081 Mt. (at a conservative value of concentration of metals—Mn = 22%, Ni = 1.0%, Cu = 0.78%, Co = 0.1%). Out of the 281.25 Mt., only 10.6–21.2% (i.e. 30–60 million tonnes) of the resource will be actually mined at a mining rate of 1.5 million tonnes/year (ISA, 2008a) and three million tonnes/year (UNOET, 1987), respectively, over a 20-year period as the life of a mine-site, with a large balance (78.8–89.4%) that could be mined for more than a century (Sharma, 2017a). Besides these minerals, deep-sea sediments have also been identified as a source for rare earth elements, as found near Minamitorishima island in the western North Pacific Ocean, with an estimated resource of 1.2 Mt. of rare earth oxide that accounts for 62, 47, 32 and 56 years of annual global demand of Y, Eu, Tb and Dy, respectively, that can be exploited as REY resource in future (Takaya et al., 2018). A study on cobalt-rich crusts on one of the seamounts in the North-Western Pacific Ocean suggests that '…the area tested around the Takuyo No. 5 Seamount contains enough cobalt to meet Japan's demand for 88 years and enough nickel to meet Japan's demand for 12 years' (www.jogmec.go.jp). An economic feasibility study has also suggested the mining of cobalt-rich crust along with its substrate rich in phosphorus as a by-product to increase the viability of the resource (Yamazaki et al., 2023).

Whereas some of the areas in which deep-sea minerals are located fall within the Exclusive Economic Zone (EEZ) of different countries and their exploration and exploitation are governed by the regulations of the respective state, large deposits of these minerals also occur across thousands of square kilometres in international waters in all the oceans of the world and are regulated by the International Seabed Authority (ISA), Jamaica. Currently, 31 contracts have been signed between different contractors and ISA for these deep-sea minerals (www.isa.org.jm).

Although fluctuating metal prices as well as reduction in demand of metals after the world wars were responsible for deferred technological developments for commercial exploitation, these deposits have always been considered important in the overall metal budget of the earth and expected to cover the twenty-first century demand for metals such as Mn, Fe, NI, Co, Cu, Mo and many others, including rare earth elements (Kotlinski, 2001), as the commercial viability of deep-sea deposits lies in their concentration and magnitude as compared to deposits on land (Lenoble, 2000).

This chapter provides an overview of deep-sea mining, dwells upon the possible environmental impacts, including on the water column, and discusses various approaches towards managing and mitigating the same.

2 Basic Characteristics of Deep-Sea Minerals

The occurrence of deep-sea minerals has been known since the discovery of polymetallic nodules from the deep seabed onboard H.M.S. Challenger expedition, during which the dredge haul of 7 March 1873 was described as 'peculiar black oval bodies about 1 inch long' and 'almost pure manganese oxide' (en.wikipedia.org/wiki/HMSChallenger). Later, Mero (1965) recognised the economic potential of these deposits and predicted that 'deep-sea mining would commence in 20 years' time, which led to a race for developing these resources as an alternative source of metals for the future.

Detailed studies on the distribution, geochemistry and mineralogy of these mineral deposits in different parts of the Pacific Ocean (Hein et al., 1979; Thijssen et al., 1981; Glasby, 1982; Usui & Moritani, 1992) and Indian Ocean (Glasby, 1972; Siddiquie et al., 1978; Frazer & Wilson, 1980; Cronan & Moorby, 1981) gave critical information on this resource. Research also focussed on their formation process, geological factors and their relationship with the sedimentary environment (Cronan, 1980; Frazer & Fisk, 1981; Glasby et al., 1982; Rao & Nath, 1988; Martin-Barajas et al., 1991). Subsequently, studies on other deep-sea minerals, such as hydrothermal sulphides (Rona, 1988; Plueger et al., 1990) and cobalt-rich ferromanganese crusts (Halbach et al., 1989; Hein et al., 1997), also helped in identifying these as potential resources for critical metals, in future.

These are known as deep-sea minerals owing to their occurrence in water depths ranging from 1000 to 6000 m, each having some common and some distinct features in terms of their occurrence, formation process, geochemistry and physical characteristics (Table 1.1). While the polymetallic nodules (or manganese nodules)

Table 1.1 Salient features of deep-sea minerals[a] (Sharma & Smith, 2019)

Type	Description	Area, thickness of the deposit	Metals and their mean concentration[b]	Principal deposits
Poly-metallic nodules	Concretions of layered iron and manganese oxides with associated metals from the water column or sediment	Up to thousands of km^2., thickness generally upto 50 cm, rarely deeper	Mn (28,4%), Ni (1.3%), Cu (1.07%), Co(2098 ppm), Mo (590 ppm), Zn (1366 ppm), Zr (307 ppm), Li(131 ppm), Pt (128 ppm), Ti (199 ppm), Y (96 ppm), REEs (813 ppm)	Clarion-Clipperton zone, Peru Basin, Central Indian Ocean and Penrhyn Basin
Seafloor massive sulfides (SMS)	Concentrated deposits of sulphidic minerals (>50–60%) resulting from hydrothermal activity on the seabed	Up to several km^2; several metres thick	Cu(0.8–17.9%), Au(0.4–13.2 ppm), ag(64–1260 ppm), Zn(2.7–17.5%), Pb(0.02–9.7%),co, as, Al, Si, REEs	Red Sea, back-arc basins, mid-oceanic ridges and other plate boundaries
Ferro-manganese crusts	Layered manganese and iron oxides with associated metals on hard substrate rock of subsea mountains and ridges	Up to several km^2; generally a few cm thick	Mn (21%), Co (6647 ppm), Ni (4326 ppm), Cu (573 ppm), Te (34 ppm), Mo (431 ppm), Zr (423 ppm), Ti (TiO_2–1.4%), Pt (0.273 ppm), W (68 ppm), REEs (1628 ppm)	Equatorial Pacific Ocean and Central Atlantic Ocean

[a]Modified from Cuyvers et al. (2018)

[b]Concentrations for sulphides from Cherkashov (2017), nodules from Hein et al. (2013), crusts from Halbach et al. (2017)

Fig. 1.1 Polymetallic nodules at the pelagic basin near the Takuyo-Daigo seamount, NW Pacific basin (JAMSTEC Cruise KR16–13#704, at 5490 m depth). Width approximately 2.0 m. (Photograph courtesy of Prof. A. Usui, Kochi University, Japan)

are layered concretions of iron and manganese oxides around a nucleus (generally a sediment particle, rock piece, shark tooth or any hard substrate) and occur as loosely lying spherical objects on abyssal plains (Kuhn et al., 2017 and the references therein), the ferromanganese crusts (or cobalt-rich ferromanganese crusts) are layered iron and manganese oxides on hard substrate of a rock (Cherkashov, 2017 and the references therein) and are generally located on underwater mountains or ridges, both containing metals precipitated from the water column and the associated substrates. On the other hand, hydrothermal sulphides (or seafloor massive sulphides) are rich in sulphide minerals deposited from hydrothermal activity on the seafloor along mid-oceanic ridges or subduction zones (Halbach et al., 2017 and the references therein).

Whereas the nodules (Fig. 1.1) are found to be spread over thousands of square kilometres, generally within the top 50 cm of the sediment (sometimes down to a few meters deeper), the crusts (Fig. 1.2) and sulphides (Fig. 1.3) could be spread over several square kilometres, with their thickness ranging from a few centimetres (for crusts) to several metres (for sulphides). Owing to different processes of formation, each of these deep-sea minerals has varying metal concentrations of major, minor and trace elements, of which some (Ni, Cu, Co, REEs and others) are

Fig. 1.2 Hydrothermal sulphide chimneys with black smokers, biological communities, and microbial mats, at the Kita-Bayonnaise Caldera, Izu-Bonin Arc (JAMSTEC Cruise NT13–05 #1494 at 1370 m depth). Width approximately 2.0 m. (Photograph courtesy of Prof. A. Usui, Kochi University, Japan)

considered critical for meeting the global demands in future (Van Nijen et al., 2018; Hein et al., 2020).

Fig. 1.3 Ferromanganese crusts near the summit of the Takuyo-Daigo seamount, NW Pacific basin (JAMSTEC Cruise KR09–02#955, at 1449 m depth). Width approximately 1.5 m. (Photograph courtesy of Prof. A. Usui, & B. Thornton)

3 Potential Sites for Deep-Sea Mining

Exploration and exploitation of some of these mineral deposits that are found to occur within the jurisdiction of some countries, that is Exclusive Economic Zones (EEZ), are governed by their respective laws. Some examples are the polymetallic nodules found within the national jurisdictions of Japan, Cook Islands and Kiribati (Hein et al., 2005; Hein et al. 2015), as well as Korea and France (Fouquet & Lacroix, 2014). Recently, Cook Islands has issued exploration licenses for mapping, sample collection and environmental studies for polymetallic nodules over an area of 254,654 km^2 within their EEZ (https://www.sbma.gov.ck). Approximately 650,000 km^2 area of the seafloor is covered by exploration contracts within EEZs of different coastal states in the Pacific Ocean (SPC, 2013; Petersen et al., 2018). Similarly, prospecting licenses to investigate hydrothermal sulphide deposits rich in silver and gold that occur along the Kermadec Volcanic Arc and on back-arc seamounts in their EEZ have been issued by the New Zealand government (Boschen-Rose et al., 2022). Recently, rich deposits have been found in the Norwegian Sea and Greenland Sea consisting of 38 million tonnes of copper and 45 million tonnes of zinc in polymetallic sulphides, as well as 24 million tonnes of magnesium, 3.1 million tonnes of cobalt and 1.7 million tonnes of cerium, a rare earth metal used in

Table 1.2 Estimated area under exploration contracts in the Area (as of December 2023)

Mineral type	Area (km^2)/contract	Total contracts[a]	Total area (km^2)
Nodules	75,000	19	1,425,000
Crusts	1000	5	5000
Sulphides	2500	7	17,500
Total		31	1,447,500

[a]www.isa.org.jm

alloys in manganese crusts, which the Norwegian government is considering developing as a future resource (https://www.reuters.com).

On the other hand, the deep-sea minerals that are located in international waters, beyond the national jurisdiction of any country, referred to as the 'Area', are regulated by the International Seabed Authority (ISA) established under Part XI of UN Law of the Sea (UNCLOS, 1982). There has been considerable interest over the last few decades in exploring these deposits, as evident from the fact that whereas only eight contracts were signed during the first decade of the century (2001–2010) by the erstwhile pioneer investors who had claimed deep-sea areas earlier (1980–2000), 23 new contracts were signed in the following decade (2011–2021) (www.isa.org.jm). Currently, there are 19 contracts for polymetallic nodules (each measuring 75,000 km^2), 5 for ferromanganese crusts (each measuring 1000 km^2) and 7 for hydrothermal sulphides (each measuring 2500 km^2) distributed over different parts of Pacific, Indian and Atlantic Oceans, totalling to 1,447,500 km^2 (Table 1.2). Although the total area appears to be large, it is critical to note that it forms just 0.45% of the area of all oceans in the world put together (Sharma, 2022).

The details of all the contracts between ISA and different contractors for each of the mineral types are given in Tables 1.3a, 1.3b and 1.3c and areas shown in Figs. 1.4a, 1.4b, 1.4c, 1.4d and 1.4e.

4 Development of Deep-Sea Mining Systems

In view of the evolving scenario of possible mining of deep-sea minerals in coming decades, several entities, including contractors and private enterprises, have initiated research and development for the design, fabrication and testing of different components as well as integrating them into pre-prototype systems that can be scaled up into full-scale commercial mining systems. Among the contractors, Korea has demonstrated a pilot mining system to collect nodules from 5000 m depth (Hong et al., 2019), India is developing a pre-prototype system for nodule collection that has been tested at 5200 m (Atmanand & Ramadass, 2017; Atmanand et al., this volume), and China has carried out a successful nodule collecting test at 500 m depth in South China Sea (https://chinadialogueocean.net). Japan has operated its prototype system for excavating sulphide ores containing Zn, Au, Cu and Pb from a depth of 1600 m off the coast of Okinawa (Kawano & Furaya, 2022) as well as a

Table 1.3a Contractors for exploration of polymetallic nodules in the Area (as of December 2023)

Contractor	Sponsoring State	General location of the exploration area under contract	Contract start date
InterOceanMetal Joint Organization	Bulgaria, Cuba, Czech, Poland, Russia, Slovakia	Clarion-Clipperton Fracture Zone (CCFZ), Pacific Ocean	29 March 2001
JSC Yuzhmorgeologiya	Russia	CCFZ, Pacific Ocean	29 March 2001
Government of the Republic of Korea	Republic of Korea	CCFZ, Pacific Ocean	27 April 2001
China Ocean Mineral Resources Research and Development Association	China	CCFZ, Pacific Ocean	22 May 2001
Deep Ocean Resources Development Co.	Japan	CCFZ, Pacific Ocean	20 June 2001
Institut français de recherché pour l'exploitation de lamer	France	CCFZ, Pacific Ocean	20 June 2001
Government of India	India	Indian Ocean	25 March 2002
Federal Institute of Geosciences and natural resources of Germany	Germany	CCFZ, Pacific Ocean	19 July 2006
Nauru Ocean Resources Inc.	Nauru	CCFZ, Pacific Ocean	22 July 2011
Tonga Offshore Mining Limited	Tonga	CCFZ, Pacific Ocean	11 January 2012
Global Sea-mineral Resources NV	Belgium	CCFZ, Pacific Ocean	14 January 2013
UK Seabed Resources Ltd. – I	UK & northern Ireland	CCFZ, Pacific Ocean	8 February 2013
Marawa Research and Exploration Ltd.	Kiribati	CCFZ, Pacific Ocean	19 January 2015
Ocean Mineral Singapore Pte Ltd	Singapore	CCFZ, Pacific Ocean	22 January 2015
UK Seabed Resources Ltd. – II	UK & northern Ireland	CCFZ, Pacific Ocean	29 March 2016
Cook Islands Investment Corporation	Cook Islands	CCFZ, Pacific Ocean	15 July 2016
China Minmetals Corporation	China	CCFZ, Pacific Ocean	12 May 2017
Beijing Pioneer Hi-Tech Development Corporation	China	Western Pacific Ocean	18 October 2019
Blue Minerals Jamaica	Jamaica	CCFZ, Pacific Ocean	16 March 2021

Table 1.3b Contractors for exploration of ferromanganese crusts in the Area (as of December 2023)

Contractor	Sponsoring State	General location of the exploration area under contract	Contract start date
Japan oil, Gas and Metals National Corporation	Japan	Pacific Ocean	27 January 2014
China Ocean Mineral Resources Research and Development Association	China	Western Pacific Ocean	29 April 2014
Ministry of natural resources and environment of the Russian Fed.	Russia	Pacific Ocean	10 March 2015
Companhia De Pesquisa de Recursos Minerais	Brazil	South Atlantic Ocean	9 November 2015
Republic of Korea	Republic of Korea	Western Pacific Ocean	27 March 2018

Table 1.3c Contractors for exploration of hydrothermal sulphides in the Area (as of December 2023)

Contractor	Sponsoring State	General location of the exploration area under contract	Contract start date
China Ocean Mineral Resources Research and Development Association	China	Southwest Indian Ridge	18 November 2011
Government of the Russian Federation	Russia	Mid-Atlantic Ridge	29 October 2012
Government of the Republic of Korea	Republic of Korea	Central Indian Ridge	24 June 2014
Institut français de recherche pour l'exploitation de la mer	France	Mid-Atlantic Ridge	18 November 2014
Federal Institute of Geosciences and natural resources of Germany	Germany	Southeast and Central Indian Ridge	6 May 2015
Govt of India	India	Central Indian Ocean	26 September 2016
Government of Republic of Poland	Poland	Mid-Atlantic Ridge	12 February 2018

Source: www.isa.org accessed on 1 December 2023

crust-excavation testing machine and collected 649 kilograms of cobalt and nickel-rich seabed crust (www.jogmec.go.jp).

Several contractors have entered into contracts with private enterprises for exploration and exploitation of seabed mineral resources in their contract areas. These include Nauru Ocean Resources Inc. (NORI, Nauru), Tonga Offshore Mining Limited (TOML, Tonga), Global Sea-mineral Resources NV (GSR, Belgium), UK Seabed Resources Ltd. (taken over by Loke Marine Minerals, Norway), Marawa Research and Exploration Ltd. (MRE, Marawa), Cook Islands Investment Corporation (CIIC, Cook Islands), China Minmetals Corporation (China), Beijing

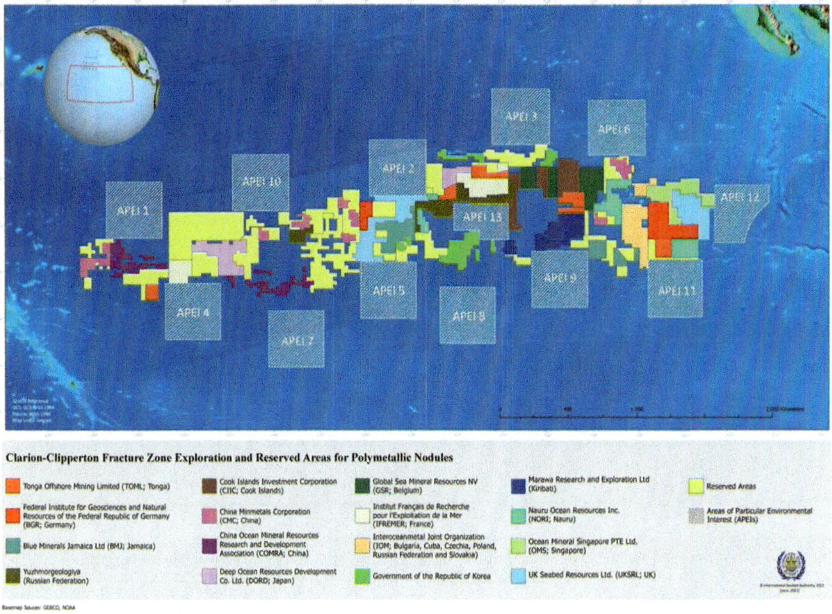

Fig. 1.4a Exploration areas for polymetallic nodules in Clarion-Clipperton Zone. (Courtesy: International Seabed Authority, Jamaica)

Pioneer Hi-Tech Development Corporation (China), Blue Minerals (Jamaica) and Ocean Mineral Singapore Pte Ltd. (OMS, Singapore) (www.isa.org.jm).

Some of the private enterprises have also started to develop and test their pre-prototype mining systems in international waters. Global Sea-mineral Resources NV (GSR) tested a 25-tonne nodule collector in the Belgian and German areas in the Clarion-Clipperton Zone (CCZ) of the Pacific Ocean over a distance of 54.3 km with a throughput of ~2000 tonnes of nodules in April 2021 (Bruyne et al., 2022). Nauru Ocean Resources Inc. (NORI—a subsidiary of The Metals Company) conducted a pilot collector test in September 2022 over 80 km of seafloor in NORI-D area and collected 4500 tonnes of nodules (of which 3000 tonnes were transported to a surface vessel) at a production rate of 86.4 tonnes per hour and plans to scale up from 1.3 to 12.5 Mtpa in future (www.tmc.com). Impossible Metals 'has successfully completed its first trial of selectively harvesting rocks in an underwater environment' in December 2022 and 'plans to have the technology ready for large scale deployment by 2026' (www.impossiblemetals.com).

Similar developments are also seen with respect to developing mineral resources within the EEZ of some countries. For example, Cook Islands Seabed Minerals Authority has signed 5-year contracts with three companies, viz. CIIC Seabed Resources Ltd., CIC Ltd. and Moana Minerals Ltd., for the development of seabed mineral resources (polymetallic nodules) within their EEZ (www.sbma.gov.ck/exploration). Odyssey Marine Exploration, which has experience in deep-sea

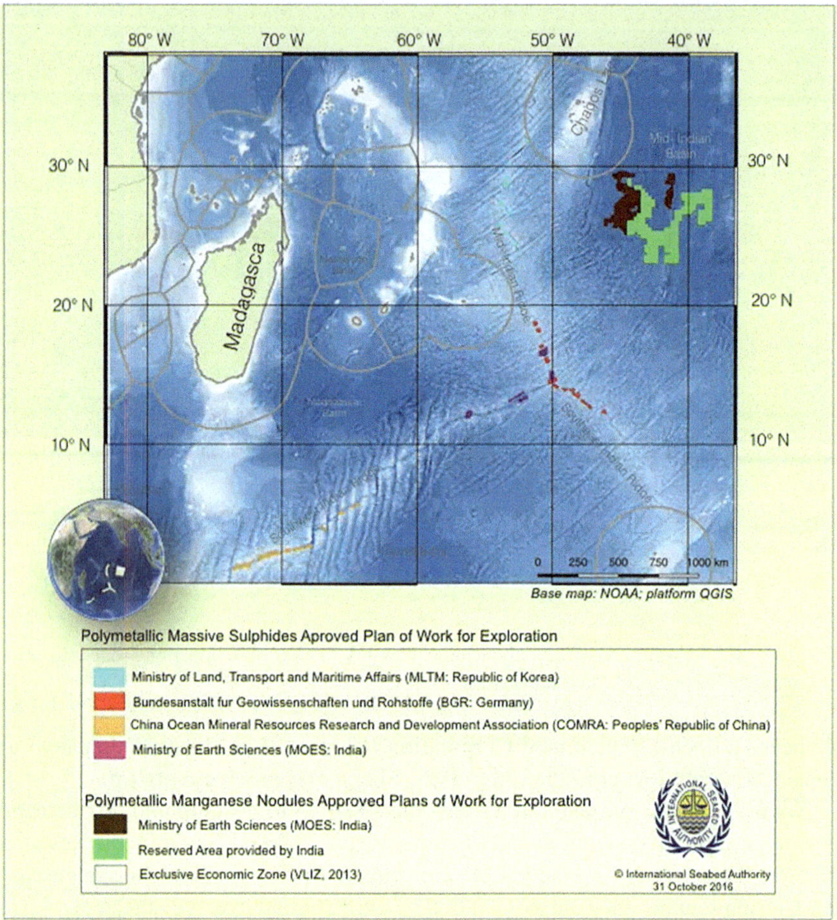

Fig. 1.4b Exploration areas for polymetallic nodules and sulphides in the Indian Ocean. (Courtesy: International Seabed Authority, Jamaica)

mineral exploration, validation and development, is involved with several government agencies interested in identifying mineral resources within their EEZs (https://www.odysseymarine.com/projects). Nautilus Minerals (taken over by Deep Sea Mining Finance Limited—https://dsmf.im) announced successful completion of seafloor production tools with an aim to mine the seafloor massive (sulphide) deposits to produce Cu, Au and Ag from 1600 m depth off Papua New Guinea (http://dsmobserver.com).

However, different activities of deep-sea mining are expected to have diverse environmental impacts (Weaver & Billet, 2019), right from the seafloor from where the minerals will be picked up to the water column through which the ores will be lifted to the mining platform, as well as at the surface where pre-processing, transfer to supply vessels and transportation of ores to land will be conducted. Hence, there is a need to evaluate the impacts on the physical, chemical and biological conditions

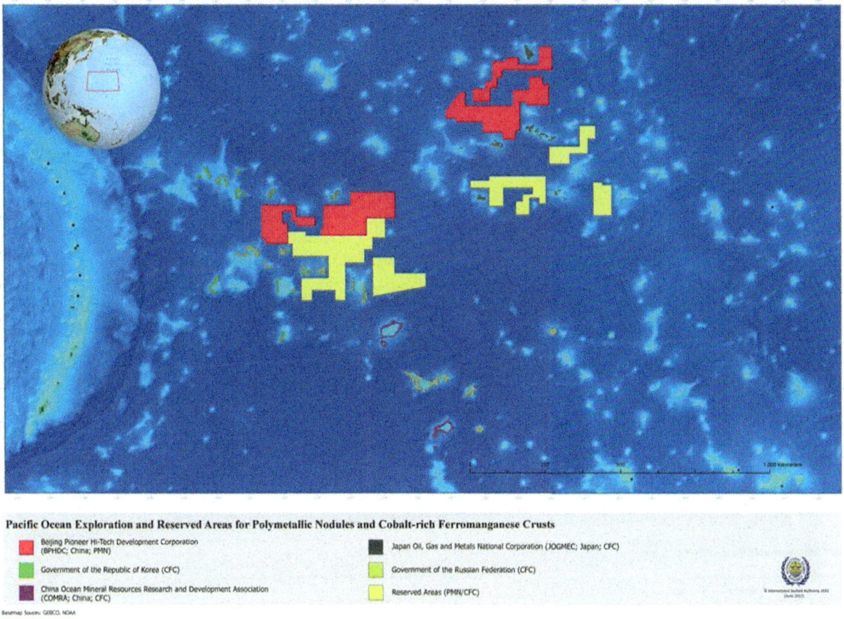

Fig. 1.4c Exploration areas for polymetallic nodules and cobalt-rich ferromanganese crusts in the Pacific Ocean. (Courtesy: International Seabed Authority, Jamaica)

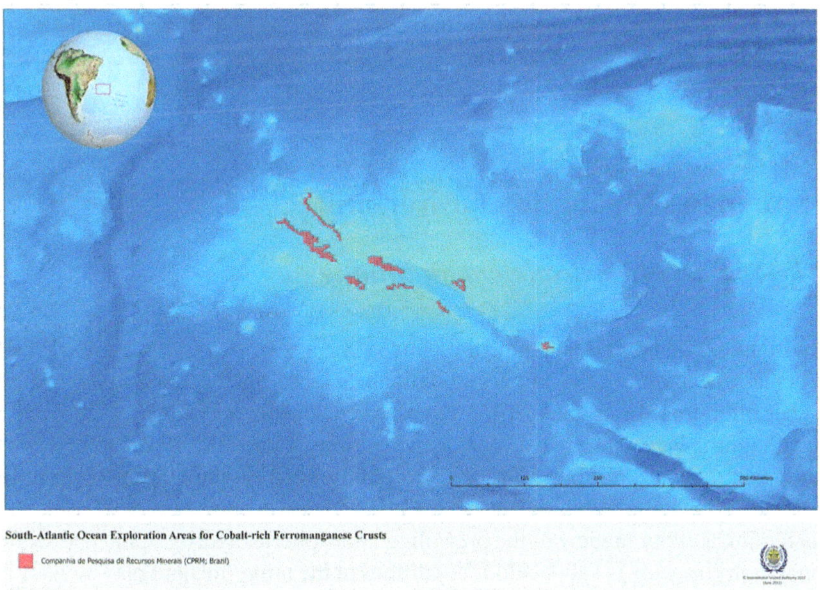

Fig. 1.4d Exploration areas for cobalt-rich ferromanganese crusts on South Atlantic seamounts. (Courtesy: International Seabed Authority, Jamaica)

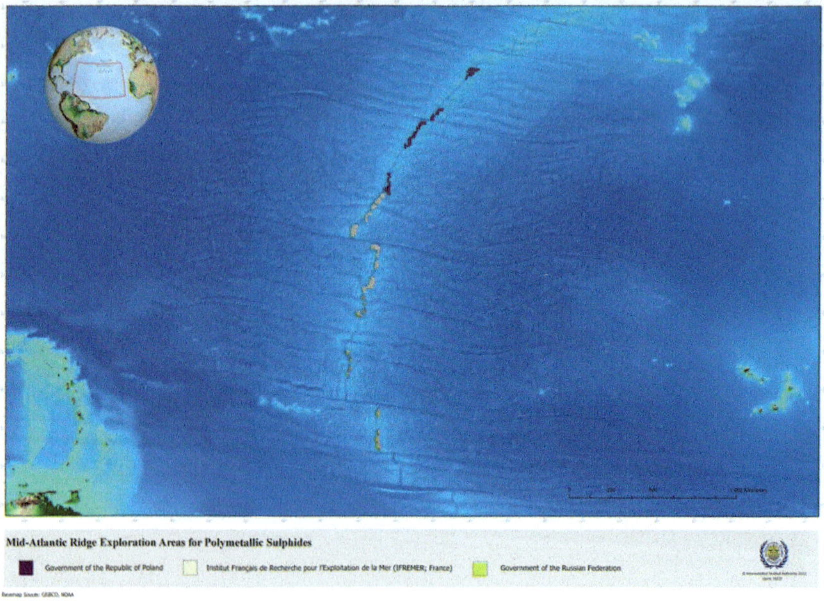

Fig. 1.4e Exploration areas for polymetallic sulphides on the Mid-Atlantic Ridge. (Courtesy: International Seabed Authority, Jamaica)

in the water column (Christiansen et al., 2020; Drazen et al., 2020), as well as the baseline conditions, and the likely impacts on the benthic environment (Sharma et al., 2001; Jones et al., 2017; Tilot, 2019; Radzeijewska et al., 2022, Fukushima et al., 2022).

5 Deep-Sea Mining and Environment

5.1 Impact of Environment on Mining

Environment and any developmental activity always have a two-way relationship. Whereas the issue of the impact of human activity on the environment attracts higher attention, the impact of the environment on the activity itself also has an equal significance. This is true for deep-sea mining as well. Whereas it is anticipated that deep-sea mining could have certain impacts on the marine environment, the environmental conditions at sea would also have different types of impacts on the mining activity in view of the prevailing atmospheric, hydrographic and seafloor topography, as well as sub-seafloor conditions at the mine site, and play a major role in the design and performance of different mining sub-systems (Table 1.4). Hence, evaluation of environmental conditions would not only help in the assessment of

Table 1.4 Influence of environmental conditions on mining system design and operation

Sr. no.	Conditions (key parameters)	Influence on mining system
1	Atmospheric (wind, rainfall, cyclone)	To determine the weather conditions for design as well as operation of the mining system during different seasons of the year, including likely shutdown periods.
2	Hydrographic (waves, currents, temperature, pressure)	Will influence operations on the platform including ore-handling as well as mining system deployment at the surface; and stability of riser system in the water column.
3	Topographic (relief, macro and micro-topography, slope angles)	Will have a bearing on the manoeuverability and stability of the mining device on the seafloor.
4	Mineral characteristics (grade, size, abundance, morphology, distribution pattern)	Important for designing the mechanism for collection, crushing and screening of mineral at the seafloor from un-wanted material before pumping the ore to the surface as well as identifying relatively richer deposits for better output.
5	Associated substrates (sediment size, composition, engineering properties; rock outcrops—Extent, elevation)	Will affect the mobility and efficiency of the collector device to be able to operate without sinking (or getting stuck) in the sediment and be able to avoid the rock outcrops for its safety.

Modified from Sharma (2017a)

impacts resulting from mining activity but also play a key role in the design and planning of the operation of the mining system itself (Sharma, 2011).

Long-term data on environmental parameters in the area of operation that need to be evaluated from local, regional or global meteorological perspectives, include seasonal variation of maximum, minimum and average for various atmospheric as well as sea-surface parameters such as sea surface temperature, wind speed, rainfall, frequency; path and intensity of cyclones (depressions, storms, hurricanes/typhoons, severe hurricanes/super typhoons); waves (season-wise frequency, direction, height and wave period); and the water column characteristics (depth wise profile of temperature, salinity) and currents (Sharma, 2019). Similarly, detailed evaluation of seafloor characteristics that will have an influence on the performance of the seabed mining device would include distribution characteristics of mineral deposits (size, coverage, abundance), substrates (sediment thickness, rock outcrops), topography, micro-topography and slopes (Sharma, 2017b, 2019).

5.2 *Impact of Mining on Environment*

Various activities associated with deep-sea mining are expected to create impacts at various levels in the marine ecosystem (Fig. 1.5). The likely impacts include those related to at-sea processing and transportation on the surface, altering the physico-chemical characteristics of the water column due to particles discharged (accidently or otherwise) during lifting of the ores, as well as on the seafloor, from where the

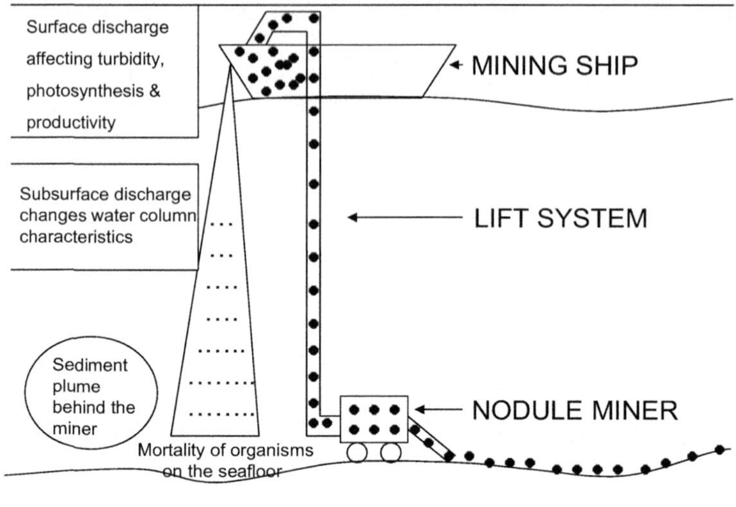

Fig. 1.5 Schematic for environmental impact of deep-sea nodule mining. (Sharma, 2017c)

minerals will be picked up and separated from the associated substrate either by scooping (in case of polymetallic nodules), scraping (in case of ferromanganese nodules) or drilling (in case of hydrothermal sulphides), leading to alteration in environmental conditions due to resuspension and redistribution of debris in the bottom water along the path of the collector device as well as in the vicinity of the mining tracks, leading to changes in diversity and abundance of benthic organisms (Pearson, 1975; Amos et al., 1977).

Various activities associated with deep-sea mining that could lead to environmental impacts include (Sharma, 2022):

(a) *Offshore activities.*

(i) *Picking up or separation of minerals and the quantity of substrate disturbed on the seafloor due to operation of the miner, crusher and discharge mechanisms.*
(ii) *Suspension of fine particles of minerals and associated sediment into the water column.*
(iii) *Resettlement of suspended particles and smothering of seafloor.*
(iv) *Impacts due to light and sound during mining operation.*
(v) *Oil spills and leakages from mining platform and transport vessels.*
(vi) *Ballast water discharge from transport vessels.*
(vii) *At-sea processing, dewatering, waste disposal including chemicals, debris.*
(viii) *Sub-system losses such as pipes, chains, tools or any other hardware.*
(ix) *Human waste such as garbage, including plastics, metals, glass and other non-biodegradable items.*

(b) *Onshore activities.*

   (i) *Pollution during loading/unloading, on land transportation of ore from the port to the processing plant.*
   (ii) *Pollution during processing (fumes, discharges) around the processing plant.*
   (iii) *Pollution after processing (dumping of slag or unwanted material) away from the processing plant.*

Impacts that are expected at various levels in the water column and on the seafloor due to deep-sea mining activities are summarised as follows (after ISA, 1999):

(a) *Potential impacts due to mining activity on the seabed.*

   (i) *mortality of organisms along the collector track,*
   (ii) *smothering of the benthic fauna away from the mining site where the sediment plume settles,*
   (iii) *clogging of suspension feeders and dilution of deposit-feeders food resources.*

(b) *Potential impacts due to discharge of tailings at mid-water depths.*

   (i) *Mortality of zooplankton species at mid-water depths.*
   (ii) *Effects on meso- and bathypelagic fishes and other nekton caused directly by the sediment plume.*
   (iii) *Effects on fish behaviour and mortality caused by the sediments or trace metals.*
   (iv) *Depletion of oxygen by bacterial growth on suspended particles.*
   (v) *Dissolution of heavy metals and their potential incorporation into the food chain.*
   (vi) *Impacts on deep-diving marine mammals.*

(c) *Potential impacts due to surface discharge and movement of vessels.*

   (i) *Trace-metal bioaccumulation leading to reduction in primary productivity due to shading on phytoplankton.*
   (ii) *Effects on marine mammals due to noise, oil spills and waste disposal.*

It is also critical to note that due to differences in characteristics and environmental settings of each of the deep-sea mineral type, there could be different kinds of impacts for mining of these on the marine ecosystem (Weaver & Billet, 2019).

5.3 Evaluating the Impact of Mining on Environment

In order to develop an understanding of the possible impacts of deep-sea mining, studies have been undertaken during the pilot mining tests under the Deep Ocean Mining Environment Study (DOMES, 1972–81) conducted by Ocean Mining Inc.

Table 1.5 Benthic Impact Experiments (BIEs) (Sharma, 2022)

Experiment	Year	Conducted by	Area	Tows	Duration	Area/ Distance	Discharge[f]
DISCOL[a]	1989	Hamburg University, Germany	Peru Basin	78	~12 days	10.8 km^2	–
NOAA-BIE[b]	1991	National Oceanographic & Atmospheric Administration, USA	Clarion Clipperton Fracture Zone	49	5290 mins	141 km	6951 m^3
JET[c]	1985	Metal mining Agency of Japan	Clarion Clipperton Fracture Zone	19	1227 mins	33 km	2495 m^3
IOM-BIE[d]	1995	InterOceanMetal – consortium of East Europen Countries	Clarion Clipperton Fracture Zone	14	1130 mins	35 km	2693 m^3
INDEX[e]	1997	National Institute of Oceanography, Govt. of India	Central Indian Ocean Basin	26	2534 mins	88 km	6015 m^3

Source
[a]Foell et al. (1990)
[b]Trueblood (1993)
[c]Fukushima (1995)
[d]Tkatchenko et al. (1996)
[e]Sharma and Nath (2000)
[f]Yamazaki and Sharma (2001)

(OMI) and Ocean Mining Associates (OMA) in the Pacific Ocean (Ozturgut et al., 1980). Subsequently, experimental mining was conducted using devices such as plough harrow and sediment suspension mechanism in the Pacific and Indian Oceans (Table 1.5) (Foell et al., 1990; Trueblood, 1993; Fukushima, 1995; Tkatchenko et al., 1996; Sharma & Nath, 2000; Theil, 2001; Sharma, 2001; Sharma, 2005). These experiments have revealed that whereas the impacts were severe immediately after the disturbance (Shirayama, 1999), the impacts were masked by natural variability over a period of time (Sharma et al., 2007). Studies have shown that while mobile and small fauna were less negatively impacted and the density and diversity of meio and mobile megafauna recovered within 1 year, some of the faunal groups returned to control conditions after two decades (Jones et al., 2017).

Analysis of the intensity of disturbance shows that the scale of these experiments was significantly small as compared to that expected during commercial mining (Yamazaki & Sharma, 2001), leading to an engineering assessment of 'experimental' deep-sea mining that suggested to 'test benthic disturbance in scale and system large enough to represent the commercial mining scale' (Chung et al., 2001). This has been evidenced in the case of Global Seabed Resources (GSR) testing its prototype nodule collector in the Belgian and German areas in the CCZ in 2021 (Bruyne et al., 2022), wherein the results have shown that the resulting sediment plume from test mining was influenced by the local seabed morphology and bottom currents (Gazis, 2022). Similarly, an environmental study conducted by 'independent scientists in the *MiningImpact* consortium (during the testing of GSR equipment) also addresses the direct effects of collecting polymetallic nodules, and also those of the sediment plume created by the process' with an aim to 'provide information on ecosystem effects of potential future mining' (https://www.deme-gsr.com) of which the results are awaited.

Observations on nanofauna, meiofauna and macrofauna at varying distances from the impact site and over different time periods from 'test' mining of hydrothermal sulphides in the Okinawa Trough by Japan have shown that different faunal groups behaved differently based on the distance, levels of resedimentation as well as toxicity. The study also showed 'the importance of sea floor topography as a driver of the spatial extent of impacts from deep-sea mining', including 'the lack of impacts at an elevation of as little as 5 m highlights the need to understand fine-scale topography when planning mining operations', suggesting that 'Mitigation of impacts may thus include mining of areas that are situated in localized seafloor depressions' (Washburn et al., 2023).

6 Estimation of Disturbance on Seafloor and in Water Column

The disturbance on the seafloor can be determined from the width of the miner, which includes locomotion tracks and the pickup device, the expected depth of penetration and the distance covered by the miner. In the case of sulphides and crusts, the design of the seafloor mining device would involve either drilling or scraping, whereas for nodules, it will be more of a scooping, sieving and sucking method. This section provides a hypothetical case for estimating the disturbance on the seafloor and the associated sediment resuspension in the water column during nodule mining. The dimensions used in this estimate are based on that of the pre-prototype nodule collector designed by the National Institute of Ocean Technology, Chennai, India (personal communication K. Gopkumar, NIOT, Chennai). However, the dimensions and scale of the commercial mining system are expected to be much larger, and necessary modifications will be required in the following calculations, in order to evaluate the disturbance of the same.

For this calculation, it is considered that the nodule collector will have four moving tracks (each of 0.75 m width), with two pickup units (each of 0.8 m width). During the trials, the machine is expected to cover a total distance of 1000 m, with a seabed penetration of a maximum of 0.3 m. Hence, the volume and weight of sediment and nodules (Table 1.6) to be disturbed are based on the following parameters:

1. Nodule collector machine with 4 moving tracks (width 0.75 m × 4 no. = 3.00 m width) and 2 pickup units (width 0.8 m × 2 no. = 1.6 m width), hence total width = 4.6 m.

Table 1.6 Estimations of area, volume and weight of sediments and nodules to be disturbed

(a) Volume and weight of sediment disturbed per m^2 of area		
(i) Volume of sediment impacted (1 m × 1 m × depth of penetration) For penetration of 0.3 m	0.3 m^3	
(ii) Weight of sediment disturbed (volume × density of sediment 1.2 kg/m^3) For penetration of 0.3 m	0.36 kgf	
(b) Volume and weight of sediment lifted within pickup width of 1.6 m (2 × 0.8 m) only		
(i) Volume of sediment lifted (width × depth × length of track) For penetration of 0.3 m (1.6 m × 0.3 m ×1000 m)	480 m^3	
(ii) Weight of sediment lifted (volume ×density of sediment 1.2 kg/m^3) For penetration of 0.3 m (240 m^3 × 1.2)	576 kgf	
(c) Area, volume and weight of sediment disturbed for entire width of collector (moving tracks + pickup unit) Collector with 4 moving tracks (0.75 m × 4 = 3 m) and 2 pickup units (0.8 m × 2 = 1.6 m) = Total width disturbed = 4.6 m		
(i) Area covered (width of miner 4.6 m × length of track 1000 m)	4600 m^2	
(ii) Volume of sediment disturbed (area × depth) For penetration of 0.3 m	1380 m^3	
(iii) Weight of sediment disturbed (volume × density of sediment = 1.2 kg/m^3) For penetration of 0.3 m	1656 Kgf	
(d) Quantity of nodules and adhered sediment expected to be lifted		
(i) Area to be swept (1.6 m width of pickup units over 1000 m)	1600 m^2	
(ii) Quantity of nodules picked (@ abundance 10 kg/m^2 × area) and at 50% nodule pickup efficiency	8000 kg	
(iii) Sediment to be lifted with nodules at 50% efficiency of nodule collector[a] If sediment is 20% by weight of nodule (sediment density 1.2 kg/m^3)	Weight 1600 kg	Volume 1333.3 m^3
[a]That will go into suspension when it is released either at the top of the pickup unit or riser pipe exit		
(e) Concentration of sediments in water after suspension considering width of dispersion = 100 m (i.e. 50 m on either side from point of discharge)		
(i) Volume of water Moving point dispersion at 80 m altitude (l × b × h = 1000 × 100 m × 80 m)	80,00,000 m^3	
(ii) For 50% efficiency of nodule collector, the sediment concentration will be	0.0167%	

2. Distance covered = 1000 m cumulative.
3. Depth of penetration = 0.3 m.
4. Density of wet sediment = 1.2 kg/m^3.

The volume of sediment expected to be impacted over 1 m^2 of seabed area for penetration of 0.3 m would be 0.3 m^3. Considering a wet density of 1.2 kg/m^3, the weight of sediment disturbed would be 0.36 kgf. Estimates also show that the volume of sediment disturbed within the pickup width of 1.6 m along a 1000 m long track would be 480 m^3 for a penetration of 0.3 m, which translates into a weight of

576 kgf. The seabed surface area expected to be disturbed by the 4-tracks' mining machine and two pickups would be 4600 m^2 over a cumulative distance of 1000 m. For this area, the volume of sediment disturbed by the mining machine would be 1380 m^3 with a corresponding weight of 1656 kgf

For a total pickup width of 1.6 m, an average nodule abundance of 10 kg/m^2, the quantity of nodules expected to be picked up is estimated to be 8000 kg at a conservative 50% nodule pickup efficiency. The estimates show that 1600 kg of sediment would be picked by the collector movement with a corresponding volume of 1333.3 m^3 if the adhered sediment is 20% by weight of nodules at 50% pickup efficiency of nodule collector. Estimates also show that in case of discharge at 80 m altitude along 1000 m track considering a width of 100 m for resettlement of sediment (i.e. 50 m on either side), the concentration of solids (sediments) would be 0.0167% of the total volume of water in the area. It must be noted that the higher the area (or width) of resettlement, the lower will be the concentration of sediment in the corresponding volume of water

In these estimates, the velocity and direction of currents at the point of discharge also need to be taken into consideration in order to estimate the lateral distribution of the suspended particles that could be either unidirectional or multidirectional depending on seasonal variability and other factors. The intensity of impact will depend on the quantity of sediment particles discharged in the water column and the lateral transportation to adjacent areas due to prevalent currents as well as the settling velocity of these fine-grained particles to the seafloor that would determine the resedimented area. At the same time, midwater discharge plume will also contain nodule fines, the size of which will depend on the separation process and also if the nodules will be crushed before pumping to the surface but expected to be within the same size class as the sediment, whereas currently there is no evidence that dissolved metals will be released from these crushed particles (Carlos Munoz-Royo et al., 2021). Moreover, results from the testing of pre-prototype nodule collector Patania II of GSR have shown that fine-grained sediment particles aggregated into larger ones facilitating relatively fast-sinking of particles that could promote the settling of most of the mobilized sediment in the vicinity of the trial site (Gazis, 2022). Observations from a small-scale experiment have shown that sediment particles could travel a few kilometers from the 'mining' site (Sharma et al., 2001), whereas model studies suggest that these could travel as far as 50–100 km from the mine site (Rolinski et al., 2001).

Modeling studies by Carlos Munoz-Royo et al. (2021) suggest that:

1. 'Recognizing that it takes about 1 year for 10 μm sediment to settle to the bottom from midwater column, over which time the sediment can readily be transported up to 1000 km in different directions by variable ocean currents throughout a 20 year commercial mining operation, the area over which the sediment and fines settle is a few million square kilometers.'
2. 'This results in a sedimentation rate on the order of 0.01 μm per year at the seabed compared to background sedimentation rates of 1 μm per year (Arrhenius, 1963).'

Further, results of sediment distribution during the test mining have shown that 'At the time of measurement, 2–8% of the sediment mass were detected 2 m or higher above the seabed and were not observed to settle over several hours, with the remaining 92–98% below 2 m and some fraction of that locally deposited. Our results suggest that turbidity current dynamics sets the fraction of sediment remaining suspended and the scale of the subsequent ambient sediment plume' (Carlos Muñoz-Royo et al., 2022).

The area of resettlement would also depend on the height of discharge, which may correspond to the altitude of the intermediate buffer where initial screening would happen, depending on the type of mechanism used for pickup. According to Allahdad et al. (2023), 'hydraulic system is the most preferred among different mechanisms for nodule mining, as it has the least interaction with seabed during collection, has higher pickup efficiency, less power consumption and better adaptability to seabed morphology'. They have also proposed a Coanda-based collector that creates a suction to dislodge the nodules from the seabed and carry them towards the collection duct. Reducing the height of the plume and increasing the settling velocity by increasing particle aggregation (Peukert et al., 2018) could help reduce the area of resedimentation on the seafloor.

Since sustained plumes become more dilute in the water column with increasing distance from the area where they form (Van Zyl et al., 2002), the impact of particles on the water column would be less than that on the sediment-water interface in the adjacent zones of DSM activities.

It is important to recognize that the farther the sediment particles travel in the water column, the concentration of suspended particles in the plume will become more diluted due to mixing with larger volumes of water masses (Van Zyl et al. 2002), and hence, lesser would be the thickness of resedimentation on the seafloor and much less the entombment on benthic organisms, minimizing the actual impact on benthic communities, in areas adjacent to or away from the mining zone.

7 Management, Mitigation and Regulation of Deep-Sea Mining Impacts

7.1 Environmental Management

Impacts on the water column due to dewatering, sediment discharge, noise, light and temperature variation (Miller et al., 2018) are likely to impact the pelagic organisms that live in this three-dimensional habitat (Drazen et al., 2020), in addition to the likely release of toxic compounds, acidification, oxygen depletion and introduction of alien species at different levels (Christiansen et al., 2020). While it is known that impacts on biodiversity are inevitable (Van Dover et al., 2017), it is important that a science-based management approach is adopted to minimize them.

Since midwater ecosystems are known to encompass >90% of the biosphere on earth and play a key role in carbon export, nutrient generation and connecting shallow and deep-water ecosystems, it is advisable to avoid any discharge in the epipelagic (0–200 m) and mesopelagic (200–1000 m) zones, but only below the mesopelagic–bathypelagic transition zone (1500–2000 m) or close to the seafloor where plume may already exist due to operation of the mining system (Drazen et al., 2020; Christiansen et al., 2020). While it is not possible to achieve 'No net loss' of biodiversity, the only viable means of reducing biodiversity loss from deep-sea mining is minimizing of impacts by focusing on a precautionary approach (Niner et al., 2018) and adopting ecosystem-based and adoptive management techniques (Thornborough et al., 2019; Cormier, 2019; Clark et al., 2022). This can be achieved by adopting scientific methodologies for evaluating local environmental conditions (Swaddling et al., 2016) as well as mitigation hierarchy to avoid, minimize, restore and offset the impacts (Miller et al., 2018; Niner et al., 2018; Clark, 2019).

It is critical to focus on trophic structure and biomass production as also list the ecological variables required for monitoring of deep-water column (Danovaro et al.; 2020) and incorporate these while the Contractors are still developing their technologies (Drazen et al., 2020) so that the requirements are known in advance. Hitchin et al. (2023) have also suggested that 'in the absence of data, thresholds can be set based on broader environmental goals, while more specific thresholds can be set as more data is available'. This is important because merely highlighting the impacts and their implications on marine ecosystems may not be enough for the contractors unless certain limits are specified for compliance that need to be met as a part of the mining activity. Moreover, a unified approach for regional environmental assessment (REA), developing a regional environmental management plan (REMP), optimizing data collection during baseline and monitoring to ensure cost-effective and robust impact assessment, as well as a quantitative prediction, could help achieve effective management of the marine environment (Sharma, 2017c; Jones et al., 2017).

An interdisciplinary approach that includes environmental, legal, economic and societal issues is required for the sustainable development of deep-sea mining ventures (Koschinsky et al., 2018), comprising of screening, scoping, impact assessment, development of EIA, EMP, stakeholder consultation and regulatory review (Durden et al., 2018). This requires a detailed ecological risk assessment (Doorn et al., 2022), determination of acceptable thresholds (Washburn et al., 2019), test mining with robust environmental assessment (Lodge & Verlaan, 2018) and identification of high-risk communities (Haughten et al., 2017).

A systematic conservation plan based on scientific data (Boschen-Rose et al., 2022) using faunal communities as indicators of impacts (Tilot, 2019), employing new techniques for standardization of environmental impact assessment (Fukushima et al., 2022), combined with environmental management practices (Billet et al., 2019), as well as quantifying the ecosystem services provided by the deep-sea (Thiele, 2019) and using traditional and social dimensions for development of legal framework (Tilot et al., 2022) could help in ensuring sustainable deep-sea mining.

Economic impacts could be on activities such as shipping or fishing, either in the mining area or en route the transportation pathways on one hand, and due to changes in employment opportunities directly (those employed in land-based mining), or indirectly (those providing services related to mining activities) associated with deep-sea mining. Hence, it would be important to outline the implications between other economic activities and deep-sea mining, examine the potential benefits of deep-sea mining to contribute to the economies in general and climate change in particular (such as battery metals) and discuss ways to mitigate the negative effects (Lebre et al., 2023) before deep-sea mining commences. A detailed social impact assessment (SIA) would also be required to be carried out that includes community profiling, identification of possible impacts and their significance on society, developing a social impact management plan (SIMP) and designing and implementation of monitoring programmes (Vanclay et al., 2015).

The deep-sea minerals found in the Area are the 'Common heritage of mankind' implying that their exploitation should be carried out for the benefit of the entire humankind (UNCLOS, 1982) for which ISA is in the process of developing a mechanism of benefit sharing (Jaeckel et al., 2016). Some of the issues, including developing a mechanism for equitable sharing and establishment of a seabed sustainability fund, have been discussed in the article by Lodge and Bourrel-McKinnon (2022). Whereas concerns over the likely impacts of deep-sea mining are being addressed, it could also become a driver for achieving the UN 2030 agenda for sustainable development (Madureira et al., 2023).

Finally, the sponsoring entities will have to take up the responsibility of protecting the marine environment and preventing serious harm (Levin et al., 2016). Different approaches to achieve this could include the requirement to comply with direct regulations, creating incentives for Contractors to reduce environmental damage, levying a cost for environmental damage, provision of environmental liability insurance and environmental bonds (Lodge et al., 2019).

7.2 *Mitigation*

Mitigation measures that could be incorporated in the design and operation of different sub-systems of deep-sea mining in order to minimise environmental impacts could be as follows (after Sharma 2015):

(a) Collector device:

- There should be minimum interaction of the collector system with the seafloor environment, to keep it as disturbance free as possible.
- The separation of minerals from sediments (or other debris) should be as close as possible to the seabed so that minimum water column is affected.

- Strip-wise (or 'patch') mining to be carried out, leaving alternate strips of undisturbed seafloor, to allow re-population by organisms from adjoining areas.

(b) Ore lifting:

- Prevention of leakage through riser pipes.
- Reduction of noise levels from pumps.
- Avoidance of breakage of riser pipes.

(c) Surface discharge:

- Sediment discharge should be the least at the surface to allow sufficient sunlight to penetrate for photosynthetic activity.
- Any surface water discharge can be sprayed over a large area so that it gets diluted without much delay.
- Discharge of bottom waters and debris should be at different levels of the water column, preferably below the oxygen minimum zone, where the faunal density is relatively lower than in the levels above.

(d) At-sea processing, ore transfer and transport:

- Proper treatment of unwanted material to be carried out before discharging.
- Biodegradable methods could be used for treatment of discharge.
- Proper care to be taken during transfer of ore from the collector to the mining platform as well as to the transport vessels to avoid spilling.
- Oil spills from these ships also should be monitored.

Some of the factors that could be taken into consideration to predict the extent of impacts of deep-sea mining could be as follows:

- The proportion of area that will be affected, with respect to total area.
- Influence of surface and subsurface currents in different seasons.
- Distance of coastal areas and inhabited areas from the area of influence.
- Existence of fishing potential or any other commercial activity in the area of influence.

7.3 Regulations and Guidelines

Deep-sea mining on a commercial scale is yet to begin in any part of the world which has given a lead time to the regulatory agencies to develop regulations for exploitation of the seabed resources. Since its establishment in 1994 by Article 156 of the 1982 United Nations Convention on Law of the Sea (UNCLOS, 1982), the ISA has issued recommendations for the assessment of possible environmental impacts from exploration of nodules (ISA, 2001) and for the establishment of environmental baselines and associated monitoring programme for exploration of polymetallic sulphides and cobalt crusts (ISA, 2005). The International Marine Minerals

Society has also prepared a code for environmental management for marine mining that provides a framework for the development and implementation of an environmental programme for a marine exploration and extraction site by marine mining companies and for other stakeholders in evaluating such programmes (www.immsoc.org). Further, several guidelines and regulations have been either developed or are in different stages of development aimed at the protection of the marine environment from the harmful effects of deep-sea mining (www.isa.org.jm), such as:

(i) Rationale and recommendations for the establishment of preservation reference areas for nodule mining in the Clarion-Clipperton Zone ISBA/14/LTC/2 (ISA, 2008b).
(ii) Environmental Management Plan for the Clarion-Clipperton Zone ISBA/17/LTC/7 (ISA, 2011).
(iii) Recommendations for the guidance of contractors for the assessment of the possible environmental impacts arising from exploration for marine minerals in the Area ISBA/19/LTC/8 (ISA, 2013) that were subsequently modified (ISA, 2019a) and also revised (ISA 2020).
(iv) Draft regulations on the exploitation of mineral resources in the Area. International Seabed Authority, Jamaica, ISBA/25/C/WP.1 (ISA, 2019b), which also includes several provisions for monitoring and management of marine environment.

Other draft regulations that are under consideration at the ISA include developing provisions for environmental impact statement and format for environmental monitoring and management plan to be furnished by the contactor while applying for mining contract; guidelines for the establishment of baseline environmental data; guidelines for environmental impact assessment process; guidelines for the preparation of an environmental impact statement; guidelines for the preparation of environmental management and monitoring plans; and guidelines on tools and techniques for hazard identification and risk assessments (www.isa.org.jm).

Besides these, the following regulatory framework of the International Maritime Organisation (www.imo.org) could also apply to all the shipping-related activities related to deep-sea mining, including the mining platform and the supply and transport vessels:

(i) Convention on the prevention of marine pollution by dumping of wastes and other matter.
(ii) International convention of preservation of marine pollution from ships (MARPOL).
(iii) Guidelines for reduction of underwater noise from commercial shipping to address adverse impacts on marine life.
(iv) The International Convention for the Control and Management of Ships' Ballast Water and Sediments, 2004 (BWM Convention).

8 Comparison of Deep-Sea Mining with Land-Based Mining

In recent times, a few studies have focussed on comparing the impacts of deep-sea mining with respect to terrestrial mining (Koschinsky et al., 2018; Hein et al., 2020; Lebre et al., 2023). As the quantity and quality of ores in terrestrial deposits are depleting, mining on land could become more expensive, as in the case of copper production, which has decreased by 25% in the last decade, whereas the energy consumption to mine it has increased by 30% (Calvo et al., 2016). Moreover, in the case of deep-sea minerals, mining of one mineral type could yield at least three or four metals with a lower carbon footprint than mining at three or four separate sites on land for one metal each. While recycling is not a viable option as yet for such quantities of metals as required to meet the future demands of humankind, the large expanses of deep-sea minerals could be the alternative source for critical metals unless new land-based resources are found or such materials are produced that can substitute the metals of interest (Hein et al., 2020).

Whereas land mining completely devastates the ecosystem in a way it can never be restored to its original state (Sonter et al., 2017), it is possible that deep-sea environment could restore itself over a period of time due to its dynamic nature (Sharma, 2022). Hein and Koschinsky (2014) concluded that 'deep-sea mining may be preferable alternative because it does not require the same excessive preparation of terrestrial mining sites, which often includes the deforestation of large areas and the dislocation of local inhabitants'. Moreover, whereas in land mining, infrastructure such as roads and electric and water lines need to be built, these are not required in deep-sea mining, hence minimizing social and environmental impacts (Hein et al., 2020).

Whereas different activities of deep-sea mining such as collection, screening, lifting, transportation and pre-processing of ores would impact different sections in the water column and on the seafloor, resulting in a dilution effect as it could be spread over a vast area, impacts due to all activities in case of terrestrial-based mining such as mining, screening, transportation, processing of ores and effluent disposal would be concentrated in and around the mining site on land (Fig. 1.6).

The relative advantages of deep-sea mining with respect to land-based mining are listed in Table 1.7

The apprehensions regarding the environmental impact of deep-sea mining are based on our experience of large-scale destruction on land due to deforestation and lack of effective implementation of reforestation programmes. This calls for robust environmental monitoring programmes to ensure compliance, data sharing and accountability on the part of the contractors in case of deep-sea mining

Fig. 1.6 Environmental impacts of mining activities on different areas (indicated in pink). (Sharma, 2022)

Table 1.7 Advantages of deep-sea mining with respect to land mining

Factors	Deep-sea mining	Land mining
Mining infrastructure	Same infrastructure can be shifted to different mining locations	Dismantling of old and construction of new infrastructure at each location
Social impact on local communities	No dislocation, relocation or influx of new communities	Relocation of local communities, influx of new communities
Land conflicts	No	Yes
Roads, electricity, water line, communication	Not required	Required
Deforestation, damage to nature	No, only local impacts, thousands of km from shore	Yes, permanent damage to ecology
Removal of overburden	No overburden removal or dumping	Yes, causing severe problems

9 Challenges in Deep-Sea Mining

While deep-sea mining is in different stages of development, some of the challenges that lie ahead are as follows:

1. Operating under extreme conditions for extended periods of time (250–300 days/year), very low temperatures (1–2 °C), high pressure conditions (300–500 bars), cross-cutting currents (at different levels within the water column), uneven topography on the seafloor (with gradients of 1–10° or more) and heterogeneous distribution of minerals (<10 to >90% coverage).
2. Upscaling from pre-prototypes to commercial mining systems as well as integration of various subsystems being developed for mining.
3. Protecting the marine environment by operating within 'acceptable' limits of impact on physical, chemical and biological conditions.
4. Estimating the timing of commencement of commercial scale mining and optimising the mining rate depending upon global metal markets.
5. Establishment of regulations for sustainable exploitation of deep-sea minerals.

10 Conclusion

Mining of deep-sea minerals is fast emerging as an alternative for meeting the growing demands of certain critical metals due to the depletion of land deposits and also limited capability for processing low-grade ores and recycling. Moreover, the monopoly of some of the countries over exporting these metals and the occurrence of such ores in conflict zones around the world are putting pressure on the developing economies to become self-sufficient for their sustained growth. As a result, several entities have identified potential mining areas within their EEZ and also claimed large areas with potential deep-sea mineral deposits in international waters and are developing infrastructure to pick up the ores from deep-sea floor and transport them to land for further processing.

It is important to recognise that conducting mining activities in the deep sea is a cost-intensive and high-risk operation for humans as well as the machinery in view of operating the equipment at depths of a few kilometres underwater at extreme temperatures, pressure and other atmospheric conditions, with limited backup and emergency services due to remoteness of mining sites (located hundreds or sometimes thousands of kilometres) from any land-based facility. At the same time, there are concerns about possible environmental impacts on marine ecosystems, as well as social, cultural and economic impacts, that need to be attended to by various national and international organisations and contractors to find ways of mitigating and regulating the activities with minimum impact.

All activities around developing deep-sea mining systems are aimed at ensuring a steady supply of metals for meeting the current and future demands of the humankind, unless the world community as a whole can decide that it does not require the metals from the deep sea and can survive with the available resources or find alternative metals and/or processes (e.g. recycling or economical processing of low-grade minerals) to meet their requirements. As in the case of any human activity, the ingenuity of the human mind is capable of finding solutions to any foreseen issues, so as to make use of the resource in the most responsible manner.

Acknowledgements The author gratefully acknowledges the permission given by the Secretary General, International Seabed Authority, Jamaica for reproducing the maps from the website. The seafloor photographs of nodules, crusts and sulphides were provided courtesy of Prof. Akira Usui of Kochi University and JAMSTEC, which are gratefully acknowledged. Inputs from K. Gopkumar, Sr. Principle Scientist, National Institute of Ocean Technology, Chennai, India for estimation of area and volume of disturbance are highly appreciated.

References

Al Barazi, S., Brandenberg, T., Kuhn, T., Schmidt, M., & Vetter, S. (2018). DERA Rohstoffinformationen 36. Kobalt. ISBN: 978-3-943566-49-9. https://www.deutsche-rohstoffagentur.de/DE/Gemeinsames/Produkte/Downloads/DERA_Rohstoffinformationen/rohstoffinformationen-36.pdf

Allahdad, S., Mehta, D., & Helmons, R. (2023). Mining of deep-seabed nodules using a Coand˘a-effect-based collector. *Results in Engineering, 17*(2023). https://doi.org/10.1016/j.rineng.2022.100852

Amos, A. F., Roels, O. A., Garside, C., Malone, T. C., & Paul, A. Z. (1977). Environmental aspects of nodule mining. In G. P. Glasby (Ed.), *Marine manganese deposits* (pp. 391–438). Elsevier Scientific Publication.

Arrhenius G. 1963—Pelagic sediments. In: Hill, M.N. (Ed.) The sea, Vol. 3, Interscience, pp. 655–727.

Atmanand, M. A., & Ramadass, G. A. (2017). Concepts of deep-sea mining technologies. In R. Sharma (Ed.), *Deep-sea mining: Resource potential, technical and environmental considerations* (pp. 305–344). Springer International Publishing AG.

Atmanand, M. A., Janardhanan, K. C., Amuda, K., Gnanaraj, A., Rajesh, S., Muthuvel, P., Kuttikrishnan, G., & Ramadass, G. A. (this volume). Contemporary technological progress in deep sea mining. In R. Sharma (Ed.), *Deep-sea mining and the water column*. Springer International Publishing AG.

Billet, D. S. M., Jones, D. O. B., & Weaver, P. P. E. (2019). Improving environmental practices in deep-sea mining. In R. Sharma (Ed.), *Environmental issues of deep-sea mining – Impacts, consequences and policy perspectives* (pp. 403–446). Springer International Publishers AG.

Boschen-Rose, R. E., Clark, M. R., Rowden, A. A., & Gardner, J. P. A. (2022). Integrated environmental management of the ecological impacts from seafloor massive sulfide mining – Perspectives from the Kermadec Volcanic Arc, New Zealand. In R. Sharma (Ed.), *Perspectives on deep-sea mining – Sustainability, technology, environmental policy and management* (pp. 373–422). Springer International Publishers AG.

Calvo, G., Mudd, G., Valero, A., & Valero, A. (2016). Decreasing ore grades in global metallic mining: A theoretical issue or a global reality? *Resources, 5*(4), 36.

Cherkashov, G. (2017). Seafloor massive sulfides: Distribution and prospecting. In R. Sharma (Ed.), *Deep-sea mining: Resource potential, technical and environmental considerations* (pp. 143–164). Springer International Publishing AG.

Christiansen, B., Denda, A., & Chrisitansen, S. (2020). Potential effects of deep-sea mining on pelagic and benthopelagic biota. *Marine Policy, 114*, 103442.

Chung, J. S., Schriever, G., Sharma, R., & Yamazaki, T. (2001). Deep seabed mining environment: Engineering and environment assessment. In *Proceedings of ISOPE – Ocean mining symposium* (pp. 8–14). International Society for Offshore and Polar Engineers.

Clark, M. (2019). The development of environmental impact assessments for deep-sea mining. In R. Sharma (Ed.), *Environmental issues of deep-sea mining – Impacts, consequences and policy perspective* (pp. 447–470). Springer International Publishers AG.

Clark, M. R., Johnson, R., & Hyman, J. (2022). Adaptive management as a tool for effective environmental management of deep-sea mining. In R. Sharma (Ed.), *Perspectives on deep-sea mining – Sustainability, technology, environmental policy and management* (pp. 339–372). Springer International Publishers AG.

Cormier, R. (2019). Ecosystem approach for the management of deep-sea mining. In R. Sharma (Ed.), *Environmental issues of deep-sea mining – Impacts, consequences and policy perspectives* (pp. 381–402). Springer International Publishers AG.

Cronan, D. S. (1980). *Underwater minerals* (p. 362). Academic.

Cronan, D. S. (2022). Deep-sea mining: Historical perspectives. In R. Sharma (Ed.), *Perspectives on deep-sea mining – Sustainability, technology, environmental policy and management.* Springer International Publishers AG. (this book).

Cronan, D. S., & Moorby, S. A. (1981). Manganese nodules and other ferromanganese oxide deposits from the Indian Ocean. *Journal of the Geological Society, 138*, 527–539.

Cuyvers, L., Berry, W., Gjerde, K., Thiele, T., & Wilhem, C. (2018). *Deep seabed mining: A rising environmental challenge* (x + 74pp). IUCN and Gallifrey Foundation.

Danovaro, R., Fanelli, E., Aguzzi, J., Billett, D., Carugati, L., Corinaldesi, C., Dell'Anno, A., Gjerde, K., Jamieson, A. J., Kark, S., McClain, C., Levin, L., Levin, N., Ramirez-Llodra, E., Ruhl, H., Smith, C. R., Snelgrove, P. V. R., Thomsen, L., Van Dover, C. L., & Yasuhara, M. (2020). Ecological variables for developing a global deep-ocean monitoring and conservation strategy. *Nature Ecology and Evolution, 4*, 181–192. https://doi.org/10.1038/s41559-019-1091-z

De Bruyne, K., Stoffers, H., Flamen, S., De Beuf, H., Taymans, C., Smith, S., & Van Nijen, K. (2022). A precautionary approach to developing nodule collector technology. In R. Sharma (Ed.), *Perspectives on deep-sea mining – Sustainability, technology, environmental policy and management* (pp. 137–165). Springer International Publishers AG.

Doorn, E. V., Laugesen, J., Haeckel, M., Mestre, N., Skjeret, F., & Vink, A. (2022). Risk assessment for deep-seabed mining. In R. Sharma (Ed.), *Perspectives on deep-sea mining – Sustainability, technology, environmental policy and management* (pp. 497–526). Springer International Publishers AG.

Drazen, J., Smith, C. R., Gjerde, K. M., Haddock, S. H. D., Cartera, G. S., Choye, C. A., Clark, M. R., Dutrieux, P., Goetze, E., Hauton, C., Hatta, M., Koslow, J. A., Leitner, A. B., Pacini, A., Perelman, J. N., Peacock, T., Sutton, T. T., Watling, L., & Yamamoto, H. (2020). Midwater ecosystems must be considered when evaluating environmental risks of deep-sea mining. *Proceedings of National Academy of Science, 117*(30), 17455–17460. www.pnas.org/cgi/doi/10.1073/pnas.2011914117

Durden, J. M., Lallier, L. E., Murphy, K., Jaeckel, A., Gjerde, K., & Jones, D. O. B. (2018). Environmental impact assessment process for deep-sea mining in the 'area'. *Marine Policy, 87*, 194–202.

Foell, E. J., Thiel, H., & Schriever, G. (1990). *DISCOL: A longterm largescale disturbance – Recolonisation experiment in the abyssal eastern tropical Pacific Ocean.* Proceedings of offshore technology conference, Houston, USA, Paper No. 6328, pp. 497–503.

Fouquet, Y., & Lacroix, D. (2014). *Deep-sea marine mineral resources*. Springer.

Frazer, J. Z., & Fisk, M. B. (1981). Geological factors related to characteristics of seafloor manganese nodule deposits. *Deep-Sea Research, 28A*, 1533–1551.

Frazer, J. Z., & Wilson, L. L. (1980). Nodule resources in the Indian Ocean. *Marine Mining, 2*, 257–291.

Fukushima, T. (1995). *Overview Japan deep-sea impact experiment = JET*. Proceedings of 1st ISOPE ocean mining symposium, Tsukuba, Japan, ISOPE, pp. 47–53.

Fukushima, T., Tsune, A., & Sugishima, H. (2022). Comprehensive understanding of seafloor disturbance and environmental impact scenarios. In R. Sharma (Ed.), *Perspectives on deep-sea mining – Sustainability, technology, environmental policy and management* (pp. 313–338). Springer International Publishers AG.

Gazis, I-Z (2022). *Deep-sea mining plume monitoring: Results from the first deep-sea trial of the Patania-II pre-prototype nodule collector and parallel plume monitoring with hydroacoustic and optic methods*. Proceedings of Underwater Mining Conference, St. Petersberg (USA), 2–7 October 2022, 21.

Glasby, G. P. (1972). Geochemistry of manganese nodules from the Northwest Indian Ocean. In D. R. Horn (Ed.), *Ferromanganese deposits on the ocean floor* (pp. 93–104). National Science Foundation.

Glasby, G. P. (1982). Manganese nodules from the South Pacific: An evaluation. *Marine Mining, 3*, 231–270.

Glasby, G. P., Stoffers, P., Sioulas, A., Thijssen, T., & Friedrich, G. (1982). Manganese nodules formation in the Pacific Ocean: A general theory. *Geo-Marine Letters, 2*, 47–53.

Halbach, P., Sattler, C. D., Teichmann, F., & Wahsner, M. (1989). Cobalt rich and platinum bearing manganese crust deposits on seamounts: Nature, formation and metal potential. *Marine Mining, 8*, 23.

Halbach, P. E., Andreas, J., & Cherkashov, G. (2017). Marine ferromanganese crust deposits: Description and formation, occurrence and distribution, estimated world-wide resources. In R. Sharma (Ed.), *Deep-sea mining; resource potential, technical and environmental considerations* (pp. 65–142). Springer International Publishing AG.

Haughton, C., Brown, A., Thatje, S., Mestre, N. C., Bebianno, M. J., Martins, I., Bettencourt, R., Canals, M., Sanchez-Vidal, A., Shillito, B., Ravaux, J., Zbinden, M., Duperron, S., Mevenkamp, L., Vanreusel, A., Gambi, C., Dell'Anno, A., Danovaro, R., Gunn, V., & Weaver, P. (2017). Identifying toxic impacts of metals potentially released during deep-sea mining – A synthesis of the challenges to quantifying risk. *Frontiers in Marine Science*. https://doi.org/10.3389/fmars.2017.00368

Hein, J. R., & Koschinsky, A. (2014). Deep ocean crusts and nodules. *Treatise Geochem, 13*, 273–291.

Hein, J. R., Yeh, H.-W., & Alexander, E. (1979). Origin of iron-rich montmorillonite from the manganese nodule belt of the north equatorial Pacific. *Clays and Clay Minerals, 27*, 185–194.

Hein, J. R., Kochinsky, A., Halbach, P., Manheim, F. T., Bau, M., Kang, J.-K., & Lubick, N. (1997). Iron and manganese oxide mineralisation in the Pacific. In K. Nicholon, J. R. Hein, B. Buhn, & S. Dasgupta (Eds.), *Manganese mineralisation: Geochemistry and mineralogy of terrestrial and marine deposits* (Geological Society special publication no. 119) (p. 123).

Hein, J. R., McIntyre, B. R., & Piper, D. Z. (2005). *Marine mineral resources of Pacific Islands – A review of the exclusive economic zones of islands of US affiliation, excluding the state of Hawaii*. US Geological Survey. http://pubs.usgs.gov/circ/2005/1286/

Hein, J. R., Mizell, K., Koschinsky, A., & Conrad, T. A. (2013). Deep-ocean mineral deposits as a source of critical metals for high- and green-technology applications: Comparison with land-based resources. *Ore Geology Reviews, 51*, 1–14.

Hein, J. R., et al. (2015). Critical metals in manganese nodules from The Cook Islands EEZ, abundances and distributions. *Ore Geology Review, 68*, 97–116.

Hein, J. R., Koschinsky, A., & Kuhn, T. (2020). Deep-ocean polymetallic nodules as a resource for critical materials. *Nature Reviews – Earth and Environment, 1*, 158–169.

Hitchin, B., Smith, S., Krögerc, K., Jones, D. O. B., Jaeckel, A., Mestre, N. C., Ardron, J., Escobar, E., van der Grient, J., & Amaro, T. (2023). Thresholds in deep-seabed mining: A primer for their development. *Marine Policy, 149*, 105505. https://doi.org/10.1016/j.marpol.2023.105505

Hong, S., Kim, H.-W., Yeu, T., Choi, J.-S., Lee, T. H., Bae Dae, S., & Lee, J.-G. (2019). Technologies for safe and sustainable mining of deep-seabed minerals. In R. Sharma (Ed.), *Environmental issues of deep-sea mining – Impacts, consequences and policy perspectives* (pp. 95–143). Springer International Publishers AG.

ISA. (1999). *Deep seabed polymetallic nodule exploration: Development of environmental guidelines*. Proceedings of ISA Workshop, Sanya, China, 1–5 June 1998, The International Seabed Authority, Pub No ISA/99/02, pp. 222–223.
ISA. (2001). *Recommendations for guidance of contractors for the assessment of the possible environmental impacts arising from exploration for polymetallic nodules in the Area*. International Seabed Authority, Jamaica. ISBA/7/LTC/1 2001; pp. 11.
ISA. (2005). *Recommendations of the workshop on polymetallic sulphides and cobalt crusts: Their environment and considerations for the establishment of environmental baselines and an associated monitoring programme for exploration*. International Seabed Authority, Jamaica. ISBA/11/LTC/2 2005; pp. 26.
ISA. (2008a). *Executive summary of the International Seabed Authority's workshop on polymetallic nodule mining technology: Current status and challenges ahead*. International Seabed Authority: Chennai, India, February, 2008, pp. 20.
ISA. (2008b). *Rationale and recommendations for the establishment of preservation reference areas for nodule mining in the Clarion-Clipperton Zone*. International Seabed Authority, Jamaica, ISBA/14/LTC/2.
ISA. (2011). *Environmental management plan for the Clarion-Clipperton Zone*. International Seabed Authority, Jamaica, ISBA/17/LTC/7,
ISA. (2013). *Recommendations for the guidance of contractors for the assessment of the possible environmental impacts arising from exploration for marine minerals in the area*. International Seabed Authority, Jamaica, ISBA/19/LTC/8.
ISA. (2019a). *Recommendations for the guidance of contractors for the assessment of the possible environmental impacts arising from exploration for marine minerals in the area*. International Seabed Authority, Jamaica, ISBA/25/LTC/6.
ISA. (2019b). *Draft regulations on exploitation of mineral resources in the area*. International Seabed Authority, Jamaica, ISBA/25/C/WP.1.
ISA. (2020). Recommendations for the guidance of contractors for the assessment of the possible environmental impacts arising from exploration for marine minerals in the. *Area, ISBA/25/LTC/6/Rev*, 1.
Jaeckel, A., Ardron, J. A., & Gjerde, K. M. (2016). Sharing benefits of the common heritage of mankind – Is the deep-sea mining regime ready? *Marine Policy, 70*, 198–204.
Jones, D. O. B., Kaiser, S., Sweetman, A. K., Smith, C. R., Menot, L., Vink, A., Trueblood, D., Greinert, J., Billet, D. S. M., Arbizu, P. M., Radzeijewska, T., Singh, R., Ingole, B., Stratman, T., Simon-Lledo, E., Durden, J. M., & Clark, M. R. (2017). Biological responses to disturbance from simulated deep-sea polymetallic nodule mining. *PLoS ONE, 12*(2). https://doi.org/10.1371/journal/.pone.0171750
Kawano, S., & Furaya, H. (2022). Mining and processing of seafloor massive sulfides: Experiences and challenges. In R. Sharma (Ed.), *Perspectives on deep-sea mining – Sustainability, technology, environmental policy and management* (pp. 167–198). Springer International Publishers AG.
Koschinsky, A., Heinrich, L., Boehnke, K., Cohrs, J. C., Markus, T., Shani, M., Singh, P., Stegen, K. S., & Werner, W. (2018). Deep-sea mining: Interdisciplinary research on potential environmental, legal, economic, and societal implications. *Integrated Environmental Assessment and Management, 9999*(9999), 1–20.
Kotlinski, R. (2001). Mineral resources of the world's ocean – Their importance for global economy in the 21st century. In: *Proceedings of 4th ISOPE ocean mining symposium, Szczecin, Poland*. International Society for Offshore and Polar Engineers, pp. 1–7.
Kuhn, T., Wegorzewski, A., Ruhlman, C., & Vink, A. (2017). Composition, formation and occurrence of polymetallic nodules. In R. Sharma (Ed.), *Deep-sea mining; resource potential, technical and environmental considerations* (pp. 23–64). Springer International Publishing AG.
Lebre, E., Kung, A., Savinova, E., & Valenta.R.K. (2023). Mining on land or in the deep sea? Overlooked considerations of a reshuffling in the supply source mix. *Resources, Conservation & Recycling, 191*(2023), 106898.
Lenoble, J. P. (2000). A comparison of possible economic returns from mining deep-sea polymetallic nodules, polymetallic massive sulphides and cobalt-rich ferromanganese crusts. In *Proceedings of workshop on mineral resources* (pp. 1–22). International Seabed Authority.

Levin, L. A., Mengerink, K., Gjerde, K. M., Rowden, A. A., Van Dover, C. L., Clark, M. R., Ramirez-Llodra, E., Currie, B., Smith, C. R., Sato, K. N., et al. (2016). Defining "serious harm" to the marine environment in the context of deep-seabed mining. *Marine Policy, 74*, 245–259.

Lodge, M. W., & Bourrel-McKinnon, M. (2022). Sharing financial benefits from deep seabed mining: The case for a seabed sustainability fund. In R. Sharma (Ed.), *Perspectives on deep-sea mining – Sustainability, technology, environmental policy and management* (pp. 559–578). Springer International Publishers AG.

Lodge, M., & Verlaan, P. A. (2018). Deep-sea mining: International regulatory challenges and responses. *Elements, 14*, 331–336. https://doi.org/10.2138/gselements.14.5.331

Lodge, M., Segerson, K., & Squires, D. (2019). Environmental policy for deep-sea mining. In R. Sharma (Ed.), *Environmental issues of deep-sea mining – Impacts, consequences and policy perspectives* (pp. 347–380). Springer International Publishers AG.

Madureira, P., Squires, D., & Ribeiro, L. P. (2023). The international seabed authority and the United Nations 2030 agenda for sustainable development. *Resources Policy, 86*, 104166. https://doi.org/10.1016/j.resourpol.2023.104166

Martin-Barajas, A., Lallier-Verges, E., & Lecraire, L. (1991). Characteristics of manganese nodules from the Central Indian Basin: Relationship with sedimentary environment. *Marine Geology, 101*, 249–265.

Mero, J. L. (1965). *The mineral resources of the sea* (312 pp). Elsevier.

Miller, K. A., Thompson, K. F., Johnston, P., & Santillo, D. (2018). An overview of seabed mining including the current state of development, environmental impacts, and knowledge gaps. *Frontiers in Marine Science, 4.*, Article 418. https://doi.org/10.3389/fmars.2017.000418

Muñoz-Royo, C., Peacock, T., Alford, M. H., Smith, J. A., Le Boyer, A., Kulkarni, C. S., Lermusiaux, P. F. J., Haley Jr, P. J., Mirabito, C., Dayang Wang, E., Adams, E., Ouillon, R., Breugem, A., Decrop, B., Lanckriet, T., Supekar, R. B., Rzeznik, A. J., Gartman, A., & Ju, S.-J. (2021). Extent of impact of deep-sea nodule mining midwater plumes is influenced by sediment loading, turbulence and thresholds. *Earth and Environment, 2*. https://doi.org/10.1038/s43247-021-00213-8

Muñoz-Royo, C., Ouillon, R., El Mousadik, S., Alford, M. H., & Peacock, T. (2022). An in situ study of abyssal turbidity-current sediment plumes generated by a deep seabed polymetallic nodule mining preprototype collector vehicle. *Science Advances, 8*, eabn1219. https://doi.org/10.1126/sciadv.abn1219

Niner, H. J., Jeff, A., Ardron, E. G., Gianni, M., Jaeckel, A., Jones, D. O. B., Levin, L. A., Smith, C. R., Thiele, T., Turner, P. J., Van Dover, C. L., Watling, L., & Gjerde, K. M. (2018). Deep-sea mining with no net loss of biodiversity—An impossible aim. *Frontiers in Marine Science, 5.*, Article 53. https://doi.org/10.3389/fmars.2018.00053

Ozturgut. E., Lavelle, J. W., Steffin, O., Swift, S. A. (1980). *Environmental investigation during manganese nodule mining tests in the North Equatorial Pacific, in November 1978*. NOAA Tech.Memorandum ERL MESA-48, National Oceanic and Atmospheric Administration, USA, 50.

Pearson, J. S. (1975). *Ocean floor mining* (p. 201). Noyes Data Corporation.

Petersen, S., Kratschell, A., Augustin, N., Jamieson, J., Hein, J. R., & Hannington, M. D. (2018). News from the seabed—Geological characteristics and resource potential of deep-sea mineral resources. *Marine Policy, 70*, 175–187.

Peukert, A., Schoening, T., Alevizos, E., Koser, K., Kwasnitschka, T., & Greinert, J. (2018). Understanding Mn nodule distribution and evaluation of related deep-sea mining impacts using AUV – Based hydroacoustic and optical data. *Biogeosciences, 15*, 2525–2549. https://doi.org/10.5194/bg-15-2525-2018

Plueger, W. L., Herzig, P. M., Becker, K.-P., Deissmann, G., Schops, D., Lange, J., Jenisch, A., Ladage, S., Richnow, H. H., Schultz, T., & Michaelis, W. (1990). Dicovery of the hydrothermal fields at the Central Indian Ridge. *Marine Mining, 9*, 73.

Radzeijewska, T., Mianowicz, K., & Abramowski, T. (2022). Natural variability versus anthropogenic impacts on deep-sea ecosystems of importance for deeps-sea mining. In R. Sharma

(Ed.), *Perspectives on deep-sea mining – Sustainability, technology, environmental policy and management* (pp. 281–312). Springer International Publishers AG.

Rao, V. P., & Nath, B. N. (1988). Nature, distribution and origin of clay minerals in grain size fractions of sediments from manganese nodule field, Central Indian Ocean Basin. *Indian Journal Marine Sciences, 17*, 202–207.

Rolinski, S., Segschneider, J., & Sundermann J. (2001). Long-term propagation of tailings from deep-sea mining under variable conditions by means of numerical simulations. *Deep-Sea Research II, 48, 3469–3485.* https://doi.org/10.1016/S0967-0645(01)00053-4

Rona, P. A. (1988). Hydrothermal mineralisation at oceanic ridges. *Canadian Mineralogist, 26*, 431.

Sharma, R. (Ed.). (2001). Indian deep-sea environment experiment (INDEX): A study for environmental impact of deep seabed mining in Central Indian Ocean. *Deep-Sea Research Part II, 48*, 3295–3426.

Sharma, R. (Ed.). (2005). Indian Deep-sea Environment Experiment (INDEX): Monitoring the restoration of marine environment after artificial disturbance to simulate deep-sea mining in Central Indian Basin. *Marine Georesources & Geotechnology, 23*, 253–427.

Sharma, R. (2011). Deep-sea mining: Economic, technical, technological and environmental considerations for sustainable mining. *Marine Technology Society Journal, 45*, 28–41.

Sharma, R. (2015). Environmental issues of deep-sea mining. *Procedia-Earth and Planetary Science, 11*, 204–211.

Sharma, R. (2017a). Deep-sea mining: Current status and future considerations. In R. Sharma (Ed.), *Deep-sea mining: Resource potential, technical and environmental considerations* (pp. 3–22). Springer International Publishing AG.

Sharma, R. (2017b). Assessment of distribution characteristics of polymetallic nodules and their implications on deep-sea mining. In R. Sharma (Ed.), *Deep-sea mining: Resource potential, technical and environmental considerations* (pp. 229–256). Springer International Publishing AG.

Sharma, R. (2017c). Development of environmental management plan for deep-sea mining. In R. Sharma (Ed.), *Deep-sea mining: Resource potential, technical and environmental considerations* (pp. 483–506). Springer International Publishing AG.

Sharma, R. (2019). Environmental factors for design and operation of deep-sea mining system—Based on case studies. In R. Sharma (Ed.), *Environmental issues of deep-sea mining – Impacts, consequences and policy perspectives* (pp. 315–344). Springer International Publishers AG.

Sharma, R. (2022). Approach towards deep-sea mining—Current status and future prospects. In R. Sharma (Ed.), *Perspectives on deep-sea mining – Sustainability, technology, environmental policy and management* (pp. 13–51). Springer International Publishers AG.

Sharma, R., & Nath, B. N. (Eds.). (2000). Indian BIE : Indian deep-sea environment experiment (INDEX). *Marine Georesources & Geotechnology, 18*, 177–294.

Sharma, R., & Smith, S. (2019). Deep-sea mining and environment – An introduction. In R. Sharma (Ed.), *Environmental issues of deep-sea mining – Impacts, consequences and policy perspectives* (pp. 3–22). Springer International Publishers AG.

Sharma, R., Nath, B. N., Parthiban, G., & Sankar, S. J. (2001). Sediment redistribution during simulated benthic disturbance and its implications on deep seabed mining. *Deep-Sea Research II, 48*, 3363–3380.

Sharma, R. et al. (2007). *Do natural changes mask artificial impacts on benthic ecosystem over a period of time?* Proceedings of Underwater Mining Institute, Tokyo, Japan, 15–20 October 2007.

Sharma, R., Mustafina, F., & Cherkashov, C. (2019). Review of mining rates, environmental impacts, metal values and investments for polymetallic nodules mining. In R. Sharma (Ed.), *Environmental issues of deep-sea mining – Impacts, consequences and policy perspectives* (pp. 519–546). Springer International Publishers AG.

Shirayama, Y. (1999). *Biological results of JET Project: An Overview.* Proceedings of 3rd ISOPE ocean mining symposium, Goa, India, ISOPE, pp. 185–190.

Siddiquie, H. N., Das Gupta, D. R., Sen Gupta, N. R., Shrivastava, P. C., & Mallik, T. K. (1978). Manganese-iron nodules from the Indian Ocean. *Indian Journal of Marine Science, 7*, 239–253.
Sonter, L., Herrera, D., Barrett, D., Galford, G., Moran, C., & Soares-Filho, B. (2017). Mining drives extensive deforestation in the Brazilian Amazon. *Nature Communications, 8*(1), 1013.
SPC. (2013). Deep-sea minerals and the green economy. In: E. Baker, Y Beaudoin (Eds.), *Secretariat of the Pacific Community*, Vol. 2.
Swaddling, A., Clark, M. R., Bourrel, M., Lily, H., Lamarche, G., Hickey, C., Rouse, H., Nodder, S., Rickard, G., Sutton, P., & Wysoczanski, R. (2016). *Pacific-ACP states regional scientific research guidelines for deep sea minerals*. Noumea.
Takaya, Y., Yasukawa, K., Kawasaki, T., Fujinaga, K., Ohta, J., Usui, Y., Nakamura, K., Kimura, J.-I., Chang, Q., Hamada, M., Dodbiba, G., Nozaki, T., Iijima, K., Morisawa, T., Kuwahara, T., Ishida, Y., Ichimura, T., Kitazume, M., Fujita, T., & Kato, Y. (2018). The tremendous potential of deep-sea mud as a source of rare-earth elements. *Scientific Reports, 8*, 5763. https://doi.org/10.1038/s41598-018-23948-5
Theil, H. (Ed.). (2001). Environmental impact study for mining of polymetallic nodules from the deep sea. *Deep-Sea Research II, 48*, 3427–3882.
Thiele, T. (2019). Deeps-sea natural capital: Putting deep-sea ecosystem activities into an environmental context. In R. Sharma (Ed.), *Environmental issues of deep-sea mining – Impacts, consequences and policy perspectives* (pp. 507–518). Springer International Publishers AG.
Thijssen, T., Glasby, G. P., Schmitz, W. A., Friedrich, G., Kunzendorf, H., Muller, D., & Richter, H. (1981). Reconnaissance survey of manganese nodules from the northern sector of the Peru Basin. *Marine Mining, 2*, 385–428.
Thornborough, K. J., Juniper, S. K., Smith, S., & Wong, L.-W. (2019). Towards an ecosystem approach to environmental impact assessment for deep-sea mining. In R. Sharma (Ed.), *Environmental issues of deep-sea mining – Impacts, consequences and policy perspectives* (pp. 147–190). Springer International Publishers AG.
Tilot, V. (2019). Assessment of deep-sea faunal communities – Indicators of environmental impact. In R. Sharma (Ed.), *Deep-sea mining and environment – Impacts, consequences and management* (pp. 147–190). Springer International Publishers AG.
Tilot, V. B. Guilloux, K. Willaert, C.Y. Mulalap, T. Bambridge, F. Gaulme, E. Kacenelenbogen, A. J. de Grissac, J. M. Navas, A. Dahl, 2022 Traditional and socio-ecological dimensions of seabed resource management and applicable legal frameworks in the Pacific Island States. In: Sharma, R. (Ed.) Perspectives on deep-sea mining – Sustainability, technology, environmental policy and management, Springer International Publishers AG, 613–660.
Tkatchenko, G., Radziejewska, T., Stoyanova, V., Modlitba, I., & Parizek, A. (1996). *Benthic impact experiment in the IOM pioneer area: Testing for effects of deep-sea disturbance*. International seminar on deep sea-bed mining tech, China Ocean Mineral Resources R&D Assoc., Beijing, China, C55–C68.
Trueblood, D. D. (1993). *US cruise report for BIE— II Cruise*. NOAA technical memo OCRS 4, National Oceanic and Atmospheric Administration, Colorado, USA, pp. 51.
UNCLOS. (1982). *UN Convention on Law of the Sea*. https://www.un.org/depts/los/convention_agreements/texts/unclos/UNCLOS-TOC.htm
UNOET. (1987). *Delineation of mine sites and potential in different sea areas* (27 pp). UN Ocean Economics and Technology Branch and Graham & Trotman Limited.
Usui, A., & Moritani, T. (1992). Manganese nodule deposits in the Central Pacific Basin: Distribution, geochemistry and genesis. In B. H. Keating & B. R. Bolton (Eds.), *Geology and offshore mineral resources of the Central Pacific Basin, Earth Science series* (Vol. 14, pp. 205–223). Springer.
Van Dover, C. L., Ardron, J. A., Escobar, E., Gianni, M., Gjerde, K. M., Jeckel, A., Jones, D. O. B., Levin, L. A., Niner, H. J., Pendleton, L., Smith, C. R., Thiele, T., Turner, P. J., Watling, L., & Weaver, P. P. E. (2017). Biodiversity loss from deep-sea mining. *Nature Geoscience, Printed online, 26 June 2017.*
Van Nijen, K., Passel, S. V., & Squires, D. (2018). A stochastic techno-economic assessment of seabed mining of polymetallic nodules in the Clarion Clipperton Fracture Zone. *Marine Policy, 95*, 132–141.

Van Zyl, D., Sassoon, M., Digby, C., Fleury, A. M., & Kyeyune, S. (2002). *Mining for theFuture. Appendix A: Large volume waste working paper* (No. 31). International Institute for Environment and Development and World Business Council for Sustainable Development. https://www.iied.org/sites/default/files/pdfs/migrate/G00883.pdf

Vanclay, F., Esteves, A. M., Aucamp, I., & Franks, D. M. (2015). *Social impact assessment: Guidance for assessing and managing the social impacts of projects*. International Associaiton of Impact Assessment.

Washburn, T. W., Turner, P. J., Durden, J. M., Jones, D. O. B., Weaver, P., & Van Dover, C. L. (2019). Ecological risk assessment of deep-sea mining. *Ocean and Coastal Management, 176*, 24–39.

Washburn, T., Iguchi, A., Yamaoka, K., Nagao, M., Onishi, Y., Fukuhara, T., Yamamoto, Y., & Suzuki, A. (2023). Impacts of the first deep-sea seafloor massive sulfide mining excavation tests on benthic communities. *Marine Ecology Progress Series, 712*, 1–19. https://doi.org/10.3354/meps14287

Weaver, P. P. E., & Billet, D. S. M. (2019). Environmental impacts of nodule, crust and sulphide mining – An overview. In R. Sharma (Ed.), *Deep-sea mining and environment – Impacts, consequences and management* (pp. 27–62). Springer International Publishers AG.

Yamazaki, T., & Sharma, R. (2001). Estimation of sediment properties during benthic impact experiments. *Marine Georesources and Geotechnology, 19*, 269–289.

Yamazaki, T., Nakatani, N., Arai, R.. (2023). *An economic feasibility analysis for combined mining of cobalt-rich crusts and phosphorous ores in North-west Pacific*. IEEE Underwater Technology (UT23), 6–9 March 2023, Tokyo, Japan.

Websites Accessed

en.wikipedi.org/wiki/HMSChallenger. Information on HMS Challenger expedition. Accessed 29 May 2021.

https://chinadialogueocean.net/10891-china-deep-sea-exploration-comra/. Accessed 24 May 2021.

https://www.deme-gsr.com/news/article/news-deep-seabed-miningexploratory-mission-begins-today. Accessed 6 Apr 2021.

http://dsmobserver.org/2017/07/nautilus-png-submerged-trials. Accessed 19 July 2017.

https://www.immsoc.org/environmental-management-code/Code-for-environmental-management-of-marine-mining. Accessed 24 Mar 2023.

https://www.imo.org/en/publications/Pages/CurrentPublications.aspx. Accessed 24 Mar 2023.

http://www.impossiblemetals.com/eureka-impossible-metals-reveals-successful-proof-of-concept-for-sustainable-seabed-mining-of-critical-minerals. Accessed 7 Jan 2023.

https://www.isa.org.jm. Accessed 1 Dec 2023.

https://www.jogmec.go.jp – NEWS RELEASE – JOGMEC Conducts World's First Successful Excavation of Cobalt-Rich Seabed in the Deep Ocean – August 21, 2020.

http://www.mineralsindepth.org/Minerals in Depth – The Story of Seabed Minerals. Accessed 1 May 2022.

https://www.odysseymarine.com/projects. Accessed 7 Jan 2023.

https://single_market_economic.ec.europa.eu/sectors/rawmaterials. Accessed 24 Mar 2023.

https://www.reuters.com/markets/commodities/norway-finds-substantial-mineral-resources-its-seabed. Accessed 27 Jan 2023.

https://www.tmc.com/TMC and Allseas Achieve Historic Milestone: Nodules Collected from the Seafloor and Lifted to the Production Vessel – Oct 12, 2022. Accessed 7 Jan 2023.

https://www.sbma.gov.ck/exploration. Accessed 3 Aug 2022.

https://www.usgs.gov/news/national-news-release/us-geological-survey-releases-2022-list-critical-minerals. Accessed 8 Apr 2023.

Dr. Rahul Sharma retired as Chief Scientist from the CSIR-National Institute of Oceanography in Goa, India with a career spanning more than 40 years in the field of deep-sea mineral exploration and exploitation. He has led a multi-disciplinary group on 'environmental studies for marine mining'. He has a master's degree in geology and a doctorate in marine science. His professional interests include application of exploration and environmental data to deep-sea mining.

He has edited three special issues of journals, published 38 scientific papers, authored 30 articles and 41 technical reports and presented more than 60 papers at national and international conferences. He has also edited three books, viz. *Deep-Sea Mining* (2017), *Environmental Issues of Deep-Sea Mining* (2019) and *Perspectives on Deep-Sea Mining* (2022), all published by Springer International Publishers, containing 18–23 chapters contributed by experts from around the world.

His international assignments include: visiting scientist to Japan, visiting professor to Saudi Arabia, member of the UNIDO mission 'to assess the status of Deep-sea mining technologies' in Europe, the USA and Japan and invited speaker and consultant for the International Seabed Authority, Jamaica. He has contributed to the 'World Ocean Assessment Report I' of the United Nations and has also been invited to contribute a chapter on 'Potential Impacts of Deep-Sea Mining on Marine Ecosystem' for the Oxford Encyclopedia for environmental science. In addition to his research career, he has been involved with several activities relating to science communication and outreach, as well as training programmes for international participants, professionals and students.

Currently, he is a freelance consultant, advising international agencies on various aspects of deep-sea mining.

Chapter 2
Interdisciplinary Approach to Deep-Sea Mining—With an Emphasis on the Water Column

Steinar Løve Ellefmo, Murat V. Ardelan, Siri Granum Carson, Rudy Helmons, and Svein Sævik

Abstract Deep sea mining refers to the mining of valuable mineral resources from the deep ocean floor. Given the complex and fragile nature of deep-sea ecosystems, adopting an interdisciplinary and holistic approach is crucial to ensure the sustainable and responsible development of deep-sea mining (DSM) operations. This includes work related to the assessment of potential environmental impacts where physical, chemical, and biological characteristics of the target area are studied along with potential short-term and long-term effects on the surrounding ecosystems. These effects will be mining system dependent. Stakeholder engagement is essential. There are however knowledge gaps related to the deep-sea ecosystems and their interconnectedness, biodiversity, ecosystem dynamics and both the potential impacts from a single operation and cumulative impacts of the mining activities, as well as the mining systems themselves and the characteristics of the deposits. Collaboration between marine biologists, oceanographers, geologists, engineers and other relevant disciplines is essential to gain comprehensive insights. Closing these gaps would enable the development and implementation of a robust regulatory framework at both national and international levels to govern potential deep-sea mining operations. Monitoring and enforcement mechanisms must also be put in place to ensure compliance with the not-yet-developed set of standards. Multiscale adaptive management approaches where different temporal- and spatial scales are taken into consideration and where scientific knowledge, stakeholder engagement, robust regulations, and responsible practices are integrated, are the prerequisite for future responsible extraction of mineral resources from the ocean floor. This chapter gives an overview of topics relevant and needed for a proper multiscale marine

S. L. Ellefmo (✉) · M. V. Ardelan · S. G. Carson · S. Sævik
Norwegian University of Science and Technology, Trondheim, Norway
e-mail: steinar.ellefmo@ntnu.no

R. Helmons
Norwegian University of Science and Technology, Trondheim, Norway

Delft University of Technology, Delft, The Netherlands

R. Sharma (ed.), *Deep-Sea Mining and the Water Column*,
https://doi.org/10.1007/978-3-031-59060-3_2

mineral management. Its focus on the water column is restricted to vertical transportation and the impact of plume resettlement on biogeochemical processes.

Keywords Adaptive management · Responsible mining · Holistic marine mineral resource management · Vertical transportation · Biogeochemical processes

1 Introduction

1.1 Setting the Scene

Recently, large potentials of deep-sea minerals, such as three-dimentional seafloor massive sulphides and the laterally extensive manganese-rich crusts, have been documented on the extended Norwegian continental shelf (OD, 2023). The Norwegian government has completed a regional environmental impact assessment (OED, 2022) and has suggested to open portions of the extended Norwegian continental shelf for "mining activities" (OED, 2023). Cook Island has recently presented its first JORC-compliant (JORC, 2012) estimate and classification of the nodule resources inside its national jurisdiction (Tay et al., 2023). In a similar situation, having signed 31 contracts for different seabed mineral deposits in the "Area,", that is, areas beyond the national jurisdictions or the international waters, the International Seabed Authority (ISA) has decided to accept submission of applications for deep-sea mining (DSM) (DW, 2023). However, this does not automatically mean that exploitation permits will be given and that mining operations can start. There will be decision gates that are tailored to ensure that the mining will be done responsibly. There are several questions that still need to be taken into consideration by the respective regulating agencies (either national or international), the contractors as well as different stakeholders. Examples of such questions are: What is responsible mining? What is sustainable mining and what does mining *for sustainable development* mean? What is and how can serious harm be defined? What is adaptive management and what are the spatial and temporal scales that adaptive management should act on? What parameters must or should be monitored to assess "serious harm"? How often and where should we monitor with what technology at what temporal- and spatial scales? Do we need deep-sea mining? Do we need minerals at all, or can recycling meet the future demand? What is really the life-cycle cost of some potential future deep-sea mining operation vs. some onshore mining operation?

Exploration activities around the world oceans have shown that the resource potential of deep-sea minerals is significant. It is argued that more research is required to close knowledge gaps associated with exploiting these deposits. Moreover, the issue of climate change is critical, and international consensus states that human activity has contributed to this change (EU, 2023; United Nations, 2023). Mining of minerals and metals from the deep sea may constitute a part of the solution along with onshore mining, more recycling, designs that enable repairing and environmental restoration and arguably a global change in the consumption pattern. One thing to consider is whether it is most responsible to allow for publicly funded scientists to make a first evaluation on whether deep-sea mining can be performed responsibly, or whether we should allow for public–private joint ventures to accelerate data collection and analysis to find necessary solutions. In this context, it is important to note that the interdisciplinary nature of deep-sea mining requires scientists across disciplines to cooperate. The impact is dependent on both the mining system and the ecosystem.

This chapter deals with parameters and approaches for assessment of environmental impact of deep-sea mining in the water column in a holistic marine mineral resource management perspective.

1.2 Scope, Aim and Rationale

This contribution aims to present the interdisciplinary activities performed at the Norwegian University of Science and Technology (NTNU) within the broad field of deep-sea mining with an emphasis on activities associated with the water column. This includes (1) vertical transportation and, specifically, vortex-induced vibrations and the simulation and the modelling thereof and (2) environmental impact from a sediment plume and the biogeochemical processes at the sediment-water interface. The part on vertical transportation is a review of Thorsen et al. (2019). This interdisciplinary approach is needed to properly elucidate the above-mentioned questions and thereby assess the future of deep-sea mining. Saying a "yes"or "no" to deep-sea mining is not a matter of technology alone, but also of the challenges associated with formal, natural and social sciences.[1] In this context, the deep-sea mining pilot programme was established at NTNU in 2015, and the vision was "*to develop new solutions for evaluation, exploration and extraction of deep-sea minerals under societal responsibility for the environment and the international heritage of mankind*".

[1] With *formal science* it is meant knowledge that deals with abstract concepts, logical reasoning and mathematical principles rather than direct observations of nature and natural processes. Unlike *natural sciences* (such as geology, physics or biology) that investigate the physical universe or *social sciences* (such as psychology or sociology) that study human behaviour and the societies they live in, formal sciences focus on theoretical frameworks and formal models that are independent of specific empirical observations but play an important role in practical applications, like mining.

1.3 Components of Deep-Sea Mining

The deep-sea mining value chain consists of a series of working processes including ore evaluation and assessment, prospecting and exploration, ore development, ore production on the deep ocean floor including thorough and extensive environmental monitoring, vertical transportation, dewatering, loading and transport, off-loading of ore at shore, mineral processing, refinement and, finally, sales and logistics of data, equipment and personnel to name a few. The value chain as it is presented here is technology focused, but equally important for a responsible sourcing of minerals are disciplines within the natural, social, and formal sciences. There is a demand for minerals and metals in society, where electrification and reduction of poverty, as well as reducing carbon footprint, are a few of the important driving forces. With the activities defined in the *deep-sea mining pilot programme*, connected with associated projects like the EU-funded projects Blue Mining and Blue Nodules and MarMine (Ludvigsen et al., 2016), supported by the Norwegian Research Council, and the NTNU-funded TripleDeep project (Ingulstad, 2023), NTNU has attempted to address challenges related to mining minerals from the deep sea by including ethical, social and environmental aspects.

The pilot programme was initially organized and established standing on six pillars to address the challenges: (1) mineral exploration, (2) resource geology, (3) ethics and environment, (4) historical and legal aspects, (5) platform development and (6) subsea systems and processes (Fig. 2.1).

Based on these pillars, the pilot programme has since its beginning, in addition to the activities presented in sections 2 and 3 of this chapter, completed projects on exploration and resource geology (Dumais & Brönner, 2020; Juliani, 2019; Juliani & Ellefmo, 2018a, b; Lim et al., 2019; Pryadunenko et al., 2022; Ryan, 2023; Ryan et al., 2023; Snook et al., 2018; Sture et al., 2019), on exploitation (Sevillano & Sangesland, 2022; Knudsen et al., 2016; Lesage et al., 2018; Lesage, 2020; Perera et al., 2022; Perera & Nilsen, 2022; Fard & Tedeschi, 2017, 2018; Eidsvik & Schjølberg, 2016a, b), on logistics and mineral processing (Kowalczuk et al., 2018, 2019; Solheim et al., 2020, 2022) and within the humanities and social sciences (Stabell & Steel, 2018; Meyer, 2018; Stabell, 2020, 2021; Nilsen, 2020).

2 Vertical Transportation

2.1 Aspects of Structural Dynamics Relevant for Hydraulic Lift Risers

There are several uncertainties and challenges related to the commercialization of deep-sea mining, where environmental issues resulting from the extraction of the targeted resources may become critical (Miller et al., 2018). The technical challenges include mineral processing, energy supply, extraction and vertical

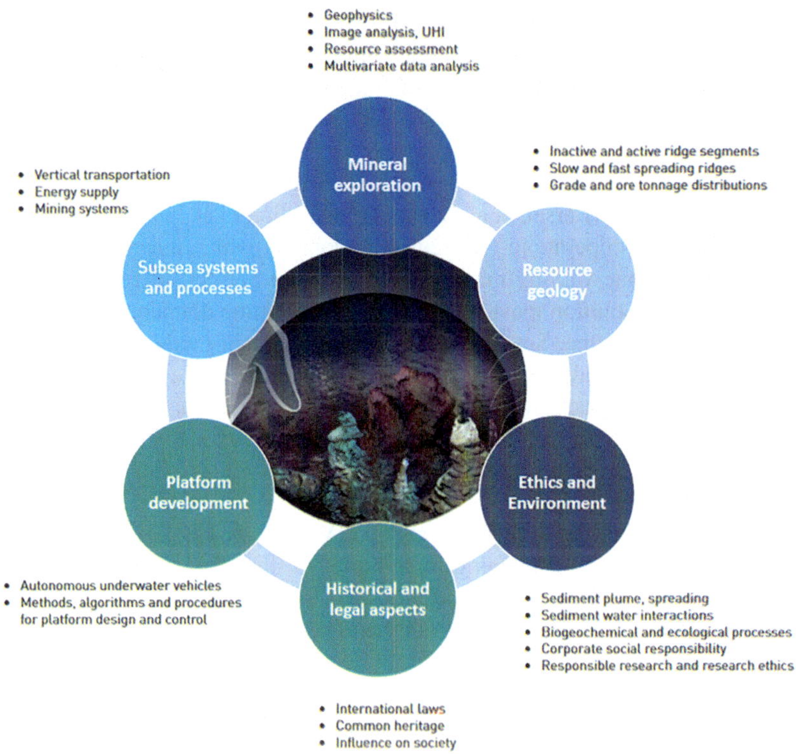

Fig. 2.1 The six original pillars of the Deep-Sea Mining Pilot programme at NTNU (Dumke et al., 2018)

transportation. With respect to the latter, suggested alternatives include mechanical bucket systems, container lifting and hydraulic lifting using slurry pumps (Laugesen et al., 2021), the latter eventually combined with airlift obtained by injecting compressed air into the liquid/solid mixture (Schulte, 2013). The hydraulic lift methods require a vertical pipe, that is, a riser, to transport the ore/water mixture. The riser will be exposed to current and wave loads that can cause fatigue failure. To avoid such failures in risers, which are extensively used in the offshore industry for transportation of oil and gas, much research has been carried out regarding their dynamic response (Patel & Seyed, 1995).

The response related to vortex-induced vibrations (VIV) is known to be associated with large uncertainties, representing a safety risk and a major design issue. VIV occurs due to the oscillating fluid forces caused by flow separation and shedding of vortices in the wake of the cylinder when the riser is exposed to current. The research on VIV has been considerable, as reflected by several reviews (Bearman, 1984; Williamson & Govardhan, 2004; Wu et al., 2012). The basic vortex shedding process for stationary cylinders is reasonably well understood (Sumer & Fredsøe, 1997); however, the complexity increases as the cylinder is allowed to oscillate, which may lead to unsteady and chaotic response that may be detrimental to the fatigue life (Modarres-Sadeghi et al., 2010, 2011). A flexible cylinder will in general oscillate with a moderate amplitude both in the crossflow (CF) and in-line (IL) directions. However, the fatigue damage may be significant due to high modes of vibration at a relatively high frequency (Baarholm et al., 2006). Prediction of fatigue damage caused by VIV is an important part of riser design where the current state of art applied by the offshore industry relies on empirical models such as SHEAR7, VIVA and VIVANA (Larsen et al., 2005; Triantafyllou et al., 1999; Vandiver & Li, 2005). These tools operate in the frequency domain, which means that non-linear effects such as time-varying mass and stiffness are not considered. To deal with these effects, the solution must be carried out in the time domain, using a step-by-step integration scheme. Several empirical time domain methods have been proposed, such as the wake-oscillator models (Kurushina et al., 2018), the artificial neural network approach (Mainçon, 2011) and the phase-coupled synchronization model (Thorsen & Sævik, 2017). A common challenge with all empirical models is to find suitable parameters for different flow regimes and real-life riser configurations.

Although most of the knowledge about VIV from the offshore industry can be directly applied to deep-sea mining, an important difference is related to the internal flow, which may include heavy particles and density variations (van Wijk et al., 2016). It is, therefore, of interest to study how such variations affect riser VIV and fatigue.

The published literature on VIV response in deep-sea mining risers is limited. Wang et al. (2018) studied vessel motion-induced VIV of a free-hanging riser, however, excluding the effects from internal flow or density variations. Verichev et al. (2011) discussed various excitation mechanisms relevant for mining risers, however, focusing on the instability that occurs when the slurry velocity exceeds the critical value. Boomsma and Blanken (2014) describe recent developments in vertical transportation models; however, no quantitative results were presented. Knudsen et al. (2016) carried out numerical simulations of riser VIV, accounting for the time-varying inertia and centrifugal forces caused by an internal flow of water and solids, demonstrating that the internal flow can change the maximum curvature of the riser. Fatigue was not part of the investigation. Thorsen and Sævik (2018) assumed a sinusoidal density variation and used a simplified analytical model to predict the wavelength that would give the largest disturbance on VIV; however, the model is not capable of quantifying the effect on fatigue life from the internal density wave.

As part of the Centre for Research–based Innovation MOVE, financially supported by the Norwegian Research Council and the consortium partners, Thorsen et al. (2019) presented a framework for non-linear time-domain dynamic analysis of deep-sea mining riser systems including both VIV and flow loads. The framework is based on a full two-way coupling between the structural and flow models, that is, the effect from both riser motion and flow dynamics can be fully accounted for in one analysis. This means that phenomena related to flow transients (e.g., start/stop of flow) and non-linear structure behaviour when exposed to combined loads from currents, waves and floater motions can be accounted for in a single analysis.

Here, a brief description of the models developed and reported in Thorsen et al. (2019) is given and some key results are commented upon with a focus on how the flow may influence the VIV response and fatigue behaviour of riser systems relevant for deep sea mining.

2.2 Numerical Models

2.2.1 Riser Dynamics and Hydrodynamic Loads

The structural analysis is based on the finite element programme RIFLEX (Fylling et al., 1995) for dynamic analysis of slender marine structures. The riser is modelled by beam elements, formulated using a co-rotated reference frame to account for large displacements and rotations in 3D space. The non-linear dynamic analysis is based on the incremental form of the dynamic equilibrium equation, which is solved by the Newmark-β method and a Newton-Raphson iteration procedure. The external hydrodynamic loads (per unit length) at any point along the riser are calculated as:

$$\begin{aligned}\mathbf{f} = C_M \rho \frac{\pi D^2}{4} \dot{\mathbf{u}}_n - (C_M - 1)\rho \frac{\pi D^2}{4} \ddot{\mathbf{x}}_n + \frac{1}{2}\rho D\ C_D |\mathbf{v}_n| \mathbf{v}_n \\ + \frac{1}{2}\rho D\ C_{v,y} |\mathbf{v}_n| (\mathbf{j}_3 \times \mathbf{v}_n) \cos\phi_{\text{exc}}\end{aligned} \quad (2.1)$$

The three first terms make up Morison's equation (Faltinsen, 1999), while the final term represents the oscillating force due to vortex shedding, as proposed by Thorsen and Sævik (2017). C_M and C_D are the inertia and drag coefficients, while $C_{v,y}$ determines the strength of the vortex shedding force. Furthermore, $\dot{\mathbf{u}}_n$ is the normal component (i.e., perpendicular to the cylinder axis) of the fluid particle acceleration, $\ddot{\mathbf{x}}_n$ is the normal component of the cylinder acceleration and $\mathbf{v}_n$ is the normal component of the relative fluid velocity. The relative flow velocity is given as $\mathbf{u}_n - \dot{\mathbf{x}}_n$, where $\boldsymbol{u}_n$ is the incoming flow velocity and $\dot{\mathbf{x}}_n$ is the velocity of the cylinder cross-section. ρ is the seawater density and D is the outer diameter of the cross-section. $\boldsymbol{j}_3$ is a unit vector pointing in the direction of the cylinder axis, which means that the cross product in Eq. (2.1) ensures that the vortex shedding force is acting perpendicular to the instantaneous relative velocity. There will also be vortex shedding forces in the flow direction, causing in-line VIV. Therefore, an in-line excitation term was later added to Eq. (2.1) by Ulveseter et al. (2018), which was further improved by Kim et al. (2021). However, in the present study, only the crossflow term was considered as described in some detail.

The oscillations of the vortex shedding force are described through the time-varying excitation force phase ϕ_{exc}, where the evolution in time is given by Eqs. (2.2) and (2.3):

$$\frac{d\phi_{\text{exc}}}{dt} = 2\pi \frac{\hat{f}_{\text{exc}} \left| \mathbf{v}_n \right|}{D} \tag{2.2}$$

$$\hat{f}_{\text{exc}} = \begin{cases} \hat{f}_0 + \left(\hat{f}_{\max} - \hat{f}_0 \right) \sin\left(\phi_{\dot{y}_{\text{rel}}} - \phi_{\text{exc}} \right), \left(\phi_{\dot{y}_{\text{rel}}} - \phi_{\text{exc}} \right) \geq 0 \\ \hat{f}_0 + \left(\hat{f}_0 - \hat{f}_{\min} \right) \sin\left(\phi_{\dot{y}_{\text{rel}}} - \phi_{\text{exc}} \right), \left(\phi_{\dot{y}_{\text{rel}}} - \phi_{\text{exc}} \right) < 0 \end{cases} \tag{2.3}$$

This makes it possible for the vortex shedding force to vary its instantaneous (dimensionless) frequency between $f_{\min}$ and $f_{\max}$, upon watching the phase difference $\phi_{\dot{y}} - \phi_{\text{exc}}$ between the structural velocity and VIV load and by that locking on to the frequency of vibration. For more details, see Thorsen and Sævik (2017).

2.2.2 Internal Flow Loads

The internal flow model adopted is a one-dimensional two-fluid model originally developed for gas-liquid two-phase flows, which is based on a finite volume Lagrangian slug tracking model concept first tested by Nydal (2012) and later extended and modified (Smith, 2017). In the model, solid particles and water phases are treated as a part of the liquid mixture.

The effect of the internal flow on the structural dynamics is presented in the following. First, the weight of the internal fluid will contribute to the effective submerged weight, w_s of the pipe:

$$w_s = w_d + w_f - \rho g A_e \tag{2.4}$$

where w_d is the dry weight of the empty pipe, w_f is the weight of fluid inside (including water and particles), and the last term is the buoyancy.

Due to the VIV response, the riser will be exposed to curvature deformation that changes over time, forcing the inside fluid/particles to follow a time-varying curve path. To illustrate this, consider the deformed beam element in Fig. 2.2.

By assuming that the pipe deformations are small in the co-rotated reference frame, the time-dependent position of an internal fluid particle can be expressed as:

$$\mathbf{r}_f(t) = Vt\mathbf{j}_3 + u\mathbf{j}_1 + v\mathbf{j}_2 \tag{2.5}$$

where V is the internal slurry velocity, while u and v are the lateral displacements of the beam in the local x- and y-directions, respectively. $\boldsymbol{j}_1$,$\boldsymbol{j}_2$ and $\boldsymbol{j}_3$ are orthogonal unit vectors as shown in Fig. 2.2, and t is some arbitrary instant of time. The velocity of a fluid particle is found as the material derivative of the position:

$$\mathbf{v}_f = \frac{D\mathbf{r}_f}{Dt} = \frac{\partial \mathbf{r}_f}{\partial t} + V\frac{\partial \mathbf{r}_f}{\partial z} = \left(\frac{\partial u}{\partial t} + V\frac{\partial u}{\partial z} \right)\mathbf{j}_1 + \left(\frac{\partial v}{\partial t} + V\frac{\partial v}{\partial z} \right)\mathbf{j}_2 + V\mathbf{j}_3 \tag{2.6}$$

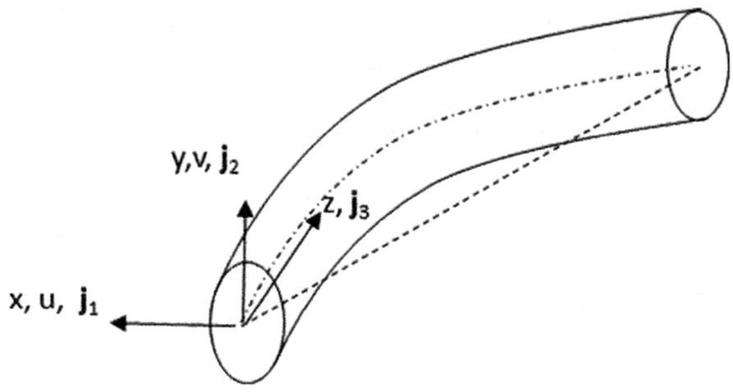

Fig. 2.2 The riser is curved during vortex-induced vibrations and the deformation changes over time forcing the inside particles to follow a time-varying curve path. This is illustrated considering a beam element in a deformed state

The fluid particle acceleration is found through similar differentiation of the velocity:

$$\mathbf{a}_f = \frac{D\mathbf{v}_f}{Dt} = \frac{\partial \mathbf{v}_f}{\partial t} + V\frac{\partial \mathbf{v}_f}{\partial z} = \left(\frac{\partial^2 u}{\partial t^2} + 2V\frac{\partial^2 u}{\partial t \partial z} + V^2\frac{\partial^2 u}{\partial z^2}\right)\mathbf{j}_1 + \left(\frac{\partial^2 v}{\partial t^2} + 2V\frac{\partial^2 v}{\partial t \partial z} + V^2\frac{\partial^2 v}{\partial z^2}\right)\mathbf{j}_2 \tag{2.7}$$

Hence, the following three force components will act on the pipe because of the internal fluid (Païdoussis & Li, 1993):

$$\mathbf{f}_{\text{inertia}} = -m_f\left(\frac{\partial^2 u}{\partial t^2}\mathbf{j}_1 + \frac{\partial^2 v}{\partial t^2}\mathbf{j}_2\right) \tag{2.8}$$

$$\mathbf{f}_{\text{Coriolis}} = -2m_f V\left(\frac{\partial^2 u}{\partial t \partial z}\mathbf{j}_1 + \frac{\partial^2 v}{\partial t \partial z}\mathbf{j}_2\right) \tag{2.9}$$

$$\mathbf{f}_{\text{centrifugal}} = -m_f V^2\left(\frac{\partial^2 u}{\partial z^2}\mathbf{j}_1 + \frac{\partial^2 v}{\partial z^2}\mathbf{j}_2\right) \tag{2.10}$$

where m_f is the internal slurry mass per unit length. The inertia term is independent of the internal flow velocity and is accounted for in the numerical model by adding the mass, m_f, to the element mass matrix, whereas the Coriolis- and centrifugal forces are added to the element load vector.

2.2.3 Effect of Internal Flow on Structural Dynamics of a Mining Riser

The numerical model described above was used to study the VIV and fatigue of an example mining riser at a water depth of 2350 m corresponding approximately to the depth at Loki's Castle, which is an active hydrothermal vent field located along the Arctic Mid-Atlantic Ridge (see Thorsen et al., 2019 for details). Compared to traditional oil and gas risers, the essential difference is the internal flow where a hydraulic lift method was assumed consisting of water and ore particles. Focus was on density variations in the internal slurry flow and the effects this may have on riser fatigue, which has previously received little attention. A case study was carried out, where the example riser was subjected to VIV caused by a steady current, along with different types of internal slurry flows. The internal flows considered were a steady state flow with a constant mixture density and time-varying flows generated by turning the inflow of particles on and off with periods ranging from 10 to 100 s. The time-varying flows were associated with density waves of different wavelengths moving up the riser. A sheared current profile was further assumed as the basis for studying displacements, stresses and fatigue along the riser.

The study confirmed that internal density waves in the conveyed slurry may influence VIV of deep-sea mining risers, in terms of both increasing and decreasing riser fatigue depending on the wavelength of density variations versus the VIV mode number excited for a given environmental condition. It is, therefore, difficult to give general guidelines other than performing numerical simulations supported by experiments on a case-to-case basis.

3 The Potential Impact of Plume Resettling on Biogeochemical Processes at Sediment–Water Interface

3.1 Background

Most of the expected environmental impact of DSM due to plume development, transport and resettling on the surface sediment in marine environments are based on conjectures extracted from limited knowledge and data. Those data and knowledge are insufficient to reduce the existing epistemological uncertainties. Besides the lack of data, heterogeneity of seabed and biota adjacent to potential DSM zones in seafloor massive sulphides (SMS), manganese nodules and cobalt crusts regions make the challenge even more prominent.

There is still a great deal of uncertainty among researchers and decision-makers in identifying essential variables needed in models to predict, to evaluate the potential severity of DSM activities and to mitigate the effects of the proposed operation on the marine environment and ecosystem (Beaulieu et al., 2017; MIDAS, 2016).

When we consider the inevitable limitation of resources for monitoring and acting in deep-sea environments, the recommended adaptive strategy for environmental management with a prioritized hierarchy for the mitigation measures (Durden et al., 2017) reveals the necessity of prioritizing also to study impacts according to the expected severity of the effects that may occur during DSM activities.

Regarding the priority, it is tempting to evaluate and discuss whether the primary focus of studying the potential impact of DSM should be on the sediment–water interface or water column. Undoubtedly, the water column will be affected directly and indirectly through disruption of benthic-pelagic coupling; therefore, some baseline parameters, such as oxygen, sulphide, turbidity, bio-essential and toxic elements and zooplankton migration (Collins et al., 2013), must be studied. In addition, coagulation/flocculation of natural particles exported from the epipelagic zone and their fluxes (i.e., part of the process of carbon export from the surface to the deep ocean and seafloor) should be studied. However, since most of the suspended particles are expected to settle rapidly to the vicinity of DSM zones, and the remaining suspended particles be carried away to adjacent regions for a more extended period near the seafloor, it is hard to believe that water with excess density due to resuspended sediment from any mining activity on the seafloor and plume from dewatering will move vertically up (Bashir et al., 2012) and create significant impact in the upper water column. The plumes usually affect the sediment-water interface and a portion of the low-lying water column just 25–50 m above the seabed where the excess water and sediment are discharged (Muñoz-Royo et al., 2022). Since sustained plumes become more dilute in the water column with increasing distance from the area where they form (Van Zyl et al., 2002), the impact of particles on the water column would be less than that on the sediment–water interface in the adjacent zones of DSM activities. The effect of DSM on the water column biogeochemistry and biota might be more pronounced due to the disruption of elemental fluxes between the sediment surface and overlying water column, which is central for benthic–pelagic coupling, and makes the sediment–water interface more critical.

Even though the plume production by resuspension of sediment from dredging and excavation and sediment–water discharge after preliminary (subsea) separation process of potential DSM activity are the major concerns for environmental impact, there are still very few research activities focusing on the effects of resettling of resuspended sediment on crucial biogeochemical processes at the sediment–water interface on the adjacent seabed zones of prospected DSM regions, probably due to logistical difficulties.

This section aims to focus on the aspects and parameters related to plume dispersion and resettling in sediment–water interface, factors affecting how far the plume will spread over both proximal and distal zones from DSM areas, to enhance the spreading zones and to identify essential variables and the knowledge gaps for understanding the potential environmental impacts on vital processes at the sediment–water interface. All the aspects and parameters listed and discussed below should be evaluated when developing models of sediment dispersion and consequent resettling, blanketing and their impact on biogeochemical processes at the sediment–water interface. This could be eventually used to decide which variables should be prioritized and introduced to the model to examine the plume dispersion and consecutive effects on the biogeochemical process at the sediment–water interface.

3.2 *Physical, Chemical Characterization of Suspended Sediment and Physical Oceanographic Data*

Size, structure, porosity and elemental composition of resuspended materials are important information for understanding the impacts and developing models of sediment dispersion and resettling and its effects on marine biogeochemistry and ecosystem. To predict the impacts, it is crucial to evaluate parameters such as the plume direction and longevity, the strength, direction of bottom current and water stability, as well as the physical character of resuspended particles (Bashir et al., 2012). Besides these, the condition of turbidity current developments such as sudden density changes in the water column and topography of the seabed are critical factors that affect dispersion of the plume. A recent study (Muñoz-Royo et al., 2022) suggests that turbidity current dynamics were overlooked in previous modelling attempts. Turbidity currents (or seafloor sediment flows) are among the volumetrically most important yet least documented sediment transport processes in oceanic systems (Paull et al., 2018). The topography of the different potential deep-sea mining regions varies greatly. Sites with sharp slopes may have contributed to the formation of turbidity currents (Birney et al., 2006). Experiments in the deep seabed polymetallic nodule zone, which is a relatively flat seabed, showed that a disturbance might create a low-lying, laterally spreading turbidity current (Muñoz-Royo et al., 2022). If DSM sites have higher elevations relative to the surrounding area, as in the case of ferromanganese crusts and hydrothermal sulphides, which are located on seamounts and ridges, the potential of triggering turbidity current because of the plume formation will be greater and the extent of dispersal of such plumes might be increased.

It is known that sediment carried by turbidity flows can create completely new habitats in the deep sea by removing existing communities, uncovering or triggering new resources and transforming aerobic communities into anaerobic ones (Bigham et al., 2020). Formation of turbidity current due to plumes of DSM activities may jeopardize the recommended buffer zone in the vicinity to facilitate recolonization after the mining operation (Birney et al., 2006). Therefore, the assessment of the potential impacts of plumes by modelling should have a risk analysis of the plume initiating a turbidity current downslope.

3.3 *Sedimentation Rate*

Most of the zones which might experience resettling of sediment from the plume produced by DSM activities are expected to be in the deep-sea region. Natural sedimentation rate of these areas varies from 0,0001 to 0,03 mm year^{-1}(Lyle, 2015), which are very low. Even relatively higher sedimentation rates in proximal active hydrothermal zones and hemipelagic surface sediment are between 0,8- and 1-mm year^{-1} (Lyle, 2015). In the case of DSM activity, sedimentation rate due to resettling of plume materials in the adjacent zone will be much higher than the background sedimentation flux. This extreme sedimentation will affect some critical biogeochemical processes such as oxygen penetration and connected redox chemistry, nutrient flux, organic and inorganic carbon fluxes and coupled activity of microbial fauna.

Decomposing of organic matter and inorganic C assimilation by heterotrophic and autotropic bacterial activity, respectively, and C sequestration in the sediment will be affected by resettling of resuspended materials on the surface sediment carried by plumes. Besides this, transporting excessive sediment may have harmful effect on sessile organisms due to its clogging effects and its toxic elemental content.

In one of the few on-site tests, a dredge was pulled over the seabed in the Clarion Clipperton Zone, showing that the sediment redeposition thickness can reach up to 9 mm directly next to the dredging tracks and 0.07 mm in about 300 m away from the dredging centre (Purkiani et al., 2021) during the test period. These sedimentation rates are almost >100–1000 times higher than the background annual sedimentation rates in deep-sea environment. These tests were however conducted with a dredge, which, in a technical sense, is not comparable to any likely mining equipment. Bed disturbance and flow conditions around the source would be different, effectively making a comparison for such near-field scale challenging. Another test has been conducted with a nodule collector system in Malaga Bight (300 meters water depth), sediment traps installed at 50- and 250 m distance from the trajectory followed by the nodule collector. Sedimentation rates of 200–800 mg/m^2/h were observed at 2.5 metre above (sea-) bed (mab), which was the lowest altitude at which the sediment traps could be installed (Haalboom et al., 2023).

Sessile organisms are part of the higher benthic trophic level in deep-sea sediment which being adapted to very low sedimentation rates would be killed by as little as a few millimetres of sediment burial (Bigham et al., 2020). In the deep-water ecosystem, semiochemicals are critical for ecological functions including defence and competition for resource; it is also possible that mining-related sediment resuspension and eventual blanketing could impact species or the function of semiochemicals and prevent future use of genetic resources (Levin et al., 2020) in deep-sea water column, especially in bathypelagic (1000–4000 m), abyssopelagic (>4000 m) zones and the deep-sea benthos.

3.4 Rate of Change in the Fluxes of Bio-Essential and Toxic Elements

Fluxes, production and consumption of dissolved and particulate organic carbon (DOC and POC), Dissolved and particulate inorganic carbon (DIC & PIC), nutrient, bio-essential and toxic elements at sediment–water interface are essential to define the ongoing biogeochemical conditions of sediment prior to any anthropogenic impact. These fluxes are expected to change simply due to sudden accumulation of resettled sediment from the plume from DSM activities.

Sulphide minerals contain potentially toxic elements such as Hg, Cd, As, Ni, etc. (Koski, 2012). To evaluate the impact of these contaminants on the biota living in the surface sediment, under the effect of resettlement of plumes, it is necessary to have information on (i) geochemical and ecological baselines (Collins et al., 2013; Koski, 2012), (ii) the natural variations in elemental concentrations in the surface environment prior to mining and (iii) transport and transformation process of the mobilized elements.

During resuspension of sediment, both redissolution and mobilization of some elements and/or removal of existing elements by adsorption on the surface of suspended oxyhydroxide particles will happen (Lyle, 2015). As bioavailability and potential toxicity are functions of the physical and chemical forms of elements (speciation), specific transportation and transformation pathways, such as adsorption on the suspended particles and ingestion of or bacterial colonization on particles, are equally important to follow.

3.5 Microbial Community Responses

Benthic microbial ecosystem plays a crucial role in carbon, nitrogen, phosphorous and other elemental cycles and ecosystem functioning. While benthic heterotopic bacteria decompose organic carbon, autotropic bacteria can convert dissolved inorganic carbon into biomass at rates that are as high as the bacterial assimilation of POC-DOC (Sweetman et al., 2018). Any changes in the flux of POC/DOC and DIC into sediment will affect bacterial activities and microbial community structure. How the deep-sea bacterial community will react to the sudden and extreme sedimentation rates and potential blanketing effects of resettling of suspended sediment is not yet clear. Changing the flux of essential elements for biological activities will affect microbial community structure and this change eventually will influence the redox condition of the sediment and in turn new biogeochemical settings will be established.

3.6 Redox Condition of Surface Sediment

Due to changes in redox condition after DSM activities resulting in plume transport and resettling of suspended sediment, the solubility and mobility of some redox-sensitive metals (such as Fe, Mn, Cr, etc.) might change. These changes will have critical effects on other particle reactive toxic metals (such as Hg and Pb) mobilities (Kohler et al., 2022) and toxicities. It is therefore necessary to monitor the oxidation-reduction conditions, and other redox condition proxies, such as organic matter, NO_3, NH_4, PO_4, Fe and Mn concentrations in surface sediment and on the sediment–water interface.

4 Multiscale Marine Mineral Resource Management Framework

Building on an established management framework (Durden et al., 2017), a future deep sea mining operation must be managed through a smart, integrated and holistic multilevel and multi-perspective mineral resource and environmental management system. Two perspectives must arguably be considered: (1) public governance and (2) business management. Both perspectives are linked to the regional environmental management plan (REMP) and the environmental management and monitoring plan (EMMP) in the sense that they are developed by the public regulatory body and put constraints on or limitations on how the mining operation can and should be executed. Modelling and prediction techniques are needed to model both passive and active biotic and abiotic substances at temporal and spatial resolutions of months to years and from hundreds of km and down to a few hundred metres, respectively. Relevant monitoring capacities for ground-truthing must also be able to work on these scales. For example, in the case of modelling of the larvae, the way the plume will affect the larvae dispersal and the interactions on the sediment–water interface should guide the temporal and spatial scale of the connectivity processes in the REMP and EMMP. There are, to the authors' knowledge, no models available that simultaneously perform adequately across all temporal and spatial scales. It is, therefore, essential that the near-field models in the future management systems are properly translated to relevant source terms in the large-scale models, based on a realistic description of the mining operation and the stressors it causes.

Given the international consensus that active hydrothermal sites should be avoided, areas beyond the active vent sites at different ridge settings must also be handled in a future system. A model must be established that quantifies how and to what extent a plume affects the dispersal of bathyal species. Monitoring activities must contribute to integrating chemical and textural (grain size and shape) modules to the plume. Further, it must model and simulate the plume at scales less than 200 m and down to the immediate surroundings of the mining tool, effectively describing the source term of a disturbance generated by the equipment, which can be excavation, separation processes or manoeuvring. The composition of the sediment brought into suspension, for example, sediment type, particle size and shape and extracellular polymeric substance (EPS), are of utmost importance. These properties influence the settling velocity and distribution of the sediment plume. Together with typical flow properties such as momentum, concentration, and the resulting mixture densities, they are relevant to enable appropriate modelling on the larger scales. Such multi-resolution modelling and simulation should cover a potential mining site and beyond. Predictive habitat modelling must combine existing cross-disciplinary data from biology, microbiology, geochemistry, geology and oceanography in an integrated spatial analysis of multidisciplinary and multiscale datasets. Predictive habitat mapping of physical and biogeochemical settings of biological communities and ecosystem functions should be integrated into the spatial management framework. This would guide science-based spatial planning and management of mining operations, EMMPs and REMPs and make sure that future monitoring actions are implemented to ensure that models are appropriate (e.g., having the right spatial scale) for the development of relevant policies.

As touched upon in Sect. 3, the assessment of ecotoxicological environmental hazards posed by sediment plumes is also of importance for any management system. Given a mineral resource and economic model, a preferred mining method, the REMP and the EMMP and the monitoring procedures and analysis methods and techniques, long- and short-term mine plans will have to be developed for each mining project. The monitoring procedures will contribute to maximize resource utilization or any other objective while minimizing the environmental impact. A thorough understanding of concepts such as "serious harm" constitutes a core in such developments.

Like in any extractive industry, any mined area will be significantly impacted if exploitation takes place. The generated environmental impact should be kept as localized as reasonably possible. Containing environmental effects closer to the mining site will leave larger unaffected areas that could serve as refugia (Wittmann, 2022) for the fauna and significantly reduce impacts outside of the contract area. It would also provide more options for locating Preservation Reference Zones (PRZs) and lead to less impacts on adjoining contract areas (Weaver et al., 2022).

The data collected during a mining operation must be fed back into both the development and the reconciliation and update of the mine plan, the EMMP and the REMP in an adaptive manner. Modules and procedures for such updates by developing multilevel and multi-perspective feedback loops are needed. Further, a link to thresholds and how to assess impact must be developed (Hitchin et al., 2023; Washburn et al., 2019; Weaver et al., 2022). In effect, modules and procedures in a future management system will define the interface between the public and the private governance perspective, and it will be an integrated part of an adaptive management system (Clark et al., 2022; Hyman et al., 2022) that constantly will be better at quantifying the link between environmental pressure and stressors and the environmental impact.

5 Discussion and Way Forward

A total of 31 exploration licenses have been issued by the International Seabed Authority in the Area. Nineteen of these licenses are for polymetallic nodules in the Clarion-Clipperton Fracture Zone, the Central Indian Ocean Basin and in the Western Pacific Ocean. Seven licenses are for polymetallic sulphides in the Southwest Indian Ridge, Central-Indian Ridge and Mid-Atlantic Ridge. The five remaining licenses are for cobalt-rich crusts in the Western Pacific Ocean. The global mineral resource potential on the deep ocean floor is huge (Mizell et al., 2022). International and national authorities are assessing the possibilities to open mining activities on the deep ocean floor.

Although the quantity is disputed (Moana et al., 2022), the green energy transition will increase the demand for minerals and metals (European Commission Joint Research Centre, 2023; IEA, 2021). Can and should the (deep) ocean floor be a part of the solution? There are many unanswered questions associated with this, including: (1) How can local consequences and global benefits be balanced? (2) How can we compare the impact of deep-sea versus terrestrial mining? (3) How can we proceed without complete knowledge in accordance with the precautionary principle? (4) Is it possible to extract without an environmental impact? (5) How should governments develop incentives in a world of uncertainty?[2]

Adaptive management is necessary and must be implemented. The environmental stressors described in Sect. 3 will have to be considered as a constraint in the development of future mine plans and iterative loops must be defined to ensure thorough and effective updates of the REMPs and the EMMPs. The effect of the bathymetry on the potential development of the turbidity flows should also be considered as a controlling factor in the mine plan development. Various technology concepts are under development, challenging the environmental performance of current technology. Whether these concepts are eventually viable options to reduce the environmental footprint needs to be seen. It is important that the industry is challenged to improve upon its environmental footprint, leading to stricter requirements on the permissible environmental impacts. Incorporation of new technology is only likely if the technology concept is robust and has been proven to work appropriately at higher technology readiness levels. Therefore, pilot testing to ground truth results from, for example, the simulation approach presented in Sect. 2, must be encouraged and permitted.

Data collection, data analysis and data management are part of the core in adaptive management. Data must be distributed and shared to ensure well-informed decisions both by national and international regulating and managerial bodies and commercial entities to questions such as: (1) How should the flow of data be set up to make sure that the right data arrive correctly at the receiving end? (2) How do we make sure that all relevant data are taken into consideration when optimizing the mine plan? (3) How can we convince the society that the published data are reliable? (4) Are we as human beings capable of understanding the complexity of these systems and the needed interdisciplinary nature? NTNU Oceans has aimed at an interdisciplinary approach on the topic as a part of the pilot programme on deep-sea mining. Over the years, many challenges have been resolved, but the more one knows, the more questions arise.

Our society needs minerals and metals. Interdisciplinary and holistic thinking is the key to making sound future decisions about undertaking deep-sea mining and making sure that the first deep-sea mining operation is not the last due to improper public and private environmental and resource management.

---

[2] Some of these are addressed in the TripleDeep-project (Ingulstad, 2023).

Acknowledgement The authors acknowledge NTNU Oceans and project- and proposal partners for financial and scientific support. The section on multiscale marine mineral resource management framework is in particular based on interdisciplinary cooperation and valuable cross-boundary discussions over the last decade, for example, during the development of the DeepSeaWatch proposal.

References

Baarholm, G. S., Larsen, C. M., & Lie, H. (2006). On fatiguedamageaccumulation from in-line and cross-flowvortex-inducedvibrationson risers. *Journal of Fluids and Structures, 22*(1), 109–127. https://doi.org/10.1016/j.jfluidstructs.2005.07.013

Bashir, M. B., Kim, S. H., Kiosidou, E., Wolgamot, H., & Zhang, W. (2012). *A concept for seabed rare earth mining in theEastern South Pacific* (p. 138). University of Southampton. https://www.southampton.ac.uk/assets/imported/transforms/content-block/UsefulDownloads_Download/7C8750BCBBB64FBAAF2A13C4B8A7D1FD/LRET%20Collegium%202012%20Volume%201.pdf

Bearman, P. W. (1984). Vortex shedding from oscillating bluff bodies. *Annual Review of Fluid Mechanics, 16*(1), 195–222. https://doi.org/10.1146/annurev.fl.16.010184.001211

Beaulieu, S. E., Graedel, T. E., & Hannington, M. D. (2017). Shouldwe mine thedeep-seafloor? *Earth's Future, 5*(7), 655–658. https://doi.org/10.1002/2017ef000605

Bigham, K. T., Rowden, A. A., Leduc, D., & Bowden, D. A. (2020). Review and syntheses: Turbidityflows—Evidence for effects on deep-sea benthic community productivity is ambiguous but the influence on diversity is clearer. https://doi.org/10.5194/bg-2020-359

Birney, K., Griffin, A., Gwiazda, J., Kefauver, J., Nagai, T., & Varchol, D. (2006). *Potential Deep-sea mining of seafloorsulfides: A case study in Papua New Guinea* (Master of Environmental Science and Management: Master's group project). University of California. https://bren.ucsb.edu/projects/potential-deep-sea-mining-seafloor-sulfides-case-study-papua-new-guinea. Accessed 17 Apr 2023

Boomsma, W., &Blanken, A. (2014). *Overview and assessment of vertical transport systems.* Presented at the Day 4 Thu, May 08, 2014, OTC. https://doi.org/10.4043/25272-ms.

Clark, M. R., Johnson, R., & Hyman, J. (2022). Adaptive management as a tool for effectiveenvironmental management of deep-sea mining. In R. Sharma (Ed.), *Perspectiveson deep-sea mining* (pp. 339–371). Springer International Publishing. https://doi.org/10.1007/978-3-030-87982-2_13

Collins, P. C., Croot, P., Carlsson, J., Colaço, A., Grehan, A., Hyeong, K., et al. (2013). A primer for the environmental impact assessment of mining at seafloor massive sulfide deposits. *Marine Policy, 42*, 198–209. https://doi.org/10.1016/j.marpol.2013.01.020

Dumais, M.-A., & Brönner, M. (2020). Revisiting Austfonna, Svalbard, with potential field methods—A new characterization of the bed topography and its physical properties. *The Cryosphere, 14*(1), 183–197. https://doi.org/10.5194/tc-14-183-2020

Dumke, I., Nornes, S. M., Purser, A., Marcon, Y., Ludvigsen, M., Ellefmo, S. L., et al. (2018). First hyperspectral imaging survey of thedeep-seafloor: High-resolution mapping of manganese nodules. *Remote Sensing of Environment, 209*, 19–30. https://doi.org/10.1016/j.rse.2018.02.024

Durden, J. M., Murphy, K., Jaeckel, A., Van Dover, C. L., Christiansen, S., Gjerde, K., et al. (2017). A procedural framework for robust environmental management of deep-sea mining projects using a conceptualmodel. *Marine Policy, 84*, 193–201. https://doi.org/10.1016/j.marpol.2017.07.002

DW. (2023, April 1). UN to start allowingdeep-sea mining applications from July – DW – 04/01/2023. dw.com. https://www.dw.com/en/un-to-start-allowing-deep-sea-mining-applications-from-july/a-65202212. Accessed 29 Apr 2023.

Eidsvik, O. A., & Schjølberg, I. (2016a). Determination of Hydrodynamic Parameters for RemotelyOperatedVehicles. In *Volume 7: Ocean Engineering* (p. V007T06A025). Presented at the ASME 2016 35th international conference on ocean, offshore and arctic engineering, Busan, South Korea: American Society of Mechanical Engineers. https://doi.org/10.1115/OMAE2016-54642.

Eidsvik, O. A., & Schjølberg, I. (2016b). Time domain modeling of ROV umbilical using beam equations. *IFAC-Papers Online, 49*(23), 452–457. https://doi.org/10.1016/j.ifacol.2016.10.447

EU. (2023). *Causes of climate change.* https://climate.ec.europa.eu/climate-change/causes-climate-change_en. Accessed 29 Apr 2023.

European Commission Joint Research Centre. (2023). Supply chainanalysis and material demand-forecast in strategictechnologies and sectors in the EU: A foresightstudy. LU: Publications Office. https://data.europa.eu/doi/10.2760/386650. Accessed 17 Mar 2023.

Faltinsen, O. M. (1999). *Sea loadsonships and offshore structures* (1. Paperback ed., repr. transferred to digital printing). Cambridge University Press.

Fard, R. N., & Tedeschi, E. (2017). Investigation of AC and DC powerdistributions to seafloor mining equipment. In *OCEANS 2017 – Aberdeen* (pp. 1–7). Presented at the OCEANS 2017 – Aberdeen, Aberdeen, United Kingdom. IEEE. https://doi.org/10.1109/OCEANSE.2017.8084903

Fard, R. N., & Tedeschi, E. (2018). Power system design considerations for a seafloor mining vehicle. In *2018 IEEE Energy Conversion Congress and Exposition (ECCE)* (pp. 1164–1171). Presented at the 2018 IEEE Energy Conversion Congress and Exposition (ECCE), Portland, OR, USA. IEEE. https://doi.org/10.1109/ECCE.2018.8558004

Fylling, I., Larsen, C., Sødahl, N., Ormberg, H., Engseth, A., Passano, E., & Holthe, K. (1995). *Riflextheory manual* (Manual No. STF70 F 95219: 53). SINTEF.

Haalboom, S., De Stigter, H. C., Mohn, C., Vandorpe, T., Smit, M., De Jonge, L., & Reichart, G.-J. (2023). Monitoring of a sediment plumeproduced by a deep-sea mining test in shallow water, Málaga Bight, Alboran Sea (southwesternMediterranean Sea). *Marine Geology, 456*, 106971. https://doi.org/10.1016/j.margeo.2022.106971

Hitchin, B., Smith, S., Kröger, K., Jones, D., Jaeckel, A., Mestre, N., et al. (2023). Thresholds in deep-seabed mining: A primer for theirdevelopment. *Marine Policy, 149*, 105505. https://doi.org/10.1016/j.marpol.2023.105505

Hyman, J., Stewart, R. A., & Sahin, O. (2022). Adaptive management of deep-seabed mining projects: A systems approach. *Integrated Environmental Assessment and Management, 18*(3), 674–681. https://doi.org/10.1002/ieam.4395

IEA. (2021). *The role of critical minerals in clean energy transitions* (World energy outlook special report) (287). Paris. https://www.iea.org/reports/the-role-of-critical-minerals-in-clean-energy-transitions

Ingulstad, M. (2023, April 5). *Triple deep—The deep dilemmas: Deep sea mining for the new deep transition – NTNU*. TripleDeep. https://www.ntnu.edu/sustainability/tripledeep. Accessed 5 Apr 2023.

JORC. (2012). *Australasian code for reporting of exploration results, mineral resources and ore reserves*. Joint Ore Reserves Committee: The Australasian Institute of Mining and Metallurgy, Australian Institute of Geoscientists and Minerals Council of Australia. http://www.jorc.org/docs/JORC_code_2012.pdf

Juliani, C. (2019). Automateddiscrimination of faultscarpsalong an Arctic mid-oceanridgeusing neural networks. *Computers & Geosciences, 124*, 27–36. https://doi.org/10.1016/j.cageo.2018.12.010

Juliani, C., & Ellefmo, S. L. (2018a). Probabilistic estimates of permissive areas for undiscovered seafloor massive sulfide depositson an Arctic mid-ocean ridge. *Ore Geology Reviews, 95*, 917–930. https://doi.org/10.1016/j.oregeorev.2018.04.003

Juliani, C., & Ellefmo, S. L. (2018b). Resource assessment of undiscoveredseafloor massive sulfidedepositson an Arctic mid-ocean ridge: Application of grade and tonnagemodels. *Ore Geology Reviews, 102*, 818–828. https://doi.org/10.1016/j.oregeorev.2018.10.002

Kim, S. W., Sævik, S., Wu, J., & Leira, B. J. (2021). Prediction of Deepwater riser VIV with an improved time domainmodelincluding non-linear structuralbehavior. *Ocean Engineering, 236*, 109508. https://doi.org/10.1016/j.oceaneng.2021.109508

Knudsen, T. H., Sævik, S., & Thorsen, M. J. (2016). *Numerical analysis of combined VIV and slug flow in time domain*. Presented at the Volume 2: CFD and VIV, American Society of Mechanical Engineers. https://doi.org/10.1115/omae2016-54891

Kohler, S. G., Heimbürger-Boavida, L.-E., Petrova, M. V., Digernes, M. G., Sanchez, N., Dufour, A., et al. (2022). Arctic Ocean'swintertimemercuryconcentrationslimited by seasonal loss onthe shelf. *Nature Geoscience, 15*(8), 621–626. https://doi.org/10.1038/s41561-022-00986-3

Koski, R. (2012). Metal dispersion resulting from mining activities in coastal environments: A pathways approach. *Oceanography, 25*(2), 170–183. https://doi.org/10.5670/oceanog.2012.53

Kowalczuk, P. B., Manaig, D. O., Drivenes, K., Snook, B., Aasly, K., & Kleiv, R. A. (2018). Galvanic leaching of seafloor massive sulphides using $MnO2$ in $H2SO4$-NaCl media. *Minerals, 8*(6), 235. https://doi.org/10.3390/min8060235

Kowalczuk, P. B., Bouzahzah, H., Kleiv, R. A., & Aasly, K. (2019). Simultaneous leaching of seafloor massive sulfides and Polymetallic nodules. *Minerals, 9*(8), 482. https://doi.org/10.3390/min9080482

Kurushina, V., Pavlovskaia, E., Postnikov, A., & Wiercigroch, M. (2018). Calibration and comparison of VIV wake oscillator models for lowmass ratio structures. *International Journal of Mechanical Sciences, 142–143*, 547–560. https://doi.org/10.1016/j.ijmecsci.2018.04.027

Larsen, C. M., Vikestad, K., Yttervik, R., & Baarholm, G. S. (2005). *VIVANA- theory manual version 3.4*. MARINTEK.

Laugesen, J., Aasly, K., Ellefmo, E., Steinar, Volkmann, S., & Knodt, S. (2021). Teknologirapport havbunnsmineraler (No. 2020–1218, rev. 2, 1231487) (p. 128). Oslo. https://www.npd.no/globalassets/1-npd/fakta/havbunnsmineraler/teknologirapport-havbunnsmineraler-oppdatert-13102021.pdf

Lesage, M. (2020). *A framework for evaluatingdeep-sea mining systems for seafloor massive sulphides deposits* (Doctoraltheses at NTNU; 2020:280). Norwegian University of Science and Technology, Trondheim. Retrieved from https://ntnuopen.ntnu.no/ntnu-xmlui/handle/11250/2677911

Lesage, M., Juliani, C., & Ellefmo, S. (2018). Economic block model development for mining seafloor massive sulfides. *Minerals, 8*(10), 468. https://doi.org/10.3390/min8100468

Levin, L. A., Amon, D. J., & Lily, H. (2020). Challenges to thesustainability of deep-seabed mining. *Nature Sustainability, 3*(10), 784–794. https://doi.org/10.1038/s41893-020-0558-x

Lim, A., Brönner, M., Johansen, S. E., & Dumais, M. (2019). Hydrothermal activity at the ultraslow-spreading Mohns ridge: New insights from near-seafloor magnetics. *Geochemistry, Geophysics, Geosystems, 20*(12), 5691–5709. https://doi.org/10.1029/2019GC008439

Ludvigsen, M., Aasly, K., Ellefmo, S., Hilario, A., Ramirez-Llodra, E., Søreide, F., et al. (2016). *MarMine Cruise report Arctic Mid-Ocean Ridge (AMOR) 15.08.2016–05.09.2016* (No. 1) (p. 120). NTNU.

Lyle, M. (2015). Deep-sea sediments. In *Encyclopedia of marine geosciences* (pp. 1–20). https://doi.org/10.1007/978-94-007-6644-0_53-2

Mainçon, P. (2011). A Wiener-Laguerre model of VIV forces given recent cylinder velocities. *Mathematical Problems in Engineering, 2011*, 1–43. https://doi.org/10.1155/2011/414702

Meyer, T. (2018). *Elisabeth Mann Borgese deep ideology* (PhD Thesis). Norwegian University of Science and Technology, Trondheim. Retrieved from https://ntnuopen.ntnu.no/ntnu-xmlui/handle/11250/2581916.

MIDAS. (2016). *Managing impacts of deep sea resource exploitation research highlights*. https://www.eu-midas.net/sites/default/files/downloads/MIDAS_research_highlights_low_res.pdf

Miller, K. A., Thompson, K. F., Johnston, P., & Santillo, D. (2018). An overview of seabed mining including the current state of development, environmental impacts, and knowledge gaps. *Frontiers in Marine Science, 4*, 418. https://doi.org/10.3389/fmars.2017.00418

Mizell, K., Hein, J. R., Au, M., & Gartman, A. (2022). Estimates of metals contained in abyssal manganese nodules and ferromanganese crusts in the global ocean based on regional variations and genetic types of nodules. In R. Sharma (Ed.), *Perspectiveson deep-sea mining* (pp. 53–80). Springer International Publishing. https://doi.org/10.1007/978-3-030-87982-2_3

Moana, S., Aponte, F., & Wiebe, K. (2022). *The future is circular – Circular economy and critical minerals for the green transition*. SINETF.

Modarres-Sadeghi, Y., Mukundan, H., Dahl, J. M., Hover, F. S., & Triantafyllou, M. S. (2010). The effect of higherharmonicforcesonfatiguelife of marine risers. *Journal of Sound and Vibration, 329*(1), 43–55. https://doi.org/10.1016/j.jsv.2009.07.024

Modarres-Sadeghi, Y., Chasparis, F., Triantafyllou, M. S., Tognarelli, M., & Beynet, P. (2011). Chaoticresponse is a genericfeature of vortex-induced vibrations of flexible risers. *Journal of Sound and Vibration, 330*(11), 2565–2579. https://doi.org/10.1016/j.jsv.2010.12.007

Muñoz-Royo, C., Ouillon, R., El Mousadik, S., Alford, M. H., & Peacock, T. (2022). An in-situ study of abyssal turbidity-current sediment plumes generated by a deep-seabed polymetallic nodule mining preprototype collector vehicle. *Science Advances, 8*(38), eabn1219. https://doi.org/10.1126/sciadv.abn1219

Nilsen, H. R. (2020). Stayingwithinplanetaryboundaries as a premise for sustainability: On theresponsibility to addresscounteractingsustainabledevelopment goals. *Etikk i praksis – Nordic Journal of Applied Ethics, 1*, 29–44. https://doi.org/10.5324/eip.v14i1.2863

Nydal, O. J. (2012). Dynamic models in multiphase flow. *Energy & Fuels, 26*(7), 4117–4123. https://doi.org/10.1021/ef300282c

OD. (2023). *Ressursvurdering havbunnsmineraler* (p. 132). https://www.npd.no/globalassets/1-npd/fakta/havbunnsmineraler/publikasjoner/2023/ressursvurdering-havbunnsmineraler-20230127.pdf

OED. (2022). *Konsekvensutredning – undersøkelse og utvinning av havbunnsmineraler på norsk kontinentalsokkel* (119). Oslo. https://www.regjeringen.no/contentassets/dbf5144d0fb-c42b5a4db5fc7eb4fa312/horingsdokument-konsekvensutredning-for-mineralvirksomhet-pa-norsk-kontinentalsokkel-l1415388.pdf

OED. (2023). *Havbunnsmineraler gir nye muligheter på havbunnen*. Regjeringen.no. Taleartikkel, regjeringen.no. https://www.regjeringen.no/no/aktuelt/havbunnsmineraler-gir-nye-muligheter-pa-havbunnen/id2986484/. Accessed 21 July 2023.

Païdoussis, M. P., & Li, G. X. (1993). Pipes Conveying Fluid: A Model Dynamical Problem. *Journal of Fluids and Structures, 7(2), 137–204.* https://doi.org/10.1006/jfls.1993.1011

Patel, M. H., & Seyed, F. B. (1995). Review of flexible riser modelling and analysis techniques. *Engineering Structures, 17*(4), 293–304. https://doi.org/10.1016/0141-0296(95)00027-5

Paull, C. K., Talling, P. J., Maier, K. L., Parsons, D., Xu, J., Caress, D. W., et al. (2018). Powerful turbidity currents driven by dense basal layers. *Nature Communications, 9*(1), 4114. https://doi.org/10.1038/s41467-018-06254-6

Perera, A., & Nilsen, R. (2022). Online identification of six-phase IPMSM parameters using prediction-ErrorSensitivities to model parameters. In *2022 international power electronics conference (IPEC-Himeji 2022- ECCE Asia)* (pp. 1225–1232). Presented at the 2022 international power electronics conference (IPEC-Himeji 2022- ECCE Asia), Himeji, Japan. IEEE. https://doi.org/10.23919/IPEC-Himeji2022-ECCE53331.2022.9807264

Perera, A., Nilsen, R., & Haugan, T. (2022). *Investigation of open-loop predictor implementation methods for online parameter estimation of IPMSM*. Presented at the PCIM Europe 2022 – International exhibition and conference for power electronics, intelligent motion, renewable energy and energy management. https://doi.org/10.30420/565822214

Pryadunenko, A., Nilsson, L. P., & Larsen, R. B. (2022). Geochemical characteristics of the Feragen ultramafic body, central Scandinavian Caledonides: Implications for petrogenesis and tectonic significance. *Norwegian Journal of Geology*. https://doi.org/10.17850/njg101-4-4

Purkiani, K., Gillard, B., Paul, A., Haeckel, M., Haalboom, S., Greinert, J., et al. (2021). Numerical simulation of deep-sea sediment transport induced by a dredge experiment in the North-eastern Pacific Ocean. *Frontiers in Marine Science, 8*. https://doi.org/10.3389/fmars.2021.719463

Ryan, E. (2023). Onshore to offshore correlation of ore-forming hydrothermal systems: Geological investigations of the Reinfjord Ultramafic Complex and Southwestern Barents Sea (PhD Thesis). Norwegian University of Science and Technology, Trondheim, Norway. Retrieved from https://hdl.handle.net/11250/3057271

Ryan, E. J., Sørensen, B. E., Fichler, C., Larsen, R. B., Gresseth, J. L., & Bjørlykke, A. (2023). Fault linkage on south-eastern Bjørnøya: Implications for structural interpretations surrounding fertile ore-forming fault systems offshore. *Norwegian Journal of Geology*. https://doi.org/10.17850/njg102-4-5

Schulte, S. A. (2013). *Vertical transport methods for deep sea mining* (Master thesis). TU Delft.

Sevillano, L. C., & Sangesland, S. (2022). Assessment of power requirements for alternative vertical transportation system for deepsea mining. In *Volume 4: Ocean space utilization* (p. V004T05A010). Presented at the ASME 2022 41st international conference on ocean, offshore and arctic engineering, Hamburg, Germany. American Society of Mechanical Engineers. https://doi.org/10.1115/OMAE2022-80149

Smith, I. E. (2017). *A 7-field Lagrangian slug capturing and slug tracking model with higher order methods* (PhD thesis). Norwegian University of Science and Technology, Trondheim, Norway. Retrieved from http://hdl.handle.net/11250/2455762

Snook, B., Drivenes, K., Rollinson, G. K., & Aasly, K. (2018). Characterisation of mineralised material from the Loki's castle hydrothermal vent onthe Mohn's ridge. *Minerals, 8*(12), 576. https://doi.org/10.3390/min8120576

Solheim, A. V., Lesage, M., Asbjørnslett, B. E., & Erikstad, S. O. (2020). *Deep sea mining: Towards conceptual design for underwater transportation*, 10.

Solheim, A. V., Pettersen, S. S., Agis, J. J. G., Brett, P. O., Asbjørnslett, B. E., Erikstad, S. O., & Ellefmo, S. L. (2022). Early stage decisions in marine systems design for deep-sea mining. *International Journal of Maritime Engineering, 164*.

Stabell, E. D. (2020, February). Hard environmental choices: Comparability, justification and the argument from moral identity. https://doi.org/10.3197/096327119X15678473651009

Stabell, E. D. (2021). Hard environmental choices: Comparability, justification and the argument from moral identity. *Environmental Values, 30*(1), 111–130. https://doi.org/10.3197/096327119X15678473651009

Stabell, E. D., & Steel, D. (2018). Precaution and fairness: A framework for distributing costs of protection from environmental risks. *Journal of Agricultural and Environmental Ethics, 31*(1), 55–71. https://doi.org/10.1007/s10806-018-9709-8

Sture, Ø., Snook, B., & Ludvigsen, M. (2019). Obtaining hyperspectral signatures for seafloor massive sulphide exploration. *Minerals, 9*(11), 694. https://doi.org/10.3390/min9110694

Sumer, B. M., & Fredsøe, J. (1997). Hydrodynamics around cylindrical structures. *World Scientific*. https://doi.org/10.1142/3316

Sweetman, A. K., Smith, C. R., Shulse, C. N., Maillot, B., Lindh, M., Church, M. J., et al. (2018). Key role of bacteria in theshort-term cycling of carbon at the abyssal seafloor in a low particulate organic carbonflux region of the eastern Pacific Ocean. *Limnology and Oceanography, 64*(2), 694–713. https://doi.org/10.1002/lno.11069

Tay, S., Browne, R., &Berrtoli, O. (2023). *Cook Islands polymetallic nodule deposit – Technical report and supporting documentation for The Cook Islands polymetallic nodules resource estimate* (Resource estimation and -classification) (p. 219). Cook Island. https://www.sbma.gov.ck/s/CIMinRscRevt_final_b.pdf. Accessed 21 Apr 2023.

Thorsen, M. J., & Sævik, S. (2017). *Simulating riser VIV in current and waves using an empirical time domain model*. In Volume 2: Prof. Carl Martin Larsen and Dr. Owen Oakley Honoring Symposia on CFD and VIV (p. V002T08A011). Presented at the ASME 2017 36th International Conference on Ocean, Offshore and Arctic Engineering, Trondheim, Norway: American Society of Mechanical Engineers. https://doi.org/10.1115/omae2017-61217

Thorsen, M. J., & Sævik, S. (2018). *An analytical model of the effect of internal density waves in risers subjected to vortex shedding* (p. ISOPE-I-18-701). Presented at the the 28th international ocean and polar engineering conference. Accessed 4 Apr 2023.

Thorsen, M. J., Challabotla, N. R., Sævik, S., & Nydal, O. J. (2019). A numerical study on vortex-induced vibrations and the effect of slurry density variations on fatigue of ocean mining risers. *Ocean Engineering, 174*, 1–13. https://doi.org/10.1016/j.oceaneng.2019.01.041

Triantafyllou, M., Triantafyllou, G., Tein, Y. S. D., & Ambrose, B. D. (1999). *Pragmatic riser VIV analysis*. Presented at the All Days, OTC. https://doi.org/10.4043/10931-ms

Ulveseter, J. V., Thorsen, M. J., Sævik, S., & Larsen, C. M. (2018). Time domainsimulation of riser VIV in current and irregularwaves. *Marine Structures, 60*, 241–260. https://doi.org/10.1016/j.marstruc.2018.04.001

United Nations. (2023). *Causes and effects of climate change*. United Nations. https://www.un.org/en/climatechange/science/causes-effects-climate-change. Accessed 29 Apr 2023.

van Wijk, J. M., Talmon, A. M., & van Rhee, C. (2016). Stability of verticalhydraulic transport processes for deepocean mining: An experimental study. *Ocean Engineering, 125*, 203–213. https://doi.org/10.1016/j.oceaneng.2016.08.018

Van Zyl, D., Sassoon, M., Digby, C., Fleury, A. M., &Kyeyune, S. (2002). *Mining for the future. Appendix A: Large volume waste working paper* (No. 31). International Institute for Environment and Development and World Business Council for Sustainable Development. https://www.iied.org/sites/default/files/pdfs/migrate/G00883.pdf

Vandiver, J. K., & Li, L. (2005). *SHEAR7 V4.4 program theoretical manual*. Department of Ocean Engineering, Massachusetts Institute of Technology.

Verichev, S., Metrikine, A., Plat, R., &Hendrikse, H. (2011). *Dynamics of thevertical hydraulic transport system for deep sea mining*. Presented at the volume 4: Pipeline and riser technology, ASMEDC. https://doi.org/10.1115/omae2011-49464

Wang, J., Fu, S., & Baarholm, R. (2018). Evaluation of vortex-inducedvibration of a steelcatenary riser in steady current and vessel motion-induced oscillatory current. *Journal of Fluids and Structures, 82*, 412–431. https://doi.org/10.1016/j.jfluidstructs.2018.07.018

Washburn, T. W., Turner, P. J., Durden, J. M., Jones, D. O. B., Weaver, P., & Van Dover, C. L. (2019). Ecological risk assessment for deep-sea mining. *Ocean & Coastal Management, 176*, 24–39. https://doi.org/10.1016/j.ocecoaman.2019.04.014

Weaver, P. P. E., Aguzzi, J., Boschen-Rose, R. E., Colaço, A., de Stigter, H., Gollner, S., et al. (2022). Assessingplumeimpactscaused by polymetallicnodule mining vehicles. *Marine Policy, 139*, 105011. https://doi.org/10.1016/j.marpol.2022.105011

Williamson, C. H. K., & Govardhan, R. (2004). Vortex-induced vibrations. *Annual Review of Fluid Mechanics, 36*(1), 413–455. https://doi.org/10.1146/annurev.fluid.36.050802.122128

Wittmann, F. (2022). The landscape role of river wetlands. In *Encyclopedia of inland waters* (pp. 51–64). Elsevier. https://doi.org/10.1016/B978-0-12-819166-8.00191-2

Wu, X., Ge, F., & Hong, Y. (2012). A review of recent studies on vortex-induced vibrations of long slender cylinders. *Journal of Fluids and Structures, 28*, 292–308. https://doi.org/10.1016/j.jfluidstructs.2011.11.010

Dr. Steinar Løve Ellefmo is a professor at the Department of Geoscience and Petroleum, Norwegian University of Science and Technology (NTNU). His research and teaching interests are within the broad field of mineral resource management and mining engineering, specifically geostatistics, 3D geometric ore body modelling, mine planning and mineral resource potential assessment. With colleagues at NTNU, he initiated the MarMine-project on exploitation technologies for deep-sea mining and has since 2013 led the interdisciplinary pilot programme on deep-sea mining at NTNU. Dr. Ellefmo is a member of the Australasian Institute of Mining and Metallurgy (AusIMM) and the Society of Economic Geologist (SEG).

Dr. Murat V. Ardelan is a professor at the Department of Chemistry at the Norwegian University of Science and Technology (NTNU). Dr. Ardelan studies marine biogeochemical cycles of bio essential, toxic elements and organic carbon, mainly in Antarctic and Arctic waters. In recent years, his research has focused on the potential impact of CO_2 capture and storage, aquaculture and deep-sea mining activities on the biogeochemistry of the sediment–seawater interface and the water column. He is also working on automatizing clean sampling to detect the biogeochemical signature at the water column and sediment surface.

Dr. Siri Granum Carson is Director of NTNU Oceans, a strategic research area at the Norwegian University of Science and Technology (NTNU). She chairs the Norwegian UN Ocean Decade National Committee and is part of the leader group of the Norwegian Marine University Consortium (NMU). She is a professor of applied ethics, and her main research areas are corporate social responsibility and responsible innovation.

Dr. Rudy Helmons is an associate professor in offshore and dredging engineering at TU Delft and adjunct associate professor for deep sea mining at NTNU. Dr. Helmons' research focusses on the handling of solids in a marine environment, specifically for dredging and mining applications. His research involves the excavation, transport, (in situ) separation and deposition of large solid particles and fine sediments. He has led the Blue Harvesting project on the environmental footprint of hydraulic nodule collection and initiated the PLUMEFLOC project, studying the effect that natural flocculation might have sediment dispersion.

Dr. Svein Sævik is a professor in marine structures at the Department of Marine Technology, Norwegian University of Science and Technology (NTNU). Dr. Sævik has a broad background from engineering, applied research and academia. His research and interest include development of methods for response analysis and monitoring of slender marine structures. This includes the development of several non-linear finite element codes for global and local response analysis of pipelines, flexible pipes, power cables and umbilicals, which are used by industry worldwide. Dr. Sævik has published many journal and conference articles within various topics and received several awards for his work, including the Equinor Research award. He is a member of the Norwegian Academy of Technological Sciences.

Part II
Engineering Concepts for Deep-Sea Mining, Ore Handling and Processing

Chapter 3
Contemporary Technological Progress in Deep-Sea Mining

M. A. Atmanand, K. C. Janardhanan, K. Amuda, A. Gnanaraj, S. Rajesh, P. Muthuvel, Gopkumar Kuttikrishnan, and G. A. Ramadass

Abstract This chapter provides an overview of the technological advancements in the field of deep-sea mining. The chapter commences with an introductory section that explores the initial technologies developed for deep-sea mining. It serves as a foundation for the subsequent discussions on the advancements made by different entities in the field. The chapter then presents a historical perspective related to deep-sea mining and then transitions to a discussion of present-day efforts and phased execution strategies undertaken. This section highlights the evolution of deep-sea mining endeavours. Subsequently, the chapter outlines various subsystems and infrastructure required for activities related to deep-sea mining. These components are critical to the successful execution of deep-sea mining operations. The chapter then discusses the operation of collector systems, crushers, and nodule transport systems, with a focus on local pumping techniques. This section sheds light on the crucial equipment and processes involved in deep-sea mining.

The narrative progresses to describe the successful sea trials conducted at a depth of 5270 meters in the Central Indian Ocean Basin (CIOB). These trials assess the manoeuvrability of a crawler in deep-sea conditions, a pivotal aspect of the mining program. The chapter also covers the different aspects of environmental disturbances and their technological remedies. This chapter helps us understand the importance of responsible and sustainable practices in deep-sea mining.

Keywords Deep-sea mining · Mining machine · Deep-sea crawler · Slurry pumping · Sub-sea umbilicals · Manganese nodules · Deep-sea minerals

M. A. Atmanand (✉)
Indian Institute of Technology, Madras, Chennai, Tamil Nadu, India
e-mail: atmanandma@hotmail.com

K. C. Janardhanan · K. Amuda · A. Gnanaraj · S. Rajesh · P. Muthuvel · G. Kuttikrishnan · G. A. Ramadass
National Institute of Ocean Technology, Chennai, Tamil Nadu, India

R. Sharma (ed.), *Deep-Sea Mining and the Water Column*,
https://doi.org/10.1007/978-3-031-59060-3_3

1 Introduction

The advent of green technology has paved the way to develop newer electrical drives for vehicles and other systems. This, in turn, is spurning the need to get more metals like cobalt, nickel, etc. from the deep sea, as the terrestrial resources are limited (Meinert et al., 2016). The availability of minerals on the sea bed varies as per their association with seafloor features (Fig. 3.1). However, the regulations for environmental impact assessment are in the process of discussion at the International Seabed Authority (ISA) (Blanchard et al., 2023). This chapter briefly describes the history of deep-sea mining, the technologies being developed by different contractors and entities, and the mining methods adopted by India towards manganese nodule mining starting with the mining system design, its launching and retrieval, and sea trials.

1.1 Why Deep-Sea Mining?

The demand for cobalt (Co), nickel (Ni), copper (Cu), and rare earth elements (REE) has increased as a result of the global push towards green technologies and renewable energy (Cronan, 2005) (Fig. 3.2). These metals are terrestrially available

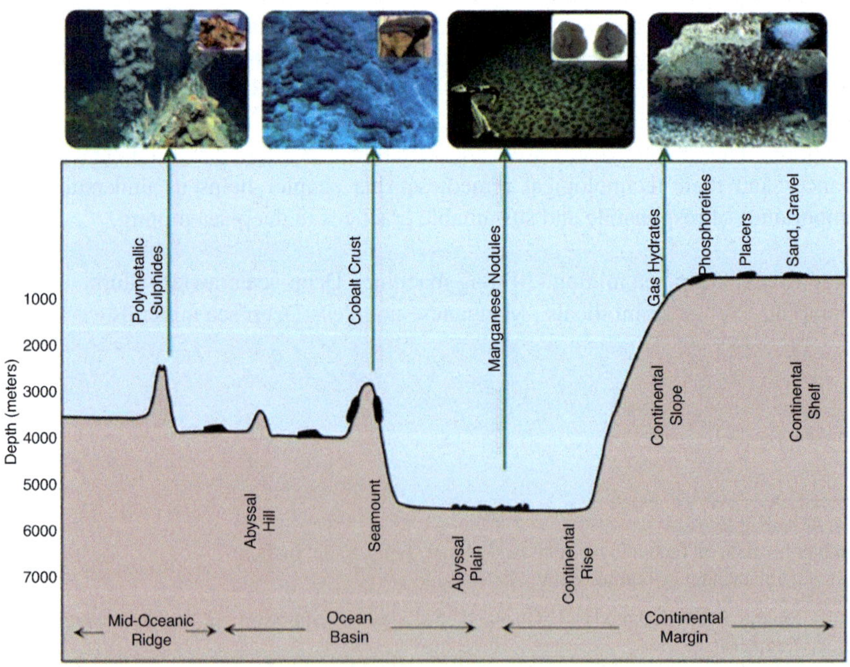

Fig. 3.1 The occurrence of mineral resources on deep ocean bed. (Source: Social and Economic Impact of Earth Sciences. Springer, Singapore. https://doi.org/10.1007/978-981-19-6929-4_17)

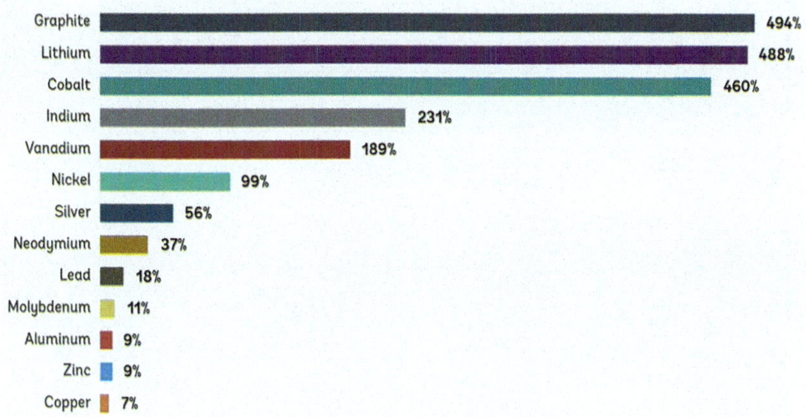

Fig. 3.2 Projected mineral demand by 2050 as compared to 2018 production levels. (Source: WBG Energy)

in sufficient quantities only in a few places. The exponential growth in demand anticipated over the following 20–30 years would not be met even by advanced recycling and the current initiatives to create a circular economy. The prospect for growth of the downstream industries and the high-value segment of electric vehicles, high-power batteries, and related technology is provided by the metal extraction of cobalt, nickel, copper, and manganese with manganese present in huge quantities. After the land, the deep oceans are the alternate source of these metals. All of these metals have a significant economic influence on India and are vital for development, national security, and strategic depth. All of these metals are in short supply to the nation, except a few REEs that are found in monazite sands along the southern beaches.

Deep-sea mining (DSM) is the process of excavating mineral deposits from the deep seabed**.** Currently, the mining technology developmental works for harvesting polymetallic nodules are in advanced stages, having had precedence through successful demonstrative trials during the late 1970s by many consortia of firms (Knodt et. al., 2016) from the USA, Canada, Europe, and Japan. ISA has entered into 15-year contracts with 31 contractors for the exploration of polymetallic nodules (PMN), polymetallic sulfides (PMS), and cobalt-rich ferromanganese crusts (CFC) in the deep seabed. India has also been allocated a site in CIOB by the ISA for the exploration of deep-sea polymetallic nodules (Fig. 3.3).

While exploratory mining to test equipment has occurred at a small scale, deep-sea mining has not yet been undertaken commercially. However, some contractors and mining companies have a plan to commence seabed mining operations in the coming years.

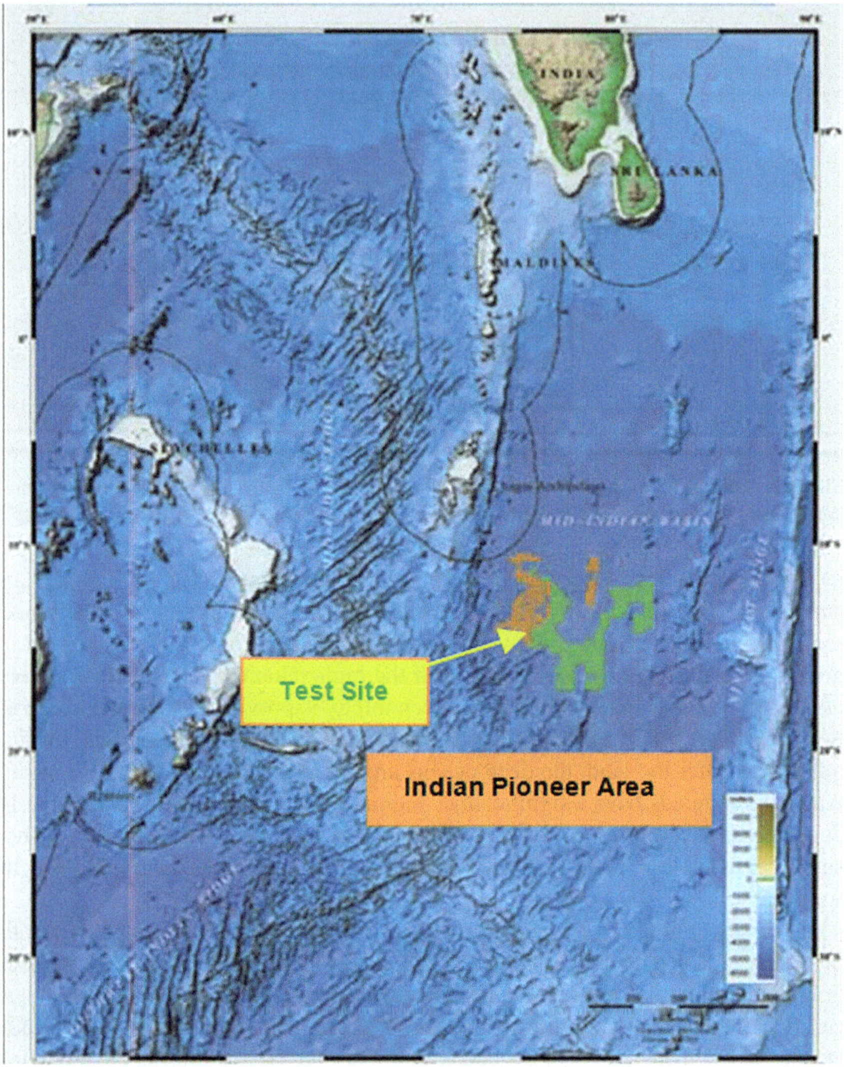

Fig. 3.3 Representation of Pioneer Area allocated for PMN exploration by the ISA. (Source: International Seabed Authority)

2 Overview of Deep-Sea Mining

2.1 History of Deep-Sea Mining Technologies

In the last five decades, various countries have started developing technologies for mining of polymetallic nodules from the World Ocean bed. But deep-sea deposit formation is a rather specific sector with its peculiarities. The major complexity consists in the fact that these mineral deposits are generally located in deep waters

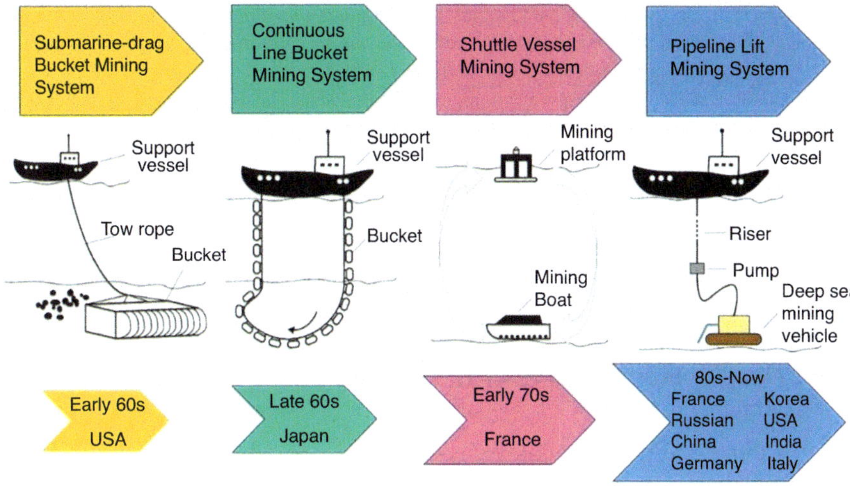

Fig. 3.4 Schematic representation of deep-sea mining technologies. (Wang et al., 2022)

(4000–6000 m). At this level of depth, the pressure is very high and the environmental conditions are adverse. In order to mine the nodules from the ocean bed, there is a need for innovative technologies that will overcome all the complexities connected with it.

The research on deep-sea mining for collecting deep mineral resources began in the 1950s (Kang & Liu, 2021). For deep-sea mining, different mining methods were proposed and applied. Among them, four major mining methods are the trawl mining system, continuous line bucket system, marine shuttle mining system, and hydraulic system, as shown in Fig. 3.4 (Deepak et al., 2001, 2007; Kuntz, 1980; Masuda et al., 1971). In the trawl mining system, the collecting box is attached to the mother ship and dragged so that mineral particles enter the collecting box. In the continuous-line bucket system, multiple buckets are installed on a looped rope belt. The buckets lift the mineral particles from the seabed to the mother ship using the circular motion of the belt. In the shuttle mining system, the shuttle sinks to the seabed, collects the mining vehicle, and comes back to the mother ship.

Deep-sea mining technology has evolved over several decades, driven by the increasing demand for minerals and metals, and advancements in underwater exploration and mining equipment. Here is a brief history of the development of deep-sea mining technology:

1960s—Early exploration:	The concept of deep-sea mining gained attention in the 1960s when geologists discovered the economic value of mineral deposits on the ocean floor, including polymetallic nodules, polymetallic sulfides, and cobalt-rich ferromanganese crusts.
1970s—Initial research:	Initial interest started to develop in the field of mining technology and plans to initiate laws to regulate and manage mineral-related activities in international waters.
1980s—Development of mining technology:	The 1980s saw the development of remotely operated vehicles (ROVs) and autonomous underwater vehicles (AUVs), which enabled researchers to explore and sample deep-sea mineral deposits.

1990s—First commercial ventures:	The first commercial deep-sea mining venture, Ocean Minerals Company (OMCO) tested the mining equipment for polymetallic nodule mining in the late 1980s and early 1990s.
2000s—Advancements in mining tools:	Significant technological advancements, including improved ROVs and seafloor mapping capabilities made deep-sea mining more feasible. Companies like Nautilus Minerals and Neptune Minerals began developing plans for mining operations.
2010s—Growing interest and concerns:	The interest in deep-sea mining continued to grow, with several countries and private companies securing exploration licenses.

2.2 Development of Deep-Sea Mining Systems

In 1970, the first testing of nodule collection systems based on the hydraulic method was carried out on the Florida coast in the Atlantic Ocean at a depth of 1000 meters (Cheng et. al., 2023a, b). The testing was organised by Deep Sea Ventures using a specially equipped cargo vessel called the Deep Sea Miner. Nodules were collected with the help of a pneumatic system and conveyed up to the vessel of 6750 tons displacement, equipped with a 25-meter derrick-crane and a central 6 × 9 m tank. In the late 1970s, three American consortiums conducted a series of tests using hydraulic mining systems. Figure 3.5 shows the recovered nodules by Sedco 445 in 1977 and the Clementine II seafloor crawler.

Since then, several deep-sea mining systems have been developed over the last few decades as shown in Fig. 3.6. In 2012, Apollo II field trials- a 60-minute pilot collection run, approximately 14 metric tons of nodules were collected during a sweep of approximately 150 m, transported to the surface using a 4.3-km riser system, and placed into the hold at a water depth of 130 m. Harbour trials were conducted in the year 2022 with the upgradation in the system. South Korea developed the "MineRo I" and "MineRo II" mining vehicles from 2007 to 2013, and completed the polymetallic nodule collection simulation test at a water depth of 1370 m (blue-nodules.eu).

Fig. 3.5 Recovered nodules by Sedco 445 in 1977 in the North Pacific Ocean, Depth = 5300 m (left) and Clementine II seafloor crawler. (Knodt et al., 2016; Spickermann, 2012)

Fig. 3.6 Representation of mining systems developed by different countries/organisations. (Chung et. al., 1997; Charles et. al., 1990; Cheng et. al., 2023a, b) (Sources: https://doi.org/10.3390/su15054572; https://newatlas.com/technology/seafloor-mining-nodule-apollo2-blue-harvesting/; https://www.deme-group.com/; https://metals.co/)

In 2021, under the organization of the China Oceanic Association, the Changsha Research Institute of Mining and Metallurgy, the Changsha Research Institute of Mines, Central South University along with other research institutes and universities jointly developed a 3500 m water depth, 1/10 commercial mining scale mining vehicle and 1000 m water depth test lifting pump and pipeline system. The same system was tested down to 3500 m water depth and the locomotion of the machine at 1300 m water depth was completed. Also, Global Sea Mineral Resources (GSR) conducted the sea trial of the pre-prototype miner in the Belgian contract area and completed the locomotion and collection test of the miner on the seabed in May 2021 (Cheng et. al., 2023a, b).

In 2022, Nauru Ocean Resources Inc. (NORI), a wholly owned subsidiary of DeepGreen and Canada's The Metals Company (TMC), developed a polymetallic nodule collection, transportation, and water surface system, where the ore collector adopts a crawler driving mechanism and a hydraulic ore-collecting mechanism. The deep-sea mining test and environmental impact monitoring were conducted at 4400 m of water depth (Source: https://www.sec.gov/Archives/edgar/data/1798562/000121390021033645/fs42021a2ex96-1_sustainable.htm).

India has conducted the world's deepest underwater locomotion trials for an indigenously developed seabed crawler-based mining machine in the Central Indian Ocean Basin (CIOB). For this sea trial, the crawler was lowered to the ocean's depth from the ship's deck using a custom-built umbilical cable, deep-sea winch, and the stern-mounted A-frame. The descent to a depth of more than 5000 m took more than 4 h while the ship moved dynamically with the motion of the waves. The trial team meticulously monitored and recorded every aspect and parameter of the operation. During the deployment, the system was periodically activated and tested to ensure its proper functioning and integrity throughout the descent. The crawler made contact with the seabed at a depth of 5270 m, and the entire operation was observed in real-time through live video feeds at the control station on the ship. The robotic system was operated remotely from the ship via a 6500-meter-long umbilical cable connection. The seabed locomotion continued at various speeds within the planned mining speed range, covering a distance of over 120 m on the seabed for more than 2 h. All systems performed as anticipated, and all relevant data were collected in accordance with the objectives. The following sections describe the subsystems developed or under development for deep-sea mining at the National Institute of Ocean Technology, Chennai, India.

3 Deep-Sea Mining Components

3.1 Nodule Collection System

The nodule collection system is interfaced on the deep-sea mining system. The nodule collection system consists of two sub-systems, namely pickup system and conveyor system. The pickup system collects the polymetallic nodules during the mining operations and transfers them to the conveyor system. The conveyor system feeds the polymetallic nodules to the crusher. The nodule crushed in the required size will be pumped to the mother ship through a flexible riser system. The nodule collection system testing is shown in Fig. 3.7.

Fig. 3.7 Lab-based nodule collection system testing in bentonite soil. (Janarthanan et al., 2022)

The development of lightweight nodule collection system of the sub-sea mining machine was the prerequisite for reducing the overall underwater weight and volume of the deep-sea mining machine (Janarthanan et al., 2022). The existing nodule collection system configuration was redesigned for weight and volume reduction without compromising the functional capabilities of the nodule collection system. The weight reduction was done with a systematic approach by individual component level design modification and material modifications. The sub-components and assemblies were tested individually on the material interfaces with various operational conditions. The 3-D solid models of nodule collection system versions are shown in Fig. 3.8.

3.2 *Nodule Crushing System*

The underwater mining system has to be very compact, lightweight, and reliable for continuous operations. For efficient operation of the pumping system, the size reduction of the polymetallic nodules has to be done before conveying to the pumping system. The inner diameter of the transport hose is 111 mm. The nodules from the deep seabed are in the range 20–70 mm (Vineesh et al., 2009). For the hydraulic transport of grain-type solids, there is a guiding rule that the grain diameter should be one-third of the hose diameter to avoid choking and clogging. Therefore, in the current scenario, the maximum allowed grain size should be less than 30 mm. The pick-up device in the underwater mining machine is used to collect the nodules with a diameter ranging from 10–100 mm. The requirement of a crusher prior to the nodule grains being sent to the slurry pump is inevitable. Therefore, size reduction of the polymetallic nodules plays an important role in deciding the overall

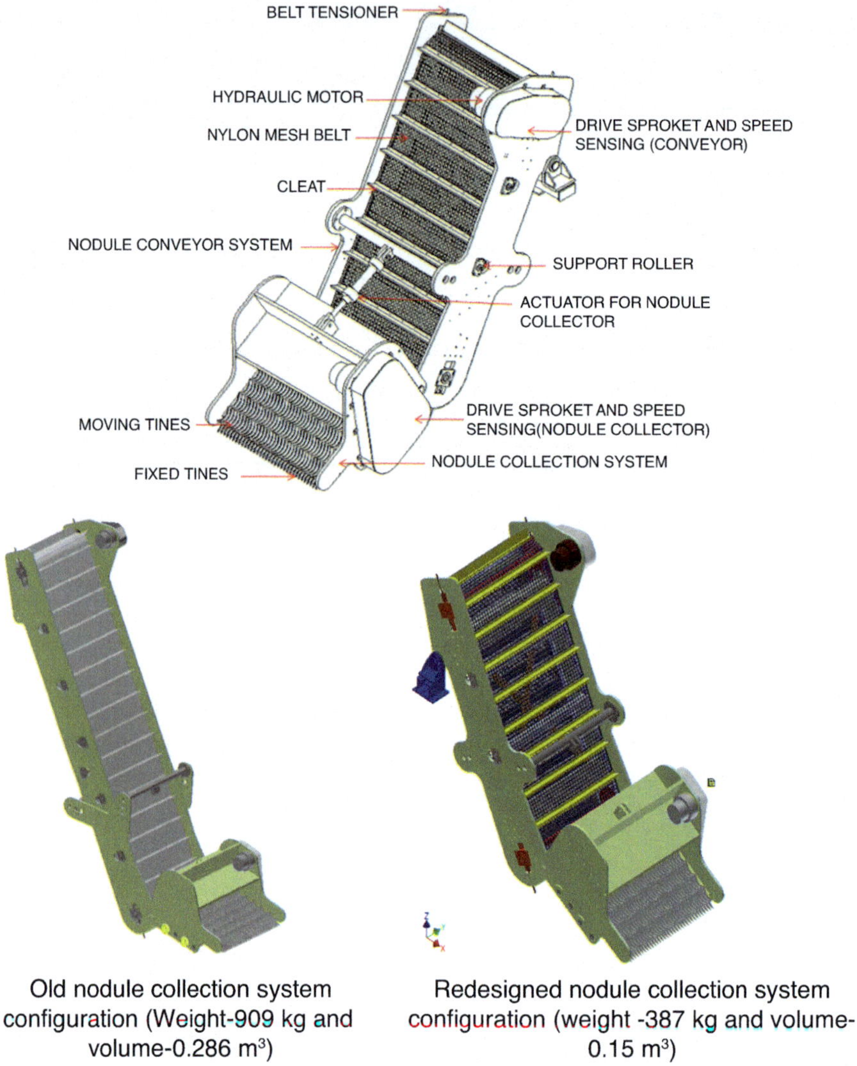

Fig. 3.8 Representation of old and new mechanical-based nodule collection systems. (Janarthanan et al., 2022)

efficiency of the system and so it must be addressed. Moreover, the crusher controls the nodule size distribution to enhance the pumping operations over the vertical riser system. The impact crusher system is light in weight with minimum rotating parts for reliability and also work continuously for a longer duration without any discrepancy. The behaviour of the particles differs with respect to the operating conditions. The nodule crusher can be structured by three different sets of features, which are related to the material flow, the machine, and the system itself. The

Table 3.1 Parameters influencing the design of the crusher system

Feed parameters	Design parameters	Operation parameters
Particle: – Material – Diameter – Shape	Surface: – Number – Material – Shape	Surface: – Displacement – Velocity – Condition Particle: – Moisture – Temperature Particle/surface: – Contact locus
Crushing parameters		
Product parameters	**Interaction parameters**	
Particle: – Fragment size distribution – Fragment shape distribution	Particle: – Probability of breakage Surface: – Contact force – Contact time – Energy consumption – Wear	

machine changes the dispersion of the particulate material flow. Therefore, the product shows different features, such as the diameter, shape, and size distributions of the outgoing particles. The machine transfers energy and applies forces to the particles by contact surfaces which are described by their geometry and material. Furthermore, the operating conditions are determined by the conditions of the contact surfaces either on land or in sub-sea conditions under which the particles are loaded. Assessments of particles in different conditions have been studied in detail (Amudha et al., 2009) showing that during the operation, the particles can behave differently due to drag in sub-sea conditions. The parameters that influence the design of the crusher system, such as the feed, design and operational factors, and the crushing parameters are shown in Table 3.1. The 3-D design model of the crusher unit is shown in Fig. 3.9.

4 Slurry Transportation System

4.1 Slurry Pumping System

In the mining system, the slurry pumping system conveys the solid particles/nodules from the crusher end to the mother ship on the principle of suction. The suction action of the pump causes the solids from the crusher output to be drawn into the system. In the current scenario, we have adopted the positive displacement (PD) pumping system. In the event of a high concentration of solids entering the system,

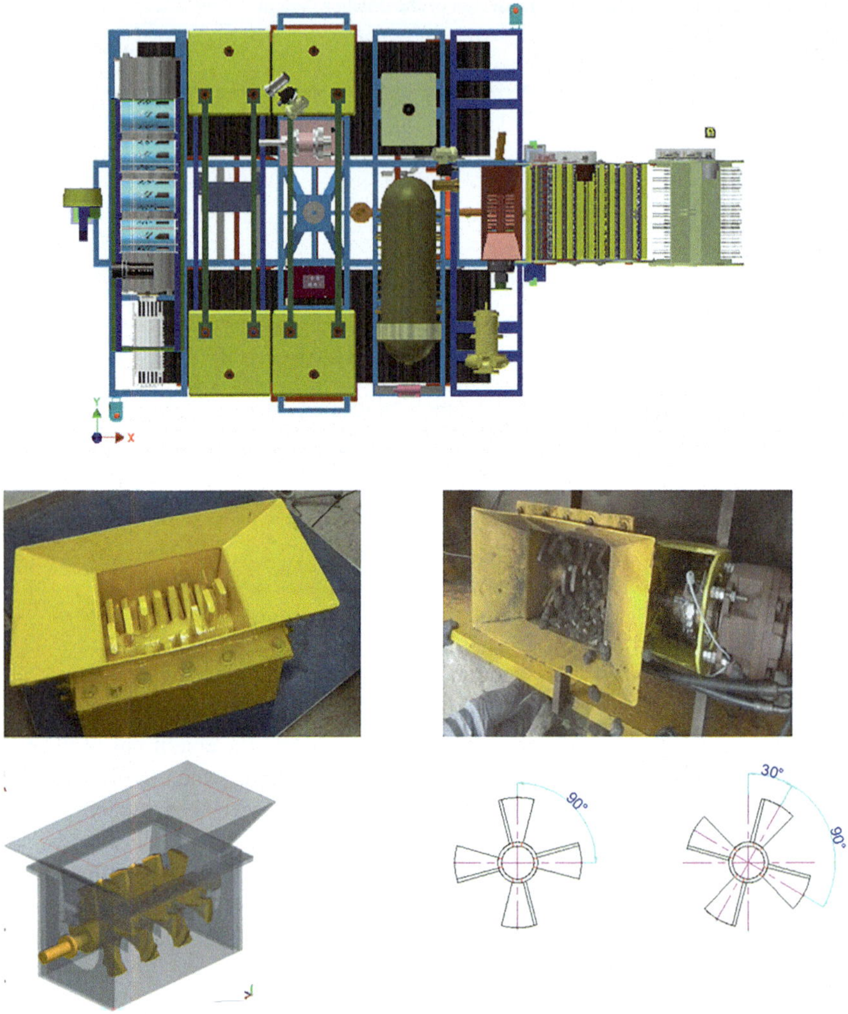

Fig. 3.9 Representation of the crusher unit and its integration with the underwater mining machine

either the collector is lifted up or the crawler speed is reduced to prevent nodule inflow till the system stabilizes. In the pumping design, all the materials of the piston, cylinders, and other parts of the pump that are directly in contact with the solids are manufactured with suitable materials to ensure trouble-free operation under extreme conditions of wear and corrosion. The flow rate of the pump has been calculated for a hose size of 90 mm internal diameter (ID) to be equal to a velocity of 2 m/s. A hydraulically actuated drain valve has been proposed at the bottom of the hopper to facilitate dumping of solids in the event of plugging or power and communication failure.

4.2 Local Pumping System

The coarse polymetallic nodules thus collected by the mining machine are crushed to the required size for transportation to the vertically suspended high-pressure positive displacement pump. A dedicated mechanical collector/pickup device attached to the mining machine picks up the nodules in the size range 1–10 cm. Further, the collected nodules are crushed to particle size less than 3 cm to form nodule slurry. The size reduction is dictated by the size of the hose used for slurry transport to avoid clogging. The crushed nodules slurry will be transported/pumped to the height of 150–200 m from the seabed.

As in a local pumping system, the rotary airlock valve is used to convey material continuously while maintaining a good airlock condition. Though this pump is not a new concept and is used in certain industries, such as mining, food, plastics, and chemicals; for material transport, these have not been used in sub-sea conditions. The valve consists of a rotor that turns at a given rpm in close clearance to its casing, thus maintaining a uniform rate of material flow providing a complete seal and it is normally operated pneumatically. The rotary valve or rotary feeder provides an airlock in the pressure conveying system for land-based applications.

4.3 Deep Water Vertical Slurry Transport

The positive displacement (PD) pumps transport the slurry from the pump system to the mother ship. The critical design challenge of such a system is that it should be compact in size and weight, and consume low power. The vanes that arrest the leakage have to withstand the pressure at a particular water depth with proper sealing to maintain the desired head. The work intends to design and develop a system to achieve a controlled uniform feed of polymetallic nodules at 6000 m water depth. The system has to operate for a continuous mining operation in sub-sea with ultra-high-pressure applications (600 bars) by incorporating suitable seals and bearings. The most interesting feature of the system is that it only requires several vanes which provide sealing without leakage.

Pump Specifications	
Pump type:	Hydraulic Double piston pump with 'S' transfer tube type or equivalent configuration for solids handling with hopper and top plate.
Delivery pressure:	130–160 bar (Normal: 130 bar maximum: 160 bar)
Delivery rate:	Variable from 10 to 60 m^3/h
Medium	
Conveyed medium:	Manganese nodules with water
Structure:	Screened, non-metallic
Average particle size:	30 mm

Particle density of manganese nodules:	1.6–2.5 t/m^3
Concentration:	10% by volume (max)
Rheological behaviour:	Abrasive
Ph value:	6.5–7.5
Temperature:	+2 °C to +25 °C
General	
Operating hours/day:	10 h (continuous)
Pressure class of the delivery line:	To suit the delivery pressure
Ambient temperature:	+2 °C to +40 °C

5 Deep-Sea Umbilical Cable

5.1 *Electrical Power Distribution*

The underwater mining machine is operated by hydraulic power. However, the high voltage high-power sub-sea electric motors consume majority of the electric power and act as prime movers for the hydraulic pump of the underwater mining machine. The electrical system describes the power distribution methodology used for the Integrated Mining System (IMS), which can be classified as:

- Ship-based systems.
- Sub-sea systems.

In the proposed scheme, as the mining machine is being operated onboard the ship, the systems/sub-systems that support the operation of the IMS from the ship are classified as ship-based systems that include power distribution systems, launching and retrieval systems, motor control and drives, and control and monitoring systems. The sub-sea systems comprise the mining machine, the cable, and the slurry transportation hose.

5.2 *Data and Video Communication*

The Integrated Mining Machine for 6000 m has various electrical and instrumentation systems for operation, data acquisition, telemetry, for operation areas monitoring control, and sub-systems operation. The following sections broadly detail the data acquisition and control, and telemetry and software systems of IMS. The Data Acquisition and Control System (DACS) will be a modular type with a stand-alone controller, which is more versatile for system configuration.

5.2.1 Data Acquisition and Control System (DACS)

The DACS used for the mining system is divided into two main parts,

1. Sub-sea DACS
2. Ship-side DACS

5.2.1.1 Sub-Sea DAC System

The sub-sea part of the system has various underwater transducers and control devices interfaced with the Compact Reconfigurable Input Output (CRIO) stand-alone controllers-based modular data acquisition system. The DACS system, that is, the I/O modules and the software of the same company will be used to have tight integration. The system will be assembled in a cylindrical rack and will be housed in pressure enclosures which are rated for an operational depth of 6000 m shown in Fig. 3.10. All the underwater sensors and devices will be connected to the electronic enclosure through underwater cables and bulkheads which penetrate through the dome.

5.2.1.2 Ship-Side DAC System

The data acquisition and control system of the mining machine will be designed, configured, and mounted in an air-conditioned-containerized control console. It is also used to interface with the sub-sea DAC system, with the user interface control devices and the CCTVs that are used for monitoring and controlling the sub-sea and

Fig. 3.10 A sub-sea enclosure with Bulkheads

ship side (deck) operations of mining machine in real time. Both the sub-sea and ship-side DAC systems will be interconnected through TCP/IP (Ethernet) connectivity to have real-time communication between the ship and sub-sea system. Single-mode fibre optic communication is used for this purpose.

5.3 Selection of Umbilical Cable

The umbilical cable is used to launch and retrieve underwater systems like the mining system. Its outer layer supports the weight of the system. It also provides the required power and signal communication to the oceanographic/hydrographic equipment from onboard ORV Sagar Nidhi during the entire operation. A brief specification of the umbilical cable is shown in Table 3.2. The construction of umbilical cable generally has three major sections: (a) inner core, (b) strength member, and (c) outer sheath, as depicted in Fig. 3.11.

Table 3.2 Specification of the umbilical cable

Description	Details
Outer diameter	40 mm
Length	7000 m
Safe working load	105 kN
Minimum bending diameter	1600 mm
Armour strength member	Aramid yarns
Power core	$3 \times 8\ m^2$
Power	120 kVA
Position	Mounted on deck strengthened area

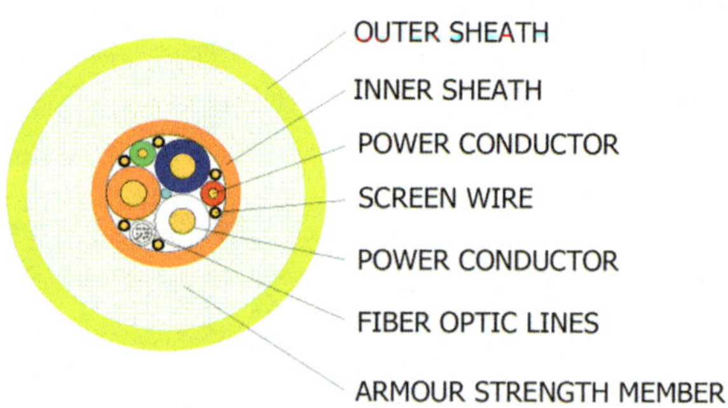

Fig. 3.11 Schematic representation of the cross-section of deep-sea umbilical cable

5.3.1 Inner Core

The inner core consists of power conductors and fibre optic cables. The cables are tightly buffered for sub-sea application for greater depths of operation. The cables are arranged to form a uniform circular cross-section and fillers are used to fill up the void spaces. The cross-section of the deep-sea umbilical cable is shown in Fig. 3.11. The fibre optic cables are usually drawn within a gel-filled stainless tube. The cables are wound in helical layers to have a specific layout to accommodate cable elongation during loading and to counteract tensional forces.

5.3.2 Strength Member

They form the armour for the cable, and it can be either metallic or synthetic material. It provides the cable with the required strength and stiffness, particularly when it needs to handle the weight and in a vertically suspended mode of operation. They are generally woven as a mesh around the inner core/cable, and in the case of synthetic yarns, they are woven in multiple layers to provide the required strength as per the user requirements. In the case of metal strands, they are wound in a helical pattern often changing the lay direction for torque balancing. The strands are wound at 20° to 23° angle. Nowadays, synthetic armours are being used widely due to their non-corrosive properties.

5.3.3 Outer Sheath

They form the outer covering for the cable and protect against abrasion. The inner core has a sheath over which the strength members are woven. Generally, the cable with metallic armour does not have an outer sheath. The sheath material will be elastomeric.

Based on the power transmission, signal communication, and safe handling; the load capacity requirement of the umbilical cable was selected for the deep-sea application. Very few proven manufacturers are available in offshore industries to supply these umbilical cables for deep-sea conditions. Sub-sea mining equipment and systems that are carried/installed onboard the mobile offshore mining unit and exclusively used for sub-sea mining operations are to be designed, constructed, and tested in accordance with the applicable standard requirements like ABS, DNV, etc. In case there does not exist any recognized national, international, or industrial codes/standards, the designer has to submit a document outlining the design basis. The design basis is to cover the technical and operational specifications including the intended service, functionality, and the expected operational conditions and limitations. Such design basis is to be used as the manufacturer's standard and is to be complied with.

6 Different Types of Disturbances Likely from Deep-Sea Mining

Deep-sea mining, an industry poised to meet the growing global demand for minerals and metals, faces a unique set of environmental challenges that require innovative solutions. Among these challenges are dewatering, sediment disturbances, light, noise, and the introduction of foreign materials into the deep ocean environment. Each of these factors presents complex issues that demand careful consideration in order to minimize their impact (Leal Filho et al. 2021).

6.1 Dewatering

Dewatering is a critical aspect of deep-sea mining operations, as it involves the removal of water from near the seafloor of the mining sites. The challenge lies in managing the discharged water, which can contain fine sediments, chemicals, and other contaminants. To address this, advanced dewatering systems and treatment technologies are employed to ensure that the discharged water meets stringent environmental standards. This minimizes the risk of waterborne pollution and ecosystem disruption.

6.2 Sediment Disturbances

The extraction of minerals from the deep ocean floor invariably disturbs sediments, potentially releasing sediment plumes into the water column. These plumes can smother marine life and impact the surrounding environment. Scientists and engineers engage in sediment disturbance studies, utilizing the models and monitoring equipment to predict, assess, and mitigate the consequences of sediment plumes. Responsible mining practices and real-time monitoring are required to be employed to minimize the disturbances and protect the sensitive ecosystems.

6.3 Light and Noise Pollution

Deep-sea mining equipment, such as remotely operated vehicles (ROVs) and subsea machinery, emit light and generate noise. These factors can disrupt the natural behaviours and habitats of deep-sea organisms, many of which are adapted to the darkness of the abyss. To mitigate these disturbances, low-impact lighting and noise-reduction technologies are required to be developed and integrated. These solutions will help strike a balance between operational needs and minimizing the ecological footprint of mining activities.

6.4 Foreign Material Introduction

The introduction of foreign materials, including plastics and debris, into the deep ocean poses a significant threat to marine life. Stringent waste management practices are essential to prevent pollution, and recovery systems are employed to retrieve any accidental foreign materials. Recycling and proper disposal are fundamental to maintaining the pristine nature of the deep-sea environment.

7 Conclusion

Deep-sea mining technologies are at a crucial juncture in developing prototype systems that can be upscaled and integrated into commercial-scale systems in the future. Given the extreme operating conditions (such as hurricanes and high waves, high underwater pressures upto 600 bars, low temperature of 2–3 °C, cross-cutting currents, hundreds to thousand-kilometre distance from shore, tight maintenance schedules while operating for 250–300 days per year, unforeseen system malfunction, damages, losses, etc.), developing deep-sea mining systems faces numerous challenges that need to be overcome in order to achieve the expected outcomes of supplying the required minerals and in turn metals to meet the demands of future generations as well as in conserving the marine environment. Addressing issues of dewatering, sediment disturbances, light and noise pollution, and the introduction of foreign materials is imperative for the responsible and sustainable development of deep-sea mining systems. By combining innovative technologies, stringent environmental regulations, and a commitment to safeguarding the unique ecosystems of the deep ocean, the industry can aspire to harmonize resource extraction with ecological preservation. As this field continues to evolve, the delicate balance between meeting resource demands and environmental conservation remains at the forefront of deep-sea mining endeavours.

References

Amudha, K., Rajesh, S., Ramesh, N. R., Babu, M., Abraham, R., Deepak, C. R., & Atmanand, M. A. (2009). Development and testing of remotely operated artificial nodule laying system at 500-m water depth. In *Proceedings of the ISOPE ocean mining symposium* (pp. 233–238).

Blanchard, C., Harrould-Kolieb, E., Jones, E., & Taylor, M. L. (2023). The current status of deep-sea mining governance at the International Seabed Authority. *Marine Policy, 147*, 105396. https://doi.org/10.1016/j.marpol.2022.105396

Charles, C., Herrouin, G., Mauviel, F., & Bernard, J. (1990). Views on future nodule technologies based on IFREMER-GEMONOD studies. *Materials Science, 14*(3/4), 299–326.

Cheng, Y., Dai, Y., Zhang, Y., Yang, C., & Liu, C. (2023a). Status and prospects of the development of deep-sea polymetallic nodule-collecting technology. *Sustainability., 15*(5), 4572.

Cheng, Y., Dai, Y., Zhang, Y., Yang, C., & Liu, C. (2023b). *Status and prospects of the development of deep-sea polymetallic nodule-collecting technology*. Special issue Frontiers in Deep Sea Mining: Current Status and Future Prospects. MDPI.

Chung, J. S., Park, I.-K., & Min, K. S. (1997). *An R&D plan for deep-ocean manganese nodule production technology-development of a TIS program* – 2nd ISOPE ocean mining symposium, 24–26 November 1997, Seoul, ISOPE-M-97-001.

Cronan, D. S. (2005). Sedimentary rocks | oceanic manganese deposits. In *Encyclopedia of geology* (pp. 113–120). Elsevier. https://doi.org/10.1016/B0-12-369396-9/00499-8

Deepak, C. R., Shajahan, M. A., Atmanand, M. A., Annamalai, K., Jeyamani, R., Ravindran, M., Schulte, E., Handschuh, R., Panthel, J., Grebe, H., & Schwarz, W. (2001). *Developmental tests on the underwater mining system using flexible riser concept*. Fourth ISOPE Ocean Min. Symposium.

Deepak, C. R., Ramji, S., Ramesh, N. R., Babu, S. M., Abraham, R., Shajahan, M. A., & Atmanand, M. A. (2007). *Development and testing of underwater mining systems for long term operations using flexible riser concept. Seventh ISOPE Ocean Min. Symposium.*

Janarthanan, C., Chandran, V., Sundaramoorthi, V., Vishwanath, B. O., Venkatesan, K., Muthuvel, P., Muthukrishna Babu, S., Gopkumar, K., & Ramadass, G. A. (2022). Light weight nodule collection system for deep sea mining machine. In *OCEANS 2022 – Chennai* (pp. 1–7). IEEE. https://doi.org/10.1109/OCEANSChennai45887.2022.9775130

Kang, Y., & Liu, S. (2021). The development history and latest progress of deep-sea polymetallic nodule mining technology. *Minerals, 11*, 1–17. https://doi.org/10.3390/min11101132

Knodt, S., Kleinen, T., Dornieden, C., Lorscheidt, J., Bjørneklett, B., & Mitzlaff, A. (2016). *Development and engineering of offshore mining systems – State of the art and future perspectives*. https://doi.org/10.4043/27185-MS

Kuntz, G. (1980). Advantages of submersible motor pumps in deep-sea mining. *Journal of Petroleum Technology, 32*, 2241–2246. https://doi.org/10.2118/8551-pa

Leal Filho, W., Abubakar, I. R., Nunes, C., Platje, J., Ozuyar, P. G., Will, M., Nagy, G. J., Al-Amin, A. Q., Hunt, J. D., & Li, C. (2021). Deep seabed mining: A note on some potentials and risks to the sustainable mineral extraction from the oceans, geological oceanography. *Journal of Marine Science and Engineering, 9*(5), 521.

Masuda, Y., Cruickshank, M. J., & Mero, J. L. (1971). Continuous bucket-line dredging at 12,000 feet. In *Proceedings of the annual offshore technology conference* 1971-April, 837–841. https://doi.org/10.4043/1410-ms

Meinert, L. D., Robinson, G. R., & Nassar, N. T. (2016). Mineral resources: Reserves, peak production and the future. *Resources*. https://doi.org/10.3390/resources5010014

Spickermann, R. (2012). *Rare earth content of manganese nodules in the Lockheed Martin Clarion-Clipperton Zone exploration areas*. Proceedings offshore technology conference, Houston Texas.

Vineesh, T. C., Nagender Nath, B., Banerjee, R., Jaisankar, S., & Lekshmi, V. (2009). Manganese nodule morphology as indicators for oceanic processes in the Central Indian Basin. *International Geology Review, 51, 27–44.* https://doi.org/10.1080/00206810802622773

Wang, R., Zhu, Z., Su, X., Mianowicz, K., Jia, H., Wu, K. (2022). Slurry pumps in deep-sea mining: A review of numerical and experimental studies. *Ocean Eng, 251, 111150.* https://doi.org/10.1016/j.oceaneng.2022.111150.

Dr. M. A. Atmanand has 38 years of experience in the field of ocean technology and retired as the Director of the National Institute of Ocean Technology Chennai. He is currently working as a Visiting Professor at Indian Institute of Technology, Madras, Chennai. He received the Ministry of Earth Sciences (MoES) National Award of Excellence in Ocean Technology, in view of pioneering work done in the area of deep-sea technologies in 2020, Life time achievement award from IEEE Madras for 2021, the Dr. D. Srinivasan Endowment Award for the year 2021, instituted by the Ocean Society of India (OSI), for outstanding contributions in the areas of Ocean Engineering and Technology, the International Society for Ocean and Polar Engineers award in 2020. the Institute of Electrical and Electronics Engineers (IEEE) Oceanic Engineering Society, Presidential Award in 2016. Dr. Atmanand has published more than 160 papers including international journals, international conferences, book chapters, national conferences, and others. He has delivered more than 300 invited lectures in conferences. He is the founder Chair of IEEE Oceanic Engineering Society in India.

Mr. K. C. Janardhanan, Scientist "E", is a design engineer working in deep-sea mining programme, India. He completed his Master's degree at PSG College of Technology, Tamil Nadu. He has done design and dynamic analysis of deep-sea mining systems in India, such as the development of tracked vehicle design for underwater mining machines and the development of subsystems for deep-sea mining. He has published more than 18 papers including international journals and international conferences.

Ms. K. Amuda is Scientist "E" at NIOT in deep-sea mining programme, India. She has done pioneering work in the area of deep-sea technologies, such as underwater deep-sea mining system design and simulations, especially for crusher and nodule collection systems, and is also involved in the design of mooring systems. She has published many research papers in reputed international journals and presented her research work in many national and international conferences.

Mr. A. Gnanaraj, Scientist "E", is a mechanical engineer working in deep-sea mining programme, India. He completed his Bachelor's degree in mechanical engineering and Master's degree in MBA-Operational from the University of Madras, Chennai. He is involved in developing integrated mining system protocols and associated systems analyzing deep-water flexible riser systems and launching and recovery systems of mining systems. He is involved in design and realisation of remotely operated in situ deep water soil tester (6000 m) and pressure-rated and compensated enclosures for more than 6000 m depth operations. He also maintains Hyperbaric Test Facility for pressure testing and qualification of sub-sea components/equipment. He has published a few research papers and has five patents to his credit.

Mr. S. Rajesh is Scientist F, mechatronics system design, and is pursuing PhD at IIITDM. He is in charge of hydraulic systems for underwater applications and slurry pumping system (IMS) for deep-sea mining programme, India.

Dr. P. Muthuvel received BE degree in Instrumentation and control from the University of Madras in 1998; ME degree from the College of Engineering, Guindy, Anna University, Chennai, in 2009; and PhD degree in electrical engineering from the Indian Institute of Technology (IIT) Madras, Chennai, India, in 2021. He has 25 years of experience in the field of Instrumentation and control and underwater autonomous system development. He is currently working as a scientist with the National Institute of Ocean Technology, Chennai, India. He has done pioneering work in the area of deep-sea technologies in India. He received the Young Scientist Award from the Ministry of Earth Sciences, India, in 2005. He has published more than 40 papers in international journals and national and international conferences. He has been granted two patents and has filed for more than six patents.

Cdr. Gopkumar Kuttikrishnan is Scientist "G" and Project Director for deep-sea mining programme India. He completed his Masters degree from University of Pune. He has done pioneering work in the area of deep-sea technologies in India, such as the development of underwater mining machines and the development of sub-systems for deep-sea mining. He has published more than 20 papers including international journals and national and international conferences.

Dr. G. A. Ramadass, Director, National Institute of Ocean Technology, Chennai, India, holds a PhD degree (Physics) from Indian Institute of Technology, Madras, India. As a chief scientist of Deep-Sea Technologies group, he has been involved in the development of technology for remotely operable vehicle, deep-water drilling system, gas hydrate exploration, and mining of polymetallic nodules. He has four patents (two international, two national) and has published several scientific papers in international journals as well as national and international conferences.

Chapter 4
Handling of Bulk Solids in a Marine Environment, from Seabed to Shore

Dingena Schott, Edwin de Hoog, Jort van Wijk, and Rudy Helmons

Abstract In this chapter, the work that is being done in relation to machine-cargo interactions relevant to deep-sea mining is elaborated. To ensure safe and efficient operations across the entire mining value chain, it is important to be aware of the implications of certain design and process decisions. An overview of mechanisms influencing the mechanical response of the bulk materials and the main effects leading to (mechanical) degradation of the ore, e.g., fragmentation, abrasion, is presented. Although the concepts are applicable to each of the deep-sea deposits, our focus is on polymetallic nodules in a riser-based concept.

Results are discussed of experiments in which nodules are fragmented due to particle-particle collisions and collisions with different handling equipment, such as the seabed harvester, riser pipe, pump impeller, pipe bends, etc. Next to fragmentation, degradation due to abrasion occurs due to particles rolling, sliding, and collid-

D. Schott (✉)
Department of Maritime and Transport Technology, Faculty of Mechanical Engineering, Delft University of Technology, Delft, The Netherlands
e-mail: D.L.Schott@tudelft.nl

E. de Hoog
Department of Maritime and Transport Technology, Faculty of Mechanical Engineering, Delft University of Technology, Delft, The Netherlands

Royal IHC, Kinderdijk, The Netherlands
e-mail: e.dehoog@royalihc.com

J. van Wijk
Independent Consultant Hydraulic Transport System Engineering, Sliedrecht, The Netherlands

Geotechnical Advisory Services, IQIP, Sliedrecht, The Netherlands

R. Helmons
Department of Maritime and Transport Technology, Faculty of Mechanical Engineering, Delft University of Technology, Delft, The Netherlands

Institute of Geoscience and Petroleum, Norwegian University of Science and Technology (NTNU), Trondheim, Norway
e-mail: R.L.J.Helmons@tudelft.nl

R. Sharma (ed.), *Deep-Sea Mining and the Water Column*,
https://doi.org/10.1007/978-3-031-59060-3_4

ing with each other, resulting in the generation of nodule fines. Based on the same set of nodules, mechanical bulk properties of dry and wet nodules are studied. The obtained results provide relevant insights for the design of the nodules handling equipment. Furthermore, modelling approaches applied to other fragile bulk materials where the breakage and generation of fines play an important role are outlined. It is described how these modelling approaches can assist in the design of handling equipment, and recommendations are given for next steps to further optimize their design.

Keywords Polymetallic nodules · Degradation · Fragmentation · Friction

Abbreviations

BPM	Bonded particle model
DEM	Discrete element method
FBM	Fast-breakage model
MSV	Mining support vessel
PmN	Polymetallic nodules
PRM	Particle replacement model

1 Introduction and Overview of Handling Steps

Deep sea mining has emerged as a promising solution to meet the growing demand for critical metals and minerals. Figure 4.1 shows a schematic of the steps involved in transport from seabed to shore, where the manganese or polymetallic (PmN) nodules will be processed. The nodules that will be picked up from the seabed, subsequently handled by the seabed harvester, and transported vertically to the mining support vessel (MSV). Various vertical transport concepts exist, such as hydraulic, airlift and skip-lift concepts. At the MSV, handling of bulk materials takes place optionally with processing steps such as dewatering and filtering of fines. The excess of water and sediments is to be discharged subsea.

While the MSVs work far offshore, the material will be transhipped to vessels that operate between the MSV and a location on shore. Once the material is onshore it might be first handled and stored at a bulk storage terminal before being transported to the nodules processing plant. At this moment, such a supply chain is not yet practically established, although a first trial focussing on the first steps in Fig. 4.1 resulted in the collection from the seabed to the MSV of about 4500 tons PmN (Allseas, 2022). It goes without saying that for an optimal, sustainable and energy-efficient supply chain, the focus should not only be on developing the initial steps, but efforts are required to develop the logistics system including all the handling steps from seabed to shore.

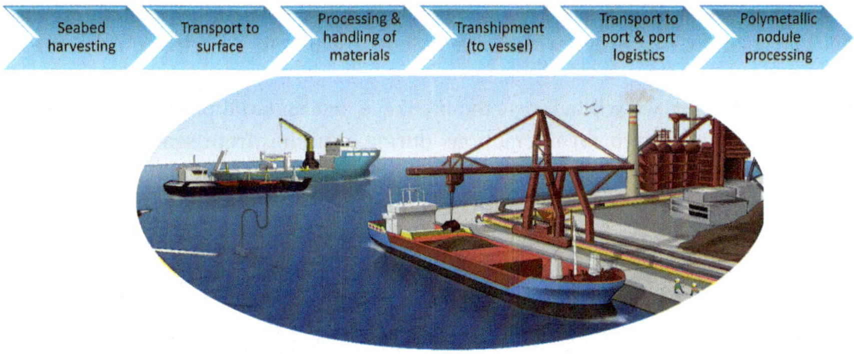

Fig. 4.1 Deep-sea mining value chain and bulk handling steps from MSV to processing facility

Starting from the right side in Fig. 4.1, the demands of the polymetallic nodule extraction plant in terms of (1) required material quality/properties (form, size and distribution) depending on the extraction processes and (2) infeed capacity are crucial to design the logistic system. These demands are posed to the previous steps upstream in Fig. 4.1.

For the mining capacity in general, if the processing capacity is too low relative to the mining capacity, there may be a bottleneck in the supply chain as the raw material accumulates and waiting times for processing increase. This can lead to increased costs and delays in the supply chain. On the other hand, if the processing capacity is too high relative to the mining capacity, the supply chain may be underutilized, and there may be excess capacity in downstream operations, leading to wasted resources and increased costs.

For the material quality, the requirement depends on the type of extraction process, which can be divided into two main streams (Monhemius, 1980), pyro–hydrometallurgical routes and hydrometallurgical processes, where drying of the wet nodules is not necessary. Randhawa et al. (2016) describe different processing routes in detail and indicate crushing and grinding steps for all processes. The ideal material properties for the extraction processes and state of the material (wet or dry) from downstream should be considered in the design of the logistics systems and its handling components. In every handling step, due to the interaction of the material with the equipment, breakage of the material might happen.

It is important to reduce the breakage of nodules during the handling steps from seabed to shore in deep-sea mining operations for several reasons. Firstly, nodules are relatively fragile and can easily break apart during the handling steps and transport. This can result in a significant loss of valuable material, reducing the overall yield of the operation and potentially impacting its economic viability. Secondly, the broken nodules can generate fine particles or dust, which can be difficult to handle and may cause problems during processing or transportation or are economically unattractive to separate at the MSV. Fine particles can cause problems such as clogging of pipes and increased wear and tear on machinery and can even be hazardous to workers' health, if inhaled. Lastly, broken nodules can also have a negative impact on the marine environment as the fine particles generated by breakage can get dispersed over a wider area, potentially affecting marine ecosystems and biodiversity.

In this chapter, we will address the breakage and material properties in the context of material equipment interactions during handling from seabed to shore. In Sect. 2, we address degradation mechanisms in relation to hydraulic transport processes and provide quantitative results of laboratory tests. In Sect. 3, the mechanical response of the bulk and its machine cargo interactions will be reported. Issues related to particle breakage, degradation and the creation of fines are also common in the biomass handling field. Hence, Sect. 4 will cover the experimental and numerical methods used in research on the handling of biomass pellets specifically. Section 5 concludes this chapter by summarizing the key findings and discussing their implications for the outlook of deep-sea mining.

2 Fragmentation of Polymetallic Nodules Due to Impacts from Seafloor to MSV

From the bulk processing and mining industries, a method is available to quantify and measure the wear or degradation of minerals and ores (Beekman et al., 2002). However, in the context of deep-sea mining, we first need to approach this problem from the viewpoint of hydraulic transportation as it is the most considered transportation option among deep-sea mining concepts. Since the reduction of the size of the nodules and generation of nodule dust determine the design choices, of processing equipment on the MSV to a large extent, the degradation mechanisms associated with the transport processes need to be evaluated to enable further quantification.

This section gives an overview of work done to assess the magnitude of polymetallic nodule degradation during deep-sea mining. First, we explain degradation mechanisms and relate them to hydraulic transport processes in typical pipeline concepts. Second, we quantify a bandwidth of degradation rates and effects, based on laboratory tests. Each test is dedicated to investigating one specific degradation mechanism, and the results are summarized.

2.1 Degradation Mechanisms

Vertical transport of polymetallic nodules to the sea surface can be realized by a number of different technologies, of which mechanical lifting, airlifting and centrifugal pump-based hydraulic transport are the most prominent candidates (Ma, 2019). Airlift technology is proven on full-scale, however, controllability and sufficiently high production rates are a challenge (Ma et al., 2017). Hydraulic transport, by means of a riser system with centrifugal pump booster stations, has the potential of stable, large production rates, however at the cost of more technology components being below sea level (van Wijk, 2016).

The actual particle size distribution (PSD) of the mixture leaving the riser is a key design parameter for the processing equipment (both on the mining support vessel and on land). The degradation of nodules to particle size fractions smaller than those that can be effectively recovered during the processing of the slurry on board the mining vessel means loss of valuable ore and aggravates the environmental pressures related to the discharge of fines in the marine environment. It is, therefore, important to have predictive models and data that allow for quantification of nodule degradation.

Particle degradation occurs in all multiphase flows involving solids in a carrier fluid. Depending on the force, magnitude and direction, the modes of degradation are attrition, fragmentation, abrasion, and chipping (see Fig. 4.2) (Laarhoven et al., 2012).

The degradation process is closely linked to the hydraulic transport process. The degradation process is complex, involving both impact fracturing as well as abrasion due to friction. Impact fracturing mainly occurs in centrifugal pumps, while abrasion mainly occurs in the pipeline (Worster & Denny, 1955). Bergh and Grima (2006) have reported on degradation experiments with limestone and coarse gravel in a test circuit and on the degradation of quartz, sand and rock salt during dredging. These four material types showed both impact fracturing by the centrifugal pump and abrasion due to the horizontal hydraulic transport, but the relative contribution of one mechanism over the other differs between the cases. Especially rock salt and coarse gravel were influenced by impact fracturing occurring in the centrifugal pump.

Yamazaki and Sharma (2001)experimentally studied the degradation of subsea sediments from the CCZ zone. They used a small-scale circulating flow loop, including one pump and a few metres of pipeline, operated for 30 min. The resulting PSDs point at the production of significant amounts of fines. The maximum sediment diameter in their test reduced from about 200 μm to 50 μm. This indicates that the already fine seabed sediment will become even finer upon transport. Yamazaki et al. (1991) show the PSD of abraded nodules after passing through the pump and

Magnitude	Direction	
	Normal	Tangential
Low (wear dominant)	Attrition 	Abrasion 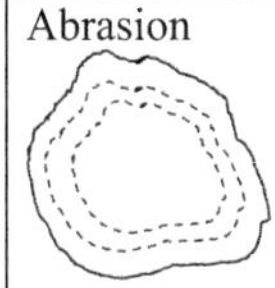
High (fracturing dominant)	Fragmentation 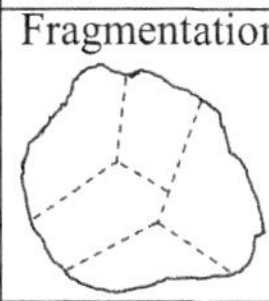	Chipping 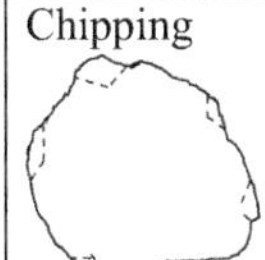

Fig. 4.2 Particle degradation depends on the applied force and its direction based on terminology by Beekman et al. (2002). (Figure from: Laarhoven et al. (2012), Copyright Elsevier 2012, reprinted with permission)

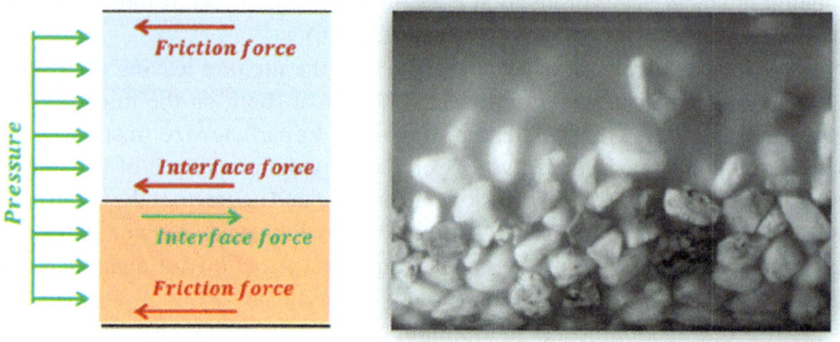

Fig. 4.3 A sliding bed of particles with shear layer and its associated force balance. This situation typically occurs in the flexible jumper hose. (From: de Hoog (2016), reprinted with permission)

riser pipe. Unfortunately, no initial PSD is known, so no conclusions about degradation rates could be drawn. A closer look at the individual transport processes is necessary to quantify the relative contributions of the different degradation mechanisms.

The slurry first enters the SMT pump, where it is subjected to mainly impact fracturing in the pump. The lazy-wave configuration of the jumper hose will mainly result in heterogeneous hydraulic transport. When particles are too large to be suspended by turbulence, they will form a sliding bed layer at the bottom of the pipe (Vlasak et al., 2013) (Ravelet et al., 2013) (de Hoog et al., 2017; Spelay et al., 2016) (see Fig. 4.3). Such a sliding bed of nodules and sediments is predominant in the inclined and horizontal parts, above which a shear layer will develop. This sliding bed transport regime will mainly cause attrition and abrasion of the nodules by sliding over the pipeline bottom, by Coulombic friction. In addition, attrition and abrasion are also caused by particle layers sliding and tumbling over each other in the shear zone.

In the riser segments, there are two mechanisms involved. The first is particle abrasion by particles sliding along the riser segments. The second mechanism is degradation by particle interaction in the riser segments. Different particle fractions have different transport velocities (van Wijk et al., 2016), and their relative velocities give rise to collisions and probably mild impact fracturing or attrition. Particle–wall interaction will have velocities in the order of the pipeline bulk velocity, that is several meters per second. Particle–particle interaction will have velocities in the order of the hindered settling velocity, which is an order of magnitude smaller than the bulk velocity (van Wijk, 2016). In the case of vertical transport with centrifugal pump booster stations, the bends and pumps in the booster give rise to degradation by impact fracturing.

Summarized, Table 4.1 provides an overview of all degradation mechanisms expected during hydraulic transport.

Table 4.1 Main transport mechanisms and their associated degradation mechanisms

VTS component	Transport mechanism	Degradation mechanism
Jumper	Heterogeneous flow, shear layer	Abrasion by friction (sliding)
Riser	Pseudo-homogeneous flow	Abrasion with riser wall
Riser	Pseudo-homogeneous flow	Inter-particle collision
Pumps, boosters	Highly turbulent, many machine–nodule interactions	Impact fragmentation

2.2 Fragmentation of Nodules Due to Impact

In Sect. 2.1, we discussed the different transport processes occurring in the SMT and VTS, and we related these transport processes to degradation mechanisms. The next step is the quantification of the different degradation mechanisms. In this section, we first investigate impact fragmentation.

The impact fracturing of polymetallic nodules in centrifugal pumps shows a complex interplay between particles and the pump. The role of centrifugal pumps in the degradation of coal was investigated in the 1980s (Shook et al., 1979; Gilles et al., 1982). Despite the many similarities with hydraulic transport for deep-sea mining, the work on coal transport is too specifically dedicated to horizontal transport under atmospheric conditions to be representative and of direct use for vertical transport from the deep sea.

Especially the very large ambient pressure in the deep sea is an important difference that needed further investigation. To this end, a special centrifugal pump test setup was constructed in the high-pressure basin at the Royal Netherlands Institute for Sea Research, NIOZ, The Netherlands (van Wijk et al., 2019).

A polymetallic nodule is formed as various minerals precipitate from the surrounding water on a nucleus, typically, an older nodule, shark tooth, volcanic rock or shell. As the nodule grows, a layered internal structure is formed, commonly with local sediments trapped between the layers.

These sediments compromise the strength of the nodule and cause breakage typically to occur along these weak points. Larger nodules consist of more layers and weak points; therefore, they are usually weaker than smaller nodules as is evident by compressive strength tests (Dreiseitl, 2017) (van Wijk & de Hoog, 2020). Some nodules are an agglomeration of smaller nodules that grew together over time.

The test setup at the NIOZ was constructed to accommodate the largest nodules that can reasonably be expected during normal VTS operation, for example $d/D_i \approx 3/5$. From a sample of UK and Belgian nodules from the CCZ (fully saturated with fresh water at 5°), the 16–32 mm fragments were chosen to fit the test setup. In contrast to the rounded appearance of intact nodules, these fragments were mostly of irregular and angular shape. The fragments have been formed due to the disintegration of intact nodules along the internal weakest parts. The fragments themselves could therefore be relatively strong compared to the original intact nodules.

The test setup is depicted in Fig. 4.4. After the passage of the pump, the nodule enters a hydro cyclone where the fine grit and coarser fragments are separated to enable detailed analysis of the particle size distributions. In the tests, nodules were passed through the setup applying different pump velocities (750, 1200 and 1800 rpm, respectively) and ambient pressures of 5, 95, 189, 283, 378 and 500 bar. Nodules and nodule grit recovered after each test run were wet-sieved over a series of sieves with logarithmically increasing mesh size from 0.063 to 32.00 mm. The retained fractions were freeze-dried and weighed. Fractions below $d < 63$ μm were measured with a laser particle analyzer. Next to single runs carried out under one particular combination of pump speed and pressure, tests were carried out simulating the passage of nodules through the full series of pumps of the VTS positioned at different water depths. While running the pump at different velocities, nodules and nodule grit were passed multiple times through the pump, consecutively at 500, 378, 283, 189, 95 and 5 bar. It should be noted that the fine-grained material retrieved from the water contained abundant chips of blue paint abraded from the interior of the hydrocyclone, providing evidence of abrasion occurring within the hydrocyclone, adding fine-grained material to the fractions produced in the pump.

Figure 4.5 shows the particle size distributions produced at different ambient pressures but similar pump speeds of 750 and 1800 rpm, respectively. This figure shows that the distribution of material after impact fracturing is hardly influenced by the ambient pressure. The results at 1800 rpm show a similar trend, but far more material has fractured and moved into smaller size classes. The single passage experiments show hardly any influence of the high ambient pressure conditions and a large influence of nodule velocity or impeller speed. The sequential experiments, in which the same batch of nodules was pumped at decreasing ambient pressure, with six stages in total, are therefore not expected to be influenced by the changing ambient pressure. Pump speed however does have a large influence on nodule degradation, so for the sequential experiments, more degradation is expected at higher pump speeds as well. Figure 4.6 shows the particle size distributions after six pump passages at 750 rpm and 1800 rpm. The consequence of the increased pump speed now is very clear, with only 5% of nodules left in the original size class at 1800 rpm compared with 40% at 750 rpm. The increase in the $63 < d < 0$ μm size class even amounts to 20% at 1800 rpm. Abrasion in the hydrocyclone is expected to have greatly contributed to the production of fines since the sequential experiments took much more time than single passage experiments, thereby promoting the abrasion process.

The impact fragmentation experiments in the pressure tank at NIOZ showed that for ambient pressures in the range of $5 \leqslant p \leqslant 500$ bar, the ambient pressure has no significant influence on the nodule fragmentation. The experiments showed that nodule degradation in the pump.

strongly relates to the nodule velocity and the speed of the impeller, as expected due to the large energy of the impeller.

After the experiments at NIOZ with nodule fragments, there was a need to better understand the actual impact mechanics and degradation rates of intact nodules. Therefore, atmospheric impact experiments were conducted at well-controlled

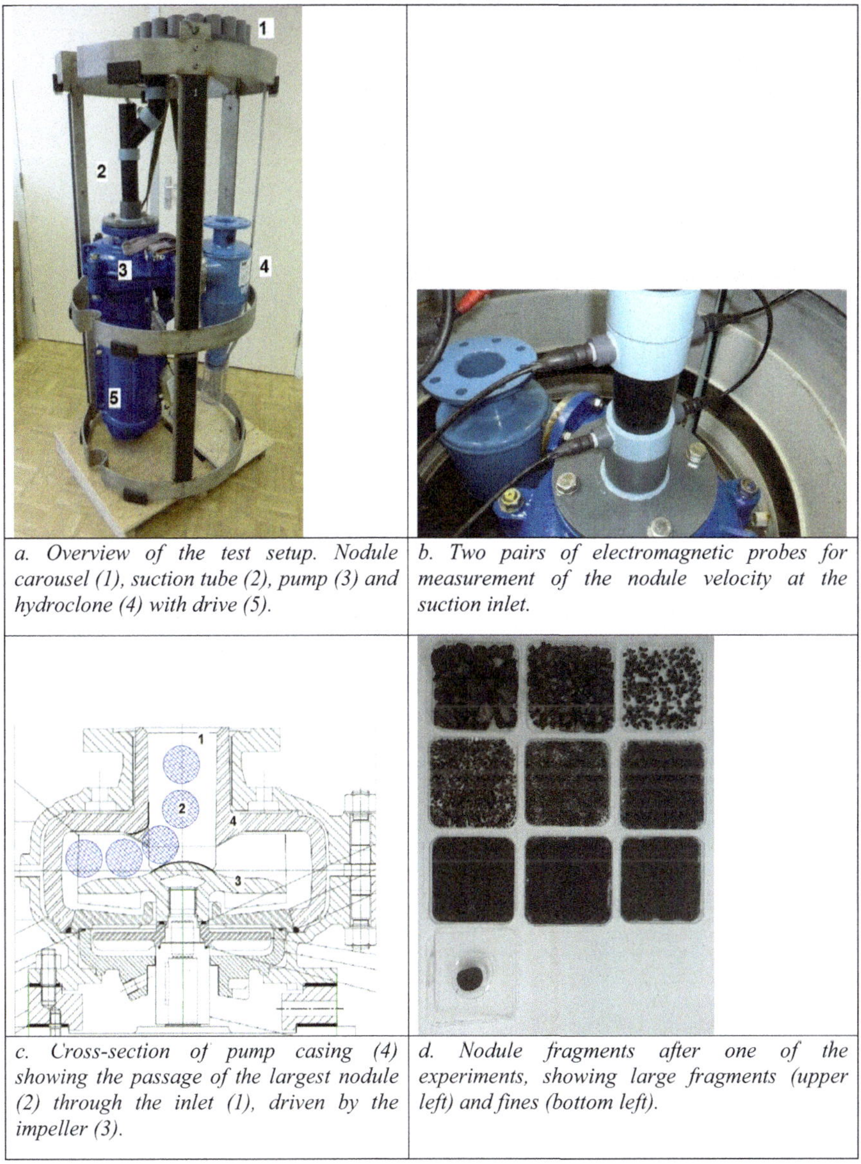

a. Overview of the test setup. Nodule carousel (1), suction tube (2), pump (3) and hydroclone (4) with drive (5).

b. Two pairs of electromagnetic probes for measurement of the nodule velocity at the suction inlet.

c. Cross-section of pump casing (4) showing the passage of the largest nodule (2) through the inlet (1), driven by the impeller (3).

d. Nodule fragments after one of the experiments, showing large fragments (upper left) and fines (bottom left).

Fig. 4.4 Centrifugal pump test rig at NIOZ and typical nodule fragments. (**a**) Overview of the test setup. Nodule carousel (1), suction tube (2), pump (3) and hydroclone (4) with drive (5). (**b**) Two pairs of electromagnetic probes for measurement of the nodule velocity at the suction inlet. (**c**) Cross-section of pump casing (4) showing the passage of the largest nodule (2) through the inlet (1), driven by the impeller (3). (**d**) Nodule fragments after one of the experiments, showing large fragments (upper left) and fines (bottom left). (From: van Wijk et al. (2019), Copyright 2020 Elsevier, reprinted with permission)

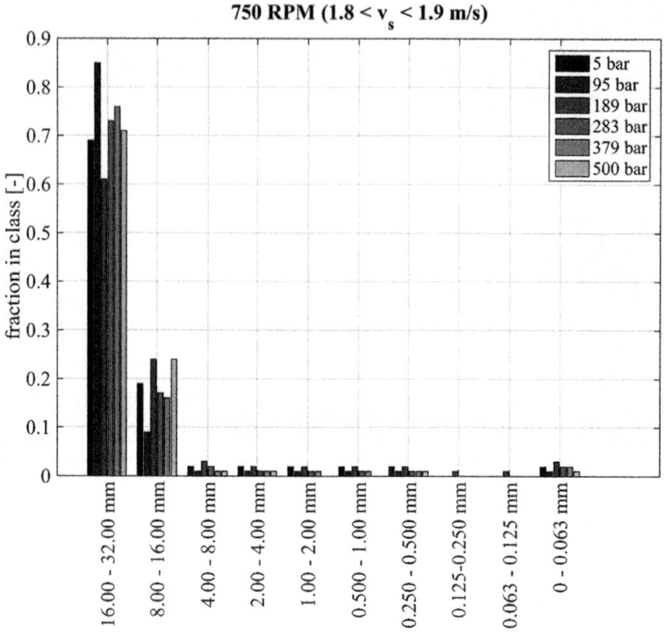

a. *750 RPM test.*

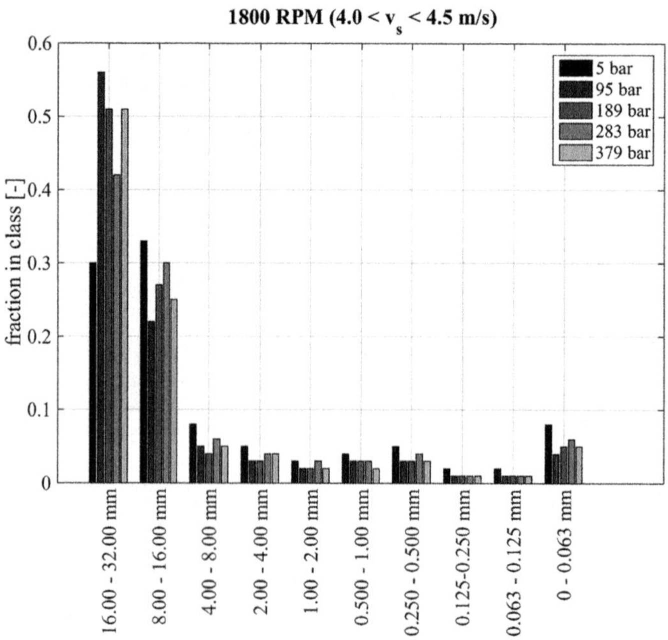

b. *1800 RPM test.*

Fig. 4.5 Comparison of PSD at different ambient pressures at two different pump speeds and associated fluid velocities in the suction pipe. The ambient pressure hardly has an influence on nodule degradation, while pump speed does have an influence. (**a**) 750 RPM test. (**b**) 1800 RPM test. (From: van Wijk et al. (2019), Copyright 2020 Elsevier, reprinted with permission)

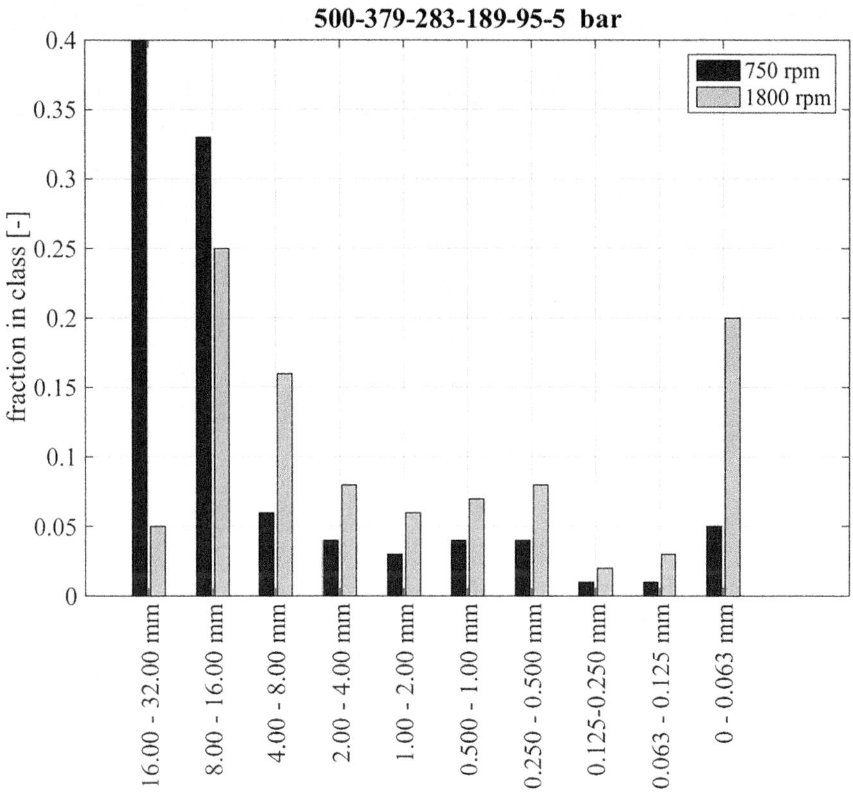

Fig. 4.6 Particle size distributions of two experiments at two velocities, in which a sequence of pump passages was simulated that a nodule would experience upon transport through the VTS with a decreasing ambient pressure from 500 to 5 bar. Note the stronger effect of degradation due to increased pump revolutions. Also note the vast increase in fines (0–0.063 mm), which is caused by a large increase in abrasive wear in the hydrocyclone. (From: van Wijk et al. (2019), Copyright 2020 Elsevier, reprinted with permission)

impact velocities v = 4, 6 and 8 m/s (van Wijk et al., 2019; van Wijk & de Hoog, 2020). The experiment consisted of a perpendicular impact of a nodule with a steel plate, on top of which a thin layer of water was applied, as shown in Fig. 4.7. This water has the function to mimic the presence of water during submerged impact, where the water acts as a buffer that dampens the impact.

The nodule diameters were in the range 16 < d < 90 mm and prior to testing the compressive strength was measured with a UCS test setup, which was correlated to tensile strength. A decreasing tensile strength was measured with increasing particle size, with more than a factor of ten difference in tensile strength between the largest and smallest nodules.

The particle size distributions of the nodules after the impact and the peak-stress at impact as a function of nodule diameter and impact velocity were measured. Particle size distributions are shown in Fig. 4.8. Chipping (i.e. small debris, but the

Fig. 4.7 Video stills of a polymetallic nodule colliding with a wet steel plate. The nodule breaks into a few large pieces, and there is a limited amount of debris. (From: van Wijk et al. (2019), Copyright 2020 Elsevier, reprinted with permission)

main nodule remains intact) was observed in the size range $30 < dm < 60$ mm and full fragmentation for the size class $dm > 60$ mm. Below $dm = 30$ mm no significant damage was observed. The results suggest a lower limit in particle size for the nodules being susceptible to impact fragmentation under the investigated conditions.

With the atmospheric experiments, it was possible to demonstrate that the production of fine material due to pure fragmentation is substantially smaller than

Fig. 4.8 Original, and measured and predicted PSD after single impact. (**a**) Nodule fragment sizes after single impact with v = 4 m/s. (**b**) Nodule fragment sizes after single impact with v = 6 m/s. (**c**) Nodule fragment sizes after single impact with v = 8 m/s. (From: van Wijk and De Hoog, (2020), Copyright Elsevier 2020, reprinted with permission)

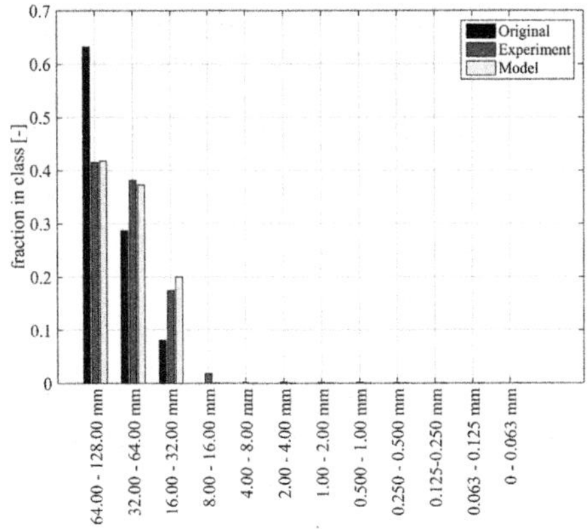

a. Nodule fragment sizes after single impact with v = 4 m/s.

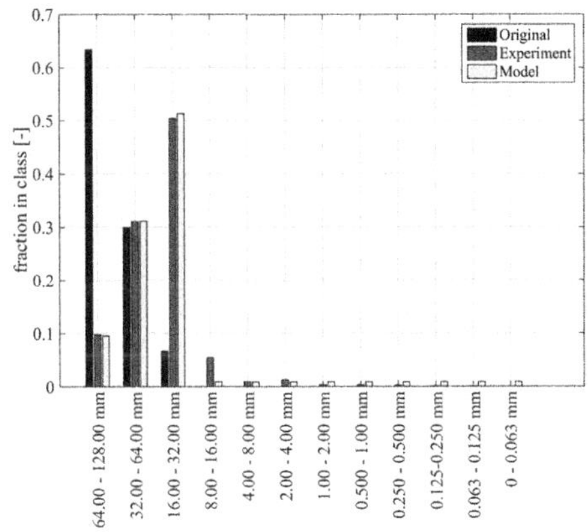

a. Nodule fragment sizes after single impact with v = 6 m/s.

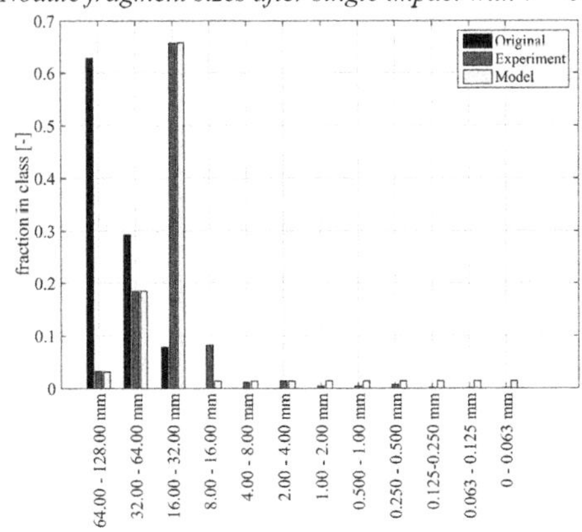

a. Nodule fragment sizes after single impact with v = 8 m/s.

found at NIOZ. The atmospheric experiments resulted in more rapid degradation of the nodules compared to the NIOZ experiments, which is thought to be related to the use of intact nodules instead of fragments, an argument supported by the relation between strength and nodule size.

2.3 *Abrasion and Chipping in the Jumper and Riser Segments*

While impact fragmentation is the dominant degradation mechanism in pumps, abrasion and chipping are the main causes of degradation in the jumper hose and riser system. To quantify these effects, dedicated abrasion and attrition experiments are needed (de Hoog et al., 2020).

The jumper, which is associated with a sliding bed with shear layer and thus abrasion and attrition of the nodules, leads the slurry into the VTS. In the riser, two mechanisms dominate the degradation process: (i) abrasion by sliding along the riser segments, which will mainly affect nodules close to the wall, and (ii) particle–particle interaction.

Different particle fractions with different sizes have different transport velocities and their relative velocities give rise to collisions and probably mild impact fragmentation, chipping and attrition. Particle–wall interaction will have velocities in the order of the mixture bulk velocity (several meters per second) in the axial direction, resulting in sliding contact. Radial particle velocities will be very limited. Particle–particle interaction will have velocities in the order of the hindered settling velocity. Typical hindered settling velocities range between 0 and 1 m/s for polymetallic nodules of tens of millimetres size, which is an order of magnitude smaller than the bulk velocity.

Abrasion takes place due to interaction between the asperities on the sliding surfaces and the effective contact area. In the contact area, abrasion is dominated by plastic deformation as a result of the normal load on the surfaces. This process is independent of the relative velocity of the moving bodies but fully dependent on the normal force and sliding distance.

The relation between particle degradation and horizontal hydraulic transport has been studied for the transport of coal. Worster & Denny (1955) made an important contribution by relating the degradation of coal to the velocity and transport distance of the coal in mainly stratified flow regimes with a sliding bed with saltation of coal lumps on top of it. Degradation relates to the combined effect of abrasive wear in the bed layer and attrition in the shear layer (particle–particle interaction). They report a strong relation between degradation and mixture velocity ($\sim v_m^3$) and a weaker relation with sliding distance ($\sqrt{s}$), pointing at a dominant chipping process even in sliding bed flow with saltating particles.

De Hoog et al. (2020) were able to quantify the specific wear rates (in conjunction with Archard's abrasive wear theory) of nodules sliding on PVC by using a continuously tilting test tube setup, as shown in Fig. 4.9, which is deemed a good proxy of jumpers and riser systems with liners. When calculating sliding distances and contact forces between nodules and the pipeline, Archard's theory then allows

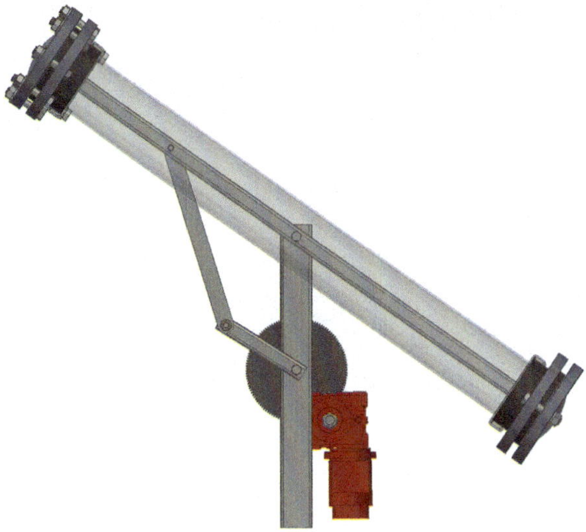

Fig. 4.9 Front view of the tilting pipe setup. A transparent PVC pipe, an electric motor and a crankshaft system to ensure a continuous tilting motion. (From: de Hoog et al., 2020. Copyright Elsevier 2020, reprinted with permission)

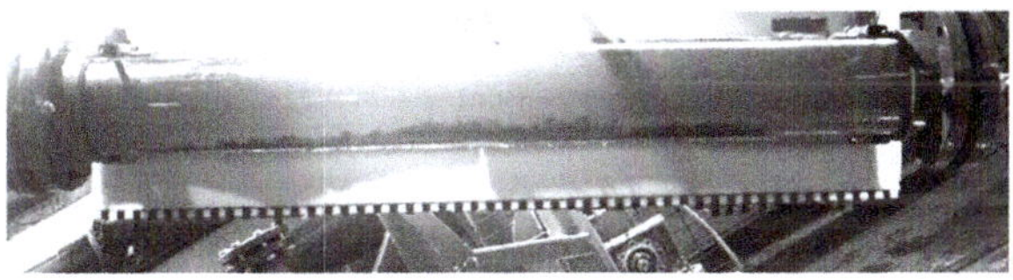

Fig. 4.10 Sliding of the batch of nodules inside the experimental setup. High-speed videos were used to track the position of the particles. (From: de Hoog et al., 2020. Copyright Elsevier2020, reprinted with permission)

for calculating the overall production of fines due to abrasion. The titling pipe batch experiment degradation process includes, on top of particle-wall abrasion, particle–particle abrasion and attrition as particles saltate and roll over each other, as shown in Fig. 4.10. This increases the number of collisions, as nodules are practically in constant contact with each other, with a higher degradation rate as a result. However, production of the finest fractions, as shown in Fig. 4.11, is attributed to abrasion only. A more detailed view on the abrasion of a single nodule is given in Fig. 4.12, which shows the ratio of mass loss due to abrasion over the normal force and the particle density, plotted as a function of sliding distance.

The tilting pipe batch experiments showed that degradation causes accumulation of material in the size of $0.1 < d < 2.0$ mm, as apparent in Fig. 4.13. These classes tend not to degrade further, as they become suspended by the fluid, or trapped between the pores of the sliding nodule bed. As such, if hypothetically degradation would continue to a point where the sliding bed is fully degraded into suspended

Fig. 4.11 The nodule batch after the experiments (left), the developed fines due to degradation (right). (From: de Hoog et al., 2020. Copyright Elsevier 2020, reprinted with permission)

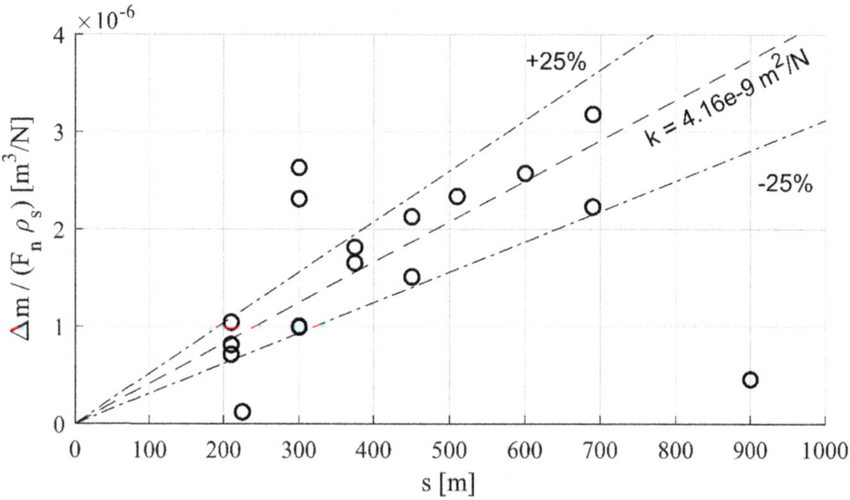

Fig. 4.12 Ratio of mass loss over the normal force and the particle density, plotted as a function of sliding distance. This mass loss is caused by pure abrasive wear, tested on a single nodule in the tilting pipe setup. (From: de Hoog et al., 2020. Copyright Elsevier 2020, reprinted with permission)

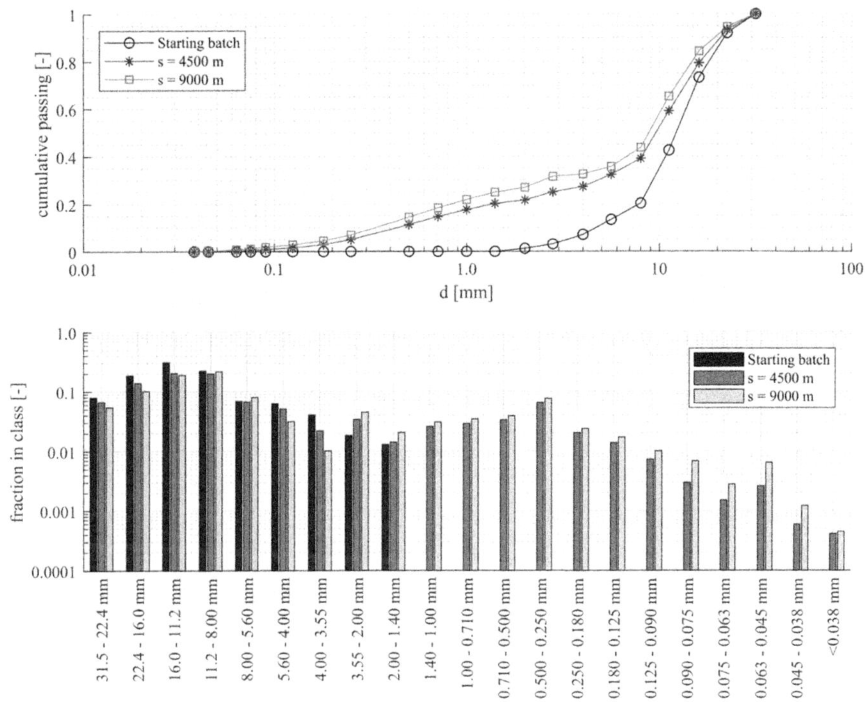

Fig. 4.13 The initial batch PSD and the remaining PSD after a sliding distance of 4500 m and 9000 m, displayed in PSD (top) and bar plot (bottom). (From: de Hoog et al., 2020. Copyright Elsevier 2020, reprinted with permission)

particles, it would be expected to have the final PSD to be similar to the distribution of the $0.1 < d < 2.0$ mm classes depicted in Fig. 4.13. Only a small portion of this is smaller than 63 μm, considered economically unattractive to recover.

Vertical fluidization experiments were then conducted to mimic interactions between nodules as they would occur in the VTS under representative conditions, since the relative velocities are governing the chipping process rather than the mixture bulk velocities. In these experiments, a bed of nodules was fluidized and kept at a more or less stable position in the fluidization column at a volumetric concentration comparable to those occurring in the VTS. In this way, it was possible to avoid the degradation of nodules due to the centrifugal pump, while maintaining representative relative particle velocities. The test setup is depicted in Fig. 4.14.

Results show combined abrasion and attrition through particle-wall and particle-particle collisions. The amount of collisions was less frequent than in the tilting pipe saltating degradation process. This was due to the nature of the fluidization experiment, where particles are spaced farther apart than in a saltating flow. The overall degradation was significantly higher in the vertical flow experiment than in the sliding experiment, suggesting that degradation in the vertical flow experiment is dominated by attrition and chipping due to particle–wall and particle–particle collisions,

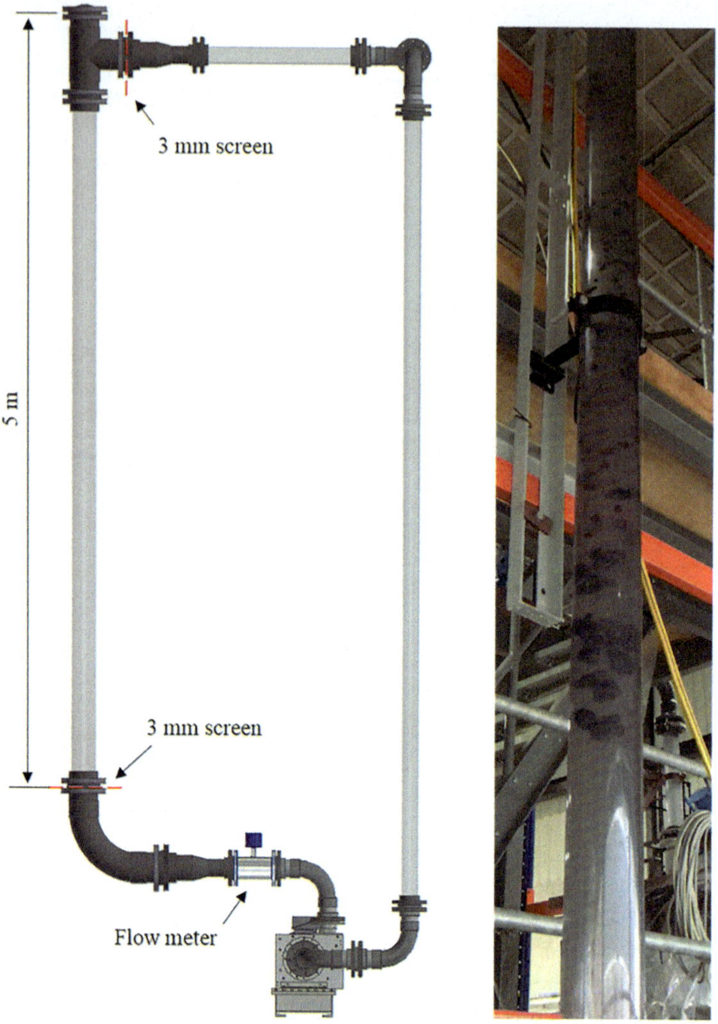

Fig. 4.14 The vertical fluidization flow loop. (From: de Hoog et al., 2020. Copyright Elsevier 2020, reprinted with permission)

and there is very little abrasion. Figure 4.15 gives a general picture of the nodule fragments and grit produced during these experiments. Similar to the tilting pipe batch experiments, degraded particles tend to accumulate in the $0.1 < d < 2.0$ mm size classes (see Fig. 4.16), because these sizes become suspended in the fluid and practically do not contribute to the degradation process.

Fig. 4.15 Fragmented nodules and fines. (From: de Hoog et al., 2020. Copyright Elsevier 2020, reprinted with permission)

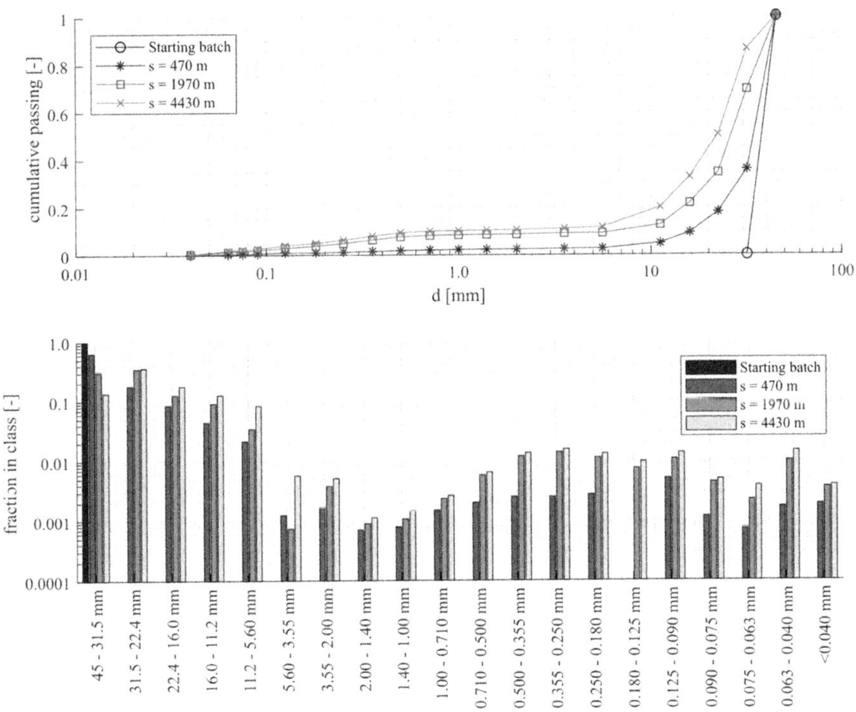

Fig. 4.16 Change of particle size in the vertical fluidization experiment starting with nodules of the 45–31.5 mm class and three consecutive experiments, displayed in PSD (top) and bar plot (bottom). (From: de Hoog et al., 2020. Copyright Elsevier 2020, reprinted with permission)

Table 4.2 Overview of hydraulic transport degradation mechanisms and their effect on the initial nodule PSD

VTS component	Degradation mechanism	Effect on PSD
Pumps/bends	Impact fragmentation	Strong reduction of d_{50}
Jumper	Abrasion/attrition	Creates mainly new fractions 0.1 < d < 2.0 mm. Small amount of fines d < 63 μm
Riser	Abrasion/attrition	Moderate decrease of d_{50}. Creates new fractions 0.1 < d < 2.0 mm. Small amount of fines d < 63 μm

2.4 An Integral Assessment of Nodule Degradation during Hydraulic Transport

Having quantified the response of CCZ nodules to impact fragmentation, abrasion and chipping under conditions comparable to those during operation, it is possible to make an integral assessment of the development of the PSD during transit. Regarding impact fragmentation, a mathematical framework of a predictive model was provided in Van Wijk and de Hoog (2020), to aid in the design of a deep-sea mining operation.

Table 4.2 summarizes the degradation mechanisms encountered during vertical transport and their relative effect on the PSD of the nodules. The relative contribution of each mechanism depends on the configuration of the transport system. Knowledge of the different flow regimes in the various segments of the transport system, the associated degradation process and dedicated calculation models for the flow process, fragmentation, abrasion and chipping are required to make this integral assessment.

These results are based on tests conducted with CCZ nodules from the Belgian License area. Nodules with different mechanical properties will have different strengths and will result in different degrees of degradation. The framework provided in Van Wijk and de Hoog (2020) will allow for the quantification of degradation rates when actual nodule samples from the envisaged license area are available. Fortunately, if the size and density are somewhat comparable, the trends in Table 4.2 will be similar since this mainly depends on flow-related mechanisms.

3 Mechanical Response of Dry Bulk Material

The nodules that remained from the tests described in the previous sections have been utilized to study the mechanical response of the nodules as a bulk material. As generally known, the response of the bulk behaviour highly depends on particle shape and size distribution. Therefore, it is strongly advised to conduct comparable tests and analyses resembling realistic mining conditions. The particle properties of

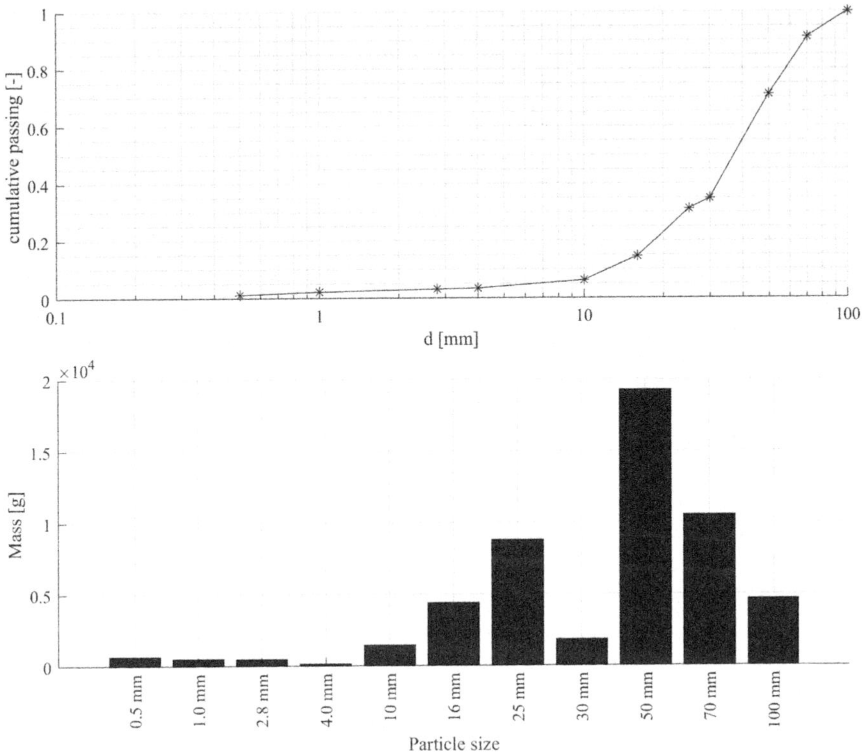

Fig. 4.17 PSD of nodules used for dry bulk behaviour. (Doorn, 2022. Reprinted with permission)

the nodules arriving at the vessel might differ significantly based on nodule characteristics at the bed, collection method and means of vertical transport, as discussed in Sect. 2. Here, an overview of the type of tests that were conducted to study the behaviour of the nodules in dry conditions is presented (Doorn, 2022).

Based on the nodule materials remaining after the fragmentation tests described in sect. 2, approximately 50 kg of nodules were available. Nodules were sorted for size based on sieving, the larger nodules were sorted by hand to avoid fragmentation. The resulting PSD is shown in Fig. 4.17. The exact history of these nodules is not known. The quality of storage conditions is questionable, especially regarding whether the nodules were stored wet. In our experiments, the nodules were again stored in water to saturate them. Then the saturated density of the nodules was measured as 1911 ± 35 kg/m^3. After oven drying for 24 h at 105 °C, the dry density was 1341 ± 46 kg/m^3.

The angle of repose represents the shear stress of bulk material under the force of gravity. The angle of repose results are helpful to categorize the flow granular flow properties. In this research, the angle of repose is measured following the ledge test procedure, often referred to as 'shear box' and 'rectangular container test' (cf. Fig. 4.18). During a ledge test, a container was filled with granular material, after

Fig. 4.18 Shear box test setup (Doorn, 2022). Reprinted with permission

which one side of the container was opened for the material to slide and roll out until an equilibrium was reached. The angle of repose was the resulting angle between the horizontal plane and the free surface. Due to the quantity of nodules and the size of the container, only the batch of nodules at a size of 10–16 mm was thoroughly tested. The angle of repose, based on 18 tests, is 49.5° ± 1.46° and 49.6° ± 1.1°, respectively, for saturated and dry nodules. Dreiseitl (2017) measured substantially lower angles of repose of 30° and 33° for intact nodules recovered from box-cores (Fig. 4.18).

An inclined surface tester (cf. Fig 4.20) was used to measure the sliding and rolling friction of particles along various materials commonly employed in bulk handling equipment, such as steel (S235) and rubber (Ethylene Propylene Diene Monomer). The tester employs a motor to adjust the inclination of the bottom surface, where the steel or rubber plate is mounted. The angle between the inclined surface and the horizontal is referred to as the inclination angle. The sliding and rolling response of individual nodules depends significantly on how constrained their motion is due to neighbouring particles. Therefore, several tests were conducted as follows (cf. Figure 4.19):

1. Single nodule.
2. Free movement of batch of nodules.
3. Confined movement of batch of nodules.
4. Added mass rolling configuration.
5. Added mass sliding configuration. Nodules are confined in a mould, restricting the nodules from rolling and forcing them to slide.

3.1 Single Nodule

In the 'single nodule' test, it was observed that when using rubber as the bottom surface, the primary mode of movement was rolling. On the other hand, when employing a steel plate, the primary mode of movement was sliding. No significant

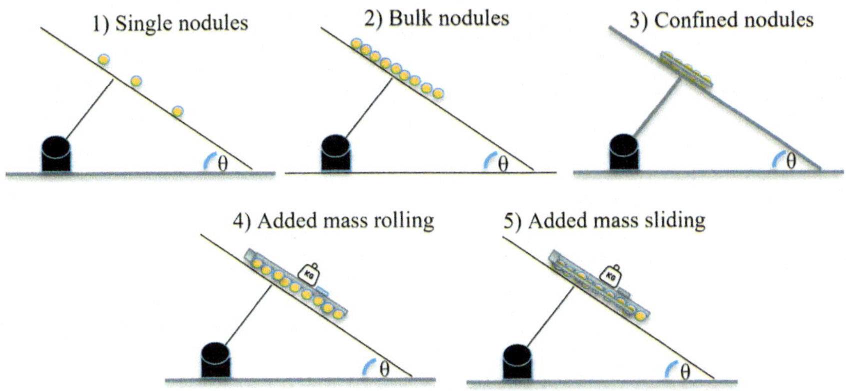

Fig. 4.19 Conducted sliding and rolling friction for single and bulk nodules. (Doorn, 2022. Reprinted with permission)

Table 4.3 Recorded angle at which the nodule starts to move

Rubber		**Dry**	**Saturated**
Count [−]		18	16
Mean [°]		30.4	29.2
Std		2.7	2.7
Steel	**Dry smooth**	**Dry corroded**	**Saturated**
Count [−]	39	23	16
Mean [°]	19.8	28.4	27.7
STD [°]	0.8	1.5	1.7

Doorn (2022). Reprinted with permission

difference was observed between dry and saturated nodules. However, it was noted that the response of a single nodule on the steel plate was affected by whether the plate was corroded or not. As a result, tests were conducted with freshly polished steel plates and with corroded plates resulting from the experiments with the fresh plates. The surface roughness of the plates was measured with a profilometer, indicating roughness values of 0.46 ± 0.07 and 1.16 ± 0.12 μm for the smooth and corroded plates, respectively (see Table 4.3) (Figs. 4.20 and 4.21).

3.2 Layer of Nodules

Similar tests were conducted for a single layer of particles. Here, it was tested as to at what angle the majority of the particles started to slide or roll. This typically occurs with many of them moving altogether, behaving mostly like a granular flow. Additional tests were conducted, with the particles being constrained to remain together by means of a water-resistant paper wrapped around the batch of nodules. Results from both sets are shown in Table 4.4.

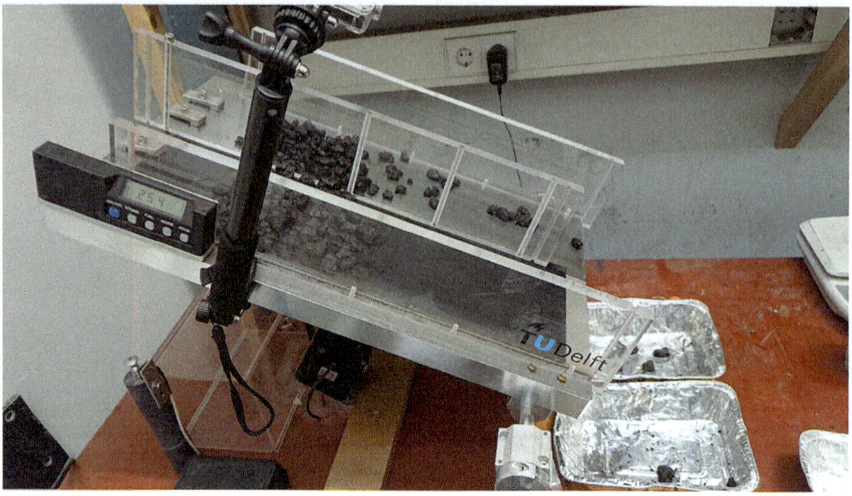

Fig. 4.20 Tilting table at the initiation of sliding/rolling of a set of nodules. (Doorn, 2022. Reprinted with permission)

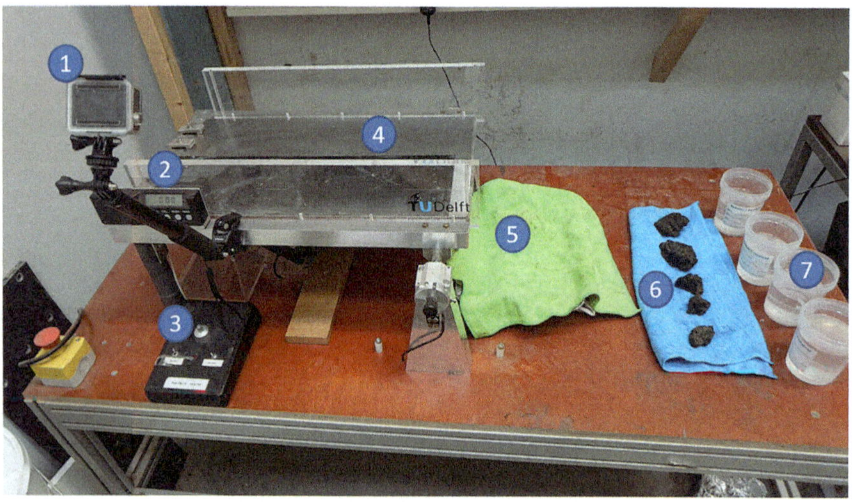

Fig. 4.21 Tilting table setup. (Doorn, 2022. Reprinted with permission)

3.3 Layer of Nodules under Normal Load

Finally, a set of experiments was conducted where a normal force was applied on top of the particles, simulating a layer of particles above. As depicted in Fig. 4.19, one set of tests comprised a sledge mounted on top of the nodules with an additional mass, increasing the applied normal force on the nodules. The other set of tests utilized a matrix of nodules embedded in plaster, effectively interlocking the layer

Table 4.4 Inclination angle at which bulk starts to slide/roll Doorn (2022). Reprinted with permission

Rubber	**Dry**	**Dry (constrained)**	**Saturated**	**Saturated (constrained)**
Count	8	13	8	14
Mean [°]	30.1	32.9	30.2	31.6
95% confidence	2.7	1.1	1.6	1.0
Steel	**Dry (corroded)**	**Dry (constrained & corroded)**	**Saturated**	**Saturated (constrained)**
Count	8	17	8	14
Mean [°]	26.2	30.6	30.2	33.7
95% confidence	1.0	0.8	2.1	1.3

Table 4.5 Friction angle found under added normal force. Doorn (2022). Reprinted with permission

Rubber	**Dry**	**Saturated**
Count	18	18
Mean [°]	32.48	30.54
STD [°]	4.88	3.25
Steel	**Dry**	**Saturated**
Count	18	18
Mean [°]	28.28	30.07
STD [°]	3.27	3.76

of nodules on the bottom surface. The mass added on top of both the nodules cast and the sledge was 3 kg, equivalent to a mass of approximately 330 kg/m^2. Once again, tests were carried out for both dry and saturated nodules, and the results are presented in Table 4.5.

3.4 Findings

It has been observed that the behaviour of saturated nodules differs slightly from that of dry nodules. When nodules are in contact with steel equipment (e.g. hopper), dry nodules tend to exhibit a lower friction angle of approximately 2°. Conversely, when in contact with rubber (e.g., conveyor belt), dry nodules tend to have a higher friction angle compared to saturated nodules. Furthermore, during sliding along the steel plate, nodules experience sliding wear, which was not observed in the case of rubber. The exact reasons for these differences in material response are not yet fully understood and could be related to (1) the particle density of the nodules and (2) the hardness of the contact surface and, consequently, the effective contact area. Further research into the micro-level phenomena at the contact interface between the nodules and equipment surface may provide insights into the underlying mechanisms.

4 Approaches to Breakage and Fines Generation in Biomass Handling

Deep-sea mining is not the only industry facing challenges related to the handling of fragile materials and issues concerning particle breakage, degradation and fines generation. These challenges are also prevalent in the biomass handling sector, although in a drier environment, with degradation mechanisms similar to those outlined in Table 4.1. An analogous handling chain to PmN can be found in biomass pellets, where it is crucial to prevent the undesirable generation of fines during handling. Additionally, the internal structure of the particles is inhomogeneous, and different degradation mechanisms occur depending on the handling method.

This section focuses on the experimental and numerical methods used in researching the handling of granular material and in particular biomass pellets, and how they can offer valuable insights for analyzing and designing handling solutions in the deep-sea mining industry.

4.1 *Biomass Pellets*

Though the formation processes of PmNs and man-made biomass pellets differ significantly, both internal structures exhibit inhomogeneities that act as weak spots susceptible to breakage.

Duliu et al. (1997) and Rizescu et al. (2001) utilized computer tomography to reconstruct PmNs (see Fig. 4.22a), revealing the inhomogeneity within the nodules' internal structure. They observed variations in nucleus shape and origin, primary and secondary growth surfaces, highly mineralized areas, and the presence of a fracture network.

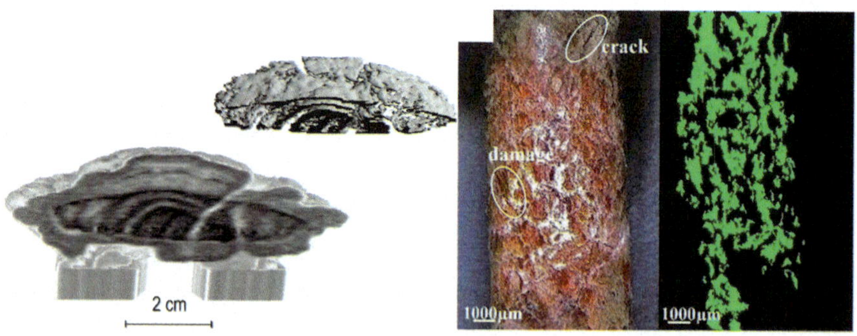

Fig. 4.22 Left Image of PmN structure (From: Rizescu et al. (2001) Copyright Elsevier 2001, reprinted with permission) and right, Digital images and corresponding determined damage area of wood pellet. (From: Cutz et al. (2021). Copyright 2021 Springer Nature, reprinted with permission)

The process of creating biomass pellets involves grinding and passing dried biomass through a pellet dye at high temperature and pressure. This results in the melting of lignin within the biomass, binding the particles together. After cooling, the particles reform into solid cylindrical-shaped pellets with uneven ends. As an example, Cutz et al. (2021)also employed CT to analyze the internal structure, uncovering fractures both on the surface and inside the particles (see Fig. 4.22b).

4.2 *Experiments to Assess Breakage*

Experimental techniques to determine the tendency of particles to break and evaluate their breakage characteristics include:

- Compression testing: Involves subjecting particles or specimens to compressive forces by uniaxial or triaxial tests to evaluate their resistance to breakage and analyzing the resulting particle size distribution.
- Impact testing: Evaluates the susceptibility of particles to breakage under impact or collision forces by analyzing the generated fragments and assessing fragmentation patterns (Sect. 2.2).
- Shear testing: Particularly useful for cohesive particles or materials with interparticle bonding, where particles are subjected to shear forces to simulate sliding or movement. Changes in particle arrangement and breakage are analyzed.
- Tumbling tests: Simulates continuous or repetitive tumbling and rolling motion to assess particle breakage, measuring changes in particle size distribution or fines generation (Sect. 2.3).
- Attrition testing: Evaluates particle tendency to undergo attrition or abrasion by subjecting particles to abrasive conditions and quantifying particle wear, loss of material and changes in size or shape.
- Particle size distribution analysis: Indirectly assesses particle breakage by comparing particle size distribution before and after a treatment or testing, using techniques like sieve analysis, laser diffraction or image analysis.

The choice of test should be relevant to the actual conditions of the handling process. In the case of biomass wood pellets, Gilvari et al. (2019) analyzed reported tests in the literature and concluded that the enormous variety of often tailor-made tests inspired by other domains makes it difficult to compare results due to the absence of a prescribed overall standard.

Encouraging the development of standards is important to have unified tests adopted by all parties in the handling chains. Furthermore, using standards and obtained experimental data under representative operational conditions allows for necessary validation data for modelling of breakage in the supply chain.

4.3 Biomass Handling Experiments

To gain insights into the various handling stages of the biomass handling chain, a comprehensive analysis was conducted (Gilvari, 2021). This analysis involved multiple testing methods, including compression testing of individual particles(Gilvari et al., 2021a, b), tumbling tests (Gilvari et al., 2020), impact testing and laboratory-scale bulk handling tests under different operational conditions (cf. Fig. 4.23) (Gilvari et al., 2022) and industrial-scale testing and analyses of fines generation using particle size distribution analysis (Gilvari et al., 2021a, b). These testing approaches were systematically examined and integrated to provide a comprehensive understanding of the biomass handling chain at different scales (Gilvari, 2021).

The industrial-scale tests to evaluate the generation of fines and dust during handling processes were conducted on an existing handling chain at a port. Pellets were discharged from a seagoing vessel through a hopper onto a conveyor system, which transported the material to a silo. In addition to the continuous sampling system

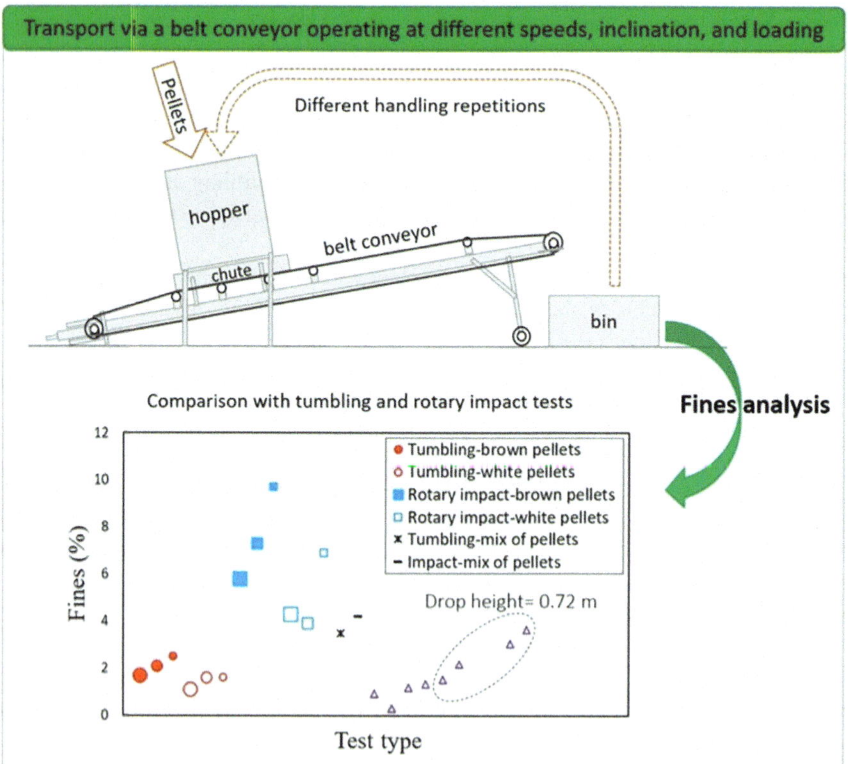

Fig. 4.23 Schematic of the laboratory setup and results of analysis of the effect of repetitions on biomass pellet fines generation under operational conditions. (From Gilvari et al. (2022). Copyright, 2022, Elsevier, reprinted with permission)

installed for quality control, samples were collected at multiple points in the chain. The primary finding was that at a transfer point between two conveyors, the proportion of small particles (<5.6 mm) increased from around 9% to 14%. This highlights the significance of equipment design and operation in relation to material degradation.

These particle-level investigations also served as a basis for developing numerical models using the discrete element modelling approach, laying the groundwork for further studies on pellet breakage in handling systems.

4.4 Modelling Approaches

The modelling of individual particle breakage can be accomplished using various methods. For example, finite element modelling allows the material to be described as a continuum and can incorporate fracture criteria to simulate particle breakage under different loading conditions. However, due to the granular nature of the material and the necessity to model the entire system (equipment and granular material interactions), discrete element modelling (DEM) is a more suitable approach. DEM, first proposed by Cundall and Strack (1979), represents individual particles as discrete elements and considers their interactions based on contact mechanics principles. DEM models can include parameters like particle strength, fracture toughness, and critical stress for breakage simulation. Moreover, DEM naturally allows fragmented parts to continue as smaller particles.

By tracking particle movements and forces, DEM can offer insights into individual particle breakage patterns, overall bulk material behaviour and the performance of the entire system, including forces, power, wear, and material flow. After defining input parameters and calibrating and validating with experimental data under representative operational conditions, DEM becomes a powerful tool for designing material handling equipment. For example, Schott et al. (2021) introduced a design framework successfully used for novel granular material handling equipment, along with full-scale validation. To develop such models describing particle-particle and equipment interactions, defining input parameters is crucial. This requires experimental data at the interface between material and equipment, as discussed in Sects. 2 and 3.

Figure 4.24 (Jiménez-Herrera et al., 2018) illustrates three ways of modelling particle breakage in DEM simulations. The bonded-particle model (BPM) shows bonds that can break when stressed beyond a critical level. The fast-breakage model (FBM) employs polyhedral elements and is a particle-replacement model, potentially powerful for simulating large-scale comminution systems, provided that the models describing the breakage probability and distribution offer the necessary degrees of freedom. In the particle replacement model (PRM), the simulation removes the spherical parent particle when the fracture occurs and replaces it with a set of progeny particles

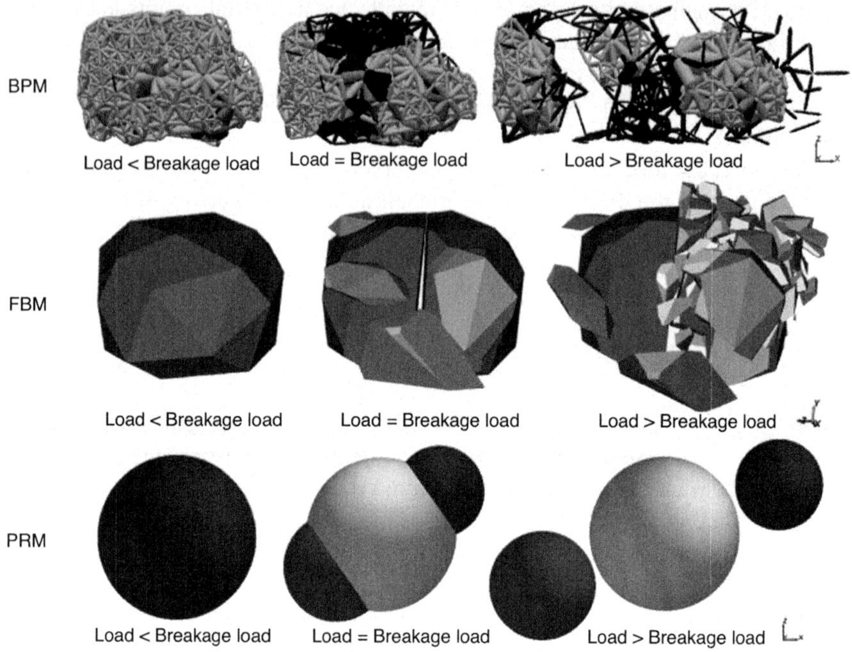

Fig. 4.24 Simulation stills of breakage of single particles: In the BPM, the particle is represented by means of their bonds (light grey: unstressed; gray: stressed; black: broken). In the FBM, primary breakage occurs when the stressing energy surpasses the breakage energy and secondary afterwards. In the PRM, only primary breakage is illustrated, with an additional moment after that. Load applied in the vertical direction. (From: Jiménez-Herrera et al. (2018). Copyright Elsevier 2018, reprinted with permission)

In the comparison of these three models, Jiménez-Herrera et al. (2018) concluded that each model possesses unique strengths. The particle-replacement method (PRM) demonstrated an excellent fit for the size distribution of the progeny from both single-particle breakage and particle bed experiments, making it an appealing alternative to previous models, especially when simulating large particle populations.

Tavares (2022) presents a critical analysis of his comprehensive mathematical breakage model suitable for describing the breakage of populations of brittle materials in conjunction with the discrete element method, either as part of particle replacement schemes or coupled to it in microscale population balance models. This breakage model has been extensively used in advanced models of crushers and various mills, enabling the prediction of degradation during material handling, including iron ore pellets. To be effective in such models, it is crucial to understand the particle response under various conditions encountered in a system that can cause mechanical degradation. These conditions encompass diverse particle sizes, stressing intensities, geometries, rates of stressing and modes of stress application. Furthermore, he notes that challenges persist in estimating breakage parameters for fine particles, despite models accounting for essential variables. Additionally, models describing surface breakage play a crucial role in predicting mechanical degradation during handling, demanding increased attention due to being in an early stage of development and validation.

4.5 Biomass Pellets

In the modelling of individual biomass pellets' breakage, Gilvari et al. (2021a, b) employed a bonded-particle model with the bonded contact based on the Timoshenko-Ehrenfest beam theory. Each pellet was represented by a varying number of 960–8000 spheres. Initially, uniaxial and diametrical compression tests were conducted on seven different types of biomass pellets in an experimental setup. The results revealed variations between pellets of different types and even within pellets of the same type due to heterogeneity in the pellet structure. This heterogeneity could be attributed to differences in the particle size distribution of the raw materials, heterogeneous porosity, presence of micro-cracks and other factors.

The maximum stress at failure for the tested pellets ranged from 8 to 32 MPa. The numerical outcomes demonstrated that the mechanical strength of various types of pellets could be represented by using the number of spheres and the coordination number as variables, in addition to the bond parameters. The calibrated model successfully predicted the stress-strain curves and modulus of elasticity of individual pellets and reasonably estimated their breakage behaviour under uniaxial and diametrical compressions.

However, it's worth noting that the amount and size of particles released from the pellets during compression tests were excluded both in the experimental and numerical aspects of the study. Building on Jimenez-Herrera et al.'s (2018) work and based on Gilvari's findings, it is recommended that fragmentation should be incorporated in the next stages of the modelling process. Moreover, considering the computational expenses associated with using the bonded-particle model, depending on the number of spheres used for a single particle, it may be unfeasible for modelling industrial scale systems.

4.6 Next Modelling Steps

To assess the performance of nodule handling equipment with respect to the breakage of nodules, generation of fines and consumption of energy, the DEM would be a suitable method to simulate the processes at the scale of the handling equipment. A set of standardized experiments to assess the relevant material characteristics of the nodules is necessary to allow for proper comparison of equipment and operations. Experiments like those presented in Sect. 2 provide a basis for such a standardized set. Such experiments should provide valuable information to allow for calibration of the particle properties in DEM, which is necessary to mimic the breakage behaviour of PmN. To effectively model breakage behaviour in the context of deep-sea mining, with a specific focus on material behaviour and degradation, it is recommended to employ DEM in conjunction with a particle replacement method to accurately capture fragmentation. However, it is also important to take into account the development of models that describe surface breakage to further enhance the accuracy and comprehensiveness of the simulations. A promising approach is Tavares' mathematical breakage model suitable for describing the breakage of populations of brittle materials in conjunction with the discrete element method, again under the condition of availability of experimental data as elaborated in this chapter.

PmN is handled in a more complex environment, potentially requiring dedicated breakage models for submerged, moist and dry conditions. Similar aspects hold for the volumetric concentration of nodules in each of the handling steps, ranging from collision-dominated to contact-dominated material handling. Further research regarding the bulk behaviour of nodules, ranging from submerged to dry condition, is needed to allow for the generation of the standardized nodule granular and breakage properties. These, combined with calibrated models in DEM, will enable for optimization of the key handling steps and equipment.

5 Conclusions and Recommendations

For a sustainable (green, energy and cost-efficient) supply chain, it is important to carefully balance the processing capacity with the mining capacity and the expected demand for the final products to optimize the efficiency and effectiveness of the supply chain. This requires careful planning and coordination between all stakeholders involved in the deep-sea mining supply chain, including the mining and processing companies, logistics providers and end customers.

Hydraulic transport is one of the key technologies that is being considered for conveying the polymetallic nodules from the seafloor to the mining support vessel. Degradation of the polymetallic nodules is closely related to the hydraulic transport regime. Experiments have shown that, in the jumper hose predominantly abrasion and chipping occurred, in the riser attrition and abrasion, and in the centrifugal pumps predominantly impact fragmentation was observed.

These mechanisms have been studied in more detail with Polymetallic Nodules from the Belgian and UK CCZ license areas under various conditions. Research into impact fragmentation on polymetallic nodules fragments indicates that the pump impeller speed has a major influence on the degradation process, and no significant dependency on the ambient pressure (in the range of 5–500 bar) was found. These findings were confirmed in successive impact experiments, where nodules were subjected to perpendicular collision with a steel plate under atmospheric conditions at impact velocities up to 8 m/s. In general, impact was found to cause fragmentation of nodules down to mm scale without significant production of finer particles.

The tilting pipe batch experiments showed that degradation by sliding and particle collision causes accumulation of material in the size range of $0.1 < d < 2.0$ mm, after which further degradation stops as debris becomes suspended in the fluid or settles within the pores of the sliding bed. A limited portion of the debris was found to be smaller than 63 μm.

The degradation during the fluidization experiment was significantly larger than in the sliding bed experiment, suggesting that degradation in the vertical flow experiment is dominated by attrition and chipping due to particle–wall and particle–particle collisions. Nodule fragments and grit tend to accumulate in the $0.1 < d < 2.0$ mm size classes, similar to the tilting pipe experiment, because these sizes become suspended in the fluid, and practically do not contribute to the degradation process. These results indicate only very little abrasion to take place during the actual vertical transport in the riser.

The mechanical response of nodules has been extensively studied through various tests and analyses. The behaviour of the bulk material is influenced by factors such as particle shape and size distribution, which vary based on nodule characteristics at the bed, collection method, and vertical transport means. The conducted experiments have provided valuable data on the angle of repose, sliding and rolling friction, and the effect of normal force on nodules. Understanding the differences between saturated and dry nodules in contact with different equipment surfaces calls for further research to uncover the underlying mechanisms at the micro-level interface.

A range of experimental techniques have been employed to determine particle breakage tendencies and evaluate breakage characteristics in various handling processes. These include compression testing, impact testing, shear testing, tumbling tests, attrition testing, and particle size distribution analysis. Encouraging the development of standardized tests is essential for reliable comparison of results and validation of breakage models from seafloor to mining support vessel. As an example from the context of biomass handling, comprehensive analyses integrating multiple testing methods at different scales have provided valuable insights into equipment design and material degradation, paving the way for further studies and numerical modelling of breakage in any handling system.

Various modelling approaches for particle breakage exist, with discrete element method being well-suited for granular materials and equipment interactions. DEM, along with particle replacement model, provides valuable insights into breakage patterns and bulk material behaviour, aiding in material handling equipment design. Challenges remain in estimating breakage parameters for fine particles and developing surface breakage models. The Tavares model is the state of the art describing the breakage of populations of brittle materials in conjunction with the discrete element method, either as part of particle replacement schemes or coupled with microscale population balance models. It is crucial to acknowledge that for particles to be effectively utilized in such models, their response must be understood across a range of conditions that they may encounter in real conditions. These conditions typically involve diverse particle sizes, stressing intensities, geometries, rates of stress and modes of stress application. To provide relevant input for these models, standardized experiments to characterize nodules and bulk responses at these conditions should be developed. Once this is in place, the models can be used for optimizing handling steps and equipment in deep-sea mining.

Acknowledgements The authors would like to thank Floris Doorn for the extensive number of experiments he conducted during his master's in Offshore and Dredging Engineering at TU Delft, Paul Vercruijsse from DEME-GSR for providing us with nodules.

A very special thanks needs to go to Cees van Rhee, full professor in Dredging Engineering. As a pioneer, he was the first to introduce us to the exciting topic of deep-seabed mining. Cees unexpectedly passed away at the age of 64. We will remember him as a dear colleague and inspiring mentor.

References

Allseas. (2022). *Deep-sea polymetallic nodule collection*. Allseas.

Beekman, W., Meesters, G., Scarlett, B., & Becker, T. (2002). Measurement of granule attrition and fatigue in a vibrating box. *Particle & Particle Systems Characterization, 19*(1), 5–11.

Cundall, P., & Strack, O. D. L. (1979). A discrete numerical model for granular assemblies. *Géotechnique, 29*(1), 47–65.

Cutz, L., et al. (2021). Microstructural degradation during the storage of biomass pellets. *Communications Materials, 2*(1).

de Hoog, E. (2016). *Coarse particle slurry flow in inclined pipes and vertical S-bends*. MSc thesis TU Delft.

de Hoog, E., In't Veld, J., van Wijk, J. & Talmon, A. (2017). *An experimental study into flow assurance of coarse inclined slurries*. Prague, Czech Republic, 18th international conference on the transport and sedimentation of solid particles.

de Hoog, E., van Wijk, J., Wijnands, J., & Talmon, A. (2020). Degradation of polymetallic nodules during hydraulic transport under influence of particle-wall and particle-particle interaction. *Minerals Engineering, 155*, 106415.

Doorn, F. (2022). *The effect of moisture on the granular characteristics of polymetallic nodules in the context of bulk handling*. TU Delft.

Dreiseitl, I., 2017. *About geotechnical properties of the deep seabed polymetallic nodules*. Prague, Czech Republic, 18th international conference on the transport and sedimentation of solid particles.

Duliu, O., Tufan, M., & Szobotka, S. (1997). Computer axial tomography investigation of polymetallic nodules. *Marine Geology, 138*(3–4), 303–311.

Gilles, R., Haas, D., Husband, W. & Small, M. (1982). *A system to determine single pass particle degradation by pumps*. Johannesburg, South Africa, 8th international conference on the hydraulic transport of solids in pipes.

Gilvari, H. (2021). *Degradation of biomass pellets during transport, handling and storage: An experimental and numerical study*. PhD thesis TU Delft.

Gilvari, H., de Jong, W., & Schott, D. L. (2019). Quality parameters relevant for densification of bio-materials: Measuring methods and affecting factors – A review. *Biomass and Bioenergy, 120*, 117–134.

Gilvari, H., De Jong, W., & Schott, D. L. (2020). The effect of biomass pellet length, test conditions and torrefaction on mechanical durability characteristics according to ISO standard 17831-1. *Energies, 13*(11).

Gilvari, H., et al. (2021a). Large-scale transportation and storage of wood pellets: Investigation of the change in physical properties. *Particuology, 57*, 146–156.

Gilvari, H., de Jong, W., & Schott, D. L. (2021b). Breakage behavior of biomass pellets: An experimental and numerical study. *Computational Particle Mechanics, 8*(5), 1047–1060.
Gilvari, H., et al. (2022). Fragmentation of fuel pellets during transport via a belt conveyor: A design of experiment study. *Particuology, 66*, 29–37.
Jiménez-Herrera, N., Barrios, G. K., & Tavares, L. M. (2018). Comparison of breakage models in DEM in simulating impact on particle beds. *Advanced Powder Technology, 29*(3), 692–706.
Laarhoven, B. V., Schaafsma, S., & Meesters, G. (2012). Toward a desktop attrition tester; validation with dilute phase pneumatic conveying. *Chemical Engineering Science, 73*, 321–328.
Ma, W. (2019). *Sustainability of deep sea mining transport plans*. PhD thesis TU Delft.
Ma, W., Van Rhee, C., & Schott, D. (2017). Technological and profitable analysis of airlifting in deep sea mining systems. *Minerals, 7*(8).
Monhemius, A. (1980). The extractive metallurgy of deep-sea manganese nodules. In A. Burkin (Ed.), *Topics in non-ferrous extractive metallurgy – Critical reports on applied chemistry* (pp. 42–69). Blackwell Scientific Publication.
Randhawa, N., Hait, J., & Jana, R. (2016). A brief overview on manganese nodules processing signifying the detail in the Indian context highlighting the international scenario. *Hydrometallurgy, 165*(Part 1), 166–181.
Ravelet, F., Bakir, F., Khelladi, S., & Rey, R. (2013). Experimental study of hydraulic transport of large particles in horizontal pipes. *Experimental Thermal and Fluid Science, 45*, 187–197.
Rizescu, C. T., Georgescu, G. N., Duliu, O. G., & Szobotka, S. A. (2001). 3-D dual gamma-ray computer axial tomography investigation of polymetallic nodules. *Deep Sea Research Part I: Oceanographic Research Papers, 48*(11), 2529–2540.
Schott, D., et al. (2021). esign framework for DEM-supported prototyping of grabs including full-scale validation. *Journal of Terramechanics, 96*, 29–43.
Shook, C., Haas, D., Husband, W., & Small, M. (1979). *Degradation of coarse coal particles during hydraulic transport.* Kent, UK, 6th international conference on the hydraulic transport of solids in pipes.
Spelay, R., Gillies, R., Hashemi, S., & Sanders, R. (2016). Effect of pipe inclination on the deposition velocity of settling slurries. *The Canadian Journal of Chemical Engineering, 94*(6), 1032–1039.
Tavares, L. M. (2022). Review and further validation of a practical single-particle breakage model. *Kona Powder and Particle Journal, 39*, 62–83.
van den Bergh, C., & Alvarez Grima, M. (2006). Tbilisi, Georgia, 13th international conference on transport and sedimentation of solid particles.
van Wijk, J. (2016). *Vertical hydraulic transport for deep sea mining; A study into flow assurance.* TU Delft.
van Wijk, J., & de Hoog, E. (2020). Size reduction of CCZ polymetallic nodules under repeated impact fragmentation. *Results in Engineering, 7*, 100154.
van Wijk, J., Talmon, A., & van Rhee, C. (2016). Stability of vertical hydraulic transport processes for deep ocean mining: An experimental study. *Ocean Engineering, 125*, 203–213.
van Wijk, J., et al. (2019). Impact fragmentation of polymetallic nodules under deep ocean pressure conditions. *Minerals Engineering, 134*, 250–260.
Vlasak, P., Chara, Z., Konfirst, J., & Kysela, B. (2013). *Experimental investigation of coarse-grained particles in pipes.* Rostock, Germany, 16th international conference on the transport and sedimentation of solid particles.
Worster, D., & Denny, D. (1955). *Hydraulic transport of solid material in pipes.* s.l., Proceedings of the Institution of Mechanical Engineers.
Yamazaki, T., & Sharma, R. (2001). *Preliminary experiment on powderization of deep-sea sediment during hydraulic transport.* Poland, 4th ocean mining symposium.
Yamazaki, T., Tsurusaki, K., & Handa, K., (1991). *Discharge from manganese nodule mining system.* Edinburgh, UK, First international offshore and polar engineering conference.

Dingena Schott is full professor of Machine Cargo Interaction Engineering at TU Delft's department of Maritime and Transport Technology. She obtained her PhD in 2004 from TU Delft on the homogenization of bulk materials in mammoth silos. She is fascinated by the complex behaviour of granular materials and the interaction with equipment operating in a logistic context. In 2007, she started the GranChaMlab@TUDelft to characterize, model, calibrate and validate granular materials for enabling simulation-supported design for cargo handling equipment on an industrial scale. Since then she has worked on developing calibration frameworks and modelling particle-based systems in various design contexts, including terminal designs for particulate materials, as well as an award-winning new grab design. Her main research interests include: machine-cargo interfaces, simulation-supported design, biomass materials and energy transition–driven handling and logistics.

Edwin de Hoog received his MSc in offshore and dredging technology from the Delft University of Technology in 2016. After that, Edwin started working as a research and development engineer at Royal IHC and a part-time PhD researcher at the Department of Maritime and Transport Technology, section of Dredging Engineering, Delft University of Technology. He specializes in hydraulic transport of sand, gravel and manganese nodules and works on many related aspects of deep-sea mining. He mainly worked on flow assurance issues in jumper hoses and vertical pipelines and degradation of manganese nodules during hydraulic transport, as part of the EU Blue Mining and Blue Nodules programmes. As part of his PhD research, he developed 1D CFD technology to simulate pump-pipeline systems and the effect of slurry dynamics and density wave amplification on flow assurance, applicable for deep-sea mining and dredging pipelines.

Jort van Wijk received his MSc and PhD from Delft University of Technology, where he specialized in offshore engineering. The subject of his PhD was flow assurance of vertical hydraulic transport for deep-sea mining applications. Jort started his career in 2008 at the corporate R&D department of Royal IHC, manufacturer of specialized offshore, dredging and mining vessels and equipment, where he worked in different specialist and managerial roles on dredging and (deep-sea) mining technology. He was closely involved in the EU FP7 Blue Mining programme and the EU H2020 Blue Nodules programme, two major research programmes aimed at developing knowledge and technology for deep-sea mining. After joining IQIP in 2020, his area of interest expanded to offshore geotechnics and soil-structure interaction problems for foundation installation. Next to this, as an independent consultant, Jort has a continued interest in deep-sea mining technology.

Rudy Helmons obtained his MSc in Mechanical engineering in 2011, after which he started as a research engineer at Royal IHC. He obtained his PhD 'cum laude' from TU Delft in 2017 on the topic of rock excavation in underwater conditions. After that, he started as an assistant professor in offshore and dredging engineering at TU Delft, and was Adjunct Associate Professor for deep-sea mining at NTNU from Feb 2020–Feb 2024. He is currently associate professor and focuses his research on the handling of solids in a marine environment, for example excavation, transport, separation and deposition. He is the project coordinator of various (deep) seabed mining research projects.

Chapter 5
Considerations for Using Polymetallic Nodules as Alternative Metal Extraction Resource: Focus on Energy-Related Applications

P. K. Sen and Sanjay Kali

Abstract The worldwide imperative for transitioning to sustainable energy depends on the availability of essential minerals such as copper, lithium, nickel, cobalt, and rare earth elements. Some minerals and metals are especially susceptible to supply shortages and have been classified as "critical" due to the dangers connected with their availability and accessibility. Certain key metals possess diverse applications that extend beyond their utilization in clean technology.

Polymetallic nodules (PMN) are being explored as a substitute source for essential materials like Co, Ni, Mn, Cu, etc., which are needed for the clean energy transition in the electrical mobility sector. These minerals are mostly used for energy storage purposes rather than energy generation. Unlike land-based resources that are processed individually to recover each metal, PMN contains multiple essential metals that can be simultaneously retrieved.

Extracting battery metals from seabed resources would mitigate supply vulnerabilities associated with terrestrial resources. While there is currently no evidence of terrestrial resource depletion, recent initiatives in Australia involving alternative ore exploitation and the use of nickel pig iron from laterite ores in China and Indonesia for battery-grade nickel sulfate production indicate a growing interest in utilizing terrestrial resources.

This chapter compares the energy needed to create battery metal precursors and metal from polymetallic sea nodules with the energy required to produce these precursors from land-based resources as part of the cathode mix. These values exhibit a favorable comparison. This comparison enables the allocation of energy costs in MJ/km for the energy consumed during battery production. Consequently, the emissions associated with the assigned energy during a life cycle study of a vehicle uti-

P. K. Sen (✉)
Indian Institute of Technology, Kharagpur, India
e-mail: prodip.sen@gmail.com

S. Kali
CSIR-Institute of Minerals and Materials Technology, Bhubaneswar, Odisha, India
e-mail: ksanjay@immt.res.in

R. Sharma (ed.), *Deep-Sea Mining and the Water Column*,
https://doi.org/10.1007/978-3-031-59060-3_5

lizing a particular battery capacity can be approximated as grams of carbon dioxide per kilometer (gm CO_2/km). The superiority of polymetallic nodules is increased when energy allocation is specifically directed toward the production of saleable manganese product.

India has recently been researching the utilization of PMN resources in the energy storage sector. This research has focused on developing flow sheets, determining the energy requirements for production, and assessing GHG emissions. The gross energy needs (GER) for hydrometallurgical operations, ranging from 9000 to 15,600 MJ/T, and pyrometallurgical processes utilizing CNG, which is 10,000 MJ/T, while recovering manganese bearing products, are translated to energy requirements for battery use in MJ/km and emissions. Utilizing renewable energy sources in the established processes might further decrease the emissions generated during the process.

While it seems promising to utilize sea bed nodule resources for battery precursor production with reasonable energy use and emissions per kilometer for specific cases, further research is required to actualize this potential. This entails recalculating the net present value of products obtained from sea nodule processing based on additional experiments.

Keywords Polymetallic nodules · Energy transition · Battery production energy · Flowsheet development · Renewable energy use

1 Introduction

Currently, there is a substantial transition occurring in the global energy sector, with a strong movement toward more environmentally friendly sources. The need to significantly and promptly decrease greenhouse gas emissions in light of increasing evidence of global climate change is widely recognized. This dedication is evident in the ever-lofty goals established by numerous nations. The efforts of an increasing number of countries and companies to reach a state of zero net emissions require the extensive implementation of various clean energy technologies. Several of these technologies rely on essential minerals such as copper, lithium, nickel, cobalt, rare earth elements, and others.

The worldwide shift toward renewable energy is accelerating, propelled by a combination of government policies, technological progress, and economic considerations. As of November 2022, a total of 87 countries and the European Union have pledged to attain carbon neutrality within the current century. The aforementioned commitments encompass more than 85% of global emissions and 85% of the world's gross domestic product, as reported by the IEA in 2023 (IEA, 2023). Significant declarations made since 2021 encompass the People's Republic of China's objective to attain carbon neutrality by 2060, India's aspiration to achieve net-zero emissions by 2070, and Indonesia's ambition to reach net-zero emissions by 2060.

If all these promises and ambitions are achieved as intended, they will have the ability to restrict the global temperature rise to approximately 1.7 °C by the year 2100. To do this, it is necessary to transition to more advanced procedures and technology that reduce carbon emissions. Nevertheless, achieving these goals requires particular minerals and metals in quantities that are far greater than what is already obtainable from existing resources.

Some minerals and metals may face supply limitations during the transition to cleaner energy. These resources are considered "critical" because their availability and accessibility pose hazards to the implementation of clean energy transition operations. The broad concept of criticality surpasses these immediate issues. A critical mineral is a type of element, either metallic or non-metallic, that has two important qualities: it is absolutely necessary for the proper functioning of modern technologies, economies, or national security, and there is a potential risk that its supply chains could be interrupted.

Essential minerals are fundamental to the production of advanced technology, such as mobile phones, computers, fiber-optic cables, semiconductors, banknotes, and their use in defense, aerospace, and medical industries. Several of these minerals are essential components in environmentally friendly technology, such as electric vehicles, wind turbines, solar panels, and rechargeable batteries. At the same time, certain elements are essential for everyday items such as stainless steel and electronics. Within the domain of energy applications, both the methods for generating energy and storing it heavily depend on essential metals.

Nations globally have articulated ambitious strategies for imperative energy transition endeavors, involving both energy production and storage. The effectiveness of these methods depends on several metals that are considered crucial due to their limited availability on Earth and/or worldwide supply concerns. The focus on metals such as copper, cobalt, nickel, lithium, specialty metals like tellurium, silver, and rare earths like neodymium and dysprosium, which are crucial for renewable energy technologies, could result in a possible scarcity of terrestrial resources (Teske et al., 2016). Offshore mining businesses are currently considering the profitable extraction of resources from the deep-sea bed to provide materials for energy storage throughout the energy transition. Although rare earths and specialty metals may not currently be the main factors influencing prospective deep-sea mining projects, their projected future demand is estimated to be substantial.

The presence of dominant countries or firms in mineral production or processing might pose risks to vital mineral supply chains, potentially leading to limited availability. Other potential hazards encompass the underdevelopment of the market, governmental determinations, civil disturbances, environmental catastrophes, mining mishaps, geological insufficiency, widespread diseases, and armed conflicts. Hence, the notion of metal criticality in the context of clean energy use encompasses supplementary aspects that require deliberation about the resource-enhancing capacity of deep-sea polymetallic nodules, specifically for energy-related purposes.

2 Minerals, Clean Energy Transition and Polymetallic Nodules Resource

The explicit significance of minerals in the processes of clean energy transformation has been highlighted in many recent publications (IEA, 2021; Azevedo et al., 2022). It is expected that the shift to renewable energy will require a far greater amount of minerals compared to the production of power from fossil fuels. It is essential to comprehend the worldwide increase in the need for minerals to meet the need for renewable energy and storage technologies. Copper, aluminum, chromium, manganese, molybdenum, and nickel are essential for a range of low-carbon technologies, making them crucial elements for attaining a low-carbon future (Hund et al., 2020). Copper, chromium, and molybdenum are versatile minerals that are used in at least eight different technologies. Copper, in particular, plays a crucial role in all energy generating and storage technologies (Hund et al., 2020).

In the Sustainable Development Scenario, the International Energy Agency (IEA, 2021) expects a significant increase in the demand for key materials by 2040 compared to 2020. This is due to the high mineral intensity of clean energy technologies and their expected widespread use. When analyzing the need for battery-related minerals, renewables, and network-related minerals, it is estimated that the demand for lithium alone will increase by nearly 42 times. Additionally, the need for rare earth elements for low-carbon networks is likely to increase sevenfold (IEA, 2021). There is a significant increase in demand for minerals, especially in energy storage technologies such as batteries. Graphite is expected to have a 25-fold increase in demand, while cobalt and nickel are forecast to see a 19–21-fold increase. It has been proposed that the G20 take a proactive stance and create a G20 vital Minerals Security Partnership. This partnership would aim to guarantee a consistent, secure, and sustainable supply of vital metals for the worldwide transition to clean energy (Narula et al., 2023).

Moreover, it is crucial to acknowledge that more than 60% of metals that are deemed vital for the energy transition and digital technologies are manufactured as co-products or by-products (Bellois & Ramdoo, 2023). This has important consequences since increasing demand could result in challenging supply reactions, given that these metals are not extracted directly for their own sake but rather as partners to host metals. Companion metals are frequently disposed of in waste rocks and tailings during the concentration process for reasons related to economics and other issues, leading to the possibility of wasting valuable resources.

Extracting low-grade metal-bearing resources from the Earth without taking into account the economic extraction of companion metals, some of which may be crucial, presents difficulties. Untapped mineral resources can be exported, and insufficient datasets on trade movements provide challenges in evaluating these supply streams. The lack of data on the volume and value of all minerals sold or exported is an issue, particularly with the export of major concentrates without additional testing for minor metals. Without a comprehensive database encompassing all material flows, the task of evaluating supply issues and measuring criticality becomes arduous.

The supply dynamics of companion items such as cobalt and key inputs like lithium, which are utilized in specific energy storage systems like EV batteries, pose distinctive issues. The recent surge in prices for lithium and cobalt has raised concerns regarding the long-term availability of these resources due to varying patterns in their supply. Most of the global lithium supply is obtained as a main product, but cobalt is mainly generated as a secondary product. This raises concerns regarding transparency and the dynamics of the supply chain, particularly in the Democratic Republic of Congo (DRC) (Azevedo et al. 2018).

The European Commission's suggested methodology for judging criticality provides valuable insights in the context of discussions on critical materials. The evaluation process takes into account the economic significance (EI) and supply vulnerability (SR) as crucial factors, assessing more than 80 raw materials according to their utilization in economic sectors, stability of the source country, substitutability, and recyclability. A material is considered significant when both the EI and SR indicators exceed predefined numerical criteria (set at 2.7 for EI and 1.0 for SR). The technique, comprehensively examined by the ITRE Committee in 2022, offers a systematic strategy for assessing the importance of raw materials within the framework of the clean energy transition (ITRE Committee).

2.1 *Metals Requirement Estimates for Clean Energy Applications: Role of Polymetallic Sea Nodules*

The metal needs for energy transition applications have been thoroughly established, as demonstrated by a previous study (Teske et al., 2016). When examining wind, solar PV, CSP, hydro, and electric vehicles, it becomes apparent that copper, silver, and specialty metals/rare earths are mainly used in applications other than electric vehicles.

A recent study conducted by researchers from the University of California, Irvine, and MIT highlights that the existing stockpiles of minerals, including aluminum, copper, manganese, and silver, are sufficient to facilitate the development of energy transition technologies associated with wind and solar power. These technologies are crucial for achieving climate objectives (Wang et al., 2023). Although there is an increasing demand for materials, it does not exceed the total amount available in geological reserves. However, there may be a significant requirement for large growth in the annual production of certain metals such as neodymium (Nd), dysprosium (Dy), tellurium (Te), as well as fiberglass and solar-grade polysilicon. The study investigates whether the anticipated future demand for primary metals would result in enough production of specialty metals, which are by-products of mines primarily focused on extracting commodities such as copper or zinc. These specialty metals include cadmium, gallium, indium, selenium, and tellurium.

The extraction of metals from polymetallic nodules seems to be more suitable for E-mobility applications because of the particularity of the metals involved, namely nickel, cobalt, copper, and manganese. The ITRE Committee published a study in

2022 that provides a comprehensive analysis of the material requirements for E-mobility (ITRE Committee, 2022). The paper focuses on enhancing the security of supply for products that use critical raw materials, with the aim of supporting the green transition and decarbonisation efforts. The primary material requirements for the initial significant component of electric motors are defined as boron and rare earth elements (RREs), specifically dysprosium, neodymium, and praseodymium. The second significant element of the battery supply chain comprises lithium, cobalt, natural graphite, manganese, copper, and nickel.

When comparing electric vehicles (EVs) to internal combustion engine (ICE) cars in the context of electric motors, a significant distinction lies in the greater utilization of copper and NdFeB permanent magnets in EVs. An average electric vehicle (EV) typically necessitates approximately 80 kilograms of copper, which is roughly four times the amount needed in internal combustion engine (ICE) cars (ITRE Committee, 2022). The rotor of the engine, which weighs between 1.7 and 3 kg, is mostly composed of neodymium (0.25–0.50 kg per vehicle) and other rare earth elements (0.06–0.35 kg per car), together with copper, iron, and boron. Regarding battery components, Chile and Argentina are the leading countries in terms of refined lithium capacity from brine operations, with Chile accounting for 32% and Argentina accounting for 20%. The Democratic Republic of the Congo is responsible for 54% of global cobalt mine production. In addition, China dominates the processing of cobalt, accounting for 46% of global refined cobalt production. According to the IEA (2021), electric cars require six times more mineral inputs than conventional gas-powered cars, which is worth mentioning.

2.1.1 Specific Metal Recovery Focus for Meeting Metal Demands

In their recent paper, Paulikas et al. (2020) explored the potential of using metals found in deep-sea bed polymetallic deposits for applications in E-mobility. The analysis indicates that these deposits fortuitously possess metal concentrations that are well-suited to the needs of electric vehicle (EV) battery and assembly makers, specifically nickel, cobalt, manganese, and copper. The surveyed Clarion-Clipperton Zone (CCZ) is considered to have substantial amounts of these metals, making it a possible primary supplier of battery metals in the future decades. This advancement is expected to result in decreased greenhouse gas emissions during the production phase of conventional electric vehicle (EV) batteries, due to reduced energy consumption.

The study removes lithium, graphite, and aluminum from consideration, despite their significance in EV battery materials. The primary focus of EV batteries lies in materials like lithium, cobalt, natural graphite, and manganese, as well as other base metals such as aluminum, copper, and nickel. Certain exclusions are established on the assumption that certain metals should only be linked with resources found on Earth. Therefore, any conversations about the energy needs for extracting and the resulting emissions are exclusively focused on nickel (Ni), cobalt (Co), and manganese (Mn) obtained from deep-sea bed polymetallic deposits.

Table 5.1 Metal demand predictions from battery energy demand

No.	EV Metals Demand	Year 2020 World	Forecast for 2030 World	Remarks for links at table bottom
1.	Total energy supply, GWh	164.61 million	184 million (refer page 1985 in the pdf)	2020 link [1] 2030 link [2]
2.	Energy consumed for transportation, GWh	29 million	36.63 million	2020 link [3] 2030 link [4] (page 1/11 fig. 8–2. Converted to GWh)
3.	EV battery capacity demand, GWh	560	4300	2022 and 2030-link [5]
4.	kT of Ni, Co, Li, Mn for EV batteries	80 kT Ni 21 kT Co 20 kT Li 25 kT Mn	1567 kT Ni 358 kT Co 257 kT Li 246 kT Mn	2020 & 2030 link [6] (fig: Total mineral demand from new EV scenario)

1. https://yearbook.enerdata.net/total-energy/world-energy-production.html
2. https://www.researchgate.net/publication/343380608_A_critical_review_of_comparative_global_historical_energy_consumption_and_future_demand_The_story_told_so_far
3. https://www.iea.org/data-and-statistics/charts/energy-consumption-in-transport-by-fuel-in-the-net-zero-scenario-2000-2030
4. https://www.eia.gov/outlooks/ieo/pdf/transportation.pdf
5. https://www.mckinsey.com/industries/automotive-and-assembly/our-insights/battery-2030-resilient-sustainable-and-circular
6. https://www.iea.org/data-and-statistics/charts/total-mineral-demand-from-new-ev-sales-by-scenario-2020-2040

Table 5.1 refers to some quantitative estimates made regarding metal demands worked out from battery energy demand predictions.

It would thus be interesting to examine if part of the metal demands for E mobility (other than Li shown in the table) can be met from deep sea bed polymetallic nodules resource.

2.1.2 Metal Supply Status for Specific Critical Metals

The differentiation between mineral reserves and resources is crucial, as reserves denote the economically and efficiently extractable quantity of a mineral, whereas resources cover the projected overall quantity accessible, regardless of economic viability. Regarding cobalt (Shanon, 2023), the U.S. Geological Survey reports that there are roughly 7.6 million tons of cobalt deposits worldwide. Nevertheless, the overall cobalt reserves greatly exceed this amount, as estimates indicate that there are 25 million tons accessible on land and 120 million tons on the ocean floor.

As the value of a commodity rises, the quantity of reserves and resources generally increases. Previously unprofitable resources become feasible, and intensified endeavors in cartography and exploration lead to the expansion of documented resources. From 2012 to the present, there has been an increase in lithium reserves from 13 million to 22 million tons, and estimated lithium resources have more than doubled, rising from 40 to 88 million ton.

A study conducted by Heijlen et al. (2021) suggests that the existing process for creating new mines on land would face difficulties in satisfying the expected increase in the need for nickel and cobalt, particularly in situations that match with the emission targets set by the 2016 Paris Agreement. One possible approach entails a substantial augmentation of the installed capacity of current mines. The International Energy Agency (IEA) predicts that there will be around 45 million electric vehicle (EV) sales each year by 2030. This would necessitate the use of around 1.5–2 million tons of nickel annually, assuming an average of 40–50 kg of nickel per vehicle. The projected nickel production for 2030 is estimated to be 3.8 million tons, while the International Energy Agency (IEA) forecasts a demand of up to 6.2 million tons for the same year.

Furthermore, the development and expansion of alternate pipelines could potentially enhance the future supply of battery metals, alongside the growth of terrestrial mining. The recycling of nickel and cobalt is becoming increasingly important as studies emphasize their major contribution to future metal supplies. Reprocessing old mine tailings is seen as a possible means of augmenting metal production in the future.

2.1.3 Evaluation of Supply Risks

Obtaining an abundance of metals specifically for electric vehicles (EVs) is not a simple task due to the combination of main and co-products and the need to optimize land resources. The difficulty may not reside in the worldwide presence of essential minerals, but rather in their attainability and prompt extraction in suitable locales. The global distribution of minerals is characterized by an uneven pattern, where certain regions have a significant dominance in production. Notably, the Democratic Republic of Congo is a major producer of cobalt, China is a leading producer of rare earths, and Australia is known for its lithium production. Assessing the supply risk index (Blengini et al., 2017) at the government level becomes crucial from this standpoint.

The proposed utilization of polymetallic nodules for the extraction of battery metals, as indicated by Paulikas et al. (2020), entails possible advantages such as the mitigation of greenhouse gas emissions in battery production through the utilization of marine resources. The feasibility of this procedure in relation to land-based resources requires careful examination, ensuring that the comparatively lower energy output from oceanic nodules is in line with the objective of minimizing the energy demands of battery production. Implementing a lower-energy polymetallic nodule processing technology, if successful, could reduce supply risks caused by the unequal geographic distribution of metals.

In order to tackle issues connected to supply, the initiation of deep-sea mining from sources such as polymetallic nodules might provide a reliable source of minerals and metals like cobalt, nickel, and copper that are in high demand. By combining terrestrial supplies of lithium, graphite, and aluminum with underwater resources of nickel, cobalt, and copper, we can potentially establish a new approach for utilizing oceanic resources in battery manufacturing. By using terrestrial resources for specific battery metals, it would be advantageous to supplement battery production that relies on marine resources, ultimately mitigating supply vulnerabilities.

3 Energy Used, Emissions for New Resource Characterization

The inherent efficiency of the EV itself does not tell the whole story about the energy used in the manufacturing of the batteries related to metal extraction and related processes (Fig. 5.1).

The production of electricity requires energy consumption both during the extraction process (Input 1) and during its utilization (Input 2). The energy is subsequently stored in batteries, with additional energy being expended during the production process of these batteries. Inputs 1 and 2 pertain to the extraction of primary energy and its subsequent conversion into an energy carrier. However, the diagram does not depict the energy consumed during the production of the storage device. The battery manufacturing process yields the specified usable energy of the battery, which is measured in kilowatt-hours (kWh).

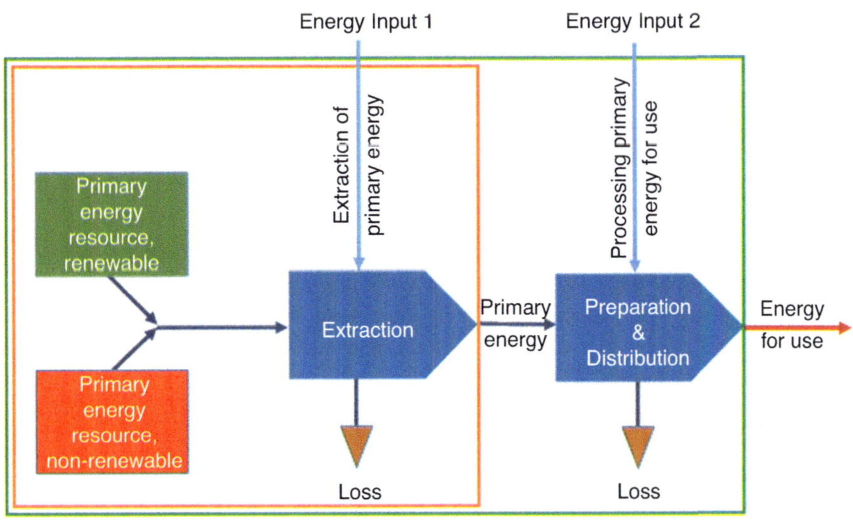

Fig. 5.1 Upstream energy processes in the manufacturing of the batteries are related to metal extraction and related processes. It is better understood by realizing that the battery of an EV is an energy storage device, and the eventual energy delivery process is related to the various upstream energy processes

Equation 5.1 determines the net electrical energy output of the battery. The equation accounts for the usable portion of the gross energy output, which is denoted by the energy carrier units less the losses that occur during the transfer of the energy carrier to the storage device. It is crucial to acknowledge that the energy consumed during the construction of the storage device is eventually recovered as recycled product energy at the end of its lifespan.

$$\textit{Net}\,\text{Energy} = \text{Gross Energy Output}\begin{pmatrix}\text{energy used}\\ \text{for product}\\ \text{making} + \text{energy}\\ \text{carrier}\\ \text{from storage}\\ \text{device}\end{pmatrix} - \Sigma\begin{pmatrix}\text{Energy}\\ \text{Investments}\\ \text{for product}\\ \text{making} + \text{energy}\\ \text{transmission}\\ \text{losses}\end{pmatrix} \tag{5.1}$$

The net energy available needs to be moderated by accounting for the product energy over the lifecycle of the battery to arrive at the energy cost incurred during the lifecycle.

The net energy available from any energy source can generally be characterized by an evaluation of energy return on energy invested (EROI). The energy delivered by a technology is divided by the energy invested to make such delivery possible to arrive at the EROI ratio. The "energy investment" is defined as the energy used to harvest an energy resource from the environment and convert it into a usable energy carrier. Thus,

$$\text{EROI} = \text{Gross Energy Output} / \Sigma\text{Energy Investments} \tag{5.2}$$

Whereas the EROI ratio is applicable to any energy source, the output carrier energy used to drive a vehicle may be further subjected to EROI analysis of the energy carrier (electricity/fuel) pertaining to energy inputs used for electricity generation. With known estimates of EROI for the input energy used (electricity/fuel), the additional energy inputs used for energy generation need to be added to other energy costs. This is explained subsequently for different vehicle types.

Tesla's design of the 75-kWh lithium-ion battery is particularly energy-intensive, mainly because of the metal extraction procedures used in the early stages. The extraction and processing of raw minerals, such as lithium, cobalt, copper, and nickel, have a substantial impact on the overall energy intensity. The production of the battery cell, namely the "drying" of the anode and cathode slurry, contributes significantly to the high energy consumption of the process.

The total energy consumption associated with the creation of lithium-ion batteries has been a topic of discussion, with estimates varying between 900 and 1800 mega joules per kilowatt-hour (kWh). There is an increasing agreement that the overall energy expenditure is included within this spectrum, taking into account both the production process and the extraction of raw materials. The objective of reevaluating manufacturing methods and upstream material selection is to minimize primary energy use, specifically for the cathode. The selection of the active material and the optimization of the production process are recognized as essential elements in reducing environmental consequences.

The energy consumption in battery manufacture can be categorized into two distinct components: energy required for acquiring raw materials and energy utilized throughout the battery manufacturing process. During the past 5 years, there have been notable reductions in energy use during the manufacturing process, namely in the operation of "dry rooms" utilized for slurry drying. However, achieving additional energy reduction depends on minimizing the energy needed for the processing of raw materials.

The influence of battery manufacturing on the overall energy efficiency is demonstrated by analyzing the energy expenditure per kilometer. Although electric cars (EVs) clearly demonstrate greater energy efficiency compared to conventional fuel-efficient sedans, it is crucial to consistently decrease the energy expenditure per kilometer in order to minimize emissions (Fulton 2023). The debate highlights the significance of minimizing the energy expenditure associated with the extraction and processing of cobalt, copper, lithium, nickel, and manganese.

The analysis includes the energy return on investment (EROI), which quantifies the increased energy expenses per kilometer associated with varying EROI levels. The primary objective is to minimize energy expenses in order to reduce the impact of greenhouse gas (GHG) emissions and improve the long-term viability of electric vehicles.

When energy costs in manufacturing are converted to GHG emissions, the following Table 5.2 provides the indicative ranges of emissions during different stages of the manufacturing of the batteries (Romare & Dahllöf, 2017).

Table 5.2 Emissions per kWh for battery-grade material production

	kg CO_2-eq/kWh battery		
Component	Raw material mining and refining	Battery-grade material production (including mining and refining)	Manufacturing (component and cell + battery assembly)
Anode	2 11	7 25	
Cathode	7–18	13–20	
Electrolyte	4.00	4–13	
Separator	<0.5	~1	
Cell case	<0.1	~1	
Battery case	4–13	10–25	
Cooling	0–3	2–6	
BMS	<1	4–30	
Total	18–50	48–121	20–110
Most likely value[a]		60–70	70–110

After Romare and Dahllöf (2017)

[a]Based on the assessment of transparency and scientific method done in the report

For minimizing GHG emissions, it is necessary to reduce energy costs related to manufacturing, which includes the stages of raw material mining and refining, material production, and other manufacturing processes. It is important to assess as to what extent the choice of a resource can affect the energy costs of production.

3.1 Polymetallic Sea Nodules as a New Resource for EV Battery Production

When evaluating polymetallic nodules as a prospective supplier of battery metals for electric vehicle (EV) manufacture, it is essential to comprehensively analyze the resource in terms of its energy expenditure for battery manufacturing. It is noteworthy that the Ni:Cu:Co ratio in sea nodules worldwide is very consistent, which makes it highly suitable for the development of high-nickel batteries, such as NMC 811 (as indicated in Table 5.3). This factor is not relevant when utilizing terrestrial sources of nickel, copper, and cobalt.

The conversation explores the potential of utilizing both marine and terrestrial resources for the creation of batteries. However, it highlights that using such a combination of methods could potentially heighten the susceptibility of the overall production process to supply concerns associated with input metals. The research specifically focuses on the creation of electric vehicle (EV) batteries using sea resources. It compares the energy requirements of this process to the existing methods that mainly depend on land resources.

The research proposes that sea resources, which are good for NMC batteries, could also be appropriate for LMO batteries by considering the nodules as a low-grade manganese ore. The narrow focus of this study enables a thorough analysis of the energy implications associated with using sea resources for battery manufacturing. It provides valuable insights into the feasibility and potential benefits or obstacles of this approach, as compared to the current reliance on land-based resources.

The specific compositional dependence of the batteries is given in Table 5.3 below.

Table 5.3 Battery cathode chemistry and composition (Mittal & Anand, 2019)

Type	Use	Chemistry	Li (%)	Co (%)	Ni (%)	Mn (%)	P (%)	Fe (%)
LCO	Portable electronics	$LiCoO_2$	11	89	–	–	–	–
NCA	Tesla model S (EV)	$LiNiCoAlO_2$	11	14	73	–	–	–
LMO	Nissan leaf (EV)	$LiMn_2O_4$	6	–	–	94	–	–
LFP	Popular in China but swithing to NMC	$LiFePO_4$	7	–	–	–	33	60
NMC (111)	Most popular in EV	$LiNiMnCoO_2$	12	30	30	28	–	–
NMC (433)	Most popular in EV	$LiNiMnCoO_2$	11	27	36	26	–	–
NMC (532)	Most popular in EV	$LiNiMnCoO_2$	11	18	45	26	–	–
NMC (622)	Most popular in EV	$LiNiMnCoO_2$	11	18	54	17	–	–
NMC (811)	Most popular in EV	$LiNiMnCoO_2$	11	9	72	8	–	–

4 Energy Requirements for Processing Terrestrial and Polymetallic Nodules Resources for EV Battery Precursors

4.1 Terrestrial Resources

The Gross Energy Requirement for metal extraction from terrestrial resources has been examined by several researchers. The following Table 5.4 provides representative values (Nuss & Eckelman, 2014; Rankin, 2014).

It is also interesting to observe from the table that the Gross Energy Requirement values reported for terrestrial mines depend on the energy requirements associated with the different phases of extraction, such as mining, beneficiation, and processing as shown in Table 5.4, bottom half (Rankin, 2014).

4.1.1 Energy Consumption for Terrestrial Nickel-Based Resources

The study conducted by Wei et al. (2020) offers significant insights into the primary energy demands of various nickel products, including nickel metal, nickel oxide, ferronickel, and nickel pig iron. The energy requirements for manufacturing these goods vary from 60 to 369 GJ/t alloy (including Ni). Ferronickel and nickel pig iron are considered unsuitable for battery-grade input because they require extra treatment, notably sulfidization, to provide an acceptable nickel input for battery precursors. The processing of sulfide ore typically uses less energy and results in a pure final product, but the treatment of laterite ores requires more energy to produce a mixed oxide product. This mixed-oxide product then needs to undergo additional processing to be used as a precursor for batteries (Dry et al., 2019).

Table 5.4 Gross Energy Requirement for metal production with stage wise energy demands

Nuss and Eckelman (2014)							
Gross energy requirement (MJ/kg metal production)	**Al**	**Cu (hydro)**	**Cu (pyro)**	**Steel**	**Ni (pyro)**	**Ni (hydro)**	
Mining	0.18	16.59	7.91	0.11	11.52	9.30	
Beneficiation	0.18	1.58	11.69	0.45	19.23		
Chemical processing I	30.58	46.29	10.33	2121	33.33	184.47	
Chemical processing II	180.57			0.93	49.44		
Embodied energy	211.51	64.46	33.02	22.70	113.52	193.77	
Rankin (2014)							
Gross energy requirement (GJ/ton metal production)	**Cu**	**Ni**	**Pb**	**Zn**	**Al**	**Steel**	**Cement**
	64.5	193.08	19.6	48.4	211.5	22.7	5.6
	6.16	16.08	2.07	4.61	22.7	2.19	0.9
	125	351	14.8	29.3	4.5	2.4	–

The acknowledgment of Tsingshan's strategy (Sherritt International Corporation 2021) to provide the lithium-ion battery industry with nickel matte obtained from nickel pig iron (NPI) is recognized. Nevertheless, it is specified that this plan presents specific obstacles and factors to be taken into account, such as substantial capital investment, supplementary processing expenses, potential environmental ramifications, redistribution rather than overall augmentation of nickel supply, potential constraints on Class I premium, and the potential for causing supply shortages in other industries dependent on Class II nickel.

The talk also addresses the minimum energy consumption required for incorporating nickel into the battery cathode mixture of NMC 811. According to Wei et al. (2020), the analysis indicates that the energy value per battery (600 MJ/kg Ni) might exceed 33.5 GJ. It is stressed that this value must be augmented by the energy consumption for additional metals (such as copper, cobalt, and manganese) as well as other raw material constituents. The energy demand per unit mass of a metal is examined in connection with the ore origin being processed, with a specific emphasis on lateritic nickel-based resources. According to the analysis, the most preferable approach for processing a resource to recover metal is to have a larger proportion of nickel in the product, which results in reduced energy consumption per ton of resource. This choice takes into account the related conversion costs.

4.1.2 Additional Energy Usage for Other Metals in EV Batteries

Batteries contain other metals in addition to nickel, such as lithium, cobalt, manganese, and copper for connectivity. Additional energy consumptions, not discussed in this section, are foreseen for these metals.

4.2 *Polymetallic Nodule Resources*

The conversation highlights the distinct difficulties and factors involved in the processing of polymetallic nodule resources for battery production. The main focal elements emphasized are:

(a) Low-Grade Manganese Resource: Polymetallic nodules are predominantly regarded as a resource with a relatively low concentration of manganese. Although the nickel and cobalt levels are important for determining battery-grade suitability, it is necessary to efficiently handle the surplus manganese in the nodules.
(b) Copper Content: Polymetallic nodules include copper as well as an excess amount of manganese. The discussion proposes that copper, which is connected to certain processes, can be employed in a range of low-carbon energy generation methods.

(c) Metal Recovery Techniques: The methods used to recover nickel, cobalt, and copper from marine nodules vary. Precursor compounds are used to recover nickel and cobalt, whereas copper is recovered in its metallic state.
(d) Excessive amounts of manganese found in sea nodules give rise to inquiries regarding its utilization. The discussion explores the potential for reclaiming surplus manganese as a ferroalloy component in the production of steel.
(e) Process Flow Sheet: A customized process flow sheet is required to effectively use sea nodules as a producer of battery metals and copper metal, taking into account their distinct composition and challenges.
(f) Energy Sources: Metal recovery techniques employ both hydrometallurgical and pyrometallurgical methods, utilizing a variety of fuel-based energy sources. The energy balance encompasses both primary and secondary energy sources, with primary energy being classified as either non-renewable or renewable.
(g) Energy Computation: The computation of energy inputs includes fundamental energy sources, such as fuel oil, diesel oil, and syngas, as well as major energy carrier streams, such as electricity. The strategy takes into account the concepts of circular economy, while also considering the energy required for obtaining fresh reactants from upstream sources, as explained in an earlier publication (Sen, 2022).

The slurry pulp density is a crucial factor in predicting energy requirements in hydrometallurgical flowsheets, especially in processes involving ammoniacal leaching where the energy needed for ammonia recycling must be taken into account.

The primary objective is to create a detailed and enduring process flow sheet that not only tackles the specific difficulties of sea nodule processing but also conforms to the concepts of circular economy and effective energy utilization.

4.2.1 Energy Requirements for Hydrometallurgical Options of Treatment

Hydrometallurgical techniques utilize direct leaching of the input material to specifically dissolve the desired metals. The energy needed for leaching, taking into account only the heat that can be measured, is greatly affected by the density of the leach pulp and the temperature at which it is operated. Pressure-acid leaching procedures generally require elevated temperatures for effective leaching. Utilizing a reductant, such as a mixture of ammonia and sulfur dioxide, has the potential to significantly decrease the temperature at which leaching occurs (Das et al., 1998). Within this particular framework, it is imperative to take into account the supplementary energy required for the recycling of the leach reactant (ammonia) in order to minimize the expenses associated with the input reagents. Hence, apart from the energy needs specified in Table 5.4, it is essential to consider additional energy for the recycling of chemicals, particularly for compounds such as ammonia. During a steam stripping process to remove ammonia, the ratio of steam to ammonia is around 5–7 tons of steam per ton of ammonia. The leach has a pulp density of 20%. As a result, there may be additional energy expenses of up to 3000 MJ per ton of ore. Hence, the aggregate energy demand for slurry heating (1500 MJ per ton of ore, with a pulp density of 20%) and reagent recycling may amount to around 4500 MJ per ton of ore, contingent upon the particular leaching conditions.

4.2.1.1 Extraction of Metals from Leach Solutions

Electrowinning is commonly used to extract metal from leach solutions. The typical energy demand for a sea nodule operation is roughly 70–75 kilowatt-hours (kWh), which is equivalent to about 900 megajoules (MJ) per metric ton of ore in terms of fuel oil.

4.2.1.2 Options for Treating Leach Residue

Hydrometallurgical activities produce a residue containing manganese that needs further processing to be converted into a valuable co-product. Additional pre-treatments are required for the leach residue due to its unfavorable Mn/Fe ratio, in order to make it suitable for direct usage in a silicomanganese alloy manufacturing machine. The residual material obtained by leaching, which contains a substantial amount of manganese (specifically, 26% manganese, 10% iron, and 16% silica), cannot be used directly for the process of silicomanganese smelting because its Mn/Fe ratio (2–3) is too low. The process of pyrometallurgical upgradation, aimed at improving the Fe/Mn ratio, necessitates an approximate energy consumption of 200 kWh per ton of ore. This energy requirement is calculated to be comparable to 2400 MJ per ton of ore when using fuel oil for residue treatment.

4.2.1.3 Total Energy Requirements for Flow Sheets Utilizing Reagent Recovery

The total energy demand for an ammoniacal-based process is around 7800 MJ per metric ton of ore (sea nodules), excluding the energy required for the initial input of fresh ammonia. Given an initial energy input of 30 gigajoules per metric ton of ammonia, there is an expected additional energy input of 6000 megajoules per metric ton of ore, as indicated by the data from the ammoniacal flowsheet.

4.2.1.4 Total Energy Requirement Excluding Reagent Recovery

Studying the energy needs of the process without using a recovery procedure that requires a lot of energy is of interest. This particular variation, demonstrated by a high-pressure leaching process using sulfuric acid without the need for an extra reducing agent, entails directly treating the input material in a water-based solution. This approach necessitates a distinct analysis of energy usage. Terrestrial laterite processing utilizes different flowsheet approaches, including the Moa Bay/Amax and Australian concepts of Murrin-Murrin, Bulong, and Cawse plant operations (Mayze, 1999). The high-pressure acid leaching process, operating at a leach temperature of 250 °C and a L:S ratio of 2:1, necessitates a thermal input of 2100 MJ per metric ton of feed. Using high-pressure steam, the conversion rate is 0.67 tons of steam per ton of ore. An H2SO4 generation facility integrated with sulfur can fulfill around 20–30% of the steam need. The projected fuel oil need is 50 kg per ton of feed, accounting for 70% of the fuel oil needed for steam generation. The overall energy demand fluctuates based on the heat load, peaking at 3060 MJ per metric ton of ore at a L:S ratio of 3:1.

4.2.1.5 Calculation of Total Energy Requirement Considering Manganese Dissolution

To decrease the amount of steam required for heating, one can employ a lower temperature in acid leaching by introducing extra reductants such as SO_2 (Khalafalla et al., 1981) or organic substances to dissolve manganese. At a leach temperature of 110 °C, the heating load is approximated to be 1500 MJ per metric ton of ore. Introducing more reductants leads to a significant presence of manganese in the liquid phase, enabling the investigation of alternate manganese products. Electrolytic manganese dioxide (EMD) necessitates approximately 2 kilowatt-hours of energy per kilogram of EMD. By utilizing a modest amount of manganese (23–25%) and achieving an 80% recovery rate, the manufacturing of electrolytic manganese dioxide (EMD) yields an impressive energy output of 9300 MJ per ton of nodules. This exceeds the usual requirements of the electric vehicle (EV) battery industry. When assessing the viability of a complete manganese dissolution procedure, it is important to take into account the existing manganese markets. This is because the demand from the electric vehicle (EV) battery industry may not be enough to accommodate the dissolved manganese content extracted from sea nodules.

4.3 Observations on Energy Requirements for Hydrometallurgical Processing Options

The aforementioned debates yield the following observations:

(a) The main sources of energy consumption are steam, electricity, and chemical reagents, together with their recycling processes. The overall energy expenditure for thermal loads, which includes reagent recycling, amounts to approximately 4500 MJ per ton. In contrast, the subsequent utilization of electrical energy to manufacture copper, nickel, and cobalt metallic products, as well as a marketable manganese-bearing product for the ferroalloy industry, requires 3300 MJ per ton. The energy demand for producing a marketable manganese-containing product from leach residue is substantial and would rise considerably if a ferrosilicon manganese product were to be produced. If the lean grade slag formed cannot be easily sold to a ferroalloy factory, the extra electrical energy consumption would be increased even more. An extra 650 kWh of electrical smelting energy (equivalent to 7800 MJ) must be supplied per ton of nodules, resulting in a twofold increase in the total energy demand to 15,600 MJ per ton of nodules.

(b) The inclusion of extra substances in a hydrometallurgical process, like ammoniacal leaching, increases the amount of energy needed at the beginning of the process. This additional energy demand would be added to the total energy consumption of 7800 MJ per ton of nodules when reagent recycling is taken into account. The addition of new ammonia results in an input energy of 6000 MJ per ton of feed. Investigating alternative energy-efficient techniques for producing ammonia from waste product streams and utilizing renewable sources for steam generation, instead of relying on fossil fuels, has the potential to result in a decrease in overall energy consumption. While it is believed that sulfuric acid is produced within the plant from elemental sulfur, the sulfur that is added to generate sulfuric acid contributes to an extra energy input of 1000 MJ per ton of ore. This additional energy input, which is not considered in the current analysis, is assumed to consume 0.35 T acid per ton of ore.

(c) When comparing a sulfuric acid leach system to an ammoniacal-SO2 leaching system that uses renewable organic reactants, it is found that the thermal energy required during leaching is similar, amounting to 1500 MJ per ton of ore. Nevertheless, the energy demands of the reductant would decrease if the source were renewable. The parameters for low-temperature reductive acid leaching have been established (Aishvarya et al., 2013). This technique leads to the dissolution of manganese and the creation of a distinct product containing manganese, such as electrolytic manganese dioxide (EMD), which requires additional investigation before being used on a big scale. The energy needed to make EMD (about 650 kWh per ton of feed) is calculated to be 7800 MJ per ton of feed. In addition, it is necessary to include the electrical energy needed to produce metals from the leach solution once the dissolved manganese has been separated. This results in a total energy need that surpasses 9000 MJ per ton of feed.

(d) When comparing options (a) and (c) mentioned earlier, it may be necessary to restrict the metal production possibilities for recovery in option (a) to only copper electrowinning and precipitating a mixed nickel-cobalt combination. This limitation is due to the possibility of the total process requiring a significant amount of energy. The remaining dissolved elements should be treated to provide battery precursors that comprise pure nickel and cobalt, rather than metals that were used for energy requirement calculations. This phase may not need a significant amount of energy. While option (c) has a lower overall energy consumption than option (a), it does have the drawback of producing an excessive amount of manganese-bearing product, which may not have readily available markets. The majority of the worldwide manganese market is currently focused on the manufacturing of ferroalloys, with only modest demand for electrolytic manganese metal and EMD production. The surplus manganese generated is unlikely to be utilized for significant amounts of LMO manufacture, as the existing demand is restricted. The available choices, for instance (c), involve either manufacturing mixed metal precursors for commercial purposes or, following copper electrowinning, precipitating nickel, cobalt, and any impurities in the solution to create a mixed hydroxide compound. The utilization of this substantial amount of manganese-containing material remains ambiguous unless additional research is conducted.

The flow sheets pertaining to options (a) and (c) discussed are shown in the Figs. 5.2 and 5.3 below.

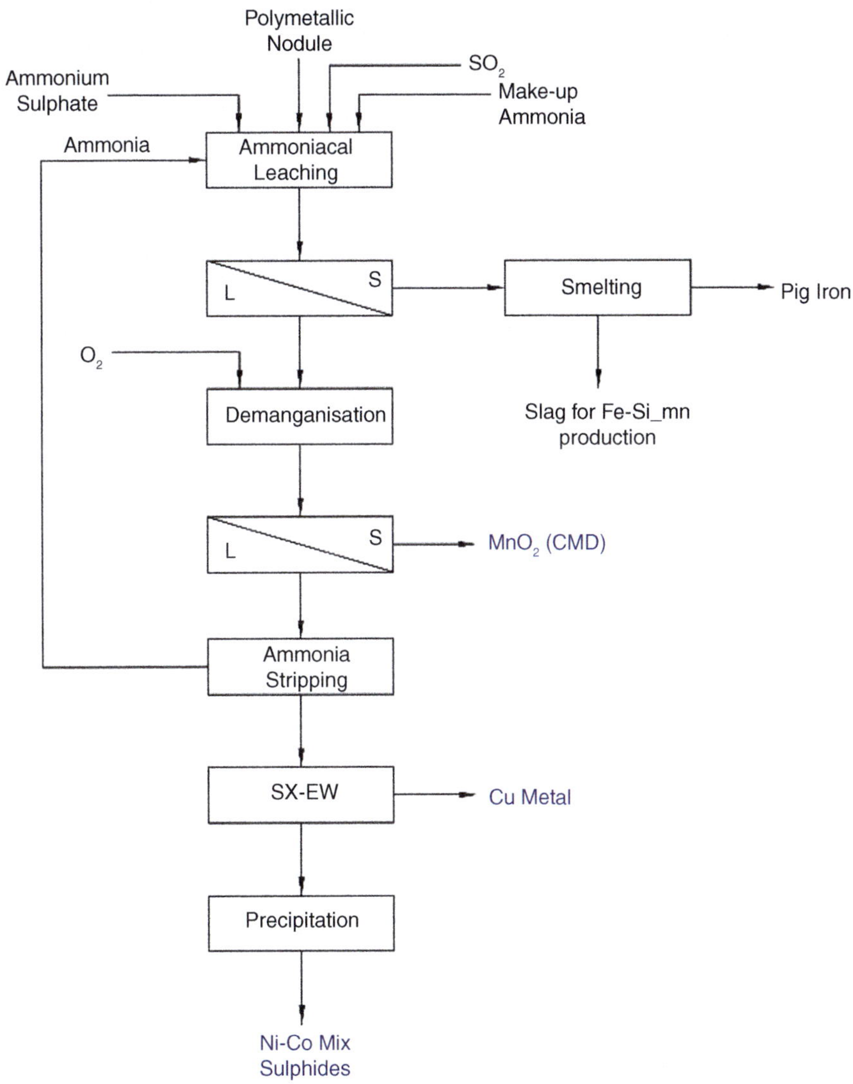

Fig. 5.2 (Case a) Bulk Sulfide precipitation with copper metal and Ferro silico manganese

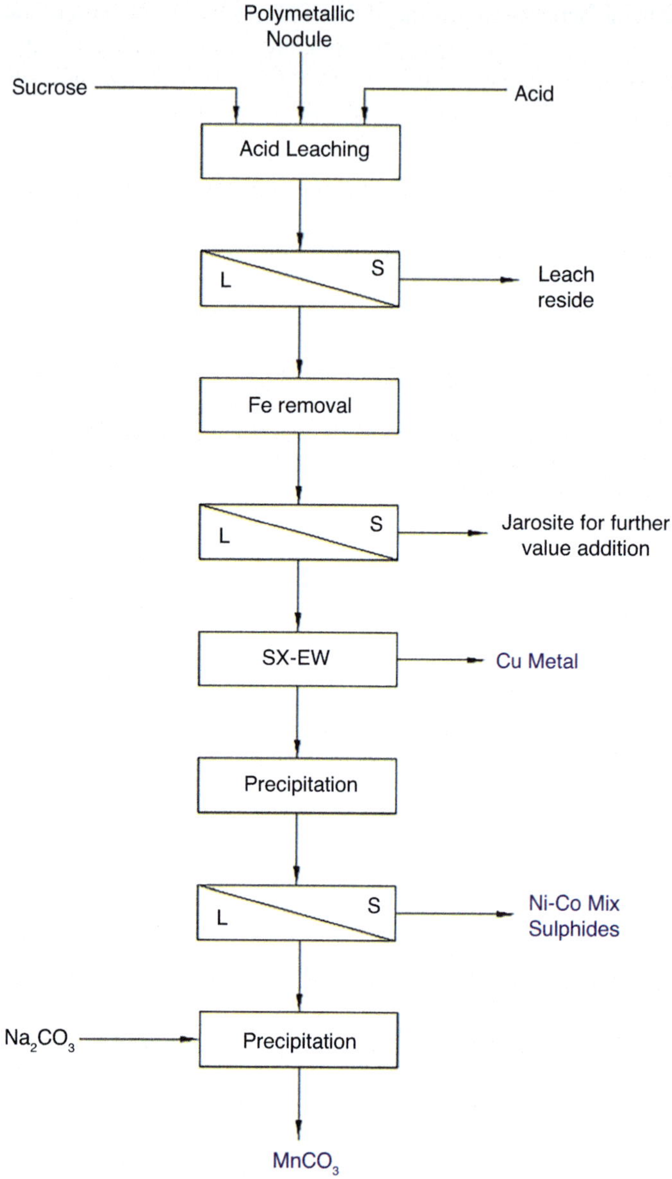

Fig. 5.3 (Case c) Bulk sulfide precipitation with manganese carbonate for conversion to EMD

A possible flow sheet version (Fig. 5.2) based on earlier work carried out at the Institute of Minerals and Materials Technology (Bhubaneshwar, India) can result in the production of battery precursor bulk sulfides from which copper can be electro won after bulk sulfide dissolution. The resultant nickel-cobalt containing solution can be further processed to separate nickel and cobalt battery precursors or removed as a mixed hydroxide for external processing. For case (c), in addition to the production of battery precursor bulk sulfides, which contain nickel and cobalt, different manganese-bearing products can be generated (Fig. 5.3). For example, EMD can be readily obtained from manganese carbonate.

4.4 *Energy Requirements for Pyrometallurgical Options of Treatment*

4.4.1 Pyrometallurgical Pretreatment Process for Sea Nodules

Pyrometallurgical options for processing sea nodule ores resemble laterite treatment processes, which eventually produce ferronickel after an ore drying/calcination/pre-reduction step. Pure hydrogen (0.56 kmol/100 kg of limonitic laterite ore, 0.125 Nm^3/kg ore) has been used (Elliott & Pickles, 2017), to study the reducibility of laterite, but the reducing agent requirement is likely to be different than an ore similar to sea nodules because of different iron and manganese contents. This pretreatment/reduction step based on laterite ores has been estimated earlier (Roorda & Hermans, 1981) to consume about 5300 MJ/T equivalent to around 0.5 Nm^3 hydrogen/kg ore. In any case, the supply of this energy will be beneficial if the hydrogen supplied is generated as green hydrogen, which can well serve as a reducing agent.

Natural gas, although not renewable, has a low emission coefficient. Laterite ores have been reduced by methane by a number of researchers, such as Pickles and Anthony (2018b) and Li et al. (2018). Pickles and Anthony (2018b) indicated methane consumption of 0.03–0.1 Nm^3/kg for saprolitic laterite, whereas the indicated consumption by Pickles and Anthony (2018a) for limonitic ores was up to 0.05 Nm^3/kg. If the entire heat consumption of 5300 MJ/ton of laterite (Roorda & Hermans, 1981) including reduction, drying, etc., a methane consumption of 0.13 Nm^3/kg is to be foreseen. It is difficult to compare the consumption values of laterite because of different raw materials, but the values indicate the ranges of gas consumption for acceptable nickel recoveries. However, preliminary experiments indicate that this value may be enhanced to 6500 MJ/T sea nodules.

4.4.2 Energy Requirement for Processing Reduced Mass

The reduced mass is melted and then the alloy phase is sulfidized and subjected to further iron removal, if required. The mixed sulfides are then subjected to oxidative pressure leaching to generate a mixed metal sulfate solution for further purification and metal recovery. The slag phase generated after melting is expected to be an appropriate raw material for ferroalloy manufacture.

Estimates indicate that 200 kWh energy would be required for melting and 75 kWh for metal recovery processes. The total energy required for processing the reduced mass is then 3300 MJ.

4.4.3 Total Energy Requirement for Present Approach

This is estimated at 10,000 MJ/t of sea nodules, if this approach is followed. Since the energy requirements are much higher than hydrometallurgical processing approaches, it is necessary to produce value added precursors for battery metals (Ni and Co) and also part of the dissolved manganese to produce precursors for both Li-Mn and NMC batteries. This is in contrast to the approach adopted for hydrometallurgical processing, where the final battery precursor metals may be in a mixed state but a ready stock for use by external parties.

4.4.4 Current Approach Being Researched

With respect to attaining a viable solution to using sea nodules for battery precursor inputs using the pyrometallurgical route, the approach being researched at the Indian laboratory is briefly discussed.

This work pertains to the development of a technology based on gaseous reduction-melting route for the recovery of Cu, Ni, Co, and Mn from polymetallic nodules. The process involves gaseous reduction of polymetallic nodules in the form of pellets of 2–3 mm size using compressed natural gas (CNG) at a temperature of 850 °C followed by melting of the reduced nodule pellets at about 1600 °C for the production of Cu-Ni-Co-Mn-Fe alloy, while generating a slag with enhanced Mn:Fe ratio >6 and free from any carbon content. The process flowsheet is shown in Fig. 5.4. During gaseous reduction of polymetallic nodule pellets, controlled reduction is carried out so that the oxides of Cu, Ni, and Co are reduced to their respective metallic phases while allowing partial reduction of oxides of Fe and Mn to join the alloy phase and rest of Fe as well as Mn content to join the slag phase during melting operation. The melting operation, carried out subsequent to the CNG-assisted gaseous reduction stage, is characterized as a zero-carbon emission process as it only facilitates the separation of metallic phase from the oxides by forming Cu-Ni-Co-Mn-Fe alloy and a slag with enhanced Mn:Fe ratio. The composition of the alloy obtained from the melting operation is presented in Table 5.5.

As shown in Fig. 5.4, the Cu-Ni-Co-Mn-Fe alloy is being treated further for downstream processing for the dissolution of alloy, the removal of Fe, solvent extraction-electro winning of Cu, and the production of $MnSO_4$, $CoSO_4$, and $NiSO_4$, which can be used for battery applications. The Mn-enriched slag with Mn:Fe ratio >6 has been tested for its suitability to be used as a feed material for the production of silicomanganese alloy.

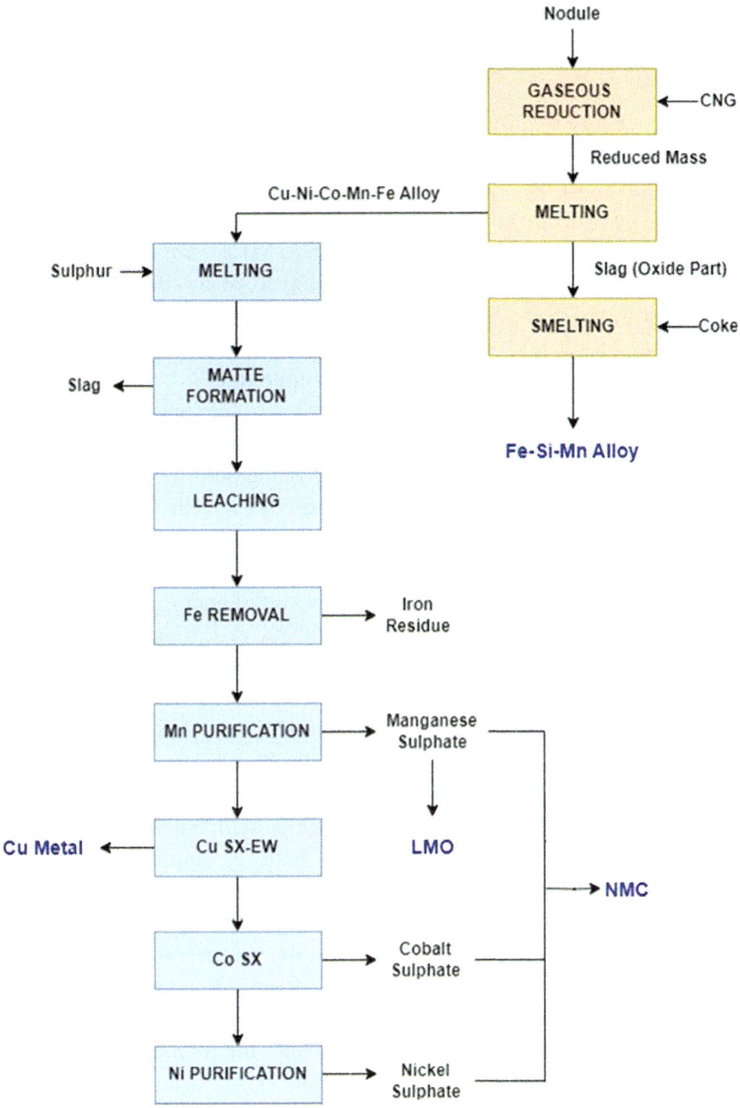

Fig. 5.4 Approach for producing battery precursors using pyrometallurgical treatment with methane reduction

Table 5.5 Composition of Cu-Ni-Co-Mn-Fe alloy produced through gaseous reduction process

Element	Cu	Ni	Co	Mn	Fe
Wt (%)	5–6	6–7	0.8–1	10–18	65–70

4.5 *Suitability of Sea Nodules for Making Battery Precursors, Considering Energy Expended*

A study of the energy requirements linked to hydrometallurgical and pyrometallurgical processing methods emphasizes important factors to consider when evaluating the appropriateness of deep-sea nodules as battery precursors in the cell manufacturing phase. The hydrometallurgical methods for processing sea nodules require an energy consumption ranging from 7800 to 15,600 MJ per ton of sea nodules. On the other hand, the pyrometallurgical method is expected to consume approximately 10,000 MJ per ton of nodules. These values are obtained by the process of electrowinning copper, nickel, and cobalt metals, as well as the manufacture of manganese-containing slag, ferroalloys, and manganese compounds.

Hydrometallurgical processes have difficulties in maximizing the utilization of products containing manganese due to uncertainty surrounding the application of manganese products. As mentioned earlier, the manufacturing of ferroalloy tailored for specific markets using hydrometallurgical methods leads to the most significant energy usage per ton of nodules (see Fig. 5.2). In contrast, significant amounts of manganese-containing goods produced in specific hydrometallurgical procedures (see Fig. 5.3), which require a considerable amount of energy, may have difficulties finding rapid market applications.

In order to evaluate the significance of energy consumed in these processes, it is crucial to examine the documented information on energy utilized in the production of battery systems as reported in several publications (Amarakoon et al., 2013; Romare & Dahllöf, 2017; Ambrose & Kendall, 2016). The recorded energy consumption for cell manufacturing shows significant variability, spanning from 316 to 2318 MJ/kWh, with a probable average of 960 MJ/kWh. The articles measure energy consumption as primary energy use by battery component (MJ/kWh capacity), including different components such as cathode paste, anode paste, substrates, anode and cathode, cell containers, battery management systems, and pack manufacture (Romare & Dahllöf, 2017). For example, the overall energy content of a pack (1920 MJ/kWh) translates to an energy consumption rate of around 650 MJ/kWh for the cathode paste that includes battery precursor metals. Dai et al. (2019) have indicated 1126 MJ/kWh for NMC 111 where the total energy powder energy is indicated approximately as 35% of this value.

Within the scope of this investigation, the precursor metal compounds derived from sea nodules would include the cathode paste, which additionally encompasses graphite and lithium (Li) for conventional batteries such as NMC. By assuming a moderate energy value of 650 MJ/kWh for the precursor powders, a comparison is conducted using NMC 811 composition as an illustrative example. The composition of this substance consists of 7.36% lithium (Li), 46.01% nickel (Ni), 5.4% manganese (Mn), 6.28% cobalt (Co), and 34.95% oxygen (O). Given a Li/kWh ratio of 0.12 kg and a 75 kWh battery pack requiring around 9 kg of Li, the total amount of nickel used per pack is approximately 56 kg. The energy consumption of lithium in the battery pack is calculated by considering the energy needed to recover 4700 MJ for every 1 ton of lithium carbonate. (Kelly et al. 2021). This results in an energy need of 230 MJ for the lithium in the battery pack, equivalent to 3.1 MJ per kilowatt-hour (kWh). The process of extracting and purifying nickel from 4.48 tons of sea nodules for the battery pack requires an energy input of around 590–600 MJ/kWh for the cathode pack, which also includes the energy needed for lithium extraction and refining. Given that this accounts for around 35% of the total energy of the pack, the overall energy need for the pack is approximately 1700 MJ/kWh. Significantly, a fraction of the energy needed for the battery pack would be provided by copper metal generated during the processing of sea nodules.

To summarize, it seems that both hydrometallurgical and pyrometallurgical methods investigated for extracting metals from marine nodules could be energetically compatible with lithium-ion batteries. However, there are difficulties in using manganese-bearing products in hydrometallurgical processes. Additional improvement of these processes is necessary to customize energy needs for the manufacture of metal precursors (namely Ni, Co, and Mn) rather than the recovery of fully formed metallic metals, which was the initial purpose of these processes.

4.6 *Effect on Economics on Processing Sea Nodules to Precursors for Battery*

In order to streamline the process of comparing resources based on revenue, it is possible to simplify by turning the quantities of all metals to be produced into nickel equivalents. For this estimation, metals such as nickel, cobalt, and copper are assumed to be acquired in their pure metallic states. On the other hand, manganese is found in reduced forms like ferroalloy or enhanced products such as slag and manganese dioxide, each having its own specific pricing per unit. This approach facilitates a further comparison between deep-sea mining and land-based lateritic nickel ore mining. The production quantity of a selected metal, defined in terms of nickel equivalents, indicates the amount of nickel needed to generate equal revenue as the sale of the metal in question, taking into account a certain nickel price and price ratios for other metals relative to nickel. Hence, it is necessary to comprehend the price correlation between the chosen metal and nickel in order to accurately compute nickel equivalents, as elaborated further below.

The function Ni Equivalent (Ni Eqv t/h) can be defined as:

$$\mathrm{NiEqv, t / h} = \sum_{i=1}^{4} \frac{P_i M_i}{P_{\mathrm{Ni}}} \tag{5.3}$$

where,

P_i denotes price of metals/kg for recovered metal from nodules
M_i is Metals recovered, ton/h
P_{Ni} is Price of the Nickel/kg

(Index "i" refers to the four metals considered in this study, namely Mn, Cu, Ni, and Co.)

The nickel equivalent function is thus a direct measure of the sales revenue of processing scheme employed for metal recovery from a similar resource and would form the basis of revenue comparison with a land-based resource (Biswas et al. 2009).

When the final products are not metals, the revenue accrued per ton of nodules can be visualized using the following expression:

$$\text{Revenue} = \Sigma W_i \frac{W_n}{W_i} P'(n) \tag{5.4}$$

W_i denotes the weights of metals, whereas W_n and $P'(n)$ denote the weights of actual products and price of the actual product. To arrive at a notional price of the metal, the ratio $\frac{W_n}{W_i} P'(n)$ needs to be evaluated. Depending on the actual product price and the quantity of product generated, which would be higher than the metal quantity that would have been produced, the notional metal generated revenue may be higher than the actual revenue. In this case, when converted to nickel equivalents, the value may be lower. This would lead to a less total revenue and a lower rate of return for a given capacity. To arrive at an acceptable internal rate of return, the nodule throughput may need to be increased as compared to earlier cases. One may need to foresee building of larger capacity plants than the referenced value of four metal recovery processes.

Required equivalent nickel production rate for an acceptable profit of operation and an internal rate of return may be estimated using the following expressions (Biswas et al., 2009).

$$E_{Ni} = C_{\text{opmin.}} + C_{\text{opmetal}} - \sum_{i=1}^{i=N} P_i * R_i + \frac{C_{\text{capexmin}}}{\text{POP}} + \frac{C_{\text{capexmetal}}}{\text{POP}} \tag{5.5}$$

$$\text{PAT} = \left(E_{Ni} + \sum_{i=1}^{i=N} P_i * R_i - C_{\text{opmin.}} - C_{\text{opmetal}} \right) * \text{Ni}\left(\frac{\text{kg}}{\text{year}} \right) * \left(1 - T_{\text{rate}}\right) \tag{5.6}$$

$$\text{NPV} = \text{PAT} * \frac{(1+i)^n - 1}{i.(1+i)^n} - \text{INV} \tag{5.7}$$

C_{opmin}, C_{opmetal}, and C_{capex} are operating cost for mining and metallurgical processing in \$/kg nickel, C_{capexmin} represents capital cost for mining and transportation (converted to \$/annual kg nickel based on POP), $C_{\text{capexmetal}}$ represents capital costs for processing in (converted to \$/annual kg nickel based on POP). E_{Ni}, P_i and R_i relate, respectively to, the price of nickel, the price of other ith component and recovery ratio values of the considered component with respect to nickel, T rate is the prevailing tax rate, I is the required interest rate, *n* is the lifetime of the investment, INV is the total capital investment, NPV is the net present value. Because of nickel and cobalt product being recovered as compounds also for purposes of nickel equivalent comparison with metal producing plants, the metal production-based nickel equivalent which when multiplied with nickel metal processed would yield the total revenue generated when appropriate correction factors are used in the definition of metal equivalent. For example, the correction for nickel would be given as:

$$F_{\text{Ni}} = \frac{W_{\text{Ni}}\,\text{product}}{W_{\text{Ni}}\,\text{metal}} \cdot \frac{P_{\text{Ni}}\,\text{product}}{P_{\text{Ni}}\,\text{metal}} \tag{5.8}$$

5 Correlation of Energy Used in Manufacturing with Emissions

The main goal of assessing the manufacturing energy needed for producing battery precursors from sea nodules is to determine the influence of the manufacturing process on emissions. As previously mentioned, the production of cathode powder is expected to account for around 30–35% of the total energy consumption in the manufacturing process. Evaluating the influence of battery production on the total emissions of electric vehicles (EVs) is a multifaceted topic. Recent research has unveiled a wide range of findings, demonstrating discrepancies ranging from 56 to 494 kg of carbon dioxide emitted per kilowatt-hour (Hall & Lutsey, 2018).

In order to acquire a more profound comprehension of these values, it is imperative to carry out thorough life cycle analyses. Although electric vehicles do not directly release greenhouse gases, they rely on electricity, which is primarily produced from fossil fuels in numerous locations globally. Emissions are generated as a result of the energy consumption involved in making the vehicle, especially the battery. The emissions are classified into three categories: emissions from the fuel cycle (including oil production, transportation, refining, and electricity generation), emissions from manufacturing the non-battery components of the vehicle, and a cautious estimation of emissions from manufacturing the battery. Table 5.6 (Hall & Lutsey, 2018) shows that there is significant variability in emissions, including those associated with battery manufacture. This is because energy is used during the manufacturing process, leading to emissions when considering the entire life cycle of the battery.

The main objective of evaluating the energy consumption in the creation of battery precursors from sea nodules is to determine the extent to which the manufacturing process contributes to emissions. As mentioned before, around 30–35% of the energy consumed in the manufacturing process is allocated to the production of cathode powder. The collective influence of battery manufacturing on emissions from electric vehicles (EVs) is complex, as evidenced by recent research that has

Table 5.6 Relative proportions of battery manufacturing emissions and other emissions

	Emissions in g/km (Nissan Leaf: 40 kWh)			
	Tailpipe Emissions	Fuel Cycle Emissions	Other Manufacturing	Batteries Manufacturing
Average euro (conventional)	165	47	46	
EU average		44	38	27–120.8
France		7	38	27–84.2
Germany		65	38	27–142
Netherlands		55	38	27–55
UK		29	38	27–106
USA		65	38	27–145

yielded a broad spectrum of findings, ranging from 56 to 494 kg CO_2 per kWh (Hall & Lutsey, 2018).

Across the globe, the primary source of emissions from both electric and conventional vehicles is the operation of the vehicles, which includes emissions from the tailpipe and the fuel cycle, rather than the manufacturing process. Countries such as Norway or France are notable outliers, as they primarily rely on low-carbon sources such as hydroelectric or nuclear power for their electrical generation. Regrettably, this report lacks precise information regarding the calculation of energy-related emissions specifically relevant to industry.

Prior research conducted by Ishihara et al. (2001) has already documented the associations between manufacturing energy and estimated emissions. Their investigation determined that there were little disparities in energy input and CO2 emissions between Ni/Co-system and Mn-system lithium-ion batteries for electric vehicle (EV) applications. The energy input and CO_2 emission per 1 kWh of battery energy capacity were documented to be around 360 Mcal/kWh and 75 kg-CO_2/kWh, respectively, which is similar to nickel-metal hydride batteries. It is important to highlight that the emission units in this report are different from those in the preceding Table 5.6.

In order to enhance the utility and comparability of variables, it is necessary to integrate them with functional components, such as the kilowatt-hour (kWh) measurement of a battery's capacity. The primary units used to measure both key variables are kWh (with MJ/kWh used to quantify the embodied energy in battery manufacture) and kg CO_2 equivalents/kWh (which represents the emissions converted into CO_2 equivalents required to produce the battery's capacity). Additional functional components could encompass the overall mass of the battery (expressed in kilograms of carbon dioxide equivalent per kilogram) or the emissions produced by the battery for each kilometer driven (measured in grams of carbon dioxide equivalent per kilometer) when utilized in a vehicle.

The energy demands of battery manufacturing operations, measured in MJ/kWh, are used as a foundation for further emission calculations. Various studies utilize either a top-down methodology, where energy consumption for a standard plant is assigned to different production stages, or a bottom-up technique, where energy usage for each process is calculated or estimated and then allocated per battery. The majority of life cycle evaluations focus on the "cradle-to-gate" perspective, which examines the environmental impact from the extraction of materials until the point at which the battery is prepared for sale. The omission of recycling in these assessments underscores the necessity for data that is specific to each circuit. The cumulative energy consumption for battery production exhibits significant variation, ranging from 500 to 2000 MJ/kWh, contingent upon the specific battery type (Peters et al., 2017).

After determining the energy needs for extracting and refining a battery metal precursor, it becomes feasible to calculate the emissions produced by this process. Direct emissions, derived from the mass-energy balance, encompass emissions occurring during the operation of the circuit, whereas indirect emissions arise from the production of reagents and utilities used in the circuit. Precise data pertaining to energy sources, such as electricity and heating fuel, are required for precise calculations. The emission values for different circuits used in processing nodules are determined using published references for both hydrometallurgical and pyrometallurgical methods.

The 1 kWh pack (Ishihara et al. 2001) has a total energy value of around 1504.8 MJ, which results in emissions of 75 kg CO_2/kWh when converting energy to emissions using natural gas. These correlations can be applied to source materials that undergo extraction and refining processes. Emissions from the battery pack exhibit significant variation, ranging from 20 to 120 kg/kWh for cobalt (Co) and nickel (Ni), depending on the specific terrestrial mineral sources containing Ni and Co. Mining and refining assumes nickel and cobalt are refined to nickel and cobalt sulphate for conversion to precursor with inputs like Nickel Sulphide, Ferronickel and Ni Pig Iron (Riggall, 2020).

When marine nodules are regarded as a source of nickel, cobalt, manganese, and copper, there is reduced variability in energy per metric ton of source material. Converting these energy values to MJ/kWh is a straightforward process, but estimating emission levels relies on the specific energy source utilized. The predicted energy consumption per kilowatt-hour (kWh) for the manufacturing of precursors is consistent with the numbers documented in the literature for an NMC 811 battery. Furthermore, this energy consumption includes the production of copper metal and a marketable product containing manganese. The text acknowledges that it does not provide a breakdown of total energy usage for various product components. However, it implies that if energy values were allocated using an economic allocation approach often employed in life cycle analysis (LCA), the resulting values would probably be lower than the anticipated 600–700 MJ/kWh for the cathode mix. The emission values per kilowatt-hour (kWh) can be determined by analyzing the characteristics of the energy sources used in the hypothetical flowsheet for extraction and refining.

5.1 Further Lowering of Energy Values for Metal Precursor Production from Sea Nodules

There are several options available for using renewable hydrogen as a reducing agent in the pyrometallurgical process, as seen in Indian practices. The worldwide accessibility of renewable energy capacity appears encouraging when combined with water electrolysis technology for the creation of hydrogen. However, additional examination is necessary to ascertain the favored origin of sustainable electricity. At present, the capacity of electrolyzers does not appear to be limited by the expense or accessibility of renewable energy. However, there is a lack of sufficient electrolyzer capacities to fully integrate with a sea nodules facility that has a capacity of 1.5 MTPA.

To achieve the calcination/reduction stage using renewable hydrogen, a hydrogen input of 0.5 Nm^3/kg of source material is needed. In order to operate at full capacity throughout the year, the total electrolyzer capacity required is anticipated to be 370 MW. This capability surpasses the existing scope of commercially tested PEM-based electrolyzers. According to Siemens literature, the Silyzer 300 and larger capacities (10–100 MW) are expected to be available from 2018 to 2023, with projected capacity expansion after 2023. Modifications to the capacity of the electrolyser may be required depending on differing assessments of the amount of hydrogen being inputted.

After the calcination/pre-reduction process in the pyrometallurgical flowsheet, the next steps involve melting/smelting the reduced mass in electric arc furnaces to create a byproduct of manganese. This is followed by the electro-winning of the other metals from the metallized residue after leaching. The primary energy inputs for this stage are derived from the consumption of electricity. The emission factor for electricity is a distinctive feature of the supply grid, where European countries generally exhibit lower grid emission factors in contrast to Asian countries such as India, primarily owing to a greater amount of renewable electricity in the grid. By relying solely on hydropower for the supply of electrical energy, emissions associated with smelting can be substantially diminished. This choice is viable for a sea nodule facility, particularly if it is located near water supplies.

5.2 Auxiliary Considerations for Use of Renewable Energy Resources

Renewable electricity plays a crucial role in the production of renewable hydrogen and other unit processes in a pyrometallurgical flowsheet. According to Martin et al. (2022), the factors that influence the selection of technology for generating renewable electricity are:

(a) Energy Return on Investment (EROI): This factor quantifies the effectiveness of energy generation by evaluating the ratio of useful energy produced to the energy consumed during the production process. The energy return on investment (EROI) can have a significant influence on both the financial expenses and the levels of emissions.
(b) Life Cycle Greenhouse Gas (GHG) Emissions: An exhaustive assessment of greenhouse gas (GHG) emissions throughout all stages of energy production is crucial for a thorough comprehension of emissions, especially when evaluating renewable technologies. The consideration of the life cycle perspective is essential due to the disparities in emissions among various renewable electricity systems. Hydropower and wind energy typically have lower emissions, whereas bioenergy technologies frequently have higher emissions.
(c) Measurement of Essential Resources: It is crucial to take into account the elements that are essential for building energy-related infrastructure, as previously mentioned in the document. Prior to commencing a renewable energy supply project, it is essential to carefully evaluate material supply factors.
(d) Sociopolitical Acceptance: Local opposition and political concerns can significantly impact the acceptance and viability of renewable technologies. The success of renewable energy projects is influenced by factors such as public support, environmental concerns, land use limits, and other quality of life difficulties (Bailey et al. 2011). Public perceptions of technologies such as marine energy are shaped by their perceived economic advantages and negligible negative consequences.

Recent discussions have arisen regarding the potential effects of sea mining on marine biodiversity, particularly in relation to marine-based technologies like deep-sea nodules processing for lithium-ion battery metals. Although the use of tidal waves for energy generation has not faced significant problems in this regard, it is worth mentioning that incorporating metals from sea nodules into lithium-ion batteries might greatly improve the capacity for renewable energy utilization while keeping emissions to a minimum.

(e) Rates of Acquisition for Emerging Technologies: The notion of learning rates, which are linked to the experience curves of emerging technologies, results in a gradual reduction in manufacturing costs, hence enhancing the economic viability of the technology. Promoting the adoption of novel technology, potentially through the provision of financial assistance, becomes imperative.

To summarize, the integration of renewable hydrogen and renewable electricity into pyrometallurgical processes involved in smelting and refining might greatly decrease the energy needed for the creation of battery precursors. Research laboratories should prioritize the development of renewable energy-powered material flow processes, in addition to the use of renewable hydrogen, in order to improve sustainability and decrease emissions in battery material production.

5.3 Role of EROI in the Use of Renewable Energy

As explained in a previous section with the definition of EROI, the following equations are useful to evaluate EROI for a proposed renewable resource used for generating electricity for smelting and refining operations.

$$\text{Net Energy} = \text{Gross Energy Output} - \sum \text{Energy Investments} \tag{5.9}$$

$$\text{Net to Gross Energy Ratio}(\text{NTG}) = \frac{\text{Net Energy}}{\text{Gross Energy Output}} = \left(1 - \frac{1}{\text{EROI}}\right) \tag{5.10}$$

Assessing an acceptable energy return on investment (EROI) number entails taking into account many errors in measurements. In general, energy sources with higher energy return on investment (EROI) are regarded as more economically and environmentally favorable, as they need less energy to obtain. Research data by Martin et al. (2022) demonstrates a correlation between higher energy return on investment (EROI) values and reduced greenhouse gas (GHG) emissions. Solar photovoltaic (PV) technologies demonstrate a reasonably high energy return on investment (EROI) and a relatively low susceptibility to sociopolitical obstacles. Another critical feature that influences the future prospects of solar cells is their capacity to be readily and cost-effectively manufactured, as evidenced by high learning rates. Nevertheless, the availability of raw materials presents a possible limitation for solar photovoltaic (PV) cells, and specific sub-technologies might exhibit comparatively elevated greenhouse gas (GHG) emissions.

International initiatives are currently in progress to tackle limitations in the availability of raw materials for battery metal precursors, with a focus on promoting the use of renewable energy and developing innovative technologies. The Sunrise Battery Materials Project, undertaken by Clean TeQ in Australia, seeks to generate an ample supply of top-notch nickel and cobalt to sustain the manufacturing of electric vehicles (EVs).

Pyrometallurgical reduction methods utilizing renewable hydrogen offer a promising alternative to decreasing process emissions without a significant decrease in process energy. This decrease in emissions can have a beneficial effect on the performance of battery emissions. The operation of a 1.5 MTPA sea nodule processing plant may necessitate the use of 370 MW of renewable energy for the purpose of hydrogen generation and its related activities. The provision of renewable energy for lower-temperature hydrometallurgical processes powered by steam can pose difficulties because of the substantial initial investment required for concentrated solar power to produce high-temperature steam.

It is expected that large-scale advancements in technology would enable the utilization of renewable energy in high-temperature operations while minimizing greenhouse gas emissions. Continual progress in this area is vital for improving the sustainability and environmental efficiency of producing battery precursors from sea nodules.

6 Conclusions

1. Polymetallic Nodules (PMN) as a Resource:

Polymetallic nodules are being considered as a valuable alternative resource for critical materials like Co, Ni, Mn, and Cu, vital for the clean energy transition, particularly in the production of batteries for electric vehicles. With considerations of terrestrial resources, reserves, future metal demands, and ongoing efforts to explore new resource avenues, metals from sea nodules could emerge as a viable option to mitigate supply risks for these crucial metals.

2. Significance of Sea Bed Resources:

The utilization of seabed resources gains significance due to the unique composition of polymetallic nodules. Unlike land-based resources, where metals are processed individually, PMNs comprise multiple critical metals necessary for clean energy processes. They can be recovered together as battery precursor products, potentially reducing energy costs, with manganese as a valuable by-product. This approach aligns with price-based energy allocation procedures, as seen in Life Cycle Assessment (LCA) studies.

3. Energy Demand Estimation:

The energy demand for processing sea nodule resources, expressed in MJ/kWh, can be estimated based on the composition of the resources and extrapolated amounts of precursor metals intended for use in batteries of specified capacity. The energy demand for cathode powder preparation compares favorably with the energy used for processing separate metal-bearing terrestrial resources. Alternatively, the energy demand for batteries expressed in MJ/km can be estimated based on supplied energy use (in MJ/km, using the battery's design attribute in Wh/km), adjusted for Energy Return on Investment (EROI) values for the supply source. Additional energy use for manufacturing a battery over its lifecycle needs to be considered, and the energy used can be converted to emissions/km.

4. Renewable Energy in Pyrometallurgical Processing:

The use of renewable energy in producing hydrogen for pyrometallurgical processing of sea nodules can significantly lower process emissions. A 370 MW capacity electrolyser for hydrogen generation in pyrometallurgical processing, along with renewable energy during smelting operations and downstream pregnant solution processing, contributes to environmental sustainability.

5. Future Work and Considerations:

While processing sea nodules presents a strong case to mitigate the supply risks of battery metals, additional work is needed. This includes flowsheet development incorporating renewable hydrogen, recovery of battery metal precursors with energy integration instead of metal electro-winning, and reworking net present value (NPV) for a business venture, considering a specific plant capacity for battery metal precursors as products, along with a manganese-bearing product for sale. Ongoing research and development are crucial for optimizing processes and ensuring the economic viability of such ventures.

Acknowledgments One of the authors (P. K. Sen) acknowledges the useful discussions he had with CSIR-IMMT scientists, Dr. Chinmaya Kumar Sarangi and Dr. Abdul Rauf Sheikh, during manuscript preparation on different aspects covering metal demands, flowsheet conceptualization and process concepts. Dr. K. Sanjay (coauthor) is grateful to his laboratory director for allowing him to participate in the manuscript preparation process.

References

Aishvarya, V., Mishra, G., Ghosh, M. K., & Subbaiah, T. (2013). Reductive acid leaching of polymetallic manganese nodule with polyol: Preliminary study. In *XIII international seminar on mineral processing technology* (Vol I, pp. 825–829).

Amarakoon, S., Smith, J., & Segal, B. (2013). *Application of LCA to nanoscale technology: Li-ion batteries for electric vehicles*. EPA 744-R-12-001, United States Environmental Protection Agency, pp. 1–120.

Ambrose, H., & Kendall, A. (2016). Effects of battery chemistry and performance on the life cycle greenhouse gas intensity of electric mobility. *Transportation Research Part D, 47*, 182–194.

Azevedo, M., Campagnol, N., Hagenbruch, T., Hoffman, K., Lala, A., & Ramsbottom, O. (2018). *Lithium and Cobalt-A tale of two commodities* (pp. 1–20). McKinsey and Company.

Azevedo, M., Baczynska, M., Bingoto, P., Callaway, G., Hoffman, K., & Ramsbottom, O. (2022). *The raw-materials challenge: How the metals and mining sector will be at the core of enabling the energy transition* (pp. 1–10). McKinsey and Company.

Bailey, I., West, J., & Whitehead, I. (2011). Out of sight but not out of mind? Public perceptions of wave energy. *Journal of Environmental Policy & Planning, 13*(2), 139–157.

Bellois, G., & Ramdoo, I. (2023), *Searching for critical minerals? How metals are produced and associated together*. IGF, Inter Governmental on mining, minerals metals and sustainable development, pp. 1–25.

Biswas, A., & Chakraborty, N., Sen, P. K. (2009) Use of process optimization and cost model for metal recovery from manganese nodules: The role of manganese recovery. In *Proceedings of the eighth ISOPE ocean mining symposium*, Chennai, India, September 20–24, 2009, ISBN 978-1-880653-75-3.

Blengini, G. A., Nuss, P., Dewulf, J., Nita, V., Talens Peirò, L., Vidal-Legaza, B., Latunussa, C., Mancinia, L., Blagoev, D., Pennington, D., Pellegrini, M., Van Maercke, A., Solar, S., Grohol, M., & Ciupage, C. (2017). EU methodology for critical raw materials assessment: Policy needs and proposed solutions for incremental improvements. *Resources Policy, 53*, 12–19.

Dai, Q., Kelly, J. C., Gaines, L., & Wang, M. (2019). Life cycle analysis of lithium-ion batteries for automotive applications. *Batteries, 5*(48), 1–15.

Das, P. K., Anand, S., & Das, R. P. (1998). Studies on reduction of manganese dioxide by (NH4)2SO3 in ammoniacal medium. *Hydrometallurgy, 50*, 39–49.

Dry, M., Vaughan, J., & Hawker, W. (2019). *Environmental evaluation of making nickel sulphate*. Presented at Alta 2019 NCC, Perth, Australia, May 2019.

Elliott, R., & Pickles, C. A. (2017). Thermodynamic analysis of the selective reduction of a nickeliferous limonitic laterite ore by hydrogen. *High Temperature Materials and Processes, 36*(8), 835–846.

Fulton, R. E. (2023). https://getjerry.com/car-repair/tesla-model-3-battery-size#tesla-model-3-battery-size-range-and-charging-specs. Date accessed 15th June 2023.

Hall, D., & Lutsey, N. (2018). *Effects of battery manufacturing on electric vehicle life-cycle greenhouse gas emissions*. https://theicct.org/publication/effects-of-battery-manufacturing-on-electric-vehicle-life-cycle-greenhouse-gas-emissions/

Heijlen, W., Franceschi, G., Duhayon, C., & Van Nijen, K. (2021). Assessing the adequacy of the global land-based mine development pipeline in the light of future high-demand scenarios: The case of the battery-metals nickel (Ni) and cobalt (Co). *Resources Policy, 73*, 1–14.

Hund, K., La Porta, D., Fabregas, T., & P., Laing, T., Drexhage, J. (2020). *Minerals for climate action: The mineral intensity of the clean energy transition.* World Bank.

IEA. (2021). *The role of critical minerals in clean energy transitions.* International Energy Agency (www.iea.org).

IEA. (2023). *Energy technology perspectives.*International Energy Agency (www.iea.org).

Ishihara, K., Kihira, N., Terada, N., & Iwahori, T. (2001). *Environmental burdens of large lithium-ion batteries developed in a Japanese National Project.* https://www.electrochem.org/dl/ma/202/pdfs/0068.pdf.

ITRE Committee. (2022). *Study report on strengthening the security of supply of products containing critical raw materials for the green transition and decarbonisation.* European Parliament.

Kelly, J. C., Wang, M., Dai, Q., & Winjobi, O. (2021). Energy, greenhouse gas, and water life cycle analysis of lithium carbonate and lithium hydroxide monohydrate from brine and ore resources and their use in lithium ion battery cathodes and lithium ion batteries. *Resources, Conservation & Recycling, 174*(105762), 1–12.

Khalafalla, S. E., Pahlman, J., & E. (1981). Selective extraction of metals from Pacific Sea nodules with dissolved sulfur dioxide. *Journal of Metals, 33*, 37–42.

Li, B., Ding, Z., Wei, Y., Zhou, S., & Wang, H. (2018). Reduction of nickel and iron from low-grade nickel laterite ore via a solid-state deoxidization method using methane. *Materials Transactions, the Japan Institute of Metals and Materials, 59*(7), 1180. to 1185.

Martin, N., Madrid-L'opez, C., Villalba-M'endez, G., & Talens-Peiró, L. (2022). Overlooked factors in predicting the transition to clean electricity. *Environmental Research: Infrastructure and Sustainability, 2 021005*, 1–14.

Mayze, R. (1999). *An engineering comparison of the three treatment flowsheets in WA nickel laterite projects.* Alta 1999 nickel/cobalt pressure leaching & hydrometallurgy forum. https://www.altamet.com.au, pp. 1–17.

Mittal, N., & Anand, S. (2019). *Techno-economic perspective on processing of polymetallic ocean nodules in environmental issues of deep-sea mining* (R. Sharma, Ed.).

Narula, K. et al. (2023). *Ensuring sustainable supply of critical minerals for a clean, justand inclusive energy transition.* T20 Policy Brief, May 2023, Observer Research Foundation, pp. 1–20. https://www.orfonline.org/research/ensuring-sustainable-supply-of-critical-minerals-for-a-clean/

Nuss, P., & Eckelman, M. J. (2014). Life cycle assessment of metals: A scientific synthesis. *PLoS One, 9*(7), 1–12.

Paulikas, D., Katona, S., Ilves, E., & Saleem, H. A. (2020). Life cycle climate change impacts of producing battery metals from land ores versus deep-sea polymetallic nodules. *Journal of Cleaner Production, 275*, 1–20.

Peters, J. F., Baumann, M., Zimmermann, B., Braun, J., & Weil, M. (2017). The environmental impact of Li-Ion batteries and the role of key parameters – A review. *Renewable and Sustainable Energy Reviews, 67*, 491–506.

Pickles, C. A., & Anthony, W. (2018a). A thermodynamic study of the reduction of a limonitic laterite ore by methane. *High Temperature Materials and Processes, 37*(9–10), 909–919.

Pickles, C. A., & Anthony, W. (2018b). Thermodynamic modelling of the reduction of a saprolitic laterite ore by methane. *Minerals Engineering, 120*, 47–59.

Rankin, W. J. (2014). *Minerals, , metals and sustainability, meeting future material needs* (pp. 1–422). CRC Press.

Riggall, S. (2020). *A focus on nickel in electric vehicle batteries-understanding cost and the carbon footprint.* Clean TeQ: BMO 29th Global Metals & Mining Conference, 23–26 Feb 2020, pp. 1–18.

Romare M., & Dahllöf, L. (2017). *The life cycle energy consumption and greenhouse gas emissions from lithium-ion batteries: A study with focus on current technology and batteries for light-duty vehicles.* IVL Swedish Environmental Research Institute Ltd, Report No. C243.

Roorda, H. J., & Hermans, J. M. A. (1981). Energy constraints in the extraction of nickel from oxide ores (II). *Erzmetall, 34*(4), 186–190.

Sen, P. K. (2022). Exploring the use of renewable resources for processing of deep-sea minerals. In R. Sharma (Ed.), *Perspectives on deep-sea mining* (pp. 219–261). Springer.

Shanon, O. (2023). Minerals are crucial for electric cars and wind turbines. Some worry whether we have enough. *The Washington Post*. https://www.washingtonpost.com/climate-environment/2023/02/02/critical-minerals-run-out-shortage/Washngton-post-NA-15. Date accessed 1st June 2023.

Sherritt International Corporation. (2021). *Does matte matter? Is nickel pig iron the answer to EV battery demand?* q4cdn.com https://s2q4cdn.com>files>doc downloads PDF Sunrise Battery Materials Complex –Sunrise (sunriseem.com)

Teske, S., Florin, N., Dominish, E., & Giurco, D. (2016). *Renewable energy and Deep Sea mining: Supply, demand and scenarios*. Report prepared by ISF for J.M.Kaplan Fund, Oceans 5 and Synchronicity Earth, July 2016.

Wang, S., Hausfather, Z., Davis, S., Lloyd, J., Olson, E. B., Liebermann, L., Nunez-Mujica, G. D., & McBride, J. (2023). Future demand for electricity generation materials under different climate mitigation scenarios. *Joule, 7*, 309–332.

Wei, W., Samuelsson, P. B., Tilliander, A., Gyllenram, R., & Jönsson, P. G. (2020). Energy consumption and greenhouse gas emissions of nickel products. *Energies, 13*(5664), 1–22.

Dr. P. K. Sen is a graduate of the Indian Institute of Technology, Kharagpur. After a stint of teaching at a University in Kolkata, India, he was inducted as a CNRS fellow in hydrometallurgy in France, working on diverse hydrometallurgical projects. He subsequently joined Engineers India Limited, New Delhi, where he had worked for more than 25 years, heading Research & Development, Environmental Engineering, and Information Technology as Group Activity Director. He has spearheaded the development of several technologies that were scaled up as well as commercialized. He subsequently joined the Indian Institute of Technology as a professor, heading research activities in process metallurgy in diverse areas. He has authored more than 80 publications, served as UNIDO expert, panel judge for the Prime Minister's Trophy related to Steel Plant performance evaluation, member of national expert committees, and Chairman of the Steel Technology Center at IIT Kharagpur. Nominated earlier as Ministry of Steel Chair Professor, he continues to be associated with developmental work in ferrous and non-ferrous metallurgy at IIT as a former professor.

Dr. Sanjay Kali, Chief Scientist, is heading the Department of Hydro and Electrometallurgy at the CSIR Institute of Minerals and Materials Technology (CSIR-IMMT), Bhubaneswar, India. He has 30 years of experience in the field of extractive metallurgy, particularly in the development of process flow sheets to recover non-ferrous metals from low-grade ores and secondaries, utilization of industrial effluents, computational fluid dynamics, and design & scale-up of hydrometallurgical plants. Dr. Sanjay also serves as a professor for the Academy of Scientific and Innovative Research (AcSIR) at the Faculty of Engineering Sciences. He is leading the CSIR-IMMT's Polymetallic Nodules Metallurgy program.

Part III
Approaches to Sea Surface and Water Column Monitoring

Chapter 6
Understanding Deep-Sea Turbulence for Environmental Impact Assessments

Yasuo Furushima, Mamoru Tanaka, Mehul Naresh Sangekar, Dhugal John Lindsay, Tatsuo Fukuhara, and Masayuki Nagao

Abstract Both the diffusion of turbid water (plume) and the redeposition of suspended particles associated with deep sea mining are expected to have a negative impact on the surrounding environment. In order to understand the actual conditions of the mixed environment near the seafloor, it is necessary to evaluate these environmental impacts.

In a review of guidelines for environmental impact assessment related to deep-sea mining published by the International Seabed Authority (ISA), the necessity of quantifying turbulence throughout the water column and vertical eddy diffusivity was suggested. Therefore, it is important to elucidate the dynamics of turbulence in the water column and directly above the seabed in the vicinity of deep-sea mining operations. However, the reality of deep-sea turbulence remains largely unexplored, and measurement techniques for deep-sea turbulence immediately above the seafloor have only recently been established. Consequently, while measurements of turbulence from the ocean surface to just above the seafloor have become possible, there are still challenges in terms of operation and cost.

In this chapter, we present the results of our field survey of deep-sea turbulence intensity in different geographical environments and describe a method for estimating deep-sea turbulence (patent pending in Japan in 2021, application number: 2021–140,907) that we developed to solve the problems of deep-sea turbulence measurements. In addition, we have described a turbulence visualization technique

---

Y. Furushima (✉) · M. N. Sangekar · D. J. Lindsay
Japan Agency for Marine-Earth Science and Technology (JAMSTEC), Yokosuka, Kanagawa, Japan
e-mail: furus@jamstec.go.jp

M. Tanaka
Tokyo University of Science, Tokyo, Japan

T. Fukuhara
KANSO TECHNOS Co., Ltd., Tokyo Branch, Tokyo, Japan

M. Nagao
National Institute of Advanced Industrial Science and Technology (AIST), Tsukuba, Ibaraki, Japan

R. Sharma (ed.), *Deep-Sea Mining and the Water Column*,
https://doi.org/10.1007/978-3-031-59060-3_6

using a shadowgraph-based imaging system with a high-resolution camera in order to understand the dynamics of suspended particles in situ, such as just above the deep-sea floor.

Keywords Turbulence proxy · Modelling · In situ observations · Imaging · EIA methodologies

1 Introduction

1.1 Ocean Turbulence

Turbulence is a phenomenon of fluid flow, therefore occurring both in seawater and air. It fluctuates irregularly in time and space and exists everywhere in the natural world around us. Turbulence has elements of diffusive, dissipative, and eddy motion. Smoke rising from chimneys and wind passing through gaps between buildings are examples of turbulent flows (Tennekes & Lumley, 1972).

Ocean turbulence also has characteristics in common with other naturally occurring fluids and flows handled in civil engineering, hydraulics, etc. (Thorpe, 2007). Ocean turbulence occurs due to irregular fluctuations in currents, temperature, and salinity over various spatial and temporal scales from the ocean surface to the seafloor (Osborn, 1980). The scale varies from small eddies of less than a few centimeters to large oceanic eddies of several hundred kilometers in extent. Sub-meter scale turbulent motion in the ocean stretches and rolls up packets of seawater, promoting the diffusion of water temperature, salinity, and any dissolved materials (Smyth & Moum, 2019), as well as homogenizing suspended particles of biological and non-biological origin suspended in the seawater. Diffusion by turbulent eddies also influences the formation of large-scale oceanic structures (Ferrari & Wunsch, 2009; Munk, 1966).

In addition, deep-sea turbulence near the seafloor is induced by the complex bottom topography unique to the bottom and the influence of hydrothermal plumes (Thorpe, 2007), which provides a driving force for vertical mixing dynamics such as the diffusion of turbid waters (plumes) and transport and re-deposition of suspended particles. It also contributes to the development of bottom mixed layers.

1.2 Importance of Deep-Sea Turbulence

In recent years, deep-sea mining surveys for resource development have been actively conducted around the world. Mining using excavators causes disturbance of the seafloor, diffusion of turbid water, and redeposition of suspended particles, which may have negative impacts on the surrounding ecosystem and environment

(Furuichi et al., 2016). Moreover, deep-sea mining will generate sediment plumes near the seabed as well as in the midwater during the course of operations (Drazen et al., 2020). Depth layers expected to be affected are expected to range from the surface to 5000 m with the midwater sediment plume expected to be ejected at 100–1200 m depth (Drazen et al., 2020). While such plumes will be advected horizontally due to currents, their vertical transport will be affected by turbulent mixing (Thorpe, 2005). Therefore, in order to evaluate the impact of deep-sea mining on the environment, it is necessary to understand the fundamentals of fluid structure, such as the intensity of vertical mixing, or "turbulence intensity," from just above the bottom to near the seafloor, and the development of the bottom mixed layer (Fig. 6.1).

Physical environment measurements in the vicinity of the deep-sea floor have been mainly based on average flow measurements using a current meter such as an Acoustic Doppler Current Profiler (ADCP) (Furushima & Yamamoto, 2015). However, it has been difficult to grasp the actual conditions of the mixing environment (i.e., turbulent environment) from such flow measurement results.

In general, the turbulent kinetic energy dissipation rate (ε) is used as an index of turbulence intensity. So far, studies directly observing the turbulent kinetic energy dissipation rate (ε) have been conducted in the coastal areas of Japan, including the Ariake Sea (Saita et al., 2008), Hiroshima Bay (Hashimoto & Takasugi, 1998), Seto Inland Sea (Nagao et al., 2004), Higo Bay (Nagao et al., 2005), and Tokyo Bay (Higashi & Maki, 2012). However, studies that have observed ε near the deep-sea floor have been quite rare due to the unavailability of instruments capable of measuring turbulence from the ocean surface to just above the deep-sea floor.

Recently, the Expendable Vertical Microstructure Profiler (VMP-X) has been developed (Shang et al., 2017), capable of measuring turbulence from the ocean surface down to just above the deep-sea floor (Furushima et al., 2019a, b, c). We have been introducing this equipment and continuously measuring deep-sea

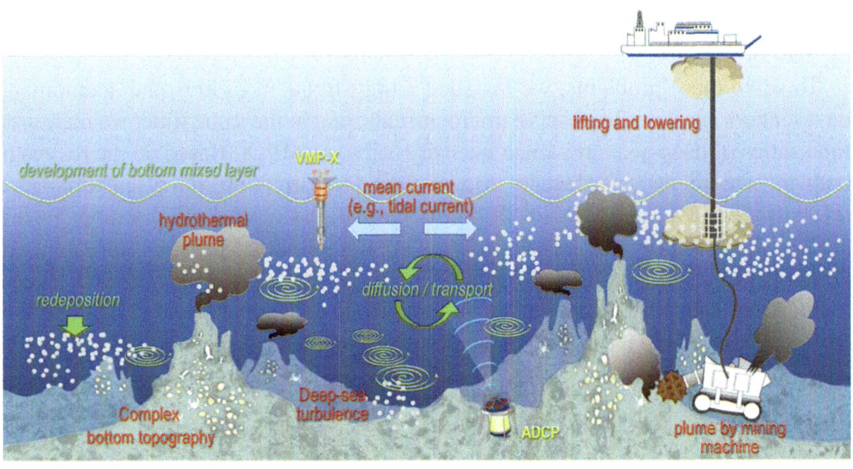

Fig. 6.1 Schematic view of mixing environment near the seafloor affected by deep-sea mining

turbulence at the study sites where deep-sea mining will be conducted to accumulate baseline data necessary for environmental impact assessments. Furthermore, we have demonstrated the importance of deep-sea turbulence by publishing the results of deep-sea turbulence measurements through side events held at ISA general meetings and through academic conferences and papers.

1.3 Necessity of Deep-Sea Turbulence for EIA

The International Seabed Authority (ISA) guidelines emphasize the importance of conducting comprehensive environmental impact assessments for seabed resource development projects. In 2013, ISA published guidelines for environmental impact assessments related to deep-sea mining with recommendations for baseline studies (ISA, 2013; ISBA/19/LTC/8). In addition, a revision published in 2020 suggested the need to quantify turbulence intensity and vertical eddy diffusion rates of the water column, which is a key parameter influencing the dispersion of particulate plumes (ISA, 2020; ISBA/25/LTC/6/Rev.1). The guideline recommended building a numerical model in order to predict sediment and plume dispersal, and this model would need to be validated by comparison with actual observational data. Therefore, it is important to understand the dynamics of turbulence in the water column and deep-sea turbulence just above the seafloor, where the mining development is expected to occur.

We have been actively conducting turbulence measurements near the deep-sea floor and obtaining highly interesting results. However, because deep-sea turbulence fluctuates largely due to currents near the bottom and complex bottom topography, the single-shot turbulence observation data we usually obtain may not be representative for the entire time period. Therefore, to ensure the availability of deep-sea turbulence data, it is desirable to measure turbulence several times over tidal cycles. However, during a side event held at the ISA General Assembly, it was pointed out that there are issues due to the operational aspects of this equipment and the cost of the surveys.

To solve these problems, we installed a high-frequency ultrasonic anemometer on a seafloor platform to observe microcurrents, and at the same time, we measured time series of deep-sea turbulence intensity using a VMP-X. Based on the results of both measurements, we developed a method for estimating deep-sea turbulence intensity (Tanaka et al., 2021a, b) and obtained a patent for this method. In this article, we present the results and characteristics of actual field measurements of deep-sea turbulence intensity, for which data have been scarce. We also introduce a method for efficiently estimating deep-sea turbulence intensity in the vicinity of the deep-sea bottom and the development of technology for visualizing the dynamics of deep-sea turbulence in the field.

2 Actual Conditions of Deep-Sea Turbulence Based on In-Situ Observations

The direct measurement of deep-sea turbulence in the vicinity of the seafloor is introduced as a new measurement technique for environmental impact assessments in deep sea mining (Furushima et al., 2019a, b, c). In this section, we will present the results of in-situ measurements of ocean turbulence conducted using a VMP-X, with a particular focus on the actual conditions of deep-sea turbulence intensity in the vicinity of the deep-sea floor and its relationship with the surrounding environment.

The turbulent motion of seawater is characterized by chaotic flow, making it difficult to fully understand. However, because it dominates the mixing of suspended particles, direct observation and understanding of deep-sea turbulence are important elements in environmental impact assessments for seabed mineral resource development. This is because the turbid water (plume) generated during seabed mining and lifting processes can result in the diffusion (entrainment) and re-deposition of suspended particles, potentially causing negative impacts on the surrounding environment and ecosystems. The International Seabed Authority (ISA) published guidelines on environmental impact assessment related to deep-sea mining in 2013, which included recommendations for baseline surveys on physical, chemical, biological, and geological observations. Furthermore, a revision published in 2019 suggested the need to quantify the turbulence and vertical eddy diffusivity of the entire water column (ISA, 2020). Therefore, technological development to assess the impacts of phenomena such as the diffusion (entrainment) and re-deposition of suspended particles in the vicinity of the seabed is important.

In order to understand the nature of deep-sea turbulence near the seafloor, we conducted turbulence measurements using the Expendable Vertical Microstructure Profiler (VMP-X) from the ocean surface to just above the seabed.

2.1 *In-Situ Observation of Turbulence*

Turbulence measurements using a VMP-X were conducted at 11 sites in Sagami Bay, the Izu-Ogasawara Trench, and the Okinawa Trough from 2016 to 2020 (Fig. 6.2). The 3-letter labels in Fig. 6.2 represent the observation points. Information on the names, water depths, observation methods using VMP-X, and the number of measurements for each observation point is summarized in Table 6.1.

Sagami Bay, where single turbulence observations were conducted at four locations using the VMP-X, is one of Japan's prominent open bays adjacent to the western side of Tokyo Bay, with depths exceeding 1000 m in some areas (Fig. 6.2a). The Izu-Ogasawara island arc, approximately 850 km in length, runs mostly north-south from the Boso Peninsula through the Ogasawara Islands, and turbulence measurements were conducted at the hydrothermal vent sites of the Bayonnaise knoll and

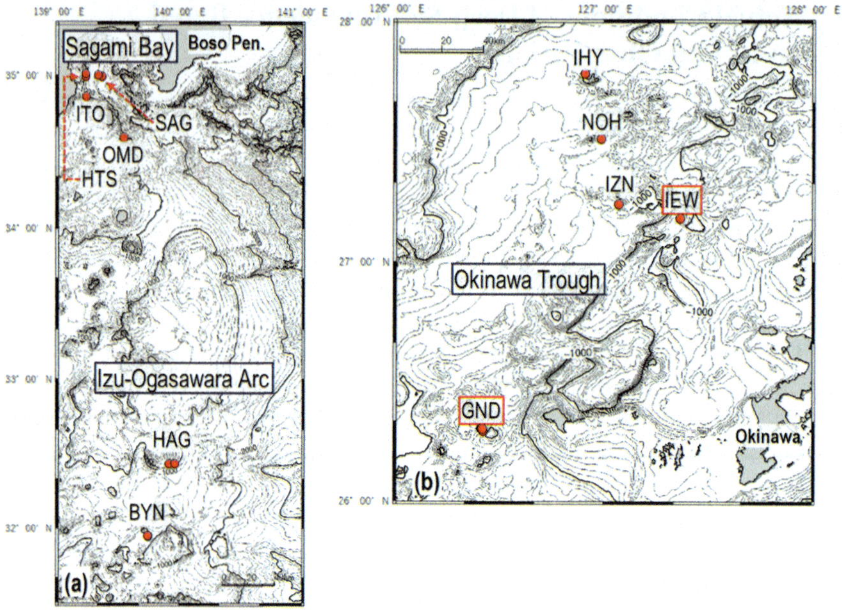

Fig. 6.2 Observation points for deep-sea turbulence measurements using the VMP-X (Expendable vertical microstructure profiler) from 2016 to 2020. (**a**) Observation sites in Sagami Bay and the Izu-Ogasawara Arc. (**b**) Observation sites in Okinawa Trough. The two red boxes in the map indicate points of continuous observation by VMP-X

the Higashi-Aogashima caldera along this arc (Fig. 6.2a). The Okinawa Trough, a depression extending about 1000 kilometers in length off the northwest coast of the Ryukyu Islands, had single turbulence observations carried out at five hydrothermal vent sites (Fig. 6.2b). Additionally, at the GONDO Site (GND) and Iejima west site (IEW) within the Okinawa Trough, time-series observations were conducted to ensure the usefulness of turbulence measurements obtained by VMP-X.

In this study, we focused on the turbulence intensity (turbulence kinetic energy dissipation rate: ε) in the upper 100 m above the seafloor at each observation point and discussed the characteristics of turbulence intensity and its influencing factors.

2.2 *Expendable Vertical Microstructure Profiler (VMP-X)*

The VMP-X (Expendable Vertical Microstructure Profiler) is capable of measuring ocean turbulence from the surface to just above the seafloor. It consists of the VMP-X main body, which serves as the data logger, and the expendable sensor MicroXM integrated with a weight (Fig. 6.3). The MicroXM has two shear sensors and a water temperature sensor. It is connected to the data logger (VMP-X) by an optical fiber.

Table 6.1 Observation points for deep-sea turbulence measurements using the VMP-X

Site	Station name	Area name	Observation date	Depth (m)	VMP-X observation (number)[a]
Sagami Bay	ITO	off Ito	2016/6/30	1000	Individual (1)
	SAG	center of the Sagami bay	2015/12/18	1500	Individual (2)
	HTS	off Hatsushima	2015/12/18 2021/10/24	1000	Individual (2)
	OMD	Ohmuro-dashi	2016/6/29 2016/10/1 2017/3/13	850	Individual (3)
Izu-Ogasawara Arc	HAG	Higashi Aogashima	2020/12/23	800	Individual (2)
	BYN	Bayonnaise	2015/12/16 2017/3/10	800	Individual (3)
Okinawa Trough	IHY	Iheya-north	2016/11/10	1000	Individual (1)
	NHO	Noho	2016/11/13	1600	Individual (1)
	IZN	Izena	2018/1/16	1600	Individual (2)
	IEW	Iejima-west	2018/8/30–8/31 2018/9/5–9/6	980	Continuity (14)
	GND	Gondoh	2017/6/24 2017/6/29	1400	Continuity (9)

[a]Number of observations and observation sites by 2020

Observations prior to 2018 were conducted under the "Next-Generation Technologies for Ocean Resource Exploration" under the Strategic Innovation Promotion Program of the Council for Science, Technology, and Innovation

The VMP-X is dropped off the ship and measures the current shear, water temperature, and pressure from the ocean surface layer to the bottom. When VMP-X reaches the seafloor, the burn-wire that secured MicroXM is automatically disconnected. This releases the sensor unit and the main body, and the VMP-X automatically floats to the sea surface. The descent velocity of the VMP-X is approximately 0.6–0.8 ms^{-1} and the ascent velocity after detaching the MicroXM is about 1.0–1.2 ms^{-1}. Therefore, after deploying the VMP-X, the approximate ascent time can be estimated based on the average round-trip velocity of the instrument (0.7 ms^{-1}) and the water depth at the survey site. It is worth noting that the VMP-X is equipped with three different mechanisms for releasing the MicroXM (weight), ensuring a reliable recovery (Shang et al., 2017).

After the VMP-X surfaces, a radio beacon is activated and indicates the surfacing direction. An iridium beacon also transmits the GPS coordinates to a preassigned e-mail address. In addition, a flashing mini-LED light is attached to the top of the VMP-X to aid recovery operations at night. The data acquired during turbulent flow measurements were processed using the ODASMatlab Library version 4.0.

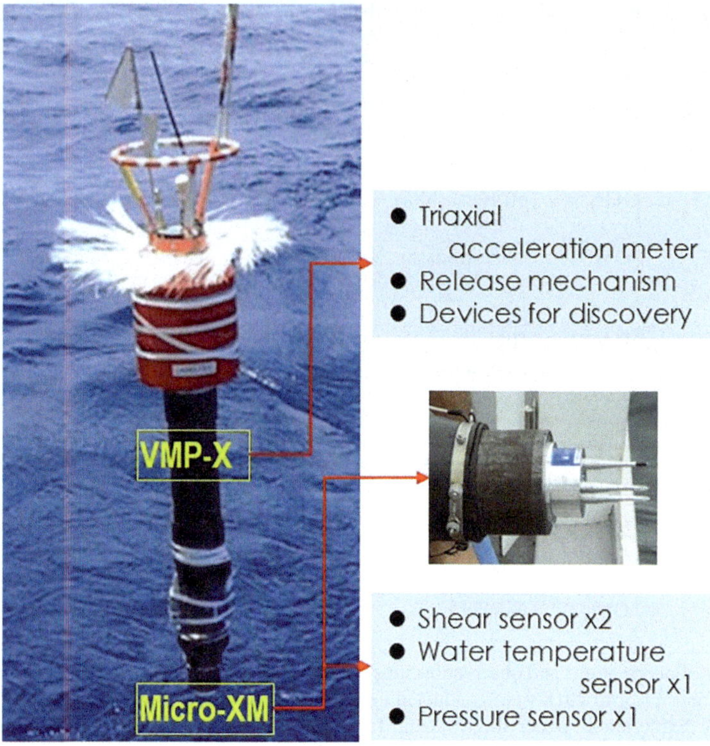

Fig. 6.3 Schematic view of the VMP-X expendable vertical microstructure profiler system

2.3 Direct Measurements of Near-Bottom Turbulence in Deep Ocean

2.3.1 In-Situ Observation at NHO (Noho Site) in Okinawa Trough

The vertical distribution of turbulence intensity (ε) and water temperature obtained at the Noho site in the Okinawa Trough is shown in Fig. 6.4a presents the vertical distribution of ε and water temperature from the ocean surface to just above the seafloor. The red and blue lines represent ε values obtained from the two shear sensors equipped on the VMP-X, while the green line indicates the distribution of water temperature.

Focusing on the vertical distribution of deep-sea turbulence intensity and water temperature from just above the seafloor to the upper 100 m layer (indicated by the red hatched region in Fig. 6.4a, it is evident that near the seafloor, ε values reach high values of around 10^{-7}(W kg^{-1}), comparable to the turbulent environment observed at the ocean surface. Additionally, an increase in ε is associated with a slight rise in water temperature near the seafloor (Fig. 6.4b).

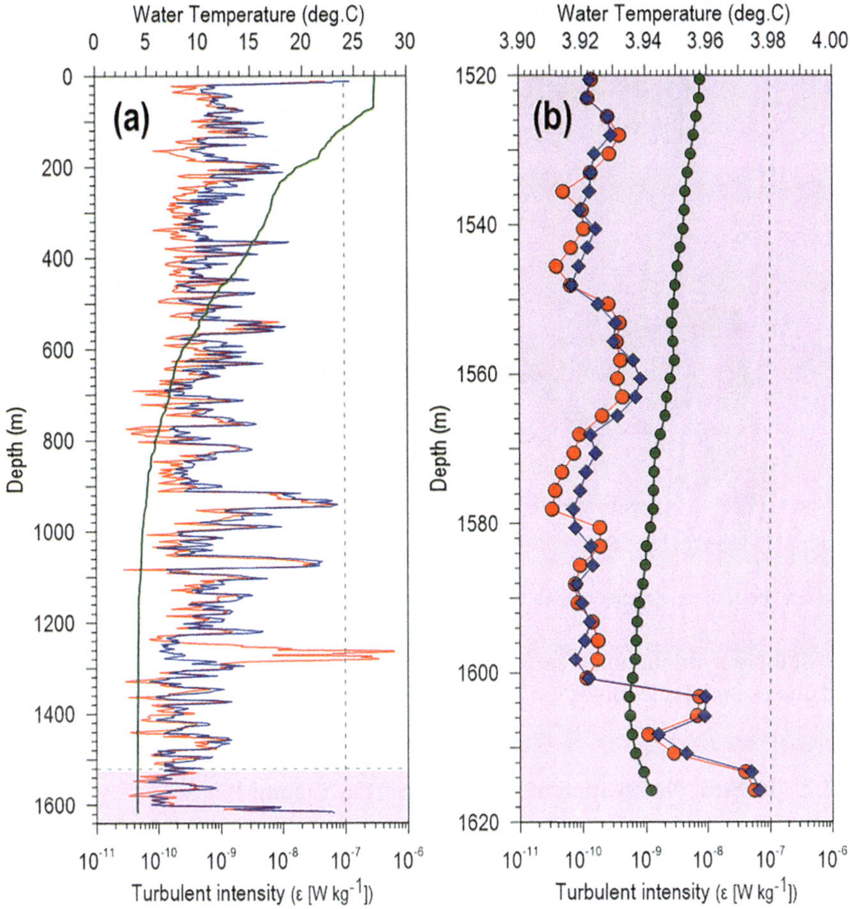

Fig. 6.4 Vertical profiles of turbulence intensity (ε) and water temperature at NHO. The blue and red lines show ε obtained from the two shear sensors. The green line in the figures shows the vertical fluctuation in the water temperature. (**a**) Vertical fluctuations of ε and water temperature from the surface to just above the seafloor. (**b**) Vertical fluctuations of ε and water temperature from just above the seafloor to the upper 100 m layer (enlargement of the pink hatched region in (**a**))

These measurement results suggest that due to the presence of numerous hydrothermal vent areas at the Noho site, it is conceivable that mixing near the seafloor is intensified by the influence of temperature and density variations associated with hydrothermal venting, leading to a possible increase in water temperature. Furthermore, the existence of hydrothermal vent areas is often associated with a highly complex seafloor topography (Fig. 6.5), indicating that besides hydrothermal activity, the influence of seafloor topography may also play a significant role in shaping the formation of a strong turbulent environment.

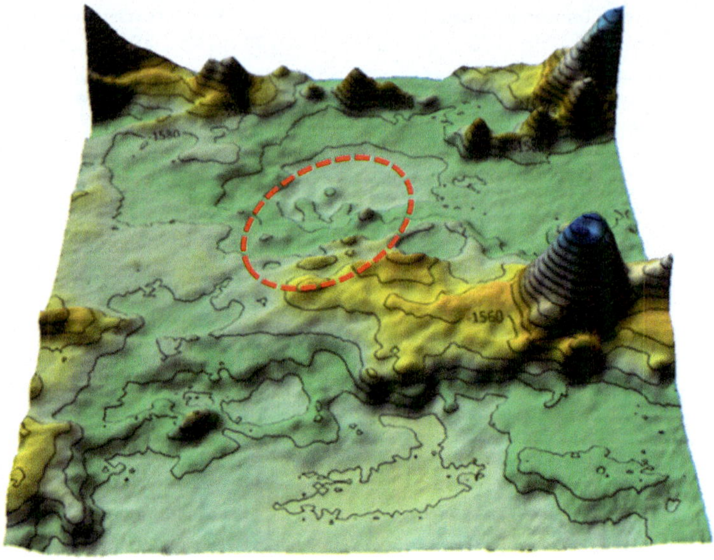

Fig. 6.5 Bottom topography map of the Noho site

Multiple hydrothermal vent areas are confirmed near the locations indicated by red circles and dashed lines.

2.3.2 In-Situ Observation at ITO (off Itoh) in Sagami Bay

Figure 6.6 shows the vertical distribution of turbulence intensity (ε) and water temperature obtained near the shelf slope off Ito in Sagami Bay. In the vicinity just above the seafloor, ε values of approximately 10^{-6} (W kg^{-1}) were obtained (Fig. 6.6b). This value is comparable to the turbulence intensity observed in the mixed layer of the ocean surface (Itoh, 2012; Saita et al., 2008). Sagami Bay is known as one of Japan's prominent open-type bays and is influenced by the inflow of Kuroshio-derived offshore water, which is affected by the variations in the Kuroshio current pathway (Iwata & Matsuyama, 1989, etc.). Therefore, the flow conditions in Sagami Bay were investigated when strong turbulence intensity was observed at Ito.

Figure 6.7 presents the time series of flow velocity, flow direction, and backscatter intensity obtained from the ADCP (Acoustic Doppler Current Profiler) deployed at the "Real-time Deep Seafloor Observatory off Matsushima Islan (Fig. 6.7d)" located approximately 10 km north of Ito at a depth of 1176 m since 1993 (Iwase et al., 2001; Momma et al., 1998).

Approximately 6 h before conducting turbulence measurements using VMP-X, a strong northeastward flow with a half-day period was recorded (Fig. 6.7a, b).

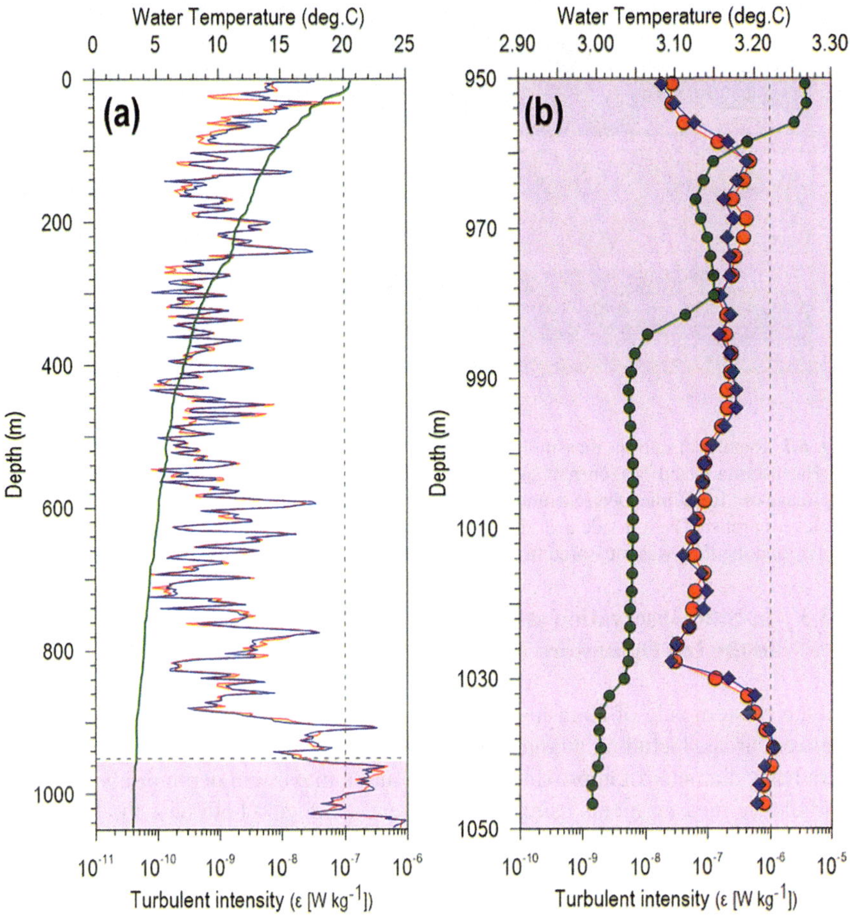

Fig. 6.6 Vertical profiles of turbulence intensity (ε) and water temperature at ITO. The blue and red lines show ε obtained from the two shear sensors. The green line in the figures shows the vertical fluctuation in the water temperature. (**a**) Vertical fluctuations of ε and water temperature from the surface to just above the seafloor. (**b**) Vertical fluctuations of ε and water temperature from just above the seafloor to the upper 100 m layer (enlargement of the pink hatched region in (**a**))

Additionally, variations in echo intensity, an indicator of turbidity caused by the flow, were shown, but no specific changes were observed (Fig. 6.7c). This northeastward flow follows the continental shelf slope in the western area of Sagami Bay. When passing over the continental shelf slope, velocity gradients may occur, suggesting the possibility of increased turbulence intensity near the seafloor. However, the data presented here are from a one-time observation, and further continuous in-situ observations will be necessary to better understand the dynamics of deep-sea turbulence in this region.

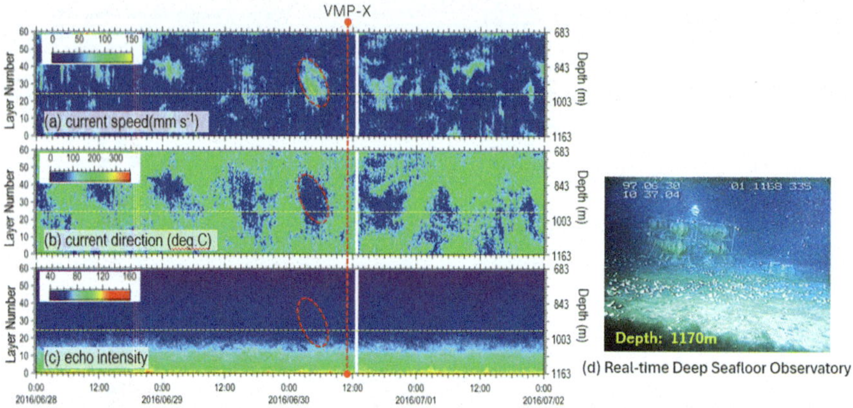

Fig. 6.7 Results of current measurement using ADCP at "Real-time Deep Seafloor Observatory off Hatsushima Island." (**a**) current speed (mm s^{-1}), (**b**) current direction (deg.), (**c**) echo intensity, (**d**) image of "Real-time Deep Seafloor Observatory off Hatsushima Island." The right axis shows the layer number. The left-side axis shows the depth from the bottom. The layer thickness was set at 8 m, and the flow was measured from the bottom in 60 layers

2.3.3 In-Situ Observation at HAG (Higashi Aogashima) in the Izu-Ogasawara Arc

On the western side of Higashi-Aogashima caldera (Fig. 6.8), there are hydrothermal vent areas, including a large chimney several tens of meters in height. The caldera has a diameter of approximately 8 km and a maximum depth of about 850 m. Turbulence measurements using VMP-X were conducted both near the hydrothermal areas within the caldera (at a depth of approximately 800 m) and on the outside of the caldera. The depth of the caldera's rim, which is located between the inner and outer measurement points, is approximately 500 m (Fig. 6.8).

The vertical distributions of ε and water temperature obtained at the inner (HAG-in) and outer (HAG-out) observation points of the Higashi-Aogashima caldera (HAG) are shown in Fig. 6.9. At depths around 350 m and 400 m in HAG-in, ε showed strong values of about 10^{-7}(W kg^{-1}) (Fig. 6.9a). Similarly, at depths around 280 m, 350 m, and 450 m in HAG-out, ε of about 10^{-6}(W kg^{-1}) was observed (Fig. 6.9c). The vertical distribution of water temperature at both locations exhibited similar variations up to a depth of approximately 450 m. However, at a depth of 500 m, there was an observed temperature difference of about 1 °C, with HAG-out having a temperature of around 12.5 °C and HAG-in having a temperature of 11.5 °C (Fig. 6.9b, d). Furthermore, at a depth of approximately 630 m, the water temperatures at both HAG-in and HAG-out reached about 8 °C, with the temperature near the seabed at HAG-in decreasing to around 7 °C (Fig. 6.9b). On the other hand, HAG-out showed a decrease in water temperature to about 5 °C (Fig. 6.9d). These results suggest that the water temperature within the caldera is higher than outside due to the influence of hydrothermal venting. Additionally, the ε values near

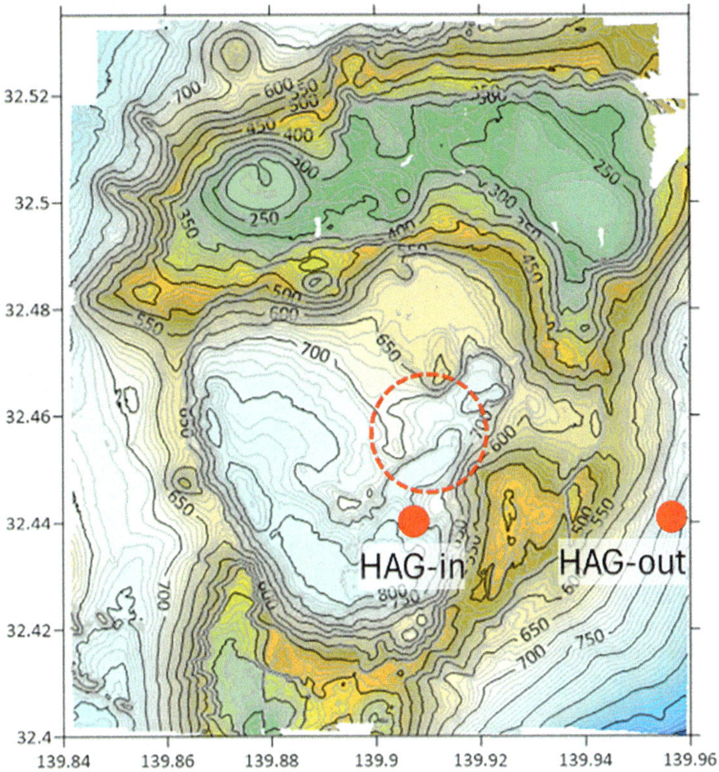

Fig. 6.8 Bottom topography map of the Higashi-Aogashima caldera. Hydrothermal vents were confirmed in the area marked by the red dotted line

the seafloor inside the caldera were greater than 10^{-7}(W kg^{-1}), while outside the caldera, they were about 10^{-9}(W kg^{-1}) (around 10^{-8}(W kg^{-1}) just above the seafloor) (Fig. 6.9b, d).

Based on the above, the strong mixing (turbulence) environment within the Higashi-Aogashima caldera is likely caused by a combination of turbulence generation due to the closed and complex seafloor topography and the mixing resulting from the temperature difference (density difference) between the surrounding seawater and the hydrothermal venting. On the other hand, the mixing environment near the seafloor outside the caldera is believed to be generated by the influence of offshore currents and topography.

2.3.4 Time Series Measurements of Turbulent Flows

In Sect. 2.3, we have presented the results of deep-sea turbulence measurements conducted on a single-shot in Sagami Bay, the Izu-Ogasawara Arc, and the Okinawa Trough. The deep-sea turbulence intensity values obtained at each site were very

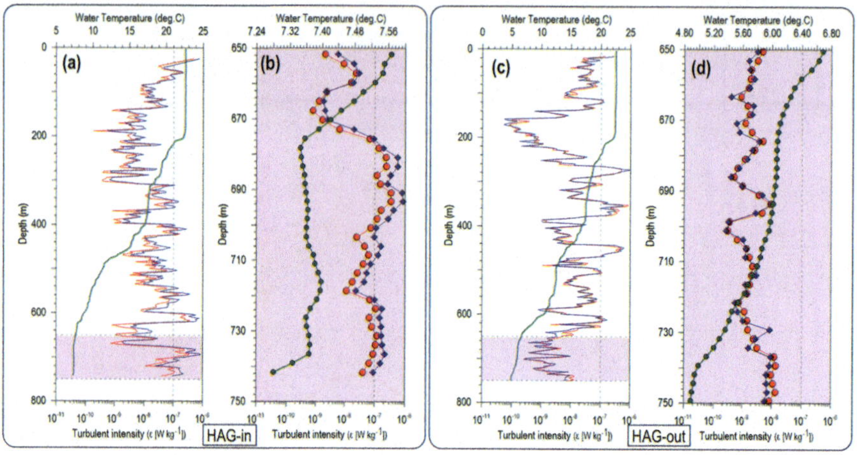

Fig. 6.9 Vertical profiles of turbulence intensity (ε) and water temperature at HAG. The blue and red lines show ε obtained from the two shear sensors. The green line in the figures shows the vertical fluctuation in the water temperature. (**a**) ε and water temperature from the surface to just above the seafloor at inside of HAG (**b**) ε and water temperature from just above the seafloor to the upper 100 m layer at inside of HAG (enlargement of the pink hatched region in (**a**)). (**c**) ε and water temperature from the surface to just above the seafloor at outside of HAG. (**d**) ε and water temperature from just above the seafloor to the upper 100 m layer at outside of HAG (enlargement of the pink hatched region in (**c**))

interesting. However, since the vertical (turbulent) mixing at the near bottom varies significantly with tides, the single-shot turbulence observations were considered to have limited temporal representativeness. Therefore, to ensure the usefulness of the observational data, continuous turbulence measurements at the same location were deemed necessary (Furushima et al., 2017; Higashi et al., 2017). Consequently, in this section, we will introduce the results obtained from the continuous observations conducted at the Gondo site (GND) in the Okinawa Trough using the VMP-X.

Continuous turbulence observations using VMP-X at GND were conducted on June 24th, 2017, and June 29th, 2017. Here, we present the results from June 24th. During the observation period, a "low head platform" owned by KANSO TECHNOS CO., LTD. was deployed on the bottom at GND, allowing for simultaneous environmental monitoring in the vicinity of the seafloor.

This platform was equipped with various instruments, including CTD (RINKO-Profiler, JFE Advantech Co., Ltd.), ADCP (Teledyne RD Instruments), LISST (Deep Ocean Laser Diffraction Particle Size Analyzer, Sequoia Scientific, Inc.), and Vector (acoustic 3D current meter, Nortek AS) (Furushima et al., 2017, 2019a, b, c).

The vertical distributions of ε and water temperature obtained from the time-series observations of ocean turbulence conducted using VMP-X at GND on June 24th, 2017, are presented in Fig. 6.10. Within the figure, the red, blue, orange, and green lines respectively depict the vertical profiles of ε and water temperature at four specific times: 08:03, 10:31, 12:56, and 15:34 (Fig. 6.10a–c) provide the vertical distributions of ε and water temperature at each time interval from the bottom to

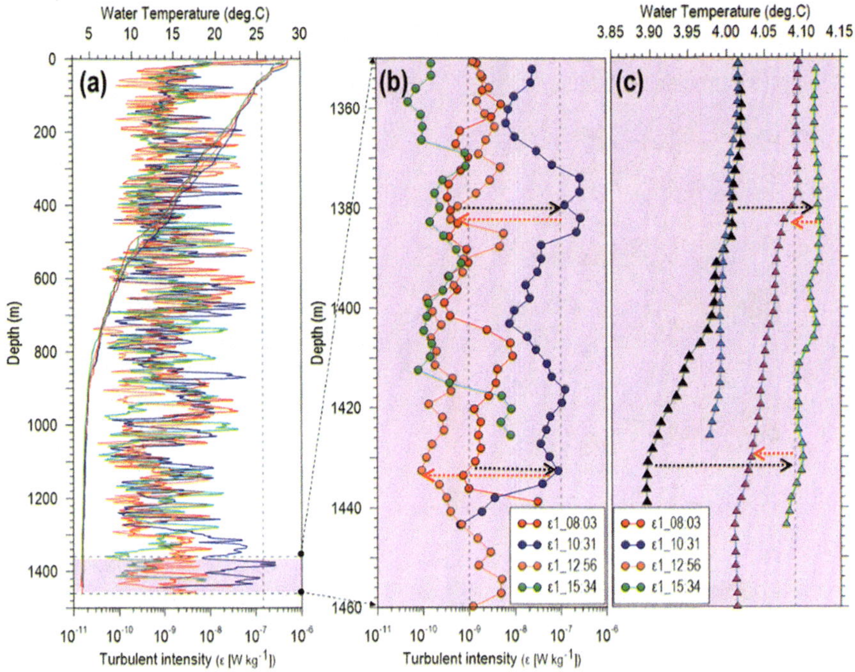

Fig. 6.10 Vertical profiles of ε and water temperature on June 24, 2017 at GND in the Okinawa Trough. The red, blue, orange, and green lines show the ε value obtained from the time series observations at 08:03, 10:31, 12:56, and 15:34, respectively, using the VMP-X. (**a**) Vertical profiles of ε and water temperature from the sea surface to the seafloor. (**b**) Vertical profiles of ε 100 m above the seafloor (enlargement of the pink hatched region in (**a**)). (**c**) Vertical profiles of water temperature 100 m above the seafloor (enlargement of the pink hatched region in (**a**))

100 m above the seafloor. It should be noted that the data obtained from shear sensor 1 (Fig. 6.3) are utilized here.

The ε values in the water column (from the surface to just above the seafloor) at GND varied within the range of approximately 10^{-10}(W kg^{-1}) to 10^{-7}(W kg^{-1}), as shown in Fig. 6.10a.

Focusing on the vertical distributions of ε and water temperature near the depths of 1430 m and 1380 m, at 8:03, the ε value was around 10^{-9}(W kg^{-1}), but by 10:31, it had intensified to about 10^{-7}(W kg^{-1}), accompanied by an increase in water temperature of about 0.1 to 0.2 °C (Fig. 6.9b, c), indicated by the black dotted arrow). Subsequently, by 12:56, the ε value had dropped to approximately 10^{-10}(W kg^{-1}), and water temperature had decreased correspondingly (Fig. 6.10b, c), indicated by the red dotted arrow). Therefore, in the vicinity of the seafloor at GND, there is a tendency for water temperature to slightly increase as mixing becomes stronger.

The isopleths of ε and water temperature in the water column at GND, obtained from continuous observations by VMP-X on June 24th, are shown in Fig. 6.11. In the upper 100 m just above the bottom, a strong mixing environment with ε values

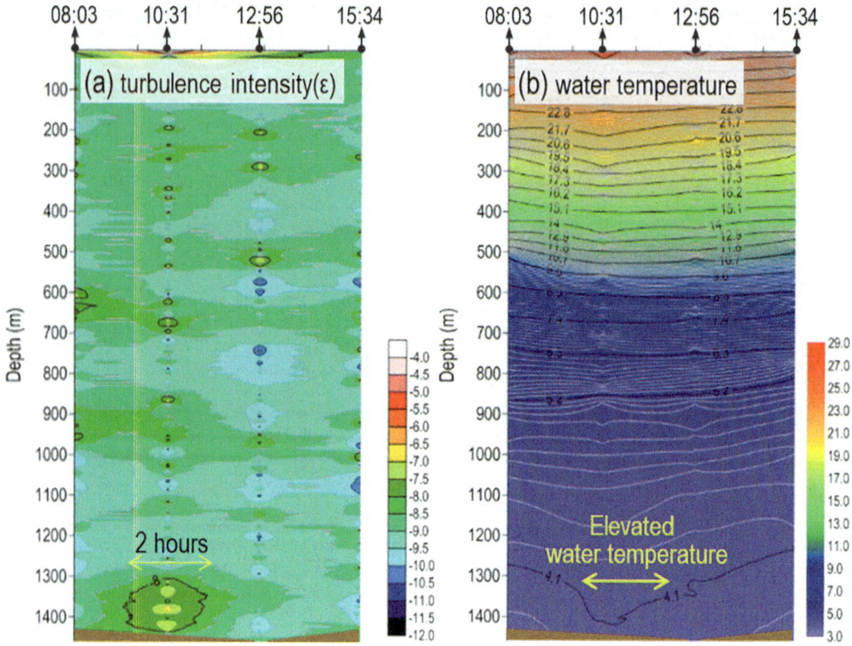

Fig. 6.11 Isopleths of ε and water temperature in the water column at GND obtained on June 24th, 2017. (**a**) Isopleth of ε, (**b**) Isopleth of water temperature

of approximately 10^{-7}(W kg^{-1}) was formed for about 2 h, centered around 10:31 (Fig. 6.11a). Simultaneously, the sea water temperature near the seafloor also increased (Fig. 6.11b). This phenomenon suggests that the hydrothermal vent fluids emitted from the seafloor at GND are mixing with the surrounding seawater under the influence of intense turbulence.

In the time series plot of environmental data obtained from the "low head platform" deployed on the seafloor at GND from June 23rd to June 25th (Fig. 6.12), during the turbulent observation period conducted using VMP-X, significant fluctuations in water temperature were not observed. However, fluctuations in both water temperature and turbidity exhibited a 24-h periodicity (Fig. 6.12a, b). The peaks of this periodic fluctuation were observed during the transition from high tide to the higher of the two daily low tides, known as "lower high water" (Fig. 6.12c). At this time, water temperature increased by approximately 0.1 °C, and there was a prevailing southward flow with a velocity of around 10 cm/s (Fig. 6.12g). Furthermore, during the period from 8:03 to 10:31 on June 24th when the maximum deep-sea turbulence intensity (ε) was observed, a southwestward flow with a velocity of around 8 cm/s was recorded. After 10:31, the flow direction shifted to a northward flow (Fig. 6.12g, h). The hydrothermal vent area within GND was confirmed to be located about 1 km southeast of the turbulence measurement point (Fig. 6.13). Additionally, at this measurement point, a dominant residual southward

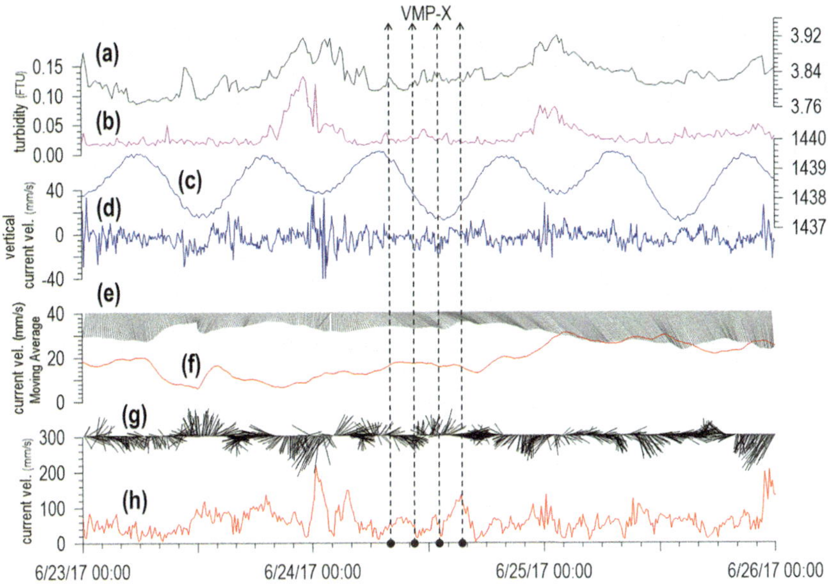

Fig. 6.12 Fluctuations in the time series of environmental data near the seafloor on June 23 and 25, 2017. The dotted black arrows show the observation time on the VMP-X. (**a**) Water temperature. (**b**) Turbidity. (**c**) Depth. (**d**) Vertical component of the current velocity at the 5th layer. (**e**) Residual current direction at the 5th layer. (**f**) Residual current velocity at the 5th layer. (**g**) Fluctuations in the current direction at the 5th layer. (**h**) Current velocity at the 5th layer

flow prevails (Fig. 6.12e, f), and considering the presence of an eastern sloping seafloor topography, it can be inferred that a counterclockwise circulation system along the slope exists (Fig. 6.13). Therefore, it can be speculated that fluctuations in deep-sea turbulence intensity and water temperature are influenced by this circulation system.

2.4 *Relationship Between Deep-Sea Turbulence Intensity and Geographic Environmental Factors*

The minimum, maximum, and mode values of ε obtained from turbulent measurements using VMP-X have been compiled in Table 6.2. The ε values are based on data acquired within the water column from just above the bottom to the upper 100 m. Additionally, the surrounding geographical characteristics have also been summarized in Table 6.2. The mode values of ε near the seafloor exhibit a range of 10^{-10} to 10^{-6} (W kg^{-1}), which corresponds to the range of ε values observed in the open ocean (Tanaka et al., 2021a, b). Furthermore, the mode values of ε within the hydrothermal vent area vary within the range of 10^{-7} to 10^{-8} (W kg^{-1}), suggesting the presence of a relatively strong mixing environment near the seafloor. This

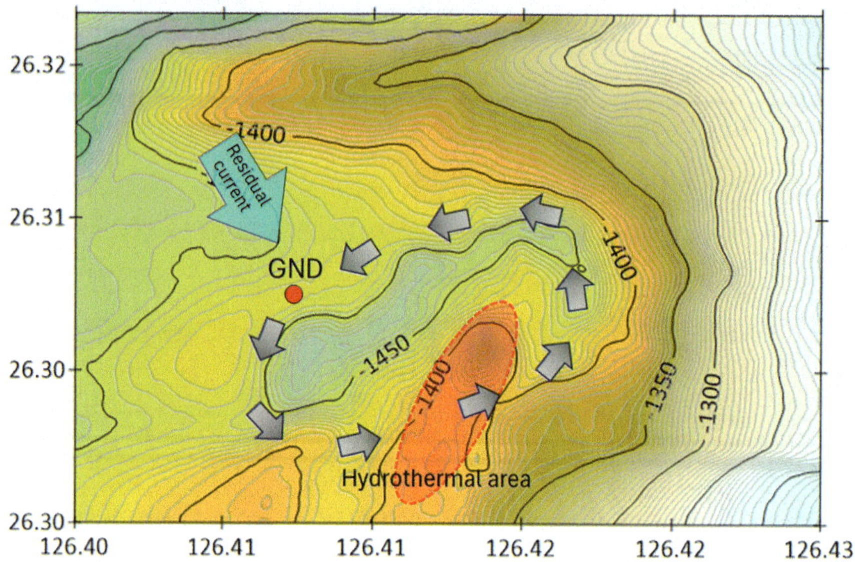

Fig. 6.13 Schematic of the predicted circulation pattern, estimated based on flow observation data around GND. Near the seafloor at observation site GND, a "low head platform" was installed, enabling flow measurements using ADCP and microscale flow measurements using ADV. Data from ADV were utilized in estimating deep-sea turbulence intensity

intense mixing environment can be attributed to the complex bottom topography unique to hydrothermal vent areas and the influence of discharged hydrothermal fluids.

Single-shot observations of ocean turbulence by VMP-X have made it possible to actually measure the intensity of deep-sea turbulence just above the seafloor, the actual conditions of which had not been revealed until now. However, due to the scarcity of actual field data on deep-sea turbulence, continuous measurements will still be necessary in the future to establish a baseline understanding of the mixing environment near the seafloor. Notably, the various characteristics observed in deep-sea turbulence intensity due to geographical and bottom environmental differences have yielded intriguing findings.

On the other hand, vertical (turbulent) mixing near the seafloor fluctuates significantly due to tides. To ensure the characterization and usefulness of temporal variations in deep-sea turbulence intensity, continuous observations using VMP-X are undoubtedly necessary. However, conducting such observations frequently presents considerable challenges both in terms of human expertise (requiring specialized skills) and financial resources (due to the high cost of sensors).

Therefore, in the subsequent section, we will introduce a method for estimating deep-sea turbulence intensity (patent obtained) that has been developed due to the possibility of field measurements of deep-sea turbulence intensity near the seafloor. This development allows us to capture the temporal dynamics of deep-sea turbulence intensity, a feat that was previously difficult to achieve.

Table 6.2 Basic statistical parameters (maximum, minimum, mode) of ε (W kg^{-1}) obtained at observation sites for ocean turbulence using VMP-X, along with geographical characteristics. Data for ε were used from just above the bottom to 100 m above the seafloor

Site	Station name	Area name	Depth (m)	Maximum value of ε	Minimum value of ε	Most frequent value of ε	Geographical characteristics	Remarks (Possible phenomena)
Sagami Bay	ITO	off Ito	1000	1.15×10^{-6}	1.42×10^{-8}	1.00×10^{-7}	Open inland Bay (open)	Tidal, Kuroshio branching current influence?
	SAG	center of the Sagami bay	1500	4.59×10^{-8}	4.29×10^{-11}	5.00×10^{-9}	Open inland Bay (open)	Influence of Kuroshio branch currents, tides…?
	HTS	off Hatsushima	1000	3.84×10^{-8}	1.02×10^{-10}	5.00×10^{-9}	Seepage site (open)	Influence of Kuroshio branch currents, tides, seep…?
	OMD	Ohmuro-dashi	850	1.32×10^{-7}	7.66×10^{-11}	2.50×10^{-8}	Hydrothermal area (Near land shelf slope)	Influence of Kuroshio branch currents, tides…?
Izu-Ogasawara Arc	HAG (in)	Higashi Aogashima	800	8.39×10^{-7}	7.80×10^{-11}	1.00×10^{-7}	Hydrothermal area (closed)	hydrothermal water
	HAG (out)	Higashi Aogashima	800	1.37×10^{-8}	2.60×10^{-10}	2.50×10^{-9}	Open ocean (outside Hydrothermal area)	Influence of Kuroshio…?
	BYN	Bayonnaise	800	4.97×10^{-7}	1.91×10^{-10}	5.00×10^{-8}	Hydrothermal area (semi-closed)	Hydrothermal water

(continued)

Table 6.2 (continued)

Site	Station name	Area name	Depth (m)	Maximum value of ε	Minimum value of ε	Most frequent value of ε	Geographical characteristics	Remarks (Possible phenomena)
Okinawa Trough	IHY	Iheya-north	1000	8.83×10^{-8}	2.71×10^{-10}	1.00×10^{-8}	Hydrothermal area (semi-closed)	Hydrothermal water, Complex bottom topography
	NHO	Noho	1600	6.52×10^{-8}	3.26×10^{-11}	1.00×10^{-8}	Hydrothermal area (open)	Hydrothermal water, Complex bottom topography
	IZN	Izena	1600	6.51×10^{-8}	3.82×10^{-11}	1.00×10^{-8}	Hydrothermal area (semi-closed)	Hydrothermal water, Complex bottom topography
	IEW	Iejima-west	980	6.56×10^{-8}	4.34×10^{-11}	1.00×10^{-8}	Hydrothermal area (semi-closed)	Hydrothermal water, Complex bottom topography
	GND	Gondoh	1400	3.18×10^{-7}	2.21×10^{-11}	5.00×10^{-8}	Hydrothermal area (semi-closed)	Hydrothermal water, Complex bottom topography

Finally, the turbulence observations conducted by VMP-X from 2015 to 2018, as presented in this section, were carried out as part of the "Next-Generation Technologies for Ocean Resource Exploration" initiative under the Strategic Innovation Promotion Program of the Council for Science, Technology, and Innovation.

3 Establishment of a New Method for Estimating Deep-Sea Turbulence Intensity

In this section, we mainly present the method for estimating deep-sea turbulence intensity (turbulent kinetic energy dissipation rate) using an ultrasonic current meter that JAMSTEC and AIST applied for domestic patent application (Application No. 2021-140907) in 2021.

Acoustic Doppler velocimeters (ADV) are frequently used to estimate turbulent kinetic energy dissipation rate (ε). Such techniques is widely used for shallow coastal regions (<20 m depth) but not optimized for the deep sea (>200 m depth). This primarily depends on that the deep sea typically exhibits low levels of sound scatterers, which lead to high levels of noise on velocity data. Here we present a new method for estimating near-bed ε in the deep sea, based on an ADV fixed on the bottom in 1400 m of water. While turbulent velocity spectra generally exhibit the "–5/3" power law in the inertial sub-range, we found the "–1" power law in our velocity data, suggesting that spectral power distribution was modulated by the presence of bed. We calculated ε from low-wavenumber ranges, which are low enough to avoid the effect of acoustic noise. The ε obtained with ADV was compared with those obtained with the conventional method, a free-falling microstructure profiler. A good agreement with the free-falling profilers suggests our method provides plausible estimates of ε near the deep-sea bed.

3.1 The Necessity of Turbulence Intensity Estimation Methods near Deep-Sea Bottom

The traditional, established method to quantify turbulence intensity is free-falling turbulent microstructure profilers equipped with velocity shear probes (Lueck et al., 2002). Fitting the theoretical curve to the observed shear spectra provides the most reliable estimates of turbulence intensity (i.e., turbulent kinetic energy dissipation rate ε) among the existing methods (Gregg, 1999). Even though fixed platforms, such as vessels, are required to deploy the profilers into the ocean, such a system is suitable for taking "snapshots" of turbulent profiles in the water column. Nevertheless, for a variety of reasons, we frequently need to continuously observe turbulence intensity for a long term (e.g., over tidal cycles). While acoustic velocimeters were originally designed to observe tidal currents, those are frequently used to quantify in situ turbulence intensity. For example, ADV records in situ turbulent velocity in a sampling volume of several cm^3 at a high sampling frequency (typically >16 Hz). The ADVs can directly quantify the variance level of in situ turbulent velocity, from which we can calculate plausible estimates of ε under various conditions, such as energetic shallow coasts (Walter et al., 2014) and calm inlets (Sato et al., 2014). In addition to ADV, acoustic Doppler current profilers (ADCP) are also available to estimate ε for marine and freshwater systems (Wiles et al., 2006; Lorke, 2007), observing the second order structure function of turbulent velocity.

The use of an acoustic velocimeter could be a simple, inexpensive solution for observing turbulence relative to traditional microstructure profilers. But there are still challenges in the application for the deep sea. The acoustic velocimeters, first of all, record Doppler shifts of reflected echo from particles suspended in the water (e.g., phyto- and zooplankton, detritus, and inorganic matters), allowing us to quantify water velocity. Because of the measurement principle, the level of noise inherent in the recorded velocity depends on the density of particles in the water—lower noise level is expected for a turbid water. Hence, acoustic-based turbulence estimations are applicable for the surface layer (<200 m depth), where high levels of biological production are typically observed, like in previous studies (Guerra & Thomson, 2017). Conversely, such acoustic-based turbulence estimations, to our best knowledge, have not been applied for the deep sea (>200 m depth), where the density of suspended particles is typically low (Wishner, 1980). Thus, there is no standard method for turbulence observations near the deep-sea bed.

Observing near-bed turbulence is especially crucial for deep-sea mining, which started in the middle twentieth century (Mero, 1965), which has a great potential to disturb natural habitat (Halfar & Fujita, 2007; van Dover, 2011). While deep-sea mining still has various challenges in technical, economical, and commercial aspects (Sharma, 2011), the International Seabed Authority (ISA), in 2013, published statements for environmental impact assessment, including physical oceanographic surveys to evaluate particle dispersal due to eddy diffusion (ISA, 2013). Additionally, recent studies suggested a substantial impact of local turbulence on marine ecosystem, especially among sub-centimeter microorganisms (Kiørboe, 2008; Koehl, 2007; Tanaka et al., 2021a, b). Given the potential importance of turbulence in local ecosystems, we developed an ADV-based technique to quantify near-bed turbulence intensity applicable to the deep sea.

3.2 Development of a New Method for Estimating Deep-Sea Turbulence

3.2.1 Field Campaign

Our experiment was made near the bottom in 1400 m of water in Okinawa Trough, the northwestern North Pacific (26°18′N, 126°28′E) during baseline survey for environmental impact assessment nearby (~1 km) a cluster of hydrothermal vents.

While the western boundary current, the Kuroshio, substantially affects environmental condition in the water column, bottom waters exhibit a typical temperature range of 4–5 °C (Furushima et al., 2020). An ADV (6-MHz Vector Current Meter, Nortek AS) was mounted on a compact platform consisting of frames, buoys, external batteries, and other environmental sensors, which has dimensions of around 2 m × 2 m × 2 m (KANSO Co., LTD., Japan). The ADV recorded in situ velocity at a sampling frequency of 32 Hz during 90-s burst measurements repeated every 300 s. ADV's sampling volume was set to 2.6 cm^3 and located about 2 meters above the bottom (hereafter MAB). Data recording lasted for a month (June 4th to July 1st, 2017).

3.2.2 Power Law in Velocity Spectrum

For fully turbulent unbounded flows, the wavenumber spectrum of turbulent velocity (ϕ) exhibits the "–5/3" power law for the inertial subrange (Thorpe, 2005):

$$\phi(k) = a \cdot \varepsilon^{2/3} \cdot k^{-5/3}, \tag{6.1}$$

where a is a constant (~0.5), and k is cyclic wavenumber (cpm). Previous methods simply back-calculated ε from Eq. (6.1), based on $\phi(k)$ observations away from the boundaries (Guerra & Thomson, 2017; Sato et al., 2014). Such methods could potentially assume ε = *constant* for the calculation process. However, such an assumption is not applicable for bottom mixed layer, where ε is a function of distance from the bottom h (m) (Thorpe, 2005):

$$\varepsilon = \frac{u_*^3}{\kappa h}, \tag{6.2}$$

where u_* is friction velocity (m s^{-1}), and κ is von Karman constant (~0.41). We here assume that we are observing $\phi(k)$ at a height h_{obs} within the bottom mixed layer with thickness of H (i.e., $h_{obs} < H$). The wavenumber spectrum $\phi(k)$ at a wavenumber range $H^{-1} < k < h_{obs}^{-1}$ is expected to have a spectral form different from Eq. (6.1), since turbulent eddies larger than h_{obs} could be interfered with the bottom. Such spectral form can be obtained by substituting Eq. (6.2) for Eq. (6.1) and assuming $k = h^{-1}$ (Nikora, 1999):

$$\phi(k) = A \cdot u_*^2 \cdot k^{-1}, \tag{6.3}$$

where A is a constant ($=a \cdot \kappa^{-2/3} = 0.9060$). The Eq. (6.3) suggests two things: First, level of $\phi(k)$ is determined with friction velocity u_*, not ε. Second, $\phi(k)$ follows "–1" power law, unlike Eq. (6.1). Those characteristics have been introduced in previous studies (Katul & Chu, 1998; Nikora, 1999).

3.2.3 Data Process

While we can estimate near-bed ε from Eqs. (6.2) and (6.3), the ADV originally obtained the frequency spectra of turbulent velocity $\psi(f)$, where f is frequency (Hz). The $\psi(f)$ was calculated from vertical component of turbulent velocity for each 90-s burst measurement. The conversion from frequency domain to wavenumber domain was made based on horizontal velocity U (m s^{-1}) during the burst measurements. Namely, $\phi(k) = \psi(f) \cdot U$ and $k = f/U$. To obtain u_* from the observed $\phi(k)$, we deformed Eq. (6.3) as follows:

$$u_* = A^{-1/2} \cdot \left[k \cdot \phi(k)\right]^{1/2}. \quad (6.4)$$

A representative value for $k \cdot \phi(k)$ was obtained by averaging $k \cdot \phi(k)$ over an arbitral wavenumber range $k_1 < k < k_2$. The lower wavenumber edge k_1 should be larger than H^{-1} and was set to be $k_1 = 1/7$ cpm, reasonable for this energetic sea (Furushima et al., 2020). A critical wavenumber, at which the transition of power law occurs, is suggested to be $k = h_{obs}^{-1}$ (Katul & Chu, 1998; Nikora, 1999). Hence, the value of k_2 was set to $k_2 = 1/2$ cpm, considering our ADV's location $h_{obs} = 2$ m. Then, the ε values were estimated from Eq. (6.2) at arbitral locations h.

3.2.4 Validation

Our velocity records exhibited numerous spikes, likely contaminated with acoustic noise due to low levels of particle density in the deep waters (Furushima et al., 2020). Most of our data (98%) was rejected by the manufacturer's quality control test, based on signal-to-noise ratio (SNR) and acoustic correlation. Instead of the use of the quality control test, we had an original test based on instrument's noise level. The effect of noise is typically serious for high frequency ranges but could be negligible for low-frequency ranges (Doroudian et al., 2010). As our estimation method is based on low-frequency turbulent fluctuations, the noise level should be determined from low frequency ranges, not from the entire range. We first calculated the variance level δ (m^2 s^{-2}), by integrating each $\psi(f)$ from 10^{-2} to 10^{-1} Hz. Given the typical range of 0.03–0.09 m s^{-1} for U, the frequency range for integration reasonably includes the wavenumber range between k_1 and k_2. The smallest variance level (δ_{min}) found during our experiment was determined as the bulk noise level. The ε values were passed, when the δ value was 3 dB (about twice) larger than $\delta_{\min}$. This process rejected 0.52% of the estimated ε.

Additionally, we had another test based on U. Under an extremely small U, our calculation wavenumber range $k_1 < k < k_2$ does not cover k ($= f/U$). Hence, we only used ε values estimated with U larger than 0.03 m s^{-1}. This process rejected 15.08%.

3.2.5 Comparison with Microstructure Profiler

To validate the ε values estimated from ADV, we deployed a turbulent microstructure profiler VMP-X (Shang et al., 2017) during our experimental period nearby (<1 km) the mooring site. The VMP-X was equipped with an expendable sensor package that carried two shear probes. The VMP-X free-falls at a rate of 0.7 m s^{-1}, recording vertical shear of horizontal velocity at a frequency of 1250 Hz until the sensors reach the bottom. We calculated ε values from the recorded shear within 2-s bins half-overlapped with the next bins (Wolk et al., 2002). Given that our method estimates ε values from velocity variance between $k_1 = 1/7$ cpm and $k_2 = 1/2$ cpm, we speculate that the comparison between ADV and VMP-X should be made at a representative spatial scale of $h = 2$ to 7 m. Hence, the ε values obtained from VMP-X were averaged over $h = 2$ to 7 m (hereafter ε_{VMP}). Similarly, the ADV-based ε was averaged over $h = 2$ to 7 m (hereafter ε_{ADV}). Because of the intermittency of dissipation rates, the representability of each ε_{ADV} value would be low (Wolk et al., 2002). Thus, the ε_{ADV} values were averaged over 1-h bin set around each VMP-X deployment for the comparison.

3.3 *Biophysical Interaction with Deep-Sea Turbulence Intensity Estimation Method*

Pressure data obtained with the ADV suggested diurnal and semidiurnal oscillations in water level (Fig. 6.14a). Northward and eastward velocity components averaged over the burst period of 90 s also exhibited the diurnal and semidiurnal oscillations (Fig. 6.14b). Magnitude of velocity frequently increased up to 5 cm s^{-1} (Fig. 6.14b) and reached over 10 cm s^{-1} (Fig. 6.15a). The mean and median of u_* were $u_* = 0.0017$ m s^{-1} and $u_* = 0.0013$ m s^{-1}, respectively (Fig. 6.15b). The minimum and maximum were $u_* = 0.00015$ m s^{-1} and 0.014 m s^{-1}, respectively. Those are about an order of magnitude smaller than shallow (<100 m) coastal beds disturbed with surface waves (Drost et al., 2018). The ε values averaged over $h = 2$ to 7 m (Eq. 6.2) ranged from $\varepsilon = 10^{-11}$ to 10^{-7} W kg^{-1} (Fig. 6.15c). The ε averaged over the observation period was $\varepsilon = 1.1 \times 10^{-8}$ W kg^{-1} (Fig. 6.15c).

The dissipation rates obtained with the ADV (ε_{ADV}) were consistent with those obtained with the VMP-X (ε_{VMP}) (black dots in Fig. 6.16). While ε_{ADV} exhibited fluctuations over hours to minutes, the ε_{VMP} are consistent with ε_{ADV}, suggesting the validity of our method based on the ADV (Fig. 6.16). The root-mean-square error in logarithmic scale was 0.44, suggesting the error level is about 3 dB.

Sub-centimeter organisms, which comprise the majority of biomass at local regions, are greatly affected by turbulence (Franks et al., 2022; Kiørboe, 2008). For example, turbulence affects prey-predator interactions (Gilbert & Buskey, 2005),

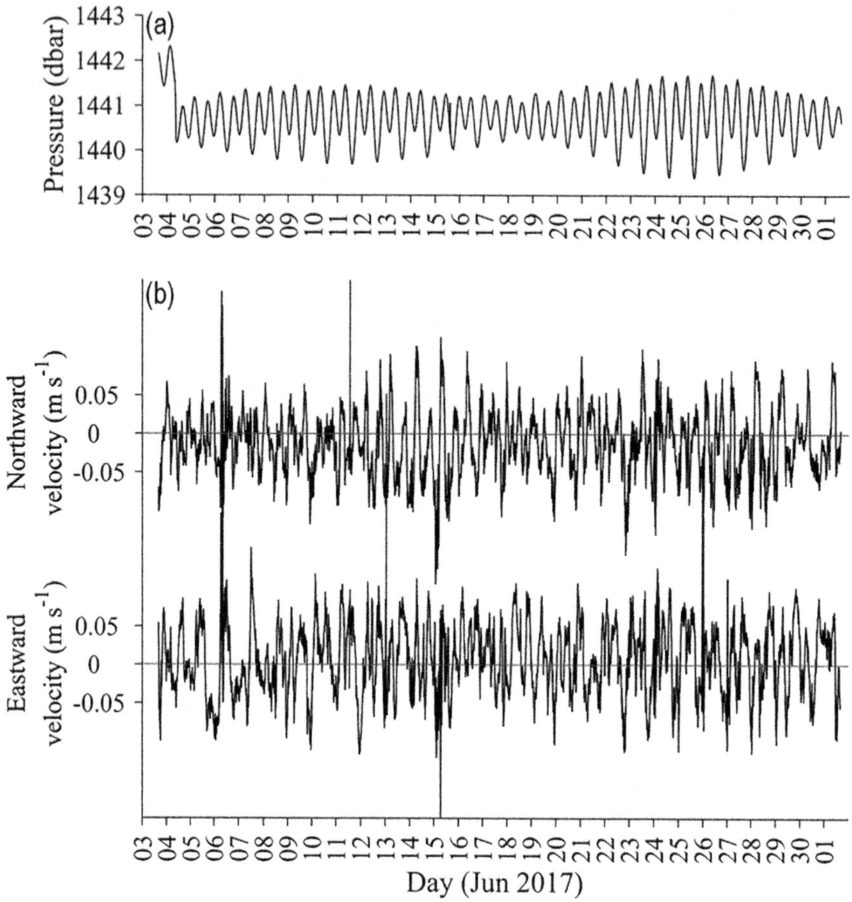

Fig. 6.14 (**a**) Pressure and (**b**) northward and eastward components of velocity, averaged over the burst period (90 s)

feeding (Pécseli et al., 2019; Rothschild & Osborn, 1988), and mating (Yamazaki et al., 1991), inducing turbulence avoidance behavior of individual organisms (Franks, 2001; Maar et al., 2006; Tanaka, 2019; Tanaka et al., 2021a, b; Visser et al., 2001). Additionally, larval settlement onto the bed, which is a key biological process governing the dynamics of bottom-dwelling organisms, is substantially hindered by high levels of turbulence (Koehl, 2007). While the effects of turbulence on abyssal ecosystems are still elusive, our method to estimate turbulent kinetic energy dissipation rates based on ADV helps understand biophysical interactions in deep-sea organisms.

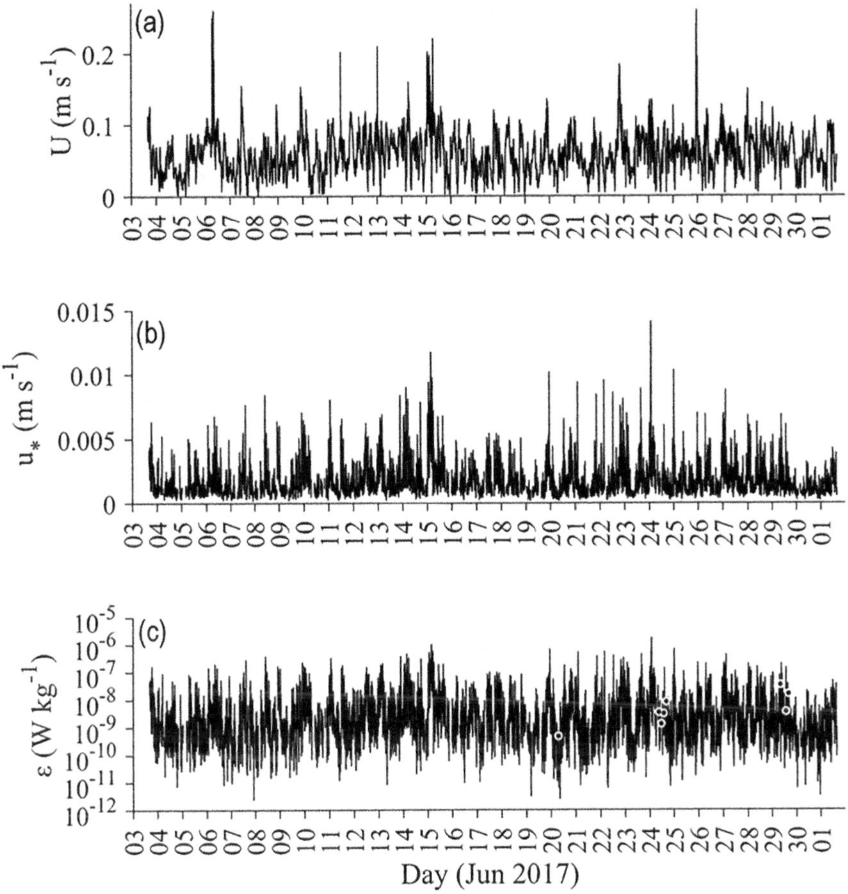

Fig. 6.15 Time series of (**a**) absolute velocity *U*, (**b**) friction velocity *u*_*, and (**c**) turbulent kinetic energy dissipation rate ε_VMP. The black dots in (**c**) denote turbulent kinetic energy dissipation rates obtained with the VMP-X

4 Visualization of Deep-Sea Turbulence

Ocean turbulence refers to the phenomenon where seawater moves in irregular patterns, and generally, it is difficult to visualize its movement. However, with the development of high-resolution cameras and illumination techniques in recent years, it has become possible to capture information about the in situ movement of particles and, therefore, the water in which they occur. Here we introduce field survey results employing an imaging system designed not only to visualize particles and their movements but also to use light refraction at density boundaries to directly image thermoclines/pycnoclines themselves. Future applications of this technique, with a focus on surveys for environmental impact assessments in the water column and near the seafloor, are introduced.

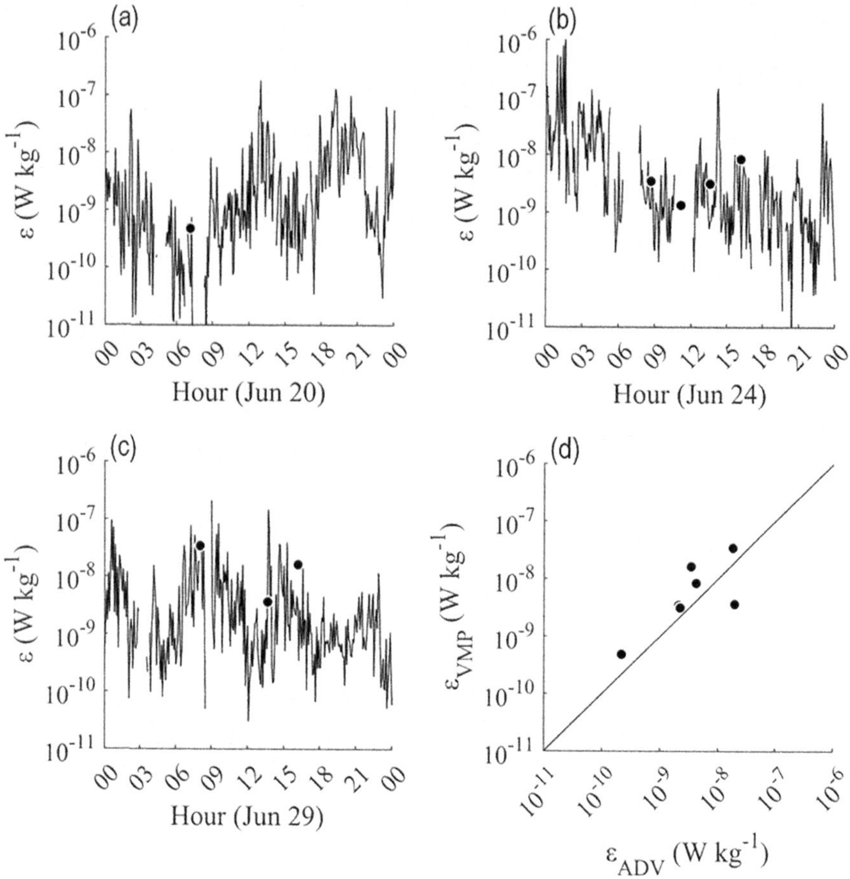

Fig. 6.16 (**a–c**) The same time series with Fig. 6.2c but zoomed in on June 20th (**a**), 24th (**b**), and 29th (**c**). (**d**) Scatter plot of turbulent kinetic energy dissipation rates obtained by ADV ("εADV") and VMP-X ("εVMP"). The solid line denotes the one-to-one relationship. An "εADV" datum (around the noon on June 24) is not available because of removal due to the quality control during this period

4.1 Comparison of Shadowgraphy-Based vs Normal Imaging Systems and Design Rationale

Shadowgraphy is an imaging technique where a camera faces a point light source (i.e., transmitted light) and collimating lenses are positioned before both the camera lens and the light source to produce parallel beams of light. The majority of imaging techniques suffer from the limitation where the imaged object appears larger in size when it is positioned closer to the camera and appears smaller when positioned further from the camera, making size calculations problematical (Durden et al., 2016; Lindsay et al., 2013, 2014). Instruments such as the UVP (Underwater Video

Profiler; Gorsky et al., 1992) or the DeepPIV (Katija et al., 2020) endeavor to overcome this limitation by using a thin sheet of light projected perpendicular to the imaging camera lens and positioned at a distance equal to the focus plane. This produces thin optical slices of a small volume of water and the particles within it, where the imaged area and thickness of the light sheet are known, and the size of an object scattering the light can therefore be calculated. Since an object's imaged size is not correlated to its distance from the camera lens when parallel beams of collimated light are used to photograph its shadow and the in-focus volume is also much larger than when the object itself is the target of the imaging, shadowgraphy has the potential to greatly increase the detection of rarer organisms or other particles such as larvacean houses.

Shadowgraph systems currently being deployed in the oceans generally have low frame rates (e.g., 2 Hz in the Zooglider—Ohman et al., 2018; 1 Hz in the DPI-ISIIS—Brisbin et al., 2020). The exceptions are generally based on line-scan cameras that have the advantage of capturing a continuous strip of ocean along the tow path (Greer et al., 2021) and therefore each object is imaged only once and double-counting of particles is avoided. However, in our application, information was also being sought on the movement of particles in the water, necessitating the use of an area-scan camera with a high frame rate in order to visualize particle movement and therefore detect turbulence directly through imaging techniques rather than with free-falling turbulent microstructure profiler. Because shadowgraph images allow visualization of density boundaries due to light refraction and can also image the size and type of particle, they should be particularly useful in areas around seafloor massive sulfide deposits where active venting of hydrothermal fluid is still taking place, at oceanographic boundaries such as thermoclines or pycnoclines, and for surveys of both the near-bottom and midwater sediment plumes produced during manganese crust or nodule mining operations. The latter plumes can be expected to have both a positively buoyant component comprised of the seawater that has heated up during mineral extraction processes on board the mining vessel at ambient air temperatures, and a negatively buoyant component comprised of sediment grains and other solid material.

Our shadowgraph system was designed to not only directly image these components, to assess whether particles were living organisms, fecal pellets or other marine snow particles, or granular-looking suspended sediments, but also to visualize particle movements in order to quantify plume transport and to assess the effects of sediment loading on organisms entrained in the plumes. Information on particle size composition would furthermore allow better interpretation of acoustic reflectance data, since different acoustic wavelengths are reflected differently based on particle size. Information on particle movement allows assessments of whether particles are sinking, being laterally transported or are entrained in turbulent flow. Since prey and other particle encounter rates will be affected by water movement within the organisms' habitat, such information will be useful for modeling the environmental impact on the organisms.

4.2 Description of the ShadowCam Shadowgraphy-Based Camera System

Power is provided via a battery (Li-Po 6S, 22.2V, 6000 mAh) deployed in the LED light containing housing with the battery charger (HiTech Balance Charger X1 NANO Premium) also in the same housing. This charger can be accessed via a Bluetooth connection using an app on a smartphone to check battery status. To prevent battery discharge during periods of storage, the charger circuitry is decoupled from the battery via a loop in the circuitry. Concretely, an underwater cable must be attached to two of the external underwater connectors to complete the loop during charging. A condenser lens of diameter 100 mm and focal point 150 mm, deployed in both housings, with the light source being white light passing through a pinhole, produced by a white LED. This light is collimated by the condenser lens and results in a circular light beam of diameter 100 mm. The housing window is only 60 mm in diameter, so the imaged diameter is 60 mm. The 100 mm diameter light field was selected so that future iterations of the camera imaging larger cross-sectional areas would not need the internal support structures to be redesigned and also to reduce optical artifacts associated with lens edge distortion effects. Images are captured by the camera (See3CAM_CU135m: 1/3.2" CMOS sensor AR1335M, Onsemi Inc.) using an 8 mm S-mount/M12 fixed-focus lens at 4096 X 3042 pixel resolution with a 10-bit color depth at 20 Hz by a Jetson NANO PC. Particles are imaged at a resolution of one pixel being 0.03 mm. Exposure time of the camera shutter was set to milliseconds to reduce blurring. Files are split automatically and recorded onto the 1 TB memory card (SanDisk Extreme PRO microSDXC A2) as 10-minute-long videos (.mpg codec) with the following filename format: SG_A_YYYYMMDD_HHMMSS_sss_+9 (nb. when capturing data in the Japan Standard Time timezone, the UTC time has a suffix appended to capture local time information), etc. A web browser-based user interface can be accessed by connecting to the web server being run on the Jetson Nano PC inside the camera housing via its wireless hotspot. Application Programming Interface (API) calls can also be made to this server. In the web browser interface, it is possible to sync the time on the Jetson Nano to the same time as the user's PC. The browser window automatically refreshes every 3 s, providing visual confirmation on system status, such as whether the camera is currently on, recording, in standby mode, etc. System time for both the Jetson Nano and the user's PC is also displayed, as is also the current camera control settings: gain and exposure. The name of the video currently being recorded is displayed along with the current file size, allowing the user to confirm that file size is increasing and therefore the system is recording correctly. The total number of video files that have been recorded during the current deployment is also displayed. The main webpage also has buttons to shut down the camera, open the camera test interface, and view all video files as a list of filenames in a new tab that also provides download (single file or zip archive) and delete functionality.

The current prototype camera system is rated at 1500 m depth, weighs 24.5 kg in air and 20 kg in water, and has dimensions of 1500 mm × 200 mm × 200 mm. Preliminary processing of shadowgraph video is performed to remove the background noise and detect objects within the frames, after import of CTD and other metadata to a relational database. The CTD data can then be extracted via API calls to get CTD data at a specific datetime or between datetimes or to get CTD between specified depths. For the image analysis, a background image is generated for each 60 s segment of video by taking the average brightness (0–255 value) of each pixel for that video segment, thereby removing any dust particles, scratches, illumination artifacts etc. from the image to be analyzed. A first-pass visualization of the dataset can be done by calculating the number of "dark" pixels per image (inputting a threshold value) or analyzing the change in brightness of an image versus the background image in order to detect large objects, mixing/turbulence layers, and/or particle plumes and extracting them to a "first-pass analysis" folder. The time-consuming process where the background image is then compared with all captured video frames to separate particles from the background is usually run post-cruise on a more powerful computer. The detected particles are then clustered into objects using 8 neighboring connected pixel methods and cropped from the video with a cleaned background. These ROIs are saved using the following name format: SG_A_YYYYMMDD_HHMMSS_sss_+9_nnnnnn, where nnnnnn is the frame number, to folders containing background-cleaned (cleaned) and background-retained (original) images, respectively. API calls to the server can get particle counts versus datetimes or depth ranges, produce particle size histograms, and graph data in various formats for visualization, including timelines, pie-charts, and bar charts. These charts can be exported as publication-ready figures (.png format) and image-metadata pairs can be exported to tools such as Morphocluster, EcoTaxa or Squidle+ for further downstream analysis.

4.3 Relationships Between Organisms and Turbulence

Field tests of the prototype system were carried out in Suruga Bay and at the Myojin Knoll hydrothermal vent site in December 2022 during R/V Kaimei cruise KM22-15-1 and in Suruga Bay and the Higashi-Aogashima caldera in May 2023 during R/V Mirai cruise MR23-03. The shadowgraph camera was deployed on a CTD rosette system during both cruises (Fig. 6.17). Many organisms as well as the pycnocline were able to be imaged, although a slight adjustment in the alignment of the two pressure housings would have resulted in images with a more evenly gray background (Fig. 6.18). Nevertheless, organisms were able to be detected (Fig. 6.19) and the backgrounds removed as planned (Fig. 6.20). Particles could also be detected and counted within each image (Fig. 6.21). Although vertical profiles of particle abundances and sizes were able to be acquired as planned, the up-down movement of the CTD rosette as the ship rolled precluded discovery of any relationship between our collected data and turbulence values due to the turbulence produced by the heaving CTD rosette itself. Disturbed organisms exhibiting an escape response were imaged in large numbers during deployments. Deployment of this system on a free-falling or free-floating platform may be necessary, although we plan to test deployment on an ROV and "flying" slowly forward as an alternative data collection method.

Fig. 6.17 Shadowgraph camera attached to the lower section of the CTD Rosette

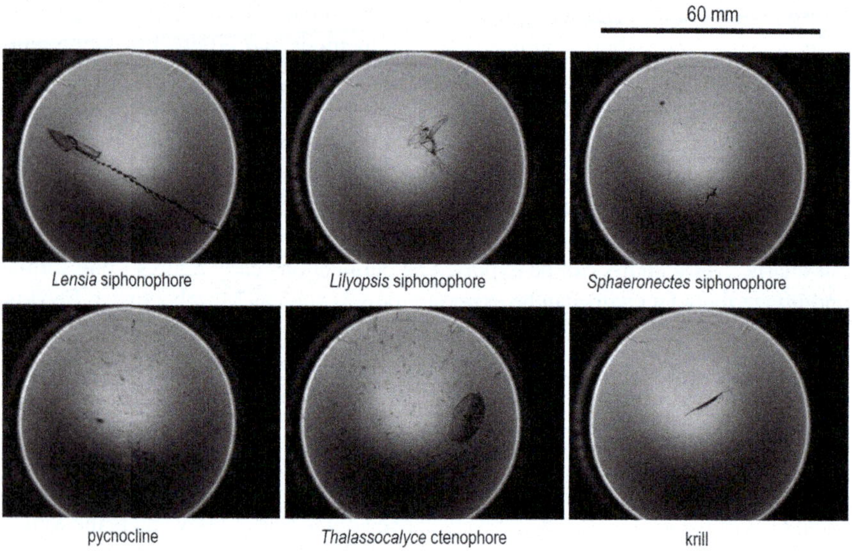

Fig. 6.18 Raw shadowgraph images from the first deployment

4.4 Prospects for Future Studies

The current design of having two pressure housings facing each other can be operationally challenging if the housings are not properly aligned and fixed to the support beam, as even a slight movement in the angle of attachment of either of the housings can lead to a black image. Designs that incorporate both camera and light source in a single housing, using mirrors to direct the light beam, are currently being investigated. Although the current prototype is easily deployable on a CTD rosette system, a smaller, single housing-based system would allow for deployment on a remotely operated vehicle (ROV) for highly accurate surveys in 4D space and time. Depending on the targets of interest, the ShadowCam camera lens can be replaced, enabling

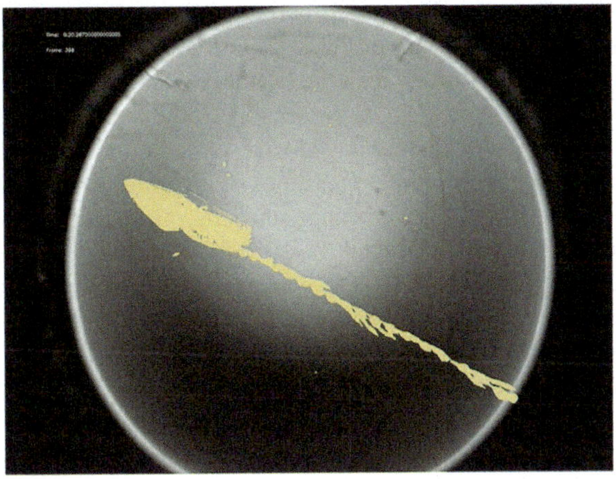

Fig. 6.19 A detected *Lensia* sp. siphonophore at a depth of 224.8 m in Suruga Bay

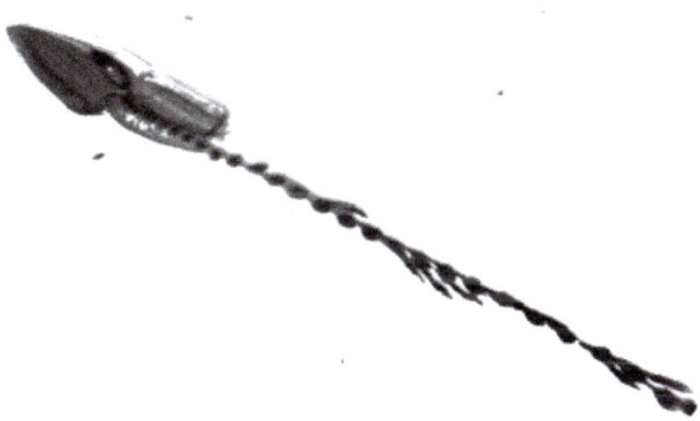

Fig. 6.20 *Lensia* sp. after background removal

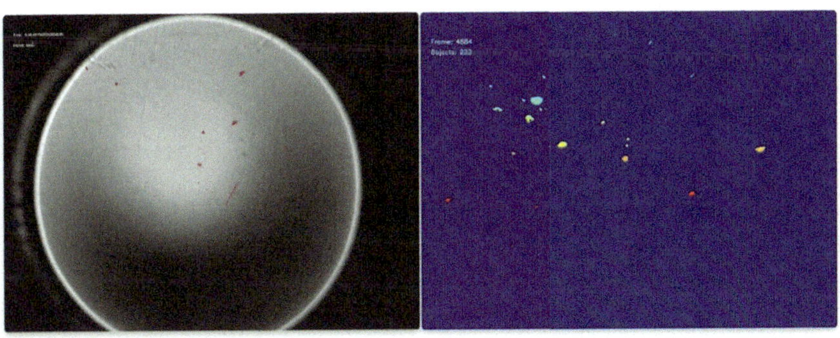

Fig. 6.21 Particle detection after background removal

optimal measurements of various sized particles. For example, if suspended sediments are found to be much finer or the near-bottom plankton community is found to be dominated by micro- rather than mesozooplankton, a narrower angle lens can be deployed to increase the imaged resolution of the smaller particles, although obviously there is a trade-off with imaged water volume.

Real-time monitoring of mining operations will require rapid analysis and interpretation of the acquired data. An image analysis, visualization, and publication pipeline is currently being developed, based on FAIR data principles (Lamprecht et al., 2020; Schoening et al., 2022). The domain transform filter (Gastal & Oliveira, 2011), as implemented in the Python library OpenCV as the "detail enhancement" function, is used to sharpen images before they are extracted as .png regions of interest (ROIs) using OpenCV and being exported/imported to Morphocluster (Schroeder et al., 2020) for unsupervised clustering based on ROI image features. Clusters are labeled during human validation and used to train a CNN model for the implementation of a YOLOv5 detection and tracking algorithm. The original captured image format is a video file (.mp4), which can be loaded into SqCapture (Sangekar et al., 2023) and processed using SquidJam (Sangekar et al., 2023), thereby allowing easy export to the Ocean Biogeographic Information System (OBIS) or other portals using the API interface.

5 Conclusion

5.1 Application of Estimation Methods for Deep-Sea Turbulence Intensity in Deep-Sea Mining

In this section, we first introduced the actual measured turbulence intensity near the deep-sea floor and its regional variation characteristics. Next, we showed a method for estimating deep-sea turbulence intensity that was developed based on actual measurements of deep-sea turbulence and microfluidic measurements. In addition, we introduced a technique to visualize particle dynamics due to turbulence and the basic technology to obtain the information necessary for environmental impact assessment in the offshore areas where deep-sea mining is conducted.

For the environmental impact assessment of deep-sea mining, it is essential to predict not only the topography and environmental baseline data of the area to be developed but also to use numerical models. In summarizing this section, we described examples of numerical model simulations to assess environmental impacts and the use of model predictions when considering the dynamics of deep-sea turbulence intensity presented in this section.

As part of a deep-sea mining project in Japan, an environmental baseline survey and the development of an environmental impact prediction model were implemented prior to the performance test of a mining vehicle for seafloor massive sulfides (SMS) conducted by JOGMEC in the waters around Okinawa in 2012. The environmental baseline survey focused on benthic community characteristics, water quality, sediment, flow direction, and velocity, which could be directly or indirectly affected by mining, and data for most of the environmental items listed in the ISA Guidelines (2013, ISBA/19/LTC/8) were obtained. In addition, a diffusion and redeposition model of suspended particles generated by prospecting and movement of the mining vehicle was developed as one of the models for predicting environmental impacts. In this model, the distribution of thickness of redeposition due to prospecting and the maximum concentration of suspended particles were estimated using the assumed mining conditions and the current data obtained from the environmental baseline study (Narita et al., 2015).

The concentration and particle size distribution of suspended particles and the thickness of redeposition around the test area were observed (Matsui et al., 2018) during the pilot test of mining that recovered seafloor massive sulfides (Okamoto et al., 2019), and during a small-scale disturbance test conducted around the Okinawa Sea area. In addition, simulations of suspended particle dispersion models prior to testing suggested that serious environmental impacts would not have occurred (Yamaji et al., 2019).

Thus, in assessing the environmental impact of deep-sea mining, it is important to understand the horizontal and vertical concentration and extent of diffusion of suspended particles due to seafloor disturbance, as well as the thickness of redeposition. We had been using the following observation methods, for example (Matsui et al., 2018; Furushima et al., 2019a, b, c; Yamaoka et al., 2023; Washburn et al., 2023):

(i) The observations by using turbidity sensors and particle size analyzers (e.g., LISST-Deep; Sequoia Scientific, Inc.) mounted on ROVs and other submersible platforms.
(ii) The collection of particles associated with redeposition by water sampling, sediment traps, etc.
(iii) The observation of turbidity plumes and the seafloor by underwater cameras mounted on ROVs and other submersible platforms (e.g., Edokko Mark 1; Okamoto Glass Co., Ltd.)
(iv) The observation of deep-sea currents using ultrasonic current meters

Information on the parameters needed for the model can be observed using these methods, and data on the particle size distribution of suspended particles and the amount of redeposition in test mining can be fed back into the model to make the simulation more closely reflect an actual situation. Observational data acquired in new surveys and findings from newly introduced equipment may also improve the accuracy of predictions. It is necessary to conduct more substantial observations and develop models for promoting sustainable deep-sea mining with reduced environmental impact.

Deep-sea turbulence, the topic we focused on in this section, can now be observed from near the sea surface to just above the deep-sea floor using the VMP-X (Rockland Scientific Inc.). New knowledge has been obtained on the intensity and spatial distribution of deep-sea turbulence in areas where seafloor massive sulfides are found, as well as data collected on their temporal variability (Furushima et al.,

2017, 2019a, b, c). Observations using the VMP-X have been a costly technique, but the method for estimating deep-sea turbulence intensity presented in this section solves this problem and can efficiently supplement future in-situ surveys. However, handling deep-sea turbulence requires advanced knowledge, and the parameters to be incorporated into the model are expected to be complex. It is important to accumulate observational data on deep-sea turbulence in areas where seafloor resource development is planned, and to examine how the estimated turbulence intensity can be applied to conventional suspended particle dispersion models.

References

Brisbin, M. M., Brunner, O. D., Grossmann, M. M., & Mitarai, S. (2020). Paired high-throughput, in situ imaging and high-throughput sequencing illuminate acantharian abundance and vertical distribution. *Limnology and Oceanography, 65*, 2954–2965.

Doroudian, B., Bagherimiyab, F., & Lemmin, U. (2010). Improving the accuracy of four-receiver acoustic Doppler velocimeter (ADV) measurements in turbulent boundary layer flows. *Limnology and Oceanography: Methods, 8*, 575–591.

Drazen, J. C., Smith, C. R., Gjerde, K. M., Haddock, S. H. D., Carter, G. S., Choy, C. A., Clark, M. R., Dutrieux, P., Goetze, E., Hauton, C., Hatta, M., Koslow, J. A., Leitner, A. B., Pacini, A., Perelman, J. N., Peacock, T., Sutton, T. T., Watling, L., & Yamamoto, H. (2020). Midwater ecosystems must be considered when evaluating environmental risks of deep-sea mining. *Proceedings of the National Academy of Sciences of the United States of America, 117*, 17455–17460.

Drost, E. J. F., Lowe, R. J., Ivey, G. N., Jones, & N. L. (2018). Wave-current interactions in the continental shelf bottom boundary layer of the Australian North West Shelf during tropical cyclone conditions. *Continental Shelf Research, 165*, 78–92. https://doi.org/10.1016/j.csr.2018.07.006

Durden, J. M., Schening, T., Althaus, F., Friedman, A., Garcia, R., Glover, A. G., Greinert, J., Jacobsen Stout, N., Jones, D. O. B., Jordt, A., Kaeli, J. W., Köser, K., Kuhnz, L. A., Lindsay, D., Morris, K. J., Nattkemper, T. W., Osterloff, J., Ruhl, H. A., Singh, H., Tran, M., & Bett, B. J. (2016). Perspectives in visual imaging for marine biology and ecology: From acquisition to understanding. *Oceanography and Marine Biology: An Annual Review, 54*, 1–72.

Ferrari, R., & Wunsch, C. (2009). Ocean circulation kinetic energy: Reservoirs, sources, and sinks. *Annual Review of Fluid Mechanics, 41*, 253–282.

Franks, P. J. S. (2001). Turbulence avoidance: An alternate explanation of turbulence-enhanced ingestion rates in the field. *Limnology and Oceanography, 4*, 959–963.

Franks, P. J. S., Inman, B. G., MacKinnon, J. A., Alford, M. H., & Waterhouse, A. F. (2022). Oceanic turbulence from a planktonic perspective. *Limnology and Oceanography, 67*, 348–363.

Furuichi, N., Higashi, H., & Furushima, Y. (2016). Toward constructing a dynamic model of suspended particles in the Deep Sea – LES experiment of the current/seabed boundary layer. *Aqua, 38*(2), 145–150. (in Japanese).

Furushima, Y., & Yamamoto, H. (2015). Periodic behavior of deep sea current in the Hatoma knoll hydrothermal system. In J.-I. Ishibashi et al. (Eds.), *Subseafloor biosphere linked to hydrothermal systems: TAIGA concept* (pp. 625–637). Springer Japan.

Furushima, Y., Higashi, H., Fukuhara, T., Furuichi, N., & Yamamoto, H. (2017). Direct measurement of near-bottom turbulence in deep ocean for predicting behavior of suspended particles in Sagami Bay and the Okinawa trough, Japan. In *Proceedings OCEANS 2017 MTS/IEEE Anchorage conference exhibitions, electronic*. ISBN:978-0-6929-4690-9, Print on Demand (PoD) ISBN:978-1-5090-6429-8.

Furushima, Y., Fukuhara, T., Kondo, S., Takashima, S., Matsuda, T., Higashi, H., Furuichi, N., Yamamoto, H., & Fukushima, T. (2019a). Characteristics of deep-sea turbulent flow fluctuation near seafloor in Izena Cauldron, Okinawa Trough, Japan. In *Proceedings OCEANS 2019 MTS/IEEE Seattle conference exhibition, electronic*. ISBN:978-0-578-57618-3., Print on Demand(PoD) ISBN:978-1-7281-1452-1.

Furushima, Y., Higashi, H., Fukuhara, T., Matsuda, T., Furuichi, N., Yamamoto, H., & Fukushima, T. (2019b). Continuous observation of turbulent flow near the hydrothermal venting area in the Okinawa Trough, Japan. In *Proceedings 29th International Ocean and Polar Engineering Conference, Honolulu, ISOPE-I-19-54*, ISBN:978-1-880653-85-2.

Furushima, Y., Yamakita, T., Miwa, T., Lindsay, D., & Fukushima, & T., Shirayama, Y. (2019c). New techniques for standardization of environmental impact assessment. In R. Sharma (Ed.), *Environmental issues of Deep-Sea mining* (pp. 275–313). Springer Nature Switzerland.

Furushima, Y., Fukuhara, T., Kondoh, S., Takashima, S., Matsuda, T., Tanaka, M., Higashi, H., Furuichi, N., & Nagao, M. (2020). Relationship between time series fluctuation of turbulence intensity and surrounding environment near the deep-sea floor. In *Proceedings in the 28th ocean engineering symposium, OES28-A0059*.

Gastal, E., & Oliveira, M. (2011). Domain transform for edge-aware image and video processing. *ACM TOG (SIGGRAPH), 30*(4), 69:1–69:12.

Gilbert, O. M., & Buskey, E. J. (2005). Turbulence decreases the hydrodynamic predator sensing ability of the calanoid copepod Acartiatonsa. *Journal of Plankton Research, 27*, 1067–1071.

Gorsky, G., Aldorf, C., Kage, M., Picheral, M., Garcia, Y., & Favole, J. (1992). Vertical distribution of aggregates determined by a new underwater video profiler. *Annales de l'InstitutOceHanographique, 68*(1–2), 275–280.

Greer, A. T., Chiaverano, L. M., Treible, L. M., Briseño-Avena, C., & Hernandez, F. J. (2021). From spatial pattern to ecological process through imaging zooplankton interactions. *ICES Journal of Marine Science, 78*(8), 2664–2674. https://doi.org/10.1093/icesjms/fsab149

Gregg, M. C. (1999). Uncertainties and limitations in measuring ϵ and *T*. *Journal of Atmospheric and Oceanic Technology, 16*, 1483–1490.

Guerra, M., & Thomson, J. (2017). Turbulence measurements from five-beam acoustic Doppler current profilers. *Journal of Atmospheric and Oceanic Technology, 34*, 1267–1284.

Halfar, J., & Fujita, R. M. (2007). Danger of Deep-Sea mining. *Science, 316*, 987–987.

Hashimoto, E., & Takasugi, Y. (1998). Measurement of the vertical mixing strength in an Inner Bay by floating-MSP. *Coastal Engineering Journal, 45*, 966–970. (in Japanese).

Higashi, H., & Maki, H. (2012). Field observations on oxygen-poor water masses and vertical mixing strength in Tokyo Bay. *Journal of Japan Society of Civil Engineers Ser B2 (Coastal Engineering), 68*(2), I_966–I_970. (in Japanese).

Higashi, H., Furushima, Y., Furuichi, N., & Fukuhara, T. (2017). Performance of vertical mixing schemes based on in-situ measurement of deep-sea turbulence in Sagami Bay. *Journal of Japan Society of Civil Engineers Ser B2 (Coastal Engineering), 73*(2), I_79–I_84. (in Japanese).

ISA. (2013). *Recommendations for the guidance of contractors and sponsoring States relating to training programmes under plans of work for exploration* (ISBA/19/LTC/14). https://www.isa.org.jm/wp-content/uploads/2022/04/isba-19ltc-14_0.pdf. Accessed 19 Apr 2023.

ISA. (2020). *Recommendations for the guidance of contractors for the assessment of the possible environmental impacts arising from exploration for marine minerals in the Area* (ISBA/25/LTC/6/Rev.1). https://www.isa.org.jm/documents/isba-25-ltc-6-rev-1/.

Itoh, S. (2012). Studies on water mass variability related to dynamics of ecosystems in the Oyashio/Kuroshio region. *Oceanography in Japan, 21*(2), 33–50. (in Japanese).

Iwase, R., Mitsuzawa, K., Hirata, K., Kaiho, Y., Kawaguchi, K., Fuji, G., & Mikada, H. (2001). Renewal of "real-time deep seafloor observatory off Hatsushima Island in Sagami Bay"-toward the development of "next-generational" real-time deep seafloor observatory. *JAMSTEC Deep Sea Research, 18*, 185–192. (in Japanese with English abstract).

Iwata, S., & Matsuyama, M. (1989). Surface circulation in Sagami Bay: The response to variations of the Kuroshio Axis. *Journal of the Oceanographical Society of Japan, 45*, 310–320.

Katija, K., Troni, G., Daniels, J., Lance, K., Sherlock, R. E., Sherman, A. D., & Robison, B. H. (2020). Revealing enigmatic mucus structures in the deep sea using DeepPIV. *Nature, 583*, 78–82.

Katul, G., & Chu, C. R. (1998). A theoretical and experimental investigation of energy-containing scales in the dynamic sublayer of boundary-layer flows. *Boundary-Layer Meteorology, 86*, 279–312.

Kiørboe, T. (2008). *A mechanistic approach to plankton ecology*. Princeton University Press.

Koehl, M. R. A. (2007). Mini review: Hydrodynamics of larval settlement into fouling communities. *Biofouling, 23*, 357–368.

Lamprecht, A.-L., Garcia, L., Kuzak, M., Martinez, C., Arcila, R., Martin Del Pico, E., Dominguez Del Angel, V., van de Sandt, S., Ison, J., Martinez, P. A., McQuilton, P., Valencia, A., Harrow, J., Psomopoulos, F., Gelpi, J. L., Chue Hong, N., Goble, C., & Capella-Gutierrez, S. (2020). Towards FAIR principles for research software. *Data Science, 3*, 37–59. https://doi.org/10.3233/DS-190026

Lindsay, D. J., Yoshida, H., Ishibashi, S., Umetsu, M., Yamaguchi, A., Yamamoto, H., Nishikawa, J., Reimer, J. D., Watanabe, H., Fujikura, K., & Maruyama, T. (2013). The uROV PICASSO, the visual plankton recorder, and other attempts to image plankton. In *Proceedings of the international symposium on underwater technology 2013*. https://doi.org/10.1109/UT.2013.6519854

Lindsay, D. J., Yamaguchi, A., Grossmann, M. M., Nishikawa, J., Sabates, A., Fuentes, V., Hall, M., Sunahara, K., & Yamamoto, H. (2014). Vertical profiles of marine particulates: A step towards global scale comparisons using an autonomous visual plankton recorder. *Bulletin of the Plankton Society of Japan, 61*(1), 72–81.

Lorke, A. (2007). Boundary mixing in the thermocline of a large lake. *Journal of Geophysical Research, 112*, C09019.

Lueck, R. G., Wolk, F., & Yamazaki, H. (2002). Oceanic velocity microstructure measurements in the 20th century. *Journal of Oceanography, 58*, 153–174.

Maar, M., Visser, A., Nielsen, T., Stips, A., & Saito, H. (2006). Turbulence and feeding behaviour affect the vertical distributions of Oithonasimilis and Microsetellanorwegica. *Marine Ecology Progress Series, 313*, 157–172.

Matsui, T., Sugishima, H., Okamoto, N., & Igarashi, Y. (2018). Evaluation of turbidity and resedimentation through seafloor disturbance experiments for assessment of environmental impacts associated with exploitation of seafloor massive sulfides mining. In *Proceedings of the 28th international Ocean Polar engineering conference, Sapporo, ISOPE-I-18-719*, ISBN:978-1-880653-87-6.

Mero, J. L. (1965). Geochemistry and descriptions of manganese nodules and crusts retrieved from the open ocean. *Elsevier Oceanography Series, 1*, 312 pp.

Momma, H., Iwase, R., Mitsuzawa, K., Kaiho, Y., & Fujiwara., Y. (1998). Preliminary results of a three-year continuous observation by a deep seafloor observatory in Sagami Bay, Central Japan. *Physics of the Earth and Planetary Interiors, 108*, 263–274.

Munk, W. H. (1966). Abyssal recipes. *Deep Sea Research, 13*, 707–730.

Nagao, M., Hashimoto, E., & Takasugi, Y. (2004). Measurement of the vertical mixing strength in the Seto Inland Sea. *Coastal Engineering Journal, 51*, 946–950. (in Japanese).

Nagao, M., Hashimoto, E., Takasugi, Y., Chiba, S., & Yamagata, Y. (2005). Measurement of the vertical mixing strength in the Ago Bay. *Coastal Engineering Journal, 52*, 341–345. (in Japanese).

Narita, T., Oshika, J., Okamoto, N., Toyohara, T., & Miwa, T. (2015). Summary of environmental impact assessment for mining seafloor massive sulfides in Japan. *Journal of Shipping and Ocean Engineering, 5*, 103–114.

Nikora, V. (1999). Origin of the '-1' spectral law in wall-bounded turbulence. *Physical Review Letters, 83*, 734–736.

Ohman, M. D., Davis, R. E., Sherman, J. T., Grindley, K. R., Whitmore, B. M., Nickels, C. F., & Ellen, J. S. (2018). Zooglider: An autonomous vehicle for optical and acoustic sensing of zooplankton. *Limnology and Oceanography: Methods, 17*, 69–86.

Okamoto, N., Shiokawa, S., Kawano, S., Yamaji, N., Sakurai, H., & Kurihara, M. (2019). World's first lifting test for seafloor massive sulfides in the Okinawa Trough in the EEZ of Japan. In *Proceedings of the 29th International Ocean and Polar Engineering Conference, Honolulu, ISOPE-I-19-655*, ISBN:978-1-880653-85-2.

Osborn, T. R. (1980). Estimates of the local rate of vertical diffusion from dissipation measurements. *Journal of Physical Oceanography, 10*(1), 83–89.

Pécseli, H. L., Trulsen, J. K., Stiansen, J. E., Sundby, S., & Fossum, P. (2019). Feeing of plankton in turbulent oceans and lakes. *Limnology and Oceanography, 64*, 1034–1046.
Rothschild, B. J., & Osborn, T. R. (1988). Small-scale turbulence and plankton contact rates. *Journal of Plankton Research, 10*, 465–474.
Saita, T., Yano, S., Tai, A., Shiki, S., Shigeta, S., & Komatsu, T. (2008). Field observation of vertical mixing strength in Ariake Sea during winter. *Journal of Coastal Engineering, 55*, 421–425. (in Japanese).
Sangekar, M., Friedman, A., Hidaka, M., Hosono, T., & Lindsay, D. (2023). SquidJam: A video annotation ecosystem. In *2023 IEEE Underwater Technology (UT23)* (pp. 1–5). https://doi.org/10.1109/UT49729.2023.10103435
Sato, M., Klymak, J. K., Kunze, E., Dewey, R., & Dower, J. F. (2014). Turbulence and internal waves in Patricia Bay, Saanich Inlet, British Columbia. *Continental Shelf Research, 85*, 153–167.
Schoening, T., Durden, J. M., Faber, C., Felden, J., Heger, K., Hoving, H.-J. T., Kiko, R., Köser, K., Krämmer, C., Kwasnitschka, T., Möller, K. O., Nakath, D., Naß, A., Nattkemper, T. W., Purser, A., & Zurowietz, M. (2022). Making marine image data FAIR. *Scientific Data, 9*, 414. https://doi.org/10.1038/s41597-022-01491-3
Schroeder, S.-M., Kiko, R., & Koch, R. (2020). MorphoCluster: Efficient annotation of plankton images by clustering. *Sensors, 20*, 3060.
Shang, X., Qi, Y., Chen, G., Liang, C., Lueck, R. G., Prairie, B., & Li, H. (2017). An expendable microstructure profiler for deep ocean measurements. *Journal of Atmospheric and Oceanic Technology, 34*, 153–165.
Sharma, R. (2011). Deep-Sea Mining: Economic, Technical, Technological, and Environmental Considerations for Sustainable Development. *Marine Technology Society Journal, 45*(5), 28–41.
Smyth, W. D., & Moum, J. N. (2019). 3D turbulence. In J. K. Cochran, H. J. Bokuniewicz, & P. L. Yager (Eds.), *Encyclopedia of ocean sciences* (3rd ed., pp. 486–496). Elsevier.
Tanaka, M. (2019). Changes in vertical distribution of zooplankton under wind-induced turbulence: A 36-year record. *Fluids, 4*, 195.
Tanaka, M., Genin, A., Endo, Y., Ivey, G. N., & Yamazaki, H. (2021a). The potential role of turbulence in modulating the migration of demersal zooplankton. *Limnology and Oceanography, 66*, 855–864.
Tanaka, M., Furushima, Y., & Nagao, M. (2021b). Turbulence observation near the deep-sea bed as environmental impact assessment: Direct measurements and indirect estimates of turbulent kinetic energy dissipation rate. In *ISOPE International Ocean and polar engineering conference* (p. ISOPE-I). ISOPE.
Tennekes, H., & Lumley, J. L. (1972). *A first course in turbulence*. MIT Press.
Thorpe, S. A. (2005). *The turbulent ocean*. Cambridge University Press.
Thorpe, S. A. (2007). *An introduction to ocean turbulence* (Vol. 10). Cambridge University Press.
van Dover, C. (2011). Tighten regulations on deep-sea mining. *Nature, 470*, 31–33.
Visser, A. W., Saito, H., Saiz, E., & Kiørboe, T. (2001). Observations of copepod feeding and vertical distribution under natural turbulent conditions in the North Sea. *Marine Biology, 138*, 1011–1019.
Walter, R. K., Squibb, M. E., Woodson, C. B., Koseff, J. R., & Monismith, S. G. (2014). Stratified turbulence in the nearshore coastal ocean: Dynamics and evolution in the presence of internal bores. *Journal of Geophysical Research, Oceans, 119*, 8709–8730.
Washburn, T. W., Simon-Liedó, E., Soong, G. Y., & Suzuki, A. (2023). Seamount mining test provides evidence of ecological impacts beyond deposition. *Current Biology, 33*, 3065–3071.
Wiles, P. J., Rippeth, T. P., Simpson, J. H., & Hendricks, P. J. (2006). A novel technique for measuring the rate of turbulent dissipation in the marine environment. *Geophysical Research Letters, 33*, L21608.
Wishner, K. F. (1980). The biomass of the deep-sea benthopelagic plankton. *Deep-Sea Research Part I: Oceanographic Research Papers, 27*, 203–216.
Wolk, F., Yamazaki, H., Seuront, L., & Lueck, R. G. (2002). A new freefall profiler for measuring biophysical microstructure. *Journal of Atmospheric and Oceanic Technology, 19*, 780–793.

Yamaji, N., Okamoto, N., Shiokawa, S., Kawano, S., & Sakurai, H. (2019). Achievement for pilot test of excavating and ore lifting conducted for seafloor polymetallic sulfides (in Japanese with English abstract). *The Journal of MMAJ, 135*(6), 42–51.

Yamaoka, K., Suzuki, A., Tsukasaki, A., Suzumura, M., Zhu, Y., Ariga, T., Fukuhara, T., Kondo, S., & Minatoya, J. (2023). Water column profile around Takuyo-Daigo seamount and monitoring of seawater chemistry in the excavating test of cobalt-rich ferromanganese crusts. In *Proceedings of the Japan geoscience union meeting 2023, Chiba, Japan*. (in Japanese).

Yamazaki, H., Osborn, T. R., & Squires, K. D. (1991). Direct numerical simulation of planktonic contact in turbulent flow. *Journal of Plankton Research, 13*, 629–643.

Dr. Yasuo Furushima is studying as Deputy Group Leader (Research Technician/Scientist) at the Research Institute for Global Change (RIGC), Marine Biodiversity and Environmental Assessment Research Center (BioEnv), Marine Environmental Impact Assessment Research Group (MEIA), Japan Agency for Marine-Earth Science and Technology (JAMSTEC). He graduated from the School of Marine Science and Technology, Tokai University, in 1988 (a bachelor's degree) and the Graduate School of Oceanography, Tokai University, in 1990 (a master's degree). In 1996, he got a doctorate in agriculture from the Graduate School of Agricultural and Life Sciences, the University of Tokyo. He was employed as research scientist in JAMSTEC in 1997. His specialty is fisheries oceanology and physical oceanography. He is studying dynamics of bottom mixed layer using an expendable vertical microstructure profiler.

Dr. Mamoru Tanaka is an assistant professor at the Department of Civil Engineering, Tokyo University of Science. He obtained his PhD in flow-animal interactions in the ocean, and his work focuses on turbulence and its impact on the marine/freshwater ecosystem. After serving as a postdoctoral researcher at the University of Tokyo and the National Institute of Advanced Industrial Science and Technology, Dr. Tanaka was awarded the best paper award from the Oceanographic Society of Japan. He also started studying plastic pollution in freshwater environments.

Mehul Naresh Sangekar Marine Engineer, Japan Agency for Marine-Earth Science & Technology (JAMSTEC)

Mehul Naresh Sangekar (b. 1983) is an Ocean Technology Engineer with the Japan Agency for Marine-Earth Science & Technology (JAMSTEC) and is a visiting researcher with the Institute of Industrial Science, University of Tokyo. Dr. Sangekar's research focuses on the development of new survey tools and data analysis pipelines for investigating marine habitats and ecosystems. During his doctoral research, he developed hardware and software frameworks for the seafloor landing of an autonomous underwater vehicle (AUV). With the University of Tokyo, he has been working on tools for manganese crust exploration, including mapping and crust thickness estimation. At JAMSTEC, he has been spearheading the development of shadowgraph imaging survey hardware and software, as well as working on traditional camera image analysis tools. Dr. Sangekar has participated in over 15 research cruises, as the leading researcher and as a member of multidisciplinary teams, aboard various Japanese research vessels as well as Singaporean and Saudi Arabian ships.

Dhugal John Lindsay Marine Biologist, senior research scientist with the Japan Agency for Marine-Earth Science & Technology (JAMSTEC)

Dhugal John Lindsay (b. 1971) is a senior research scientist with the Japan Agency for Marine-Earth Science & Technology (JAMSTEC) and holds adjunct professorships at Yokohama Municipal University and Kitasato University. Dr. Lindsay's research focuses on mid-water ecology, particularly concentrating on gelatinous organisms, and on the development of new survey tools for investigating the midwater biota. He is currently involved with the Deep Ocean Stewardship Initiative (DOSI) by serving as the Head of the New Technologies Working Group. Dr. Lindsay served as a Steering Committee Member (CMarZ) and as a National Regional Implementation Committee Member (Japan) for the Census of Marine Life and has also served on the National Academies of Science Committee "Future Needs in Deep Submergence Science." In addition to his role as vice editor-in-chief for the scientific journal *Plankton and Benthos Research*, he also serves on the Editorial Board of *Scientia Marina*. Dr. Lindsay has participated in over 60 research cruises, both as Chief Scientist and as a member of multidisciplinary teams, aboard various Japanese research vessels, cruises on Australian, U.S., and Spanish vessels from the Arctic to the Antarctic, and has dived 21 times in crewed submersibles. His publication record includes over 190 scientific papers.

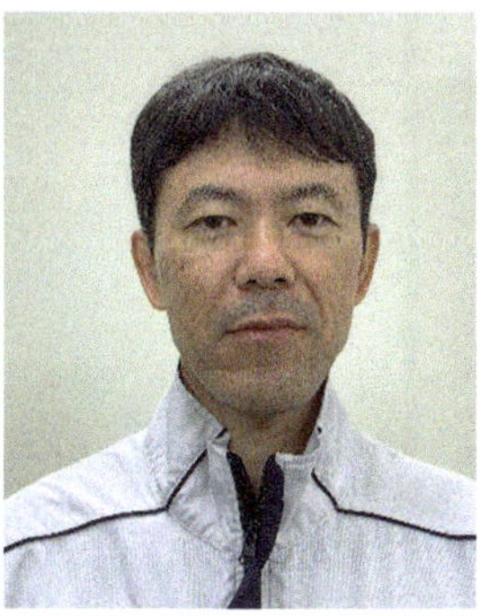

Tatsuo Fukuhara works for KANSO TECHNOS Co., Ltd, Tokyo Branch, as an environmental consultant, mainly in the field of seafloor mineral resource development. He received his M.S. in geoscience from Shizuoka University in 1997, and his specialty is geochemistry. He is a professional engineer (Applied Science). For 2 years from 2015, he was seconded to JAMSTEC, where he was involved in the establishment of a method for predicting changes in environmental impacts based on marine ecosystem observations and the international standardization of evaluation criteria. He is engaged in environmental impact assessment and oceanographic field surveys every year.

Dr. Masayuki Nagao is studying as a senior researcher in the Geological Survey of Japan (GSJ), National Institute of Advanced Industrial Science and Technology (AIST). He graduated from the Faculty of Engineering, Tohoku University in 1991 (Bachelor of Engineering) and from the Graduate School of Engineering, Tohoku University in 1993 (Master of Engineering). In 1999, he got a Doctor of Philosophy (Engineering) from the Graduate School of Engineering, Tohoku University. He was employed as a research associate in Tokyo Institute of Technology in 1995, and as a research scientist in AIST in 1999. His specialty is environmental hydraulics and physical oceanography. He is studying the physical environmental impact assessment of deep-sea mining using marine instrumentation.

Chapter 7
Turbidity at the Source: Aiming for Minimized Sediment Dispersion During Deep-Sea Mining

Rudy Helmons, Said Alhaddad, Claire Chassagne, Mohamed Elerian, Geert Keetels, Alex Kirichek, and Laurenz Thomsen

Abstract The chapter gives an overview of the sediment dispersion generated by the mining process. Within the field of dredging engineering, ample experience is available regarding equipment, turbidity generated by equipment, and sediment transport processes. High up the environmental impact mitigation hierarchy are avoidance and minimization. That is where engineering can provide (part of) the solution. It is our aim to predict and consider how we can improve the mining process and equipment. Within this context, our focus is on those processes that are likely to take place close to the seabed. On the one hand, our work focuses on the prediction and reduction of the amount of sediment that might get suspended. On the other hand, considering the conditions under which the suspended sediment might be released in the most optimal way to reduce dispersion, we have performed and analysed small-scale and full-scale laboratory experiments of a hydraulic collector design and various dynamic sedimentation experiments.

Keywords Particle aggregation · Polymetalic nodules · Sediment plume · Sediment pickup & deposition · Settling velocity

---

R. Helmons (✉)
Maritime and Transport Technology, Faculty of Mechanical, Maritime and Materials Engineering, Delft University of Technology, Delft, the Netherlands

Mineral Processing and HSE, Norwegian University of Science and Technology, Trondheim, Norway
e-mail: r.l.j.helmons@tudelft.nl

S. Alhaddad · M. Elerian · G. Keetels
Maritime and Transport Technology, Faculty of Mechanical, Maritime and Materials Engineering, Delft University of Technology, Delft, the Netherlands

C. Chassagne · A. Kirichek
Environmental Fluid Mechanics, Faculty of Civil Engineering and Geoscience, Delft University of Technology, Delft, the Netherlands

L. Thomsen
University of Gotheborg, Gothenburg, Sweden

R. Sharma (ed.), *Deep-Sea Mining and the Water Column*,
https://doi.org/10.1007/978-3-031-59060-3_7

1 Introduction

Polymetallic nodules have attracted interest from industry and scientists for several decades. As of the 1970s and 1980s of the last century, significant attempts have been made to collect nodules in an economic way. Nodules have attracted renewed interest over the last decade. Nowadays, the concern about the potential environmental impact of seabed mining has gained more attention. Industry will be challenged to work according to the best available technology and best environmental practices to minimize and mitigate the environmental impacts. It is not yet clear which technologies and methodologies will lead to the least environmental impact while still considering an economically viable mining operation.

Various solutions and technologies for exploitation are being developed. In general, any deep-sea mining system will consist of similar subsystems, i.e.:

1. Seafloor Mining Tool (SMT): Nodules need to be picked from the seabed.
2. Vertical Transport System (VTS): Nodules need to be transported to surface and excess water, sediment, and other effluents need to be transported subsea.
3. Mining Support Vessel (MSV): Nodules need to be separated from non-valuable materials. The nodules will be stored in the hold, while the non-valuable materials should be provided to the VTS to return back subsea.

In general, the material flow from the seabed to the mining vessel can be divided into four stages (Lang et al., 2019), see Fig. 7.1:

1. Seabed to seafloor mining tool
2. Seafloor mining tool to environment
3. Seafloor mining tool to mining vessel via a vertical transport system
4. Mining vessel to environment (seabed and/or water column) via a vertical transport system

Several processes occur simultaneously, influencing material flows 1 and 2. For most seafloor mining tools, it is expected that they will generate turbidity due to the machine manoeuvring on top of the seabed, e.g., sediments are suspended by the motion of the vehicle tracks. At the same time, as the SMT collects nodules, it will entrain water and sediment. Part of the entrained water and sediment will move through the collector and part of it will spill directly to the environment. Within the SMT, nodules will be separated from the excess water and sediment. The excess materials will then be expelled behind the collector vehicle, often referred to as collector or diffuser plume, see Fig. 7.2.

Various technology concepts exist, varying in their levels of maturity (Lesage, 2020). One of the tested technologies is that of The Metals Company and Allseas, as they successfully conducted an integrated field test in October 2022. Their system consists of a hydraulic (Coanda-effect-based) nodule collector, a riser-based airlift system, and the excess water and sediments on board of the MSV are discharged into the midwater column Nauru Ocean Resources Inc (2021). Similar concepts are developed by others, e.g. GSR (Ouillon et al., 2021), Royal IHC (Haalboom

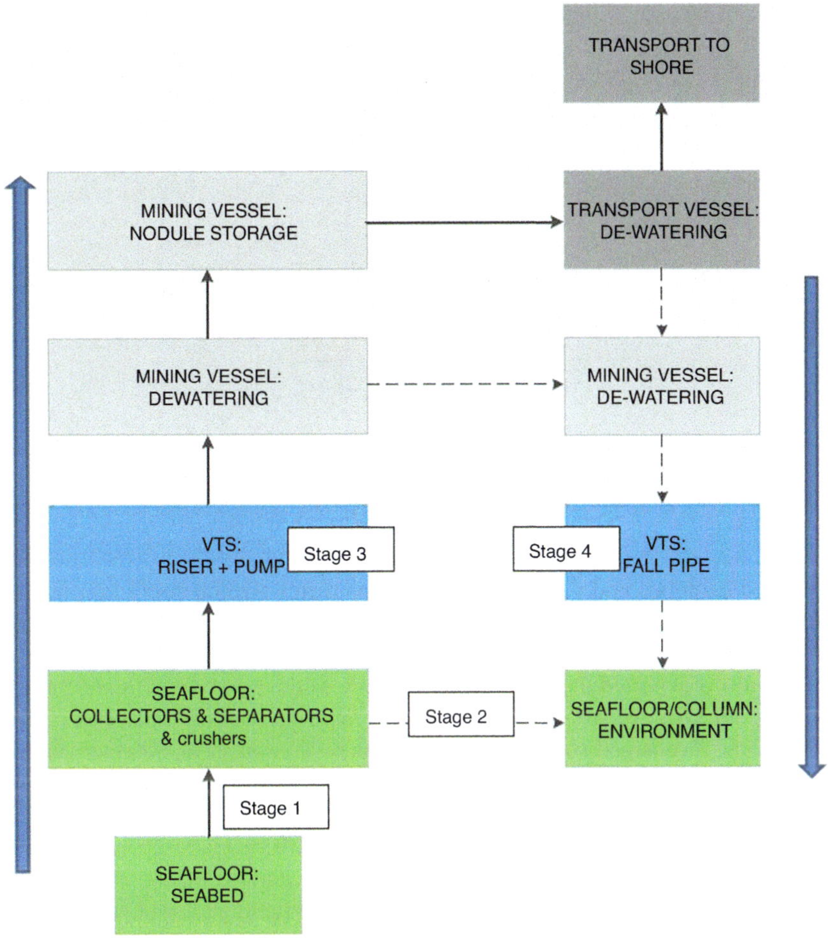

Fig. 7.1 Schematic of material flows. Solid lines represent the main flow (nodules) and dashed lines represent secondary flow (sediments, waste, and other effluants, SWOE). Blue arrows show the general direction of the flow. (Based on Blue Nodules Lang et al., 2019)

et al., 2023), based on other hydraulic concepts, e.g. KIOST (Hong et al., 2019), COMRA (Kang & Liu, 2021), or mechanical collection, e.g. NIOT (Atmanand and Ramadass, 2017). All such concepts can be grouped as continuous SMTs; that is, they are designed to effectively collect all nodules in the tool's pathway. There are other concepts under development that consider picking up individual nodules (Gillham, 2022), but such concepts are not yet demonstrated at a sufficiently high technology readiness level to be considered as available technology at the required economy of scale. Regarding VTS, most concepts are based on continuous transport systems, e.g., hydraulic (van Wijk, 2016) or airlift (NORI-D EIA, 2021), or discontinuous systems based on skip lifting of containers (van den Bosch, 2023).

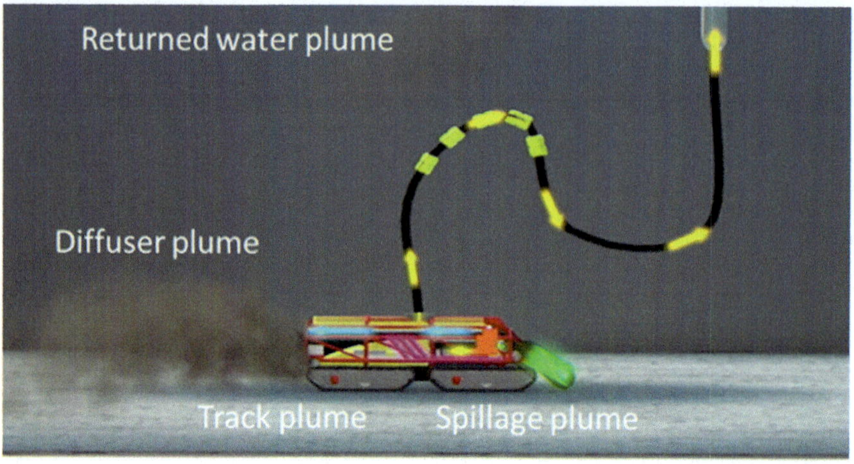

Fig. 7.2 Main source terms of turbidity generated near the bed. (Based on Blue Nodules Lang et al., 2019)

The continuous and discontinuous systems differ significantly in terms of the amount of water and potentially sediments that will be mobilized, transported to the surface, or separated to be released into the environment. For example, in hydraulic transport it is expected that approximately 12 % of the transported mixture consists of ore (van Wijk, 2016). This means that a large volume of water and sediment needs to be discharged subsea. So far, there is no consensus on whether the return flow needs to be discharged close to the seabed or whether it would be acceptable to be discharged midwater column. Some hydraulic transport concepts inherently require the return water flow to provide flow for drill-string-based systems, inherently discharging the excess water near the seabed (van den Bosch, 2023).

In this chapter, the emphasis will be on specific systems and/or physical phenomena relevant for the material flows identified in categories 1, 2, and 4 above. The experimental results that we will present are based on a variation of non-cohesive fine sediments and substitute cohesive sediments to unravel physical phenomena, allow for sufficient repetitions, and allow us to work with the limited availability of sediments collected from CCZ area. Several experiments have been executed with original CCZ sediment as well. In Sect. 2, a Coandă-effect-based nodule collection system is analysed for the physical phenomena occurring underneath the collector head. We will shed light on the pressure distribution and provide insights into the mechanisms that influence erosion of the bed. Various lab-scale tests have been conducted to investigate the near-bed release of the SWOE; these are discussed in Sect. 3. The next section discusses a combination of both simulations and experiments to further investigate the release conditions of a collector plume, in which the emphasis is on studying the effect that particle aggregation might have on the dispersion of the sediment plume, based on hydrodynamic flows such as lock-exchange as well as closed volume experiments in a sea-water column simulator.

2 Coandă-Effect-Based Nodule Collection System

Various hydraulic collection techniques were explored in the literature, including the suck-up-based approach (Zhao et al., 2019), double-jet method (Hong et al., 1999), and Coandă-effect-based method (Alhaddad et al., 2023). It is anticipated that the Coandă-effect-based collectors will cause less disturbance to sediments, as the flow from the nozzle of the collector does not directly interact with the seabed (Zhao et al., 2021). Consequently, this technology is being widely considered in the field of polymetallic-nodule mining. This section, therefore, provides an overview of the underlying physics of the collection mechanism using a Coandă-effect-based collector. Furthermore, directions towards minimizing the sediment collection by the collector are presented.

The Coandă effect, named after the Romanian scientist Henri Coandă, can be described as the tendency of a jet flow to remain attached to a nearby surface, even when this surface curves (Reba, 1966). A diagram illustrating the basic design of a Coandă-effect-based collector is depicted in Fig. 7.3. The design comprises a couple of ducts, a jet duct and a collection duct, which are formed by three curved surfaces. Due to the Coandă effect, the fast-moving water jet follows the contour of the upper curved plate. Consequently, surrounding water is entrained towards the collection duct, creating a pressure difference (suction) under the collector, kicking up nodules from the seabed and dragging them towards the collection duct. It should be noted that not only does the water jet entrain ambient water into the collection duct, but it also injects water behind the collector head, which is referred to as 'spilling water' (see Fig. 7.3).

In fact, polymetallic nodules are partially or completely embedded in the sediment of the ocean floor, meaning that sediments will unavoidably be dislodged and picked up along with the nodules during mining. Consequently, the sediments are discarded behind the hydraulic collectors, forming a sediment plume that can travel long distances and potentially disrupt aquatic ecosystems. In order to evaluate the

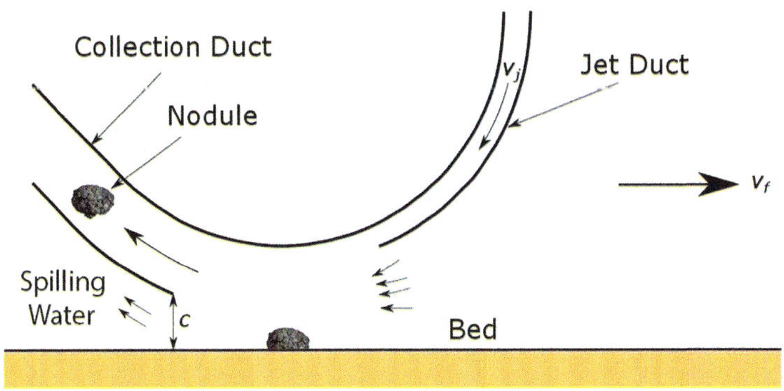

Fig. 7.3 Diagrammatic illustration of the collector head, where v_j represents the jet velocity, v_f represents the forward velocity of the collector, and c represents the clearance. The smaller arrows indicate the direction of water entrainment

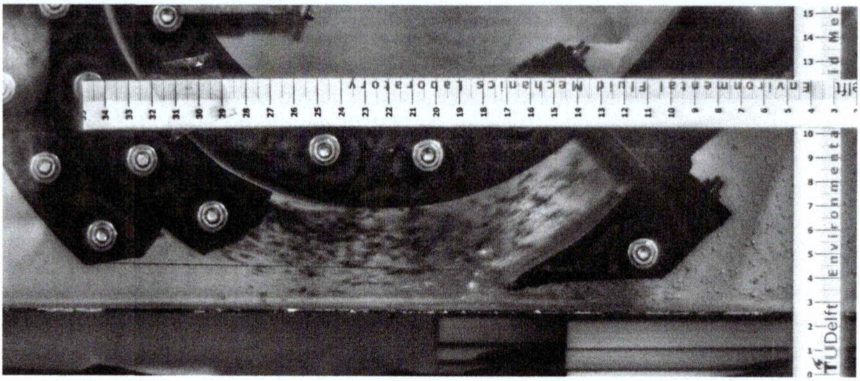

Fig. 7.4 Image obtained by a high-speed camera illustrating the flow direction including spilling water (highlighted by the presence of air bubbles) when the collector was stationary ($v_f = 0$). (Alhaddad & Helmons, 2023)

environmental impact of deep-sea mining activities, it is crucial to predict the evolution and fate of these plumes. This requires an understanding of the characteristics of the sediment-water mixture discharged from the collector. One key parameter of this discharge is the sediment concentration, which is influenced by the depth of the sediment layer picked up by the collector during operation. Therefore, gaining a thorough understanding of the interaction between Coandă-effect-based collectors and the ocean floor is essential to develop a collector with minimal environmental impact. To this end, Alhaddad and Helmons (2023) conducted a series of small-scale experiments to explore sand erosion caused by a Coandă-effect-based collector driving over a subaqueous sand bed in a water flume. Various operational conditions were explored, resulting in a set of quantitative data on sediment erosion associated with this type of collector. Their study showed that the Coandă-effect-based collector does not erode sediments primarily through flow-induced shear stress. Instead, the collector's water jet strikes the sediment bed at an inclined angle, leading to the collapse of the sediment layer. The majority of the collapsing sediment is transported with the deflecting flow into the collection duct, while the remaining portion is transported behind the collector head by the spilling water (see Fig. 7.4). The latter portion generates a sediment plume behind the collector, which brings about an additional environmental pressure besides the plume formed at the collector rear.

2.1 Operational Parameters

The key parameters controlling the pick-up efficiency of the collector in question are jet velocity v_j, collector's forward velocity v_f, and bottom clearance c (see Fig. 7.3). In the following, we will discuss the effect of these parameters on nodule pick-up efficiency and sediment erosion depth. The former is defined as the ratio between the mass of collected nodules and the mass of nodules that are available for collection, while the latter is the thickness of the sediment layer collected by the collector.

2.1.1 Jet Velocity

Results of full-scale tests conducted by Alhaddad et al. (2023) showed that higher jet velocities result in higher pick-up efficiencies. This is attributed to the fact that higher jet velocities lead to larger pressure gradients under the collector (see Fig. 7.5 left), consequently dislodging more nodules from the sediment bed. Additionally, a higher jet velocity results in more water entrainment, which enhances the pressure difference under the collector, increasing the pick-up efficiency.

On the other hand, when jet velocities are higher, they cause greater erosion depths due to the increase in the impinging force of the flow onto the sediment bed. Experimental results of Alhaddad and Helmons (2023) showed a strong logarithmic correlation between the erosion depth and the impinging force exerted by the flow onto the sediment bed (see Fig. 7.6).

2.1.2 Collector's Forward Velocity

Alhaddad et al. (2023) tested only two forward velocities, where the higher forward velocity led to a lower pick-up efficiency despite the higher pressure gradient under the collector (see Fig. 7.5, middle). This is because a lower forward velocity allows more time for the nodules to respond to the pressure gradient and be picked up. This implies that a higher pressure gradient under the collector does not always lead to a higher pick-up efficiency. In other words, the collection process is highly dependent on the available time for the nodules to respond to the pressure gradient; if this time is insufficient, even an adequate pressure gradient will fail to pick up the nodules.

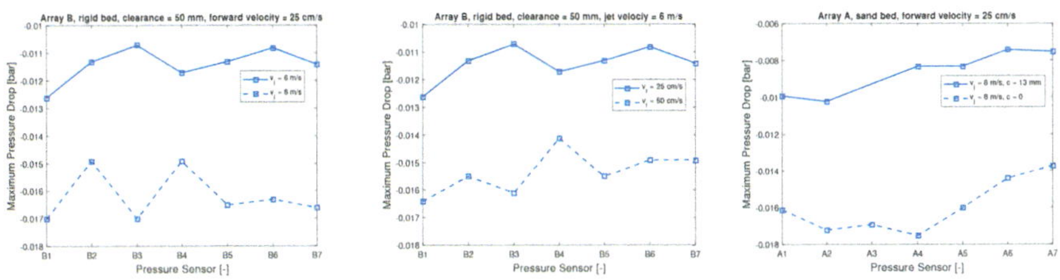

Fig. 7.5 Experimental results illustrating the effect of the jet velocity v_j (left), the forward velocity v_f (middle), and clearance c (right) on the maximum pressure drop under a Coandă-effect-based collector (Alhaddad et al., 2023). Pressure sensors B1-B7 and A1-A7 are spaced by a distance of 10 cm

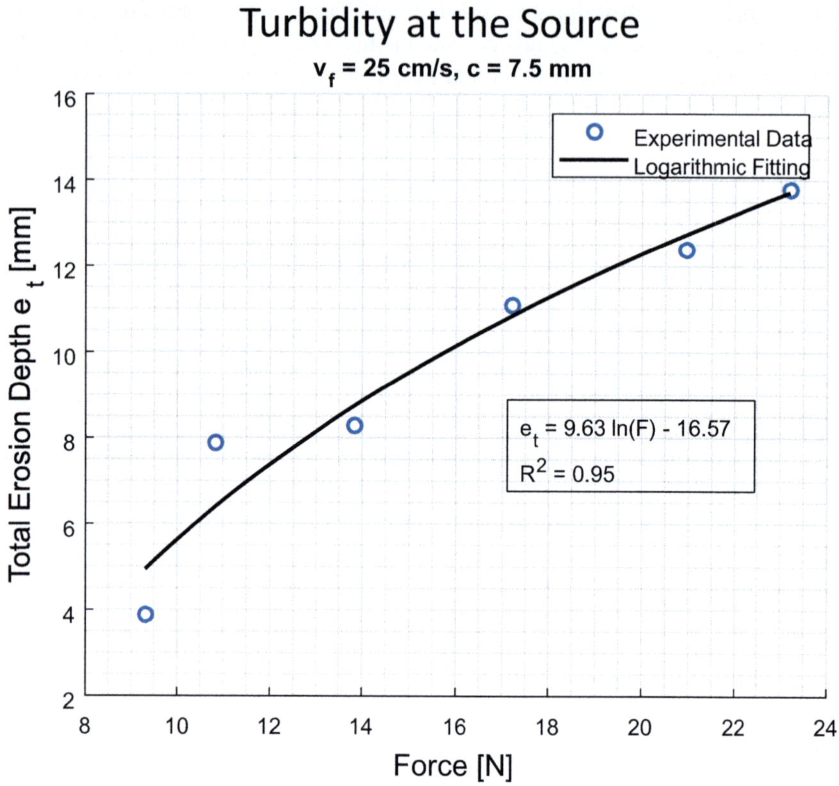

Fig. 7.6 Experimental results demonstrating a logarithmic correlation between the flow impinging force *F* and total erosion depth of sand e_t (Alhaddad & Helmons, 2023)

The sediment erosion process is significantly affected by the forward velocity, v_f, of the collector. Unsurprisingly, the experimental results of Alhaddad and Helmons (2023) clearly show that a lower v_f results in a greater erosion depth. This correlation is expected because a lower v_f exposes the sediment bed to the water flow for a longer period, resulting in the erosion of a thicker layer of sediment.

2.1.3 Bottom Clearance

The clearance under the rear cowl of the collection duct is another parameter that significantly affects the collection process. A smaller clearance results in a higher pick-up efficiency, reaching 100%, because it generates a greater suction force (see Fig. 7.5, right). Additionally, a smaller clearance implies that the centre of gravity of the nodules is closer to the underside of the collector, further increasing the pick-up efficiency.

In relation to sediment erosion, reducing the bottom clearance leads to a greater erosion depth, which is a logical correlation; when the underside of the collector is closer to the sand bed, a larger sand layer is exposed to the water flow, resulting in an increased erosion depth. To avoid the impingement of the flow onto the sediment bed, the bottom clearance should be increased above a certain threshold, which depends on the jet velocity and the forward velocity. However, this might result in a very low pick-up efficiency or even no nodule pick-up at all.

2.2 *Discussion*

In addition to pick-up efficiency, nodule collectors are also evaluated based on their environmental impact, which is a crucial performance metric. In the light of the findings described above, one can infer that if adjusting a certain operational parameter is favourable for the pick-up efficiency, it is mostly unfavourable from an environmental point of view (i.e. sediment erosion increases). This implies that the operational conditions of the collector need to be optimised to achieve the optimal pick-up efficiency and least environmental impact.

To gain further insights into the sediment erosion generated by the collector in question, a series of laboratory experiments, in which the collector drives along a cohesive sediment bed will be executed soon at Delft University of Technology.

3 Near Bed Release of Return Water Flow

The collection and separation stages of deep-sea mining will likely result in a large amount of process water that contains various types of fine sediments. Transporting this water to the surface for cleaning by filtration or hydrocyclones would presumably result in an unfeasible business case. For this reason, it is important to develop efficient strategies to return this process water in a controlled way that minimizes the impact both on the water column and the seabed. The key question is at which location in the water column should one release this water? High in the water column, to shorten the vertical transport distance of the return flow, would be economically attractive. On the other hand, this would affect a large volume of ambient water and could easily result in dispersion way beyond the designated mining area due to barotropic and baroclinic effects. For that reason it is also important to consider the release of process water in the proximity of the seabed, where the sediment was collected in the first place, and some degree of disturbance is unavoidable anyway by the act of mining vehicles for collection or excavation.

The challenge is then to let the sediment settle close to the impingement point and prevent the formation of turbidity currents that possibly extend outside the designated mining area. For this reason, it is important to understand the dynamics of sediment-laden flows close to the impingement point. Fig. 7.7 gives an overview of

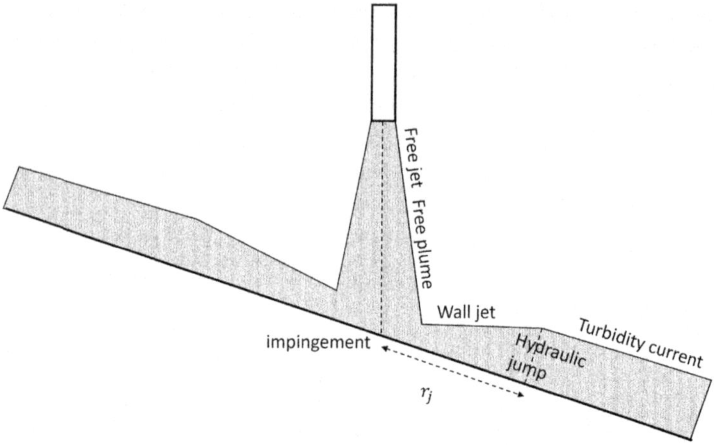

Fig. 7.7 Three different regions of mining return flow: free jet/plume region, impinging region, and turbidity current region

the different regions. Initially, the flow is controlled by the momentum flux from the nozzle; this is called the free jet region. At some point, the flow dynamics can be characterized as a free plume, and the flow can further accelerate by (negative) buoyancy associated with the sediment load. After impinging the plate the flow turns in a parallel direction relative to the seabed. The flow in this region can be described as wall-jet like (Chowdhury & Testik, 2015). After the formation of an internal hydraulic jump, a gravity (turbidity/density) current develops.

This section presents an experimental exploration of this problem and identification of scientific challenges in the prediction of the underlying near-bed sediment transport processes.

3.1 Experimental Exploration

Figure 7.8 shows the experimental set-up to explore the effect of impinging return flows Grunsven et al. (2018). The set-up consists of a large tank of 5 × 2 × 2 m and a 1 × 1 × 1 m mixing tank to prepare the jet mixture. A table is positioned inside the large flume chamber. The tabletop is mounted at 0.5 m from the flume floor. The advantage of this approach is that it limits side effects from the flume side walls as the density current tends to drop from the table and fill the volume below the tabletop. A second table with an inclined top (30°) was also available in order to study the effect of an inclined sea-bed slop on the development of the turbidity current. Light panels from the sides and mounted inside the table top helped to visualize the plume and turbidity currents. Concentrations and velocities are measured using ADV, and at several locations at the table top samples are taken to measure the local concentration. The stand-off distance H, concentrations, and discharge rate are

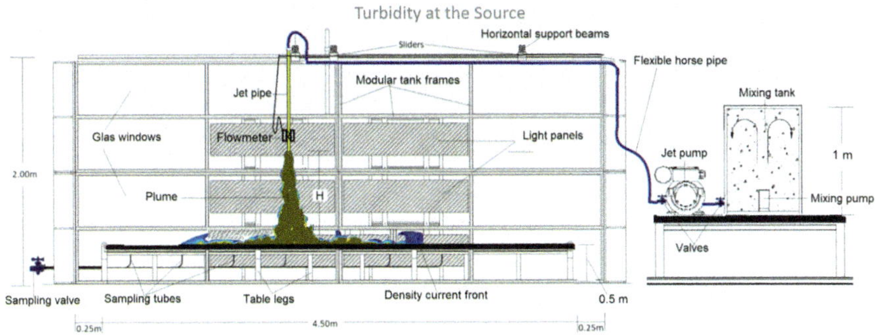

Fig. 7.8 Experimental set-up of the dredging engineering group of Delft University of Technology

Fig. 7.9 Impression of impinging return flow on inclined floor

variable. The sediment type was Millisil silica flour (or quartz powder) with a $D_{10} = 15\mu m$, $d_{50} = 36\mu m$ and $D_{90} = 70\mu m$. Figure 7.9 provides an impression of a typical impingement on an inclined floor. In the experiments, sediment deposition rings form around the impingement point and exhibit complex dynamical behaviour (see Fig. 7.10). Based on the diameter and height of these rings in the order of 10–20 cm and 1–2 cm, respectively at the end of the experiments (10 min) the total amount of sediment that is deposited during the experiments is found to be significant, i.e. 50–90% of the sediment flux from the nozzle. It is very interesting to note that even in this simple case of a smooth wall, such complex morphological patterns emerge.

To obtain some understanding of this formation process, it is helpful to analyse the vertical profiles of the radial velocity at various distances from the impingement points (see Fig. 7.11). Close to the impingement point, the velocity profile shows a sharp maximum and turns into a more gradual profile when moving to a location further downstream of the impingement point. Also the turbulent kinetic energy profile collapses, which indicates that it becomes more difficult to keep sediment

Fig. 7.10 Formation of a sediment deposition ring around the impingement point

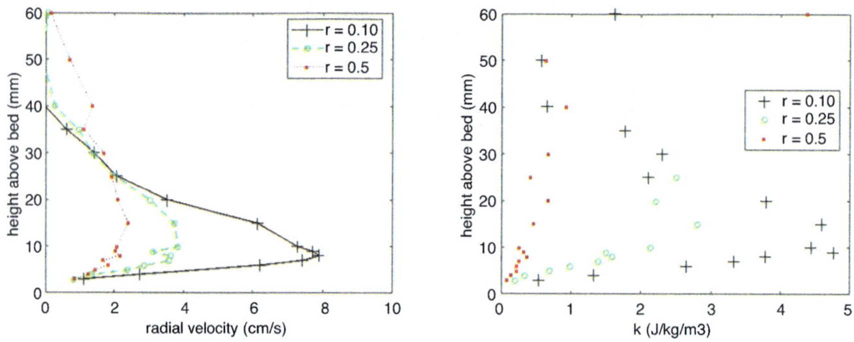

Fig. 7.11 Velocity and turbulent kinetic energy density versus height above the bed at three different radial distances r from the impingement point for the case with a stand-off distance of $H = 50$ cm, nozzle diameter 0.01 cm, and suspended sediment concentration of 20 g/l impinging on a horizontal table. Deposition rings were visible between $v_j r = 0.075$ m and $r = 0.15$ m in this experiment

suspended further downstream. Figure 7.12 shows the measured concentrations obtained with the draining systems that were mounted on the tabletop. It is seen that the sediment concentration quickly drops outside the range where the deposition rings were formed. From the velocity profiles in Fig. 7.11 at a location in this range, $r = 0.1$, and beyond this range, $r = 0.25$, it can be deduced that dilution by entrainment with ambient water is limited. The concentration mainly drops due to sediment deposition in the ring area. For $r > 0.25$, dilution by entrainment and removal of particles by settling further reduce the concentration.

In conclusion, this phenomenon is a subtle interplay between the flow in the impingement region and the turbidity current region, as indicated in Fig. 7.7. In the internal hydraulic jump that separates the wall jet region and the turbidity current region, turbulent and mean kinetic energy are strongly dissipated, which lowers the sediment carrying capacity further downstream of the jump. As a result, sediment is

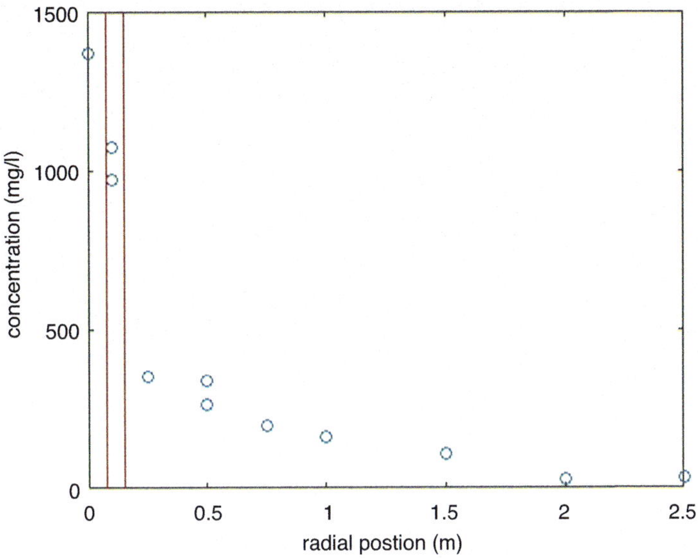

Fig. 7.12 Concentrations measured by the water tapping system versus the radial distance *r* from the impingement point for the case with a stand-off distance of $H = 50$ cm, nozzle diameter 0.01 cm and suspended sediment concentration of 20 g/l, impinging on a horizontal table. Deposition rings were visible between $r = 0.075$ m and $r = 0.15$ m in this experiment (indicated by the red vertical lines)

released from the flow and starts to form a deposition ring. In the experiments, the Shields parameter in the wall-jet region is presumably larger than the critical Shields parameter required for the initiation of motion of sediment on a smooth surface. In reality, particles that were already present on the sea bed could also be mobilized by the impinging flow, which would eventually result in a scour pit formation. At present, it is beyond the state-of-the-art of (numerical) models to describe these patterns satisfactorily. Therefore, further experimental study is required at both the laboratory and prototype scales to acquire a better understanding of the impingement process of sediment-laden return flows into the seabed. Also, high-resolution direct numerical simulations and Eulerian-Eulerian models could be helpful to further unravel the sediment transport physics near the impingement point.

4 Collector Plume Concept

The water-sediment mixture discharged from the back of the collector undergoes three distinct regimes: jet, plume, and turbidity current (Elerian et al., 2021). It is important to note that whether the flow is in the jet or plume phase depends on discharge parameters such as velocity and concentration. If the flow is

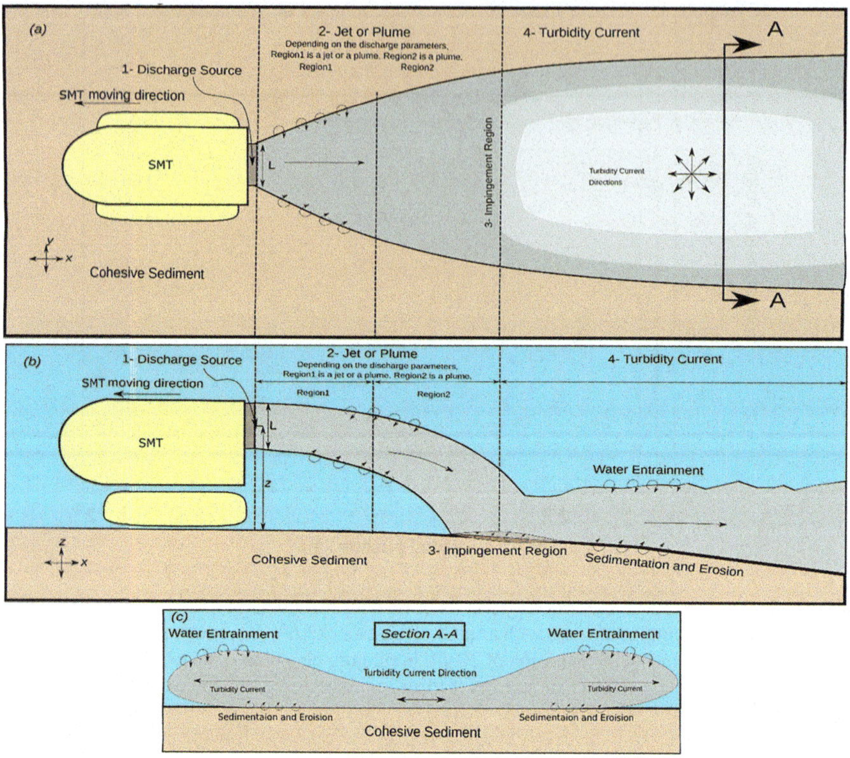

Fig. 7.13 Conceptual sketch of the evolution of the sediment-water mixture discharged by the SMT. (**a**) Top view of the collector plume, (**b**) side view of the collector plume, and (**c**) cross-section A-A indicating the sideways propagating turbidity current, after. (Elerian et al., 2022)

momentum-dominated, it is considered a jet, while if it is buoyancy-dominated, it is considered a plume. Once the mixture enters the plume phase, the density difference dominates the flow, and the mixture eventually reaches the seabed in a region known as the impingement region. At this point, the turbidity current begins to form, and a sideways propagating turbidity currents develop almost perpendicular to the direction of the collector movement (see Fig. 7.13).

The sideways turbidity current propagates due to the density difference between the denser water-sediment mixture and the lighter ambient fluid. Studying turbidity currents in the field is challenging due to the high cost of field experiments. However, scaled laboratory experiments are a widely used alternative for investigating such currents.

In this section, the study will primarily focus on the sideways propagating behaviour of turbidity currents in the near-field region, as these currents exhibit the longest time and length scales among the mixture phases previously mentioned. The investigation is divided into two main sections, which encompass experimental and numerical modelling, respectively.

4.1 *Experimental Modelling of Non-cohesive Sediments*

To investigate the influence of the particle size experimentally, lock-exchange experiments were conducted in the dredging lab of Delft University of Technology to test different sediment types with varying particle sizes.

4.1.1 Lock Exchange Experiments

Lock-exchange experiments are a type of laboratory experiment used to study the behaviour of turbidity currents. In these experiments, a tank is divided into two compartments by a gate or barrier. One compartment is filled with a denser fluid, such as water-sediment mixture, while the other compartment is filled with a lighter fluid, such as clear water. The gate is then removed, allowing the two fluids to mix and form a turbidity current (Fig. 7.14a).

In our experiments, we tested three non-cohesive sediment types: Type 1 with a particle size range of $d_p = 65 - 105\mu m$, Type 2 with a particle size range of $d_p = 4 - 60\mu m$, and Type 3, which was a 50-50 mix (in mass) of Type 1 and Type 2. In addition, we tested different initial concentrations ranging from 0.1% to 3% volumetric concentration for each sediment type to investigate their influence on the propagation of the turbidity current.

We recorded videos of each experimental run using a high-speed camera. To analyse the concentration of the turbidity current, we used a newly developed pixel-concentration method that allows us to detect concentration maps at any time of the

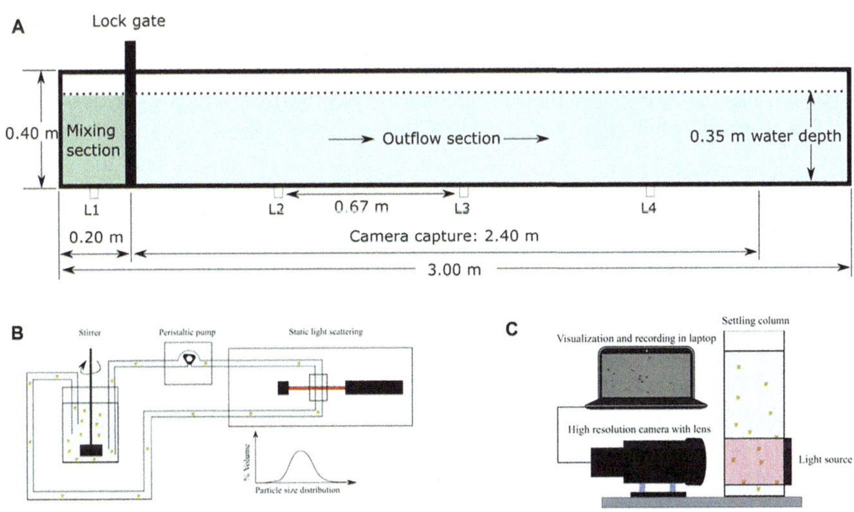

Fig. 7.14 Schematic representation of the set-ups: (Top) Lock exchange. The samples are taken at L1, L2, L3, and L4 locations. (Bottom left) Particle sizer. (Bottom right) FLOCCAM

entire turbidity current. Additionally, we were able to determine the forward velocity of the current from the recorded videos.

Based on the literature sources Rottman and Simpson (1983) and Huppert and Simpson (1980), it has been observed that a turbidity current in a lock-exchange experiment goes through three distinct phases based on its forward velocity: the slumping phase, the self-similar phase, and the viscous-buoyancy phase. The slumping phase occurs close to the lock gate, where the denser fluid near the lock collapses, leading to the formation of the turbidity current front or head. During this phase, the current propagates at a constant speed due to a balance between inertia and buoyancy forces. Subsequently, the self-similar phase begins, during which the propagation speed of the current is no longer constant and starts to decrease. Finally, the viscous-buoyancy phase takes over as viscous forces dominate the flow, causing the current to gradually diminish.

4.2 *Findings*

From these experiments, it was found that, increasing initial concentrations inside the lock yields high front velocities independent of sediment type, meaning that the transition time to the self-similar phase increases when increasing the initial concentration. Furthermore, increasing particle size leads to low front velocities, with coarse particle, lower-concentration currents transitioning to the self-similar phase more quickly than fine-particle currents, which take much longer to transition. Increasing initial concentrations neutralizes the particle size effect on current propagation. With higher-concentration currents, coarse particles have little to no effect on the motion of the current compared to fine-particle currents with the same initial concentration, while in lower-concentration currents, coarse particles settle and affect the forward velocity of the current. In the case of finer-particle, lower-concentration currents, low local concentration distribution values were observed.

For a more comprehensive understanding of these experiments, including the associated results and analysis, it is recommended that the reader read the work of Elerian et al. (2022).

5 Effect of Flocculation on Sediment Dispersion

Particle size distribution (PSD) controls how far deep-sea sediment plumes spread (Gillard et al., 2019; Spearman et al., 2020). Since aggregated particles (flocs) settle quicker than individual (unflocculated) particles of smaller size, flocculation can limit plume dispersion (Manning & Dyer, 2002; Spearman et al., 2019, 2020; Smith & Friedrichs, 2011; Gillard, 2019). The turbulence caused by mining vehicles, particularly in their vicinity, leads to rapid flocculation. Moreover, based on numerical findings, the aggregation mechanism was determined to be more dominant than the break-up mechanism. As discussed by Gillard (2019), experiments on aggregation behaviour under various turbulent conditions and plume concentrations should reveal the likelihood of their deposition in the near-field area. Aside from that study, a few other publications present the hydrodynamic behaviour of original plume particles and model the impact scale at various temporal and spatial scales in the benthic boundary layer (Aleynik et al., 2017; Jones et al., 2017). The physical and chemical parameters of bottom sediments, the hydrodynamic regime (near and far field), bottom topography, and mining equipment type must be considered for this endeavour. Plume modelling of small-scale benthic impact experiments (discharge of approximately 10 kgs^{-1} along predefined tracks) showed that 90% of the suspended particles created by artificial seafloor disturbance settled within 2 km of the disturbance area (Fukushima, 1995; Nakata et al., 1997; Haalboom et al., 2022). Recent modelling of an industrial-scale, near-bottom sediment plume in the CCZ case study area with a discharge rate of 280 kg s^{-1} showed that more than 50% of suspended particles would settle within a few kilometers of the source region within 10 days. The remaining particles left the near-field area (Aleynik et al., 2017). The cohesive characteristics, aggregation potential, and settling processes of fine-grained deep-sea sediments, which affect plume sediment deposition, were, however ignored. Site-specific particle behaviour under varied plume concentrations and high-resolution bathymetric and oceanographic data are however essential to initialize and calibrate sediment transport models for the exploitation of nodule fields, where size, density, settling velocities, and strength of flocs over time are influenced by the properties of the suspending medium (salinity, organic matter content, sediment concentration, hydrodynamics) (Manning & Dyer, 2002; Mietta et al., 2009; Smith & Friedrichs, 2011; Chassagne, 2020). The deep-sea environment is, in principle, favourable for flocculation because of its high salinity and concentration of organic matter (Mewes et al., 2014; ISA, 2015; Volz et al., 2018). Flocculation has been shown to occur at very short timescales in natural environments in the presence of (microscopic) organic matter (Deng et al., 2019; Safar et al., 2019). The aim of the present work is to quantify how much flocculation can influence a plume's propagation and how the main characteristics (size, density) of the particles composing the plume evolve as function of run-out length. Two different approaches are investigated, one with artificial sediments and synthetic flocculants and one with original sediments from the Clarion Clipperton Zone.

5.1 Approach 1, Artificial Sediments

5.2 Materials and Methods

5.2.1 Clays

As illite is one of the dominant clays found in the top layer of the Clarion Clipperton Zone (CCZ) sediment where Deep Sea Mining (DSM) is performed (ISA, 2015), it was chosen for the initial experiments.

For further experiments, a lab-made artificial clay with a composition similar to CCZ clay was used and named "Artificial CCZ" (aka ACCZ). More information about the clays can be found in (Enthoven, 2021; Ali et al., 2022).

5.2.2 Flocculant

Organic matter found in the deep-sea region is expected to be the flocculating agent for the sediment plume (Jones et al., 2017; Gillard et al., 2019; Spearman et al., 2020; Jones et al., 2021). This organic matter is expected to be composed of parts of polyssacharides, and therefore, as was done in previous studies (Shakeel et al., 2020), an anionic polyacrylamide, referenced Zetag 4110 (BASF company) of medium anionic charge with high molecular weight, was used as a proxy for organic matter content.

5.2.3 Lock Exchange Set-Up

Lock exchange experiments were done in fresh as well as saline water. The saline water was produced by mixing 10 mM of $CaCl_2$ with freshwater. Three clay concentrations were used (10, 30, and 100 g/L), 30 g/L and 100 g/L with illite, and two different concentrations with ACCZ (10 g/L) and 30 g/L. Two flocculant dosages (0.25 mg/g and 0.75 mg/g) were also used. The dry clay was mixed with water in the lock exchange's mixing section for an hour before the lock gate was opened. This mixing made sure that homogeneous suspensions with a well-defined mean particle size could be obtained. In experiments with flocculant, the anionic polyelectrolyte was added and stirred for 30 s before the opening of the lock. The plume propagation was recorded with the help of a high-resolution camera. The videos were obtained with a Navitar 17mm lens on an IL5HM8512D: Fastec high-speed camera. The camera was placed 4.75 m from the lens to the front wall of the tank, and recording speed was 130 frames per second. The camera captured the 2.40 metres to the left of the lock. The camera and tank were placed beneath a black fabric to block out most of the light from outside. Samples were collected from four locations (L1, L2, L3, and L4 in Fig. 7.14 at the top) at the end of the experiments.

5.2.4 Particle/Flocs Size Distribution

Particle Size Distribution (PSD) analysis was conducted on the obtained samples using a Malvern master sizer 2000 (Fig. 7.14 bottom left), a technique based on static light scattering (SLS). With this set-up, within a few seconds a full PSD can be recorded.

5.2.5 Floc Settling Analysis

The FLOCCAM device records, using video microscopy, the settling velocity and size of particles. It was used to estimate PSD's (above 20 microns in size) and settling velocities of the different collected samples (Shakeel et al., 2021; Ye et al., 2020; Manning et al., 2007; Mietta et al., 2009). Figure 7.14 bottom right gives a schematic representation of the equipment. The size, shape, and settling velocity of the flocs are calculated from the videos using the software package Safas (MacIver, 2019).

5.3 Results

There is no substantial difference between experiments conducted in fresh and saltwater, both for illite and for ACCZ sediment when no flocculant is used. The results for 30 g/L clay are given in Fig. 7.15 Regardless of the clay concentration in either fresh and saltwater, the sediment plumes reached the lock exchange's end in those experiments. The plume head velocity scales with the square root of the plume density, as expected (Huppert, 2006). As salt-induced flocculation is slow, with flocculation times of the order of 15–30 min even under ideal shear circumstances (Mietta, 2010). It is expected that no significant effect was seen over the experimental period of the current study (less than 1 min).

When experiments with added flocculant were performed, the distance travelled in saltwater experiments for all clay types and concentrations was reduced considerably, and the sediment plumes did not reach the end of the lock exchange. In Fig. 7.16 results for 0.75 mg/g flocculant are given. The results using 0.25 mg/g flocculant can be found in (Enthoven, 2021; Ali et al., 2022), but the general observations remain the same. For the experiment done in freshwater with flocculant, the sediment plume reached the end of the lock exchange (100 g/L) or nearly half of the

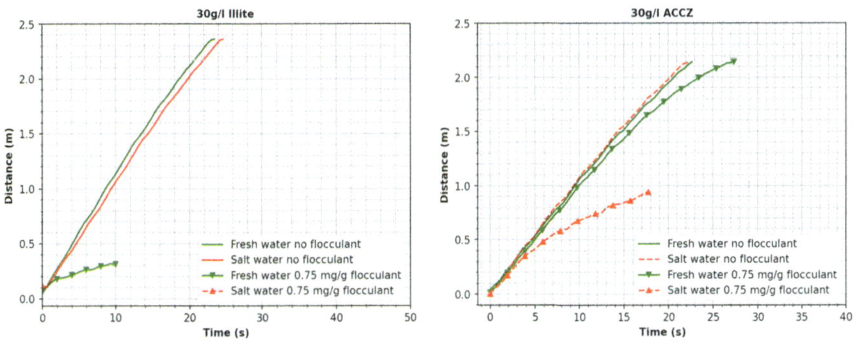

Fig. 7.15 Head propagation as a function of time for 30 g/L illite and ACCZ in different suspending media (indicated in the legends)

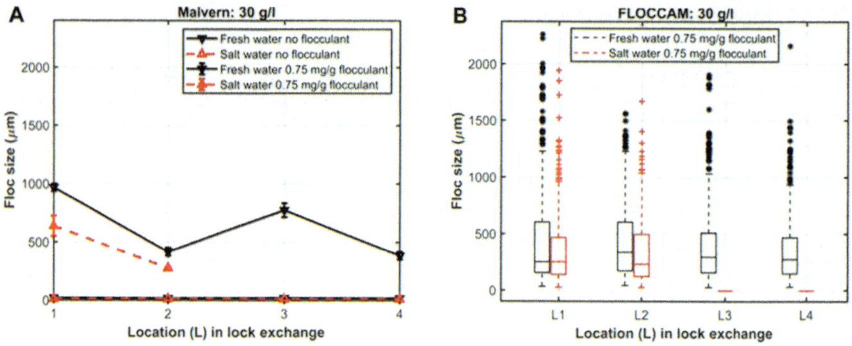

Fig. 7.16 Hydrodynamic size of ACCZ flocs at locations L1-L4 in the lock exchange. (**a**) Results obtained by SLS and (**b**) results obtained from FLOCCAM

lock exchange (10 g/L). The weak flocculation in freshwater accounts for this difference between fresh and saline water experiments. Freshwater contains enough cations to promote slight flocculation between the negatively charged clay and flocculant (Ibanez Sanz, 2018). The cation concentration in saltwater hugely promotes flocculation, because the cation chosen (Ca^{2+}) is divalent (Shakeel et al., 2020; Chassagne, 2020). For 100 g/L experiment with illite, it was found that the system did not properly flocculate and in all cases the sediment plumes reached the end of the lock with the same speed.

5.3.1 Mean Floc Size as Function of Travel Distance

Figure 7.16 displays the mean floc size of the samples taken at the various locations L1-L4. No flocculation was found to happen without the flocculant in both fresh and saltwater, as for each location, the mean particle size was found to be equal to the mean clay size. For both experiments with illite, the mean particle size was found to be around 5 microns at all locations, whereas for experiments with ACCZ, it was found to be between 12 and 20 microns. These sizes are in line with the sizes found using SLS (5 and 10–20 microns respectively). The bars given in the left figures (measured by SLS, using a Malvern ParticleSizer) represent the standard deviation around the mean floc size, whereas in the right figures (measured by video microscopy, using the FLOCCAM) the box plot represents the median particle sizes with interquartile range and outliers.

It was observed that freshwater flocs were larger than saltwater flocs in the presence of organic flocculant. Because of the electrostatic repulsion between the charged groups on the polymeric flocculant, the flocculant in freshwater is expected to be less coiled in fresh compared to saltwater. As a result of shearing during propagation, it is observed that in saltwater the size of flocs created with illite clay is reducing as function of travel distance. This difference is not observed for the ACCZ clay.

5.3.2 Settling Velocity Distributions as Function of Size and Travel Distance

Figure 7.17 displays the settling velocities and particle size ranges for 30 g/L ACCZ with 0.75 mg/g of flocculant at sites L1 and L4 (for freshwater). The settling velocities in freshwater are smaller than in saltwater at the point where the sediment plume settles down. This is due to the fact that the flocculant is less coiled in freshwater than in saltwater, leading to denser flocs with a faster settling velocity. This was confirmed by the video images. In the case of 30 g/L ACCZ, the settling behaviour and floc size for the saltwater sample did not change between L1 and L2, indicating that optimum flocculation has occurred in the mixing tank. The bridging between anionic polyelectrolyte and clay by saltwater cations is, in that case, complete. In contrast, the flocculation in the mixing tank for the freshwater sample is most probably incomplete due to the limited amount of cations. Observing the similarity between the distributions in location L1 for all experiments, the created flocs have the same features (same size, same settling velocity) as flocs formed in saltwater environments. Flocs, clay particles, and unbounded flocculant are released when the lock is opened. Freshwater containing cations comes into contact with the clay particles and unbounded flocculant. Because flocculation by polyelectrolytes is fast and of the order of seconds (Ibanez Sanz, 2018; Shakeel et al., 2020; Ali & Chassagne, 2022), these cations will act as a binding agent, inducing flocculation.

Several flocs formed during the propagation of the sediment plume in 30 g/L ACCZ experiment in freshwater are observed to be elongated, resulting in flocs with large equivalent diameters (shown by the red circle in the bottom panels of Fig. 7.17). These big flocs have a very slow settling velocity because they are formed of low-density uncoiled flocculant connected to some clay. These flocs were unable to

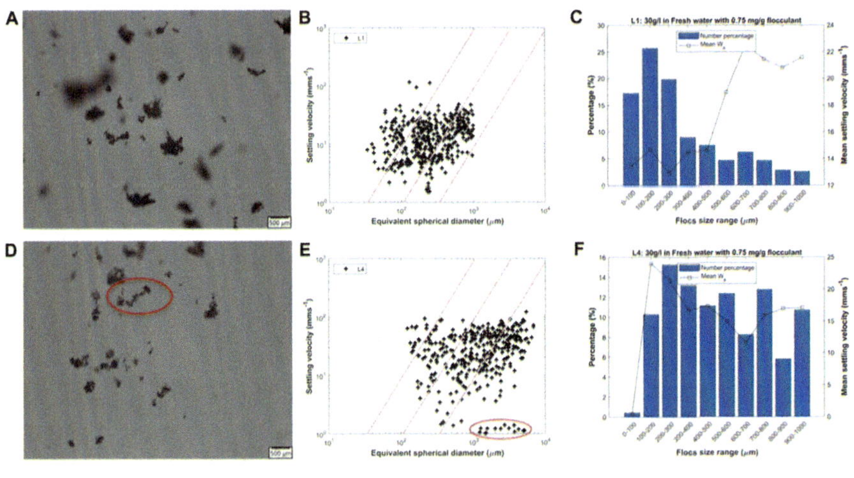

Fig. 7.17 (**a**, **d**): Snapshots of the videos; (**b**, **e**): Settling velocity and particle size analysis of the 30 g/L ACCZ with 0.75 mg/g flocculant in fresh water collected at L1 and L4. The dashed lines in (B, E) represent the effective density lines from left to right: 1600, 160 and 16 kgm^{-3}

catch more clay particles or coils due to their limited residence period in the water column. Coiling of flocs happens over longer periods of time when turbulent shear causes the polyelectrolyte's loose ends to collapse onto the floc. The flocs get rounder and denser as a result (Shakeel et al., 2020).

5.4 *Approach 2, Original Sediments*

Oceanographic data and surface sediment samples were collected using a multicorers at different locations in the eastern sectors of the CCZ, where flat abyssal plains, NW–SE trending ridge systems, and seamount rises produce fine-grained sediments at 4300 m. Roller tank, shear tank, and water column simulator incubations, as well as a settling tube, were used to analyse the particle behaviour of the sediment plumes under different turbulent conditions (Gillard, 2019). The dependence of the various concentrations of aggregated plume particles on the settling behaviour was investigated under in-situ temperature conditions for periods of 24 h. The particle size range covered original deep sea sediments which were released into the water column under high shear rates (a centrifugal pump simulating the exhaust plume behind a mining vehicle) to form large aggregates >2000 μm in size. These formed rapidly in the wake of the vehicle. For these analyses a LISTT – 100 X particle analyser and a particle camera set-up as described in (Gillard et al., 2022) were used. FIJI software (v.1.51n) was used to calculate aggregate size and shape descriptor parameters (circularity, roundness, and solidity) as described by (Shen, 2016). For mesocosm experiments visualizing the generation of fluid muds and subsequent consolidation of aggregated plume sediments, the surface sediments were dispersed using a centrifugal pump (7500 l/h) and subsequently transferred into a calibrated water column simulator (110 cm diameter, 1000 L), based on the design of (Sanford, 1997) and (Crawford & Sanford, 2001), thus allowing larger water volumes and avoiding constant shear as produced by Couette cylinders, which is intrinsically unrepresentative of natural turbulence (Arnott et al., 2021). Shear rates of 0.1 G to 1 G were simulated with such simulator. These represent flow conditions typical for slow deep sea bottom currents and elevated currents near the seafloor generated by passing eddies. Turbidity was determined using a Aquatech 310TY sensor.

5.5 *Results*

Median particle size (D50) of disaggregated surface sediments from three different sectors of the eastern CCFZ varied between 8 – 25μm. The average bulk wet density of these sediments was ≈1230 kgm^{-3} with a porosity of 87 %. Figure 7.18 presents the aggregation behaviour of two sediment plumes of different particle concentrations (1 g/L vs. 10 g/L) under two different turbulent conditions (0.1 G

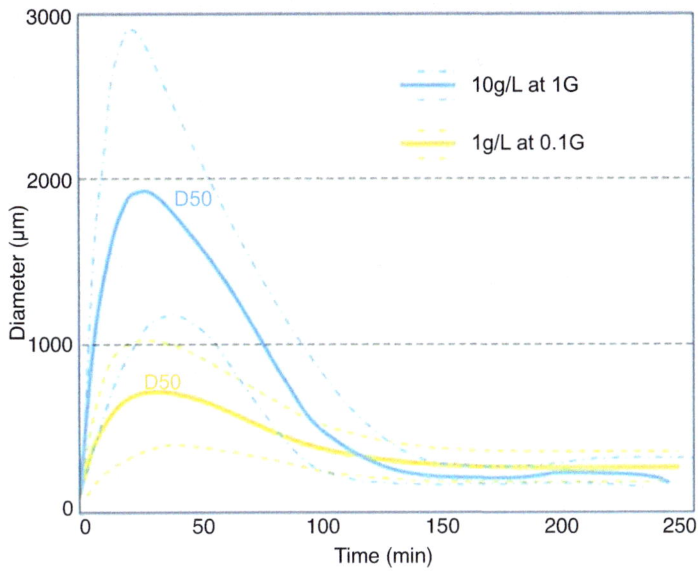

Fig. 7.18 Sediment plume aggregation dynamics in the water column simulator. d_{50} (solid lines ±STD (dashed lines) from particle camera data, analysed with Matlab

(0.1 s^{-1}, ≈ 4 cm/s flow velocity) vs. 1 G (1 s^{-1}, ≈15 cm/s flow velocity). The results show the expected spectrum of aggregate formation, from low shear/low particle concentration to high shear/high particle concentration. The results reveal that both hydrodynamic conditions result in a rapid formation of large particles within the first 30 min after release. During that time, the size distribution of primary particles of d_{50} of 8-25 µm rapidly shifts towards large aggregates with d_{50} of 700µm to almost 2000µm respectively while after 24 h particles of d_{50} of 300–400 µm still remain in suspension showing ongoing aggregation of the remaining particle loads. This would continue during a permanent release of a plume, while simultaneously these particles are settling (exported) towards the seafloor. Within the bottom boundary layer behind the mining vehicle, an export of ≈ 50 % of the initial plume concentration can be seen. After this initial phase of rapid aggregate formation and/or once the release stopped, the particle sizes decrease and smaller particles of lower settling velocities are formed which widen the general fallout area in downstream direction of the bottom currents or resulting gravity current. Most of the expelled particle mass will thus settle behind a moving mining vehicle with 1 h, thus focusing the main fallout area to a few km^2 around the exploration site. The results however also show that even after 24 h the turbidity is elevated with particle concentrations of a few mg/L and thus still one order of magnitude higher than background concentrations of ≈ 0.1 mg/L (Haalboom et al., 2023).

The settling velocities of aggregates during the first 120 min after release varied for the d_{50} size between ≈ 120 and 210 m/d under the low turbulence/low particle

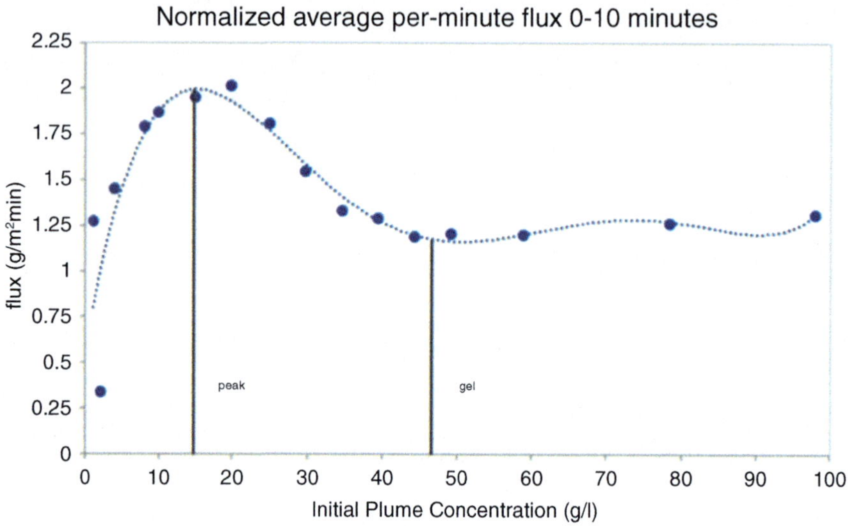

Fig. 7.19 Normalized average per-minute settling flux for CCZ sediment plume

concentration conditions and ≈ 420 to 700 m/d under the high turbulence/high particle concentration conditions. Under these conditions, 90 % of the flocs settle through a 1 m thick water column and form a fluid mud, which could be exported as gravity current. The results partly resemble the flocculation behaviour of artificial sediments and flocculants, as shown in Fig. 7.15.

The settling flux result for CCZ sediment plumes under varying particle concentrations is presented in Fig. 7.19 and should reveal the behaviour of aggregates under very high release concentrations behind the mining vehicle once settled into the fluid mud. Hindered settling is expected at particle concentrations of a few g/L (Mehta, 1986). CCZ sediment plumes behaved in three distinct modes that resembled the settling behaviour described by (Dankers & Winterwerp, 2007). Before the volumetric sediment concentration of the solids (Φ, after (Kynch, 1952) peaks at 15 g/L, the settling flux rises somehow linearly with increased sediment plume concentration. Beyond Φ_{peak}, the flux decreased showing the effect of hindered settling behaviour. Finally, the profile flattens out past Φ_{gel} (45 g/L) to a relatively constant per-minute flux, indicating that a gel-like suspension had formed at Φ_{gel}. This results in the formation of a fluid mud, which then slowly consolidates over many hours.

The experiments suggest that the fallout zones of particles behind a moving mining vehicle are relatively local. This will result in heavy blanketing of the sediments in the direct vicinity of an exploration site unless the fluid mud is carried away by a gravity current (Elerian et al., 2022). This was confirmed during the first in-situ collector test in 2021 in the CCFZ within the EU MiningImpact 2 project (Igazis et al, Nature, in review).

5.6 Findings

In this work, the influence of flocculation on turbidity currents was studied inside a lock exchange, where the current propagation time was of the same order of magnitude. It was shown that in the presence of an organic flocculating agent (anionic polyelectrolyte), flocculation was promoted. It was found that in both fresh and saltwater, flocs can be formed in a matter of seconds with the flocculant used in this study. As a result, the sediment plume was able to settle more quickly. The synthetic flocculant used is a proxy for organic matter found in marine environments (usually also negatively charged). It remains to be investigated if the type of flocculant has a significant impact on flocculation. This will be possible once the organic matter found in our area of interest (i.e. the Clarion-Clipperton Zone) has been fully characterized. The results presented in this article are generic and thus apply to a wide range of turbidity currents. We demonstrated that flocculation may occur even in freshwater, where flocculation is supposed to be difficult because of the electrostatic repulsion between organic matter and clay. This means that flocculation should be accounted for in turbidity current models. The obtained results demonstrate that flocculation is a relevant phenomenon that may already be contributing in the near field. Building experience with more conventional sediments allows us to better understand and design experiments with real CCZ sediment, which is the next step.

6 Numerical Modelling of Flocculation Effects in Near-Field

Given the time and cost-intensive nature of these experiments, it is crucial to utilize mathematical and numerical models to predict turbidity current behaviour with minimal computational effort. However, it is equally important to properly validate and calibrate these models.

6.1 The Mixture Model (Drift-flux Model)

The mixture model, which is also known as drift flux model, is a mathematical model used to describe the behaviour of multiphase flow, such as the flow of a mixture of sediment and water. In this context, each particle size fraction is considered a discrete phase, while the water is the continuous phase.

The drift-flux model treats turbidity currents as a single continuum, requiring a single momentum equation to be solved for the mixture as a whole. However, it takes into account the depositional behaviour of particles through the use of drift velocity. A clear mathematical structure for the drift-flux model is reported in detail in the work of Goeree (2018).

The multi-fraction drift-flux model was incorporated into the widely-used open-source computational fluid dynamics (CFD) software OpenFOAM. In addition, the blockMesh utility in OpenFOAM was used to generate a two-dimensional mesh that represents the lock-exchange tank. Furthermore, numerical experiments were conducted with the same initial conditions as the aforementioned experiments for the purpose of validating the newly implemented model. The measured forward velocity and concentration profiles from the experiments were compared to those obtained from the numerical simulations, revealing a high degree of agreement. This finding underscores the efficacy of the multi-fraction drift-flux model as a powerful tool for accurately predicting turbidity currents.

6.2 The Population Balance Equation (PBE)

The aforementioned multi-fraction drift-flux model has been demonstrated to be well-suited for simulating noncohesive sediment. However, the sediment in the deep sea exhibits cohesive properties in the presence of organic content. This cohesive nature allows the sediment to aggregate while in suspension, and in certain scenarios with high shear rates, the formed flocs may break up from each other. Recent research has increasingly recognized the crucial role of flocculation in deep-sea mining turbidity flows, as evidenced by studies such as those by Gillard et al. (2019), Spearman et al. (2020), Notably, Gillard et al. (2019) conducted a controlled laboratory experiment and revealed that under specific mixing conditions, the median particle size (d_{50}) of a Clarion Clipperton Fracture Zone (CCZ) sediment mixture can increase from 12 μm to 600 μm in just 7 min, indicating a rapid aggregation process among sediment particles.

The unique behaviour of cohesive sediment in suspension presents additional challenges and complexities that need to be considered in the development of accurate models for simulating turbidity currents in deep sea environments. Extensive efforts have been dedicated to the mathematical framework for describing the complex mechanisms of flocculation, resulting in the establishment of the Population Balance Equation (PBE) as a key tool in capturing the intricate dynamics of floc formation and evolution (Hounslow et al., 1988). Consequently, the multi-fraction drift-flux model was extended by incorporating the PBE to account for the aggregation and break-up processes. The six source terms of the PBE were added to the phase transport equation of the drift-flux model. The details of the coupling between the PBE and the drift-flux model are clearly reported in the work of Elerian et al. (2023) for further reference.

The aggregation of particles in shear-induced flows leads to an increase in floc size and a decrease in density, as water becomes trapped within the flocs during the aggregation process. This phenomenon directly affects the settling velocity of the flocs, which in turn impacts the accuracy of predictions for mixture hydrodynamics. To accurately account for these changes, an improved definition of settling velocity that considers both the primary particle size and the fractal dimension of the flocs is used in the model. Furthermore, the most important parameters of the model, such as the fractal dimension, floc shape coefficient, and breakage coefficient were calibrated using the experimental work of Gillard et al. (2019).

6.3 Near-Field Modelling

Following the successful validation of the multi-fraction drift flux model and calibration of the Population Balance Equation (PBE) source terms, we have gained confidence to proceed with the near-field modelling, which focuses on the area approximately 100 m behind the collector. These simulations were conducted in order to distinguish the flocculation nature and its effect in the near-field region.

Numerical simulations were conducted for a distance of 100 m behind the collector, using a realistic discharge scenario. The collector velocity was set at 0.5 m/s, the discharge velocity at 0.5 m/s, the discharge volumetric concentration at 1%, the volumetric sediment flux at 0.01 m^2/s, and the volume flux at 1 m^3/s. The numerical simulation was run for 250 s to ensure a steady state situation for the propagation of the sideways propagating turbidity current. Two numerical runs were compared: one with flocculation, where the flocculation source terms were active, and the other without flocculation, where flocculation was not considered, modelling non-cohesive sediment.

The comparison between the two cases has revealed that sediment readily forms aggregates (see Fig. 7.20), which are larger particles that settle once the turbidity current subsides, likely occurring in the far-field region. The turbulence caused by

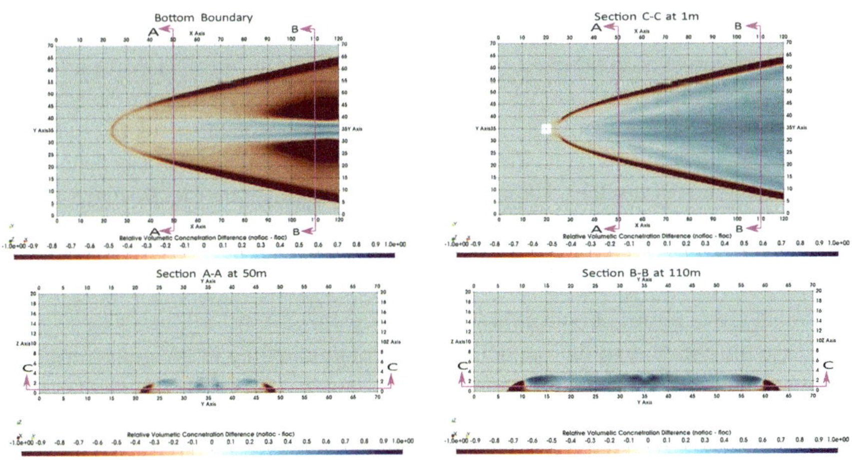

Fig. 7.20 Z and X normal sections show the relative differences between the flocculation and no-flocculation cases. The Z-normal sections are taken at the BOTTOM boundary (left) and 1 m from the bottom (right) which is section C-C, while the X-normal sections are taken at 50 m and 110 m from the INLET boundary

SMT, especially in the region close to the SMT, promotes rapid flocculation. Moreover, the aggregation mechanism was found to be more dominant than the break-up mechanism, based on the numerical findings. While the effects of flocculation can be observed in the near-field region, their significant impact on the turbidity current near the SMT is limited.

7 Conclusions

In this chapter, we have shown various essential results to allow for better prediction of sediment dispersion near the bed. Some insights have been provided into how the collector design of a Coandă-effect-based nodule collection system can influence the erosion of the seabed, differentiating between what sediment might be collected and what might be spilled into the near seabed environment, and whether and how sediment dispersion near the bed might be predicted.

The experimental explorations of the near-seabed release of SWOE open up the question of whether the formation of morphological structures could help to prevent the dispersion of sediments outside the designated mining area. The challenge is to dissipate the mean and turbulent kinetic energy of the flow and thereby reduce the sediment transport capacity of the flow. Design parameters that may influence the flow of the sediment plume will be the nozzle diameter, nozzle angle, stand-off distance, discharge rate, and discharge composition. In addition, the deposition of sediment could be promoted by using topological features or by altering the seabed, e.g creating morphological structures by the mining vehicles used for the collection process.

This kind of research should be conducted in-line with other attempts to promote the deposition rate, e.g. by flocculation.

The experiments conducted with cohesive sediments hint at several trends. Aggregation is likely to take place in the collector plume, increasing the sediment deposition rate and effectively reducing the driving mechanism of the sideways propagating turbidity currents. This means that flocculation should be accounted for in turbidity current models. The obtained results demonstrate that flocculation is a relevant phenomenon that may already be contributing in the near field. Building experience with more conventional sediments allows us to better understand the mechanisms at play. Experiments with real CCZ sediments are essential to providing quantitative information for these specific cases.

The multi-phase drift flux model for near field modelling of the collector plume shows some of the potential effects flocculation might have on the sediment composition (and thus settling velocities) already within the vicinity of the SMT. It is in this area where the design and operation of the equipment might be able to influence the collector plume release conditions, potentially improving overall settling velocities by optimizing mixing and sediment concentration. Further studies would be required to fully explore the design space available to identify under which conditions plume dispersion might be minimized. Obviously, this will only be possible under the constraint of effective nodule pick-up, enabling a viable economic production.

Acknowledgements The results presented are part of the Blue Harvesting project, which is funded by the European Institute of Innovation and Technology, EIT Raw Materials under Project Agreement 18138, Specific Grant Agreement No. [EIT/RAW MATERIALS/SGA2019/1], TREASURE project (Towards Responsible ExtrAction of SUbmarine mineral REsources), which is financially supported by NWO-STW and PLUMEFLOC project, (TMW.BL.019.004, Topsector Water and Maritiem: Blauwe route) within the MUDNET academic network.

References

Aleynik, D., Inall, M., Dale, A., & Vink, A. (2017). Impact of remotely generated eddies on plume dispersion at abyssal mining sites in the Pacific. *Scientific Reports, 7*(1), 1–14, Springer US.

Alhaddad, S., & Helmons, R. (2023). Sediment erosion generated by a Coandă-effect-based polymetallic-nodule collector. *Journal of Marine Science and Engineering, 11*(2), 349.

Alhaddad, S., Mehta, D., & Helmons, R. (2023). Mining of deep-seabed nodules using a Coandă-effect-based collector. *Results in Engineering, 17*, 100852.

Ali, W., & Chassagne, C. (2022). *Comparison between two analytical models to study the flocculation of mineral clay by polyelectrolytes*. Manuscript submitted for publication.

Ali, W., Enthoven, D., Kirichek, A., Helmons, R., & Chassagne, C. (2022). Can flocculation reduce the dispersion of deep sea sediment plumes? In *Proceedings of the world dredging conference (WODCONXXIII), Copenhagen, Denmark.*

Arnott, R. N., Cherif, M., Bryant, L. D., & Wain, D. J. (2021). Artificially generated turbulence: A review of phycological nanocosm, microcosm, and mesocosm experiments. *Hydrobiologia, 848*, 961–991.

Atmanand, M., & Ramadass, G. (2017). Concepts of deep-sea mining technologies. In R. Sharma (Ed.), *Deep-Sea mining*. Springer.

Chassagne, C. (2020). *Introduction to colloid science*. Delft Academic Press. ISBN 9789065624376.

Chowdhury, M., & Testik, F. (2015). Axisymmetric underflows from impinging buoyant jets of dense cohesive particle-laden fluids. *Journal of Hydraulic Engineering, 141*(3), 04014079.

Crawford, S. M., & Sanford, L. P. (2001). Boundary shear velocities and fluxes in the MEERC experimental ecosystems. *Marine Ecology Progress Series, 210*, 1–12.

Dankers, P. J. T., & Winterwerp, J. C. (2007). Hindered settling of mud flocs: Theory and validation. *Continental Shelf Research, 27*(14), 1893–1907.

Deng, Z., He, Q., Safar, Z., & Chassagne, C. (2019). The role ofalgae in fine sedimentflocculation: In-situ andlaboratory measurements. *Marine Geology, 413*, 71–84.

Elerian, M., Alhaddad, S., Helmons, R., & Van Rhee, C. (2021). Near-field analysis of turbidity flows generated by polymetallic nodule mining tools. *Mining, 1*, 251–278. https://doi.org/10.3390/mining1030017

Elerian, M., van Rhee, C., & Helmons, R. (2022). Experimental and numerical modelling of Deep-Sea-mining-generated turbidity currents. *Minerals, 12*(5), 558.

Elerian, M., Ziyang, H., van Rhee, C., & Helmons, R. (2023). Flocculation effect on turbidity flows generated by deep-sea mining: A numerical study. *Ocean Engineering, 277*, 114250.

Enthoven, D. (2021). *Plume dispersion of low-density clayey suspension turbidity currents created by deep-sea mining* (Master thesis), Technische Universiteit Delft.

Fukushima, T. (1995). *Overview "Japan Deep-Sea impact experiment = JET". First ISOPE Ocean mining symposium*. International Society of Offshore and Polar Engineers.

Gillard, B. (2019). *Towards Deep Sea mining-impact of mining activities on benthic pelagic coupling in the clarion Clipperton fracture zone* (PhD thesis). Universität Bremen.

Gillard, B., Purkiani, K., Chatzievangelou, D., Vink, A., Iversen, M. H., & Thomsen, L. (2019). Physical and hydrodynamic properties of deep sea mining-generated, abyssal sediment plumes in the Clarion Clipperton Fracture Zone (eastern-central Pacific). *Elementa, 7*, 5.

Gillard, B., Harbour, R. P., Nowald, N., Thomsen, L., & Iversen, M. H. (2022). Vertical distribution of particulate matter in the clarion Clipperton zone (German sector)—Potential impacts from Deep-Sea mining discharge in the water column. *Frontiers in Marine Science, 9*, 820749.

Gillham, J. (2022). Economically viable selective harvesting of polymetallic nodules. In *Proceedings of the 50th underwater mining conference, St. Petersburg, Florida, USA*.

Goeree, J. (2018). *Drift-flux modeling of hyper-concentrated solid-liquid flows in dredging applications* (PhD thesis). Delft University of Technology.

Grunsven, F., Keetels, G., & Van Rhee, C. (2018). The initial spreading of turbidity plumes—Dedicated laboratory experiments for model validation. In *Proceedings of the 47th underwater minerals conference, Bergen, Norway*.

Haalboom, S., Schoening, T., Urban, P., Gazis, I.-Z., de Stigter, H., Gillard, B., Baeye, M., Hollstein, M., Purkiani, K., Reichart, G.-J., Thomsen, L., Haeckel, M., Vink, A., & Greinert, J. (2022). Monitoring of Antrhopogeneic sediment plumes in the clarion-Clipperton zone, NE equatorial Pacific Ocean. *Frontiers in Marine Science, 9*, 699.

Haalboom, S., de Stigter, H., Mohn, C., Vandorpe, T., Smit, M., de Jonge, L., & Reichart, G.-J. (2023). Monitoring of a sediment plume produced by a deep-sea mining test in shallow water, Malaga Bight, Alboran Sea (southwestern Mediterranean Sea). *Marine Geology, 456*, 106971.

Hong, S., Choi, J.-S., Kim, J.-H., & Yang, C.-K. (1999). Experimental study on hydraulic performance of hybrid pick-up device of manganese nodule collector. In *Proceedings of the third ISOPE Ocean mining symposium*.

Hong, S., Kim, H.-W., Yeu, T., Choi, J.-S., Lee, T., & Lee, J.-K. (2019). Technologies for safe and sustainable mining of deep-sea minerals. In R. Sharma (Ed.), *Environmental issues of deep=sea mining: Impacts, consequences and policy* (pp. 95–143).

Hounslow, M., Ryall, R., & Marshall, V. (1988). A discretized population balance for nucleation, growth, and aggregation. *AICHE Journal, 34*(11), 1821–1832.

Huppert, H. E. (2006). Gravity currents: A personal perspective. *Journal of Fluid Mechanics, 554*(1), 299–322. https://doi.org/10.1017/s002211200600930x

Huppert, H., & Simpson, J. (1980). The slumping of gravity currents. *Journal of Fluid Mechanics, 99*(4), 785–799.

Ibanez Sanz, M. (2018). *Flocculation and consolidation of cohesive sediments under the influence of coagulant and flocculant* (PhD thesis). Delft University of Technology.

ISA. (2015). *A geological model of polymetallic nodule deposits in the clarion-clipperton fracture zone*. [WWW Document], Technical report 6. ISA 2019. Current status of the reserved areas with the international seabed authority. [WWW Document], Technical report.

Jones, D. O. B., Kaiser, S., Sweetman, A. K., Smith, C. R., Menot, L., Vink, A., Trueblood, D., Greinert, J., Billett, D. S. M., Arbizu, P. M., Radziejewska, T., Singh, R., Ingole, B., Durden, J. M., Clark, M. R., Stratmann, T., & Simon-lledo, E. (2017). Biological responses to disturbance from simulated deep-sea polymetallic nodule mining. *PLoS One, 12*(2), e0171750.

Jones, D. O. B., Simon-Lledó, E., Amon, D. J., Bett, B. J., Caulle, C., Clément, L., Connelly, D. P., Dahlgren, T. G., Durden, J. M., Drazen, J. C., Felden, J., Gates, A. R., Georgieva, M. N., Glover, A. G., Gooday, A. J., Hollingsworth, A. L., Horton, T., James, R. H., Jeffreys, R. M., & Huvenne, V. A. I. (2021). Environment, ecology, and potential effectiveness of an area protected from deep-sea mining (Clarion Clipperton Zone, abyssal pacific). *Progress in Oceanography, 197*, 102653. https://doi.org/10.1016/j.pocean.2021.102653

Kang, Y., & Liu, S. (2021). The development history and latest progress of deep-sea polymetallic nodule mining technology. *Minerals, 11*, 1132.

Kynch, G. (1952). A theory of sedimentation. *Transactions of the Faraday Society, 48*, 166–176.

Lang, A., Dasselaar, S., Aasly, K., & Larsen, E. (2019). *Technical report describing process flow overview*. Blue Nodules.

Lesage, M. (2020). *A framework for evaluating deep sea mining systems for seafloor massive sulphides deposits* (PhD thesis). Norwegian University of Science and Technology (NTNU).

MacIver, M. R. (2019). Development and usability assessment of a connected resistance exercise band application for strength-monitoring. *World Academy of Science, Engineering and Technology, 13*, 340–348. https://doi.org/10.5281/zenodo

Manning, A. J., & Dyer, K. R. (2002). A comparison of floc properties observed during neap and spring tidal. *Proceedings in Marine Science, 5*, 233–250.

Manning, A., Friend, P., Prowse, N., & Amos, C. (2007). Estuarine mud flocculation properties determined using an annular mini-flume and the LabSFLOC system. *Continental Shelf Research, 27*(8), 1080–1095.

Mehta, A. (1986). Characterisation of cohesive sediment proper- ties and transport processes in estuaries. In A. Mehta (Ed.), *Estuarine cohesive sediment dynamics* (Lecture notes in coastal and estuarine studies) (pp. 290–325). Springer.

Mewes, K., Mogollón, J. M., Picard, A., Rühlemann, C., Kuhn, T., Nöthen, K., & Kasten, S. (2014). Impact of depositional and biogeochemical processes on small scale variations in nodule abundance in the Clarion-Clipperton Fracture Zone. *Deep Sea Research Part I, 91*, 125–141. Elsevier. https://doi.org/10.1016/j.dsr.2014.06.001

Mietta, F. (2010). *Evolution of the floc size distribution of cohesive sediments* (PhD thesis). Delft University of Technology.

Mietta, F., Chassagne, C., Manning, A. J., & Winterwerp, J. C. (2009). In-fluence of shear rate, organic matter content, pH and salinity on mud flocculation. *Ocean Dynamics, 59*(5), 751–763.

Nakata, K., Kubota, M., Aoki, S., & Taguchi, K. (1997). Dispersion of resuspended sediment by ocean mining activity – Modelling study. In *Proceedings of the first international symposium on environmental studies for Deep-Sea mining, Tokyo, Japan* (pp. 169–186).

Nauru Ocean Resources Inc. (2021). *Collector test study environmental impact statement: Testing of polymetallic nodule collector system components in the NORI-D contract area, Clarion-Clipperton Zone, Pacific Ocean*, Technical report submitted to the International Seabed Authority.

Ouillon, R., Kakoutas, C., Meiburg, E., & Peacock, T. (2021). Gravity currents from moving sources. *Journal of Fluid Mechanics, 924*, A43.

Reba, I. (1966). Applications of the Coanda effect. *Scientific American, 214*(6), 84–93.

Rottman, J., & Simpson, J. (1983). Gravity currents produced by instantaneous releases of a heavy fluid in a rectangular channel. *Journal of Fluid Mechanics, 135*, 95–110.

Safar, Z., Rijnsburger, S., Sanz, M. I., Chassagne, C., Manning, A., Pietrzak, J., Souza, A., van Kessel, T., Horner-Devine, A., Flores, R., & McKeon, M. (2019). Characterization and dynamics of suspended particulate matter in the near field of the Rhine River Plume during a neap tide. In *Geophysical research abstracts* (Vol. 21).

Sanford, L. P. (1997). Turbulent mixing in experimental ecosystem studies. *Marine Ecology Progress Series, 161*, 265–293.

Shakeel, A., Safar, Z., Ibanez, M., Paassen, L., & Chassagne, C. (2020). Flocculation of clay suspensions by anionic and cationic polyelectrolytes a systematic analysis. *Minerals, 10*, 1–24.

Shakeel, A., MacIver, M. R., van Kan, P. J. M., Kirichek, A., & Chassagne, C. (2021). A rheological and microstructural study of two-step yielding in mud samples from a port area. *Colloids and Surfaces A: Physicochemical and Engineering Aspects, 624*, 126827.

Shen, X. (2016). *Modeling flocculation and deflocculation processes of cohesive sediments*. The College of William and Mary.

Smith, S. J., & Friedrichs, C. T. (2011). Size and settling velocities of cohesive flocs and suspended sediment aggregates in a trailing suction hopper dredge plume. *Continental Shelf Research, 31*(10 Suppl), S50–S63.

Spearman, J., Taylor, J., Crossouard, N., Cooper, A., Turnbull, M., Manning, A., & Lee, M. (2019). The measurement and modelling of plumes resulting from deep sea mining of Fe-Mn Crusts. In *Proceedings of the world dredging conference (WODCONXXII), Shanghai, China*.

Spearman, J., Taylor, J., Crossouard, N., Cooper, A., Turnbull, M., Manning, A., Lee, M., & Murton, B. (2020). Measurement and modelling of deep sea sediment plumes and implications for deep sea mining. *Scientific Reports, 10*(1), 1–14.

van den Bosch, G. (2023). *Tech transfer from – Supplier perspective, presented at GCE Ocean technology: Accelerating deep-sea exploration*, Bergen, Norway.

van Wijk, J. (2016). *Vertical hydraulic transport for Deep-Sea mining: A study into flow assurance* (PhD thesis). Delft University of Technology.

Volz, J. B., Mogollon, J. M., Geibert, W., Arbizu, P. M., Koschinsky, A., & Kasten, S. (2018). Natural spatial variability of depositional conditions, biogeochemical processes and element fluxes in sediments of the eastern Clarion Clipperton Zone, Pacific Ocean. *Deep Sea Research Part I: Oceanographic Research Papers, 140*, 157–172. Elsevier Ltd. https://doi.org/10.1016/j.dsr.2018.08.006

Ye, L., Manning, A. J., & Hsu, T. J. (2020). Corrigendum to Oil-mineral flocculation and settling velocity in saline water. *Water Research, 173*, 115569.

Zhao, G., Xiao, L., Yue, Z., Zhao, W., & Kou, Y. (2019). Investigation on characteristics of forces on spherical particles in Deep Sea hydraulic collecting. In *Proceedings of the 29th international Ocean and polar engineering conference*.

Zhao, G., Xiao, L., Yue, Z., Liu, M., Peng, T., & Zhao, W. (2021). Performance characteristics of nodule pick-up device based on spiral flow principle for deep-sea hydraulic collection. *Ocean Engineering, 226*, 108818.

Rudy Helmons (1987) obtained his MSc in Mechanical Engineering in 2011, after which he started as a research engineer at Royal IHC. He obtained his PhD 'cum laude' from TU Delft in 2017 on the topic of rock excavation under water conditions. After that, he started as an assistant professor in offshore and dredging engineering at TU Delft and as of February 2020, he is also an adjunct associate professor for deep sea mining at NTNU. Currently, Rudy is associate professor in Offshore and Dredging Engineering and he is active in the field of marine minerals. His research focusses on the handling of solids in a marine environment, e.g., excavation, transport, separation, and deposition. He is the project coordinator for various (deep) seabed mining research projects.

Dr. ir. Said Alhaddad is an assistant professor of offshore and dredging engineering at Delft University of Technology (TU Delft). The focus of his research is to gain a deeper understanding of the physical processes governing the interplay between underwater activities and the generation of turbidity flows.Dr. ir. Alhaddad carried out his postdoc in deep sea mining at TU Delft. He was awarded the degree of doctor from the Department of Hydraulic Engineering, Faculty of Civil Engineering and Geosciences, TU Delft. His PhD work has focused on breaching flow slides and the associated turbidity currents.

Claire Chassagne has been an associate professor at the hydraulic engineering department of the Technical University of Delft since 2020. She holds a PhD (cum laude) from the University of Leiden and a *Habilitation à Diriger des Recherces degree* from the Sorbonne University. Her research focusses on flocculation and sediment transport, the rheology of fluid mud, and the consolidation of slurries. She is a founding member of MUDNET, an academic platform for inter-disciplinary work about cohesive sediment (https://www.tudelft.nl/mudnet/), since 2015.

Mohamed Elerian (1990) has been a PhD candidate at the Dredging and Offshore Group in the Maritime Department at Delft University of Technology since 2019. He holds a Master of Science degree in Water Resources and Engineering Management from Stuttgart University. With expertise spanning both industry and academia, Mohamed's current research focuses on developing numerical-CFD models to improve the understanding of turbidity flows in general as well as the plumes and turbidity currents generated from deep-sea mining operations.

Geert Keetels is an associate professor in offshore and dredging engineering at Delft University of Technology. His research challenge is the numerical modelling of sediment transport in the range of dense soil and hyperconcentrated flows to dilute suspension on the scales that are relevant for seabed equipment. He obtained his PhD in applied physics at Eindhoven University of Technology.

Dr. Alex Kirichek is an assistant professor at the Department of Hydraulic Engineering, Faculty of Civil Engineering and Geosciences, TU Delft. His research interest is focused on developing sustainable solutions for port maintenance and sediment management strategies. He is one of the founders of MUDNET, which is an academic exchange platform for fine-grained sediment experts who are studying the physical processes (flocculation, settling, and consolidation), geophysical monitoring methods, and transport (plume propagation, density currents, and rheology) of sediment.

Laurenz Thomsen (1963) earned his PhD in 1992 from the University of Kiel, focusing on benthic boundary characteristics along continental margins. He was a postdoctoral researcher at the GEOMAR research centre in Kiel from 1992 to 1998, followed by 2 years as a Heisenberg stipend of DFG at the University of Washington in Seattle. He was a professor of geosciences at Jacobs University Bremen from 2001 to 2022 and has been professor of sedimentology at the University of Göteborg since then. He has also been an affiliate professor at the University of Washington's School of Oceanography in Seattle since 2004. His research focuses on the impact of humans on particle transport in the ocean, as well as marine robots.

Chapter 8
Applicability of Satellite Data in the Selection of Protected Areas Within REMP

Tomohiko Fukushima, Akira Tsune, Chisato Murakami, and Hideki Sugishima

Abstract The International Seabed Authority (ISA) is considering establishing Areas of Particular Environmental Interest (APEI) as part of the Regional Environmental Management Plan (REMP) for deep-sea mining. Ideally, the number, location, and area of APEIs should be determined based on information that considers the biodiversity network, ecosystem functions of the entire region, and other factors. However, at present, there is a significant lack of data for examination based on these features. In particular, there are issues such as accuracy discrepancies, spatial heterogeneity, and lack of time-series data. Based on these facts, the authors examined the linkage between surface and bottom ecosystems through the water column and evaluated the utility of satellite data. In the present study, the Clarion-Clipperton Zone (CCZ)-REMP and its APEIs were assumed to be the completed model, and the degree of effectiveness and limitations related to satellite data for APEI establishment were considered. As a result, it was confirmed that the satellite data captured spatiotemporal variation of the surface ecosystem, that the primary production calculated from the satellite data correlated with the amount of sedimentation of particles in the deeper ecosystem, and that the amount of sedimented particles correlated with benthic biomass. For this reason, we concluded that satellite data, which has uniform accuracy, covers the entire region, and can be expected to provide year-round data, are an effective tool for examining APEI. On the other hand, it was also found that water depth and the amount of manganese nodules, which are unrelated to the surface ecosystem, also affect benthic biomass, and because APEI study also requires biodiversity and linkages between ecosystems,

T. Fukushima (✉)
Japan Organization for Metals and Energy Security (JOGMEC), Tokyo, Japan
e-mail: fukushima-tomohiko@jogmec.go.jp

A. Tsune · C. Murakami
Deep Ocean Resources Development Co., Ltd. (DORD), Tokyo, Japan

H. Sugishima
IDEA Consultants, Inc. (IDEA), Shizuoka, Japan

R. Sharma (ed.), *Deep-Sea Mining and the Water Column*,
https://doi.org/10.1007/978-3-031-59060-3_8

satellite data alone are not sufficient. Given these strengths and weaknesses, we believe that satellite data could be utilized for primary screening for APEI determination.

Keywords Areas of Particular Environmental Interest (APEI) · Regional Environmental Management Plan (REMP) · Clarion-Clipperton zone (CCZ) · Net primary production (NPP) · Total mass flux · Benthic community

1 Introduction

The International Seabed Authority (ISA) is an autonomous international organization established under the 1982 United Nations Convention on the Law of the Sea (UNCLOS) and the 1994 Agreement Relating to the Implementation of Part XI of the United Nations Convention of the Law of the Sea (Agreement).[1] The mission of the ISA is to govern ocean mineral resources activities in the international seabed area beyond the limits of national jurisdiction for the benefit of humankind as a whole. Given this, the ISA undertakes various efforts to effectively protect the marine environment from the harmful effects of activities in the deep sea. Regional Environmental Management Plans (REMPs) are one of these efforts, and the main tools of REMPs are Areas of Particular Environmental Interest (APEIs), which are the subject of this chapter.[2] APEIs are a network of marine protected areas established in a target region that aim to protect and conserve the biodiversity and ecosystem functions of the entire area[3,4] (Smith et al., 2007). To meet the aim, the number, location, or area of APEI must be determined after comprehensively examining the spatio-temporal variation of the local environment. In CCZ-REMP, species/taxon richness, evenness and community structure, biogeography and genetic connectivity as biodiversity data, species range and the distance of APEIs and contractor area as biogeography and genetic connectivity data, SCOC (sediment-community oxygen consumption), C-cycling and C-flows, rates and depths of bioturbation, carbon burial, calcium carbonate dissolution, nutrient fluxes, and biotic abundance and biomass as ecosystem structure and functions data, in addition to POC-flux, water depth, and sediment properties as potential drivers, were referred.[5]

[1] ISA; https://www.isa.org.jm/about-isa/ (accessed 24 July 2023).

[2] ISA; https://www.isa.org.jm/protection-of-the-marine-environment/regional-environmental-management-plans/ (accessed 24 July 2023).

[3] ISA (2008). "Rationale and recommendations for the establishment of preservation reference areas for nodules mining in the Clarion- Clipperton Zone" ISBA/14/LTC/2*, International Seabed Authority, 12.

[4] ISA (2015) Biodiversity, Species Ranges and Gene Flow in the Abyssal Pacific Nodule Province: Predicting and Managing the Impacts of Deep Seabed Mining, ISA Technical Study 3, 45p.

[5] ISA (2019) Deep CCZ Biodiversity Synthesis Workshop Friday Harbor, Washington, USA, 1-4 October 2019 WORKSHOP REPORT, ISA 206p.

In practice, the central Pacific Ocean is currently the only region with sufficient data to have established APEIs as part of the CCZ-REMP,[6] and comparable spatio-temporal data are lacking for the north Mid-Atlantic Ridge (nMAR) REMP,[7] which is in draft stage, and the Indian Ocean-REMP[8] and Northwest Pacific REMP,[9] which are planning to consider their APEIs in the future (Cecchetto et al., 2023). There are several possible reasons for the delay in REMP development and APEI identification. One is the lack of comparable data due to discrepancies in data accuracy. Since REMP covers a wide regional area, data collected by a single organization are not sufficient; however, if data are provided by multiple organizations, the unification of methods cannot be guaranteed, resulting in the dilemma of nonuniform accuracy (inconsistent accuracy). The second is the spatial bias of data collection. The main data used for APEI study are the baseline data provided by the contractors who sign exploration contracts with the ISA. In that case, the data collected within the exclusive exploration area by each contractor become the main data source,[10] and the data outside the mining area are not taken into consideration (spatial heterogeneity). The third is the lack of time-series data. Most contractors conduct baseline surveys on an annual basis, resulting in insufficient seasonal data (lack of time-series data). To resolve these issues, it would be better to standardize data collection methods, investigate the target area thoroughly for each season, and then consider the number, location, and size of APEIs. However, it is easy to imagine that the road to get to that point is absurdly long. The ISA is aiming for the early establishment of exploitation regulation.[11] However, forming the regulation does not mean that development will begin immediately, but it will be possible to apply for a Plan of Work (PoW) for development in any area beyond national jurisdiction. This means the time allotted for establishing APEI is limited. Therefore, in this chapter, utilization of satellite data was evaluated as a simple method to deal with problems such as nonuniformity

---

[6] ISA (2019) Review of the implementation of the Environmental Management Plan for the Clarion-Clipperton Zone Report and recommendations of the Legal and Technical Commission, ISBA/26/C/43.

[7] ISA (2022) Regional environmental management plan for the Area of the northern Mid-Atlantic Ridge with a focus on polymetallic sulphide deposits Issued by the Legal and Technical Commission, ISBA/27/C/38.

[8] ISA (2023) Report of the Chair of the Legal and Technical Commission on the work of the Commission at the second part of its twenty-eighth session, ISBA/28/C/5/Add.1.

[9] ISA(2018) Towards the Development of a Regional Environmental Management Plan for Cobalt-Rich Ferromanganese Crusts in the Northwest Pacific Ocean. Report of an International Workshop Convened by the International Seabed Authority and China Ocean Mineral Resources Research and Development Association in Qingdao, ISA TECHNICAL STUDY NO. 23, 62p.

[10] Contractors who have an exploration contract with the ISA must submit a PoW to the ISA for approval. The PoW includes the implementation of an environmental baseline survey and provision of the results to the database (DeepData) managed by ISA. Most of the data used for the REMP study are quoted from DeepData.

[11] ISA (2023) Decision of the Council of the International Seabed Authority on a timeline following the expiration of the 2-year period pursuant to section 1, paragraph 15, of the annex to the Agreement relating to the Implementation of Part XI of the United Nations Convention on the Law of the Sea, ISBA/28/C/24, ISA, 6p.

of data accuracy, spatial nonuniformity, and lack of time-series data. Specifically, the CCZ-REMP and its APEI, which had the most abundant data on deep-sea ecosystems and were examined in detail by experts in 2007 and 2019, were assumed to be the completed model, and an attempt was made to confirm the degree of contribution of satellite data, that is, to evaluate detection of spatiotemporal variations of chlorophyll *a* and primary production that can be inferred from satellite data and verification of the linkage between the surface and bottom ecosystems through the water column.

2 Methods and Materials

The survey area was the CCZ, for which a REMP has already been completed (Fig. 8.1a). The area is located in the North Pacific Gyre, the westward flow of the North Pacific Current to the north along the Clarion fracture zone (hereafter CLA zone), the southward flow of the California Current to the west, and the eastward flow of the North Equatorial Current to the south along the Clipperton fracture zone (hereafter CLIP zone) dominate. Satellite data were collected for1 year from October 2017 to September 2018. Sample data used to compare with the satellite data were obtained from the baseline data conducted in the exclusive exploration area of Deep Ocean Resources Development (DORD west area,[12] hereafter DWA), supplemented with data from existing literature as necessary (Smet et al., 2017; Pape et al., 2017; Wilson, 2017; Hauquier et al., 2018; BGR, 2018).

2.1 *Satellite Data*

2.1.1 Data Source

The data source was Level 3 data from the moderate resolution imaging spectroradiometer (MODIS) aboard the Aqua satellite, an Earth-observing satellite operated by the National Aeronautics and Space Administration (NASA). Among the data, chlorophyll *a*, photosynthetically active radiation (PAR), and sea surface temperature (SST) were extracted. The horizontal resolution was set to 4 km (Table 8.1). Hereafter, MODIS Level 3 data is referred to as satellite data in this chapter.

[12] DORD has two exploration areas, but only the data for the DORD west area (DWA) are included in this chapter. The data were collected from the following three subareas in DWA; High Abundance Area (HAA), Preservation Reference zone (PRZ), and Japan Deep Sea Impact Experiment area (JET area) (Fig. 8.1b).

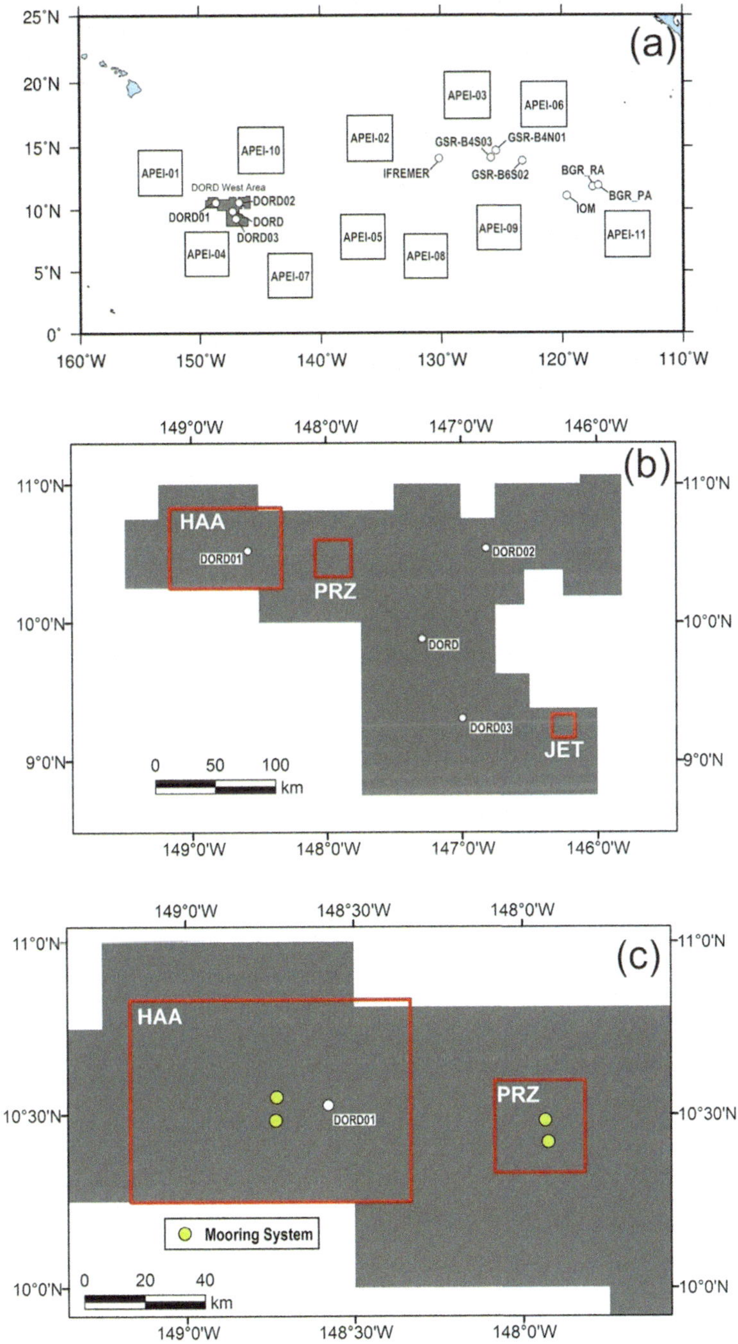

Fig. 8.1 Location maps for this study. (**a**) APEIs, DORD's exploration area, and areas of published data for reference. Published data (IFREMER, GSR-B4SO3, GSR-B6SO2, GSR-B4NO1, BGR_RA, BGR_PA, and IOM) are referred from Pape et al. (2017), Smet et al. (2017), Wilson (2017), and Hauquier et al. (2018). (**b**) HAA, PRZ, and JET sites in the DORD west area, (**c**) St. W, St. C, and St. E in the HAA and PRZ

Table 8.1 Data source

Satellite	National Aeronautics and Space Administration (NASA) Earth-observing satellite/Aqua
Sensor	Moderate Resolution Imaging Spectroradiometer (MODIS)
Collected data	Level 3 data Chlorophyll *a*; Photosynthetically Active Radiation (PAR); Sea Surface Temperature (SST)
Duration	Oct. 2017–Sep. 2018
Covered area	Longitude; 110°W to 160°W Latitude; 0°to 25°N
Resolution	Horizontal 4 km; daily

2.1.2 Estimation of Primary Production

The Vertically Generalized Production Model (VGPM) by Behrenfeld and Falkowski (1997) was adopted to estimate primary production. The VGPM estimates primary production (mgC $m^{-2}.day^{-1}$) from chlorophyll *a*, SST, and PAR and is derived from (Eq. 8.1) below.

$$PP_{eu} = 0.66125 \times P^{b}{}_{opt} \times \left[E_0 / \left(E_0 + 4.1\right)\right] \times Z_{eu} \times C_{opt} \times D_{irr} \tag{8.1}$$

PP_{eu}: Primary production in photic layer (mgC m^{-2} day^{-1})
$P^{b}{}_{opt}$: Maximum C fixation rate within a water column (mgC mg Chl^{-1} h^{-1})
E_0: Sea Surface daily PAR (mol m^{-2} day^{-1})
Z_{eu}: Physical depth receiving 1% of E_0
C_{opt}: Chl concn at $P^{b}{}_{opt}$ (mg m^{-3})
D_{irr}: Decimal hour

The values of PAR and chlorophyll *a*, obtained by MODIS, described above, were applied to E_0 and C_{op}t. D_{irr} was calculated from latitude, longitude, and time. Maximum C fixation rate ($P^{b}{}_{opt}$) is given as (Eq. 8.2) as a function of water temperature (T), so it was calculated using MODIS observed value (SST). Z_{eu} was calculated from C_{opt} using the empirical formula (Eq. 8.3) of Morel and Berthon (1989).

$$\begin{aligned} P^{b}{}_{opt} &= -3.270 \times 10-8T7 + 3.4132 \times 10-6T6 - 1.348 \times 10-4T5 \\ &+2.462 \times 10-3T4 - 0.0205T3 + 0.0617T2 + 0.2749T + 1.2956 \end{aligned} \tag{8.2}$$

$$Z_{eu} = \{ \begin{matrix} \mathbf{568.2\left(Chl_{tot}\right)^{-0.746}} \left(Z_{eu} < \mathbf{102m}\right) \\ \mathbf{200.0\left(Chl_{tot}\right)^{-0.293}} \left(Z_{eu} \geq \mathbf{102m}\right) \end{matrix}$$

$$\mathbf{Chl_{tot}} = \{ \begin{matrix} \mathbf{38.0\left(C_{surf}\right)^{0.423}} & \left(\mathbf{C_{surf} < 1.0\,mgm^{-3}}\right) \\ \mathbf{40.2\left(C_{surf}\right)^{0.505}} & \left(\mathbf{C_{surf} \geq 1.0\,mgm^{-3}}\right) \end{matrix} \tag{8.3}$$

Chl_{tot}: Accumulated chlorophyll in the photic layer
Cs_{urf}: Surface chlorophyll concentration

2.2 *Sampling Data*

Baseline studies conducted in the DWA and wider area data from existing literature to supplement them were used as sample data. The baseline study data were those collected by the mooring systems for mid-layer and bottom-layer observation, Finder-installed Deep-sea camera (FDC), Box Corer (BC), and Multiple Corer (MC).

2.2.1 Mooring Systems

Two types of mooring systems were installed at PRZ and HAA of the DWA from October 2017 to September 2018 (Fig. 8.2, Table 8.2). One was a sediment trap for middle layer observation that was set at 2000 m below sea surface, and the other was a sediment trap for bottom layer observation set at approximately 4700 m. In both cases, the turntable rotates, and a sample can be obtained twice a month. The total mass flux (total flux; $mg/m^2/day$) and total organic carbon (TOC: $mg/m^2/day$) of the obtained sample were measured.

2.2.2 FDC

The FDC is a towed seafloor observation device capable of acquiring still and video imagery data. Observations were made on three survey lines (22.66 km in total) at HAA and two survey lines (6.1 km in total) on the south side of the DWA (JET area) (Fig. 8.3, Table 8.3). The area ratio (%) of manganese nodules (nodules ratio) and the occurrence number of megabenthos (individuals/ha) were calculated from the obtained still photographs and video images. More information about the methodology of the FDC megabenthos survey conducted by DORD can be found in Fukushima et al. (2000) and Fukushima et al. (2022).

2.2.3 BC and MC

BC and MC were used for collection at the HAA, PRZ, and JET sites in the DWA. BC samples were sliced horizontally at 0–1, 1–2, 2–3, and 3–5 cm intervals and used for macrobenthos analysis. MC samples were subsampled with 1.8 cm diameter syringes, sliced horizontally at intervals of 0–0.25, 0.25–0.5, 0.5–0.75, 0.75–1.0, 1.0–1.5, 1.5–2.0, 2.0–2.5, and 2.5–3.0 cm, and were used for metazoan meiobenthos and chemical analyses. Since this chapter deals only with the number of occurrences of metazoan meiobenthos, the method of analyzing chemical samples was omitted (however, see Fukushima and Tsune (2019) for survey methods in the DWA).

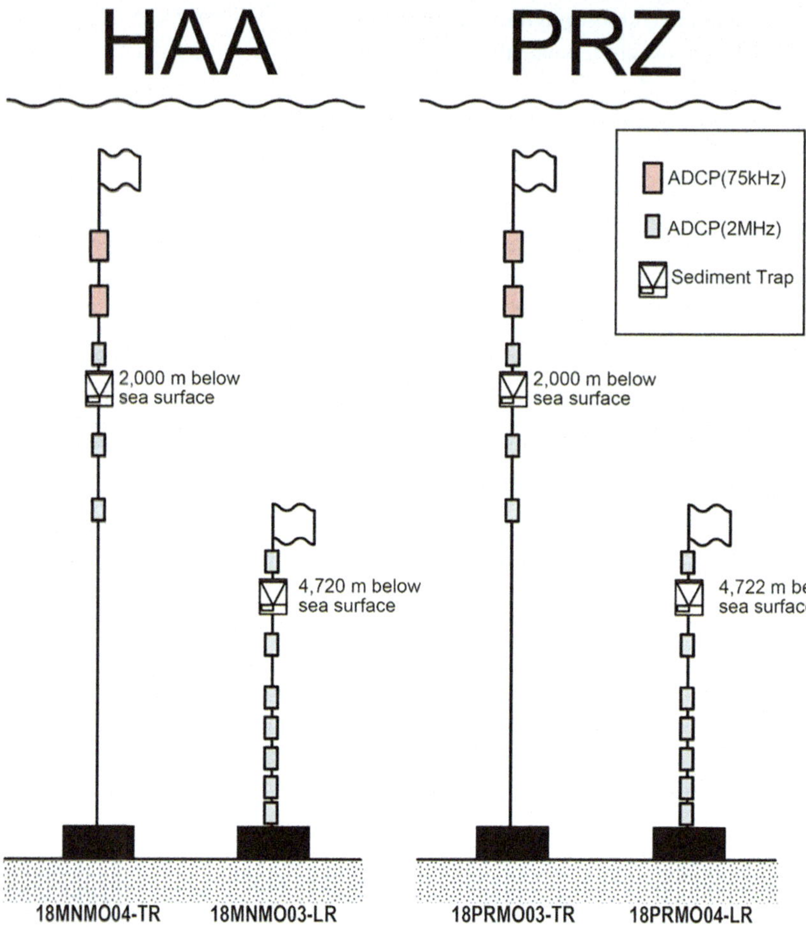

Fig. 8.2 Mooring system design used in HAA and PRZ; (left) mid-layer observation, (right) bottom-layer observation

Table 8.2 Mooring system deployment duration, position, position, and depth

Area	Mooring No	Duration		Longitude	Latitude	Water depth(m)	SD water depth (m)
HAA	18MNMO04-TR	start	2017.10.30	10°33.000′N	148°43.794′W	5227	2000
		end	2018.09.26				
HAA	18MNMO03-LR	start	2017.10.30	10°29.004′N	148°44.000′W	5230	4720
		end	2018.09.26				
PRZ	18PRMO03-TR	start	2017.10.30	10°25.227′N	147°55.498′W	5290	2000
		end	2018.09.25				
PRZ	18PRMO04-LR	start	2017.10.30.	10°28.998′N	147°55.997′W	5232	4722
		end	2018.09.27				

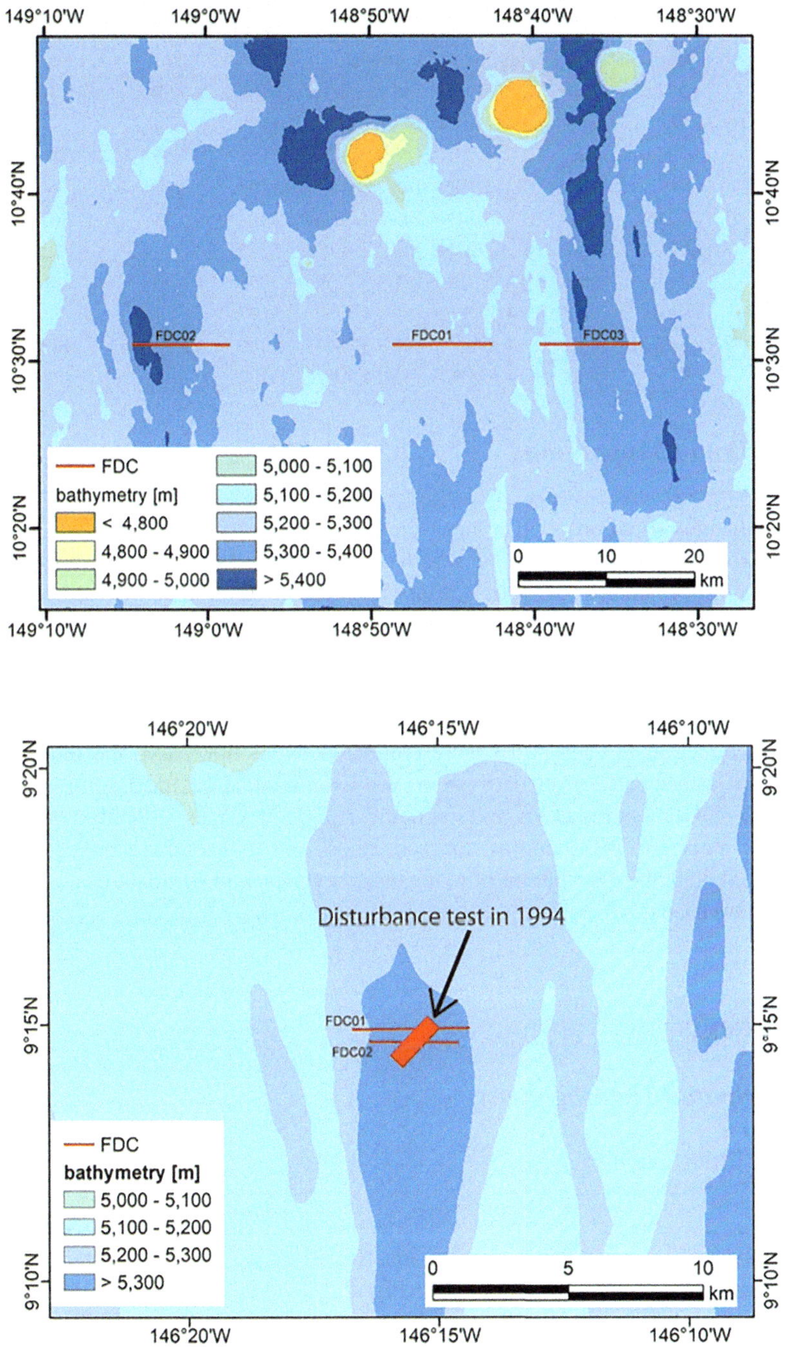

Fig. 8.3 FDC observation lines; (upper) three lines in HAA, (lower) two lines in JET area

Table 8.3 FDC tow lines

Line no	Date		Longitude	Latitude	Depth (m)	Distance (km)	Duration (min.)
HAA-FDC01	19 Nov. 2011	start	10'30.985N	148'48.667W	5268	7.60	359
		end	10'31.015N	148'42.550W	5197		
HAA-FDC02	20 Nov. 2011	start	10'30.978N	148'39.666W	5169	7.55	386
		end	10'31.019N	148'.33.428W	5294		
HAA-FDC03	21 Nov. 2011	start	10'31.011N	148'04.668W	535	6.35	361
		end	10'30.993N	148'.33.428W	5215		
JET-FDC01	7 Nov. 2011	start	9'14.923N	146'16.72W	5268	3.40	188
		end	9'14.957N	146'14.37W	5197		
JET-FDC02	15 Nov. 2011	start	9'14.699N	146'16.386W	5300	2.70	133
		end	9'14.624N	146'14.598N	5290		

2.3 *Data Comparisons*

First, changes in chlorophyll *a* and primary productivity estimated from satellite data were examined to determine whether they reflected spatiotemporal variations in surface ecosystems. For spatial variation, APEIs were divided into the CLA zone (APEI 1, 2, 3, 6, and 10) and the CLIP zone side (APEI 4, 5, 7, 8, 9, and 11), and chlorophyll *a* and primary production between the north and south were compared. At the same time, latitudinal and longitudinal variations of each APEI within the CCZ were confirmed. For time-series variation, changes in chlorophyll *a* and primary production in each APEI were compared by month. Next, the relationship between surface and bottom ecosystems was examined. Here, the baseline data collected by DORD in the DWA and the number of occurrences of benthic organisms such as metazoan meiobenthos, macrobenthos, and megabenthos collected from published data, the distribution of manganese nodules, and flux data from sediment traps were used.

3 Results

3.1 *Spatial Heterogeneity*

3.1.1 Chlorophyll *a*

The annual average chlorophyll *a* was less than 0.1 mg/m^2 in the northern APEIs along the CLA zone (APEI 1, 2, 3, 6, and 10), while it exceeded 0.1 mg/m^2 in the southern APEIs along the CLIP zone (APEI 4, 5, 7, 8, 9, and 11) except for APEI 9 (Fig. 8.4). Regarding the relationship between latitude and chlorophyll *a*, the correlation was significant ($p < 0.05$) in the CLIP zone, but not in the CLA zone (Fig. 8.5). On the other hand, chlorophyll *a* decreased slightly westward in the CLIP zone and increased slightly westward in the CLA zone, but neither was statistically

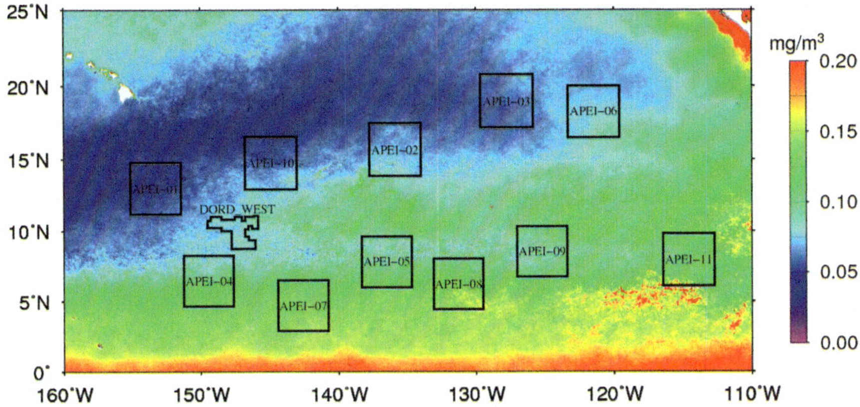

Fig. 8.4 Annual average of chlorophyll *a* distribution

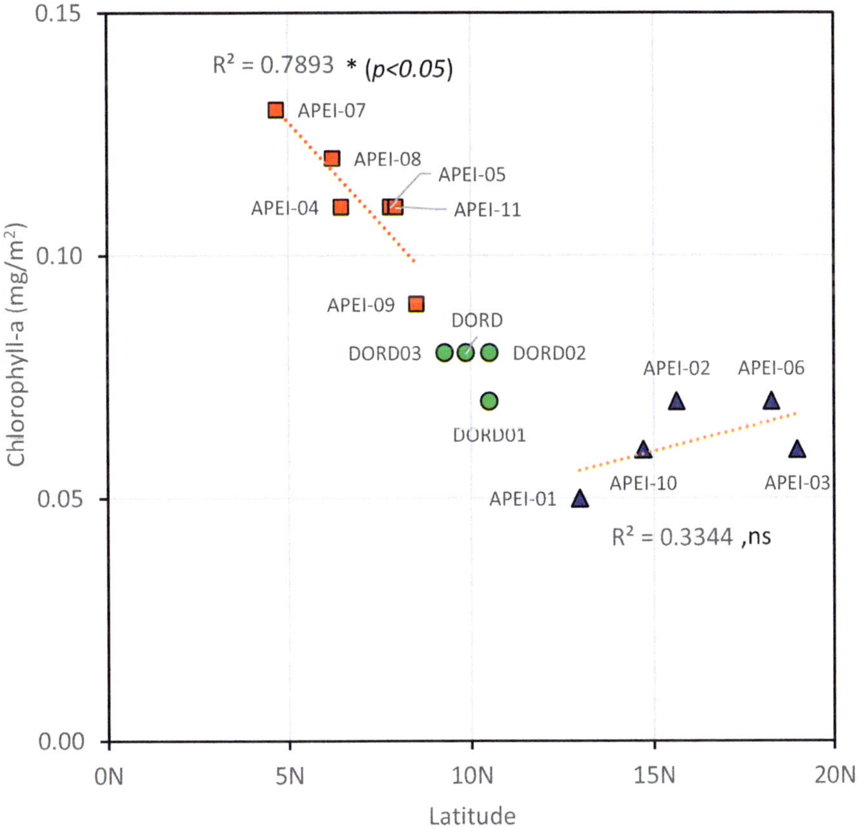

Fig. 8.5 Variation of chlorophyll *a* with latitude. Red squares indicate APEIs in the CLIP zone, green circles indicate DWA, and blue triangles indicate APEIs in the CLA zone

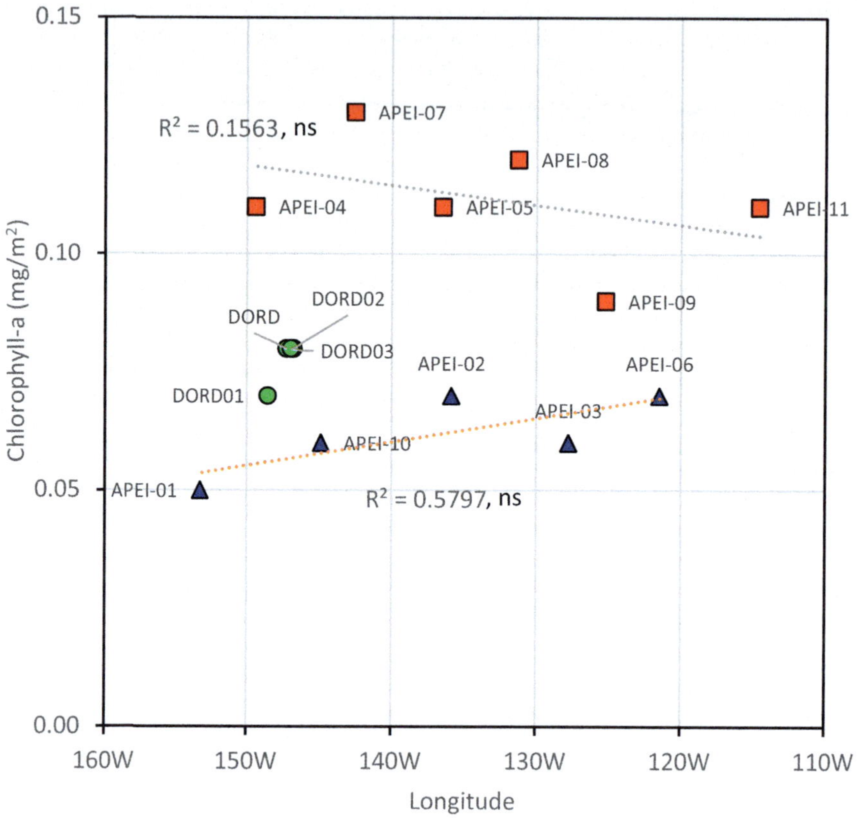

Fig. 8.6 Variation of chlorophyll *a* value with longitude. Red squares indicate APEIs in the CLIP zone, green circles indicate DWA, and blue triangles indicate APEIs in the CLA zone

significant (Fig. 8.6). This result, though not statistically clear, is consistent with Lutz et al. (2007), who showed trends with latitude and longitude of NPPs in the CCZ.

3.1.2 Primary Production

Similar to chlorophyll *a*, primary production showed significant differences between north and south every month (Fig. 8.7a, b). On the other hand, when we compared the latitude differences within each zone, the CLA zone showed increased primary production with latitude ($p < 0.05$), while the CLIP zone showed decreased primary production with latitude ($p < 0.05$) (Fig. 8.8). As for the difference with longitude, similar to chlorophyll *a*, the CLA zone showed increased primary production toward the west ($p < 0.01$), while the CLIP zone showed no definite trend (Fig. 8.9).

3.2 *Temporal Heterogeneity*

3.2.1 Chlorophyll *a*

Monthly fluctuations of chlorophyll *a* values altered synchronously in all APEIs in the CLA zone and peaked in January except for APEI 9 (Fig. 8.10). On the other hand, in the CLIP area, although there was a large difference between APEIs, a maximum value that seems to be a spring bloom appeared from February to May (Fig. 8.11).

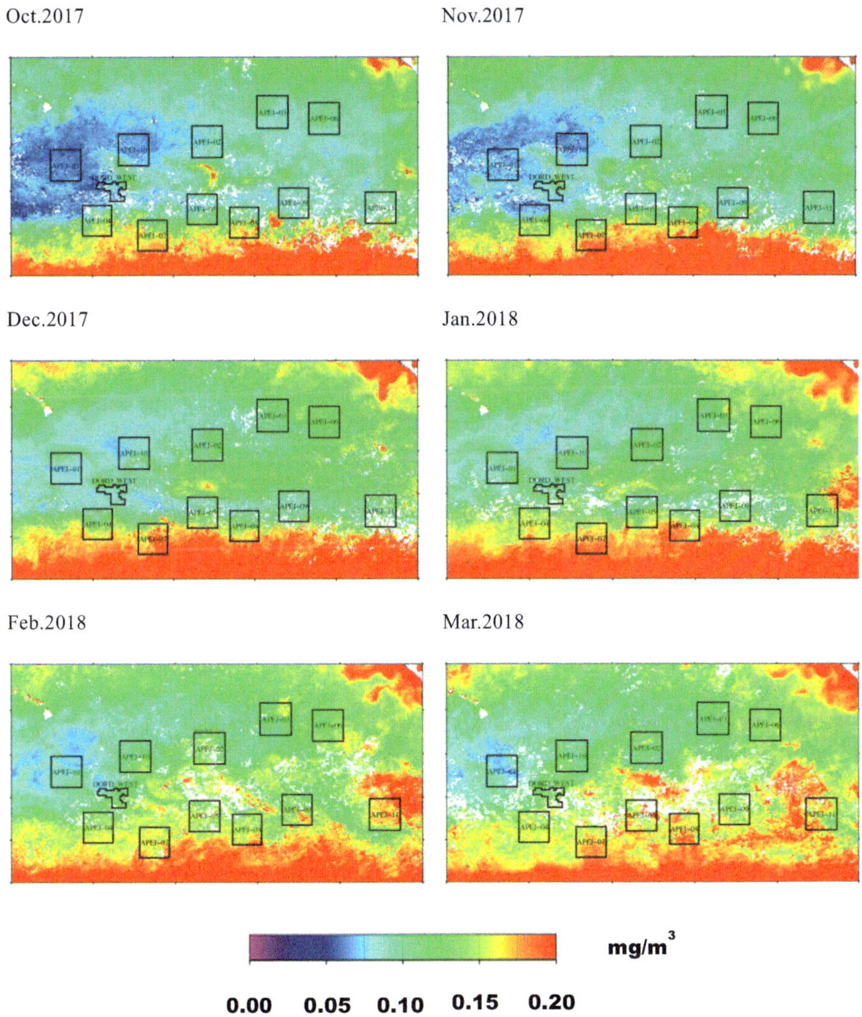

Fig. 8.7 (**a**) Monthly fluctuations in primary production (October 2017 to May 2018). (**b**) Monthly fluctuations in primary production (June 2018 to September 2018)

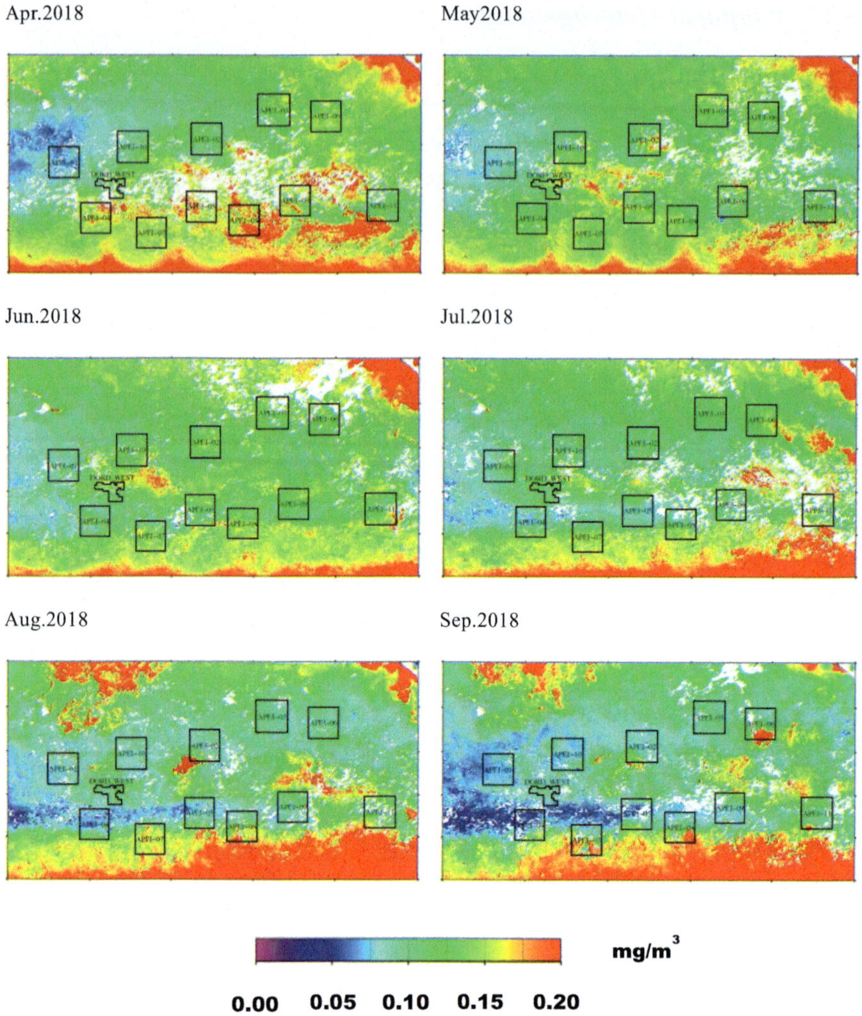

Fig. 8.7 (continued)

3.2.2 Primary Production

Temporal heterogeneity in primary production was recognized (Fig. 8.7a, b). When comparing October 2017 to March 2018 with April to September 2018, primary production in the former is clearly larger. In the CLA zone, only APEI 1 showed a different monthly fluctuation, but the other APEIs were relatively high from January to June and low in the other months (Fig. 8.12). In the CLIP zone, the difference between APEIs was large from October to January, but the difference remained at 250–300 mg/m^2/day in May–June (Fig. 8.13).

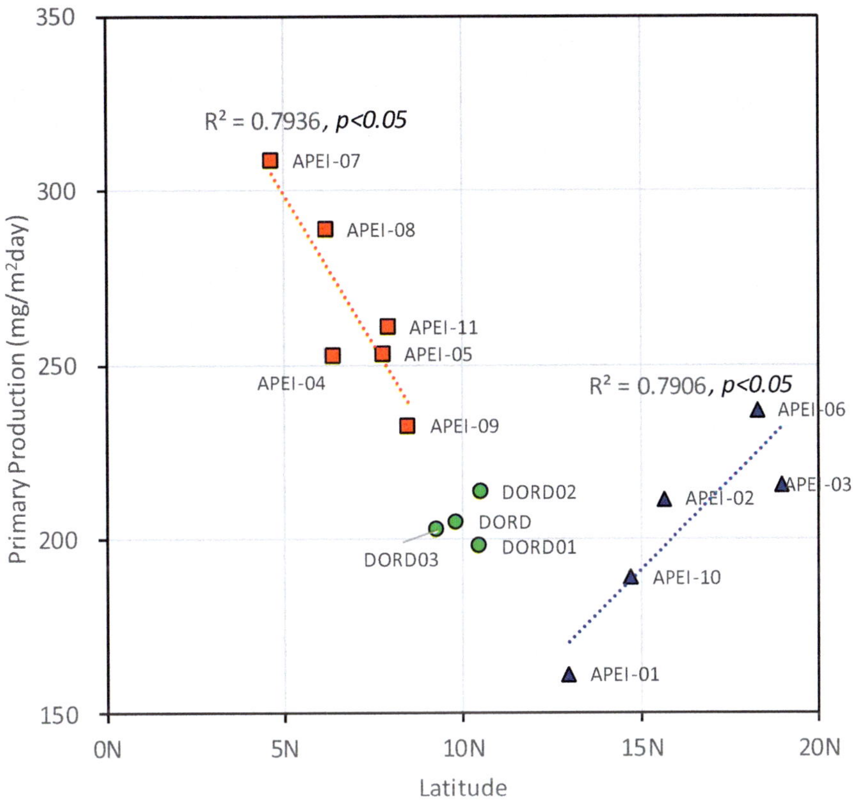

Fig. 8.8 Variation of NPP with latitude. Red squares indicate APEIs in the CLIP zone, green circles indicate DWA, and blue triangles indicate APEIs in the CLA zone

3.3 *Linkage of Upper Ecosystem*

In order to examine the linkage between the upper ecosystem and the benthic ecosystem, we investigated the correlation between net primary production (NPP) estimated from satellite data, particle flux (total flux and TOC flux), and biomass of benthic fauna (metazoan meiobenthos, macrobenthos, and megabenthos). There are many research reports on the linkage between the upper ecosystem and the bottom ecosystem (Berelson et al., 1997; Smith & Rabouille, 2002; Smith & Demoooulos, 2003; Corliss et al., 2009). This study will confirm them using our data.

3.3.1 NPP and Particle Fluxes

Monthly data of total amounts of flux and TOC flux obtained from sediment traps at 2000 m and 4700 m below the sea surface in both HAA and PRZ and monthly NPP values calculated from satellite data were compared. As the results, in both HAA

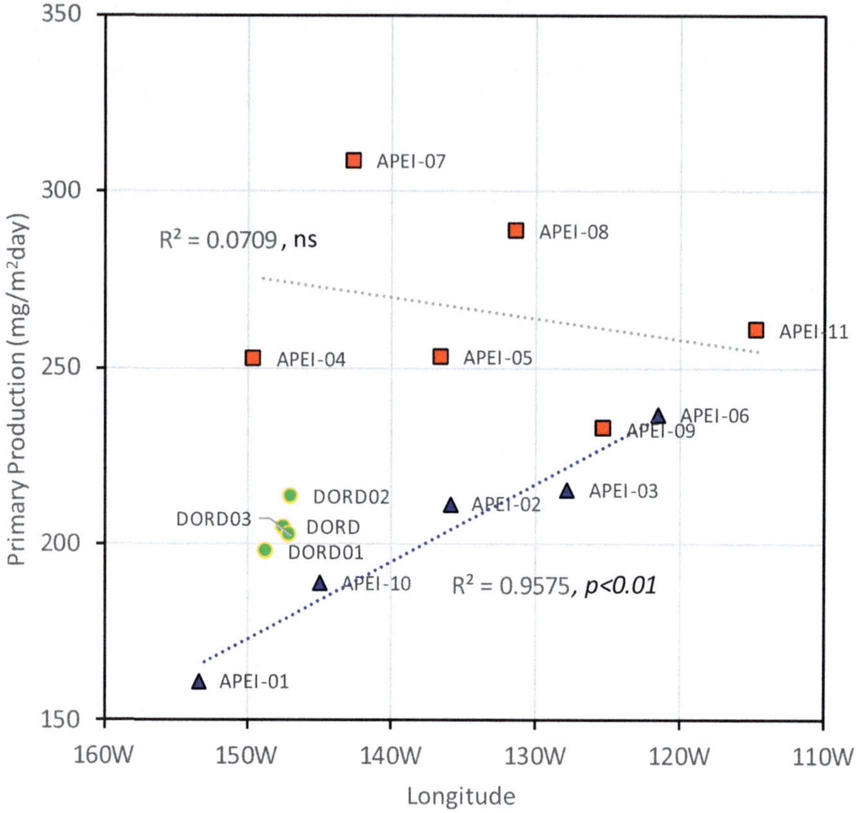

Fig. 8.9 Variation of NPP value with longitude. Red squares indicate APEIs in the CLIP zone, green circles indicate DWA, and blue triangles indicate APEIs in the CLA zone

and PRZ, NPP was statistically significantly correlated with total flux and TOC flux (Fig. 8.14).

3.3.2 Water Depth and Particle Flux

Comparing the total fluxes and TOC fluxes by depth, sediment traps at 2000 m below the sea surface collected more than those at 4700 m over the year at both HAA and PRZ (Fig. 8.15). During the observation period, 15 days of total fluxes were collected for 23 times. However, the last sample had an insufficient sampling period and was excluded from the quantitative data in Fig. 8.15.

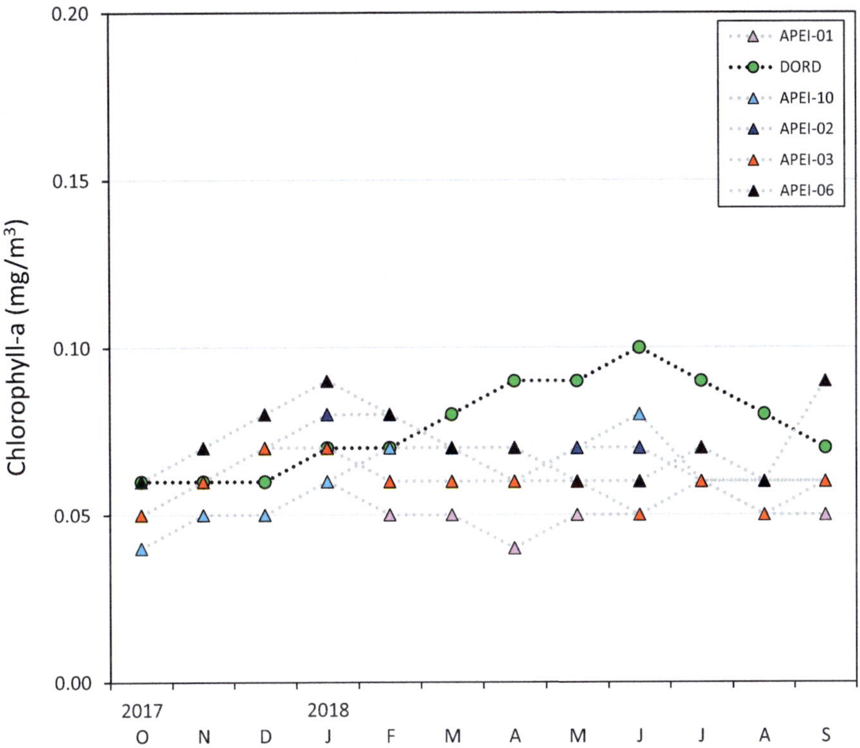

Fig. 8.10 Monthly variation of chlorophyll *a* in the CLA zone

3.3.3 Benthic Organisms and Water Depth

The relationship between benthic organisms and water depth was examined by complementing the data obtained in the DWA with existing literature values. The benthic organisms targeted here were metazoan meiobenthos and macrobenthos. For metazoan meiobenthos, 22 samples from HAA, four samples from PRZ, and 25 samples from the JET area were used. For macrobenthos, 15 samples from HAA, 2 samples from PRZ, and 18 samples from the JET area were used. Data from the existing literature that supplemented the sample data were the data groups on the east side of the collection point of the DWA (Fig. 8.1a–c). As a result, for metazoan meiobenthos abundance decreased significantly with increasing water depth in the range of 4000–5500 m ($p < 0.01$) (Fig. 8.16a). While macrobenthos abundance also showed a trend to decrease with water depth, it was not statistically significant (Fig. 8.16b). In general, the deep sea is considered an extremely food limited environment with C and N limitation increasing with increasing water depth down to 6000 m depth (Sweetman et al., 2019), then the water depth is considered to be a factor in determining biomass.

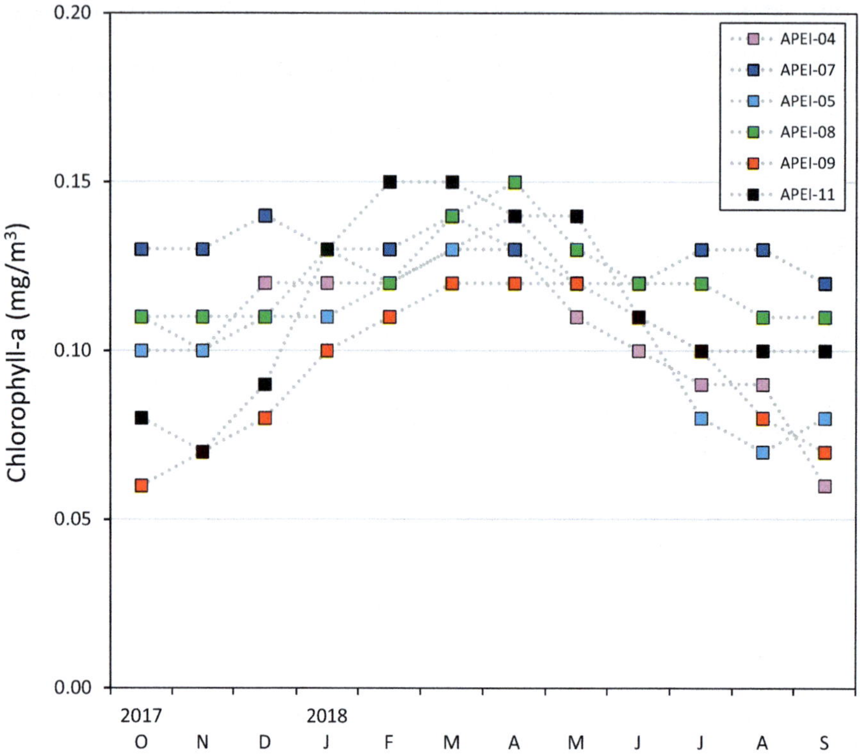

Fig. 8.11 Monthly variation of chlorophyll *a* in the CLIP zone

3.3.4 NPP and Benthic Organism

Comparing the mean of annual NPP calculated from satellite data with the abundance of benthic organisms, there was significant correlation in both metazoan meiobenthos and macrobenthos ($p < 0.05$ for metazoan meiobenthos, $p < 0.01$ for metazoan meiobenthos) (Fig. 8.17a, b). These results are consistent with that deep sea biological communities greatly depend on the transfer of carbon and nitrogen from the surface (Smith et al., 2008, 2009).

3.3.5 Benthic Organisms and Coverage Ratio of Manganese Nodules

Comparing the coverage ratio of manganese nodules observed by FDC and the abundance of megabenthos, there was a statistically significant correlation between total megabenthos and nodules ratio in 1 of 3 HAA lines ($p < 0.05$). For Porifera and Cnidaria, the correlation was not significant for both lines, but for Echinodermata, 2 out of 3 lines were significantly correlated ($p < 0.01$, <0.05) (Fig. 8.18). While in JET area, there are trends to correlations between each megabenthos and nodules

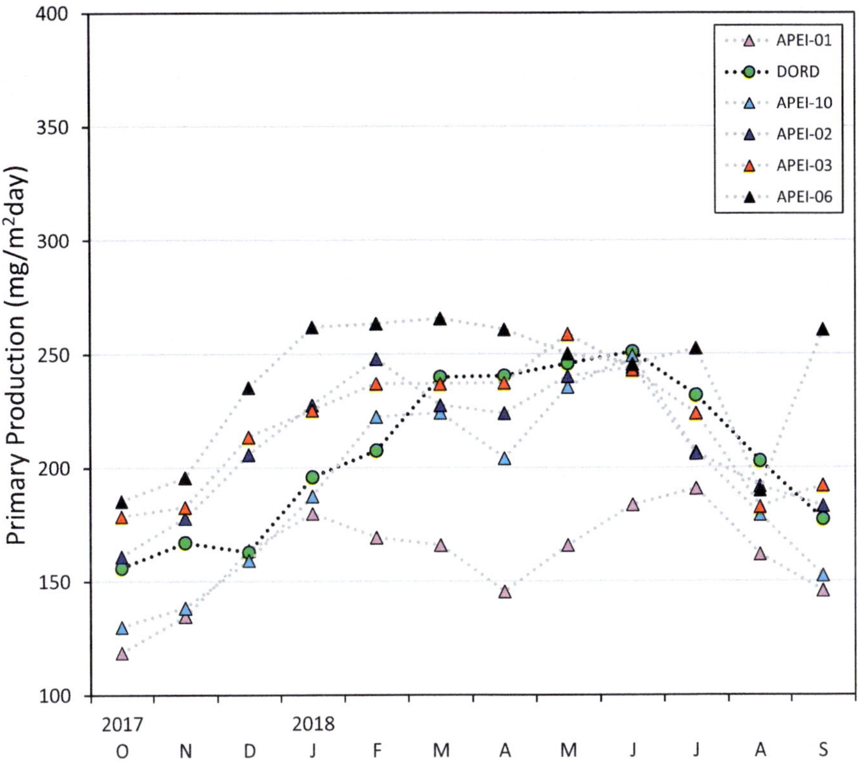

Fig. 8.12 Monthly variation of primary production in the CLA zone

ratio, but not significant (Fig. 8.19). There are already several studies on nodule densities and megabenthos abundance, all of which state that there is a close relationship between the two, although there is a difference in degree of generality Amon et al., 2016; Vanreusel et al., 2016; Simon-Lledo et al., 2019; Jones et al., 2019.

4 Discussion

The goals of this study were to evaluate the utilization of satellite data as a simple method to deal with problems such as nonuniformity of data accuracy, spatial nonuniformity, and lack of time-series data. Specifically, the CCZ-REMP and its APEI, which had the most abundant data on deep-sea ecosystems and were examined in detail by experts in 2007 and 2019, were assumed to be the completed model, and an attempt was made to confirm the degree of contribution of satellite data, that is, to evaluate detection of spatiotemporal variations of chlorophyll *a* and primary production that can be inferred from satellite data, and verification of the connection between the surface and bottom ecosystems through the water column.

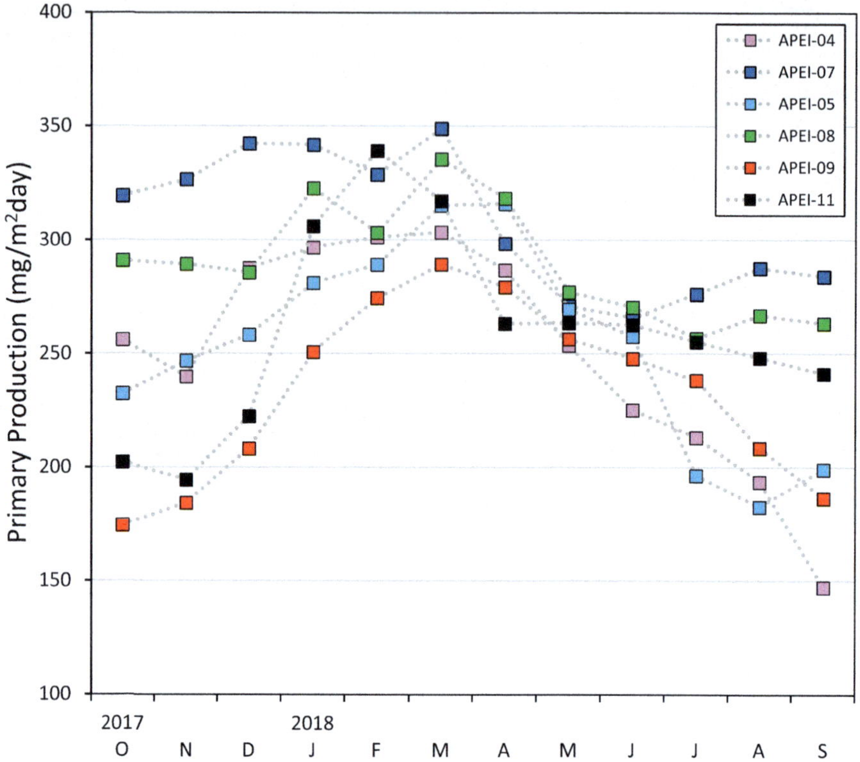

Fig. 8.13 Monthly variation of primary production in the CLIP zone

First, we examined the effectiveness of the satellite data used in this study. The amount of chlorophyll *a* and primary production calculated from satellite data reflect variations in the east-west and north-south directions of the target region, as well as annual variations such as spring blooms.

Next, the linkage between the surface ecosystem and the bottom ecosystem, which is the target of APEI, was examined. With the exception of deeper water chemosynthetic habitats, deep sea biological communities greatly depend on the transfer of organic materials from the surface have been already published and the present study confirmed those results (Smith et al., 1997; 2008). A significant correlation was observed between the primary production calculated from the satellite data and the abundance of metazoan meiobenthos and macrobenthos at the same location. The surface ecosystem and the bottom ecosystem linkage were examined from the process of material flow. As a result, a significant correlation was found between the primary production calculated from the satellite data and the total fluxes and TOC fluxes collected by sediment traps near the seafloor. In addition, a significant correlation was observed when comparing the amounts of total fluxes and TOC fluxes with the abundance of benthic organisms, including existing literature values. From these results, it is reasonable to speculate that the organic particles

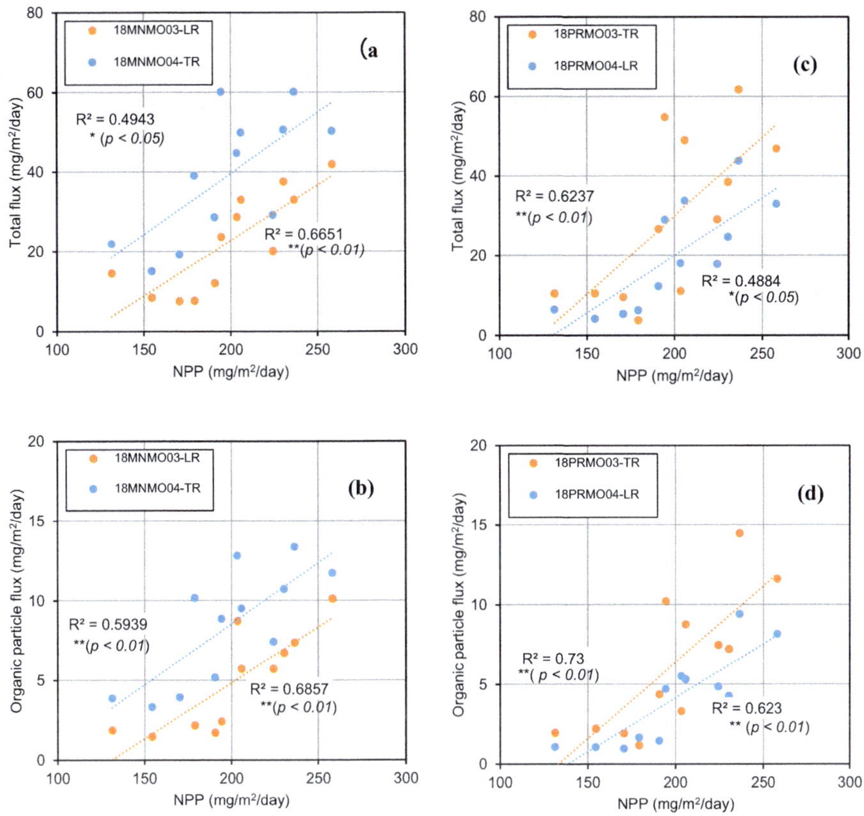

Fig. 8.14 Correlation of NPP and fluxes. (**a**) Relationship between NPP and total flux at HAA, (**b**) relationship between NPP and organic particle flux at HAA, (**c**) relationship between NPP and total flux at PRZ, (**d**) relationship between NPP and organic particle flux at PRZ

produced in the surface layer are transported vertically, becoming a food source for benthic organisms, consequently defining the benthic abundance. This assumption was reinforced by the results of separate comparisons of the correlation between water depth and the total fluxes and the correlation between water depth and benthic biomass.

Based on the above, satellite data, covers the entire region with certain precision can be expected to provide year-round data, andcan be evaluated as an effective tool for examining APEIs. The issue is the degree of effectiveness. Needless to say, surface organic matter input is not the only factor affecting benthic biomass or bottom ecosystems. The results of this study confirm that water depth and abundance of manganese nodules are correlated with benthic abundance, which is independent of variation in surface ecosystems. The former can be explained by the consumption of organic matter at intermediate depths until it reaches the seafloor. In connection with this result, Dunne et al. (2007) showed that the food supply to the seafloor,

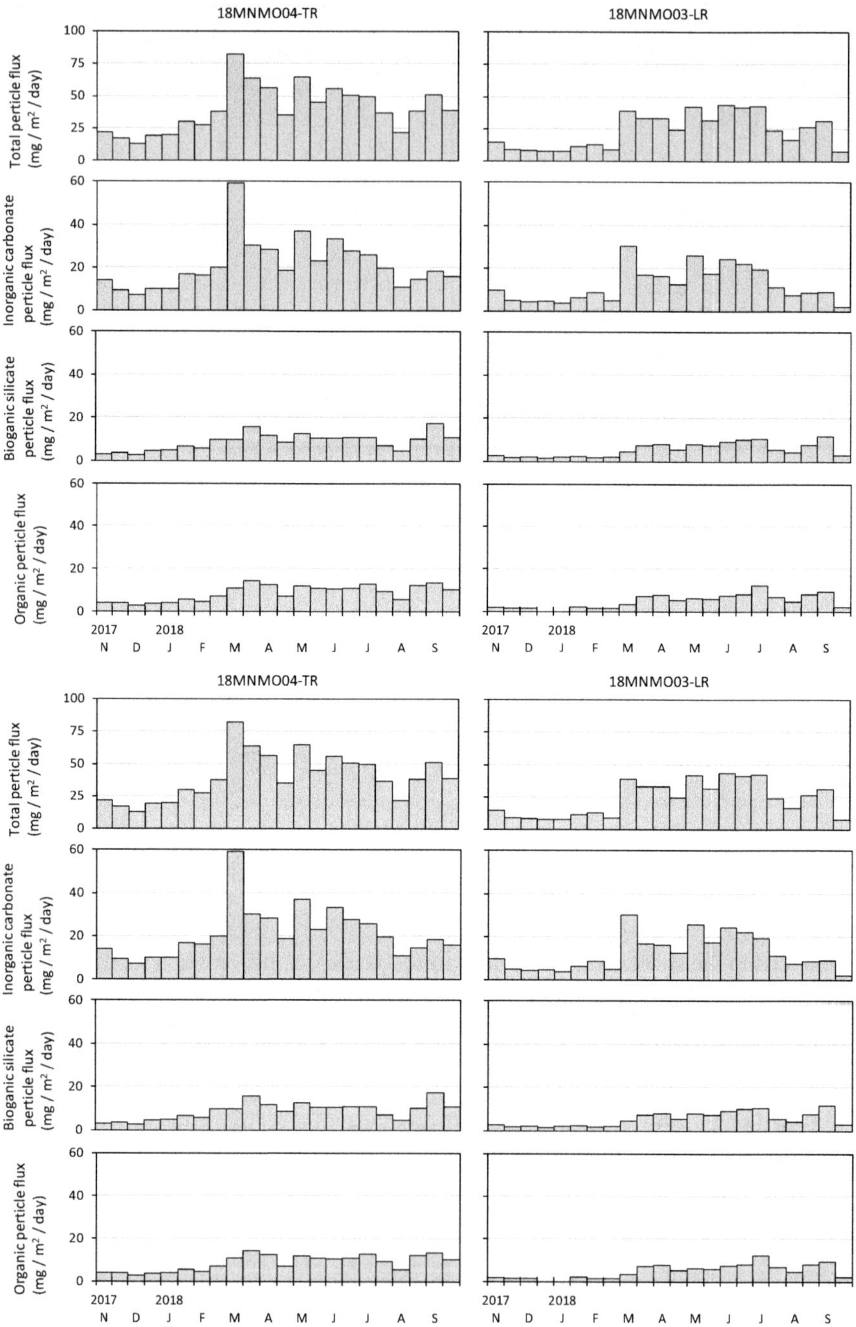

Fig. 8.15 Comparisons of monthly fluxes by water depth (upper left 2000m; upper right 4722m in HAA, lower left 2000m, lower right 4720m in PRZ)

Fig. 8.16 (**a**) Relationship between the metazoan meiobenthos abundance and water depth. (**b**) Relationship between macrobenthos abundance and water depth

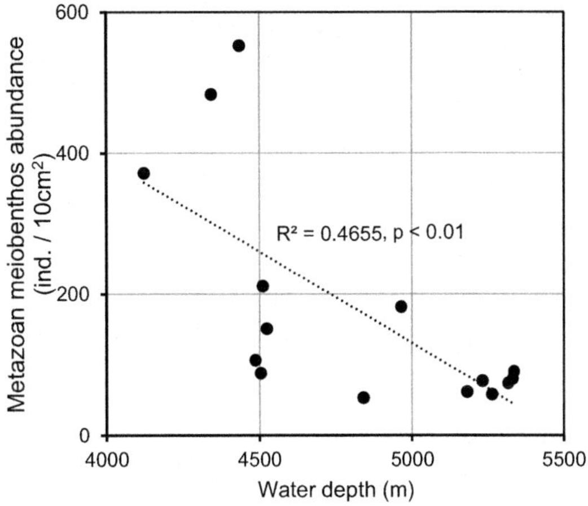

a Relationship between the metazoan meiobenthos abundance and water depth

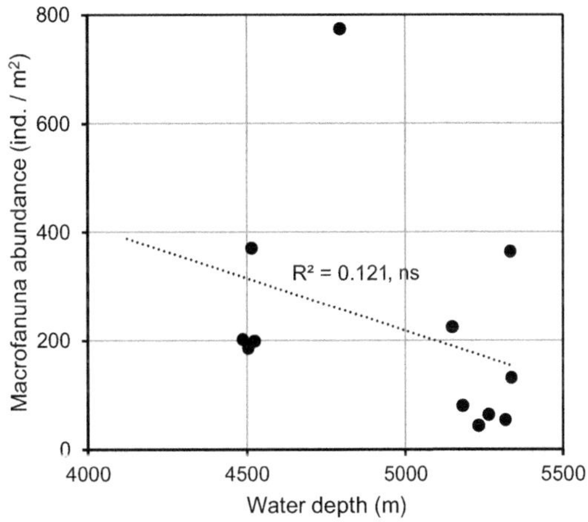

b Relationship between macrobenthos abundance and water depth

defined as the particulate organic carbon flux, often corresponds to only 0.5–2% of the total surface net primary production, and indicates the magnitude of the consumption of organic matter at intermediate depths. And the latter can be explained by the complexity of the habitat. In addition, although this study focused on the abundance of benthic organisms, APEIs also require examination of the diversity of benthic communities and the linkages between ecosystems. In either case, it is essential to acquire physical data such as seafloor topography and bottom currents, which cannot be estimated from satellite data. Furthermore, the CCZ-REMP covers only the abyssal plain, whereas the nMAR covers the mid-ocean ridge, the Indian Ocean-REMP covers the mid-ocean ridge and abyssal plain, and the Northwest

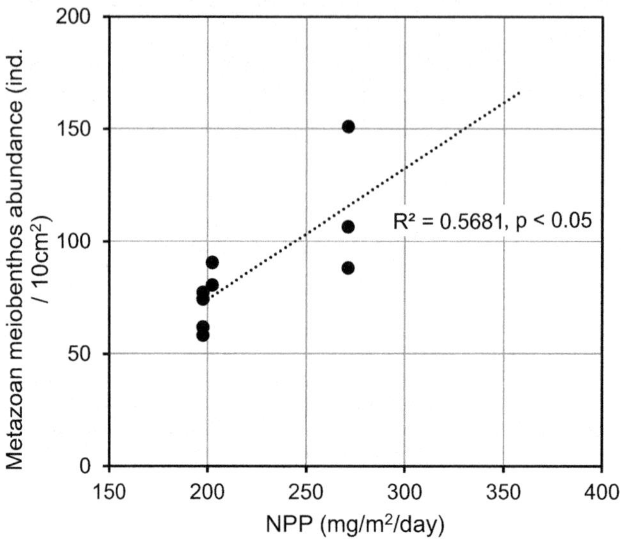

a. Relationship between metazoanmeiobenthos abundance and primary production

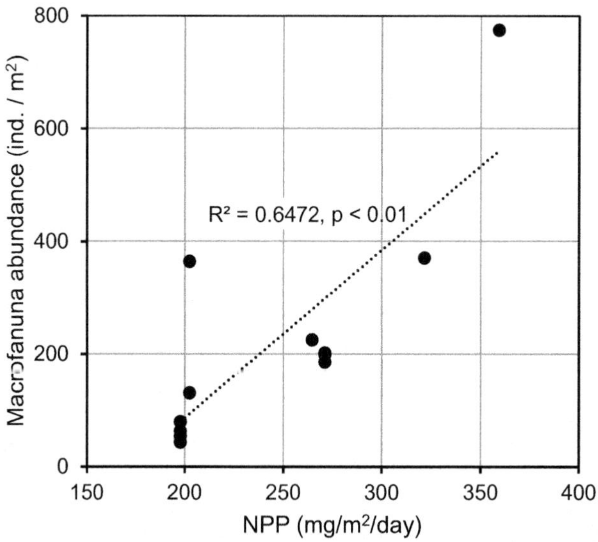

b Relationship between macrobenthos abundance and primary production

Fig. 8.17 (**a**) Relationship between metazoan meiobenthos abundance and primary production. (**b**) Relationship between macrobenthos abundance and primary production

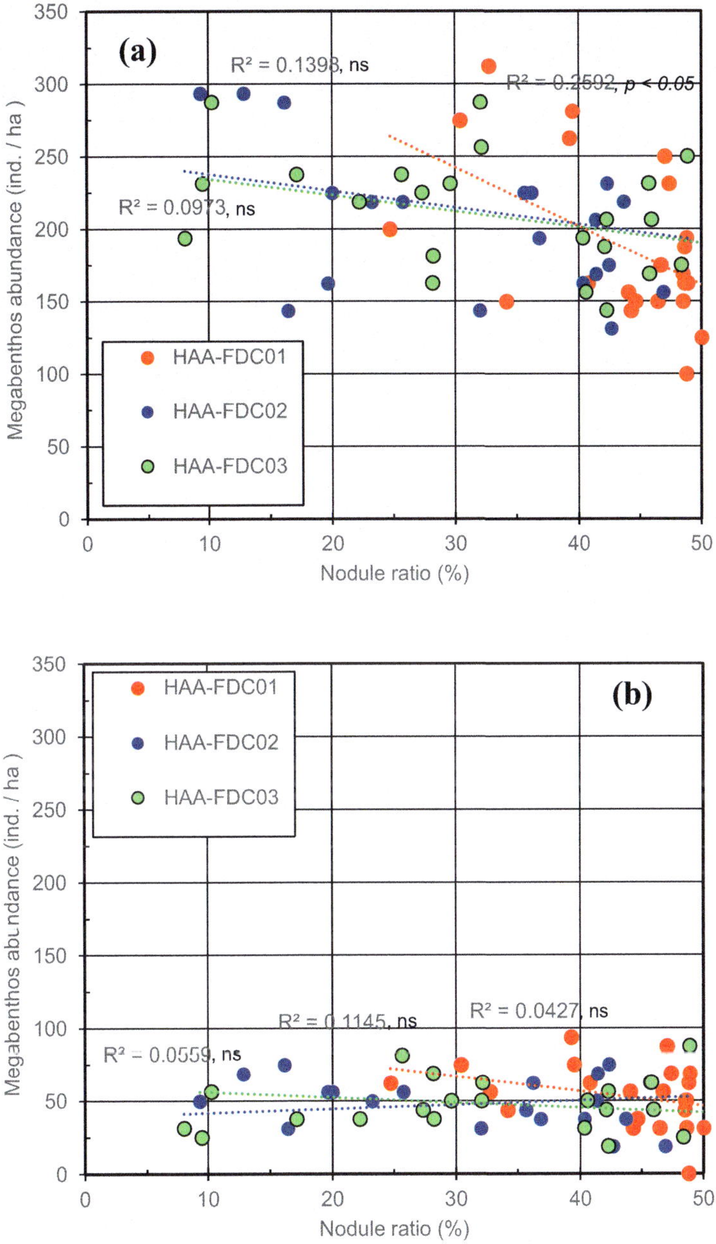

Fig. 8.18 Relationship between megabenthos abundances and nodules coverage in HAA (**a**) Total number of megabenthos, (**b**) Porifera, (**c**) Cnidaria, (**d**) Echinodermata

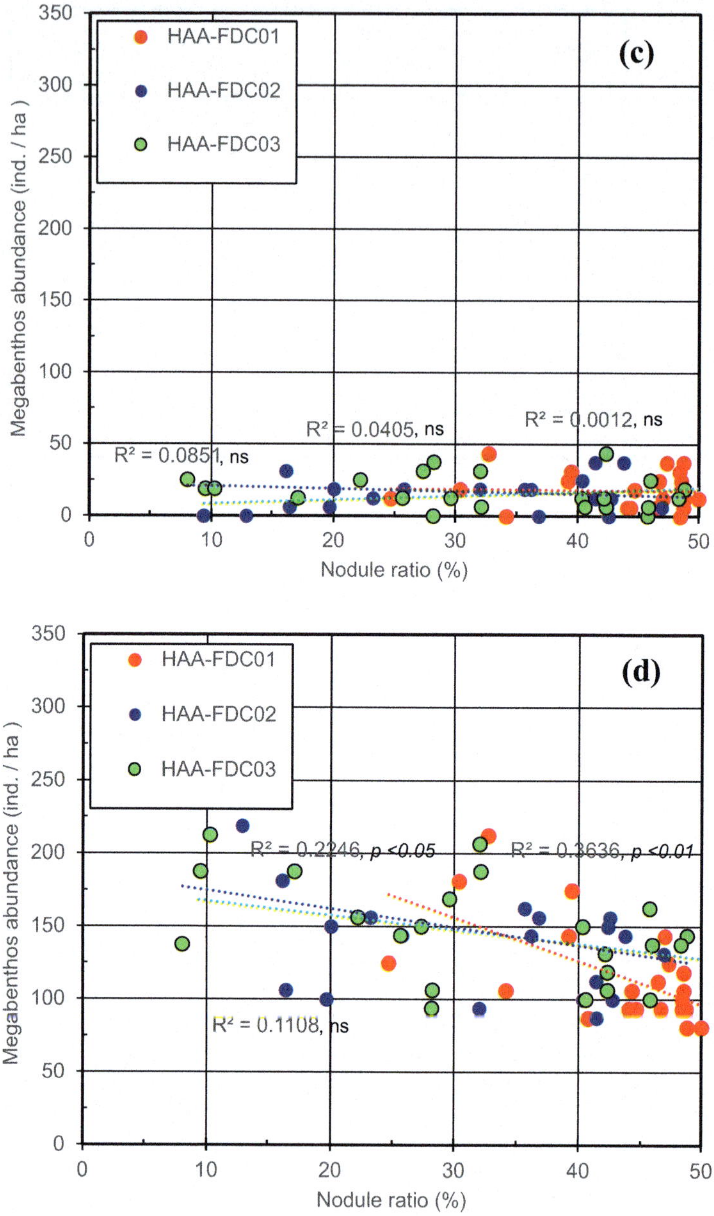

Fig. 8.18 (continued)

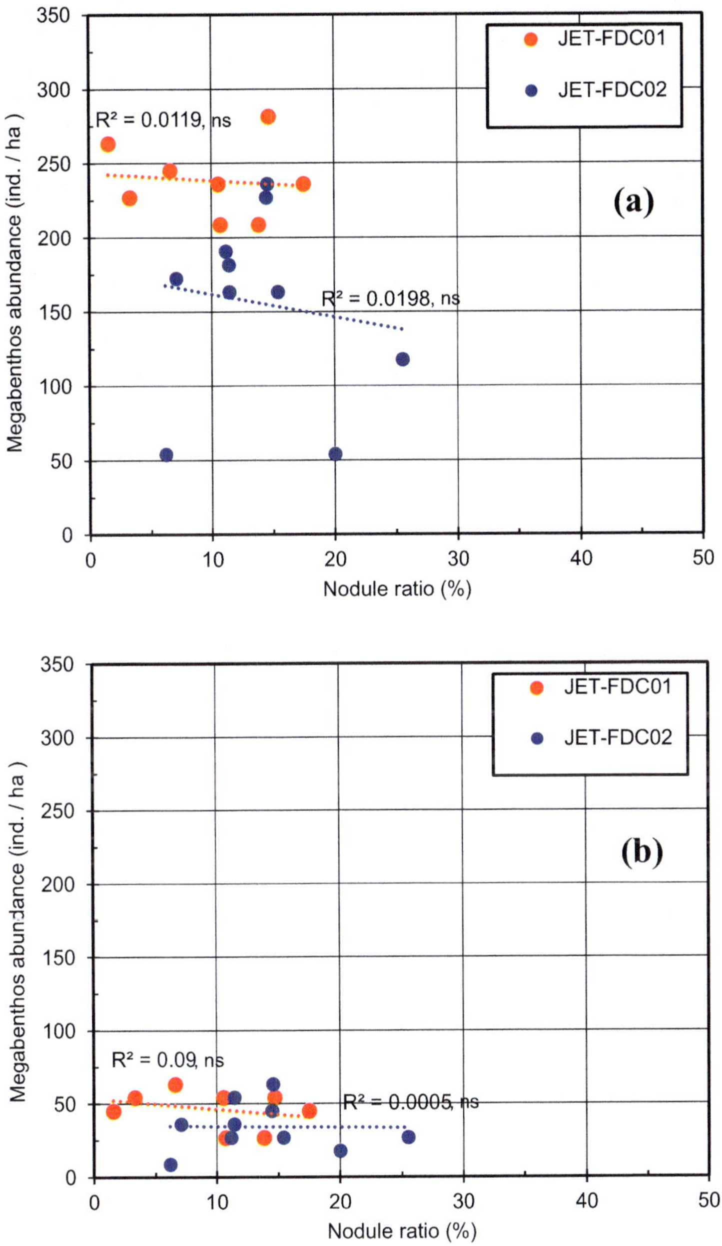

Fig. 8.19 Relationship between megabenthos abundances and nodules coverage in JET area (**a**) Total number of megabenthos, (**b**) Porifera, (**c**) Cnidaria, (**d**) Echinodermata

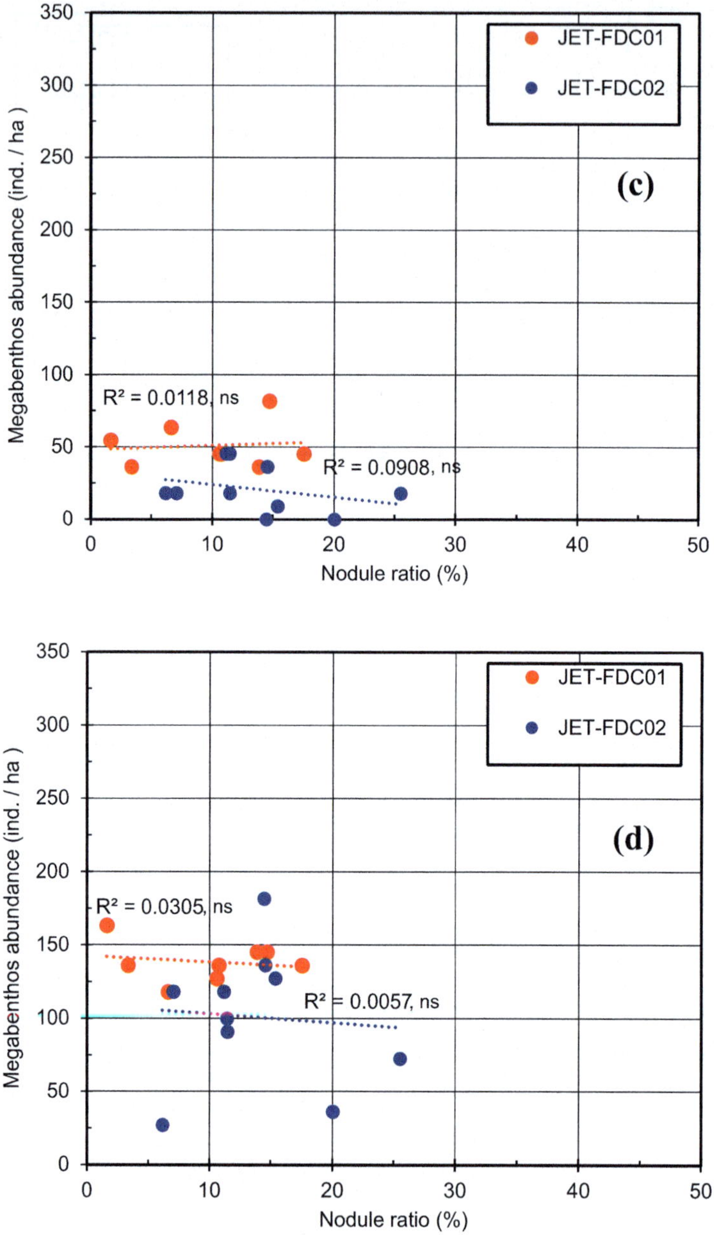

Fig. 8.19 (continued)

Pacific-REMP covers seamounts and the abyssal plain. This fact suggests that the CCZ-REMP method cannot be simply applied to other REMPs.

Considering these pros and cons, satellite data can be evaluated as effective at least as a primary screening tool in the current situation of APEI determination, where problems such as inconsistent data accuracy, spatial heterogeneity, and lack of time-series data have been pointed out. However, for a more detailed study, a separate study method must be considered. If the completion of the exploitation regulation is imminent and APEIs must be finalized as soon as possible, it will be necessary to consider what to prioritize and where to find compromises, as well as effective countermeasures based on opinions other than scientific and technological aspects.

Acknowledgments This chapter is based on results obtained from a project commissioned by the Ministry of Economy, Trade and Industry, Japan.

References

Amon, D., Ziegler, A., Dahlgren, T., Glover, A., Goineau, A., Gooday, A., Wikilund, H., & Smith, C. (2016). First insights into the abundance and diversity of abyssal megafauna in a polymetallic-nodule region in the eastern Clarion-Clipperton Zone. *Scientific Reports, 6*(1), 30492. https://doi.org/10.1038/srep20492

Behrenfield, M., & Falkowski, P. (1997). Photosynthetic rates derived from satellite-based chlorophyll concentration. *Limnology and Oceanography, 42*(1), 1–20.

Berelson, W. M., Anderson, R. F., Dymond, J., Demaster, D., Hammond, D. E., Collier, R., Honjo, S., Leinen, M., Mcmanus, J., Pope, R., Smith, C., & Stephens, M. (1997). Biogenic budgets of particle rain, benthic remineralization and sediment accumulation in the equatorial Pacific. *Deep-Sea Research Part II, 44*(9–10), 2251–2282.

BGR. (2018). *Environmental Impact Assessment for the testing of a pre-prototype manganese nodule collector vehicle in the Eastern German license area (Clarion-Clipperton Zone) in the framework of the European JPI-O MiningImpact 2 research project* (206p). BGR.

Cecchetto, M. M., Moser, A., Smith, C. R., van Oevelen, D., & Sweetman, A. K. (2023). Abyssal seafloor response to fresh phytodetrital input in three areas of particular environmental interest (APEIs) in the western clarion-clipperton zone (CCZ). *Deep Sea Research Part I: Oceanographic Research Papers, 195*, 103970.

Corliss, B. H., Brown, C. W., Sun, X., & Showers, W. J. (2009). Deep-sea benthic diversity linked to seasonality of pelagic productivity. *Deep-Sea Research Part I, 56*, 835–841.

Dunne, J. P., Sarmiento, J. L., & Gnanadesikan, A. (2007). A synthesis of global particle export from the surface ocean and cycling through the ocean interior and on the seafloor. *Global Biogeochemical Cycles, 21*, GB4006. https://doi.org/10.1029/2006GB002907

Fukushima, T., & Tsune, A. (2019). Long-term monitoring of environmental conditions of benthic impact experiment. In R. Sharma (Ed.), *Environmental issues of Deep-Sea mining* (pp. 191–211). Springer.

Fukushima, T., Shirayama, Y., & Kuboki, E. (2000). The characteristics of deep-sea epifaunal megabenthos two years after an artificial rapid deposition event. *Publications of the Seto Marine Biological Laboratory, 39*, 17–27.

Fukushima, T., Tsune, A., & Sugishima, H. (2022). Comprehensive understanding of seafloor disturbance and environmental impact scenarios. In R. Sharma (Ed.), *Perspectives on deep-sea mining* (pp. 313–338). Springer Nature.

Hauquier, F., Macheriotou, B., Bezerra, T. N., Egho, G., Martínez, A. P., & Vanreusel, A. (2018). Geographic distribution of free-living marine nematodes in the Clarion-Clipperton Zone: Implications for future deep-sea mining scenarios. *Biogeosciences Discussions, 492*. https://doi.org/10.5194/bg-2018-492

Jones, O. B. D., Simon-Lledo, E., Cuvelier, D., Durden, M., Amon, D., SêJong, J., McQuaid, K., Pape, E., Laming, S., Keitner, A., Drazen, J., Howell, K., Christodoulou, M., Martines, P., Fukushima, T., Clark, M., & Smith, C. (2019). DeepCCZ synthesis workshop -invertebrate megafauna. In *Report of the Deep CCZ biodiversity synthesis workshop* (pp. 108–133). International Seabed Authority.

Lutz, M. J., Caldeira, K., Dunbar, R. B., & Behrenfeld, M. J. (2007). Seasonal rhythms of net primary production and particulate organic carbon flux to depth describe the efficiency of biological pump in the global ocean. *Journal of Geophysical Research: Oceans, 112*, 1–26.

Morel, A., & Berthon, J. F. (1989). Surface pigments, algal biomass profiles, and potential production of the euphotic layer: Relationships reinvestigated in view of remote-sensing applications. *Limnology and Oceanography, 34*(8), 1545–1562.

Pape, E., Bezerra, T. N., Hauquier, F., & Vanreusel, A. (2017). Limited spatial and temporal variability in Meiofauna and nematode communities at distant but environmentally similar sites in an area of interest for Deep-Sea mining. *Frontiers in Marine Science, 4*, 205. https://doi.org/10.3389/fmars.2017.00205

Simon-Lledo, E., Bett, B. J., Huvenne, V. A., Schoening, T., Benoist, N. M., & Jones, D. O. (2019). Ecology of a polymetallic nodule occurrence gradient: Implications for deep-sea mining. *Limnology and Oceanography, 64*, 1883–1894.

Smet, B. D., Pape, E., Riehl, T., Bonifácio, P., Colson, L., & Vanreusel, A. (2017). The community structure of Deep-Sea macrofauna associated with polymetallic nodules in the eastern part of the clarion-Clipperton fracture zone. *Frontiers in Marine Science, 4*, 103. https://doi.org/10.3389/fmars.2017.00103

Smith, C. R., & Demopoulos, A. W. J. (2003). Ecology of the deep Pacific Ocean floor. In P. A. Tyler (Ed.), *Ecosystems of the world, volume 28: Ecosystems of the deep ocean* (pp. 179–218). Elsevier.

Smith, C. R., & Rabouille, C. (2002). What controls the mixed layer depth in the deep-sea sediments? The importance of POC flux. *Limnology and Oceanography, 47*, 418–426.

Smith, C. R., Berelson, W., Demaster, D. J., Dobbs, F. C., Hammond, D., Hoover, D. J., Pope, R. H., & Stephens, M. (1997). Latitudinal variations in benthic processes in the abyssal equatorial Pacific: Control by biogenic particle flux. *Deep Sea Research Part II: Topical Studies in Oceanography, 44*(9), 2295–2317.

Smith, C., Gaines, S., Watling, L., Friedlander, A., Morgen, C., Thurnherr, A., Mincks, S. L., Clark, M., Baco-Taylor, A., Bernardino, A. F., De Lo, F. C., Dutrieux, P., Rieser, A., Kittinger, J. N., Padilla-Gamino, J., Prescott, R., & Srsen, P. (2007). Areas of particular environmental interest (or "protected areas") for ecosystem based Management of the Clarion- Clipperton Zone: Rationale and recommendations to the international seabed authority. In *20 Expert participants in the workshop to Design of Marine Protected Areas Seamounts and the Abyssal Nodules Province in Pacific High Seas, 23–26 October*. University of Hawaii. http://www.soest.hawaii.edu/oceanography/faculty/csmith/MPA_webpage/documents/Proceedings_PEW_Workshop_MPAs_October_2007.pdf. Accessed 9Aug 2023.

Smith, C. R., De Leo, F. C., Bernardino, A. F., Sweetman, A. K., & Martinez, A. P. (2008). Abyssal food limitation, ecosystem structure and climate change. *Trends in Ecology & Evolution, 23*(9), 518–528.

Smith, K. L., Ruhl, H. A., Bett, B. J., Billett, D. S. M., Lampitt, R. S., & Kaufman, R. S. (2009). Climate, carbon cycling, and deep-ocean ecosystems. *Proceedings of the National Academy of the National Academy of Sciences of the United States of America, 106*(46), 19211–19218. https://doi.org/10.1073/pnas.0908322106

Sweetman, A. K., Wenzhöfer, F., Stratmann, T., & Cecchetto, M. M. (2019). DEEPCCZ biodiversity synthesis workshop report: Ecosystem functioning. In *Report of the Deep CCZ biodiversity synthesis workshop* (pp. 174–185). International Seabed Authority.

Vanreusel, A., Hilario, A., Ribeiro, P. A., Menot, L., & Martinez Arbizu, P. (2016). Threatened by mining, polymetallic nodules are required to preserve abyssal epifauna. *Scientific Reports, 6*, 26808. https://doi.org/10.1038/srep26808
Wilson, G. D. F. (2017). Macrofauna abundance, species diversity and turnover at three sites in the Clipperton-Clarion Fracture Zone. *Marine Biodiversity, 47*, 323–347.

Dr. Tomohiko Fukushima has served as a member of the Legal Technical Commission of the International Seabed Authority since 2023. His specialty is marine biology, and he currently belongs to the Japan Organization for Metals and Energy Security (JOGMEC). He graduated from the Faculty of Fisheries, Tokyo University of Fisheries in 1984 (bachelor's degree) and the Graduate School of Fisheries science, Tokyo University of Fisheries in 1986 (master's degree). In 2004, he was awarded the degree of Doctor of Science from Kyoto University. The title of his doctoral thesis was "Ecological Characteristics of Deep-Sea Benthic Organisms in Relation to Manganese Nodules Development Practices." After being awarded the doctoral degree, he continued research activities at the Ocean Policy Research Foundation (Research Fellow), the University of Tokyo (Associate Professor), Japan Agency for Marine-Earth Science and Technology (JAMSTEC), and Deep Ocean Resources Development Co., Ltd. (DORD) and then went on to the incumbent organization (JOGMEC).

Dr. Akira Tsune has a BSc in Earth Science from Kanazawa University in 1999 and a PhD from Kyushu University in 2005. After working in the field of geotourism, he changed jobs in 2008 and is involved in survey cruises of Massive Sulfides, Cobalt Rich Crusts, and Manganese Nodules as a researcher of Deep Ocean Resources Development., Co., Ltd. (DORD). He is now in charge of geological and resource survey in DORD.

Dr. Chisato Murakami is a marine biologist and her specialty is taxonomy of invertebrates. She graduated from the Faculty of Science, Toyama University in 1998 and got a PhD from Kyoto University in 2003. She worked in the field of environmental survey at the Marine Biological Research Institute of Japan Co., Ltd. and in the field of fishery at Japan Fisheries Research and Education Agency. She is now in charge of environmental survey at Deep Ocean Resources Development Co., Ltd. (DORD) since 2018.

Mr. Hideki Sugishima is a marine biologist. He studied at the Marine Biology Department in Tokyo University of Marine Science and Technology (formerly known as Tokyo University of Fisheries). After graduating from the university in 1995, he was employed at IDEA Consultants, Inc. From 2015 to 2018, he was transferred to the Japan Oil, Gas and Metals National Corporation (JOGMEC) as in charge of project management for environmental surveys in the areas of seafloor massive sulfides and cobalt-rich crusts. After returning to the company, he was assigned the Director of the Institute of Environmental Ecology and continues his research on environmental impact assessment for deep ocean mining.

Chapter 9
Assessment of Possible Environmental Impacts Using Flow-Cytometric Analysis of Metal Toxicity in Marine Phytoplankton

Shuhei Ota, Hiroshi Koshikawa, and Masanobu Kawachi

Abstract Mineral resources are abundant in submarine hydrothermal areas. The development of submarine resources is expected to be accompanied by metal-ore mining. This has raised concerns that mining is detrimental to marine environments because of the presence of heavy metals and arsenic. Innovations in collecting environmental baseline data and new onboard assessment methods for marine environmental impacts are required as a precautionary approach. This study aimed to evaluate the feasibility of an onboard on-chip benchtop flow cytometry (FCM) system and establish a technique to analyze the ecotoxic effects of metals on marine phytoplankton based on the FCM assay. The benchtop unit was installed on the Japan Agency for Marine-Earth Science and Technology (JAMSTEC) research vessel *Kaimei*. Experiments were conducted onboard using seawater collected onsite. The effects of heavy metals on marine cyanobacteria were evaluated by staining nucleic acids with SYBR Green I and propidium iodide (PI). PI is a stain that permeates membranes differently depending on cell damage. We found that the metal sensitivities of *Prochlorococcus* in oligotrophic waters differed depending on the ambient nutrients. A similar trend was observed in culture-based laboratory experiments.

Keywords Ecotoxicology · Cyanobacteria · Deep-sea mining · Flow cytometry · Heavy metal contaminant · Microalgae · On-chip Sort · Phytoplankton

1 Introduction

Commercial mining operations of massive seafloor sulfide deposits are considered an important source of metals, such as copper (Cu), lead (Pb), and zinc (Zn) (Hoagland et al., 2010; Petersen et al., 2016). Seawater contamination by these

S. Ota (✉) · H. Koshikawa · M. Kawachi
National Institute for Environmental Studies (NIES), Tsukuba, Ibaraki, Japan
e-mail: ota.shuhei@nies.go.jp

R. Sharma (ed.), *Deep-Sea Mining and the Water Column*,
https://doi.org/10.1007/978-3-031-59060-3_9

metals during mining is unavoidable and a concern (Boschen et al., 2013; Hauton et al., 2017; Van Dover, 2011; Wedding et al., 2015). Environmental impact assessments, baseline assessments, monitoring, mitigation, and management are important for identifying and reducing potential impacts (Jones et al., 2019). A recent study measured the leaching of metals and metalloids from hydrothermal ores in the Okinawa Trough into aerobic seawater. These findings indicate that hydrothermal ore particulates are a source of toxic pollutants that can affect primary production (Fuchida et al., 2017). Prior assessment of the leaching of hazardous elements from ores into seawater is necessary to minimize the environmental impact of mining these mineral deposits (Fuchida et al., 2017).

Cu, Pb, Zn, and arsenic (As) can be released from sulfide minerals with different chemical compositions in hydrothermal fields (Fuchida et al., 2017). The toxicity of leachate from drilled core samples was assessed onboard the Japan Agency for Marine-Earth Science and Technology (JAMSTEC) Drilling Vessel (D/V) *Chikyu* during cruise (CK16-05), using the marine cyanobacterium *Cyanobium* sp. NIES-981as the test organism(Yamagishi et al., 2018). Fuchida et al. (2018) found that heavy metals (e.g., Zn and Pb) from sulfide samples leached at an accelerated rate in a surface environment (temperate and aerobic) rather than in a seabed environment (cold and anaerobic conditions) (Fuchida et al., 2018). During the onboard experiment, copper leaching was observed in only one case from 16 experimental conditions (Fuchida et al., 2018). However, copper is problematic at low concentrations, including picomolar levels, especially in *Synechococcus* and *Prochlorococcus* cyanobacteria (Mann et al., 2002). When deep-sea mining commences, it must be approached in a precautionary and stepwise manner to integrate the new and developed knowledge (Niner et al., 2018). An onsite environmental impact assessment method is required to enable scientific evidence-based actions.

The Test No. 201 growth inhibition bioassay defined by the Organization for Economic Co-operation and Development (OECD) is an algae-based test method (OECD (Organization for Economic Cooperation and Development), 2011).This method is widely applied, but is currently limited to freshwater and test strains specified by the OECD. In addition, this method has high cultivation costs and requires at least 72 h to perform. To examine marine test strains, we screened the Microbial Culture Collection at the National Institute for Environmental Studies (MCC-NIES)to identify metal-sensitive marine strains that could be used as test organisms (Ota et al., 2020).

Delayed fluorescence (DF) is emitted by photosynthetic organisms for a short period (seconds to minutes) after exposure to light (Goltsev et al., 2009). DF intensity has been used as an alternative indicator for toxicity testing to complement Test No. 201 or other methods (Katsumata et al., 2006, 2009; Takeuchi et al., 2014; Takeuchi et al., 2020; Yamagishi et al., 2016). DF intensity correlates with algal growth inhibition and has been used as an endpoint for the rapid estimation of half-maximal effective concentration (EC_{50}) values(Katsumata et al., 2006, 2009). Although Test No. 201 takes at least 72 h, DF can be detected relatively quickly, usually within 24 h of incubation, which is an advantage in onboard operations.

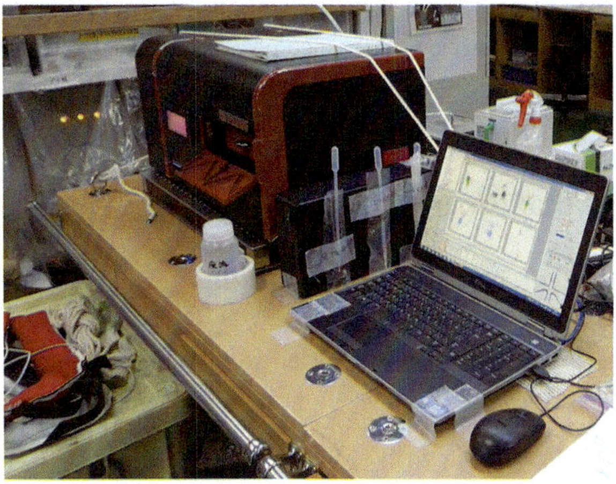

Fig. 9.1 On-chip sort system installed in the laboratory of a research vessel. The device is an On-chip Sortflow cytometer, and the laptop computer is used for analysis

However, the DF cannot be used for low-biomass oligotrophic seawater samples because of the insufficient sensitivity of the luminometer.

Flow cytometry (FCM) allows observation of individual cells rather than bulk observations. This approach helps understand the composition and interactions within microbial systems (van den Engh et al., 2019). On-chip technology allows sorting and analysis in a damage-, contamination-, and maintenance-free manner because of the disposable microfluidic chip-based system. Although generally used in medical research (Hasegawa et al., 2019; Sawada et al., 2016; Watanabe et al., 2014), this method may also be useful in onboard applications (Fig. 9.1).

The analysis of low-chlorophyll-content cyanobacteria *Prochlorococcus* with a benchtop flow cytometer underestimated cell counts because of the inadequate separation of the autofluorescence signal from background noise (Gérikas et al., 2016). One solution is to stain nucleic acids before FCM analysis (Marie et al., 1997; Ota et al., 2022). SYBR Green I (SYBR) is often used to stain nucleic acids because the stained cells can be detected using a blue laser (488 nm). However, this fluorescent dye stains both living and dead cells. If cell viability must be checked, membrane-impermeable dyes such as propidium iodide (PI), in which only dead/damaged cells are stained, should be used. Previous applications include studies on bacteria in river and marine environments (Grégori et al., 2001; Nescerecka et al., 2016; van der Merwe et al., 2014) and microalgal cultures (Gumbo et al., 2014; Nogueira et al., 2015; Suman et al., 2015; Xiao et al., 2011; Zhang et al., 2015). Several previous studies have used PI-based assays for both eukaryotic and prokaryotic algae, indicating that it is possible to assess cell health using PI even in the presence of chlorophyll autofluorescence, as in algae.

We have previously reported the efficacy of SYBR–PI double staining in cultured strains (Ota et al., 2022). Here, we performed a pilot study of SYBR–PI double staining using the portable On-chipSort flow cytometer system on board

(Fig. 9.1). Double staining enables the analysis of cell cultures and field populations to determine how the presence or absence of ambient nutrients or bacteria affects the sensitivity of marine cyanobacteria to metals.

2 Experimental Procedures

2.1 Sample Collection and Preparation

Seawater samples (500 mL) were collected from the chlorophyll maximum layer (depth of 90 m) in the Boso Peninsula (34°19.90′N, 141°24.86′E), Japan, using Niskin-X bottles attached to a conductivity–temperature–depth profiler with carousel multiple sampling (CTD-CMS) system. The collections were done on September 06, 2019, during cruise KM19-07 of the research vessel (R/V) *Kaimei* (Kawagucci et al., 2018).The seawater samples were subsampled into 500 mL polypropylene bottles. They were stored in a 25 °C incubator with a 12:12 h light/dark cycle and illumination of approximately 40 μmol photons m^{-2} s^{-1}.

2.2 Culture-Based Cell Viability Assay

The culture-based assay used *Cyanobium* sp. NIES-981 cells that were obtained from MCC-NIES (Tsukuba, Japan). NIES-981 was selected as the test cyanobacterium because it has previously been used as a marine test strain (ISO 23734), its genome has been decoded (Yamaguchi et al., 2016), and it is relatively tractable. For the experiments comparing living and dead cells, dead cells were prepared by boiling them for a few seconds.

Living and dead cells were stained with a final concentration of 0.03 mM PI. After 1.5 min, time-lapse microscopy was performed using a DFC450C digital camera mounted on a DMi8 inverted microscope (Leica Microsystems, Wetzlar, Germany). The PI-stained cells were observed using an RHOD fluorescent cube (excitation: 546/12 nm, emission: 585/40 nm). Images were obtained over 16 min at a capture interval of 1.5 min. Red channel images (512 × 512 pixels) were saved as stacked images. Twelve cells in the stack images were randomly selected, and a region of interest (ROI) was set for each cell.

We compared the detection performances of the FCM and DF. In this experiment, a 20-mL NIES-981 culture was grown under continuous illumination with approximately 50 μmol photons m^{-2} s^{-1} for 6 days at 22 °C. The 20-mL culture was centrifuged at 3000× *g* for 10 min, and the supernatant was removed and resuspended in 60 mL of 10% Pro-99 medium diluted with filtered seawater. Subsequently, 4 mL was added to a screw-top glass test tube and exposed to different Cu^{2+} concentrations for 24 h. The DF of each sample was measured using a Type-6110 high-sensitivity luminometer (Hamamatsu Photonics K.K., Hamamatsu, Japan) at 0.1 s intervals. The integrated DF (DFI) value was determined by calculating the sum of positive DF values over a period of 40 s. The same sample used in the DF assay was diluted to 5% with sterile filtered seawater and double-stained with SYBR and PI (described in Sect. 2.3) for FCM analysis.

2.3 *Onboard FCM for* **Prochlorococcus**

Niskin-sampled seawater was incubated overnight at 25 °C after adding a final concentration of 10% (v/v) Pro-99 medium (Moore et al., 2007). Pro-99 was used because of its simple nutrient composition and relatively low chelating effects compared to other media, such as f/2 or K. Cu^{2+} or Zn^{2+} exposure was performed by adding $CuSO_4$ or $ZnSO_4$stock solutions directly to the samples. After Cu^{2+} or Zn^{2+} addition, samples were incubated in a 24-well microplate for 4 h at 25 °C under illumination with approximately 150 μmol photons m^{-2} s^{-1}. After incubation, 3 μL of 0.15 mM PI in dimethyl sulfoxide was added to 96 mL of the sample, 1 μL of 100× SYBR Green I in dimethyl sulfoxide was added (final concentration of 1× SYBR Green I), and the samples were incubated for at least 15 min at room temperature (approximately 25 °C). Stained samples were analyzed using an On-chip Sort flow cytometer (On-Chip Biotechnologies, Tokyo, Japan) with 488 nm blue laser excitation. In the first step for FCM gating, FL2 (543/22 nm) and FL6 (775/46 nm) detection channels were used to discriminate the signal from background noise. The FL3 (591.5/43 nm) and FL5 (716/40 nm) detection channels were used to expand the signals into two-dimensional dot plots or histograms.

2.4 *Coculture Experiment*

A coculture experiment was performed to determine the presence of heterotrophic bacteria (HB) in populations of *Prochlorococcus*. NIES-2885 (unialgal, non-axenic culture) affected PI staining-based detection. The low number of ambient HB in the *Prochlorococcus* NIES-2885 culture was increased by transferring the culture to the nutrient-rich BF/2 medium (https://mcc.nies.go.jp/02medium-e.html). HB and NIES-2885 bacteria were pre-cultured in Pro-99 medium for 1 day to ensure the same nutritional status in this experiment. The bacteria, at five different ratios of *Prochlorococcus* to HB (100:0, 75:25, 50:50, 25:75, and 0:100), with a density of 1×10^6 cells/mL, were exposed to 2 mg/L (0.032 mM) Cu^{2+} for 4 h. The bacteria were then double-stained with PI/SYBR, as described above, and analyzed using FCM.

2.5 Onboard FCM Using a Mixed Metal Solution

An artificial complex solution of metals (abbreviated as "Cy" henceforth) was used for toxicity testing. The Cy solution was prepared based on experimental leaching data from the JAMSTEC D/V *Chikyu* CK16-05 cruise (core sample ID: C9027B 1X-CC) (Yamagishi et al., 2018). In this experiment, the Cy mixture was added to a final concentration of 10% of that of the metal leachate. The incubation conditions, staining, and FCM procedures were the same as those described above.

2.6 Culture-Based Experiment

To confirm the differences in sensitivity to Cu^{2+} in the presence or absence of nutrients, 10 mL of the stationary phase of NIES-981 culture was centrifuged (3000× *g*, 10 min). The supernatant was removed and replaced with 10 mL of sterile filtered seawater. The cultures were incubated at 22 °C under illumination with approximately 20 μmol photons m^{-2} s^{-1}. After preculturing, Pro-99 medium was added to a final concentration of 10% and incubated for 1 d. $Cu^{2+}$was added (0.5 mg/L,0.08 mM final concentrations) and incubated for 4 h. FCM analysis was performed after PI–SYBR double staining as described above.

2.7 Statistical and Data Analyses

For time-lapse imaging analysis, the intensity of light in the ROI was quantified using the plot of the Z-axis profile of ImageJ1.51 s (https://imagej.nih.gov/ij). Flow cytometry dot plots and histogram analyses were performed using OnchipFlow ver. 1.7.9.0. (http://on-chipbio.com) software. DF data from at least three independent culture experiments was used for the analysis. Best-fit curves and half-maximal concentration (EC_{50}) values were calculated using the "drc" package (Ritz et al., 2015) v.3.0.1 in R v.3.4.1 (http://www.R-project.org/) with the four-parameter log-logistic function. Data are shown as mean ± standard deviation. Significant differences were evaluated using Wilcoxon's exact test or the Student's t-test. *P*-values <0.05 were considered statistically significant.

3 Results

3.1 Cell Viability Determined Using PI Staining

Cells of *Cyanobium* sp. NIES-981 cells stained with PI were captured using time-lapse fluorescence microscopy after staining (1.5 min after preparation) (Fig. 9.2a) to confirm the appropriate time for sufficient dye permeation. Quantifying the red

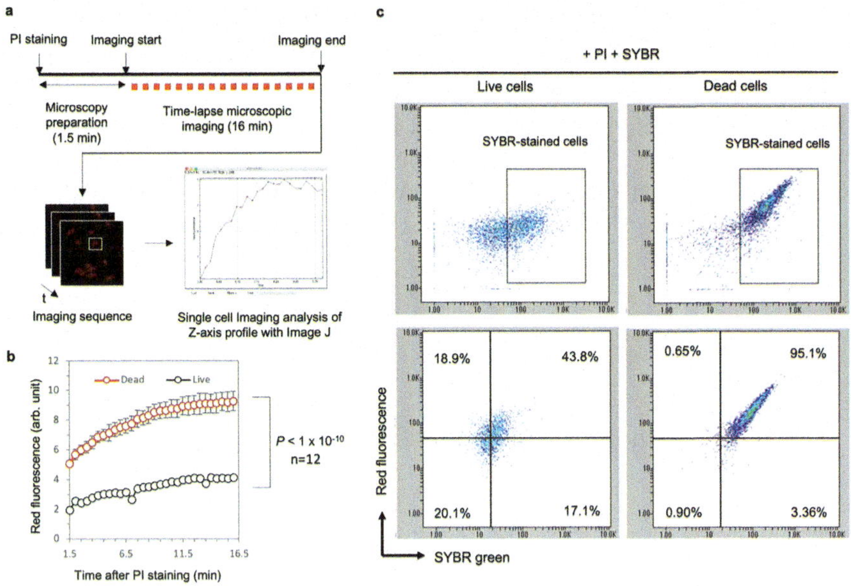

Fig. 9.2 Cell viability test using the propidium iodide staining method to evaluate metal toxicity. (**a**) The experimental flow of the time-lapse imaging analysis. (**b**) Time course changes in red intensity in living and dead NIES-981 cells due to intracellular penetration of PI. (**c**) FCM dotplots of living and dead cells stained using PI. The lower image shows the dot plots inside the gate of interest

fluorescence intensity in each cell in sequential images showed that the fluorescence intensity of dead cells was significantly higher than that of living cells and that the fluorescence intensity increased over time (Fig. 9.2b). This suggests that PI permeated dead cells. Although crosstalk and interference between the fluorescence of chlorophyll and PI were expected owing to the near spectra of chlorophyll and PI, we confirmed an increase or decrease in PI fluorescence by observing the total sum of red fluorescence (Fig. 9.2b). Because maximum light intensity was attained approximately 10 min after staining, a PI staining time longer than 10 min was adopted as the sample processing method.

On-chip Sort FCM analysis of unstained *Prochlorococcus* cells did not distinguish between cell signal and background non-cell noise. SYBR-stained *Prochlorococcus* cells distinguished both living and dead cells from the background noise. Double staining of living and dead cells using both PI and SYBR as controls revealed that SYBR-positive cells with red fluorescence exceeding 50 arbitrary units represented 95.8% (0.65% + 95.1%) of dead cells and 62.7% (18.9% + 43.8%) of living cells (Fig. 9.2c). These results show that the PI signal could be used to evaluate dead cells, even after staining with SYBR.

3.2 Evaluation of Damage to PI-SYBR Double-Stained Cells from Cu^{2+}Exposure

To test the validity of PI staining as an indicator of cell damage, the red fluorescence signal of cells detected via the SYBR-positive gate was analyzed in untreated control and Cu^{2+}-exposed cultures. More cells were detected by red fluorescence in the Cu^{2+}-exposed cultures than in the control (Fig. 9.3a, b). The gating of red-shifted

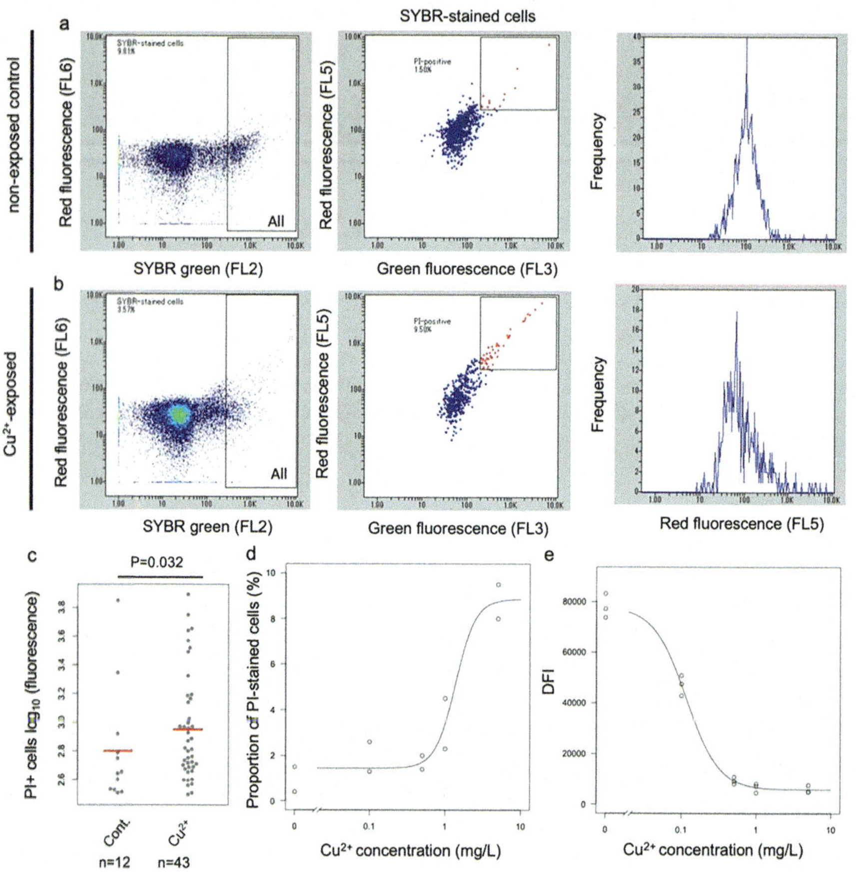

Fig. 9.3 Evaluation of damage to NIES-981 cells from metal exposure as shown using PI-SYBR double staining. (**a**) Representative pseudo-color dot plots and histograms of non-exposed controls. In this experiment, the gate of interest was set sufficiently distant from the noise area. Red dots indicate PI-positive cells. (**b**) Representative pseudo-color dot plots and histograms of Cu^{2+}-exposed cells. The dot plots in the middle and the histograms on the right show the data in the gate. The red dots in the middle diagram denote PI-positive cells. (**c**) Comparison of PI-positive cells in control and copper-exposed samples. The red line shows the mean value, where P-values are from the Wilcox exact test. (**d**) A dose-response curve estimated based on the proportion of PI-stained cells by FCM. (**e**) A dose-response curve estimated based on integrated delayed fluorescence values

cells resulted in 12 and 43 cells being counted among the unexposed and Cu^{2+}-exposed cells, respectively (Fig. 9.3a, b). The significantly higher intensity of red fluorescence detected in Cu^{2+}-exposed cells (Fig. 9.3c) suggests that red fluorescence upon PI staining is a valid indicator of damage in cells exposed to Cu^{2+}.

The proportions of PI-stained cells and DFI were measured simultaneously in the same sample to compare their detection sensitivities (Fig. 9.3d, e). In general, cells that were more intensively stained by PI were observed after incubation with higher Cu^{2+}concentrations (Fig. 9.3d). In contrast, DFI levels were generally lower during cultivation at higher Cu^{2+} concentrations (Fig. 9.3e). Higher Cu^{2+} concentrations were associated with increased cell damage using both methods (Fig. 9.3d, e). When the dose-response curves of both methods were compared, the detection level of the PI-staining method was one order of magnitude lower than that of DFI (EC_{50} of 1.37 vs. 0.12). These findings indicated the superiority of DFI for detecting Cu^{2+} exposure.

3.3 *Onboard FCM Analysis*

We examined the effects of ambient nutrients on the metals in natural seawater samples (Fig. 9.4a). When the experiments were performed with and without Pro-99 pre-incubation, there was no significant difference in the proportion of PI-stained cells in the unexposed group. However, an increase from 19.9% to 37.5% was observed in the Cu^{2+}-exposed group, and an increase from 11.8% to 24.2% was observed in the Zn^{2+}-exposed group after Pro-99 pre-incubation (Fig. 9.4b). These findings indicated that metal sensitivity increases in the presence of ambient nutrients. The histogram also revealed that the number of red fluorescent cells increased following exposure to Cu^{2+}or Zn^{2+} (Fig. 9.4c), indicating that the effect of metal stress could be assessed using the proportion of PI-stained cells.

The flowcytometry gate where *Prochlorococcus* was detected may contain heterotrophic bacteria (HB) because *Prochlorococcus* and HB cannot be completely separated using the current on-chip sorting system. We examined the effect of a mixture of these two bacterial species on the proportion of PI-stained cells. It is difficult to control the proportions of *Prochlorococcus* and HB in natural seawater. Culture-based experiments were performed using *Prochlorococcus* NIES-2885 (Fig. 9.5a). The proliferation of HB in the BF/2 medium was visually confirmed (Fig. 9.5b). Mixed cultures of *Prochlorococcus* and HB were exposed to Cu^{2+} (Fig. 9.5a). After PI and SYBR double staining, FCM analysis revealed that 35.7% of the cells stained with PI were *Prochlorococcus* alone, whereas the number of PI-stained cells gradually decreased as the proportion of HB increased, with a value of 15.6% for HB alone (Fig. 9.5c). These results indicated that detection was affected by the ratio of *Prochlorococcus* to HB (Fig. 9.5c).

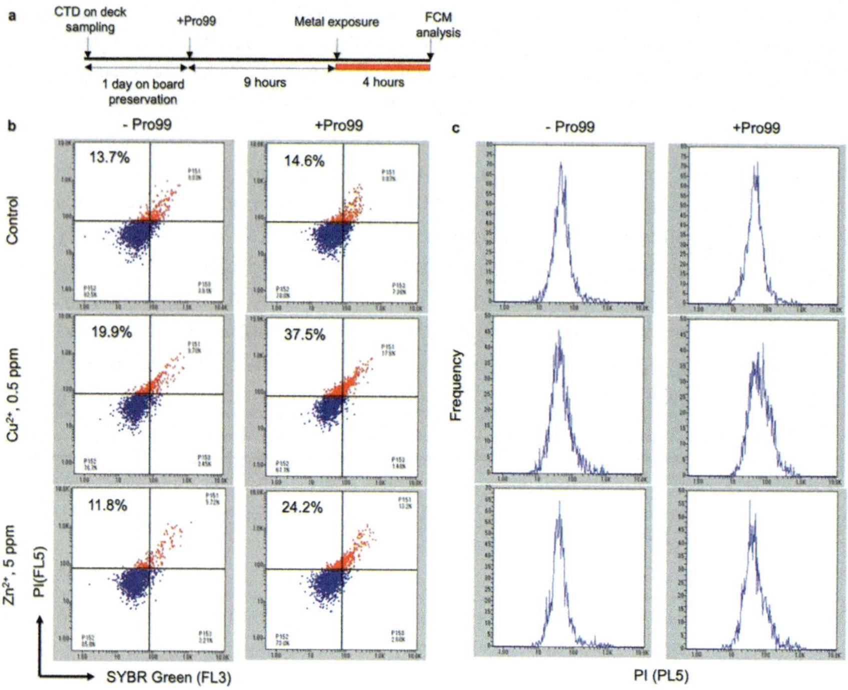

Fig. 9.4 Onboard FCM of metal-exposed *Prochlorococcus* populations. (**a**) Experimental timeline scheme from the CTD on deck to the FCM analysis. (**b**) Representative pseudo-color dot plots showing red-shifted cells (red dots) under the indicated condition. The values (%) indicate the ratio of red dots to the total cells examined. (**c**) Histogram display corresponding to the graphs in (**b**)

Using the SYBR–PI double-staining method described above, the effects of the presence or absence of complex metals and nutrients were examined (Fig. 9.6). Samples were incubated with a mixture of metals (Cy), and the red fluorescence intensity was compared after SYBR–PI double staining. The proportion of PI-stained cells in the Pro-99-treated sample was 18% for mixed metals. This value was twice as high as that of the control group (Fig. 9.6), indicating that the effects of the metals were influenced by ambient nutrients.

We investigated whether a similar effect would be observed in land-based laboratories. NIES-981 was used as a representative test strain under four different nutrient conditions (Fig. 9.7a). A significant decrease in the proportion of PI-stained cells was observed in the nutrient status upon metal exposure under oligotrophic conditions, regardless of whether the preculture was eutrophic or oligotrophic (Fig. 9.7b, c). In experiments comparing eutrophic and oligotrophic conditions or oligotrophic to oligotrophic conditions, the proportions of PI-stained cells were 55.8% and 29.5%, respectively (Fig. 9.7c). The results indicated that the field and laboratory trends were similar. Thus, it can be inferred that the ambient nutrient status affects metal sensitivity.

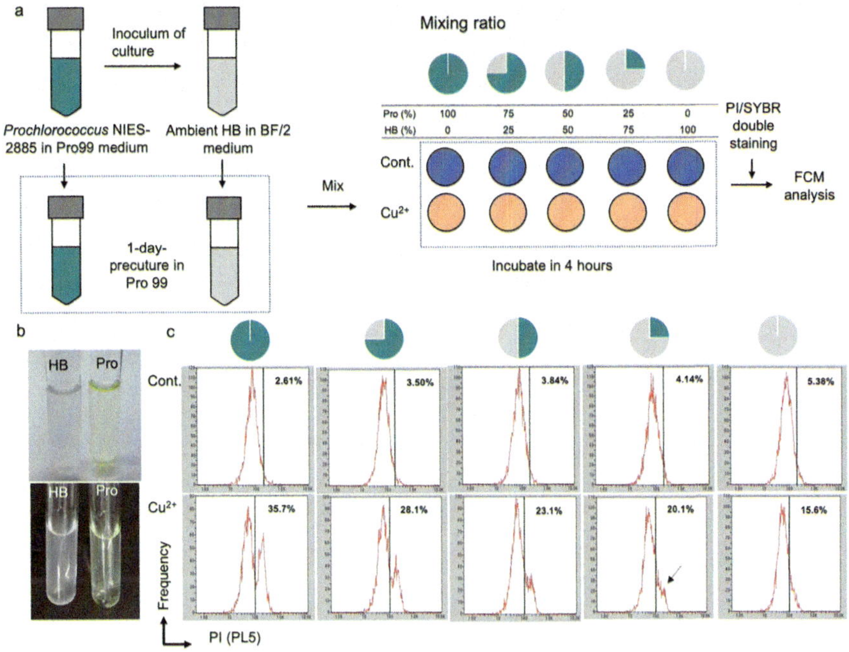

Fig. 9.5 Coculture experiments of *Prochlorococcus* and ambient heterotrophic bacteria. (**a**) Schematic flow of the mixing experiments. For each of the five mixing ratios, SYBR/PI double staining was performed 4 h after exposure to Cu^{2+}. (**b**) Photographs of *Prochlorococcus* (Pro) and ambient HB cultures (top, white background; bottom, black background). (**c**) Histograms showing the frequency of PI-positive cells with an altered population of Pro and HB. The mixing ratio is shown in the pie charts. The top row represents the control, and the bottom row represents the Cu^{2+} exposed culture. Even with 75% HB in the mixture, the proportion of PI signal could be detected (arrow). Numbers show the percentage of PI-positive cells in total cells

4 Discussion

In this study, the on-chip Sort benchtop FCM system was used for analyses on the JAMSTEC R/V *Kaimei* and in the laboratory. Few studies have compared flow cytometry experiments at sea and on land, and this is the first report of an analysis using the On-chip Sort system. FCM is a widely used method for analyzing the dynamics of plankton populations in oceanographic fields. This approach provides more accurate and rapid measurements than microscopic visual cell counting (Marie et al., 2005). A recent study used FCM sorting combined with single-cell genomics to study plankton diversity (Sieracki et al., 2019). In situ experiments require fresh cells because fixed cells cannot be used to assess toxicity. However, conventional flow cytometers are not portable. In recent years, commercially available flow cytometers have become smaller and lighter, allowing their use in onboard experiments. Conventional and benchtop flow cytometers have been reported to differ in detection sensitivity. For example, a comparison of the BD FACSCanto (a

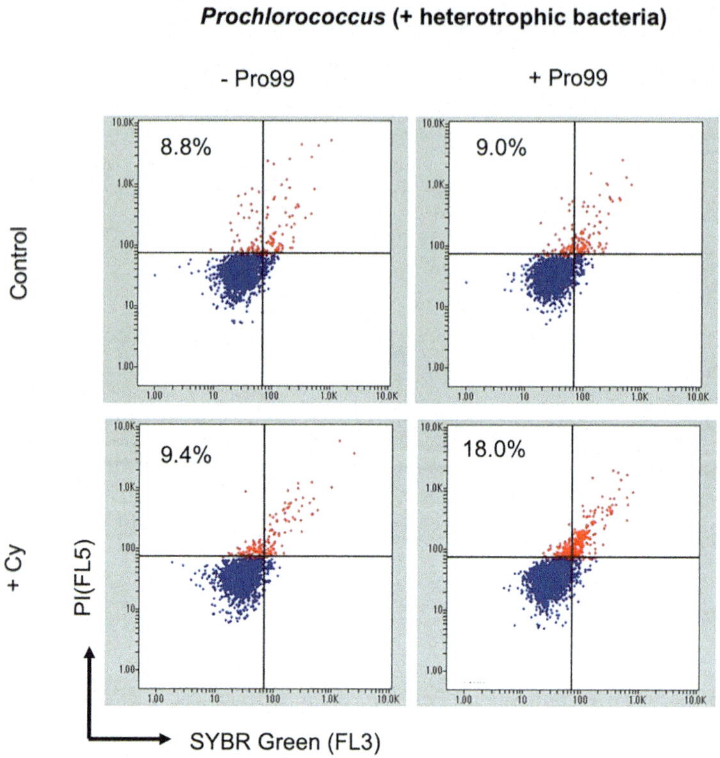

Fig. 9.6 Onboard FCM of a metal complex with or without Pro-99 culturing. Representative pseudo-color dot plots showing red-shifted cells (red dots) of the *Prochlorococcus* with other bacterial populations under the indicated conditions. The values (%) indicate the percentage of red dots relative to the total cells examined

laboratory-based instrument) and BD Accuri C6 (a benchtop instrument) showed that *Prochlorococcus* and *Synechococcus* populations were severely underestimated using the benchtop FCM (Gérikas et al., 2016). Experiments were performed using the same protocol for cultured and field populations in this study to evaluate the practicality of onboard testing. The results showed that cell viability can be determined using PI and SYBR double staining and that the major ambient nutrients, nitrogen and/or phosphorus, may affect phytoplankton sensitivity to metals.

The PI staining method was used to assess the cell's health. Algal cells exhibit autofluorescence of chlorophyll that overlaps with PI fluorescence; therefore, careful interpretation is necessary. Considering that previous studies have applied the PI method to algae (Gumbo et al., 2014; Nogueira et al., 2015; Suman et al., 2015; Xiao et al., 2011; Zhang et al., 2015) and that dead cells showed stronger red fluorescence even in cyanobacterial cells in the validation experiment (Fig. 9.2), we believe that the PI method is feasible for evaluating cell health. This study also showed that the evaluation by PI is related to DFI, which has been reported to

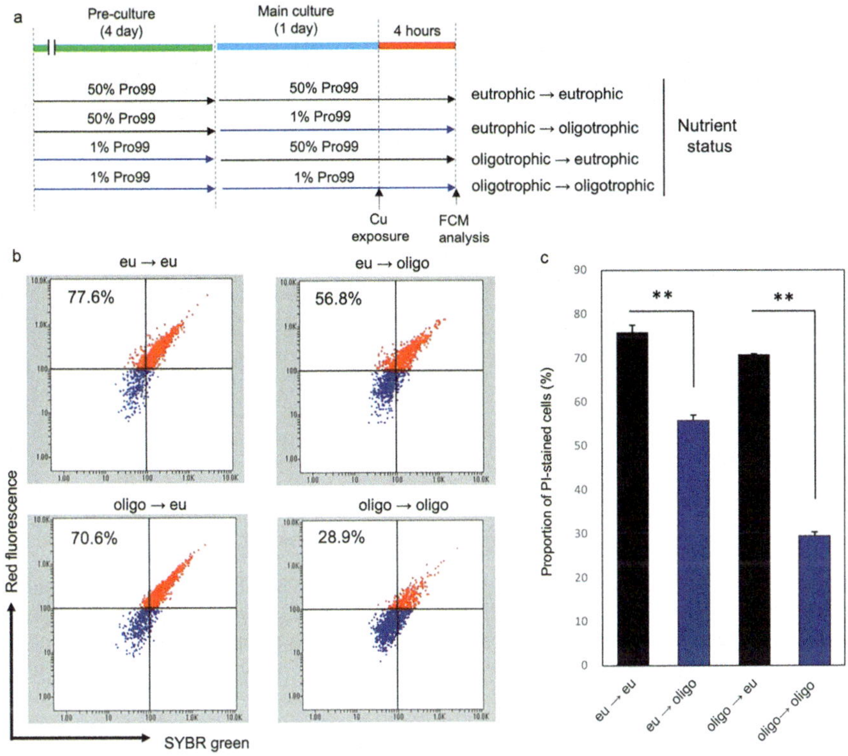

Fig. 9.7 Confirmation test of onboard experiments using a cultured sample of NIES-981 cells. (**a**) Experimental timeline scheme. (**b**) Representative pseudo-color dot plots showing the proportion of PI-positive cells in the presence of different ambient nutrients (eu: eutrophic, oligo: oligotrophic). (**c**) Graph showing the mean ± standard deviation of the proportion of PI-stained cells in four nutrient conditions. ** $P < 0.01$ by Student's t-test (triplicate biological samples)

correlate with standard growth inhibition tests (Yamagishi et al., 2016), and we believe that the PI method is useful for assessing cell health.

Although this study focused on *Prochlorococcus*, the present FCM protocol also allows for the analysis of other marine phytoplankton, such as eukaryotes and *Synechococcus*. When focusing on eukaryotes or *Synechococcus*, only PI staining is sufficient because chlorophyll or phycoerythrin autofluorescence can be used to separate them from the background noise without SYBR staining (Fig. 9.8). However, we found it difficult to separate *Prochlorococcus* from heterotrophic bacteria (HB). A laboratory-based study showed that metal sensitivity is affected by changes in the ratio of HB to *Prochlorococcus*. It was previously explained that the change in the proportion of PI-stained cells reflected the greater sensitivity of *Prochlorococcus* to Cu- or HB-derived chelation involving molecules such as siderophores (Schalk et al., 2011).This may be the case for natural samples, but it requires further investigation, and the proportion of HB should be considered in future FCM analyses.

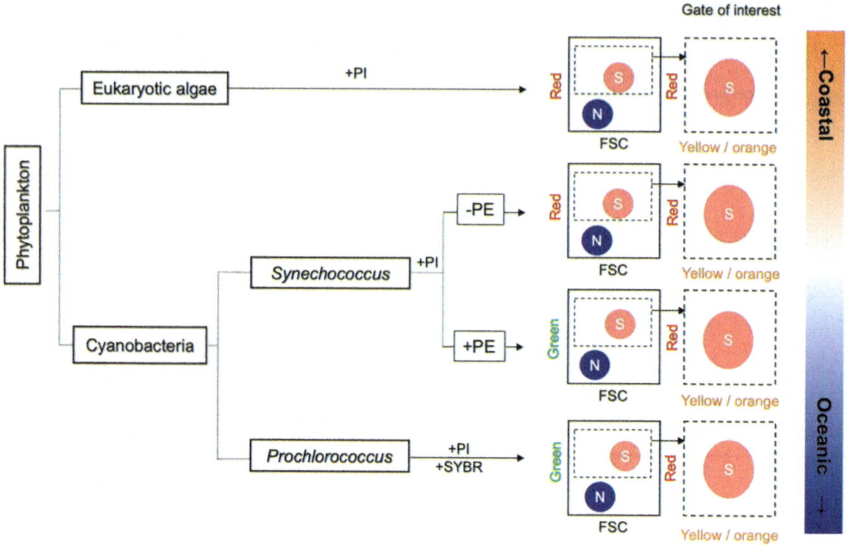

Fig. 9.8 Flow chart for gating strategy for each algal group green, yellow/orange, and red indicate fluorescence intensity of interest. Coastal and oceanic areas indicate areas where the indicated algal groups are abundant in seawater. (e.g., *Prochlorococcus* is dominant in the oceanic environment.) *PI* propidium iodide; *+PI* PI-staining; *+SYBR+PI* SYBR Green I and PI double staining, *N* noise, *S* signal, *FCS* forward scatter

This study showed that the metal sensitivity of phytoplankton is altered by ambient nutrient conditions. Changes in sensitivity have been observed in culture-based experiments and in phytoplankton populations in natural samples. It is envisaged that ores will be retrieved from deep-water environments in future commercial mining operations (Hoagland et al., 2010). Experiments comparing metal leaching from sulfide ores showed that metal release is accelerated under warmer and more aerobic conditions, such as those at the ocean surface (Fuchida et al., 2018). Deep seawater contains more nutrients than surface seawater (Sarmiento & Gruber, 2006). As a result, tailings generated and drawn up along with deep seawater are expected to be rich in nutrients. This can increase primary productivity and have a cascading trophic impact (Halfar & Fujita, 2002). Future seafloor mining operations should consider eutrophication and changes in community structure, as well as changes in phytoplankton sensitivity due to eutrophication.

We compared the detection performances of the FCM and DF assays for metal toxicity. The EC_{50} values were 1.37 or 0.12 as determined using FCM or DF, respectively (Fig. 9.3). These results indicate that the DF can be used to evaluate lower concentrations of heavy metals. However, a disadvantage of the DF method is that it cannot be used at low cell densities. For example, for NIES-981, the initial viable cell count should be 10^6 cells/mL (Yamagishi et al., 2016). Thus, anultraluminometer may not be suitable for raw natural seawater samples with a low cell density. In contrast, FCM has the advantage of being able to directly analyze low cell densities, such as that of oligotrophic seawater. Another advantage of FCM is that it can be used for population-specific analysis. For example, population-by-population analysis allows the identification of groups that are vulnerable to metals. The complementary use of FCM and DF may enable on-site bioassays and monitoring, providing multifaceted insights into phytoplankton populations and communities. Future research is needed to improve the ability of FCMs to detect metals more sensitively.

5 Conclusions

The present experiment was conducted both onboard a research vessel and, in the laboratory, to demonstrate the feasibility of onboard FCM using a double-staining method and the effect of ambient nutrients or HB on metal exposure. The sensitivity of phytoplankton to metals is increased by ambient nutrients. Previous studies have demonstrated that an increase in ambient nitrogen and/or phosphorus concentrations considerably enhances metal uptake by phytoplankton cells, such as diatoms and dinoflagellates, resulting in the bioaccumulation of metals (Wang & Dei, 2001a, b). Similarly, elevated cellular P concentrations in the cyanobacterium *Microcystis aeruginosa* significantly promoted short-term uptake of cadmium and Zn (Zeng & Wang, 2009). Although the effect of ambient nutrients on phytoplankton has only been addressed in a few studies, we speculate that intracellular metal uptake may directly affect metal toxicity. Future mining operations must consider perturbations in phytoplankton populations and communities, as well as the trophic impact on marine phytoplankton. Finally, we would like to note the gating strategy; the present FCM assay can be used for all major marine plankton. For example, because eukaryotic algae are usually more abundant in coastal areas than in the open oceans, targeting only eukaryotic algae in coastal areas may be possible. In contrast, it may be possible to target *Prochlorococcus* as the dominant phytoplankton in oligotrophic ocean waters (Fig. 9.8). The advantage of this assay system is that it can be used for a variety of groups according to the dominant phytoplankton type in the survey area.

Acknowledgments We thank the staff of the Microbial Culture Collections at NIES for providing the test strains and the master, crew, and scientific parties of the JAMSTEC R/V *Kaimei* KM19-07 cruise for their expertise and support. This work was supported by the Cabinet Office, Government of Japan, Cross-Ministerial Strategic Innovation Promotion Program (SIP), Next-Generation Technology for Ocean Resource Exploration, and NIES Research Funding (Type A).

References

Boschen, R. E., Rowden, A. A., Clark, M. R., & Gardner, J. P. A. (2013). Mining of deep-sea seafloor massive sulfides: A review of the deposits, their benthic communities, impacts from mining, regulatory frameworks and management strategies. *Ocean & Coastal Management, 84*, 54–67. https://doi.org/10.1016/j.ocecoaman.2013.07.005

Fuchida, S., Yokoyama, A., Fukuchi, R., Ishibashi, J., Kawagucci, S., Kawachi, M., & Koshikawa, H. (2017). Leaching of metals and metalloids from hydrothermal ore particulates and their effects on marine phytoplankton. *ACS Omega, 2*, 3175–3182. https://doi.org/10.1021/acsomega.7b00081

Fuchida, S., Ishibashi, J., Shimada, K., Nozaki, T., Kumagai, H., Kawachi, M., Matsushita, Y., & Koshikawa, H. (2018). Onboard experiment investigating metal leaching of fresh hydrothermal sulfide cores into seawater. *Geochemical Transactions, 19*, 15. https://doi.org/10.1186/s12932-018-0060-9

Gérikas Ribeiro, C., Marie, D., Lopes dos Santos, A., Pereira, B. F., & Vaulot, D. (2016). Estimating microbial populations by flow cytometry: Comparison between instruments. *Limnology and Oceanography: Methods, 14*, 750–758. https://doi.org/10.1002/lom3.10135

Goltsev, V., Zaharieva, I., Chernev, P., & Strasser, R. J. (2009). Delayed fluorescence in photosynthesis. *Photosynthesis Research, 101*, 217–232. https://doi.org/10.1007/s11120-009-9451-1

Grégori, G., Citterio, S., Ghiani, A., Labra, M., Sgorbati, S., Brown, S., & Denis, M. (2001). Resolution of viable and membrane-compromised bacteria in freshwater and marine waters based on analytical flow cytometry and nucleic acid double staining. *Applied and Environmental Microbiology, 67*, 4662–4670. https://doi.org/10.1128/aem.67.10.4662-4670.2001

Gumbo, J. R., Cloete, T. E., van Zyl, G. J. J., & Sommerville, J. E. M. (2014). The viability assessment of *Microcystis aeruginosa* cells after co-culturing with *bacillus mycoides* B16 using flow cytometry. *Physics and Chemistry of the Earth, 72*, 24–33. https://doi.org/10.1016/j.pce.2014.09.004

Halfar, J., & Fujita, R. M. (2002). Precautionary management of deep-sea mining. *Marine Policy, 26*, 103–106. https://doi.org/10.1016/S0308-597X(01)00041-0

Hasegawa, N., Takeda, N. I., Ueno, T., Kojima, S., Kawazu, M., Akaike, K., Okubo, T., Takagi, T., Suehara, Y., Hayashi, T., Saito, T., Kaneko, K., Mano, H., & Kohsaka, S. (2019). Detection of circulating sarcoma tumor cells using a microfluidic chip-type cell sorter. *Scientific Reports, 9*, 20047. https://doi.org/10.1038/s41598-019-56377-z

Hauton, C., Brown, A., Thatje, S., Mestre, N. C., Bebianno, M. J., Martins, I., Bettencourt, R., Canals, M., Sanchez-Vidal, A., Shillito, B., Ravaux, J., Zbinden, M., Duperron, S., Mevenkamp, L., Vanreusel, A., Gambi, C., Dell'Anno, A., Danovaro, R., Gunn, V., & Weaver, P. (2017). Identifying toxic impacts of metals potentially released during deep-sea mining—A synthesis of the challenges to quantifying risk. *Frontiers in Marine Science, 4*, 368. https://doi.org/10.3389/fmars.2017.00368

Hoagland, P., Beaulieu, S., Tivey, M. A., Eggert, R. G., German, C., Glowka, L., & Lin, J. (2010). Deep-sea mining of seafloor massive sulfides. *Marine Policy, 34*, 728–732. https://doi.org/10.1016/j.marpol.2009.12.001

Jones, D. O. B., Durden, J. M., Murphy, K., Gjerde, K. M., Gebicka, A., Colaço, A., Morato, T., Cuvelier, D., & Billett, D. S. M. (2019). Existing environmental management approaches relevant to deep-sea mining. *Marine Policy, 103*, 172–181. https://doi.org/10.1016/j.marpol.2019.01.006

Katsumata, M., Koike, T., Nishikawa, M., Kazumura, K., & Tsuchiya, H. (2006). Rapid ecotoxicological bioassay using delayed fluorescence in the green alga *Pseudokirchneriella subcapitata*. *Water Research, 40*, 3393–3400. https://doi.org/10.1016/j.watres.2006.07.016

Katsumata, M., Koike, T., Kazumura, K., Takeuchi, A., & Sugaya, Y. (2009). Utility of delayed fluorescence as endpoint for rapid estimation of effect concentration on the green alga *Pseudokirchneriella subcapitata*. *Bulletin of Environmental Contamination and Toxicology, 83*, 484–487. https://doi.org/10.1007/s00128-009-9826-1

Kawagucci, S., Makabe, A., Kodama, T., Matsui, Y., Yoshikawa, C., Ono, E., Wakita, M., Nunoura, T., Uchida, H., & Yokokawa, T. (2018). Hadal water biogeochemistry over the Izu–Ogasawara Trench observed with a full-depth CTD-CMS. *Ocean Science, 14*, 575–588. https://doi.org/10.5194/os-14-575-2018

Mann, E. L., Ahlgren, N., Moffett, J. W., & Chisholm, S. W. (2002). Copper toxicity and cyanobacteria ecology in the Sargasso Sea. *Limnology and Oceanography, 47*, 976–988. https://doi.org/10.4319/lo.2002.47.4.0976

Marie, D., Partensky, F., Jacquet, S., & Vaulot, D. (1997). Enumeration and cell cycle analysis of natural populations of marine picoplankton by flow cytometry using the nucleic acid stain SYBR Green I. *Applied and Environmental Microbiology, 63*, 186–193. https://doi.org/10.1111/j.1365-294X.2009.04480.x

Marie, D., Simon, N., & Vaulot, D. (2005). Phytoplankton cell counting by flow cytometry. In R. A. Andersen (Ed.), *Algal culturing techniques* (pp. 253–267). Academic.

Moore, L. R., Coe, A., Zinser, E. R., Saito, M. A., Sullivan, M. B., Lindell, D., Frois-Moniz, K., Waterbury, J., & Chisholm, S. W. (2007). Culturing the marine cyanobacterium *Prochlorococcus*. *Limnology and Oceanography: Methods, 5*, 353–362. https://doi.org/10.4319/lom.2007.5.353

Nescerecka, A., Hammes, F., & Juhna, T. (2016). A pipeline for developing and testing staining protocols for flow cytometry, demonstrated with SYBR Green I and propidium iodide viability staining. *Journal of Microbiological Methods, 131*, 172–180. https://doi.org/10.1016/j.mimet.2016.10.022

Niner, H. J., Ardron, J. A., Escobar, E. G., Gianni, M., Jaeckel, A., Jones, D. O. B., Levin, L. A., Smith, C. R., Thiele, T., Turner, P. J., Van Dover, C. L., Watling, L., & Gjerde, K. M. (2018). Deep-sea mining with no net loss of biodiversity—An impossible aim. *Frontiers in Marine Science, 5*, 53.

Nogueira, P. F. M., Nakabayashi, D., & Zucolotto, V. (2015). The effects of graphene oxide on green algae *Raphidocelis subcapitata*. *Aquatic Toxicology, 166*, 29–35. https://doi.org/10.1016/j.aquatox.2015.07.001

OECD (Organization for Economic Cooperation and Development). (2011). Test No. 201: Freshwater Alga and cyanobacteria, Growth Inhibition Test.

Ota, S., Yamaguchi, H., Vanel, F., Fuchida, S., Koshikawa, H., Yamagishi, T., Yamamoto, H., & Kawachi, M. (2020). Differential heavy metal sensitivity in seven algal species from the NIES culture collection based on delayed fluorescence assays. *Phycological Research, 68*, 41–49. https://doi.org/10.1111/pre.12403

Ota, S., Fuchida, S., Yamaguchi, H., Yamagishi, T., Yamamoto, H., Koshikawa, H., & Kawachi, M. (2022). Propidium iodide staining and flow cytometry-based assessment of heavy metal impact on marine phytoplankton. *Cytologia, 87*, 177–187. https://doi.org/10.1508/cytologia.87.177

Petersen, S., Krätschell, A., Augustin, N., Jamieson, J., Hein, J. R., & Hannington, M. D. (2016). News from the seabed—Geological characteristics and resource potential of deep-sea mineral resources. *Marine Policy, 70*, 175–187. https://doi.org/10.1016/j.marpol.2016.03.012

Ritz, C., Baty, F., Streibig, J. C., & Gerhard, D. (2015). Dose-response analysis using R. *PLoS One, 10*, e0146021. https://doi.org/10.1371/journal.pone.0146021

Sarmiento, J. L., & Gruber, N. (2006). *Ocean biogeochemical dynamics*. Princeton University Press.

Sawada, T., Watanabe, M., Fujimura, Y., Yagishita, S., Shimoyama, T., Maeda, Y., Kanda, S., Yunokawa, M., Tamura, K., Tamura, T., Minami, H., Koh, Y., & Koizumi, F. (2016). Sensitive cytometry based system for enumeration, capture and analysis of gene mutations of circulating tumor cells. *Cancer Science, 107*, 307–314. https://doi.org/10.1111/cas.12868

Schalk, I. J., Hannauer, M., & Braud, A. (2011). New roles for bacterial siderophores in metal transport and tolerance. *Environmental Microbiology, 13*, 2844–2854. https://doi.org/10.1111/j.1462-2920.2011.02556.x

Sieracki, M. E., Poulton, N. J., Jaillon, O., Wincker, P., de Vargas, C., Rubinat-Ripoll, L., Stepanauskas, R., Logares, R., & Massana, R. (2019). Single cell genomics yields a wide diversity of small planktonic protists across major ocean ecosystems. *Scientific Reports, 9*, 6025. https://doi.org/10.1038/s41598-019-42487-1

Suman, T. Y., Radhika Rajasree, S. R., & Kirubagaran, R. (2015). Evaluation of zinc oxide nanoparticles toxicity on marine algae chlorella vulgaris through flow cytometric, cytotoxicity and oxidative stress analysis. *Ecotoxicology and Environmental Safety, 113*, 23–30. https://doi.org/10.1016/j.ecoenv.2014.11.015

Takeuchi, A., Katsumata, M., Koike, T., Takata, Y., Itatsu, Y., & Kusui, T. (2014). Comparison of the conventional algal growth inhibition tests using cell counting and algal bioassay using delayed fluorescence: Application to industrial effluents. *Journal of Water and Environment Technology, 12*, 367–377. https://doi.org/10.2965/jwet.2014.367

Takeuchi, I., Takaichi, D., Katsumata, M., & Ishibashi, H. (2020). Succession of delayed fluorescence correlated with coral bleaching in the hermatypic coral *Acropora tenuis*. *Marine Pollution Bulletin, 154*, 111008. https://doi.org/10.1016/j.marpolbul.2020.111008

van den Engh, G., Ulloa, O., & Karl, D. M. (2019). Monitoring microbial communities in the marine environment. *Cytometry Part A, 95*, 717–721. https://doi.org/10.1002/cyto.a.23763

van der Merwe, R., Hammes, F., Lattemann, S., & Amy, G. (2014). Flow cytometric assessment of microbial abundance in the near-field area of seawater reverse osmosis concentrate discharge. *Desalination, 343*, 208–216. https://doi.org/10.1016/j.desal.2014.01.017

Van Dover, C. L. (2011). Mining seafloor massive sulphides and biodiversity: What is at risk? *ICES Journal of Marine Science, 68*, 341–348. https://doi.org/10.1093/icesjms/fsq086

Wang, W.-X., & Dei, R. C. H. (2001a). Effects of major nutrient additions on metal uptake in phytoplankton. *Environmental Pollution, 111*, 233–240. https://doi.org/10.1016/S0269-7491(00)00071-3

Wang, W.-X., & Dei, R. C. H. (2001b). Metal uptake in a coastal diatom influenced by major nutrients (N, P, and Si). *Water Research, 35*, 315–321. https://doi.org/10.1016/S0043-1354(00)00256-6

Watanabe, M., Serizawa, M., Sawada, T., Takeda, K., Takahashi, T., Yamamoto, N., Koizumi, F., & Koh, Y. (2014). A novel flow cytometry-based cell capture platform for the detection, capture and molecular characterization of rare tumor cells in blood. *Journal of Translational Medicine, 12*, 143. https://doi.org/10.1186/1479-5876-12-143

Wedding, L. M., Reiter, S. M., Smith, C. R., Gjerde, K. M., Kittinger, J. N., Friedlander, A. M., Gaines, S. D., Clark, M. R., Thurnherr, A. M., Hardy, S. M., & Crowder, L. B. (2015). Managing mining of the deep seabed: Contracts are being granted, but protections are lagging. *Science, 349*, 144–145. https://doi.org/10.1126/science.aac6647

Xiao, X., Han, Z. Y., Chen, Y. X., Liang, X. Q., Li, H., & Qian, Y. C. (2011). Optimization of FDA-PI method using flow cytometry to measure metabolic activity of the cyanobacteria, *Microcystis aeruginosa*. *Physics and Chemistry of the Earth, 36*, 424–429. https://doi.org/10.1016/j.pce.2010.03.028

Yamagishi, T., Katsumata, M., Yamaguchi, H., Shimura, Y., Kawachi, M., Koshikawa, H., Horie, Y., & Tatarazako, N. (2016). Rapid ecotoxicological bioassay using delayed fluorescence in the marine cyanobacterium *Cyanobium* sp. (NIES-981). *Ecotoxicology, 25*, 1751–1758. https://doi.org/10.1007/s10646-016-1718-7

Yamagishi, T., Fuchida, S., Katsumata, M., Horie, Y., Mori, F., Kitayama, A., Kawachi, M., Koshikawa, H., Nozaki, T., Kumagai, H., Ishibashi, J., & Tatarazako, N. (2018). Evaluation of the toxicity of leaches from hydrothermal sulfide deposits by means of a delayed fluorescence-based bioassay with the marine cyanobacterium *Cyanobium* sp. NIES-981. *Ecotoxicology, 27*, 1303–1309. https://doi.org/10.1007/s10646-018-1989-2

Yamaguchi, H., Shimura, Y., Suzuki, S., Yamagishi, T., Tatarazako, N., & Kawachi, M. (2016). Complete genome sequence of *Cyanobium* sp. NIES-981, a marine strain potentially useful for ecotoxicological bioassays. *Genome Announcements, 4*, e00736-16. https://doi.org/10.1128/genomeA.00736-16

Zeng, J., & Wang, W.-X. (2009). The importance of cellular phosphorus in controlling the uptake and toxicity of cadmium and zinc in *Microcystis aeruginosa*, a freshwater cyanobacterium. *Environmental Toxicology and Chemistry, 28*, 1618–1626. https://doi.org/10.1897/08-639.1

Zhang, H., Zhang, S., Peng, Y., Li, Y., Chen, Z., Xu, H., Yu, Z., Zheng, W., & Zheng, T. (2015). Effects of marine actinomycete on the removal of a toxicity alga *Phaeocystis globose* in eutrophication waters. *Frontiers in Microbiology, 6*, 474. https://doi.org/10.3389/fmicb.2015.00474

Dr. Shuhei Ota is an expert for the Microbial Culture Collection at the Biodiversity Division, National Institute for Environmental Studies (MCC-NIES), Tsukuba, Japan. Earlier, he served as an assistant professor in the University of Tokyo from 2011 to 2017, as well as a postdoctoral fellow at the University of Oslo (Norway) and Station Biologique de Roscoff (France). He has a PhD degree in life sciences from the Graduate School of Natural Science and Technology, Kanazawa University, Japan. He is currently involved in microbial preservation at MCC-NIES. His research interests include phytoplankton phylogeny, taxonomy, ecotoxicology, and cell biology in algae.

Dr. Hiroshi Koshikawa is working for the Regional Environment Conservation Division of the National Institute for Environmental Studies (NIES), Japan, as a head of Marine Environmental Section. His research background is in phytoplankton ecology, and he has focused on anthropogenic alterations of marine environments and their impacts on the algal production and food webs, such as eutrophication, oil pollution, and climate change issues. Since 2014, he has worked on the development of the EIA protocol, especially for the marine surface environment at the seafloor mining site, where there are risks of heavy metal contamination. He was born in 1968 and obtained the Doctor of Engineering from the Science University of Tokyo in 1997.

Dr. Masanobu Kawachi works for the Biodiversity Division of the National Institute for Environmental Studies as a group leader in the Microbial Culture Collection. His research background includes algal diversity, ecology, and applications. He has been involved in the National Bioresource Project and projects to establish appropriate methods for environmental impact assessment, particularly for the marine surface environment.

Part IV
Regional Assessment of Water Column Characteristics and Management

Chapter 10
Atmospheric and Oceanographic Characteristics of the BGR Exploration Area for Polymetallic Nodules in the Northeastern Tropical Pacific Ocean and Model Simulations of Dredge-Induced Suspended Sediment Transport

Carsten Rühlemann and Kaveh Purkiani

Abstract This chapter provides an overview on atmospheric and hydrographic conditions, and temporal variations of bottom water currents in the exploration area for polymetallic nodules of the German Federal Institute for Geosciences and Natural Resources (BGR) in the northeastern tropical Pacific Ocean. Furthermore, we compare in situ resuspension experiments with numerical model results on transport pathways and deposition patterns of the suspended sediments to assess the reliability of model predictions with respect to future deep-sea mining. The BGR area underlies seasonal climate variations due to meridional migrations of the Intertropical Convergence Zone (ITCZ) and changes in the subtropical high. In March, when the ITCZ is weakest near the equator, sea level pressure is high and precipitation and northeast winds are low, while in August and September, when the ITCZ is most northerly, pressure is low, precipitation increases, and winds are variable. The frequency of tropical storms and hurricanes peaks in June through August. Tropical storm magnitude is usually reached once a year, and hurricanes occur every other year. The two parts of the exploration area are located at the equatorward edge of the westward flowing North Equatorial Current. Water temperatures reach as high as 27 °C and salinity as low as 33.9 psu in the mixed layer with an average depth of 30 m. The thermo- and haloclines are sharp and separate the surface layers from cooler, saltier waters that are devoid of oxygen to a depth of 700 m to 900 m. These water masses overlie Pacific Intermediate waters, the deeper

C. Rühlemann (✉)
Federal Institute for Geosciences and Natural Resources, (BGR), Hannover, Germany
e-mail: carsten.ruehlemann@bgr.de

K. Purkiani
Federal Maritime and Hydrographic Agency (BSH), Hamburg, Germany

R. Sharma (ed.), *Deep-Sea Mining and the Water Column*,
https://doi.org/10.1007/978-3-031-59060-3_10

Antarctic Intermediate water, and Lower Circumpolar waters closest to the ocean bottom. Analysis of near-bottom current data recorded by moored ADCPs between 2013 and 2019 indicates low velocities of 1–5 cm s^{-1} on average and maximum values of 8–14 cm s^{-1}. Current directions are strongly influenced by semidiurnal tidal cycles and near-inertial oscillations, with predominant southeasterly to southwesterly flow. The bottom currents can be amplified and change their predominant direction by mesoscale eddies that form under the influence of Central American Gap Winds and travel several 1000 km westward. Coupled ocean-circulation and sediment-transport models, taking into account local sediment particle properties, were used to simulate the near-field and far-field transport and settling of suspension plumes on the seafloor generated by dredging experiments in the BGR exploration area and by an epibenthic sledge tow in the Peru Basin. Comparison of the observed and predicted deposition patterns showed satisfactory agreement and increased confidence in the model's ability to adequately predict the spread of suspension plumes and particle redeposition resulting from future industrial-scale deep-sea mining.

Keywords Pacific Ocean · Clarion-Clipperton Zone · Atmospheric variability · Oceanography · Bottom currents · Particle transport modeling · Deep-sea mining

1 Introduction

In 2006, the German Federal Institute for Geosciences and Natural Resources (BGR) signed a contract with the International Seabed Authority (ISA) for the exploration of polymetallic nodule deposits in the Clarion-Clipperton Zone (CCZ) of the tropical Pacific Ocean. This exploration area consists of two parts, covering 75,000 km^2 in total. The eastern area "E" of 60,000 km^2 is centered at 118°W/12°N and western area "W" of 15,000 km^2 is located approximately 2000 km to the west at 138°W/12.5°N (Fig. 10.1).

With this study, we aim to provide an overview of the oceanographic data collected in the framework of the exploration between 2008 and 2023. BGR's work program relating to environmental baseline studies within its contract areas follows the "Recommendations for the guidance of contractors for the assessment of the possible environmental impacts arising from exploration for marine minerals in the Area" as outlined in the ISA document ISBA/25/LTC/6/Rev.2. These recommendations include water column measurements in order to determine its physical and chemical characteristics before any future mining activities may take place.

Furthermore, we carried out numerical experiments on the vertical and lateral spread of suspension plumes that may be generated by potential future deep-sea mining and the deposition of resuspended sediments in water depths of 4200 m based on a sediment transport module coupled to a high-resolution hydrodynamic regional ocean model. Drift of sediment plumes and redeposition of particles on top of the seafloor are expected to exert pressure on the benthic ecosystem. Benthic

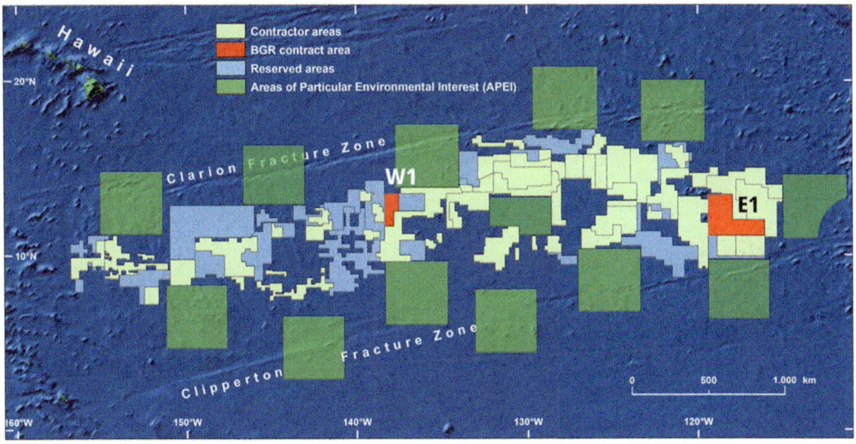

Fig. 10.1 Map of the nodule belt between the Clarion and Clipperton fracture zones in the East Pacific. The two parts W and E of the BGR exploration area are highlighted in red. The yellow areas are exploration areas of other contractors, blue-gray areas indicate "reserved areas," for which developing countries can obtain exploration contracts, and the green squares indicate Areas of Particular Environmental Interest (APEI). APEIs form a network of large no-mining zones that are intended to cover the full range of habitats, biodiversity and ecosystem functions

communities are adapted to stable environmental conditions in the deep sea and are expected to be vulnerable to sediment redeposition associated with mining-induced sediment plumes (Glover & Smith, 2003; Gollner et al., 2017; Jones et al., 2017). Therefore, the understanding of the physical behavior of the suspended sediment plume and the proposition of appropriate monitoring strategies through in-situ and numerical modeling research is crucial.

Due to the difficulties of in situ observations, including expensive logistics, limited operational research time, and unfavorable weather conditions, a thoroughly validated numerical model simulation with well-calibrated input parameters is an excellent tool for studying sediment transport and quantifying the distribution and deposition of deep-sea sediment plumes. To date, our knowledge of deep sediment transport from numerical models is limited to a few studies based on accompanying observations, for example in the Pacific (Jankowski et al., 1996; Nakata et al., 1997; Jankowski & Zielke, 2001; Schaaning et al., 2008; Roliniski et al., 2001; Schriever & Thiel, 2013; Aleynik et al., 2017; Ouillon et al., 2022; Muñoz-Royo et al., 2022) or near the Canary Islands (Spearman et al., 2020). Many commercial modeling efforts were linked to a specific resuspension experiment and developed with limited access to underlying data (Spearman et al., 2020). In addition, the role of flocculation during plume dispersion and its direct impact on the settling velocity of particles has not been considered in the past. The impact of the flocculation process on settling velocities was included in our model based on ex-situ experiments by Gillard et al. (2019). Determining the fate of sediments, the spreading of the plume and redeposition patterns on the seabed, and the modeling of disturbance experiments requires an extended variety of oceanographic data as input, such as deep-sea current properties and sediment characteristics. The validation of a numerical ocean-sediment transport model with a large sensor array dataset helps to better predict future environmental impacts of deep-sea mining. To this end, a sediment module was developed and coupled to the MITgcm hydrodynamic model to simulate the sediment distribution caused by benthic disturbances in the deep sea.

2 Methods

2.1 Atmospheric Variability and Water Mass Characteristics

Atmospheric conditions in areas W and E are based on the reanalysis of historical data (ERA-interim; Dee et al., 2011), satellite observations (AMSR-E and AMSR-2, http://www.remss.com/missions/amsr/) and hurricane post-storm analyses (HURDAT2, http://www.nhc.noaa.gov/) carried out by the Maunalani Consulting LLC (Honolulu) on behalf of the BGR (Schneider, 2016a, b). The study of Schneider (2016a) also describes the methods used in detail.

Regarding oceanographic data, the BGR acquired measurements of temperature, salinity, oxygen concentration, fluorescence, bottom currents, and turbidity through the deployment of CTDs and additional sensors, as well as moored ADCPs. In order to investigate longer time series on a monthly basis as well, ocean reanalysis based on data from the World Ocean Atlas 2013 was also carried out by the Maunalani Consulting LLC on behalf of the BGR (Schneider, 2016b). Water column parameters determined by the BGR were mainly collected in the months of April and May. During the 15 years of seagoing expeditions, we completed a total of 74 CTD deployments, 68 in area E and 6 in area W (Figs. 10.2 and 10.3, Table 10.1). Of these 74 datasets, 43 were measured using sensors mounted on a rosette water sampler, and 31 were measured using sensors mounted on a video sledge. The CTD casts have been measured with different instruments during 11 cruises with the research vessels KILO MOANA in 2008, 2013, 2014, 2016, and 2023, the old SONNE in 2010, the L'ATALANTE in 2012, the new SONNE in 2015, 2018, and 2022, and with the multi-purpose vessel ISLAND PRIDE in 2021. The data are available through the DeepData database of the ISA (https://www.isa.org.jm/deepdata-database/).

2.2 Sediment Suspension Plume Modeling

To model the spreading of sediment suspension plumes and seabed redeposition patterns associated with nodule collector mining, a sediment module was developed and coupled to the MITgcm hydrodynamic model. The model algorithm is described in detail in Marshall et al. (1997) and Adcroft et al. (2004). The following 3-D advection-diffusion equation for different size classes of particles is solved in the model.

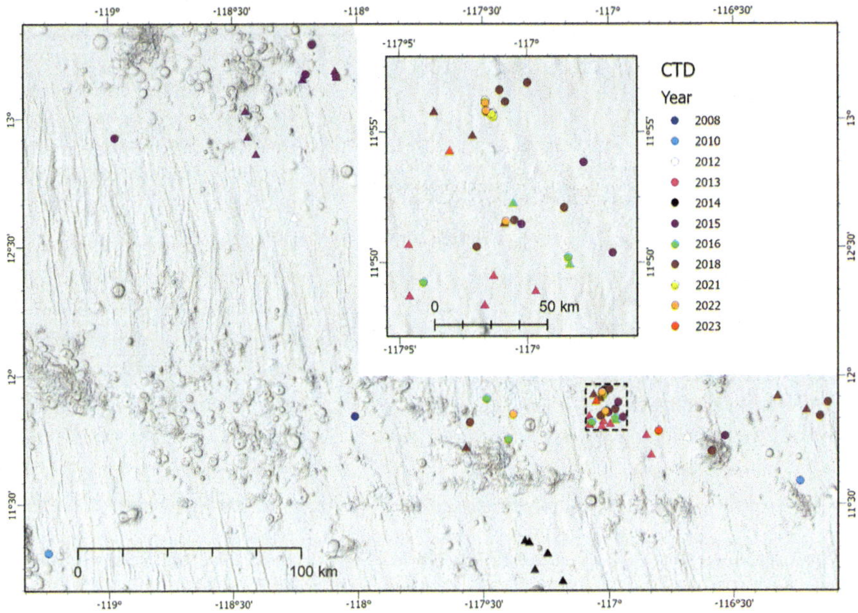

Fig. 10.2 Locations of CTD measurements from 2008 to 2023 in the eastern part of the BGR exploration area (E). Color codes indicate year of deployment, circles identify CTDs that were mounted on rosette water samplers, and triangles show sites where CTDs were attached to video sledges

$$\frac{\partial C_i}{\partial t}+u\frac{\partial C_i}{\partial x}+v\frac{\partial C_i}{\partial y}+w\frac{\partial C_i}{\partial z}+\frac{\partial\left(wS_iC_i\right)}{\partial z}$$
$$=\frac{\partial}{\partial x}\left(Kh\frac{\partial C_i}{\partial x}\right)+\frac{\partial}{\partial y}\left(Kh\frac{\partial C_i}{\partial y}\right)+\frac{\partial}{\partial z}\left(Kv\frac{\partial C_i}{\partial z}\right)+AS_{in}-D_i+E_i \tag{10.1}$$

where C_i is the suspended sediment concentration (SSC) of each sediment class (i). The model includes three size classes, which the user can define. The current velocity components are indicated as u, v, and w and the settling velocity of each sediment class is shown as ws_i. The horizontal and vertical diffusivity coefficients are presented as K_h and K_v, respectively. The same advection and diffusion schemes as for salinity and temperature are also used for the SSC. The sediment input source due to the towing of the dredge on the seafloor is given as AS_{in}. Here, A is a parameter used as a calibration coefficient to fit the observed SSC to the results of the measurements at the monitoring stations. The suspended sediment that reached the seafloor was added to the deposition layer, which is depicted as D_i in Eq. (10.1), and was removed from further suspended sediment calculations in the model. The erosion of sediment for each size class (E_i) was added to the equation, which may work as a source term in the model if the bed shear stress exceeds the critical shear stress (τ_{cb}). The simulated deposited sediment mass was converted to sediment deposition

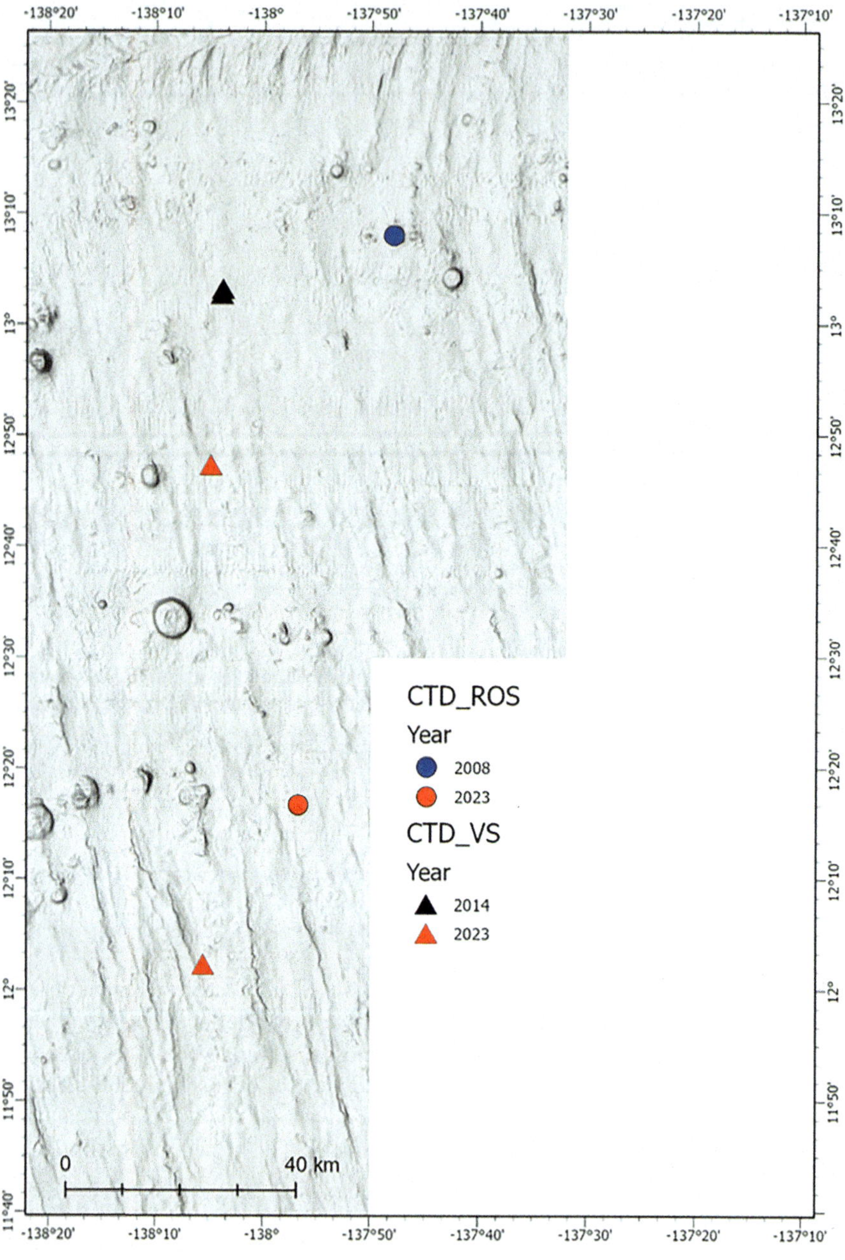

Fig. 10.3 Locations of CTD measurements in 2014 and 2023 in the western part of the BGR exploration area. Color codes indicate year of deployment, circles identify CTDs that were mounted on rosette water samplers, and triangles show sites where CTDs were attached to video sledges

Table 10.1 List of vessels used for exploration between 2008 and 2023 and instruments deployed to measure water mass properties

Year	Vessel	CTD	Oxygen	Fluorescence	Turbidity
2008	KILO MOANA	SBE 911	SBE 43	Seapoint SCF	WET Labs C-Star
2010	SONNE (old)	SBE 911	SBE 43	–	–
2012	L'ATALANTE	SBE 19plus	–	–	–
2013	KILO MOANA	SBE 19plus	SBE 43	WET Labs ECO-AFL/FL	WET Labs ECO-NTU
2014	KILO MOANA	SBE 911	SBE 43	WET Labs ECO-AFL/FL	WET Labs ECO-NTU
2015	SONNE (new)	SBE 911plus	SBE 43	WET Labs ECO-AFL/FL	Seapoint
2016	KILO MOANA	SBE 19plus	SBE 43	WET Labs ECO-AFL/FL	WET Labs ECO-NTU
2018	SONNE (new)	SBE 911plus	SBE 43	WET Labs ECO-AFL/FL	WET Labs ECO-NTU
2021	ISLAND PRIDE	SBE 911plus	SBE 43	WET Labs ECO-AFL/FL	Seapoint
2022	SONNE (new)	SBE 911plus	SBE 43	WET Labs ECO-AFL/FL	WET Labs ECO-NTU
2023	KILO MOANA	SBE 911plus	SBE 43	WET Labs ECO-AFL/FL	WET Labs ECO-NTU

thickness, which was often used to determine the resilience of benthic communities (Gollner et al., 2017; Lefaible et al., 2023). The integration takes into account the model grid cell size, a dry bulk sediment density of 200–250 kg m^{-3}, and a freshly deposited sediment porosity of 0.9.

Accurately reconstructing the flow field and its variability in a small domain with all boundaries open is a challenging yet crucial aspect when simulating the distribution of sediment plumes in the deep sea. In both case studies, the model was laterally forced by current velocity components obtained from in-situ current measurements near the dredging tracks. Temperature and salinity profiles were obtained from a CTD profile conducted near the dredging tracks. A spin-up time of 2 days was sufficient for the flow field to reach a steady state before the sediment was released into the domain at each experiment. The model was driven by ERAInterim reanalysis atmospheric forcing (Berrisford et al., 2011), which includes horizontal wind speed, air temperature, short and long-wave radiation, surface-specific humidity, and precipitation with a temporal resolution of 6 h (Purkiani et al., 2021).

In Sect. 4, we report on the results of two modeling experiments conducted to study the transport of suspension plumes and settling patterns of sediment particles for comparison with in situ dredging experiments to assess the reliability of the model predictions with respect to future deep-sea mining.

Case Study 1 describes a dredging experiment in the BGR exploration area. During the scientific expedition SO268 (MiningImpcat Project) a small-scale plume dispersion experiment in the BGR area was carried out. In this experiment, a 1 m wide chain dredge was used to generate a sediment plume. The dredge was dragged across the seafloor in 11 tracks, spanning from west-southwest to east-northeast. Each track was around 500 m in length and took between 40 and 60 min to complete. The dredging took place on April 11, 2019, and lasted 12.5 h (06:30 to 19:00 UTC) (Haalboom et al., 2022). The maximum sediment release rate, considering the size of the dredging gear, its estimated penetration depth of 5 cm into the seafloor, the towing speed of 0.2 to 0.5 ms^{-1}, the dry bulk density of the sediment of 200–250 kgm^{-3} and a surface sediment porosity of 0.9, can be estimated as 2.5 to 6.25 kg s^{-1}. The mean current speeds at 1 m above seafloor reach 4.4 cms^{-1}, with a dominant southward velocity component on the first day and a westward one during the following days (Purkiani et al., 2021). The progressive vector diagram of the deep-water current, measured at a height of 1 m above the seafloor, indicates a predominant south-to-southwest direction during the dredging experiment. At this depth, the maximum north-south water movement is approximately 5.5 km, while the maximum east-west movement is 9 km during 4 days of measurements.

The sediment release rate observed in the dredge experiment shares similarities with the experiment conducted by Gillard et al. (2019). In their lab study, they used an initial plume concentration of 35 mg l^{-1} and a shear rate of $G = 2.5\ s^{-1}$, which resulted in a relatively low gradient between current velocities observed in different bins of the ADCP measurements. Drawing from the findings of Gillard et al. (2019), we have set the settling velocities for sediment particles in our modeling study at 25, 140, and 300 m d^{-1} for aggregates categorized as class 1 (d25: 70 μm), class 2 (d50: 340 μm), and class 3 (d75: 590 μm), respectively. The sediment mass distribution was non-equally split so that classes 1, 2, and 3 have 25, 50, and 25% of total mass weight. An additional sediment class 4 with settling velocity of 2500 m d^{-1} was introduced to enable the model to simulate redeposition of sediment "lumps" that did not come into resuspension and were only dislocated by the dredging activity.

Case study 2 describes a dredging experiment in the DISCOL (DISturbance and re-COLonization Experiment) area. During expedition SO242 (MiningImpact project) in 2015, the DISCOL area in the Peru Basin was revisited for high-detail mapping and biogeochemical and biological sampling in and close to areas that were disturbed in 1989 (Greinert, 2015). A small-scale seafloor disturbance experiment was carried out to investigate the dispersion of sediment plumes by near-bottom currents. The test area is located in the Peru Basin of the southern Pacific Ocean, 900 km offshore Peru at 4150 m water depth. The area is known as the DISCOL area, where in 1989, a circular patch of about 2 nautical miles across was impacted with a plough harrow to generate a mechanical disturbance (Thiel & Schriever, 1990). The water mass near the bottom in the DISCOL is characterized by salinity of about 35 PSU and temperature < 2 °C, and currents with alternating regimes: weak (1 to 3 cm s^{-1}) and average (~5 cm s^{-1}) episodes (Klein, 1993, 1996). The topographically gently sloping area, with a lower nodule density than the surrounding area and without rocky outcrops, was chosen to avoid the failure of the plough. An epibenthic sledge (EBS) was used to generate a plume of suspended sediment. The EBS of 880 kg weight (in air), 3.6 m length, 2.4 m width, and 1.2 m height was towed for 20 min between 00:36 and 00:56 UTC on August 5, 2015, over a distance of 740 m, slightly upslope (2.8%) at a quasi-constant velocity of 0.6 cms^{-1}.

Considering the size of the EBS, the total weight of dry sediment displacement rate was estimated to 18.6 kg s^{-1}. Here we used three sediment size classes as 61, 10 and 2 μm based on the d90, d50 and d10. Near-bottom current dynamics and SPM transport were recorded with the Bottom Boundary (BoBo) lander and the Deep-sea Observation System lander (both from NIOZ and GEOMAR, institutes collaborating in the MiningImpact project).

3 Results and Observations

3.1 Atmospheric Variability

Area E is located between the subtropical sea level pressure ridge of the North Pacific High to the north and the sea level pressure trough associated with the Intertropical Convergence Zone (ITCZ) to the south. Thus, meridional migrations of the ITCZ and changes in the subtropical high influence climatic conditions in the eastern contract area on a seasonal scale (Fig. 10.4). When the ITCZ is weakest and located close to the equator in March, the contract area is dominated by high sea-level pressure, low precipitation, and northeast trades. In August, when the ITCZ is at its northernmost position, sea-level pressure in the contract area is low and precipitation is high. On interannual timescales, the El Niño-Southern Oscillation (ENSO) influences the regional atmospheric conditions, leading to an equatorward shift of the ITCZ and warming in the El Niño phase, with commensurate changes in sea-level pressure, precipitation, and winds (Fig. 10.5).

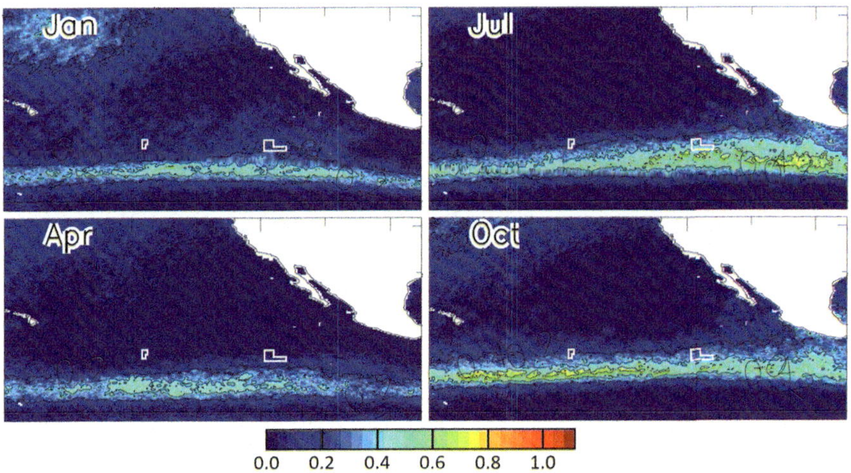

Fig. 10.4 Seasonal cycle of mean monthly precipitation in mm h^{-1} observed with the AMSRE and AMSR2 satellites. Contract areas W and E are indicated by small white polygons. (Figures from Schneider, 2016a)

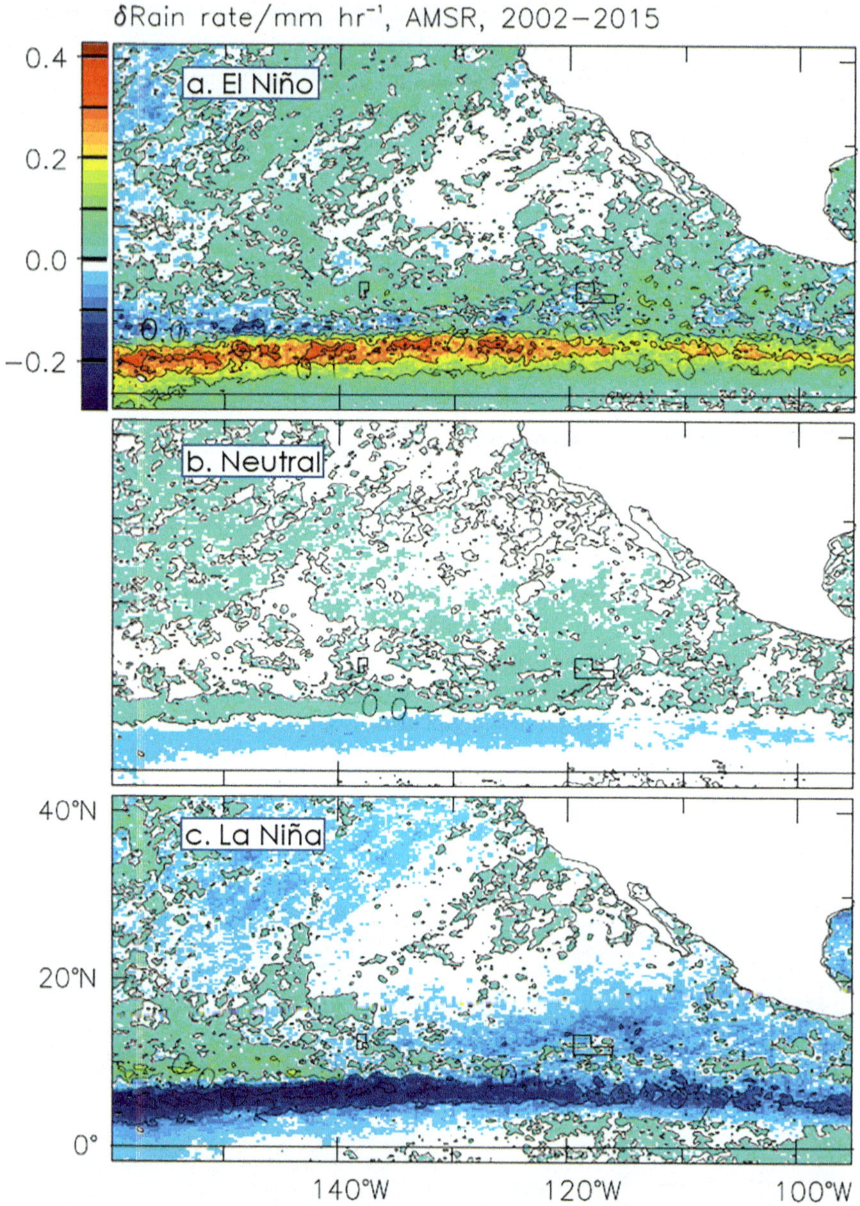

Fig. 10.5 Composite AMSR rain rate anomalies in mm h^{-1} for (**a**) El Niño, (**b**) neutral, and (**c**) La Niña conditions. El Niño and La Niña terciles are defined based on NINO3.4 index larger than one standard deviation and less than minus one standard deviation, respectively. Contract areas W and E are indicated by small black polygons. (Figures from Schneider, 2016a)

In area W, northeastern winds prevail throughout the year and are strongest in boreal winter and spring, averaging 8 m s^{-1}. In the boreal summer and fall, winds weaken to less than 4 m s^{-1} around September (Fig. 10.6). In contrast, winds in area E are highly variable and bimodal, with northeast trade winds predominating from November to June, with additional southwesterly winds occurring from July to October and dominating in September (Fig. 10.7).

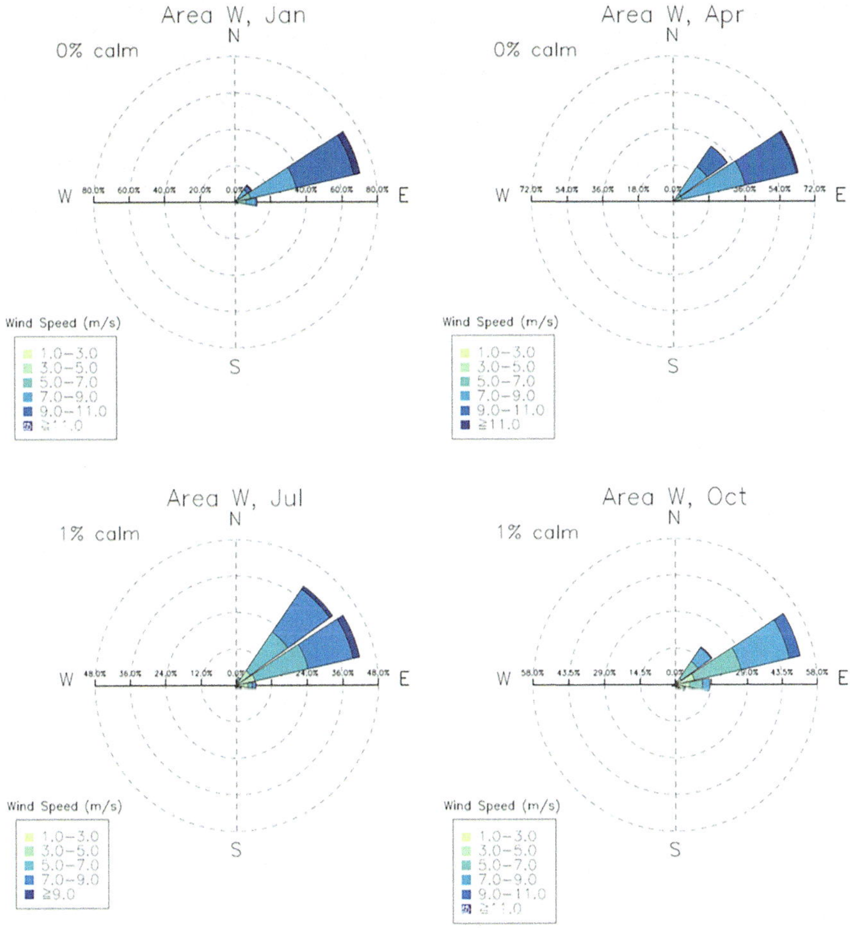

Fig. 10.6 Wind rose diagrams showing daily averaged surface winds in area W for January, April, July, and October. Shown are in 15° directional bins the wind speed frequency distribution, with percentage of occurrence in the radial direction, and speeds as indicated by the color key in the lower left of each panel. Note that wind speed bins and frequency along the radial axis are adjusted in each plot to optimally represent the data. Wind direction is defined according to meteorological convection: Easterly winds occupy the bin toward the east (E), northerlies the directional bin toward north (N) and similarly for westerlies and southerlies. Winds with speed smaller than the lowest bin are marked as calm, with the frequency of occurrence given in upper left. (Figures from Schneider, 2016a)

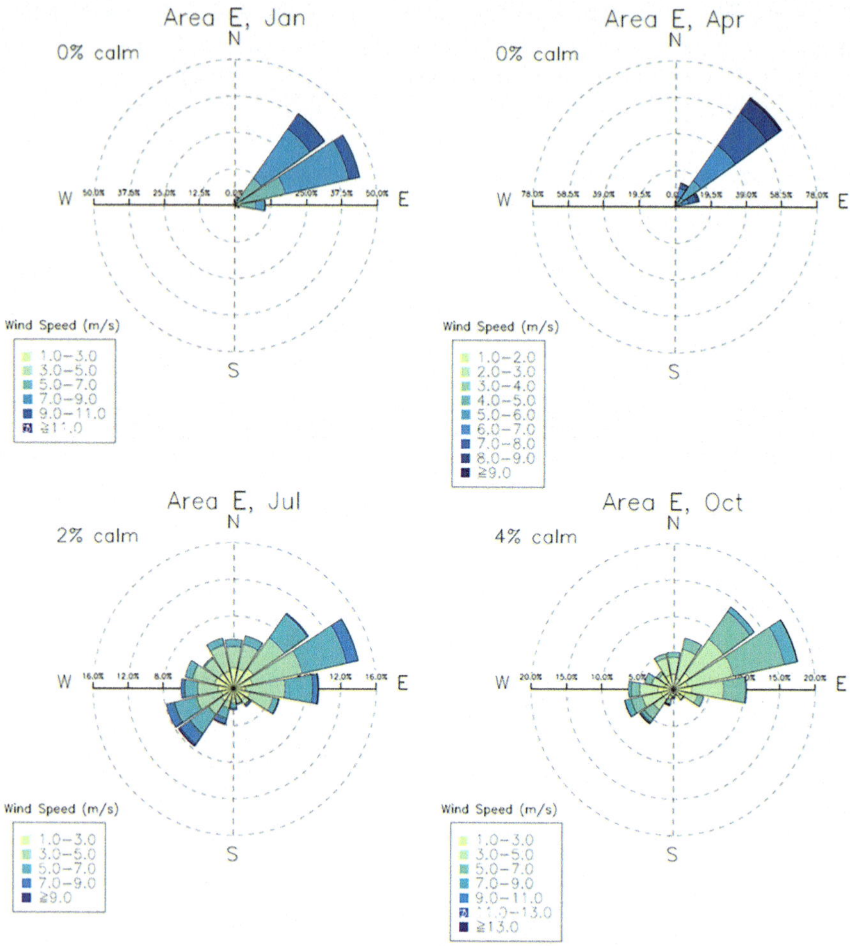

Fig. 10.7 Wind rose diagrams showing the seasonal cycle of daily averaged surface winds as in Fig. 10.6 but for area E. (Figures from Schneider, 2016a)

The eastern Pacific is the location of global maximum tropical storm and hurricane frequency (Neumann, 1993). The number of extreme wind occurrences in areas W and E are estimated by the number of tropical depressions, tropical storms (winds stronger than 35 knots) and hurricanes (winds in excess of 65 knots) (Fig. 10.8). The tropical storm season in area W commences in July with only rare occurrences of tropical depression during May and June. Activity ramps up in July, reaches its maximum in August, and diminishes in September and October. On average, area W experiences 1.4 storm events of at least tropical depression strength per season, with 0.8 of these of tropical storm strength and 0.4 of hurricane strength. In area E the hurricane seasons starts in May, quickly ramps up to reach maximum intensity in June with over 1.1 wind events per month of tropical depression strength

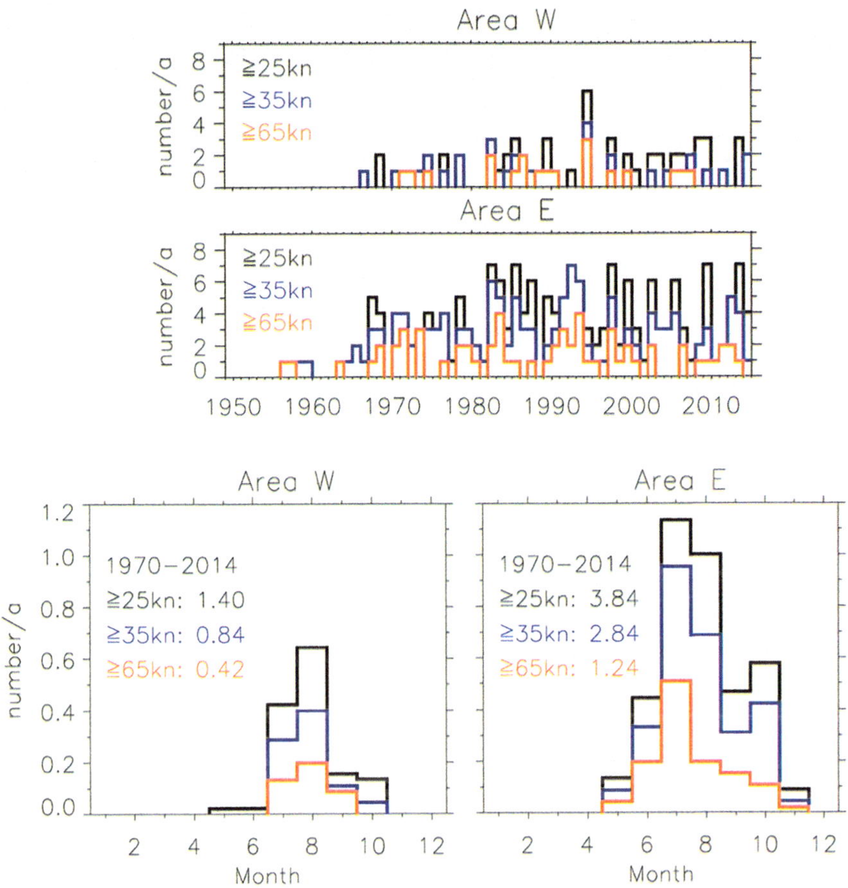

Fig. 10.8 Number of tropical depressions (black), tropical storms (blue) and hurricanes (orange) per hurricane season for areas W and E is shown in the upper panel and average numbers per year as a function of month for the two areas is shown in the lower panel. A storm is counted when its position lies within 10°N to 15°N and for area W between 140°W and 135°W, and for area E between 120°W and 115°W. Its strength is the maximum strength attained while in this area. The averaging period is 1970-2014. The numbers in the lower two panels show the average annual cumulative counts of extreme wind events for each of the three categories (Figures from Schneider, 2016a)

or stronger, of which 1.0 are at or exceed tropical storm magnitude, and, on average hurricane strength is achieved every other year. Storm counts slowly abate in September and October, but events have occurred as late as November. Event counts are highly variable from season to season, and are systematically smaller in area W compared to area E, consistent with the long-term spatial distribution of event counts.

The tropical Pacific wave climate is characterized by a multi-mode wave spectrum, dominated by swells generated in the mid-latitude storm tracks, and due to trade winds and tropical storms. In the Northern Hemisphere tropics, the dominant wave modes are due to the north-east trades and northerly swells caused by the eastward tracking mid-latitude storms during boreal winter, while during boreal summer, southerly swells from Southern Hemisphere winter storms are present (Hemer et al., 2011). Distributions of daily significant wave height (average of the largest 33% of waves) and direction in areas W and E for January, April, July and October are shown in Figs. 10.9 and 10.10, respectively. During boreal winter

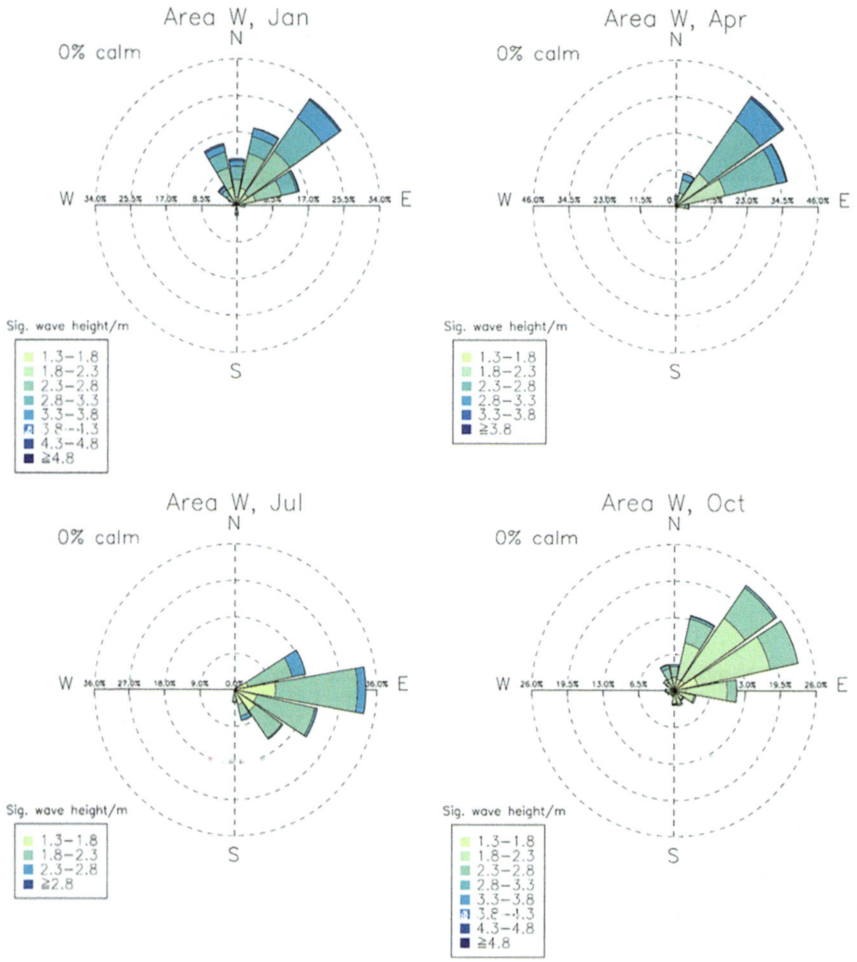

Fig. 10.9 Seasonal cycle of the distribution of daily averaged significant wave height in meters in area W. Shown are in 15° directional bins the wave direction, with percentage of occurrence in the radial direction, and significant wave heights as indicated by the color key in the lower left of each panel. Note that height bins and radial axis are adjusted in each plot to optimally represent the data. Wave direction indicate the direction of wave propagation, toward north (N), west (W), south (S), and east (E). Wave heights less the lowest bin are marked as calm, with frequency of occurrence given in upper left. (Figures from Schneider, 2016a)

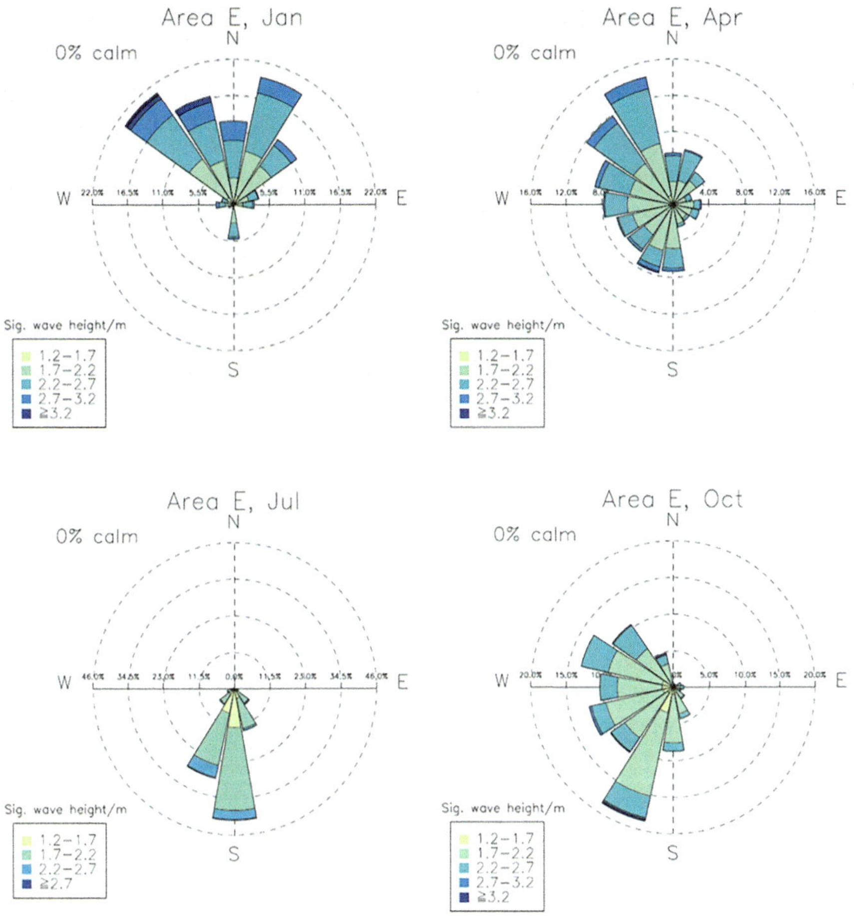

Fig. 10.10 Seasonal cycle of the distribution of daily averaged significant wave height as in Fig. 10.9 but for area E. (Figures from Schneider, 2016a)

(December to April), waves in area W are dominated by northeasterly swells with significant wave heights in excess of 2.3 m, and northerly swells in January and February. Starting in May and June, easterly and southeasterly swells with smaller significant wave heights of around 2 m occur, and peak in September when swells cover a broad fan of easterly directions. In area E, the seasonal switch is even more pronounced, with large northerly swells in November to March, with largest amplitudes in February, to a predominance of southerly swells from May to September (Fig. 10.10).

3.2 Hydrographic Conditions

The eastern and western parts of the BGR exploration area are situated within the North Equatorial Current, flowing westward under the influence of the northeast trade winds, and the westward extension of the eastern Pacific Warm Pool and just north of the ITCZ. The oceanography in these areas from the surface to the thermocline is largely a response to the wind stresses by the northeasterly trades north and the southeasterly trades south of the ITCZ. On a long-term average, seasonal variation in sea surface temperature is between 25.0 °C and 28.0 °C in area W and between 26.5 °C and 28.3 °C in area E (Fig. 10.11).

In area W, surface salinity varies from 33.6 in September to 34.6 in May, while in area E it varies from 33.2 in October to 34.1 in May. The seasonal ranges of surface salinity of ~1.0 are comparable for the two areas (Fig. 10.12), but salinity in area E is 0.4 lower on average, caused by greater freshwater flux in the east. The opposite annual variations in temperature and salinity reflect the latitudinal migrations of the ITCZ (Fig. 10.4). Mixed layer depths are shallower than 20 m in areas E and W in September, and reach to 40 m and 60 m, respectively, in March (Fig. 10.13). Throughout the year, the mixed layer depth in area W is about 10 m deeper than in area E.

As an example, for the vertical distribution of water masses and their characteristics in the BGR exploration area, we show the data for site SO295-004CTD (11°51.59′N, 117°00.84′W) measured in November 2022 from the surface to the seafloor in Fig. 10.14, and for the upper 200 m in Fig. 10.15. The temperature and salinity profiles reveal a surface temperature of 27.8 °C and surface salinity of 33.35. The mixed layer depth (MLD) was 25 m; a typical value for this season. The maximum temperature gradient was 1.1 °C m^{-1} and the temperature difference between the top and bottom of the thermocline was 13.5 °C. A pronounced pycnocline separates the warm, low salinity Tropical Surface Water (TSW) from the intermediate waters.

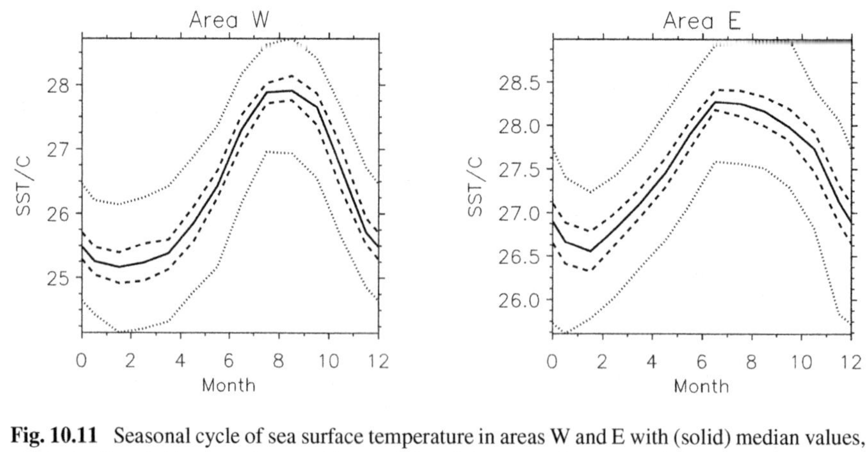

Fig. 10.11 Seasonal cycle of sea surface temperature in areas W and E with (solid) median values, (dashed) 35 and 65 percentiles, and (dotted) 5 and 95 percentile values of daily averages within each month. (Figures from Schneider, 2016a)

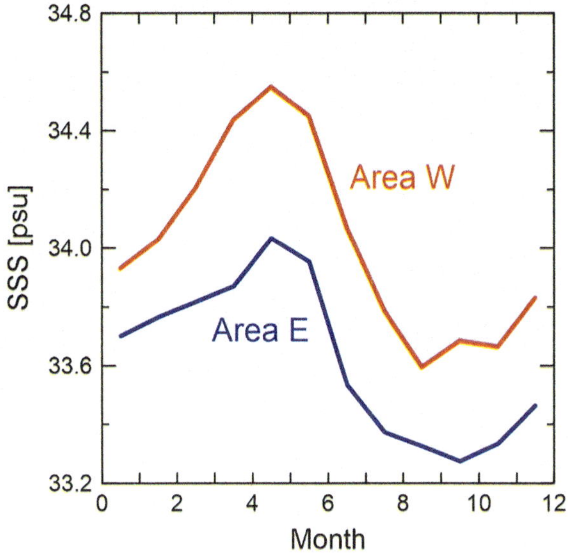

Fig. 10.12 Long-term average seasonal cycle of sea surface salinities in areas W and E based on data from the World Ocean Atlas 2018. (Garcia et al., 2019)

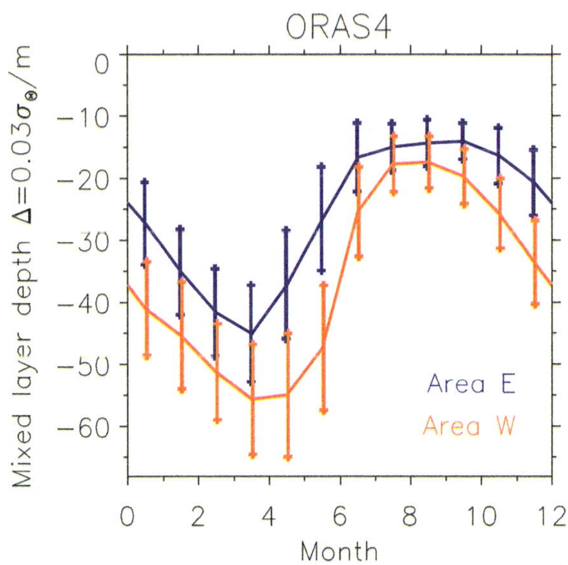

Fig. 10.13 Seasonal cycle of mixed layer depth in m, with vertical bars denoting the standard deviation, for areas W (red) and E (blue), respectively. Data are based on the monthly estimates, and from ECMWF ocean reanalysis system ORAS4. Mixed layer depths are based on a potential density σ_Θ difference to surface of 0.03 kg m^{-3}. (Figure from Schneider, 2016b)

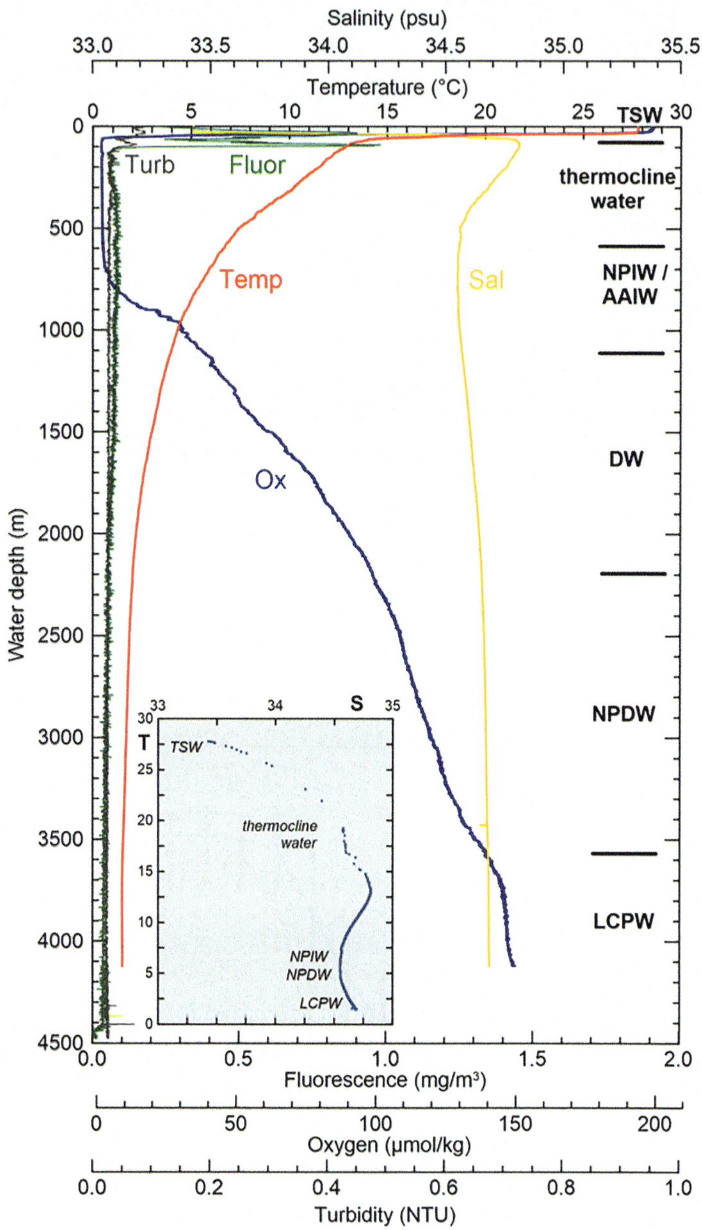

Fig. 10.14 Water mass characteristics of the BGR exploration area, exemplified by the downcast profile at site SO295-004CTD in November 2022. The vertical distribution of water masses is indicated on the right: Tropical Surface Water (TSW), North Pacific Intermediate Water (NPIW), Antarctic Intermediate Water (AAIW), Pacific Deep Water (DW), North Pacific Deep Water (NPDW), and Lower Circumpolar Water (LCPW). A temperature-salinity plot with data from the CTD cast is shown in the inset

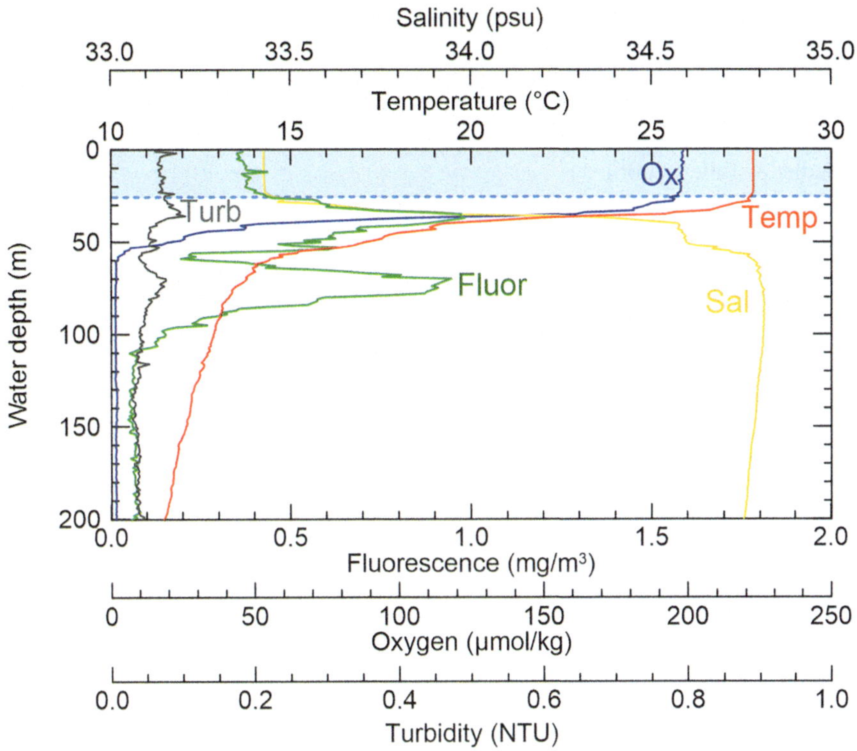

Fig. 10.15 Close-up of Fig. 10.14 showing the water mass characteristics of the upper 200 m at site SO295-004CDT in the BGR exploration area. Stippled line marks mixed layer depth

Fluorescence is mainly a measure of chlorophyll a concentration and represents the level of phytoplankton biomass. A sudden increase coincides with the lower limit of the mixed layer, below which nutrient concentrations increase significantly. A slightly less pronounced deeper fluorescence maximum occurs at 70–90 m water depth, a typical pattern that we also observed during previous cruises.

Turbidity is elevated in the mixed layer near the surface, with two maxima occurring parallel to the fluorescence maxima. Below the mixed surface layer, turbidity drops rapidly to constant, low values that extend throughout the water column. This pattern indicates increased turbidity due to biological activity near the surface and clear water below the thermocline.

The oxygen concentration drops from values of ~200 µmol kg^{-1} at 25 m water depth to 1.6 µmol kg^{-1} below 55 m. This sharp decrease within 30 m marks the beginning of the oxygen minimum zone, which extends to about 700 m. The strong increase in fluorescence concentration to maximum values at 37 m coincides with the depths where the oxygen concentrations drop sharply, indicating that primary production in the sunlit zone and subsequent remineralization of particulate organic matter under oxygen consumption at depth is an important factor in the formation of this oxygen minimum zone.

In general, subpycnocline waters in the eastern tropical Pacific are very low in oxygen. This extreme oxygen deficiency in the open ocean is mainly due to three factors: (1) high phytoplankton production at the surface; (2) a sharp permanent pycnocline that prevents local ventilation of subsurface waters; and (3) a sluggish and convoluted deep circulation and therefore the old age of subpycnocline waters (Fiedler & Talley, 2006).

Northern and southern source low-salinity intermediate (subpycnocline) water masses (between 500 m and 1000 m) meet approximately at 10 °N, somewhat south of the BGR exploration area. Antarctic Intermediate Water (AAIW) forms from the deep, winter mixed layer north of the Subantarctic Front in the southeast Pacific (Hanawa & Talley, 2001) and spreads equatorward and westward in the Pacific. North Pacific Intermediate Water (NPIW) originates from subsidence in the Sea of Okhotsk and at the Oyashio Front in the western North Pacific, with some contribution from the Gulf of Alaska (You, 2003).

The properties of the deep and bottom waters of the Pacific are determined by their distant sources in the Antarctic and North Atlantic. The deep water (DW) between 1000 m and 2500 m is relatively featureless, with a steady decrease in temperature and increase in salinity. North Pacific Deep Water (NPDW) is the 1.2–2 °C cold water mass found at depths >2500 m. The bottom water in the Pacific is Lower Circumpolar Water (LCPW), a mixture of Antarctic Bottom Water formed in the Weddell Sea and North Atlantic Deep Water formed in the northern North Atlantic. LCPW in the eastern tropical Pacific north of the equator is characterized by relatively high oxygen concentrations and is present below a water depth of ~3600 m in the BGR exploration area.

3.3 Upper Ocean Currents

Figures 10.16, 10.17, 10.18, 10.19 and 10.20 show the seasonal distribution of current speed and direction as "current roses" based on ORAS4 monthly average data from 1992 to 2015 for different water depths. In area W, the direction of surface current is west-north-west with very little variation, but with seasonal differences in speed (Fig. 10.16). Current speed is highest in October to November and lowest from April to June. In area E, surface currents are fairly constant and swift from January to March, with maximum monthly mean speeds greater than 26 cm s^{-1} (Fig. 10.17). During the fall and winter seasons, the monthly mean currents become much more variable. Comparison with currents at 100 m depth in area E suggests that winter variations are largely an Ekman response to wind forcing. In the thermocline at a depth of ~100 m, the current direction is shifted to the east, consistent with the absence of Ekman currents (Fig. 10.18). The seasonal variation in the speed distribution mirrors the surface currents, but with a more westward component and less directional variability in the October to December season. This season also shows more frequent occurrences of calm conditions, defined by speeds less than 1 cm s^{-1}. At a depth of ~500 m, currents in area E are predominantly westward from

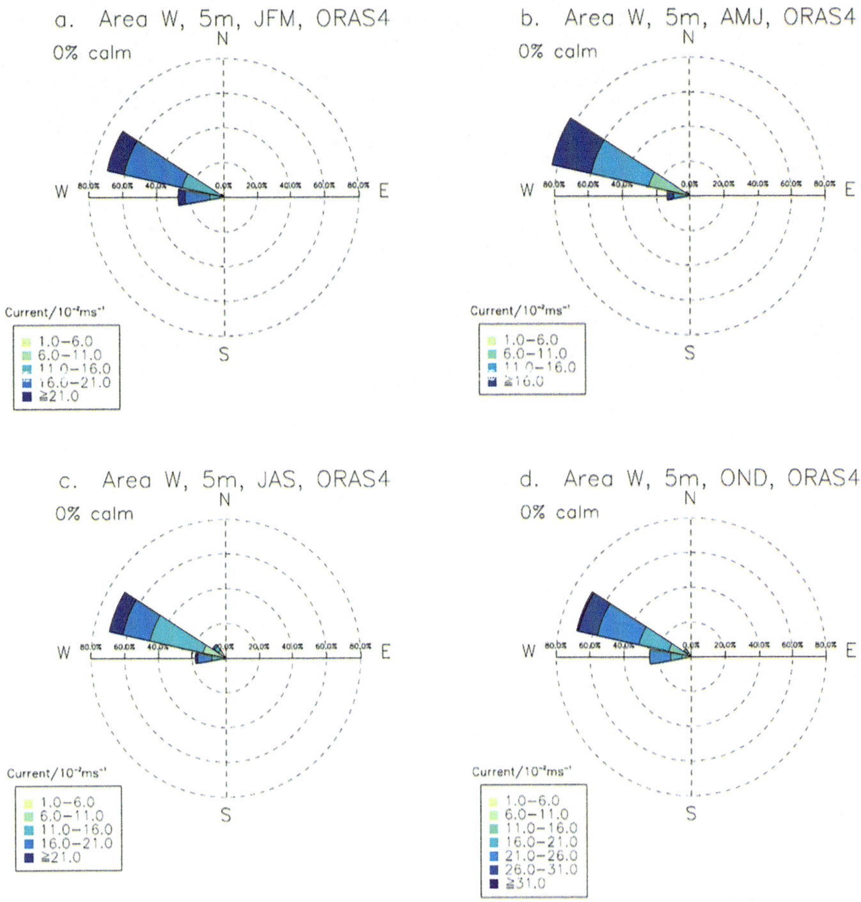

Fig. 10.16 Current rose diagrams in cm s^{-1} in area W at 5 m water depth for (**a**) January–March, (**b**) April–June, (**c**) July–September and (**d**) October–December. Data is based on monthly means of ORAS4. Note that in contrast to wind direction that depict from where winds are blowing, current direction denotes the direction towards which the water is flowing. (Figures from Schneider, 2016b)

January to March, with maximum speeds of 2.5 cm s^{-1} and calm conditions occurring about 10% of the time (Fig. 10.19). In the April to June season, currents are variable, and calm conditions prevail for 60% of the months. For the seasons from July to December, currents again are intermittent and toward the west and southwestward section with calm condition occurring in 40% of the months. At a depth of ~1000 m currents are weak throughout the year with speed of less than 1 cm s^{-1} for most of the time (Fig. 10.20).

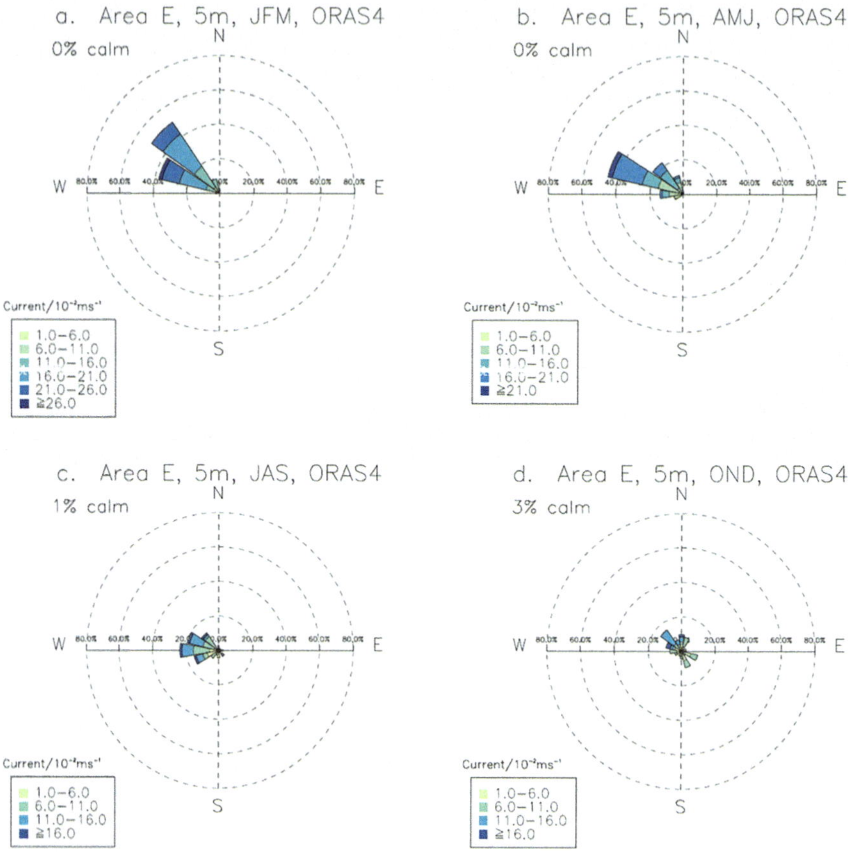

Fig. 10.17 Current rose diagrams in cm s^{-1} at 5 m water depth as in Fig. 10.16 but for area E. (Schneider, 2016b)

3.4 Near-Bottom Currents

Near-bottom current data were obtained by ADCPs and Recording Current Meters (RCMs) installed on three Ocean Bottom Moorings (OBMs) that were deployed in total from 2013 to 2016 and recovered and re-deployed in 2014 to read out the data collected over one year (Fig. 10.21 and Fig. 10.22).

In general, the current velocities and directions recorded by the ADCPs and RCMs at the three sites are in good agreement. The ADCPs indicate mean near-bottom current velocities of 3 to 4 cm s^{-1} and velocity maxima around 8 to 14 cm s^{-1} (good measurements of all collected time series derive from water depths between ~13 and 18 m above the seafloor), and thus indicating a generally weak background current (Fig. 10.21). The RCM point sensors indicate slightly lower current velocities, with averages around 2 to 2.5 cm s^{-1} and maxima around 7 cm s^{-1}. As the point sensors (2–3.5 m above the seafloor) were mounted below the ADCPs, this likely

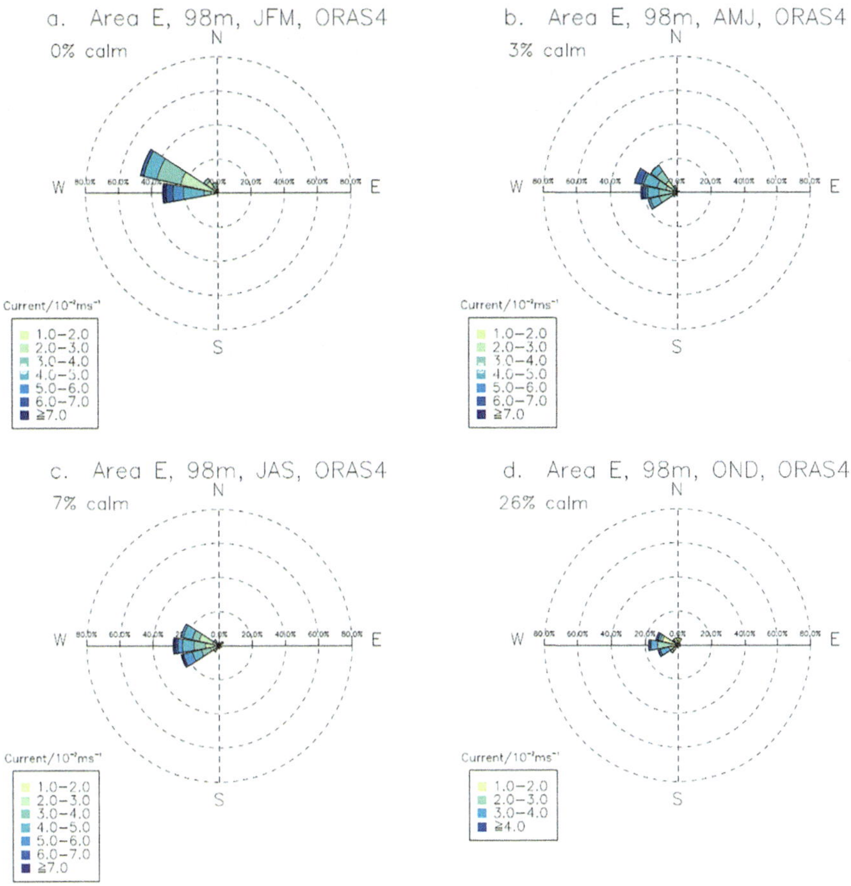

Fig. 10.18 Current rose diagrams in cm s^{-1} at 98 m water depth as in Fig. 10.16 but for area E. (Figures from Schneider, 2016b)

indicates a decreasing current velocity toward the seafloor due to bottom friction, i.e., the logarithmic decay of flow velocity toward the bottom, typical of the bottom boundary or Ekman layer.

Current directions are highly variable, with a tendency to predominant southeasterly to southwesterly directions at most observed mooring sites (Fig. 10.22). However, at some sites more northwesterly or more northeasterly to southwesterly (ADCP) or westerly (RCM) directions were observed. This shows that near-bottom currents are spatially and temporally variable (e.g., depending on topographic heights and surface ocean conditions) and not easy to predict.

Both current velocity and direction show strong cyclical variations. Spectral analyses show that the near-bottom current regime is strongly influenced by diurnal and semidiurnal tidal cycles, i.e., approximately one third of current speed variance is explained by tidal activity, with strong semi-diurnal and diurnal frequencies.

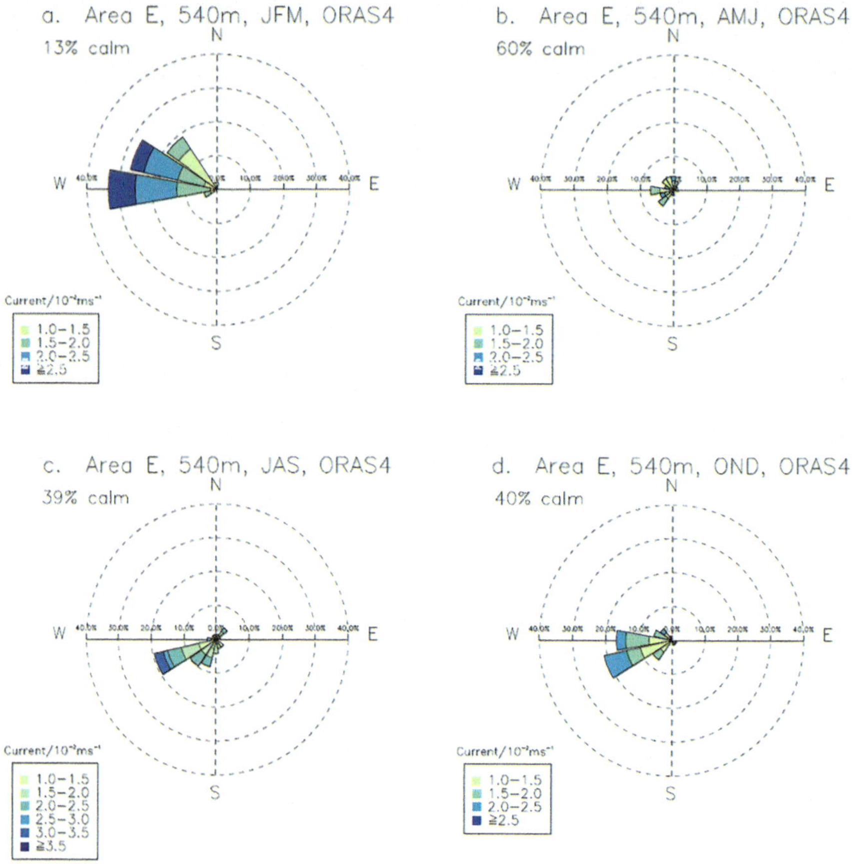

Fig. 10.19 Current rose diagrams in cm s^{-1} at 540 m water depth as in Fig. 10.16 but for area E. (Figures from Schneider, 2016b)

However, oscillating flow at near-inertial frequency (centered on a period of 38.5 h) seems to be as energetic as tidal flow (Aleynik et al., 2017). The origin of these oscillations is almost certainly the wind at the surface. This result was unexpected and means that on time scales of hours to days, currents are dominated by tides and wind-induced, near-inertial movements in equal measure. In addition, the time series shows multi-day cycles.

3.5 Mesoscale Eddies

The CCZ is regularly affected by mesoscale eddies, which originate at the western coast of central America and draw their energy from the Central American mountain Gaps Winds (Aleynik et al., 2017; Purkiani et al., 2020, 2022) (Fig. 10.23). Such

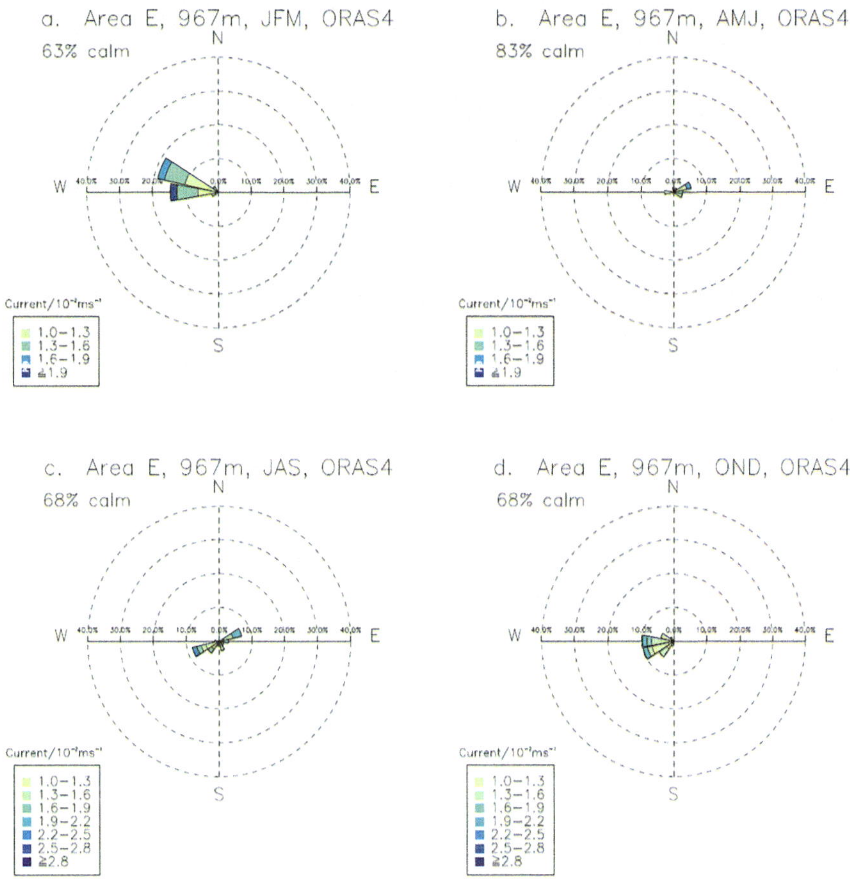

Fig. 10.20 Current rose diagrams in cm s^{-1} at 967 m water depth as in Fig. 10.16 but for area E

eddies are temporary bodies of swirling water 100-300 km in diameter that travel long distances (1000-4000 km) through the open ocean before they dissipate. On average, 0.6 cyclonic and 0.6 anti-cyclonic eddies are observed per month in area W, corresponding to 7.2 eddies per year. In area E, the number of cyclonic eddies is smaller, with typical values of 0.5 per month, or 6 cyclones per year (Schneider, 2016a). Eddy radii are usually between 50 and 150 km, with the largest reaching 300 km (Purkiani et al., 2020), and radial speeds are typically 15–20 cm s^{-1}. Eddy lifetimes are on average 100 days, and propagation westward with speeds of 10–20 cm s^{-1}.

Purkiani et al. (2022) investigated an anticyclonic eddy (characterized by clockwise rotation, a positive sea level anomaly, and warm core) that passed the area E in westward direction in April and May 2019. Corresponding anomalies in seawater properties were monitored at CTD repeat stations. The CTD profiles indicate a well-expressed deepening of the thermo-, halo-, and oxyclines by about 30 m, with more

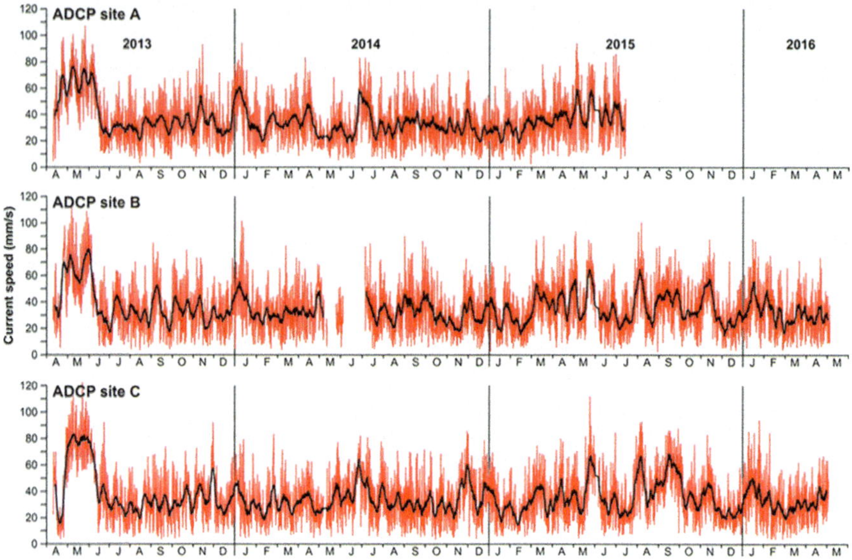

Fig. 10.21 Near-bottom current speeds exemplified by measurements of 600 kHz ADCPs at three OBM sites in PA1-West over a 3-year deployment period from April 2013 to May 2016

intense mixing of warm, nutrient-poor surface water and cold, nutrient-rich deeper water. A similar situation occurred in November 2022, when temperature and salinity profiles showed a pronounced difference in mixed layer depth between site SO295-004CTD (30 m) and sites SO295-031CTD and SO295-052CTD (15 m) that were situated only 8 km apart from SO295-004CTD (Fig. 10.24).

Despite the great water depth, near-bottom currents can be amplified during and after the passage of large mesoscale eddies. One such pronounced eddy with a diameter of ~100 km passed over the mooring deployment site during April 2013 (Figs. 10.21 and 10.22), which not only increased near-bottom current speed up to 13 cm s^{-1} but also strongly changed the current direction predominantly to the east for several weeks. Surface water velocities peaked at 40 cm s^{-1}, and near-bed velocities at 10 cm s^{-1}; the seafloor response lagged by about 10 days. However, the eddy-induced increased current speeds in May/June 2013 did not relate to any changes in seafloor turbidity, i.e. the bottom water current seemed not strong enough to cause natural erosion and resuspension of bottom sediments. This observation was corroborated by monitoring the potential effects of an eddy passing through the BGR exploration area in April/May 2019. A transect of ocean bottom moorings that were equipped with current meters and sediment traps was deployed during cruise SO268. While a slight increase in current strength and shift to more uniform current direction could be associated with the eddy passage, there was no clear indication for sediment resuspension.

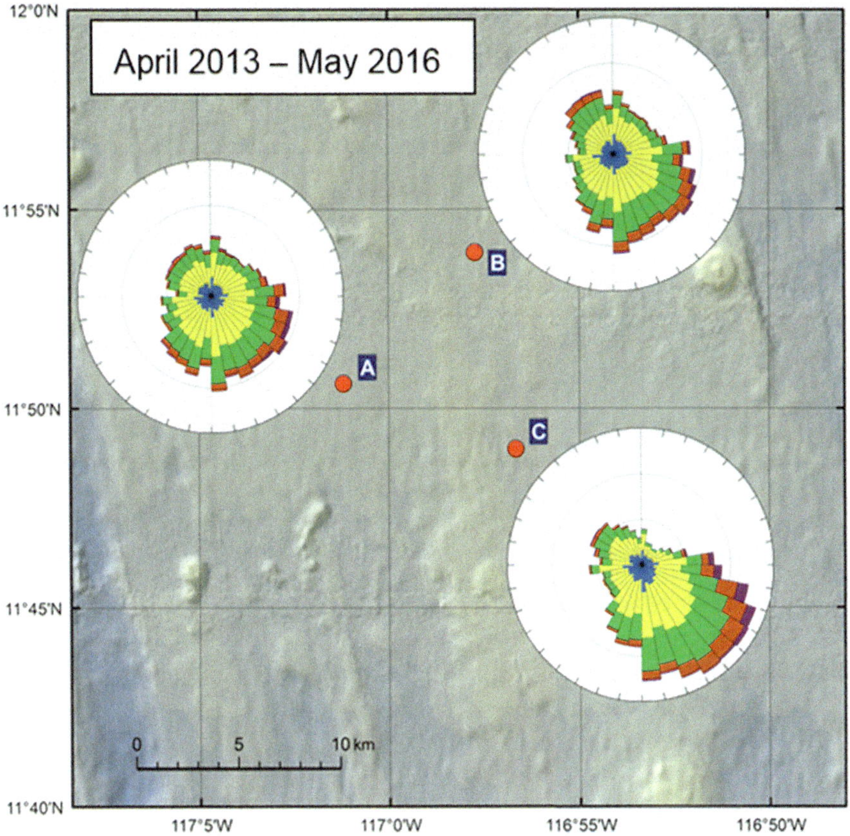

Fig. 10.22 Bathymetric map of the flat seafloor area surrounding the three moorings in the eastern part of the eastern BGR contract area. The red dots mark the locations of the moorings deployed from 2013 to 2016. Radial plots of current speeds and directions as measured by 600 kHz ADCPs at the three OBM sites A, B, and C over a 3-year period from April 2013 to May 2016 plotted as a function of their frequency of occurrence: The radius represents a relative frequency of 6%. Colors reflect increasing current speeds from <20 cm s^{-1} (blue) to >8 cm s^{-1} (purple). OBM positions have been plotted on a map of the topography of the area

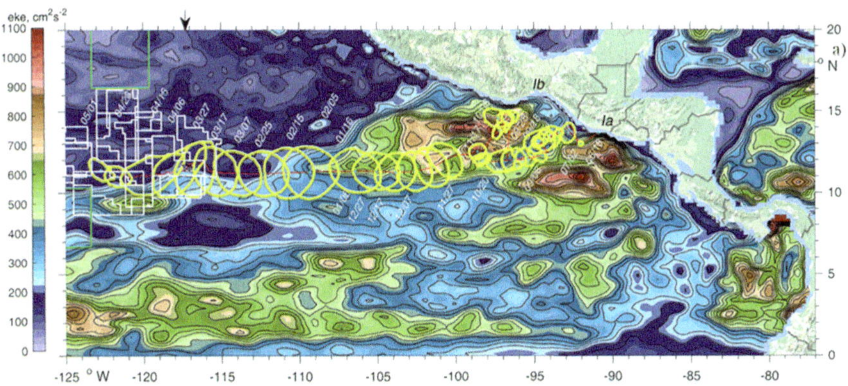

Fig. 10.23 The track of a mesoscale eddy over 318 days from the Mexican coast toward the CCZ (contract areas shown with white and APEIs with green lines) and the BGR moorings site (star). Yellow lines encircle a local maximum (>80 cm) sea surface height (SSH) anomaly (AVISO) at 10-day intervals. Colors reflect mean surface Eddy Kinetic Energy (EKE) over the period 18 June 2012 to 1 May 2013. (Figure from Aleynik et al. (2017))

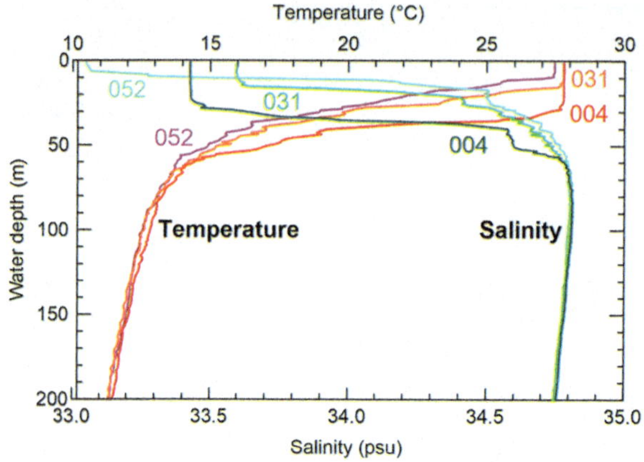

Fig. 10.24 Temperature and salinity profiles for sites SO295-004CTD (11°51.59′N, 117°00.84′W), SO295-031CTD (11°55.81′N, 117°01.61′W), and SO295-052CTD (11°56.10′N, 117°01.63′W) situated in the BGR exploration area and measured in November 2022

4 Modelling Experiments

4.1 Case Study 1: BGR Exploration Area

Figure 10.25 shows the location of the small-scale plume dispersion experiment in the BGR area. To achieve a satisfactory fit for the model results to the measurements, the sediment release rate was set at 0.1 kg s^{-1}. We estimate that this value represents approximately 4% of the maximum available sediment that could have been resuspended. The simulated suspended sediment concentration (SSC) generated by the model was validated using sensor data from two monitoring stations. According to the findings presented by Purkiani et al. (2021), the model successfully predicted the main characteristics and variations of the SSC. The simulated SSC values were interpolated at the sensor locations and closely matched the observed data. The maximum total SSC near the dredge tracks was recorded at 28 ml l^{-1} and decreased exponentially as the distance from the tracks increased toward the south. The statistical analysis revealed a correlation coefficient greater than 0.66 and a root mean square error less than 1.18 ml l^{-1}, indicating a meaningful relationship between the model predictions and the observed data.

Figure 10.26 displays the spatial and temporal variation of sediment deposition maps for particle classes 1–3, illustrating the distribution of sediment deposition over a 12-h period following the dredging activity, with each map representing a 2-h time interval. Based on our findings, the deposition pattern of sediment is influenced by both time and space. Specifically, for particle classes 2 and 3, sediment deposition began immediately, as depicted in Fig. 10.26b, c. However, for particle class 1, the initial redeposition took a longer time, occurring after approximately 4 h, as

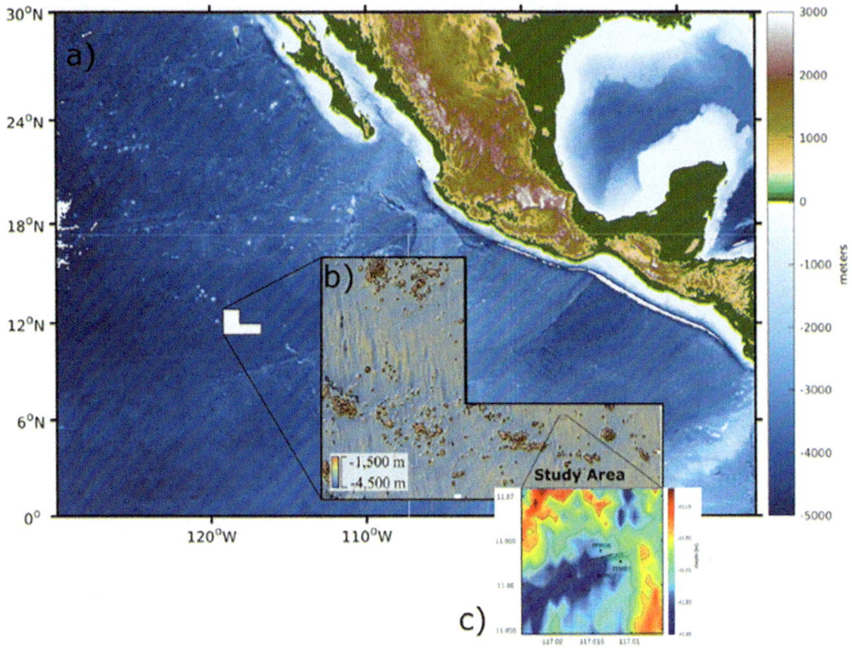

Fig. 10.25 (**a**) Location of the eastern part of the BGR exploration area in the northeastern Pacific Ocean. (**b**) Sea-floor topography. The black square in (**b**) depicts the dredging area. (**c**) Close-up of the dredging area used for the numerical simulation, with 5 m depth contours. The dredging tracks are shown as black lines. PFM02, PFM07, and PFM06 represent the locations of bottom-moored sensors for SSC and near-bed current measurements (Purkiani et al., 2021)

shown in Fig. 10.26a–d. Our observations indicate that the highest redeposition of sediment took place near the dredging tracks, particularly for particle classes 2 and 3, which have higher settling velocities. Following the prevailing south-westerly current direction, no sediment deposition was observed north of the dredge tracks. Instead, the deposition pattern extended in a south-westerly direction for all particle classes.

Based on the given bulk density of 200 kg m^{-3} for the freshly deposited sediments and considering a porosity of 90% for the newly deposited material, our calculations indicate that the maximum deposition thickness at the end of the simulation period is estimated to be 0.12 mm for sediment class 1, 0.39 mm for sediment class 2, and 0.2 mm for sediment class 3, as shown in Fig. 10.26d-l. These values represent the maximum thickness of sediment accumulation for each respective sediment class. The final accumulated sediment deposition of class 1–3 reaches up to 0.7 mm after 4 days (Fig. 10.26m). By including sediment class 4 in our calculation, which represents the remobilization of particles that do not resuspend, we have observed an improvement in the prediction of near-field deposition thickness. This did not have any significant influence on the far-field deposition pattern, as shown in Fig. 10.26n. Including sediment class 4 allows for a more accurate representation of sediment

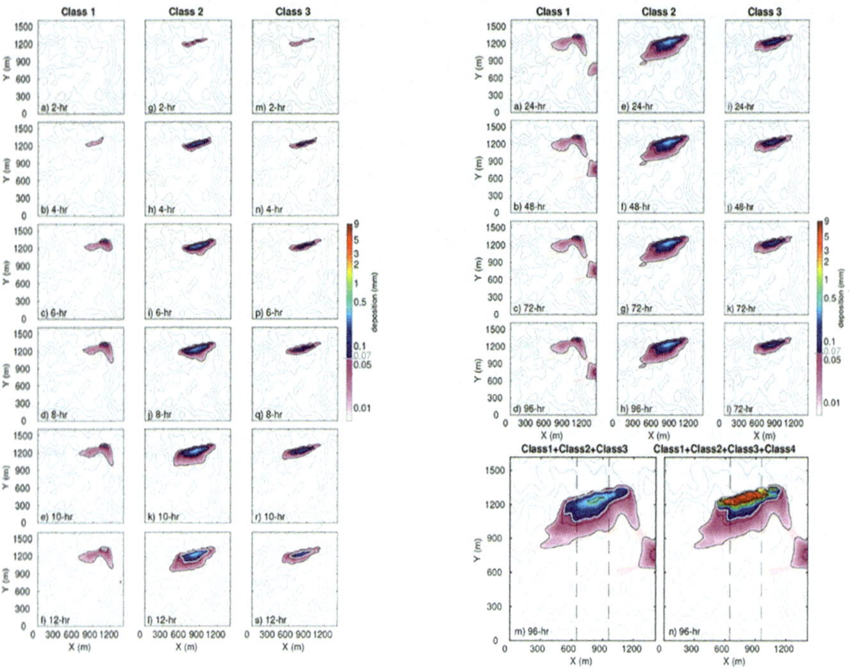

Fig. 10.26 Sediment redeposition patterns: (a) 2-h, (b) 4-h, (c) 6-h, (d) 8-h, (e) 10 h, and (f) 12 h after dredging for particle size class1. (g-s) Redeposition of particle size class 2 and class 3, respectively, for same time steps as presented for (a) to (f). The color bar shows sediment thickness ranging from 0.01 to 9 mm in logarithmic scale. Thin gray lines in the background show the seafloor topography. The 0.07 mm deposition thickness shown as thick gray contour marks the boundary between near- and far-field areas. The near- and far-field areas are customized in the color bar with different color scale. (Figure from Purkiani et al., 2021)

dynamics and enhances our understanding of the deposition process in the near-field area. Following the increase in maximum total sediment deposition in the dredging tracks to 9 mm, a significant rise in average deposition height is evident in the near-field area (Fig. 10.26n). The quantitative comparison of our results is in agreement with the previous experiments of the Japan Deep Sea Impact Experiment (JET) and the NOAA Benthic Impact Experiment (BIE) in the CCZ, which show that the maximum thickness reached 12 and 1.9 mm for much larger sediment releases of 1450 tons and 352 tons, respectively, albeit over larger regions than the dredging area in our study (Barnett & Yamauchi, 1995; Nakata et al., 1997; Trueblood, 1993).

Sediment deposition was visually inspected using video cameras on the towed Ocean Floor Observations System (OFOS). The video footage was manually annotated using the OFOP software package. The preliminary seabed categories used for annotation were 'dredge track', 'weak sediment cover (<1 mm)', 'thick sediment cover (>1 mm)', and 'no sediment cover'. A qualitative comparison between the model predictions and observed deposition patterns showed satisfactory agreement, as depicted in Fig. 10.27n. The region with a deposition of 1 mm or more covers an

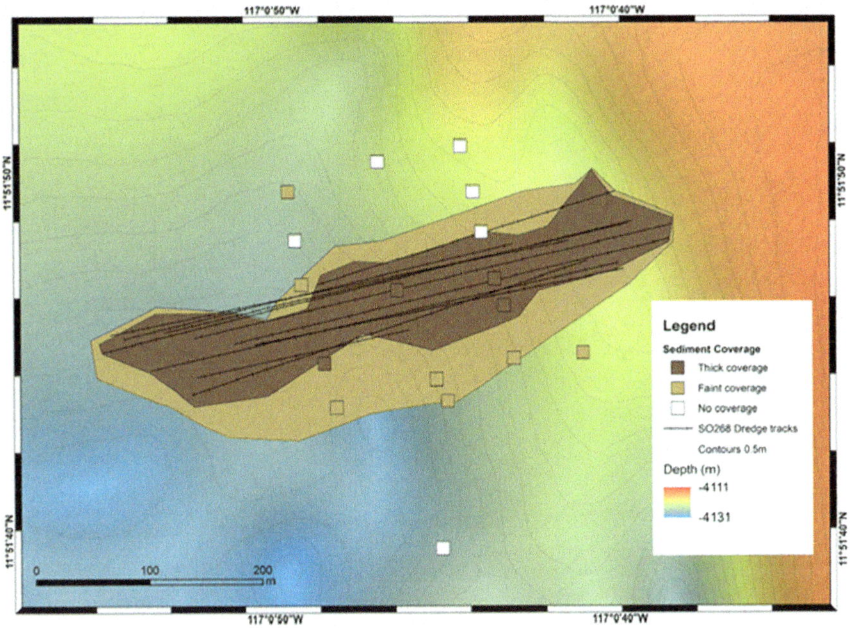

Fig. 10.27 Map showing the visually assessed sediment coverage of the nodules, distinguishing "no coverage" (white/no color), "faint coverage" (light brown), and "thick coverage" (dark brown). (Figure from Haalboom et al., 2022)

area of 33,000 m^2, which closely matches the shape and size of the area with thick sediment cover observed in photographic observations (Figs. 10.26n and 10.27).

4.2 *Case Study 2: Peru Basin*

To confirm the agreement between observed and predicted deposition patterns made for the CCZ, we also briefly report on the results obtained for the dispersion of sediment transport in the deep Peru Basin using the same numerical model (Fig. 10.28). Current velocities measured at 1 m above seafloor (mab) with the downward looking ADCP (1200 kHz) reach up to 10.5 cms^{-1}, with an average of 4.0 cm s^{-1}. The influence of tides with a large component of M2 and M4 is evident in the current data (Baeye et al., 2022). Towing EBS on the seafloor caused mobilization and partial resuspension of sediment, leading to an increase in SSC. During the first event (E), which lasted 1.7 h, there was a notable and significant increase in SSC. The SSC reached a peak value of 6.4 mgL^{-1}, which is approximately 100 times higher than the background concentration. In the second event (E2), the SSC remained relatively constant at 0.5 mgL^{-1} over a duration of 3.4 h. To understand the sediment dispersion behavior, a numerical simulation study using in-situ measurements and sediment settling velocities was performed (Baeye et al., 2022).

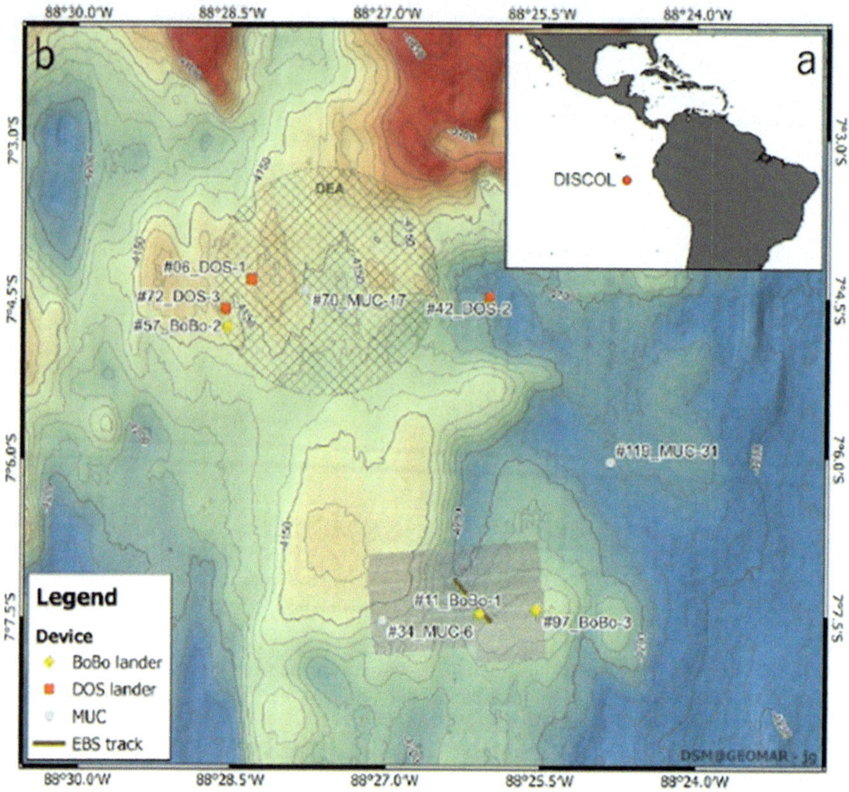

Fig. 10.28 (**a**) shows the location of the DISCOL experiment area (DEA) in the southeast Pacific Ocean) and (**b**) provides a bathymetric map of the DEA (ploughed in 1989) and surroundings. White circles show the sites of multicorer samples selected for sediment analysis. The yellow diamond symbols are the BoBo and DOS lander stations. The bathymetry was obtained during cruise SO242. (Figure from Baeye et al., 2022)

To validate the model results, a release rate of 0.142 kg s^{-1} was employed. This indicated that less than 1% of the mobilized sediment becomes resuspended. The sediment with a d90 size had a settling velocity that allowed 95% of it to be deposited within approximately 3 h (Fig. 10.29). However, the sediment with d50 and d10 sizes remained in suspension for a significantly longer time and was eventually deposited after 12 h. The E2 event and the simulated SSC in the model have a slight discrepancy in timing. The model shows a lower SSC and an offset of approximately 1 h. The plume, influenced by tidal currents, initially moved in a southward direction and then shifted toward the northeast. The influence of plume release height and seabed topography on plume dispersion was assessed by running different model simulations: plume release height at 0 to 1 mab vs. 1 to 2 mab, and actual bathymetry vs. flat bathymetry. The release height set at 1 to 2 mab produces higher SSC at 2 to 3 mab, especially during E2. With and without realistic bathymetry, the model produces differences in both SSC and timing for the second event.

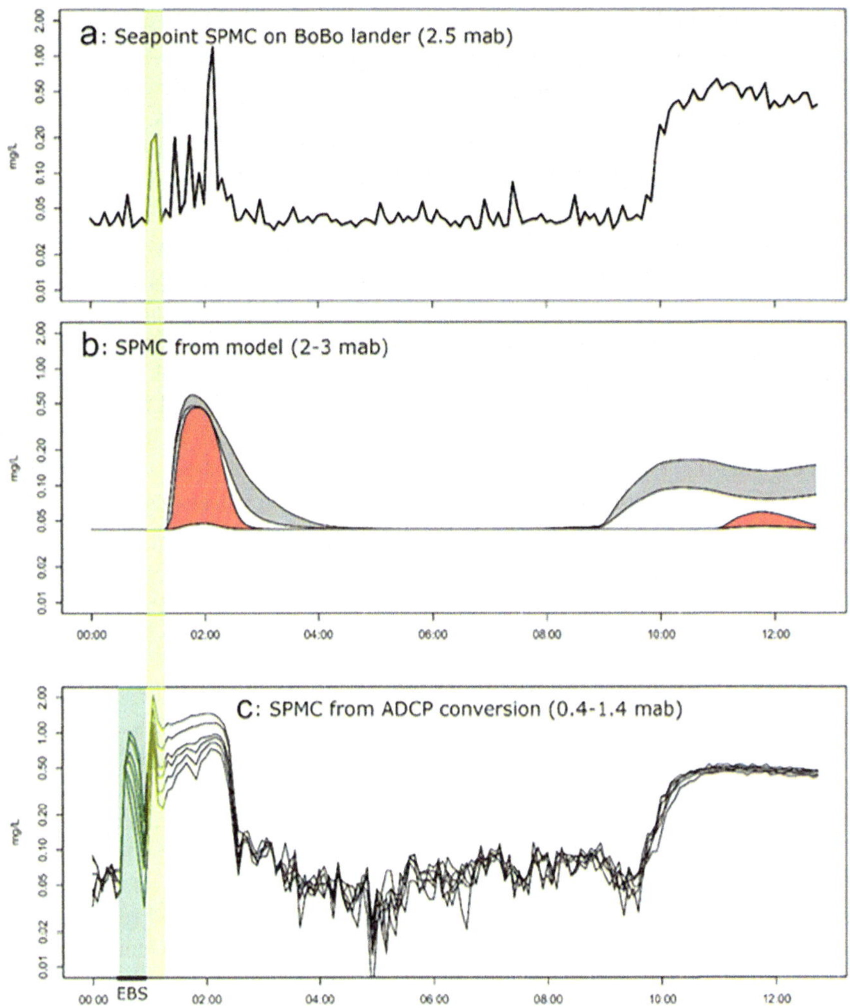

Fig. 10.29 (**a**) Time series of SSC measured at 2.5 m above seafloor (mab) by the Seapoint sensor on the BoBo lander, during the EBS experiment on August 5, 2015. The plume was detected twice (E and E2). The first peak marked by yellow shading was probably produced by the EBS towing cable sweeping over the seafloor. (**b**) Modelled SPMC time-series for the 2–3 mab vertical model grid cell corresponding to the position of the Seapoint sensor on BoBo. The gray and red shaded areas refer to the SPMC simulated while implementing the real bathymetry or assuming a horizontal seafloor, respectively. The lower and upper bounds of the shaded areas represent simulations for release heights of 0–1 mab and 1–2 mab, respectively. (**c**) Time series of SPMC derived from acoustic backscatter recorded with the downward-looking 1200 kHz RDI ADCP mounted on the BoBo lander, for 6 vertical depth bins centered, respectively, at 0.4, 0.6, 0.8, 1.0, 1.2, and 1.4 mab. (Baeye et al., 2022)

5 Conclusions

The data presented here on atmospheric variability, hydrographic conditions, and the ocean current regime show that the eastern tropical Pacific Ocean, where the BGR exploration area for polymetallic nodules is located, is a dynamic system. Surface conditions are mainly determined by the seasonal cycle of the latitudinal migration of the ITCZ. The deeper ocean shows little variation in water mass properties and current direction and strength. However, mesoscale eddies extending several 1000 km from the coast of Mexico to the BGR area cause a substantial increase in bottom water current velocities. The information presented on maximum events in atmospheric circulation (tropical storms and cyclones) and oceanic currents (eddies) at the surface and at depth can provide valuable information for planning future deep-sea mining activities and assessing their environmental impact.

Two small-scale dredge experiments were carried out in the BGR contract area for polymetallic nodule exploration in the CCZ and DISCOL areas to test the setup strategy for the monitoring of sediment plumes produced by deep-sea mining activities (Purkiani et al., 2021; Baeye et al., 2022; Haalboom et al., 2022). By conducting field observations and numerical model simulations of small-scale plume experiments, valuable insights of sediment plume dispersion and redeposition were obtained. The insights gained from these experiments play a crucial role in developing monitoring strategies that are highly effective for future projects and should consider the unique hydrodynamic and bathymetric conditions of the specific area, ensuring accurate and comprehensive monitoring of sediment plumes.

Based on the deep-sea dredge experiments, it has been determined that sediment settling velocities, bottom current characteristics, and topographic features are the key parameters influencing the transport of suspended sediment in deep-sea environments. In order to fully explore the potential impact of large-scale deep-sea mining in regions with sub-mesoscale activity, it is crucial to focus on the dynamic deployment of sensor arrays. This involves closely monitoring and responding to the retreat or advance of plume fronts, ensuring effective and efficient operations. By giving more attention to this aspect, we can better understand and optimize the deployment strategies for successful deep-sea mining endeavors. Numerical simulation needs to be further developed to adequately simulate the near-field dispersion driven by the hydraulic system of nodule-collector such as jet velocity, forward speed of vehicle, and pumping efficiency (Alhaddad et al., 2023).

Acknowledgments The authors would like to thank all colleagues who contributed to the exploration of polymetallic nodules in the BGR license area, especially Annemiek Vink and Katja Schmidt, who supported this study with CTD and mooring deployments. Our thanks also go to the crews of the research vessels SONNE, KILO MOANA, L'ATALANTE, and ISLAND PRIDE for their dedicated support during the sea campaigns.

References

Adcroft, A., Hill, C., Campin, J. M., Marshall, J., & Heimbach, P. (2004). Overview of the formulation and numerics of the MIT GCM. In *Proceedings of the ECMWF Seminar series on numerical methods, recent developments in numerical methods for atmosphere and Ocean modelling (Cambridge, MA)* (pp. 139–149).

Aleynik, D., Inall, M., Dale, A., & Vink, A. (2017). Impact of remotely generated eddies on plume dispersion at abyssal mining sites in the Pacific. *Scientific Reports, 7*, 1–14. https://doi.org/10.1038/s41598-017-16912-2

Alhaddad, S., Mehta, D., & Helmons, R. (2023). Mining of deep-seabed nodules using a Coandă-effect-based collector. *Results in Engineering, 17*, 100852. https://doi.org/10.1016/j.rineng.2022.100852

Baeye, M., Purkiani, K., de Stigter, H. C., Gillard, B., Fettweis, M., & Greinert, J. (2022). Tidally driven dispersion of a Deep-Sea sediment plume originating from seafloor disturbance in the DISCOL area (SE Pacific Ocean). *Geosciences, 12*, 8. https://doi.org/10.3390/geosciences12010008

Barnett, B., & Yamauchi, H. (1995). Deep sea sediment resuspension system used for the Japan deep sea impact experiment. In *Proceedings of the First (1995) ISOPE Ocean Mining Symposium (Tsukuba)* (pp. 175–179).

Berrisford, P., Dee, D., Poli, P., Brugge, R., Fielding, K., Fuentes, M., et al. (2011). *The Era-Interim archive version 2.0* (Era report series 1. Technical report). ECMWF, Shinfield Park.

Burns, R., Halbach, P., Friedrich, G., & Stackelberg, U. V. (1990). The manganese nodule belt of the Pacific Ocean. Geological environment, nodule formation, and mining aspects. *Geological Magazine, 127*, 93. https://doi.org/10.1017/S0016756800014394

Muñoz-Royo. C., Ouillon, R., Mousadik, S. E., Alford, M. H., & Peacock, P. (2022) An in situ study of abyssal turbidity-current sediment plumes generated by a deep seabed polymetallic nodule mining preprototype collector vehicle. *Science Advances, 8*(38). https://doi.org/10.1126/sciadv.abn1219

Dee, D. P., Uppala, S. M., Simmons, A. J., Berrisford, P., Poli, P., Kobayashi, S., Andrae, U., Balmaseda, M. A., Balsamo, G., Bauer, P., Bechtold, P., Beljaars, A. C. M., van de Berg, L., Bidlot, J., Bormann, N., Delsol, C., Dragani, R., Fuentes, M., Geer, A. J., Haimberger, L., Healy, S. B., Hersbach, H., Hólm, E. V., Isaksen, L., Kållberg, P., Köhler, M., Matricardi, M., McNally, A. P., Monge-Sanz, B. M., Morcrette, J.-J., Park, B.-K., Peubey, C., de Rosnay, P., Tavolato, C., Thépaut, J.-N., Vitart, F. (2011) The ERA-Interim reanalysis: configuration and performance of the data assimilation system. *Abstract Quarterly Journal of the Royal Meteorological Society, 137*(656), 553–597. https://doi.org/10.1002/qj.v137.65610.1002/qj.828

Fiedler, P. C., & Talley, L. D. (2006). Hydrography of the eastern tropical Pacific: A review. *Progress in Oceanography, 69*(2–4), 143. https://doi.org/10.1016/j.pocean.2006.03.008

Garcia, H. E., Boyer, T. P., Baranova, O. K., Locarnini, R. A., Mishonov, A. V., Grodsky, A., Paver, C. R., Weathers, K. W., Smolyar, I. V., Reagan, J. R., Seidov, D., & Zweng, M. M. (2019). *World Ocean Atlas 2018: Product documentation*. A. Mishonov, Technical Editor. https://www.ncei.noaa.gov/data/oceans/woa/WOA18/DOC/woa18documentation.pdf

Gillard, B., Purkiani, K., Chatzievangelou, D., Vink, A., Iversen, M. H., & Thomsen, L. (2019). Physical and hydrodynamic properties of deep sea mining-generated, abyssal sediment plumes in the clarion clipperton fracture zone (eastern-central pacific). *Elementa: Science of the Anthropocene, 7*, 1–14. https://doi.org/10.1525/elementa.343

Glover, A., & Smith, C. (2003). The deep-sea floor ecosystem: Current status and prospects of anthropogenic change by the year 2025. *Environmental Conservation, 30*, 219–241. https://doi.org/10.1017/S0376892903000225

Gollner, S., Kaiser, S., Menzel, L., Jones, D. O., Brown, A., Mestre, N. C., et al. (2017). Resilience of benthic deep-sea fauna to mining activities. *Marine Environmental Research, 129*, 76–101. https://doi.org/10.1016/j.marenvres.2017.04.010

Greinert, J. (2015). RV Sonne Fahrtbericht/Cruise Report SO242-1 [SO242/1]: JPI Oceans Ecological Aspects of Deep-Sea Mining, DISCOL Revisited, Guayaquil—Guayaquil (Equador), 28.07.-25.08.2015; GEOMAR Report; GEOMAR Helmholtz-ZentrumfürOzeanforschung: Kiel, Germany (p. 290).

Haalboom, S., Schoening, T., Urban, P., Gazis, I. Z., de Stigter, H., Gillard, B., Baeye, M., Hollstein, M., Purkiani, K., Reichart, G.-J., Thomsen, L., Haeckel, M., Vink, A., & Greinert, J. (2022). Monitoring of anthropogenic sediment plumes in the clarion-clipperton Zone, NE equatorial Pacific Ocean. *Frontiers in Marine Science, 9*, 882155. https://doi.org/10.3389/fmars.2022.882155

Hanawa, K., & Talley, L. D. (2001). Mode waters. In G. Siedler, J. Church, & J. Gould (Eds.), *Ocean circulation and climate: Observing and modeling the Global Ocean* (pp. 373–386). Academic Press.

Hemer, M., Jack Katzfey, J., & Hotan, C. (2011). *The wind-wave climate of the Pacific Ocean* (113 pp). Report for the Pacific Adaptation Strategy Assistance Program, Department of Climate Change and Energy Efficiency, The Centre for Australian Weather and Climate Research.

Jankowski, J. A., & Zielke, W. (2001). The mesoscale sediment transport due to technical activities in the Deep Sea. *Deep-Sea Research Part II, 48*, 3487–3521. https://doi.org/10.1016/S0967-0645(01)00054

Jankowski, J. A., Malcherek, A., & Zielke, W. (1996). Numerical modeling of suspended sediment due to deep-sea mining. *Journal of Geophysical Research: Oceans, 101*(C2), 3545–3560.

Jones, D., Kaiser, S., Sweetman, A. K., et al. (2017). Biological responses to disturbance from simulated deep-sea polymetallic nodule mining. *PLoS One, 12*, e0171750. https://doi.org/10.1371/journal.pone.0171750

Klein, H. (1993). Near-bottom currents in the deep Peru Basin, DISCOL Experimental Area. *Deutsche Hydrographische Zeitschrift, 45*, 31–42.

Klein, H. (1996). Near-bottom currents and bottom boundary layer variability over manganese nodule fields in the Peru Basin, Se-Pacific. *Deutsche Hydrographische Zeitschrift, 48*, 147–160.

Lefaible, N., Macheriotou, L., Purkiani, K., Haeckel, M., Zeppilli, D., Pape, E., & Vanreusel, A. (2023). Digging deep: Lessons learned from meiofaunal responses to a disturbance experiment in the Clarion-Clipperton Zone. *Journal Marine Biodiversity, 53*, 48.

Marshall, J., Adcroft, A., Hill, C., Perelman, L., & Heisey, C. (1997). A finitevolume, incompressible navier-stokes model for studies of the ocean on parallel computers. *Journal of Geophysical Research, 102*, 5753–5766. https://doi.org/10.1029/96JC02775

Nakata, K., Kubota, M., Aoki, S., & Taguchi, K. (1997). Dispersion of resuspended sediment by ocean mining activity modelling study. In *Proceedings of the symposium on environmental studies for Deep-Sea Mining (Tokyo)*.

Neumann, C. J. (1993). A global tropical cyclone climatology. Global guide to tropical cyclone forecasting, Chapter 2. http://www.wmo.int/cycloneguide/

Ouillon, R., Muñoz Royo, C., Alford, M. H., & Peacock, T. (2022). Advection–diffusion settling of deep-sea mining sediment plumes. Part 2. Collector plumes Flow 2. https://doi.org/10.1017/flo.2022.19

Purkiani, K., Paul, A., Vink, A., Walter, M., Schulz, M., & Haeckel, M. (2020). Evidence of eddy-related deep-ocean current variability in the northeast tropical Pacific Ocean induced by remote gap winds. *Biogeosciences, 17*(24), 6527–6544.

Purkiani, K., Gillard, B., Paul, A., Haeckel, M., Haalboom, S., Greinert, J., et al. (2021). Numerical simulation of Deep-Sea sediment transport induced by a dredge experiment in the northeastern Pacific Ocean. *Frontiers in Marine Science, 8*. https://doi.org/10.3389/fmars.2021.719463

Purkiani, K., Haeckel, M., Haalboom, S., Schmidt, K., Urban, P., Gazis, I. Z., de Stigter, H., Paul, A., Walter, M., & Vink, A. (2022). Impact of a long-lived anticyclonic mesoscale eddy on seawater anomalies in the northeastern tropical Pacific Ocean: A composite analysis from hydrographic measurements, sea level altimetry data, and reanalysis model products. *Ocean Science, 18*(4), 1163–1181.

Roliniski, S., Segschneider, J., & Suendermann, J. (2001). Long-term propagation of tailings from deep-sea mining under variable conditions by means of numerical simulations. *Deep Sea Research Part II: Topical Studies in Oceanography, 48*, 3469–3485. https://doi.org/10.1016/S0967-0645(01)00053-4

Schaaning, M., Trannum, H., Øxnevad, S., Carroll, J., & Blake, T. (2008). Effects of drill cuttings on biogeochemical fluxes and macrobenthos of marine sediments. *Journal of Experimental Marine Biology and Ecology, 361*, 49–57. https://doi.org/10.1016/j.jembe.2008.04.014

Schneider, N. (2016a). *Surface and atmospheric variability in the BGR exploration area for polymetallic nodules in the eastern low-latitude Pacific* (128 pp). Report by the Maunalani Consulting LLC, Honolulu, Hawaii, USA on behalf of BGR., Hannover. https://www.bgr.bund.de/DE/Themen/MarineRohstoffforschung/Meeresforschung/Downloads/schneider_n_2016_surface_atmospheric_variability.html. https://doi.org/10.25928/bq5w-nj10

Schneider, N. (2016b). *Oceanography in the BGR exploration area for polymetallic nodules in the eastern low-latitude Pacific* (p. 68). Report by the Maunalani Consulting LLC, Honolulu, Hawaii, USA on behalf of the BGR, Hannover. https://www.bgr.bund.de/DE/Themen/MarineRohstoffforschung/Meeresforschung/Downloads/schneider_n_2016_oceanography_BGR_exloration_area.html. https://doi.org/10.25928/ea0m-rw97

Schriever, G., & Thiel, H. (2013). Tailing and their disposal in deep-sea mining. In *Proceeding of the Tenth (2013) ISOPE Ocean Mining and Gas Hydrates Symposium (Szczecin)* (pp. 1–14).

Spearman, J., Taylor, J., Crossouard, N., Cooper, A., Turnbull, M., Manning, A., et al. (2020). Measurement and modelling of deep sea sediment plumes and implications for deep sea mining. *Scientific Reports, 10*, 1–14. https://doi.org/10.1038/s41598-020-61837-y

Thiel, H., & Schriever, G. (1990). Deep-sea mining, environmental impact and the DISCOL project. *Ambio, 19*, 245–250.

Trueblood, D. (1993). *US cruise report for BIE II* (NOAA technical memorandum NOS OCRM4) (pp. 1–50). NOAA, National Ocean Service.

You, Y. (2003). The pathway and circulation of North Pacific intermediate water. *Geophysical Research Letters, 30*, 2291. https://doi.org/10.1029/2003GL018561

Carsten Rühlemann is a marine geologist and head of the Marine Geology working group at the German Federal Institute for Geosciences and Natural Resources (BGR). Since 2020, he has been a member of the Legal and Technical Commission at the International Seabed Authority.

Kaveh Purkiani is a physical oceanographer. His work focuses on the numerical simulation of deep-sea sediment transport and data analysis. He has contributed to several interdisciplinary studies on the effects of sediment transport on deep-sea environments.

Chapter 11
Ocean Current Observations Throughout the Water Column in the Clarion-Clipperton Fracture Zone, Tropical North Pacific

Jon Wood

Abstract Four (4) long-term moorings were deployed in the tropical eastern North Pacific Ocean within the Clarion-Clipperton Facture Zone (CCFZ) to observe physical processes throughout the water column. High-resolution ocean current profiles were obtained within the upper 1000 m of the water column and within the lower 100 m of the near-bottom boundary layer, along with velocity data at discreet depths in the mid-water column. Many of the resulting oceanographic records spanned more than three (3) years. Currents were most energetic near the surface with a mean velocity of ~18–22 cm/sec and mean direction toward the west. The strongest currents observed in the record exceeded 165 cm/sec and occurred during Hurricane Barbara in 2019, but these peak currents were brief, decaying rapidly as the storm passed, leaving only oscillatory currents propagating at near-inertial frequencies as aftereffects. A strong density layer about 75–100 m below the surface, the top of the pycnocline, appeared to act as an effective barrier separating the near-surface layer from deeper layers. CTD casts performed at the mooring location(s) confirmed the strong gradients in the upper water column. Currents at 204 m, below the surface layer, were weaker (mean speed ~12 cm/sec) and flowed eastward much of the time. The velocity variance at this depth was about one-third the velocity variance found nearer the surface. Hurricane storm response at 204 m was much weaker also, further evidence the pycnocline limits mixing between the surface and deeper waters. Currents mid-depth of the water column—from about 1500 m down to 4000 m—were the weakest, with mean speeds of order ~3–4 cm/sec and velocity variance about 1/30th of the variance near the surface. Currents near the seabed were found more energetic, with mean flows of order ~5 cm/sec and peak speeds exceeding 20 cm/sec. Local bathymetry features likely play a role in steering near-bottom flows at each site. Some flow intensification was observed in the near-bottom boundary layer at several sites, where the mean (and peak) near-bottom currents were twice the magnitude of currents measured at the 40 m elevation. This near-bed

J. Wood (✉)
Ocean Data Technologies, Inc., Hyannis, MA, USA
e-mail: jwood@oceandatatech.com

R. Sharma (ed.), *Deep-Sea Mining and the Water Column*,
https://doi.org/10.1007/978-3-031-59060-3_11

intensification suggests a velocity shear layer which may enhance vertical mixing. Currents in the upper layer were dominated by tides and near-inertial oscillations, along with significant low-frequency(seasonal) variability. Lunar semi-diurnal tides appeared as the most significant constituent through the water column, especially at the seabed. Near-bottom M2 tides were oriented primarily in the north-south direction with little energy found in the east-west velocity component. Near-inertial oscillations with magnitudes of ~5 cm/sec were ubiquitous through the water column, and especially prominent at the seabed. Low-frequency circulation events were observed often near the seabed—some events featured peak speeds of order ~15–20 cm/sec and varied on ~monthly time scales. These low-frequency events possess properties of large-scale, westward-propagating meso-scale eddies. Comparison of low-frequency currents between moorings showed high correlation, suggesting circulation patterns of length scales greater than 7–9 km. One event estimated a translational speed of ~18 cm/sec, which was consistent with near-bottom eddy speeds described in other studies. Water properties measured in near-bottom boundary layer were extremely stable. Water temperatures varied less than 0.03 °C and salinity values ranged less than 0.005 PSU over the nearly 4-year time series.

Keywords Current observations · Deep benthic boundary layer · Water column · Meso-scale eddies · CCZ · North Pacific Ocean

1 Introduction

As potential deep-sea mining activities gain momentum, observations regarding various environmental conditions in area(s) of interest are valuable. Understanding the physical characteristics of the water column is of critical importance, not only for the different operational components of the mining process but also to assess the likely environmental impacts. In order to collect such information, an array of long-term moorings was deployed in the tropical eastern North Pacific Ocean in an area known as the Clarion-Clipperton Facture Zone (CCFZ). These mooring systems were designed primarily to observe physical processes throughout the water column, with particular emphasis on resolving ocean currents within the upper layers and the near-bottom boundary layer. Parameters of interest included variability in water temperature, salinity, pressure, and turbidity. Many of the resulting data sets spanned more than three (3) years.

This work was supported by Global Sea Mineral Resources NV (GSR), a deep-sea exploration company based in Belgium, which has generously shared these data results with the broader scientific community. Belgium holds the rights to certain discrete concessions within the CCFZ, with those investigation and exploration rights contracted to GSR for deep-sea polymetallic nodules. Within the larger Belgium concession contract area, GSR identified the B4 block and the B6 block as the focus of environmental baseline research, launching broad studies to better understand the biological, chemical, and physical environment at those locations.

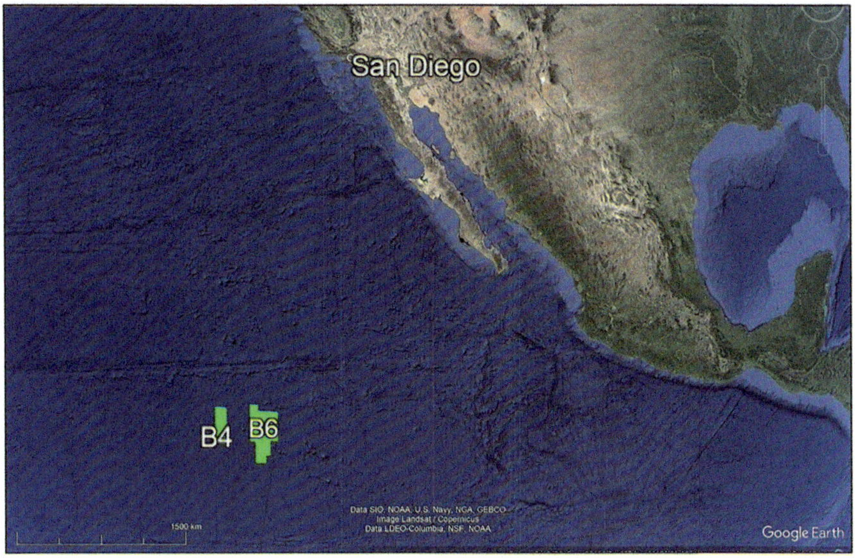

Fig. 11.1 Map of the equatorial North Pacific Ocean showing (in green) the general locations of GSR contract areas B4 and B6 approximately 3300 km (1800 nautical miles) southwest of San Diego. Moorings MOR003 and MOR008 were located in block B4. Moorings MOR007 and MOR009 were located in block B6. B4 was located about 375 km to the west of block B6

Figure 11.1 shows the location of the B4 and B6 blocks roughly 3300 km southwest of the staging port in San Diego, California and relative to the west coast of Mexico/Central America. The approximate water depths at these locations are about 4500 meters.

The B4 block is located approximately 375 km west of Block B6. The Long mooring—labeled MOR003—was deployed in the B4 block. The coordinates were 14.10198°N/125.89448°. A Short mooring—labeled MOR008—was installed also in Block B4, located about 7.1 km to the north-northwest of MOR003 (Fig. 11.2). Both moorings were located within a larger area of seafloor ridges that run approximately NNW-SSE. In block B6, the remaining short moorings were installed—MOR009 was located about 9.6 km south-southeast of mooring MOR007. These moorings were installed near several seamounts and to the southwest of a major bathymetric ridge running northwest-southeast.

During an early cruise to the B4/B6 blocks in May–June 2017, three (3) oceanographic mooring systems were installed for long-term data collection. The moorings can be described briefly as two (2) short moorings—simpler systems intended to measure current profiles in the near-bottom boundary layer only—and one (1) long mooring, which was intended to measure currents throughout the water column, providing high-resolution coverage of the near-surface and near-bottom layers. In addition, CTD casts were performed during this installation cruise, which provided instantaneous vertical profiles of temperature, salinity, density, and dissolved oxygen. During a subsequent cruise back to the area in February–March

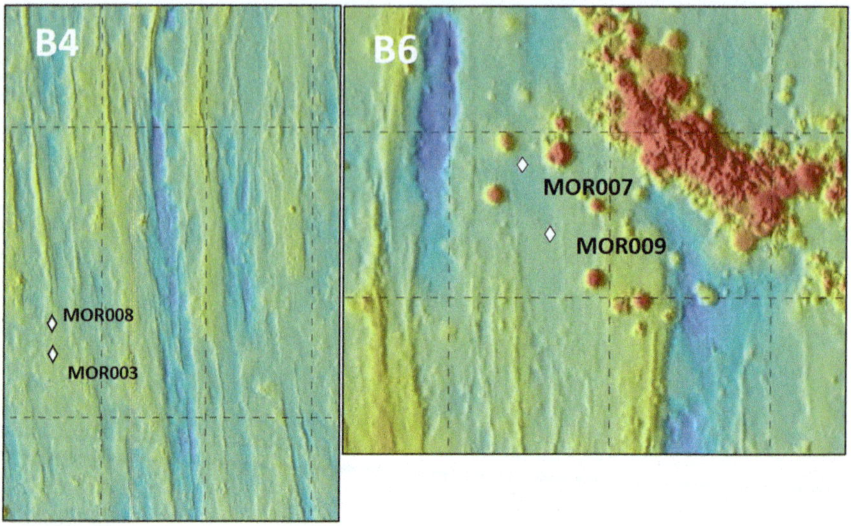

Fig. 11.2 Partial views of the B4 and B6 contract areas showing the mooring locations relative to general bathymetry features. In B4 (left side), the white diamonds depict the location of moorings MOR003 (south) and MOR008 (to the north), separated by 7.1 km. In B6 (right side), the northern white diamond represents the MOR007 mooring; the southern white diamond represents the MOR009 mooring. These two moorings are separated by 9.6 km. Maps are not to scale

2018, the two short moorings were recovered, then reinstalled at the same location, providing an early look at physical processes near the seabed. For various reasons—vessel capabilities and scheduling pressures—the long mooring remained in place at that time. A fourth mooring—also intended to measure the near-bottom boundary layer—was added during this 2018 cruise. The long mooring was recovered successfully—as were the three (3) Short moorings—during the next cruise in April/May 2021.

The data sets recorded by these moorings during the 2017–2021 time period are presented in this chapter, but with particular focus on the 3+ years of observations obtained by the MOR003 Long mooring. Near-bottom velocity profiles recorded at the other locations are also presented to provide a sense of the magnitude and variability scales of deep currents. Nearly four (4) year time series of temperature and salinity near the seabed are also presented. Numerous vertical profiles of conductivity-temperature-depth-turbidity were obtained during the various cruises and supporting evidence was provided that may help explain some of the physical processes observed during that time period.

The purpose of this document is also to describe the data sets themselves—an assessment of data quality, sensor performance, and general overview of significant behaviors, physical processes, and variability time scales identified in the data. This overview assessment is descriptive and qualitative in nature and provides to the reader an indication of the important features of the observations. We do *not* contemplate providing a fully quantitative scientific analysis of the physical oceanography in these areas, although it is hoped the information provided herein can be used as a basis for such future studies.

2 Mooring Descriptions

2.1 MOR003 Long Mooring Design and Sensor Depths

The MOR003 mooring stretched the full water column to measure vertical profiles of water velocity from the sea surface to the sea bed. A pair of Long Ranger 75 kHz ADCPs were mounted within a large buoy at the top of the mooring string at ~519 m mean depth—one ADCP upward-looking to the surface and the other ADCP downward-looking to depth, resulting in a near-continuous profile of current speed and direction in the upper 1000 m of the water column. The first valid bin was located 44 m below the surface (measurements near the very surface were rejected owing to side lobe contamination) with 8-m bin spacing down to 492 m. The deepest valid bin was 1025 m below the sea surface, with 16 m spacing up to 545 m. A ~ 42 m gap in the profile due to the ADCP blanking layers occurred around the buoy depth.

The upward-looking ADCP recorded data from June 2, 2017 to July 7, 2020, a period of 1131 days (3.1 years). The downward-looking ADCP data set ran from June 2, 2017 to September 30, 2020, a period of 1216 days (3.3 years). The upward-looking ADCP was programmed for 8-m bins, and an ensemble average recorded every 20 min. The average consisted of 20 pings every 3 s. The downward-looking ADCP was programmed for 16 m bins and 20-min ensembles—the ensemble average consisting of 12 pings every 5 s. These sample schemes were effectively a 1-min average at the start of every 20-min sample window. The standard deviation with these settings was 1.45 cm/sec and 1.82 cm/sec, respectively.

The upper buoy was relatively stable throughout the deployment—which suggests the resulting current measurements were not biased by mooring motion. Mean depth of the upper buoy was 519 m—and "blew over" as deep as 535 m. A vertical excursion of ~16 m over the deployment—given a 4000 m-length mooring—was acceptable. The buoy depth was within a half-bin depth (4 m) about 95.7% of the time, so extremely stable and upright. Two instances of higher mooring layover—when currents push the buoys laterally and downward—were observed about June 2018 and again May–June 2020 when deeper currents strengthened, increasing lateral drag forces on the string.

Single-point current meters (Teledyne RD Instrument Doppler Volume Samplers—DVS) were located at depths 1458 m, 2441 m, 3446 m, and 3948 m (corresponding to elevations of 2996 m, 2013 m, 1008 m, and 506m above the seabed). The sample interval for each unit was set to 20 min. Data return for the DVS meters were 499, 752, 968, and 642 days, respectively. Each DVS looked downward to sample currents at a range of 1.3 m from the transducer face. Each bin was 1 m in length, but only the first bin was processed—distal bins fell below the correlations/echo amplitude threshold required for valid/trustworthy data. The DVS current meter depth of 3446 m (1008 m elevation) collected valid data for 968 days—however, data after about 700 days became erratic and noisy, so a noise suppression filter was applied to the data. The number of 'noisy' points replaced by the filter procedure was mostly at the end of the record; however, the variability time scales appeared consistent with other data sets, so these data were trusted and included.

Nearest to the seabed, another pair of ADCPs—an upward-looking 150 kHz Quartermaster and a downward-looking 600 kHz Workhorse—were installed within a syntactic foam buoy positioned about 12 m above the bottom and a near-continuous vertical profile of the near-bottom boundary layer was provided (to about ~100 m above the seabed). The Quartermaster 150 kHz ADCP was programmed for 4 m bins centered 33 m up to 97 m above the seabed and sampled every 20 min (60 pings every 3 s, an accuracy of 0.69 cm/sec). This unit recorded data for about 443 days (June 2, 2017 to August 19, 2018). The downward-looking 600 kHz Workhorse ADCP was mounted at an elevation of 11.4 m and sampled also at 20 min (120 pings at 1.5 s, an accuracy of 0.99 cm/sec), similar to all other current meters. The bin size was 0.5 m. Some bins were contaminated due to acoustic interference from the mooring acoustic releases, and discarded, but otherwise, most of the bins were considered valid. The deepest valid bin was centered 2.5 m above the seabed. Data were collected for 977 days (2.7 years).

It is important to report a systemic issue affecting ADCP data quality—this issue was evident in numerous data sets. Data within the first and/or second (and sometimes third) bins of the ADCP profile had been unrealistic and therefore discarded. These near-field bins reported extremely high current speeds—which were utterly unrealistic—and therefore suspect. Consulting with engineers at Teledyne RD Instruments, manufacturer of the ADCPs, the cause of these invalid results nearest the transducers is thought to be a physical effect of excessive transducer ringing: "At those depths, the backscatter density is very low, so the weak echoes expose the ringing energy in the nearest bins." This effect has been identified for ADCPs in the upper layer (~500 m deep) as well as near the seabed. For the deep Quartermaster ADCP, the first 2 bins saw unrealistic values (speeds in excess of 50 cm/sec were considered unrealistic!) and were rejected. These first few bins were removed from the data set(s)—which had the same effect as if the blanking interval were increased.

2.2 *Short Mooring Designs*

The remaining Short moorings—MOR007, MOR008, and MOR009—were designed to measure near-bottom currents (and in some locations conductivity-temperature-pressure-turbidity). Each mooring was generally similar—using a single buoy to position ADCPs (or single-point current meters) and temperature-conductivity sensors close to the seabed. While each system had slight differences, the intent for each was similar: to measure near-bottom velocities as far up into the water column as possible given the available sensor resources.

All the Short moorings stood stable and upright at all times; there was no evidence in the pressure records suggesting the moorings were subjected to vertical/horizontal layover that could have contaminated the subsequent velocity measurements.

2.2.1 MOR008 Mooring

The MOR008 subsurface buoy was positioned about 27 m above the seabed and housed a downward-looking Workhorse 300 kHz ADCP. This mooring was deployed initially in June 2017, recovered in February 2018, and then reinstalled at the same location until recovery in 2021.

The Workhorse 300 kHz ADCP collected data successfully during both deployments. Data collection spanned from June 6, 2017 to February 7, 2018—and again from March 1, 2018 to November 13, 2020, a total data set length of 1245 days, not including the interim gap of 12 days. These data covered from 8.3 m to 18.3 m elevations—in 2 m increments—above the seabed. Again, the nearest few bins of the data were corrupted due to transducer ringing (the same manner as other ADCPs deployed at these depths). Data values recorded in the first 2 bins were simply not believable and therefore discarded. The remaining data were of high quality.

During the February 2018 turn-around, a Nortek Aquadopp single-point current meter with ancillary temperature, conductivity, and turbidity sensor was added to the bottom of the mooring. The Aquadopp recorded data successfully from March 1, 2018 until July 22, 2020—a total of 870 days. This unit was bolted in parallel to the acoustic releases, and looked downward, reasonably free of flow disturbances caused possibly by the tandem acoustic releases. These data were of good quality as well with low noise levels. The conductivity and turbidity data were also of good quality, although minor drift of the turbidity sensor was noted over time.

2.2.2 MOR007 Mooring

MOR007 was deployed also in the B6 area during February 2018. Nortek Aquadopp single-point current meters were placed on the upper part of the mooring at the discrete depths 20 m, 32 m, and 42 m. These units measured current speed and direction at a single depth only. The Aquadopp placed at 20 m also included Seabird temperature and conductivity sensors as well as a Seapoint turbidity sensor. A Teledyne Workhorse 600 kHz ADCP was mounted to a protective frame bolted parallel to the acoustic releases, 10 m above the bottom, looking downward to measure the bottom layer. This mounting technique avoided acoustic interference with releases and obtained an undisturbed profile as close as possible to the seabed.

The Aquadopp at 42 m returned 1148 days of data (3.1+ years). These data were very clean, although erratic data at the end of the record were removed as the battery power waned. The Aquadopp positioned at 20 m also returned a high-quality data set, with the data set extending 843 days (2.3+ years). The ancillary temperature, conductivity, and turbidity sensors drew power from the main battery and resulted in a shorter data set. No data were recovered from the 32 m depth level. All Aquadopps were programmed for 20-min sample intervals, based on a 60-second average at the start of each sample window. Horizontal precision was 0.7 cm/sec.

The 600 kHz Workhorse ADCP recorded for 942 days (~2.6 years) and returned a profile valid from 2.4 m to 6.9 m above the seabed at 0.5-m bin increments. Each 20-min ensemble consisted of 120 pings at 1.5-second intervals—an accuracy standard deviation of 0.99 cm/sec. These data were of very high quality; however, again, the first several bins of the profile—closest to the transducers—were corrupted and of poor quality, similar to other ADCPs performance at these deeper depths, and were discarded.

Temperature, conductivity, and pressure measurements obtained by the Aquadopp at 20 m were stable and steady, and there were no apparent problems with instrument drift or excessive noise levels.

2.2.3 MOR009 Mooring

This mooring was deployed initially in June 2017, recovered and reinstalled in February/March 2018, and recovered again during the 2021 cruise. The mooring consisted of a single syntactic foam buoy positioned about 13 m above the seabed containing back-to-back ADCPs: a 300 kHz Workhorse ADCP looking upward and a 600 kHz Workhorse ADCP looking downward. A Seabird SBE-16plus conductivity, temperature, and depth (CTD) sensor was bolted to the acoustic releases 5.5 m above the seabed. The CTD was tethered via long cables to a pair of Seapoint turbidity sensors—which were affixed to the mooring wire at elevations 5.1 m and 10.5 m.

The upward-looking 300 kHz Workhorse ADCP collected data from June 26, 2017 to February 16, 2018, and again from March 11, 2018 to January 1, 2020, a total of 919 days—more than 3 years (with a gap of 23 days). The sample rate was 20 min—each ensemble an average of 40 pings every 4.5 s (standard deviation of 0.43 cm/sec). Velocity profiles were obtained at 8 m bins from 25 m up to 73 m elevation above the seabed. All bins recorded by the ADCP appeared to be valid. The first bin was not plagued by the same transducer ringing problems experienced by other ADCPs—perhaps due to the longer blanking distance demanded by larger bin sizes. Bins beyond the range of the ADCP—more than 50 m distant—were discarded due to various suspect data quality criteria. A range of 50 m for a 300 kHz ADCP in these water conditions is considered reasonable.

The downward-looking 600 kHz ADCP collected data from Jun 26, 2017 to March 25, 2020, also with a gap of 23 days between deployments, a total of 1003 days. This ADCP was positioned 12.9 m above the seabed. The sample interval of 20 minutes consisted of 120 pings every 1.5 s (standard deviation of 0.99 cm/sec) at 0.5 m bin sizes. The first 2 bins were discarded—unrealistic data values due to transducer ringing issues described previously—as well bins surrounding the acoustic releases. During the first deployment—June 2017 to February 2018—measurements within near-bottom bins 2.4 m off the seabed were valid. However, during the second deployment, side-lobe reflections off the acoustic release and Seabird CTD contaminated all bins below 6.6 m, so no data were salvageable in the very near-bottom boundary layer. After March 2018, the measurement closest to the seabed was centered at 7.1 m. The measurement closest to the ADCP was centered 10.1 m above the seabed.

The Seabird 16plus recorded continuously for both deployments—June 2017 to April 2021—returning a data set in excess of 1381 days (almost 3.8 years) at a 20-min sample rate. Temperature, conductivity, and pressure measurements obtained by the Seabird CTD at 5 m appeared stable with low noise levels throughout the deployment. The salinity signal may have experienced a slight drift over 4 years—this drift was subtle, less than 0.003 PSU.

3 Vertical Profiling of Water Properties

Vertical profiles of oceanic parameters were obtained throughout the B4&B6 areas during the 2017 cruise using a Seabird SBE-19plus Conductivity, Temperature, and Depth (CTD) recorder mounted on a rosette frame. Cast CTD12 was chosen for display because it was the closest to the MOR003 mooring location, obtained roughly 2.2 km to the east, at 14.11183°N/125.87617°W. The CTD12 data were collected May 05, 2017. The Seabird SBE-19plus recorded the following parameters as a function of depth (or pressure): water temperature, conductivity, oxygen levels (SBE-43 sensor), and fluorescence (Seapoint Sensors). The CTD sampled continuously at 4 Hz. The descent/ascent rates for all profiles averaged about 1 meter/second. The vessel was dynamic positioned (DP), so vessel drift during a profile was minimal.

Seasoft software package (Seabird Electronics) was used to convert the raw data into engineering units, as well as calculate parameters such as seawater depth, salinity, density, and speed of sound. Subsequent data processing consisted of bin-averaging the original 4 Hz data into discrete 2 m depth bins to smooth out high-frequency variability. The bin average included both downcast and upcast data.

The thermocline in this region was observed between 75 m and 150 m (Fig. 11.3). Surface temperatures—in the mixed layer above 75 m—were relatively constant, slightly higher, and greater than 25 °C. Temperatures dropped through the thermocline to 13 °C at 150 m. Temperatures were less than 2 °C at the seabed. Salinity was a minimum in the upper mixed layer, roughly 34.2 PSU, increasing to a relative maximum (34.7 PSU) just below the thermocline (~200 m–250 m). Salinity values hit a local minimum of about 600 m, then increased gradually with depth down to the seabed to a water-column high of about 34.7 PSU. Density was also relatively

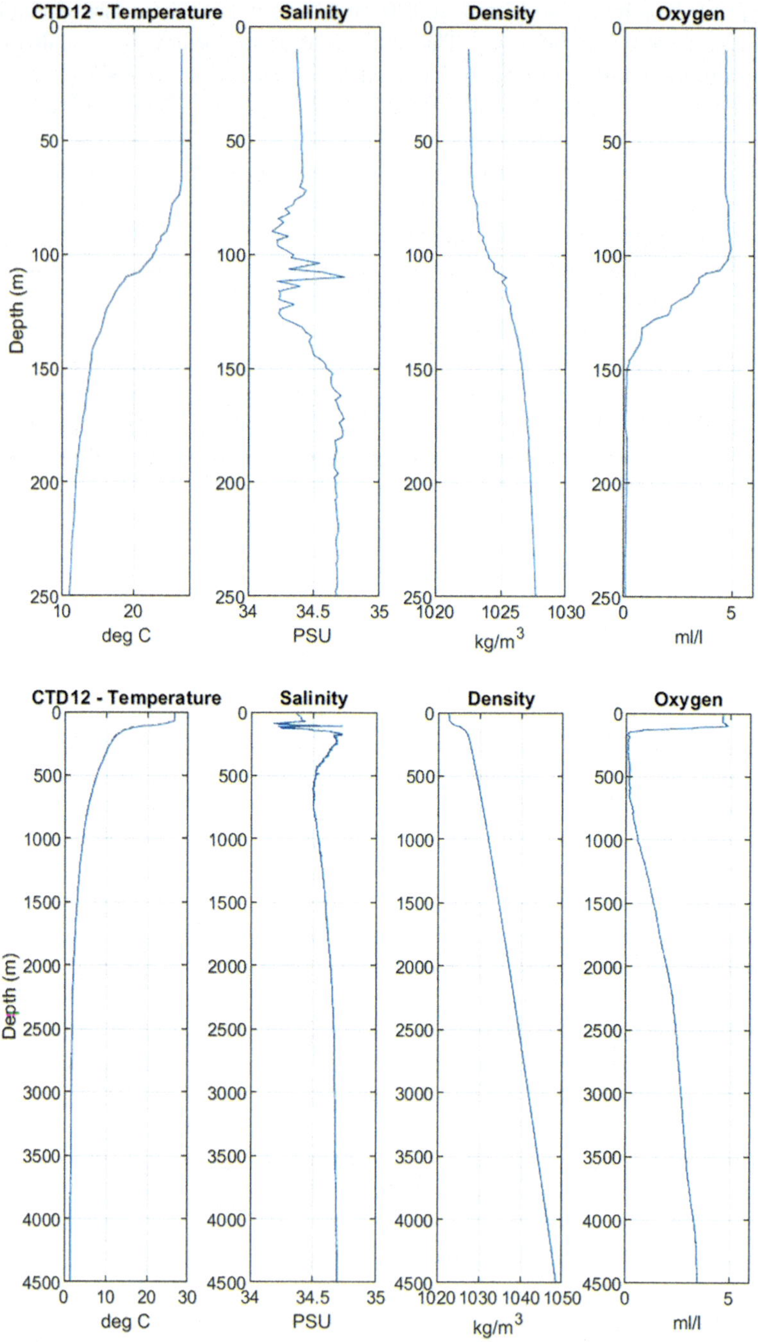

Fig. 11.3 Vertical profiles of temperature, salinity, density, and dissolved oxygen from cast CTD12 (May 05, 2017). The upper plots detail profiles from the surface to 250 m, the mixed layer, and thermocline. The lower panels represent how these parameters vary throughout the entire water column

constant in the upper mixed layer, about 1022.5 kg/m^3, increasing rapidly through the pycnocline to a value of 1026.5 kg/m^3 at 150 m. Density increased throughout the water column to a value of 1048 kg/m^3 at the seabed.

Dissolved oxygen levels were ~ 5 ml/L in the mixed layer, decreasing to a near-zero minimum at 150 m, just below the thermocline. Near-zero oxygen levels were observed from the thermocline down to about 600 m, where oxygen values began to increase steadily toward the seabed. DO levels of about 3.5 ml/L were found at the seabed.

The water properties observed during these vertical profiles are consistent with historical data in this subtropical region (Pickard & Emery, 1982). A strong pycnocline—defined as the upper layer of the water column where density increases rapidly with depth and often co-located with the thermocline—exists and presents a significant barrier to vertical mixing. Properties in the upper layer—from the sea surface to roughly 150 m depth—were nearly uniform, a result of rigorous mixing due to atmospheric winds and surface waves. Salinity minima in the upper layer suggest precipitation rates were greater than evaporation rates; however, this may have been a seasonal characteristic of early spring and therefore different during other times of the year. Oxygen levels in the upper layer were high, replenished easily due to frequent exposure to the atmosphere. Fluorescence—a proxy measure of biomass—was always a maxima at the base of the upper layer, suggesting a depth of bio-congregation.

In contrast to the well-mixed upper layer, waters below the pycnocline were quite stable. Vertical exchange between these deeper layers—even 200 m deep—and the near-surface layer is much more difficult. Low oxygen levels are a prime example of this barrier. The oxygen minimum layer has been widely reported in the eastern North Pacific. Low-oxygen layers—referred to as Oxygen Minimum Zones (OMZs)—exist in all the world's oceans and are particularly widespread in the subtropical eastern North Pacific. The CTD casts observed the OMZ in nearly all profiles, located at the bottom of the pycnocline to roughly ~600–800 m depths. Sources (Fiedler & Talley, 2006) describe the existence of these low-oxygen zones as the result of biophysical processes. Organic matter—decaying phyto- and zooplankton, detritus, waste, and other material—rains from the upper layer down toward deeper layers. Aerobic consumption—oxidation—of this biotic material by bacteria in these underlying layers reduces oxygen levels. Oxygen levels cannot be replenished from the atmosphere or by other processes (photosynthesis) via vertical mixing due to the barrier set up by the strong pycnocline. The only manner of replenishment is by horizontal circulation—oxygen advected into the layer by ocean currents. Currents in this layer are too sluggish to improve oxygen levels since the rate of addition is outpaced by the rate of consumption. Hence, a permanent deficit of oxygen is found within this layer below the pycnocline.

The low oxygen (anoxic) layer in this region of the tropical North Pacific likely leads to the lack of sound scatterers. The lack of ample sound scatterers—plankton, detritus, other ambient particles that reflect acoustic energy and are essential to the function of acoustic doppler current meters—limits severely the effective profiling range of the ADCPs.

Conditions in the deep ocean, above the seabed, were stable. Temperatures decreased with increasing depth, with minimum temperatures of order 1.5 °C. Salinity values were consistent from site to site, about 34.7 PSU at the seafloor. Water densities continued to increase with increasing depth, with maximum values of about 1048 kg/m^3. In very deep waters below 1000 m, oxygen levels rebound, suggesting the rates of oxygen consumption in deep water are lower relative to oxygen resupply rates.

4 Discussion of Flow Conditions Throughout the Water Column

The MOR003 data provides direct measurement of current velocity through the water column over an approximate 3-year time period, with particularly high-resolution coverage near the surface and near the seabed. Statistics calculated at all depths—from the surface to the seabed—are presented in Fig. 11.4. Selected statistical values comprising Fig. 11.4 are presented in accompanying Table 11.1. Note that the mean velocity and mean direction quantities were derived from vector components.

Ocean current velocities appeared dependent on the same density layers identified from the CTD cast data (Fig. 11.5). This near-surface layer, down to about 75 m depth, appeared relatively homogeneous due to the mixing effects of winds and waves –the previous section showed temperature, salinity, and oxygen properties were nearly constant within this layer. Currents in the near-surface layer were the strongest currents found in the water column—an unsurprising characteristic common to most of the world's oceans. Mean speeds in the near-surface layer were about 20–21 cm/sec and flowed to the west (~260°) on average. Flow speeds exceeded 50 cm/sec about 2% of the time; 10% of the time current speeds exceeded 35 cm/sec. Peak speeds of 165 cm/sec were observed in this layer. These peak speeds resulted from Hurricane Barbara in 2019, which passed less than 100 km south of MOR003 as a category 4 storm with wind speeds of 135 kts (more detail on this event in later sections). Variance of these upper-layer currents was more than an order of magnitude greater than the variance of currents at mid-depth (say, at 2500 m).

Below the near-surface mixed layer, the upper pycnocline about 75–150 m deep, a strong gradient in water properties was visible with increasing depth; temperature and oxygen values decreased and water density increased abruptly. This gradient also appeared to signal a transition layer for velocity, which slowed rapidly and shifted direction. Between 75 m and 150 m below the surface, mean speeds slowed from about 18 cm/sec to 12 cm/sec, and mean direction turned from the west-southwest to east-northeastward. Peak speeds in this layer varied with depth—155 cm/sec at 75 m, due to storm passage, decreasing to ~45 cm/sec at 150 m depth. The strong density gradient within the upper pycnocline layer appeared to prohibit

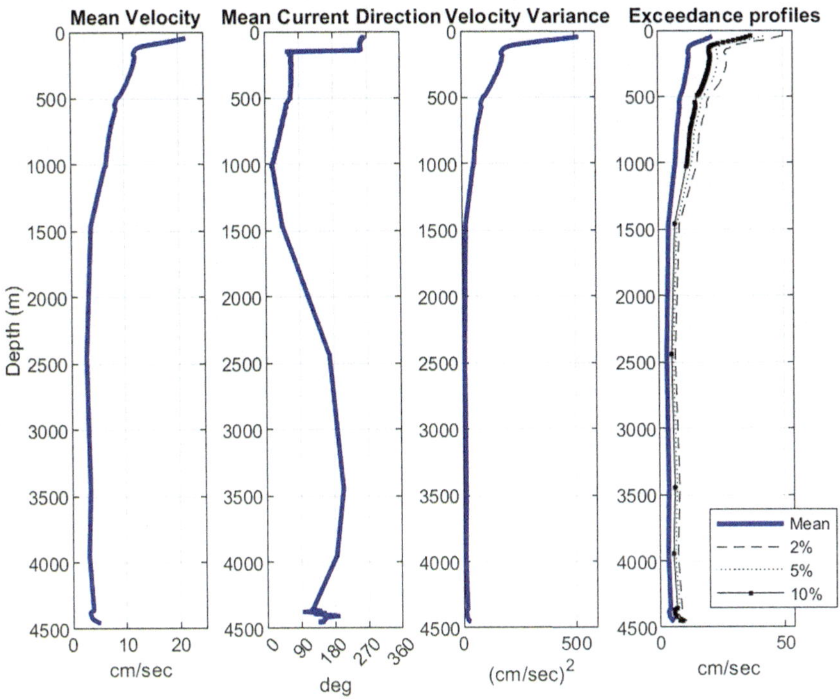

Fig. 11.4 Statistical profiles for the full water column over the entire data set duration. From left to right, (1) Mean velocity magnitude [in cm/sec], (2) Mean current direction, (3) velocity variance [cm/s^2], and (4) Speed profiles—mean and percent exceedance values through the entire water column at MOR003 location

surface wind stress from penetrating effectively to 150 m during storm events. The overall current variance decreased by half between 75 m and 150 m.

Current velocities below 150 m decreased gradually with increasing depth. The variance at 200 m—a measure of the overall energy of the flows—was about 1/3 the energy of currents in the near-surface layer. The mean speed at 200 m was approximately 12 cm/sec and directed toward the east-northeast (72°). The peak speed at this depth was about 50 cm/sec, but speeds this high were rare; during the 3+ years of observations, speeds exceeded 21 cm/sec only about 10% of the time. At 500 m depth, the mean speed was about 9 cm/sec, and flowed (on average) toward the east-northeast. The mean direction shifted slowly with increasing depth, flowing northward at 1000 m depth. Mean speeds slowed further to ~6 cm/sec. The overall variance at 1000 m was one-half the variance found at 500 m.

From 1000 m down to about 2500 m, mean current directions shifted steadily from the north to the east (at ~1800 m) and to the south at 2500 m. Mean speeds decreased from 6 cm/sec at 1000 m to 3 cm/sec at 2500 m. Mean currents from ~2500 m down to 4420 m (or about 34 m above the seabed) were 4 cm/sec. Peak

Table 11.1 MOR003 current velocity statistics

Depth(m)	Days (of data)	Mean speed	Mean direction (deg)	Peak speed	2% exceedance speed	10% exceedance speed	Variance (cm/sec)^2
44	1131	21.60	263	164.39	49.7	37.1	521.0
52	1131	20.68	260	169.29	47.2	35.4	477.0
68	1131	18.64	256	162.93	44.1	32.2	398.0
76	1131	17.53	255	154.85	42.0	30.2	360.8
92	1131	15.17	254	123.94	36.2	26.0	285.4
100	1131	14.06	254	102.63	33.5	23.9	249.0
124	1131	12.38	254	70.18	30.0	20.7	198.1
148	1131	11.99	62	45.80	26.8	20.3	182.2
204	1131	12.02	71	50.12	27.1	20.7	182.6
300	1131	11.35	72	43.19	26.7	19.8	163.5
308	1131	11.30	71	42.10	26.6	19.7	161.6
404	1131	10.27	71	38.91	24.3	18.0	133.9
492	1131	8.80	69	31.54	20.6	15.6	98.9
609	1216	8.25	54	26.80	18.8	14.4	86.2
705	1216	7.62	46	26.36	17.3	13.3	73.6
801	1216	7.16	38	23.41	16.2	12.4	64.7
913	1216	6.78	28	25.32	15.7	11.8	58.7
1025	1216	6.29	21	25.58	15.0	11.0	51.2
1458	499	3.73	45	12.51	8.1	6.3	17.1
2441	752	2.83	169	10.30	6.3	4.8	10.2
3446	968	3.40	205	13.01	7.8	5.9	14.4
3948	642	3.10	186	11.81	7.1	5.3	11.3
4357 (97 m)[a]	443	3.84	120	12.74	8.5	6.6	17.6
4365(89 m)[a]	443	3.40	139	10.51	7.3	5.8	14.3
4373(81 m)[a]	443	3.20	100	9.86	6.9	5.5	12.9
4381 (73 m)[a]	443	3.21	151	10.32	6.9	5.5	12.8
4389(65 m)[a]	443	3.25	129	10.20	7.1	5.6	13.4
4397(57 m)[a]	443	3.32	165	11.20	7.2	5.7	13.9
4405(49 m)[a]	443	3.40	188	10.49	7.3	5.9	14.6
4413(41 m)[a]	443	3.48	153	10.58	7.5	6.0	15.3
4421 (33 m)[a]	443	3.60	154	11.40	7.7	6.2	16.3
4446 (8 m)[a]	977	4.51	148	16.72	9.7	7.6	22.2
4448 (6 m)[a]	977	4.57	142	21.57	9.9	7.7	22.8
4451 (2.5 m)[a]	977	4.98	142	19.84	11.2	8.5	29.0

[a]Depth values in parenthesis denote elevation above the seabed

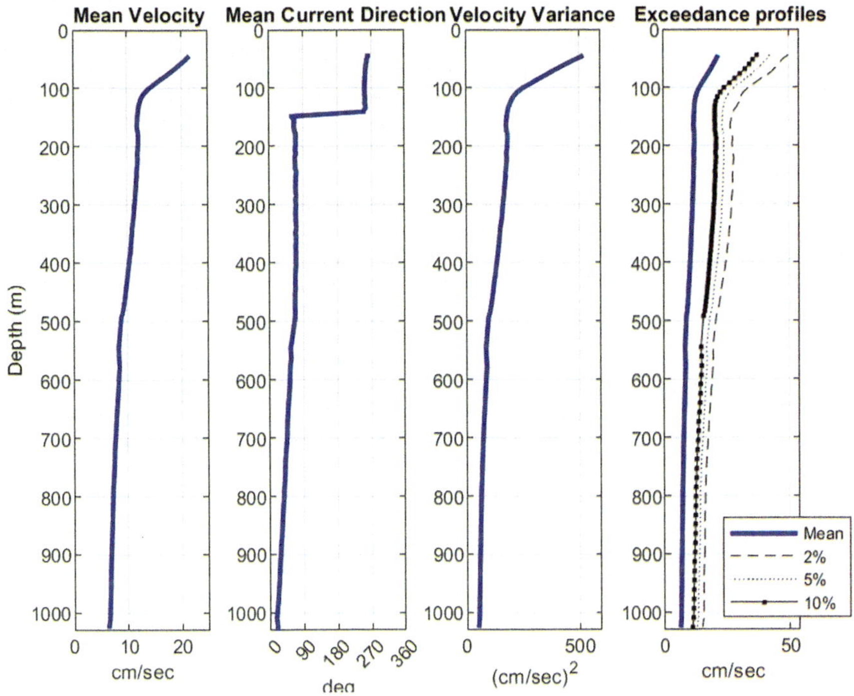

Fig. 11.5 Mean velocity, mean direction, variance, and speed exceedance profiles for the upper 1000 m of the water column. These statistical quantities were calculated over the entire 3+ year time series. Mean velocity represents the vector mean; Exceedance values represent the speed value exceeded only 2% (10%) of the time

speeds at these depths were 10–12 cm/sec. Velocity variance in these deeper layers from ~2500 m down to the near-bottom boundary layer was fairly constant—about 10-to-20 cm^2/sec^2. Variance levels at these deeper depths were about 1/30th of the energy at the surface.

Statistical quantities only tell a limited story, however—especially when the statistics are calculated over such a long time period. Another perspective is gained by assessing the time series observations at specific depths. Figure 11.6 presents color contours of the east and north velocity components for the upper 1000 m. Red/orange colors denote relatively strong flows to the east; blue colors (negatively valued) indicate flows toward the west. While there was a thin near-surface layer of light green/blue, denoting westerly flow, it is clear the predominant flow direction was eastward in this upper layer, with periodic strengthening approximately every summer (July). The northern velocity component was weaker and appeared to switch positive/negative—northeast or southeast—on ~monthly time scales.

Figure 11.7 shows the raw time series of current speed and direction at 44 m from late June 2017 to July 2020—more than 3 years. The time series confirms that speeds were generally less than 50 cm/sec much of the time, punctuated on occasion

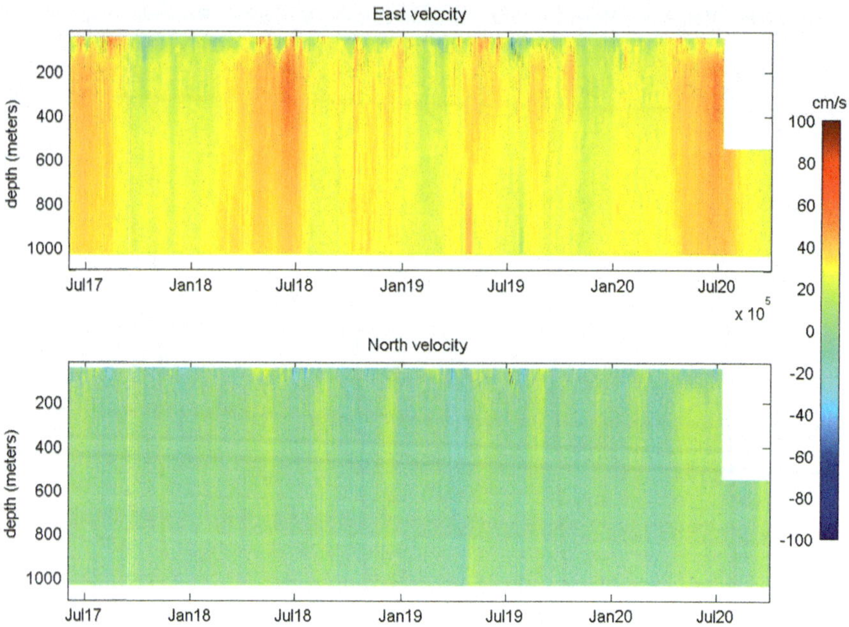

Fig. 11.6 Color contours of the east and north velocity components from late June 2017 to July 2020

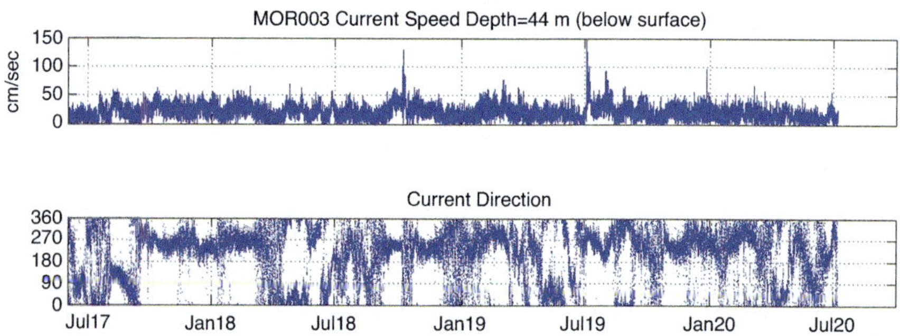

Fig. 11.7 Time series of current speed and direction 44 m below the surface from June 2017 to July 2020. Currents flowed to the southwest most of the time. The strongest currents—165 cm/sec—occurred during storms

with brief, but intense, spikes in magnitude, particularly in the middle of the plot (October 2018) and again in early July 2019. The current direction at 44 m is primarily toward the west (~270 degrees), with brief interruptions when directions reversed to the east or northeast. These periodic reversals of current directions at this depth appeared to occur in the spring of each year—approximately the May–June time frame—before resuming westward flows mid-to-late summer. Westward flows seem to persist through late winter/early spring each year.

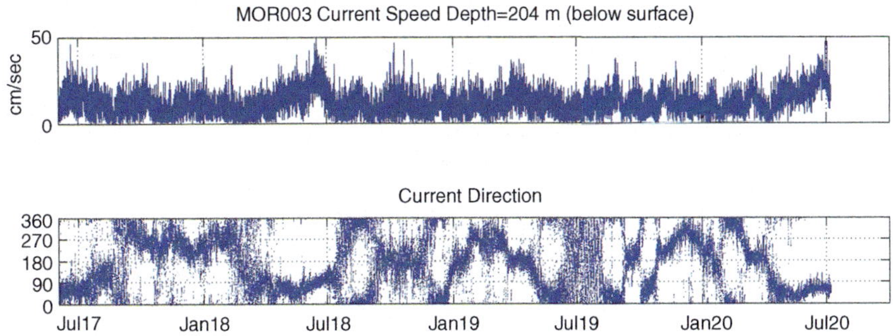

Fig. 11.8 Time series of current speed and direction 204 m below the surface from June 2017 to July 2020. Currents oscillated between the northeast and southwest most of the time

Figure 11.8 presents a time series at 204 m—below the mixed layer and steep gradient of the pycnocline. At this depth, the reversals in current directions appeared to occur on seasonal time scales—running generally eastward or northeastward from late winter (~February) into mid-summer (~July–August), then reversing toward the west or northwest late summer through winter. About September–October of 2018, currents appeared to flow southward for some time.

Figure 11.9 shows the current rose diagrams calculated at these depths. Ros diagrams are a statistical quantity—but provide a fuller representation of the flow behavior. Each 'petal' of the rose diagram represents the percentage of time currents flowed into a specific direction. The annulus rings represent 5%, 10%, 15%, and 20% occurrence—the length of each petal relative to the annulus rings indicates the percentage of time currents flowed in that direction. Further, the values in each petal are binned according to the speed scale on the left, the strongest currents in red, and the weakest currents in blue. At 44 m, most of the flow was toward the west-southwest—and the strongest flows observed at that depth were always in that direction. Flow toward the northeast was much less frequent. Below the pycnocline, at 204 m and 492 m, northeast flows were more common. These northeast flows were typically of higher magnitude than flows toward the southwest.

Velocities showed good correlation throughout the deeper water column (in the lower 1000 m). Higher-speed events identified at 3446 m depth are observed also at the 4447 m depth. Lower-frequency changes in the current direction are tracked similarly in all the deeper layers, too. Current observations in the near-bottom boundary layer are discussed in more detail in the next section.

Also of significance was a very impressive diel migration in the upper oceanic layer, as shown in the upward-looking ADCP echo amplitudes (Fig. 11.10). This daily vertical migration of plankton (sound scatterers) is a common feature in the ocean and is manifest as a daylight-dependent variability of ADCP echo amplitude through the upper water column. During nighttime hours plankton rose to the surface and during daylight descended to about ~400 m. This resulted in an abundance of scatterers at mid-depth during daylight hours—that is, higher echo

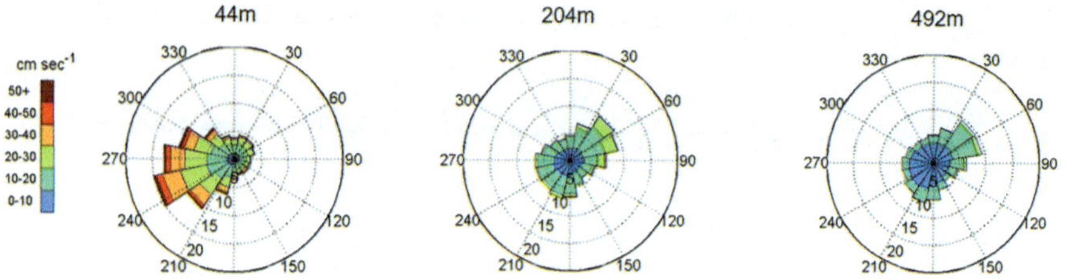

Fig. 11.9 Rose diagrams of ocean currents at the 44 m, 204 m, and 492 m depths. Flow was generally southwesterly in the near-surface layer (44 m) and northeastward at 204 m and 492 m

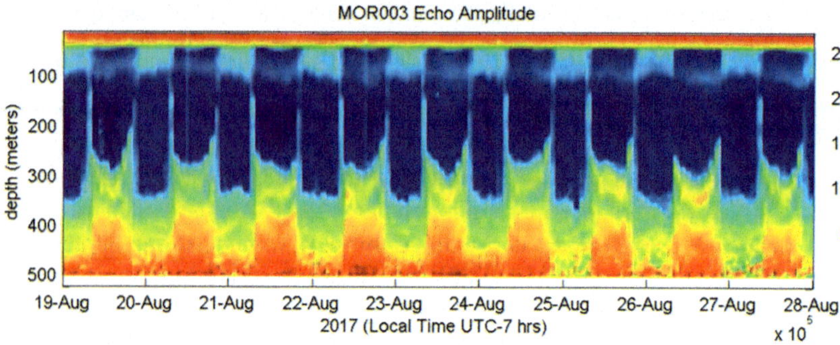

Fig. 11.10 Example of the daily vertical migration of plankton in the upper 500 meters, as observed in the echo amplitude of the upward-looking ADCP for a 9-day time period in late August 2017. Yellow/red colors indicate high echo amplitude; dark blue indicates low echo amplitude. The dark red stripe at the top of the plot indicates the strong acoustic reflection off the sea surface

amplitudes—and a corresponding dearth of scatterers at mid-depth during the night (i.e., low echo amplitude).

4.1 Variability Modes in the Upper Water Column

Velocity spectra were calculated for both east and north velocity components at every depth level, and displayed as variance-preserving spectra in both orthogonal directions. For these graphics, total variance (energy) is represented by the area under each spectral curve such that significant spikes in the spectral distribution indicate the relative energy at various time scales. Calculating the velocity spectra throughout the water column provides an idea of which time scales—and, in turn, physical processes—are responsible for the energy observed at each depth.

The spectral distribution calculation used the entire data set length at each depth—which varied by instrument from 3+ years to less than 2 years. Comparing the spectra at each depth based on differing time series lengths should not affect the results detrimentally since the data were separated—block-averaged—into 32

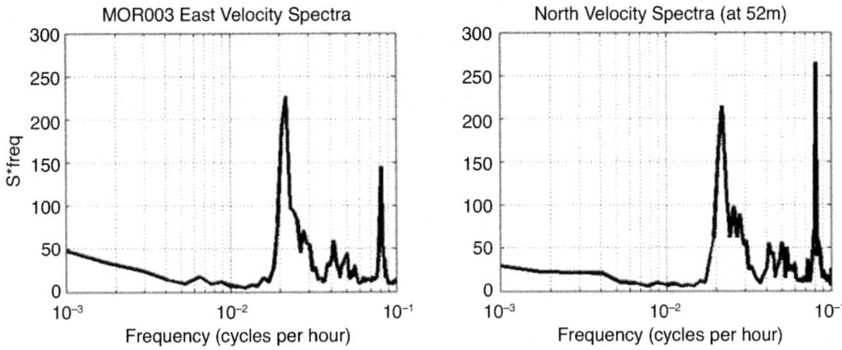

Fig. 11.11 Variance-preserving spectra for the East (left) and North (right) velocity components at 52 m depth below the surface. The spectral amplitude is depicted by the vertical y-axis, and the frequency is depicted by the horizontal (bottom) x-axis in units of cycles per hour. Large spikes at 0.02 cycles per hour and 0.08 cycles per hour represent variability at inertial and semidiurnal tidal frequencies, respectively

blocks. Since we are interested in variability modes in near-tidal bands—say, a few hours to a few days—the spectral resolution was more than sufficient to make meaningful comparisons.

Near-Inertial energy and semi-diurnal (M2 and S2) tides dominate the spectra, especially in the upper layers (see Fig. 11.11). Inertial energy is represented by the spike in the spectral curve about ~0.021 cycles/hour (~48-hour cycles), which is consistent with the theoretical inertial period at this latitude (14.1° corresponds to 49.3 hours), and is given by:

$T_{inertial} = 24\ h / (2 * \sin(\text{latitude}))$ (*see* Tolmazin, 1985).

The abrupt spike in the spectrum at this frequency appears to represent energy due to inertial motions. Note that the spike is somewhat broad—suggesting that inertial energy at this location is a relatively broad-banded physical process. The spectra also show a second significant signal representing the M2 (principal lunar semidiurnal) and S2 (principal solar semidiurnal) tides, which have periods of 12.42 (0.081 cycles/hour) and 12.00 h (0.083 cycles/hour), respectively. The spectral resolution was not fine enough—due to block averaging choices—to distinguish between these two tidal constituents. A tidal constituent analysis was performed to calculate the amplitudes of all tidal constituents.

The spectra indicate very high energy in the upper layer—particularly above the thermocline—in both the inertial and tidal bands. Energy from inertial oscillations appeared greater than tidal energy—inertial peak was high and broad-banded while tidal energy peak(s) was high but narrow-banded. Inertial energy was equivalent in the north-south and east-west directions, which is unsurprising given that inertial oscillations are orbital typically (i.e., omnidirectional). There appeared more tidal energy in the north component than the east component—almost by a factor of 2.

At 200 m, the overall variance was roughly half of the energy at 52 m—but remained distributed equivalently between east and north components. Variance in inertial bands was roughly the same in the east-west versus the north-south directions. Semidiurnal tidal energy at 204 m was substantially less than the tidal energy at 52 m. At about 404 m the overall variance was about ¼ (25%) of the variance at 52 m. Tidal energy in the north-south component was also about 2× larger than tidal energy in the east-west direction—this ratio becomes larger with increasing depth—the north-south tides were much more energetic than the east-west tides. Inertial energy at these depths remained equivalent between north and east components.

4.1.1 Tidal Variability

Tidal constituents were calculated on the velocity time series observed at the MOR003 location to determine the amplitude and phase of the major tide-generating causes—the lunar and solar tides varying at semi-diurnal and diurnal frequencies. Tidal currents in the open ocean are relatively weak generally and flow in elliptical patterns, rotating in elongated circles with a minor axis and major axis. The major tidal constituents were calculated for selected depths through the water column, and are presented in Table 11.2 below. This constituent analysis—employing a least-squares fit technique—was performed on the entire 3 + -year time series.

In the near-surface, the diurnal tide K1 (lunisolar diurnal constituent) had a higher amplitude than the lunar (M2) and solar (S2) semi-diurnal tides. The amplitudes of all tides appeared relatively equivalent in both north-south and east-west axes, suggesting tides near the surface were roughly circular in their elliptical patterns. Below the mixed layer, at 404 m, the M2 lunar semidiurnal tide had the greatest amplitude, with the north-south amplitude much greater than the east-west, suggesting a narrower ellipse pattern oriented primarily in the north-south direction. At the seabed, one can see the M2 lunar semi-diurnal tide dominated in the north-south orientation—the north-south tidal component (1.61 cm/sec) was orders of magnitude greater than the east-west tidal component (0.07 cm/sec). The diurnal constituents—O1 and K1—were weaker than the semi-diurnal tide.

Table 11.2 Velocity tidal constituents at selected depths

		52 m depth		404 m depth		4447 m depth (7 m above seabed)	
Constituent symbol	Period (hours)	North-south axis (cm/sec)	East-west axis (cm/sec)	North-south axis (cm/sec)	East-west axis (cm/sec)	North-south axis (cm/sec)	East-west axis (cm/sec)
M2	12.42	0.83	0.62	1.12	0.17	1.61	0.07
S2	12.00	0.63	0.41	0.11	0.24	0.27	0.15
K1	23.93	0.99	0.81	0.85	0.37	0.57	0.29
O1	25.82	0.20	0.77	0.49	0.50	0.52	0.15

4.1.2 Seasonal Variability

Quantifying seasonal conditions at MOR003—and in the B4 area generally—may be helpful planning future operations. Two analyses were performed in an attempt to distinguish seasonal patterns. First, a 30-day low-pass filter was applied to the data sets to smooth out higher-frequency oscillations and reveal the underlying inter-annual variability. Second, the data sets were subsectioned into seasonal groups—spring, summer, autumn, and winter—denoted by the solstice and equinox dates (March 21, June 21, September 21, and December 21)—and presented as rose diagrams to visualize more simply the directional tendencies of flows during these time periods. Note that any time period(s) could have been defined; time periods based on seasonal movements of the sun seemed a reasonable choice. These rose diagrams were created using the original data including all variability modes; the 30-day low-passed signals were not used to generate the seasonal rose diagrams.

This analysis was performed at discrete depths throughout the upper water column. As expected, seasonal conditions for the very near surface—above ~75 m—were slightly different than conditions found in deeper layers. In the near-surface layer, 52 m depth (Fig. 11.12), where flows are influenced directly by wind stress, flow was toward the west and southwest during the late summer-autumn-winter time periods, and northeast briefly during spring periods. Southwest flows seemed to be the common condition, consistent with the west-southwesterly-blowing trade winds in the tropical Pacific and the westerly-flowing North Equatorial Current (Pickard & Emery, 1982). A reversal of the main flow toward the northeast appeared

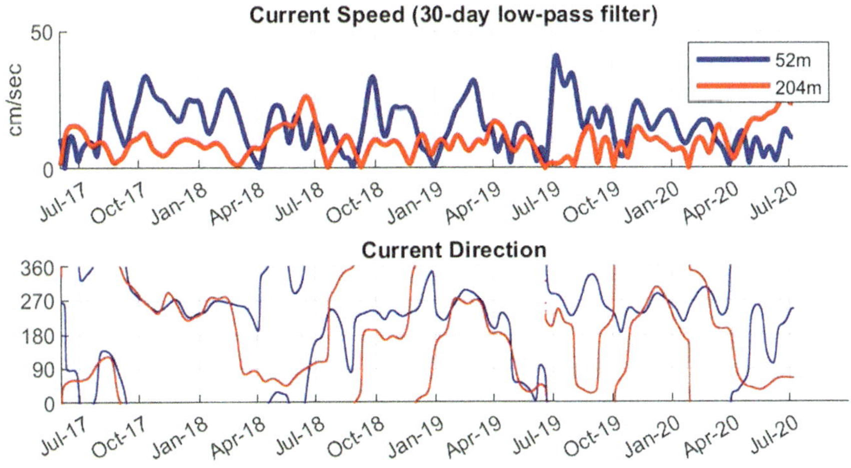

Fig. 11.12 Current speed and direction 52 m (blue) and 204 m (red) below the surface after applying a 30-day low-pass filter to flow near the surface was generally toward the west for much of the time but appeared to switch directions to the east briefly. Flow at 204 m was most often easterly or northeasterly

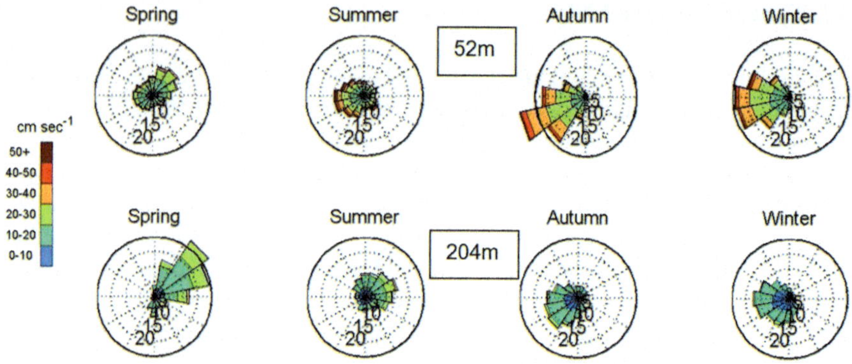

Fig. 11.13 Current speed and direction 204 m below the surface after applying a 30-day low-pass filter. Seasonal rose diagrams at this depth are also shown. Flow was generally toward the northeast and east in spring/summer, and switched to the southwest and west in the autumn/winter

to be a recurring feature—this brief reversal in the upper layer occurred about May–June each year for a short time period (a month or less).

Below the near-surface layer, for example, at 204 m depth, flow was predominantly toward the west and southwest during the autumn and winter time periods; rarely toward the east or north (Fig. 11.13). During the spring and early summer time period, the flow condition was more commonly toward the northeast. These brief flow reversals to the northeast were of order ~25 cm/sec (low-passed value)—roughly twice the speed of more common flows to the west or southwest—and accompanied by a modest strengthening, especially in late June 2018 and again late June 2020. Rose diagrams of velocities in the upper 1000 m were consistent with the example at 204 m—generally west or southwest during the autumn and winter, and generally northeast in the spring and summer.

4.1.3 Storm Events

Several significant storm events—major hurricanes—passed through the MOR003 location during the deployment and the resulting current responses were captured in high resolution. Hurricane Barbara occurred in early July 2019—taking a north-westerly track across the B4 block, with the eye passing about 100 km south of the MOR003 location at about 1200 h 03-July (Fig. 11.14). Barbara was near peak intensity as a Category 4 hurricane with winds in excess of 130 kts as it passed the mooring. Prior to the onset of the storm, nominal current speeds near the surface were of order ~20 cm/sec toward the west, typical conditions there during the summer months. Within about 8 h however current speeds (at 52 m depth) accelerated rapidly to ~165 cm/sec and shifted toward the north-northwest (Fig. 11.15). The peak was short-lived, however, and decayed just as rapidly as the storm front passed. In the storm's wake, current speeds oscillated at near-inertial frequencies—approximate 48-hour recurrences—for days afterward.

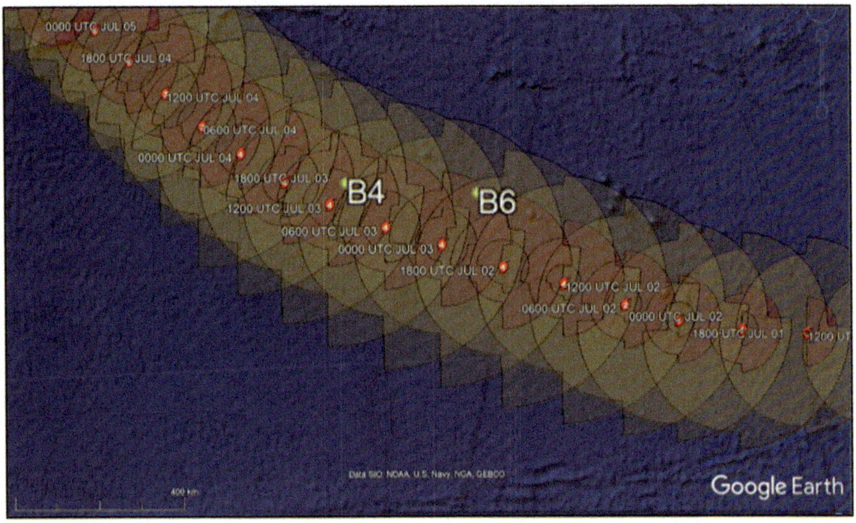

Fig. 11.14 Path of Hurricane Barbara in early July 2019. Data courtesy of National Oceanic and Atmospheric Administration National Hurricane Center. https://www.nhc.noaa.gov/data/tcr/EP022019_Barbara.pdf

Extreme wind events affect the near-surface layer via direct wind stress dragging across the water surface. The rapid increase in currents observed near the surface—the 52 m depth being an example—did not appear to penetrate downward. Currents at 204 m depth, below the well-mixed near-surface layer, did not exhibit the same intense increase/decay. Currents at 204 m were quite weak throughout the storm, never exceeding 10 cm/sec (the red signal of Fig. 11.15). This suggests the pycnocline acts to block wind-induced energy from penetrating directly into deeper layers.

Effects of Hurricane Barbara can also be visualized with color-contoured plots of the upper 1000 m. In this perspective, it was apparent that some hurricane-induced responses penetrated to deeper layers, but at some time well after passage of the storm. Figure 11.16 shows the east and north velocity components for the month of July 2019. The direct storm effects were depicted clearly in the near-surface layer as the dark blue (negatively-valued, or westward) patches in the east component and the dark red patches in the north component, and appeared confined to the very upper layer (above about 150 m). These direct effects—the bright patches—decayed after a day or two only, and appeared to affect the upper layer, but did not penetrate deeper than about 200 m. The pycnocline appeared to block these direct effects from penetrating deeper layers. However, a week or two after the storm, the color-contours showed much weaker currents as vertical light blue-and-yellow stripes, which indicate velocity oscillations in the deeper layers. These oscillations were weak inertial currents, which lingered for weeks after the storm had passed. These inertial currents—with anti-cyclonic rotations (clockwise)—were observed below the thermocline to (at least) 1000 m depth. So hurricanes affected deep currents, but

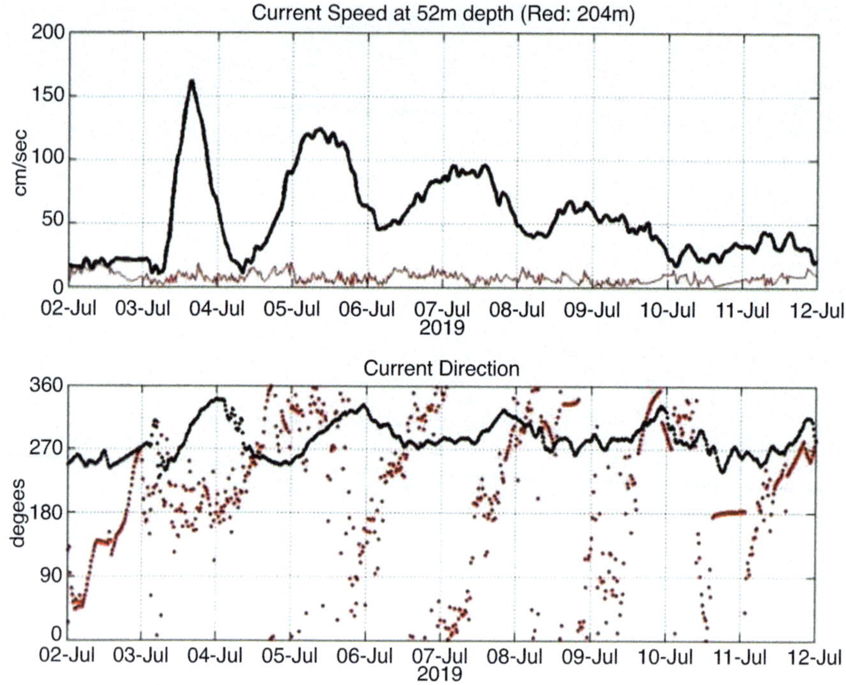

Fig. 11.15 Storm-generated current speed (top) and current direction (bottom) time series near the surface (52 m, black) and below the mixed layer (204 m, red) during Hurricane Barbara in early July 2019

those effects were much weaker and appeared delayed by a week (or two) after storm passage.

The DVS current meter positioned about 1500 m below the surface—as well as deeper DVS current meters—did not record this event due to already-depleted batteries. Evidence of this storm was not observed in the near-bottom current meter data sets.

Another hurricane event—Hurricane Sergio– was observed in October 2018. This storm passed on a west-southwesterly track less than 100 km to the north of the mooring as a category 3 hurricane with winds roughly 100–110 kts. A few days later, the storm path dog-legged abruptly to the northeast, passing MOR003 a second timesome 400 km to the north. Two current spikes—associated with the dogleg storm path—were observed in the near-surface layer, currents reached 100 cm/sec and then 135 cm/sec respectively during each pass, but both spikes decayed rapidly. Flow swirled clockwise (eddy-like) at inertial (~2-day) periods over the next 10–14 days as speeds weakened. Currents at the 204 m depth reached almost 50 cm/sec briefly, less than an hour, but otherwise held steady at ~20 cm/sec throughout the storm. Again, the pycnocline appeared to block wind energy from penetrating deeper down the water column. Similar to other extreme wind events, inertial

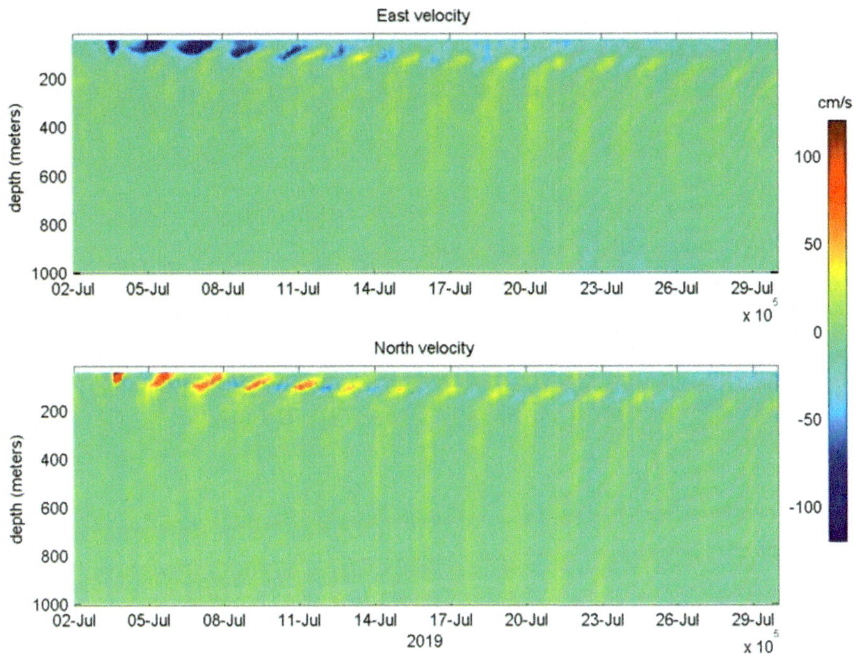

Fig. 11.16 Color-contours of the east and north velocity components in the upper 1000 m showing the response to Hurricane Barbara in July 2019. The storm affected the very near-surface layer. Deeper currents—from 200 m to 1000 m—oscillated at inertial frequencies for several weeks after the storm passed

currents were observed throughout the upper 1000 m for several weeks after the storm had passed.

Another anomalous event of interest occurred about April–May 2019 in deeper layers below the thermocline, perhaps consistent with the advection of a mesoscale eddy through the mooring location (Purkiani et al., 2022). About mid-April, currents strengthened to ~20 cm/sec to the south, then began a slow counter-clockwise rotation (cyclonic—south, then east, then north) over an approximate 3-week time period until early May. Current then normalized to 0–10 cm/sec while inertial oscillations (counter-clockwise rotations) were observed for several weeks in the wake of this event. This event was observed throughout the entire upper layer covered by the ADCPs, but the higher amplitude currents appeared most prominently in the depth layer from about ~800 m to 1000 m (Fig. 11.17). Current meters at the 1500 m and 2500 m depth levels had run out of power by this time and did not record this event. This anomalous April event was not preceded by a significant current disturbance at the surface, and so was probably not associated directly with a storm event such as a hurricane. The source of this anomalous current condition within this finite depth layer could have been remnants of a remote storm.

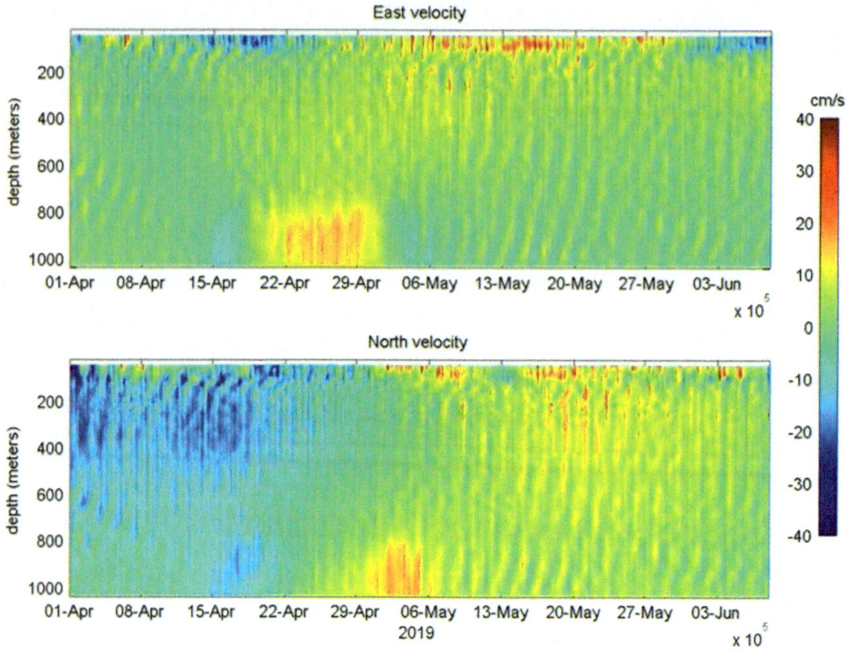

Fig. 11.17 Velocity in the upper 1000 m during the event in April–May 2019. An increase in current speed was observed throughout the water column, but most prominently about 800 m–1000 m deep

4.2 Near-Bottom Boundary Layer at MOR003

Conditions within the near-bottom boundary layer are of primary importance to DSM activities. Observed flow conditions are summarized in this section, first with a detailed look at conditions at MOR003, and also comparing to near-bottom velocity conditions measured at MOR008—about 7 km distant from MOR003.

The mean velocity closest to the seabed at MOR003 was 5 cm/sec (Fig. 11.18). Peak speeds in the near-bottom layer were ~ 20 cm/sec (at an elevation of 2.5 m), versus roughly 10 cm/sec at an elevation of 100 m above the seabed (Table 11.1). Currents exceeded 11.2 cm/sec 2% of the time and exceeded 8.5 cm/sec 10% of the time. The mean flow direction near the seabed was approximately southeast, consistent with the local bathymetric features.

Currents near the seabed appeared slightly stronger relative to flows found just ~100 m above. While the mean flow direction in the lower ~100 m remained approximately southeast (~130–160°), the mean velocity increased slightly from 3 cm/sec at ~100 m to 5 cm/sec near the seabed. Peak speeds were roughly 10 cm/sec at an elevation of 100 m above the seabed. The overall variance doubled between those depths—with near-bottom variance twice (2x) the variance at the 100 m elevation. This increase in near-bottom variance is demonstrated in Fig. 11.18 (third

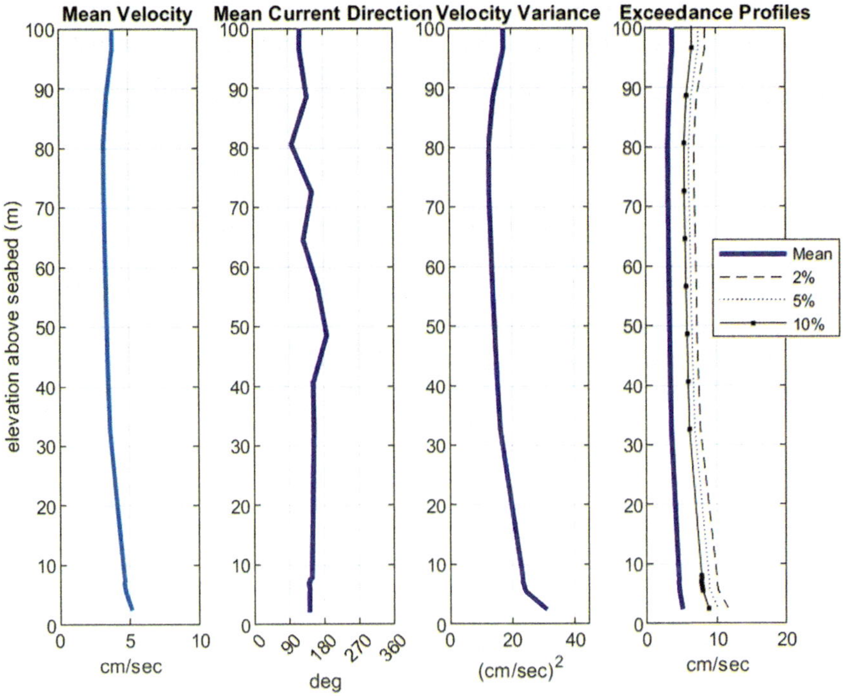

Fig. 11.18 Statistical profiles in the near-bottom layer over the entire data set duration. From left to right, (1) mean velocity [in cm/sec], (2) mean current direction, (3) velocity variance [cm/s^2], and (4) Speed profiles—mean and percent exceedance values

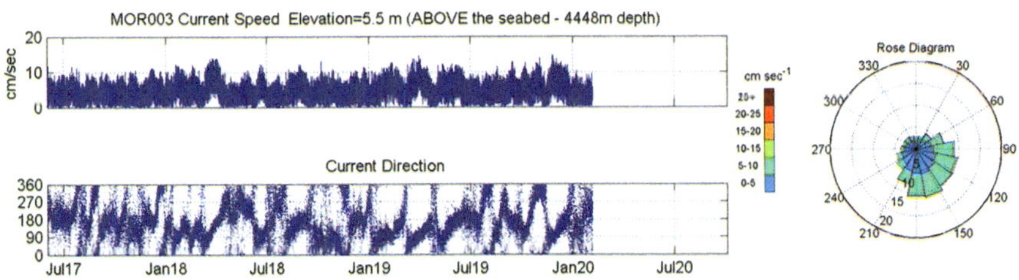

Fig. 11.19 Time series of current speed and direction at 4448 m depth (5.5 m above the seabed). The accompanying rose diagram is to the right. Current speeds of 10 cm/sec were frequent. The predominant flow direction was south-southeastward, parallel to the bathymetric ridges in that area

panel from left). The tendency for stronger flows very near the bottom relative to flows about 100 m above the seabed was observed in other data sets, too, specifically, flows observed in the B6 area (see next section).

A time series of currents at the 5.5 m elevation at MOR003 is shown in Fig. 11.19. Currents of magnitude 10 cm/sec were common—the 2% exceedance speed at this depth was 9.9 cm/sec. The 10% exceedance speed was 7.7 cm/sec. The accompanying rose diagram shows currents flowed to the south and southeast most of the time—this was evident in the direction time series as well. However, while southeasterly flow appeared the most common condition, it was not steady, rather, shifting continuously over long—approximately month-long—time scales.

Bathymetry of the area shows MOR003 located roughly southeast of what appears to be a localized rise in elevation around a seamount. This elevated bathymetric feature may cause flow steering—that is, currents running parallel to the local bathymetric contours. Near-bottom flow steering caused by local bathymetric features is a common observation in many oceanic regions.

Semi-diurnal tides were the dominant physical process at the seabed, followed by near-inertial oscillations and diurnal tides. These observations are consistent with other historical current measurements in the abyssal Pacific (Beaulieu & Baldwin, 1998). A majority of the tidal energy was found in the north-south component—very little tidal energy was found in the East velocity spectra (Fig. 11.20). The M2 lunar semi-diurnal tide dominates in the north-south orientation—the north-south tidal component (1.61 cm/sec) was orders of magnitude greater than the east-west tidal component (0.07 cm/sec). The diurnal constituents—O1 and K1—were weaker than the semi-diurnal tide.

The spectral peak in the near-inertial frequency band appeared to shift to ~0.025 cph, or about ~40-hour periods, a slightly higher frequency relative to near-inertial oscillations observed near the surface (0.021 cph, or about ~48 h).

To visualize the low-frequency motions with better clarity, these data were cleaned up by applying a 66-hour lowpass filter to the time series (Fig. 11.21). This filter will remove most of the higher-frequency motions such as tides and inertial currents. Higher-speed events exceeding 5 cm/sec were relatively frequent, occurring during times of south- or southeasterly flow. Stronger currents were observed during the early spring of 2018 and 2019 (late March-early April). Another higher-speed event occurred in October–November 2019. In total, roughly 8–10 of these events were observed, each event roughly a month or so in duration. A pattern for

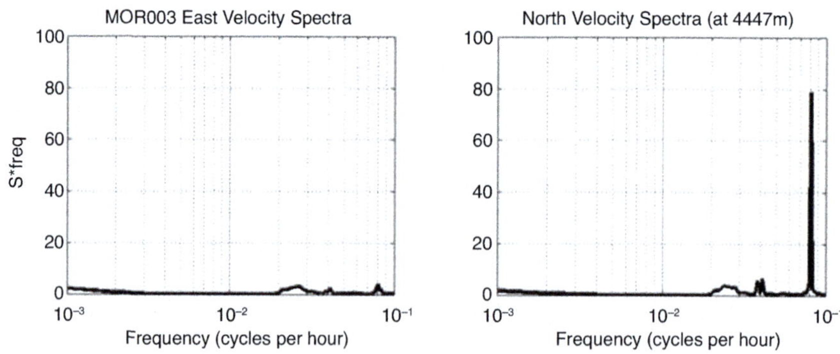

Fig. 11.20 Variance-preserving spectra 7 m above the seabed—a majority of the signal energy is represented by spikes in the spectra. At this depth, the large spike in the north velocity spectra at ~0.08 cycles per hour represents semi-diurnal tides (M2 constituent)

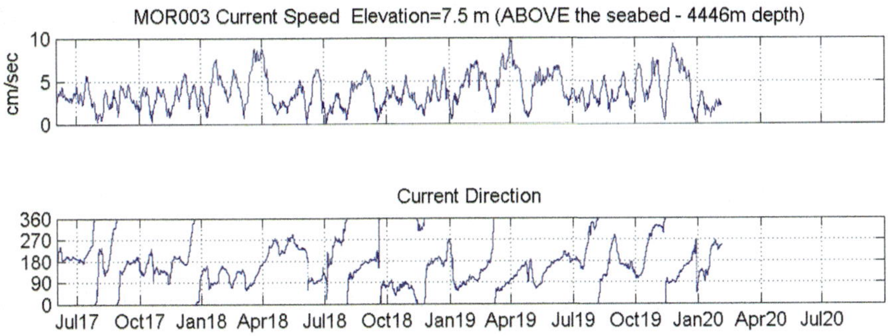

Fig. 11.21 Current speed and direction 7 m above the seabed (after cleaning up the signal with a 66-hour low-pass filter). There were numerous instances when current speeds exceeded 5 cm/sec. The strongest currents were observed in late March–April 2018, March–April 2019, and November 2019. These stronger currents flowed toward the south and east

each event emerged: currents running eastward initially, and then shifting south and then west (i.e., anti-cyclonic rotations) as speeds weakened. Most of the higher-speed events followed this clockwise rotational pattern over spans of about a month or so. The rotational patterns may be consistent with the passage of large-scale, gyre-like disturbances through the area (Aleynik et al., 2017; Chen et al., 2015).

4.3 *Near-Bottom Flow Conditions at MOR008*

MOR008 and MOR003 were located relatively close to one another, with MOR008 about 7.1 km to the north-northwest of MOR003. Figure 11.22 presents statistical profiles of these data, overlaid with the near-bottom velocity statistics from MOR003. The near-bottom flow was similar between the two sites: Mean flow direction was toward the south-southeast, about 145°. Velocities in the layer 8 m-to-18 m above the seabed, measured by the ADCP, were highly correlated. Mean velocities were 5.0–5.5 cm/sec. Peak speeds were about 18–19 cm/sec, although these higher speeds were quite rare. The 98% threshold was about 11 cm/sec, meaning currents exceeded this value only 2% of the time. About 90% of the time current speeds were less than 9 cm/sec.

At MOR008, currents measured by the Nortek Aquadopp at an elevation of 4.1 m were slightly weaker than those measured by the ADCP at the 8 m-to-18 m elevations. This was not consistent with other near-bottom velocity observations within B4, which showed a slight acceleration/strengthening of velocities very near the seabed. This same strengthening near the seabed was not observed at MOR008 (nor was near-bottom strengthening observed at the same location—MOR004—from June 2017 to February 2018, although those data were limited in vertical coverage—only 2 bins—7 m to 9 m above the seabed—were deemed valid). For this data set, the current variance at 4 m was roughly one-half the variance measured at 8 m

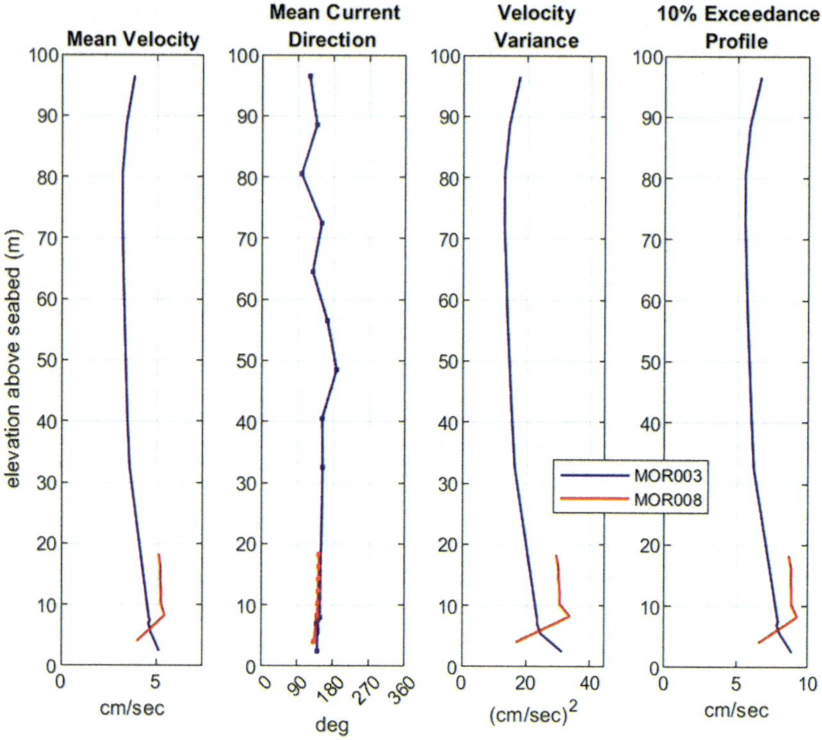

Fig. 11.22 Vertical profile of mean conditions at the B4 mooring locations. MOR003 is depicted by the blue lines; MOR008 by the red lines. Near-bottom currents at MOR008 appeared slightly stronger than those at MOR003. A slight strengthening of currents below about 40 m was observed at MOR003

and above (Fig. 11.22). Data quality from this instrument appeared sound—there was no indication of poor acoustic performance. However, this current meter was mounted next to the mooring's tandem acoustic releases, looking downward, and perhaps was affected by the presence of these large objects.

Variance-preserving spectra of MOR008 velocity at the 8 m elevation is presented in Fig. 11.23. A large amplitude spike in the north velocity spectrum was observed at the lunar semi-diurnal frequency band (0.8 cph, cycles per hour), with a much smaller bump at the diurnal frequency band (0.4 cph). The amplitude of this tidal energy was overwhelmingly manifest in the north-south component; the relative energy in the east-west direction was almost non-existent. This suggests tides oscillated primarily in the north-south direction, perhaps steered by southeast/northwest-running bathymetric ridges. Another bump in the spectrum was found at 0.024–0.025 cycles per hour (~40-hour periods)—these are near-inertial oscillations ubiquitous in these datasets. Other, much lower frequency energy was observed also at the far left of the spectra (~0.002 cycles per hour), or periods greater than ~20 days, representing the longer, slowly rotating shifts over month-long time periods.

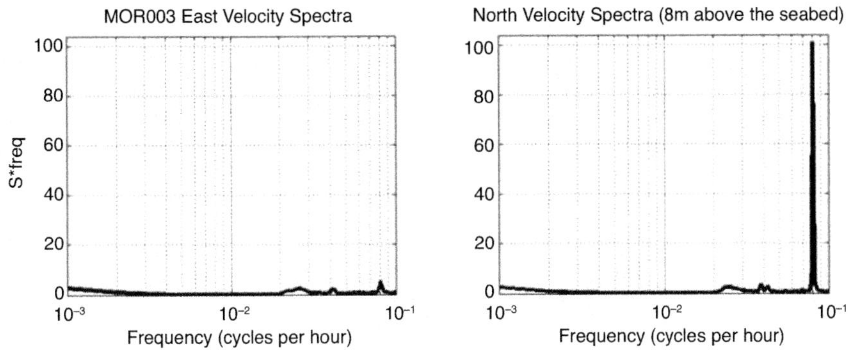

Fig. 11.23 Variance-preserving spectra for MOR008 about 8 m above the seabed

Tides were comprised generally of the major semi-diurnal (M2 and S2) and diurnal (O1 and K1) tidal constituents. The largest tidal velocity constituent was the M2—the principle lunar semi-diurnal tide with a period of 12.42 h. The M2 amplitude was 1.61 cm/sec, almost completely contained within the north velocity component. By contrast—the M2 east velocity amplitude was 0.10 cm/sec. The next largest constituent was the K1 (23.93 h)—the principle lunar diurnal tide—with an amplitude of 0.56 cm/sec. The solar tidal constituents S2 (12.00 h, 0.28 cm/sec) and O1 (25.82 h, 0.49 cm/sec) were smaller. These constituent amplitudes were nearly equivalent to amplitudes at MOR003.

Modulation of the various tidal constituents contribute to additional variability—diurnal and semi-diurnal modulations created a daily inequality where one tide was stronger than the other—as well as a fortnightly (~2 week) enveloping of spring and neap tides—spring tides featured stronger currents while the neap phases were weaker.

Near-inertial oscillations represented another physical process contributing significantly to overall observed variability and appeared roughly of the same amplitude in both the east and the north spectra. Near-inertial oscillations in the bottom boundary layer have been observed at other locations in the Pacific (Xie et al., 2022). These near-inertial oscillations were found at ~40-hour periods, different than the theoretical ~49-hour estimate for inertial periods at this latitude. An example of near-inertial velocities observed at MOR008 during the month of October 2017 is presented in Fig. 11.24. The near-inertial currents may also be in phase with tides, producing stronger-than-normal currents.

A comparison of near-bottom velocities at the two nearby locations—MOR003 and MOR008—is presented in Fig. 11.25. These curves represent the low-frequency velocity signals 8 m above the seabed, resulting from the application of a 30-day low-pass filter to reveal motions at time scales greater than about one month (removing tides and other high-frequency motions that would otherwise obfuscate the signal). Motions at these longer time scales represent circulation patterns at much larger length scales than the separation distance between the two sites, and therefore

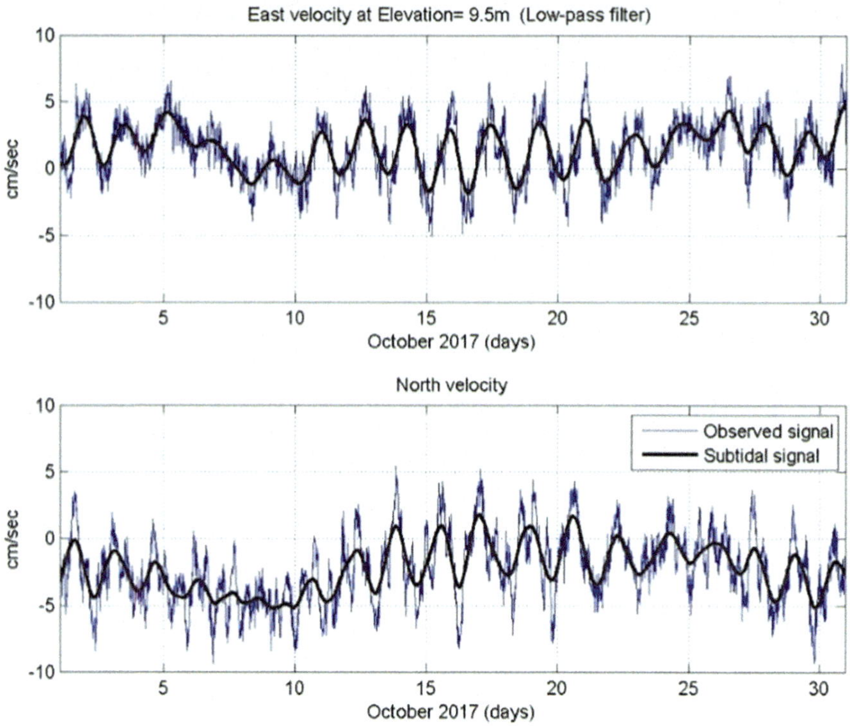

Fig. 11.24 Example of the ~40-hour velocity variability (near-inertial oscillations) observed in October 2017 at MOR008. The heavy black line represents the low-passed signal for signal clarity

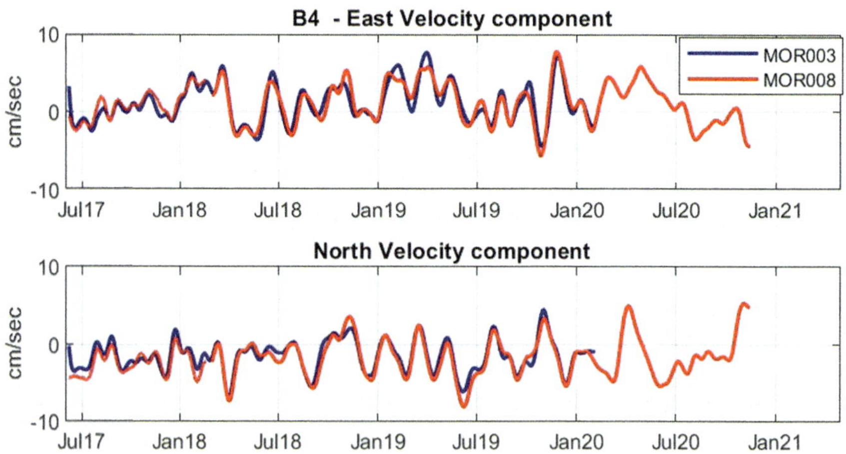

Fig. 11.25 Comparison of near-bottom velocity at MOR003 (blue) versus MOR008 (red)—after smoothing with 30-day low-pass filter. These low-frequency signals were highly correlated between locations suggesting circulation length scales much greater than 10 km

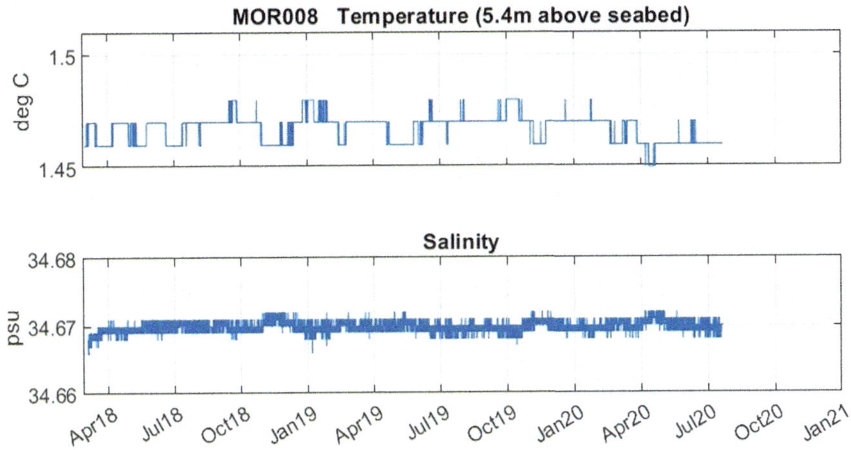

Fig. 11.26 Temperature and salinity time series from MOR008 at an elevation 5.4 m above the seabed

it is expected that velocities would be highly correlated between the two sites. These larger-scale, low-frequency motions had peak amplitudes of about 4–8 cm/sec and appeared to recur on 1-to-2-month time scales. A majority of the higher-speed events appeared to run toward the southeast or south, consistent with the direction of local bathymetric features. There appeared a delay, or phase, between the two signals that may depend on flow direction. Analysis of—and comparison between—observations at these two sites requires far more detailed analysis than is provided in this presentation.

The Nortek Aquadopp included Seabird temperature-conductivity sensors as well (Fig. 11.26). Temperatures near the seabed were steady at 1.46–1.48 °C and did not change during the 874-day time period. Salinity values were also steady, 34.67 PSU, and stable throughout the deployment, ranging less than 0.005 PSU over the 870-day record.

4.4 Near-Bottom Flow Conditions at MOR007

General flow conditions at MOR007 were consistent throughout the bottom-most 40 m of the water column. Flow directions appeared to be bidirectional—either southerly or toward the northwest (Fig. 11.27). This directional modality was a consistent feature in every data set at MOR007. The mean flow direction was toward the south (or south-southeast).

Figure 11.28 below displays the vertical profiles of east and north velocity components, speed ranges, current direction, and velocity variance averaged over the entire dataset timeframe at the MOR007 location. Mean velocities were about

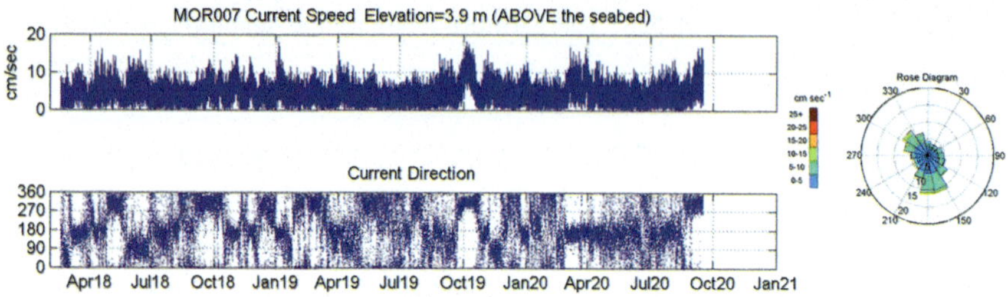

Fig. 11.27 Example of current speed and direction 3.9 m above the seabed at MOR007. Rose diagram is to the right. Flow directions varied between northwest flow and southerly flow—rarely to the east or southwest

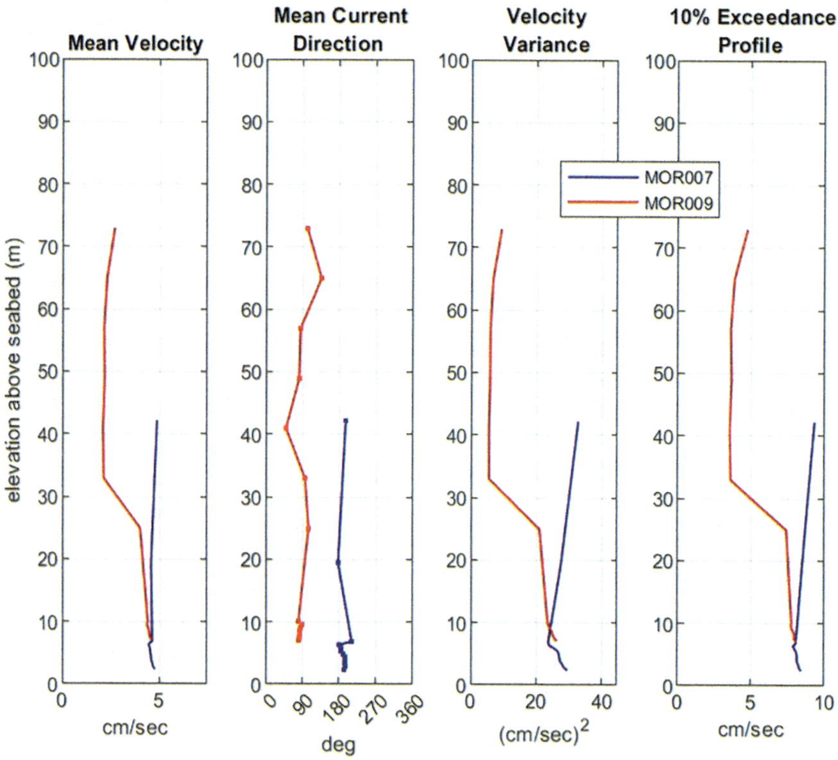

Fig. 11.28 Vertical profile of the near-bottom layer at MOR007 (blue) and MOR009 (red). Note the significant intensification of near-bottom currents at MOR009 from 25 m down to the seabed

4.5–5.0 cm/sec. Peak speeds observed near the bottom were 18–20 cm/sec. However a more useful statistic might be the exceedance value—for example, 90% of the time at this depth, current speeds were less than 8–9 cm/sec. Equivalently, current speeds exceeded 8–9 cm/sec only 10% of the time. Speeds exceeded 11 cm/sec about 2% of the time.

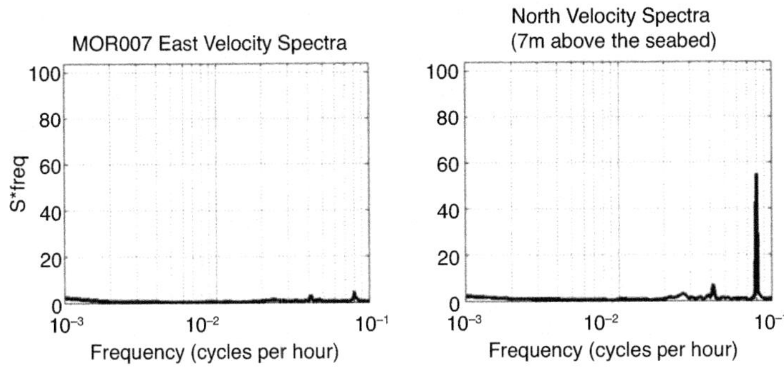

Fig. 11.29 Variance-preserving spectra 7 m above the seabed at MOR007

Current speeds at 42 m elevation were slightly greater than currents nearer the seabed (for example, those observed at 7 m). Mean speeds were 4.9 cm/sec versus 4.5 cm/sec, respectively, a small but real difference. All other statistical values—peak speeds, exceedance levels, and variance—were greater at the 42 m level relative to speeds nearer the seabed. In other locations there was an observation of flow acceleration very near the seabed—currents were greater in the near-bottom layer than those measured overhead—but that was not observed at this location—currents at 42 m were slightly stronger relative to currents near the bottom.

A majority of the current energy at MOR007 appeared at tidal frequencies—specifically, semi-diurnal tides (Fig. 11.29). Energy was present in the diurnal tidal bands as well. At all depths—tidal energy was significantly greater in the north-south direction than in the east-west direction. The M2 semi-diurnal tidal constituent was the largest—an amplitude of 1.35 cm/sec (north-south). The S2 semi-diurnal constituent was smaller, about 0.25 cm/sec. K1 diurnal constituent had an amplitude of 0.58 cm/sec, while the O1 constituent amplitude was 0.32 cm/sec. For all constituents, the north-south amplitudes were much larger than the east-west amplitudes.

Energy was also found at inertial frequencies—and also greater in the north component than in the east component. This north-south dominance of tidal and inertial motions may be due to local bathymetric features which may tend to steer flow in these preferred directions. Low-frequency energy—noted as the slight ramp of the spectral curve to the very left-hand side of the plots—also contributed to the overall variance totals. These low-frequency (also termed subtidal) processes—at frequencies less than about 0.002 cycles per hour (500 h)—revealed energy at time scales greater than ~20 days.

4.5 *Near-Bottom Flow Conditions at MOR009*

Mooring MOR009 was located approximately 9.6 km south-southeast of MOR007. A major bathymetric feature—a subsea range with an axis running northwest-southeast—lies to the northeast of the mooring. Numerous other seamounts are located near the two B6 moorings. The general flow condition was bidirectional—either toward the north-northwest or toward the southeast. This bidirectional tendency was similar to MOR007—although the directions were shifted slightly. The predominant flow directions agreed with the direction of the subsea bathymetric ridge lying to the northwest. Flow appeared to run parallel to this ridge in both directions. Overall, flow was toward the southeast generally roughly 46% of the time and toward the northwest about 29% of the time.

There appeared to be an apparent shear zone in the vertical profile separating the lower water column into two layers: a near-bottom layer and an overlying layer. There was a significant change in velocity between the two layers (red signals of Fig. 11.28). Mean velocities were of order ~4.5 cm/sec near the seabed (7 m to up 25 m) but slowed to ~2.1 cm/sec at elevations above about 33 m. Peak velocities in the near-bottom layer were ~ 20 cm/sec while peak speeds in the overlying layer were half—about ~10 cm/sec. 10% of the time currents in the near-bottom layer exceeded about 8 cm/sec; the 10% exceedance threshold in the overlying layer was only ~3.6 cm/sec. The variance profile also showed an abrupt transition between 25 m and 33 m—energy levels (variance) were 4× higher at 7 m than at 33 m.

The condition where flow accelerated in the very near-bottom layers was observed also during the first deployment from June 2017 to February 2018. This same tendency for near-bottom intensification was observed also at location MOR003, albeit less distinctly. The transition zone, or shear layer, between stronger near-bottom flow and slower overlying flow was also observed about ~30 m above the seabed at that time. This shear layer suggests enhanced potential for vertical mixing in such areas.

The spectra for the two layers—7 m and 65 m elevations—are shown in Fig. 11.30, and show the relative energy content at various frequencies for both the east and north velocity components within these two distinct layers. These graphics provide a couple of notable discussion points. First, the total energy at the 7 m elevation was much greater than the energy at the 65 m level. Second, at both elevations, the relative energy in the north component was much larger than the energy contained in the east component. Third, most of the variability (variance, or energy) appears due to semi-diurnal tides in the north component near the seabed (or, currents oscillating in the north-south direction at 7 m); tides in the east component were much weaker. The amplitude of the M2 tidal constituent was 1.74 cm/sec in the north component versus 0.38 cm/sec in the east component. Tides were much weaker at the 65 m elevation—an amplitude of 0.65 cm/sec for the north component versus 0.13 cm/sec in the east component—relative to 7 m. Near-inertial oscillations were also more energetic at the 7 m versus 65 m level. Low-frequency (subtidal) processes—variability at time scales greater than ~20 days—appeared slightly greater in the north component, with less energy in the east component. Subtidal energy at the 7 m elevation was more energetic than at 65 m.

The higher-speed, subtidal events observed at B6 (MOR007) were observed also at MOR009. Figure 11.31 compares the subtidal signals at both locations after applying a 30-day low-pass filter. Both east and north subtidal velocity signals correlate strongly between the two locations, although the two signals appear to be

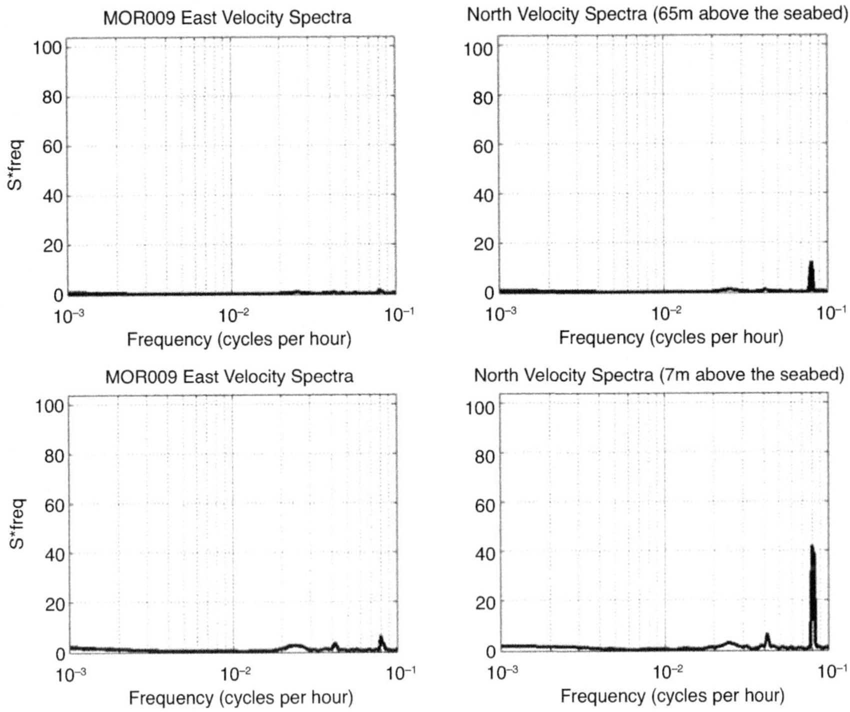

Fig. 11.30 Variance-preserving spectra 65 m (top) and 7 m (bottom) above the seabed at MOR009. Note the increase in energy (area under the spectral curves) at 7 m relative to 65 m

oriented in slightly different directions. Recall the MOR007 location between several prominent seamounts, which may tend to concentrate flow as it squeezed through the passage between these bathymetric features. Flows to the south at MOR007 appeared to turn toward the southeast at MOR009, parallel to the subsea ridge in that region. Similarly, flow direction in the opposite direction—toward the northwest or north-northwest at MOR009—may be steered to the west or west-northwest as it reaches MOR007. These differences in the subtidal flow directions were likely a result of steering and frictional effects of the bathymetric features near these mooring locations.

Numerous higher-speed events were observed—events where current speeds increased for an extended period of time (order ~weeks) before abating. These events were relatively frequent and featured peak current speeds approximately 5–10 cm/sec. The events appeared similar in behavior—flows toward the northwest for about a week while speeds increased steadily to a peak. This northwest flow would then weaken—to near zero—and turn abruptly toward the south while regaining strength. This southerly flow would persist for a week or so before weakening. This pattern—northwest flow strengthening then weakening, an abrupt turn to the south, another increase in speed followed by a decrease—was observed on several

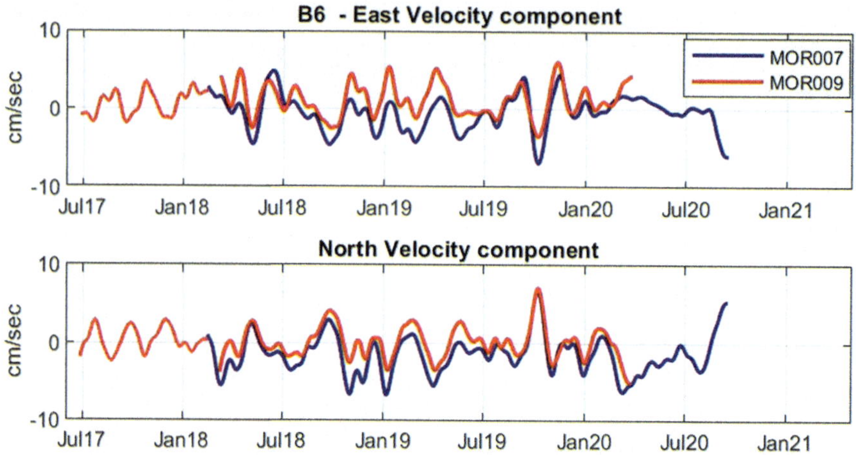

Fig. 11.31 Comparison of near-bottom velocities at the MOR007 (blue) and MOR009 (red) locations after applying a 30-day low-pass filter to the signals. East velocity components are shown in the upper panel—north velocity components in the lower panel. Correlations appear strong between the two sites

occasions: April to June 2018, December 2018 to January 2019, September to November 2019, March to April 2020, and late August through October 2020. Loosely interpreted, roughly six events were counted in the ~1150-day time series, on average one event every ~190 days. The events seemed to occur during the seasonal transitions—approximately April and October each year.

There appears evidence here of large-scale, westward-propagating eddies in all these near-bottom data sets. Previous studies indicate the propagation of large-scale eddies along the seafloor (Aleynik et al., 2017) and (Chen et al., 2015). A comparison of the low-frequency signals from B4 (Fig. 11.25) and B6 (Fig. 11.31) shows some similarities over much broader horizontal length scales. In one instance, a higher-speed event was observed at both B6 locations in October 2019—speeds increased to nearly 10 cm/sec toward the northwest. A similar increase in speed was observed at the B4 locations also, but delayed by several weeks. Both events, first observed in B6 and later at B4, had similar speed and duration attributes. Peak speed occurred in B6 09-October; the peak was observed in B4 about 02-November, roughly 24 days later. The approximate 24-day delay between B6 and B4 (~375 km distance) suggests an east-to-west propagation of a large-scale eddy at an average travel speed of 18 cm/sec. This propagation speed is comparable to the 14 cm/sec propagation speed estimates forwarded by Aleynik et al. (2017).

Temperature and conductivity measurements for the ~4-year deployment(s) are shown in Fig. 11.32. The mean temperature was 1.484 °C with a standard deviation of 0.00518 °C. There appeared a nonnegligible decrease in water temperature (~0.015 °C) in mid-October 2109 during the apparent passage of a large-scale eddy through B6. The mean salinity was 34.689 PSU, with a standard deviation of 0.00092 PSU. Salinity values increased slightly over the deployment, an increase

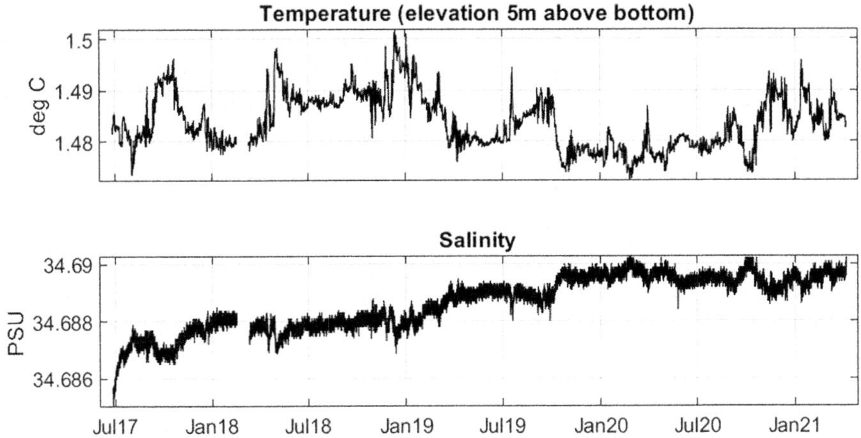

Fig. 11.32 Temperature and salinity time series from MOR009—almost 4 years—from two consecutive deployments. Temperature varied by less than 0.03 deg. C. Salinity varied by less than 0.002 PSU (practical salinity units)

suggesting sensor drift. There was no interim calibration of the conductivity cell during the February 2018 mooring turn-around, nor a post-calibration following the 2021 recovery, so we cannot provide an explanation for this signal drift over time. There was no sensor drift during this time in the MOR008 CTD data.

5 Conclusions

Environmental baseline moorings installed during this program recorded data successfully for more than 3 years at several different locations in the CCFZ, providing an outstanding and unique data set. Data coverage included the upper 1000 m of the water column down to the lower 100 m of the near-bottom boundary layer, along with other velocity data obtained at discreet depths in the mid-water column. Data return from these moorings suggested motions within these deep-sea areas were highly dynamic at a variety of time and length scales.

Currents were strongest near the surface and weakest within the mid-water column. Mean currents were ~ 18–22 cm/sec near the surface and flowed, on average, toward the west. A strong density layer about 75–100 m below the surface, the top of the pycnocline, appeared to act as an effective barrier separating the near-surface layer, which was well-mixed and responded to direct atmospheric conditions, from deeper layers, which had different flow properties. Currents within the pycnocline, below the surface layer at 204 m, were weaker (~12 cm/sec) and flowed eastward much of the time. The velocity variance at 204 m was about one-third the velocity variance found at 52 m. CTD casts at the location showed a strong gradient in temperature, salinity, and density in the upper layer.

Several major hurricanes passed the mooring location, generating the strongest currents observed in the record. Near-surface currents exceeded 165 cm/sec during Hurricane Barbara in 2019, but these peak currents were brief, decaying rapidly as the storm passed, leaving only oscillatory currents propagating at near-inertial frequencies as aftereffects. Storm response at 204 m was weaker, with peak speeds less than 50 cm/sec, further evidence the pycnocline limits mixing between the surface and deeper waters.

Currents mid-depth of the water column—from about 1500 m down to 4000 m—were the weakest, with mean speeds of order ~3–4 cm/sec. Peak speeds in the mid-depth layer reached 12 cm/sec. Mean flow was northward at 1000 m, eastward at 2000 m, and southward from 2500 m down to 4000 m. Velocity variance at mid-depth was about 1/30th of the energy near the surface.

Near-bottom currents were found more energetic. Mean current speeds within the bottom boundary layer were of order ~5 cm/sec. Local bathymetry features likely play a role in steering near-bottom flows at each site. Flow patterns in the very near-bottom layer appeared to run parallel to bathymetric ridges in B4, and around and through seamount features in area B6.

Some flow intensification was observed very near the seabed, in the lower ~20–30 m boundary layer, at several sites. An explicit example was at site MOR009, where the mean near-bottom currents at 7 m were twice the magnitude (4.6 cm/sec) of currents measured at 40 m (2.1 cm/sec). Peak speeds at 7 m were 21 cm/sec while peak speeds at 41 m were 9 cm/sec. This near-bed intensification suggests a velocity shear layer, which may enhance vertical mixing.

Currents in the upper layer were dominated by tides and near-inertial oscillations, along with significant low-frequency(seasonal) variability. Both semi-diurnal and diurnal tides were strong in the near-surface layer, but (lunar) semi-diurnal tides appeared the most significant constituent through the water column, especially at the seabed. Near-bottom M2 tides were oriented primarily in the north-south direction at B4, with little energy found in the east-west velocity component. Near-inertial oscillations were ubiquitous through the water column, and especially prominent at the seabed. Near-inertial currents with magnitudes of ~5 cm/sec were observed at MOR008.

Low-frequency circulation events were observed often near the seabed—some events featured peak speeds of order ~15–20 cm/sec. The events had rotational features—both cyclonic and anti-cyclonic patterns—and varied on ~monthly time scales. Comparison of low-frequency currents between the B4 moorings showed high correlation, suggesting circulation patterns of length scales much greater than the separation distance between the two sites (~7 km). The same held true in the B6 block—low-frequency signals were strongly correlated over the ~9 km separation distance. These low-frequency events possess properties of large-scale, westward-propagating eddies, with similar signals observed in both areas. One example—observed initially at B6 and later at B4—estimated an eddy traveled the 375 km distance between sites in about 24 days, suggesting a translational speed of ~18 cm/sec, which was consistent with near-bottom eddy speeds described in other studies.

Water properties measured in near-bottom boundary layer—temperature and salinity—were extremely stable. Mean water temperatures at MOR009 were 1.484 °C and varied less than 0.03 °C over the nearly 4-year time series. The mean salinity at this same site was 34.689 PSU, with a range of less than 0.005 PSU in that time.

Sustainable deep-sea mining operations depend on the ability of the marine environment to absorb and attenuate the impacts of nodule mining. The data presented herein provide detailed information describing conditions in this remote and largely unknown environment and hopefully prove helpful to operators, regulators, and the broader community as a whole to understand physical processes in these areas. These data sets serve as a basic building block of knowledge on which to develop further analyses and studies to help guide DSM decision-making going forward.

Acknowledgments The author would like to thank Global Sea Mineral Resources NV for their leadership and transparency throughout this program and commitment to sharing these high-quality environmental data sets with the broader DSM and scientific community. Kris van Nijen, Francois Charlet, Kris DeBruyne, Tom DeWachter, Celine Taymans, and many other individuals understood the importance of baseline monitoring and supported those efforts in full. Rick Cole, RDSea International, was instrumental in preparing and installing the moorings in 2017 and provided a wealth of constructive input which improved the program results.

References

Aleynik, D., Inall, M. E., Dale, A., & Vink, A. (2017). *Impact of remotely generated eddies on plume dispersion at abyssal mining sites in the Pacific Ocean*. Scientific Reports, December 2017. www.nature.com/scientificreports

Beaulieu, S., & Baldwin, R. (1998). *Temporal variability in currents and the benthic boundary layer at an abyssal station off Central California* (pp. 587–615). Deep-Sea Research II 45. Pergamon Press. Elsevier Science B.V.

Chen, G., Wang, D., Dong, C., Tingting, Z., HuijieXue, Y. S., Chu, X., Qi, Y., & Chen, H. (2015). Observed deep energetic eddies byseamount wake. *Scientific Reports, 5*, 17416. https://doi.org/10.1038/srep17416. www.nature.com/scientificreports

Fiedler, P. C., & Talley, L. D. (2006). Hydrography of the eastern tropical Pacific: A review. In *Progress in oceanography* (Vol. 69, p. 143). Elsevier Ltd.

Pickard, G. L., & Emery, W. J. (1982). *Descriptive physical oceanography*. Pergamon Press.

Purkiani, K., Haeckel, M., Haalboom, S., Schmidt, K., Urban, P., Gazis, I.-Z., Stigter, H., Paul, A., Walter, M., & Vink, A. (2022). Impact of a long-lived anticyclonic mesoscale eddy on seawater anomalies in the northeastern tropical Pacific Ocean:A composite analysis from hydrographic measurements, sea level altimetry data, and reanalysis model products. *Ocean Sciences, 18*, 1163–1181. https://doi.org/10.5194/os-18-1163-2022

Tolmazin, D. (1985). *Elements of dynamic oceanography*. Published by Allen and Unwin, Inc.

Xie, X., Wang, Y., Liu, Z., Liu, X., Chen, D., Zhang, D., & Wang, J. (2022). Observation of near-inertial waves in the bottom boundary layer of an abyssal seamount. *(Pending) Journal of Physical Oceanography, 53*, 635. https://doi.org/10.1175/JPO-D-22-0026.1

Jon Wood is the president/founder of Ocean Data Technologies, Inc. He has designed, installed, and successfully recovered hundreds of oceanographic mooring systems throughout the world's ocean regions—in waters from 3 feet to 3 miles deep. He holds a Bachelor of Science degree in mechanical engineering and a Master of Science degree in ocean engineering, both from the University of New Hampshire.

Chapter 12
Biogeochemistry of the South Indian Ocean—Water Masses, Nutrient Distribution, and Sinking Particulate Matter

Natalie C. Harms, Niko Lahajnar, Birgit Gaye, Tim Rixen, and Ralf Freitag

Abstract The South Indian Ocean, one of the least explored ocean regions, is dominated by the Indian Ocean subtropical gyre (IOSG), one of the five extensive oligotrophic areas in the world's oceans. Here, we present results from a comprehensive study on physical water column data (e.g., temperature, salinity and sigma-theta), nutrient and stable isotope data, and long-time sinking particulate matter collection retrieved between 2014 and 2019 by sediment trap moorings in the INDEX (Indian Ocean Exploration) area. Our research contributes to the main understanding of key processes of the N and C cycle and provides the first multidisciplinary approach of water mass interfingering, N sources and transformation processes and the quantity, quality and variability of sediment trap-based sinking particulate matter fluxes. With this knowledge, future impacts from deep-sea mining activities on persisting biogeochemical cycles in the South Indian Ocean will be better understood and evaluated. Numerous water samples along 3°S to 28°S allow detailed water mass analyses accompanied by a comprehensive study of nitrogen sources and cycling processes. We demonstrate the convergence and mixing of water masses of Antarctic and Subantarctic origin with water masses from the North Indian Ocean and track

Note: This chapter is based on the PhD thesis of Natalie C. Harms, titled *The South Indian Ocean – Biogeochemical Studies on Water Masses, Nutrient and Stable Isotope Distribution, and Particulate Matter in an Oligotrophic Ocean Region*, and was submitted to the University of Hamburg in 2020 (see reference in the reference list).

N. C. Harms (✉) · N. Lahajnar · B. Gaye
Institute for Geology, Universität Hamburg, Hamburg, Germany
e-mail: natalie.harms@uni-hamburg.de

T. Rixen
Leibniz-Centre for Tropical Research, Bremen, Germany

R. Freitag
Federal Institute for Geosciences and Natural Resources (BGR), Hannover, Germany

R. Sharma (ed.), *Deep-Sea Mining and the Water Column*,
https://doi.org/10.1007/978-3-031-59060-3_12

the changes in nutrient distribution and signatures of stable isotopes (N and O of nitrate) along these diverse water masses. This provides evidence for the injection of preformed nutrients from the Subantarctic and the Arabian Sea. Furthermore, dinitrogen (N_2) fixation evidently accounts for a significant proportion of primary production, and we calculate that ~32–34% of the nitrified nitrate is contributed by N_2 fixation in surface waters of the IOSG at ~20–24°S. Nevertheless, the IOSG is a strongly nutrient-limited ocean realm and is characterised by low primary production rates and hence very low sinking particulate matter fluxes. Sinking particulate matter considered here was collected during the 5-year sediment trap deployment experiment between 2014 and 2019 to provide new information on the major factors that control the particle export out of the biologically active zone, its transfer into the ocean's interior and its potential storage in pelagic sediments of the South Indian Ocean. Comparing particulate organic carbon (POC) fluxes to global data, the IOSG reveals the lowest fluxes worldwide (0.52 ± 0.18 mg m^{-2} d^{-1} at 500 m.a.b.), even compared to other oligotrophic areas. Preliminary estimates indicate an average POC export efficiency of $\varepsilon \approx 0.02$. Based on net primary production (NPP) data in surface waters of the IOSG only 0.17% of the POC produced at the sea surface reaches the sea floor. Due to intense degradation at the sediment–water interface, less than 0.01% (~0.02 mg POC m^{-2} d^{-1}) accumulates in surface sediments. Assuming that the IOSG, as well as comparable ocean regions, will expand under climate warming, it is of major importance to investigate POC export fluxes and its final carbon storage in the sediments in order to study the organic carbon pump and potential changes in the global C cycle. Furthermore, we present the seasonal variability of particle fluxes in the IOSG that shows a decoupling between NPP and the amount of exported particles. In the winter season, the external nutrient supply via gyre boundaries and/or subsurface waters elevate NPP and favour a complex zooplankton-dominated food web. In a first step, this enhances the organic matter production but, on the other hand, reduces the export of particles by increasing their fragmentation into free/unprotected particles, vulnerable to remineralisation/degradation. In comparison, during the low productive summer season, a foraminifera-dominated zooplankton community increases the particle export efficiency by ballast-minerals (here $CaCO_3$). Consequently, the final export of particulate matter is anti-correlated to NPP rates. Various factors, such as variations in the ocean mixed layer depth, impacts of physical mixing (surface wind stress, cyclonic eddies), changes in the plankton community structure and concomitant changes, impact the vertical transfer of particles to the sea floor. Understanding these complex mechanisms is challenging, but they have to be studied in detail to make robust statements about global climate predictions.

Keywords Indian Ocean subtropical gyre · Oligotrophic · Water masses · Nutrient-limited · N_2 fixation · Sinking particulate matter · Seasonal variability

Abbreviations

INDEX	Indian Ocean Exploration
IOSG	Indian Ocean subtropical gyre
NPP	Net primary production
OMLD	Ocean mixed layer depth
POC	Particulate organic matter

1 Introduction

1.1 Regional and Biogeochemical Setting of the Indian Ocean Subtropical Gyre

The INDEX license area is located in the southern Indian Ocean in the vicinity of the Rodrigues Triple Junction. There, the African, Antarctic and Indo-Australian tectonic plates meet, which leads, among other things, to hydrothermal fields and associated seafloor massive sulphide deposits. In contrast to other marine regions, this study area is still relatively little explored and is not only very complex because of the geological settings on the seafloor, but also very challenging scientifically. The southern Indian Ocean hosts one of the major vast oligotrophic areas in the world's oceans. In contrast to the Atlantic and Pacific Oceans, the Indian Ocean is land-locked to the north and thus developed only one subtropical gyre south of the equator—the Indian Ocean subtropical gyre (IOSG). The IOSG is an anticyclonic and counter-clockwise gyre that ranges approximately from 10° to 35°S (Baer et al., 2019; Sarmiento & Gruber, 2006; Williams & Follows, 2003). Branches of several current systems drive the circulation of the IOSG (Fig. 12.1). The South Equatorial Current (SEC) separates the IOSG from the southern equatorial Indian Ocean (Duing, 1970; Pickard & Emery, 1982; Woodberry et al., 1989) and is a broad westward current (8–20°S; Wyrtki, 1971; Godfrey & Golding, 1981; Pickard & Emery, 1982) with a maximum transport at 12–15°S (Woodbery, 1989). Additionally, the SEC is in part fed by a water mass with lower salinity signatures from the Pacific Ocean at about 10° to 25°S (e.g., Wyrtki, 1971; Woodbery, 1989; Godfrey & Golding, 1981; You & Tomczak, 1993). The South Indian Ocean Current (SIOC) marks the southern boundary of the IOSG and is located at the Subtropical Front (STF) at about 40°S (Deacon, 1933; Stramma, 1991). Together with the Agulhas Current (AC), the South East Madagascar Current (SEMC), and the West Australian Current (WAC), the SEC and the SIOC limit the IOSG and restrict its counter-clockwise rotation.

The gyre circulation results in a centre-directed (convergent) horizontal Ekman volume flux and leads to an intense downwelling of water to the gyre centre.

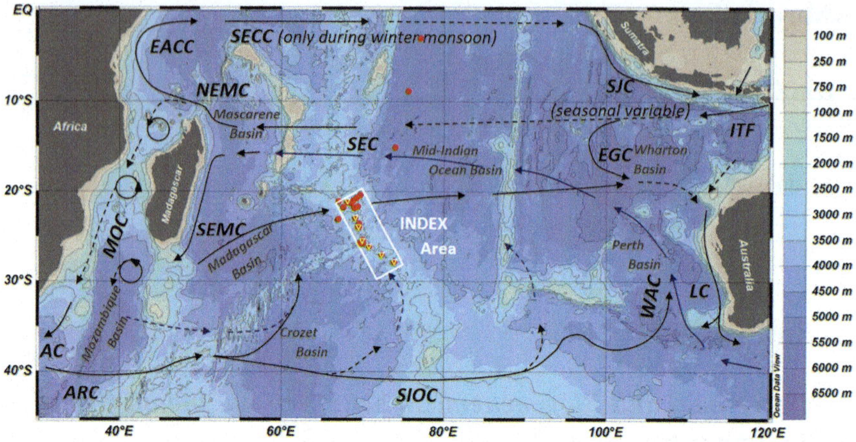

Fig. 12.1 Map of the South Indian Ocean with schematic representation of deep-sea basins and surface currents according to Stramma (1992, 1997), Schott and McCreary (2001), and Schott et al. (2009). Red dots indicate water sampling stations used for water mass analyses and yellow triangles mark the positions of sediment trap moorings deployed during INDEX expeditions in 2014–2019. The white rectangle confines the INDEX license area. Current branches are the Agulhas Current (AC), Agulhas Return Current (ARG), East African Coastal Current (EACC), East Gyral Current (EGC), Leeuwin Current (LC), Mozambique Current (MOC), Northeast Madagascar Current (NEMC), South Equatorial Current (SEC), South Equatorial Countercurrent (SECC), Southeast Madagascar Current (SEMS), South Indian Ocean Current (SIOC), South Java Current (SJC), West Australian Current (WAC), ITF indicates the Indonesian Throughflow

Consequences of the gyre-specific circulation are the deepening of thermo-, pycnocline and nutriclines, which shoal upwards the gyre margins (McClain et al., 2004; Fig. 12.2a, d). As in all subtropical gyres, this leads to strongly nutrient-depleted (oligotrophic) surface waters, where regeneration of organic matter and associated nutrients is efficient, while vertical mixing and the resulting influx of nutrients from deeper waters is reduced or absent (Ito & Follows, 2005; Knox & McElroy, 1984; Rixen et al., 2019a; Sarmiento & Toggweiler, 1984). Thus, the IOSG belongs to those extensive oligotrophic regions in the world's oceans, where the essential macronutrients, such as reactive N and phosphorus (P) are limited.

Primary producers, that is, the free-floating unicellular phytoplanktonic organisms that are responsible for photosynthesis in the global ocean, require and take up these elements (together with other macro- and micronutrients, e.g., carbon and iron) from the seawater in order to form organic matter (Sarmiento & Gruber, 2006) and couple N and C cycles in the ocean. Consequently, limited primary production in the biologically active zone (euphotic zone) is the result and leads to low sinking particulate matter fluxes. However, due to the large extent of subtropical gyres that account for ~75% of the global ocean's surface, they are responsible for half of the global organic carbon export to the ocean interior (Emerson et al., 1997). Thus, the particulate organic carbon (POC) export out of the euphotic zone, across the thermocline into the deep ocean, and the burial into the sediments constitutes the

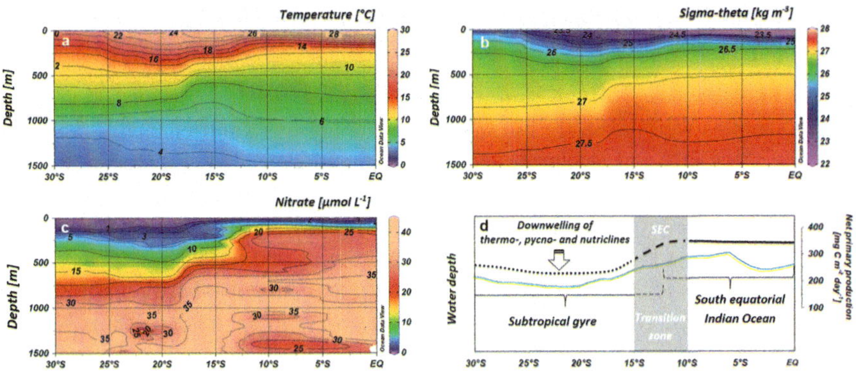

Fig. 12.2 Latitudinal cross sections from 30°S towards the equator along sampling stations (Fig. 12.1) for temperature (**a**), potential density (sigma-theta in kg m^{-3}) (**b**), and nitrate concentrations (**c**). Panel d presents the deepening of isoclines in the IOSG and their upward lift towards the southern equatorial Indian Ocean accompanied by net primary production (NNP) data in mg C m^{-2} day^{-1}. Water column and nutrient data from 2016 to 2017 are available at the PANGAEA DataBase at https://doi.pangaea.de/10.1594/PANGAEA.897503 and https://doi.pangaea.de/10.1594/PANGAEA.897504 (Harms et al., 2019). NPP data were extracted from the Ocean Productivity site (http://sites.science.oregonstate.edu/ocean.productivity/index.php) and based on the standard VGPM algorithm (Behrenfeld & Falkowski, 1997). It used data of MODIS chlorophyll-a concentrations, sea surface temperature (SST) and SeaWiFS PAR and estimates of euphotic zone depth from a model developed by Morel and Berthon (1989). Data were extracted from global 8-day averaged grids (original resolution: 1080× 2160) from 2016–2019 averaged over 65–75°E with a resolution of 1°

biological organic carbon pump (BCP) that is an essential part of the global C cycle (Volk & Hoffert, 1985). An effective BCP lowers the partial pressure of CO_2 in surface waters and thus enhances the ocean's ability to absorb CO_2 from the atmosphere (Berger et al., 1989; Lutz et al., 2007; Wefer & Fischer, 1991). Consequently, the BCP controls the partitioning of carbon between the large reservoir in the deep ocean and the relatively small atmospheric reservoir (Berger et al., 1989).

Future global warming is assumed to expand these low-productivity areas globally and to strengthen stratification in low-latitude oceans, which will lead to a further decrease of the net primary production and to corresponding changes in plankton community structure (Behrenfeld et al., 2006; Signorini et al., 2015). Our investigations on water masses, nutrient distributions and stable isotopes of nitrate in the IOSG are a baseline for this understudied ocean region and for subtropical gyres in general. It is a starting point to track the progression of C and N cycles and their budgets in a changing climate in the coming years. Particularly within the INDEX programme, our research is especially important for assessing potential deep-sea mining activities and their impact on a very complex, but also fragile, ecosystem along tectonic plate boundaries. This requires understanding not only the conditions on the seafloor but also the processes in the overlying water column, such as productivity, particle transport and initial sedimentation patterns.

1.2 *Use of Nutrients and Nitrate Isotopes to Qualify Water Mass Processes*

Water samples were taken for biogeochemical analyses of nutrients and stable isotopes of nitrate (N and O) during INDEX 2015 and 2018. Sample preparation and the analytical methods for determining NO_3^- and PO_4^{3-} concentrations, as well as the stable isotope ratios relative to a reference standard ($\delta^{15}N$ and $\delta^{18}O$) are given in detail in Harms et al. (2019). Nutrient concentrations and nitrate isotope ratios are useful indicators for N cycle processes and help to distinguish between different N sources and sinks. The dominant source of reactive N in the open ocean is N_2 fixation, the process by which atmospheric N_2 gas is converted to nitrogen forms that are useable by phytoplankton. A major sink is heterotrophic denitrification, the process during which reactive nitrogen is released back into the atmosphere by converting nitrate (NO_3^-) into gaseous N (Deutsch et al., 2001). N_2 fixation by cyanobacteria, that is, diazotrophs such as *Trichodesmium*, is observed over much of the tropical and oligotrophic subtropical oceans (Capone et al., 1997; Emerson et al., 2001; Karl et al., 1995; Michaels et al., 1996) and compensates for the loss of reactive N during heterotrophic denitrification if the ocean's marine N cycle is in a steady state (Deutsch et al., 2001).

The inputs of N through N_2 fixation are detached from inputs of P, which have their major source on land, leading to a decoupling between the nitrate (NO_3^-) and phosphate (PO_4^{3-}) pools. Deviations from the Redfield stoichiometry (N:P ~ 16:1; Redfield, 1934, 1963) are used to study rates of both N_2 fixation and denitrification (Gruber & Sarmiento, 1997; Sigman et al., 2005). Based on this, the tracer N* was established (see Deutsch et al., 2001; Gruber & Sarmiento, 1997; Harms et al., 2019) and used as an indicator for excesses or deficits in NO_3^- relative to the global NO_3^-/PO_4^{3-} ratio. However, the use of N* has limitations. First, the deviation from the Redfield stoichiometry may not always be a result of N inputs or outputs (N_2 fixation and denitrification), but may also reflect variations in uptake and remineralisation processes (Sigman et al., 2005). Second, inputs and losses partially overprint each other, when they occur simultaneously in the same water body.

Stable isotopes of nitrate are used to overcome the weakness associated with the N* approach and distinguish between sources and sinks of fixed nitrogen to study transfer processes in the N cycle (e.g., N-assimilation, denitrification, nitrification, N_2 fixation), also when they occur simultaneously. Isotope ratios are reported per mil (‰) using the δ-notation (for the formulae see Harms et al., 2019). During consumption processes of nitrate (e.g., N-assimilation and denitrification), lighter isotopes are preferred, leaving the substrate enriched in ^{15}N and ^{18}O according to intrinsic isotope effects. Several culture experiments indicate that $\delta^{15}N$ and $\delta^{18}O$ of the residual nitrate pool rise equally as consumption proceeds. Consequently, the O-to-N isotope effect ($^{18}\varepsilon$:$^{15}\varepsilon$) is close to 1 (Sigman et al., 2003, 2005; Granger et al.,

2004; Rafter et al. 2013). Nitrate consumption processes thus shift N and O isotope compositions in a nitrate pool in parallel, but nitrate production processes (N_2 fixation and nitrification) have very different effects and decouple N and O isotope compositions (Rafter et al., 2013; Sigman et al., 2005). In oxygenated environments, almost all of the ammonium generated from organic N is oxidised to nitrate. Thus, the isotope effect is negligible and the δ^{15}N-NO^{3-} depends largely on the biomass being remineralised (Rafter et al., 2013; Sigman et al., 2005). In contrast, the δ^{18}O of newly nitrified nitrate is independent of the isotopic composition of the mineralised organic matter. Significant addition to a nitrate pool by nitrification is marked by deviating values of δ^{15}N and δ^{18}O. The tracer Δ(15–18) is calculated by the deviation of δ^{15}N and δ^{18}O. It keeps track of the decoupling of N and O isotopes in the nitrate pool (Rafter et al., 2013) and thus quantifies nitrate assimilation and regeneration processes in marine environments (Casciotti et al., 2002, 2010; Sigman et al., 2005, 2009; Wankel et al., 2007).

Using these indices, we are able to maintain our water mass analyses that base mainly on physical properties (potential density, temperature and salinity). We illustrate divergent N cycle processes in various water masses and track their transport path and the original source of the water body, carrying specific nutrient-signatures into the South Indian Ocean.

2 Use of Sediment Trap Moorings to Qualify the Organic Carbon Pump

Particulate matter fluxes to the deep ocean are directly liked to surface water processes (Betzer et al., 1984; Cram et al., 2018; Eppley & Peterson, 1979; Lampitt & Antia, 1997; Suess, 1980) and particulate matter and POC fluxes collected by sediment traps are often suggested to be positively correlated to the rate of NPP (Betzer et al., 1984). Previous investigations in oligotrophic regions, however, showed contradicting findings. In the North Atlantic (e.g., Sargasso Sea at BATS stations,) positive correlations between particle flux and NPP have been found (Deuser et al., 1981; Deuser & Ross, 1980; Pabortsava et al., 2017), while in other oligotrophic regions, including the SIO gyre, no direct link or an even negative correlation have been observed (Ducklow et al., 2001; Harms et al., 2021). Thus, the export and transfer efficiency of organic matter via sinking particles through the thermocline and further into the deep ocean depends on numerous parameters, influencing the particle formation and its properties such as particle size and density, the availability of ballast material acting on particle aggregation (ballast effect), remineralisation and degradation and the plankton community structure. These effects on the export of organic matter to the deep-sea are being studied in the IOSG by using sediment trap moorings that are a reliable tool to monitor the quantity, quality and temporal variability of sinking particles (e.g., Buesseler et al., 2000; Honjo et al., 1995; Rixen et al., 2019a; Wong et al., 1998).

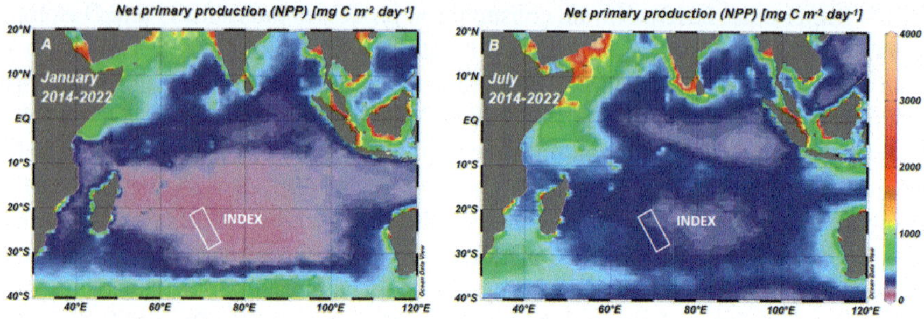

Fig. 12.3 Distribution of the net primary production (NNP in mg C m^{-2} day^{-1}) based on the standard VGPM algorithm (http://sites.science.oregonstate.edu/ocean.productivity/index.php; Behrenfeld & Falkowski, 1997) using data of MODIS chlorophyll-*a* concentrations, sea surface temperature (SST), SeaWiFS PAR, and estimates of euphotic zone depth from a model developed by Morel and Berthon (1989). Data were extracted from global monthly grids (original resolution: 1080× 2160) for January (**a**) and July (**b**; time-averaged for 2014–2022). The white rectangle marks the INDEX license area

Sediment trap moorings have been deployed for 1-year periods at seven sites in the clusters of the INDEX license area since 2014. A low annual NPP rate and minor seasonal variation characterise this region (Figs. 12.1 and 12.3). Sediment traps were moored at ~500–600 m above the bottom (m.a.b.) (2600–3500 m water depth) to collect the undisturbed vertical ("true") particulate matter that sinks through the water column and avoid the influence of resuspended material from the sea floor. An exception was sediment trap moorings in cluster 04, where an additional trap was installed each above (*Shallow* at ~700–800 m water depth) and below (*Deep* at ~3500 m water depth, i.e., ~30 m.a.b.), the trap moored at ~600 m.a.b. (*Middle*). These additional traps provide further information on water column processes, such as the degradation state of organic matter in sinking particles.

Our sediment trap moorings consist of a number of different components. In addition to the used Kiel K/MT 234 and McLane Mark 7G-21 sediment traps, float modules, acoustic releases and an anchor were attached to the mooring system. The anchor weight fixes the mooring to the ground, while the buoyant floats keep it upright in the water column. The acoustic release disconnects the anchor from the remaining system during the recovery. The sediment trap consists of a funnel (aperture size 0.5 m^2; slope angle 34°), which intercepts the vertical particle flux. A hexagonal "honeycomb" lattice covers the funnel. This serves as a current stabilizer, reduces washout and prevents large swimmers from entering the funnel. The control unit is equipped with a microprocessor for programming the rotation steps and serves as the power supply for the stepping motor. The rotation platform is placed under the funnel's orifice and is loaded with 20 HDPE (high-density polyethylene) bottles (250 ml for McLane Mark 7G-21 trap and 400 ml for Kiel K/MT 234 trap).

2.1 *Physical Properties and Water Mass Analysis of the South Indian Ocean*

Water masses in the South Indian Ocean are diverse and have specific physical and biogeochemical properties (e.g., potential density, temperature, salinity, oxygen concentration). Using Temperature-Salinity (T-S) diagrams, accompanied by an overlay of potential density surfaces (sigma-theta in kg m^{-3}), we are able to identify water masses in a region of complex water mass interfingering. The study area can be subdivided into four latitudinal sections that represent different oceanic regimes: the southern limb of the IOSG (25.66–27.80°S), the central IOSG (20.36–25.42°S), the northern boundary of the IOSG (15.08°S) and the southern equatorial Indian Ocean (2.98–8.81°S; Fig. 12.4). For each latitudinal section, we plot individual diagrams for the variables temperature (Fig. 12.4a–d), salinity (Fig. 12.4e–h) and oxygen concentrations (Fig. 12.4i–l) relative to their sigma-theta surfaces, to identify

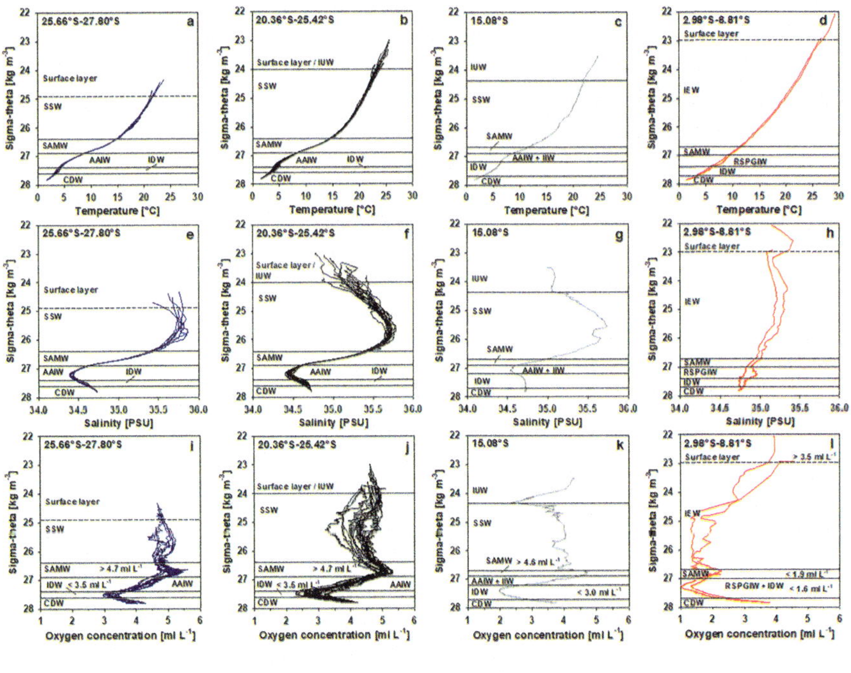

Fig. 12.4 Water mass properties from CTD measurements during INDEX expeditions between 2015 and 2019 presented as temperature vs. sigma-theta (a-d), salinity vs. sigma-theta (e-h), and oxygen vs. sigma-theta diagrams (i-l) for different oceanic regimes: the southern limb of the IOSG (25.66–27.80°S), the central IOSG (20.36–25.42°S), the northern boundary of the IOSG (15.08°S), and the southern equatorial Indian Ocean (2.98–8.81°S). The figure is modified from Harms et al. (2019). Water masses are the Indonesian Upper Water (IUW), Indian Equatorial Water (IEW), Subtropical Surface Water (SSW), Subantarctic Mode Water (SAMW), Red Sea-Persian Gulf Intermediate Water (RSPGIW), Indonesian Intermediate Water (IIW), Antarctic Intermediate Water (AAIW), Indian Deep Water (IDW) and Circumpolar Deep Water (CDW)

the occurrence of different water masses and their flow paths along the latitudinal transect. The T-S diagram in Fig. 12.5a summarises all water masses in the INDEX area. Overall, water masses in the South Indian Ocean can be subdivided into surface water masses (covered by a thin surface layer that is mainly affected by seasonal variations and weather conditions), thermocline water masses, intermediate water masses and deep water masses. Surface waters in subtropical regions of the southern hemisphere are characterised by strong evaporation rates that exceed precipitation (Schott & McCreary, 2001) and are often called Subtropical Surface Waters (SSW; see Muromtsev, 1959; Wyrtki, 1973; Schott & McCreary, 2001). The SSW of the IOSG is formed at latitudes between 25°S and 35°S (Baumgartner & Reichel, 1975) and is marked by the highest salinities (Fig. 12.4b, e–g). The SSW follows the typical gyral depression of pycnoclines and is detectable as far as ~15°S (Harms et al., 2019). Less saline water is found above the SSW at the transition from the central IOSG and its northern end and can be traced back to the Indonesian Upper Water (IUW) that is injected via the Indonesian Throughflow (ITF) from the Pacific Ocean into the gyre region (Emery, 2001). Moderate vertical changes in salinity are found in surface waters of the southern equatorial Indian Ocean (Fig. 12.4b–d, h), related to the Indian Equatorial Water (IEW). It is a mixture of subsurface water masses from the South Indian Ocean (SSW) with subsurface and thermocline waters from the Arabian Sea (Quadfasel & Schott, 1982; Schott & McCreary, 2001; Sharma, 1976; Warren, 1981; You & Tomczak, 1993).

Thermocline waters of the South Indian Ocean are ventilated south of the IOSG at 32°S by mode waters observed between surface waters (seasonally mixed and approximately isohaline) and intermediate waters (salinity minimum; McDonagh et al., 2005). These mode waters were named Subantarctic Mode Water (SAMW)

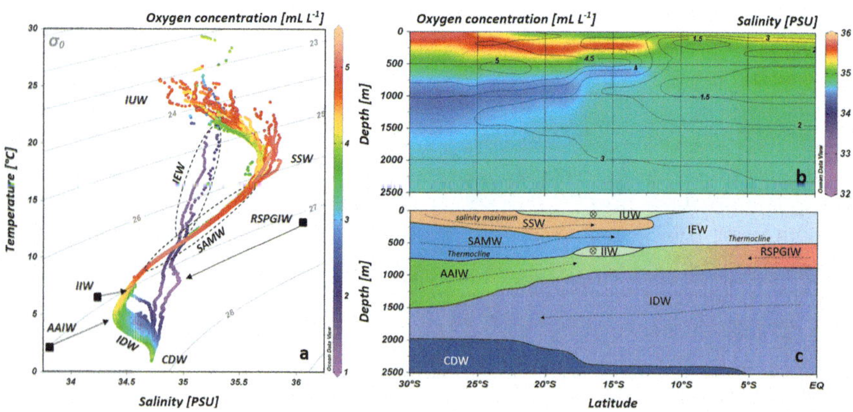

Fig. 12.5 T-S diagram of vertical CTD profiles (source properties of water masses based on Emery (2001)) accompanied by oxygen concentration in ml L^{-1} (colour coding) (**a**), salinity distribution as latitudinal section from south to north with an overlay of oxygen concentrations (contour lines) based on data collected during INDEX expeditions between 2015 and 2019 (**b**). Panel c presents an overview of the water mass distribution in the South Indian Ocean from 30°S towards the equator (reproduced from Harms et al., 2019); for water mass abbreviations see Fig. 12.7.

by McCartney (1977, 1982) and formed at latitudes between 40°S and 50°S. The SAMW injects oxygen-saturated waters into all subtropical gyres of the southern hemisphere (McCartney, 1977), as well as into the thermocline of the IOSG, where temperature and salinity vary monotonically with depth (Figs. 12.4a–c, e, f and 12.5b, c). On its transition to the north, oxygen concentrations rapidly decrease due to respiration and the absence of effective ventilation in the North Indian Ocean (Fig. 12.4i–k; Harms et al., 2019).

At intermediate water depths, a distinct salinity minimum (Fig. 12.4b, c, e, f) is associated with the Antarctic Intermediate Water (AAIW) mentioned by several authors (Bindoff & McDougall, 2000; Deacon, 1933; Fine, 1993; Schott & McCreary, 2001; Toole & Warren, 1993; Warren, 1981; Wyrtki, 1973; You, 1998). The AAIW originates in the surface waters of the Antarctic Ocean, then transported eastwards by the Antarctic Circumpolar Current (ACC), and penetrates into all three oceans. In the South Indian Ocean, the AAIW flows along a core density of 27.2 kg m^{-3} and extends towards the equator (Fine, 1993; McCartney, 1977; Piola & Gordon, 1989; Reid, 1986, 1989; Sverdrup et al., 1942; Talley, 1996; Wüst, 1935). At ~15°S a slightly divergent core density of the salinity minimum compared to the AAIW implies an additional inflow of less saline water. It is injected from the ITF into intermediate waters (Harms et al., 2019) and is called the Indonesian Intermediate Water (IIW; Emery & Meincke, 1986; Emery, 2001), which flows westwards between 10°S and 15°S (Schott & McCreary, 2001; Wyrtki, 1971; You & Tomczak, 1993).

Higher saline waters in thermocline waters of the southern equatorial Indian Ocean mark the inflow of saline water from the Arabian Sea, mainly from the Red Sea and Persian Gulf outflow. This water mass is mentioned as Red Sea-Persian Gulf Intermediate Water (RSPGIW; Warren, 1981; Emery & Meincke, 1986) and transports oxygen-depleted water from the oxygen minimum zone in the Arabian Sea towards the equator and beyond ~10°S (You, 1998). Further transport of this oxygen-depleted water into the central and southern IOSG is triggered by the Indian Deep Water (IDW; Fig. 12.4i, j). It originates in the northern basin of the Arabian Sea comprising the Circumpolar Deep Water (CDW) and the overlying RSPGIW (Talley, 2013) and flows southwards just below the density level of the AAIW. The CDW is a mixture of the North Atlantic Deep Water (NADW) and the Antarctic Bottom Water (AABW) and enters the Indian Ocean passing the Agulhas Basin, the Crozet Basin and the southern Australian Basin (Emery, 2001; Toole & Warren, 1993).

To conclude, data from INDEX expeditions from 2015 to 2018 across the INDEX area permit a detailed analysis of the water masses in the South Indian Ocean from ~3°S to 28°S. We illustrate the provenance of water masses of Antarctic and Subantarctic origin that converge and mix with water masses from the southern equatorial Indian Ocean and the Arabian Sea (Harms et al., 2019). In addition, Pacific waters from the ITF are injected into the surface and intermediate depths of the South Indian Ocean. A complete profile of the water mass distribution in the South Indian Ocean is shown in Fig. 12.5c, which is based on our previous work in Harms et al. (2019).

3 Nutrient Distribution and N Cycle Processes

Water masses from the Southern Ocean and from the North Indian Ocean import the characteristic nutrient signatures and isotope fingerprints of reactive N into the IOSG. Based on the water mass analyses, we are able to reconstruct the main factors that lead to the observed nutrient distribution and to the detected variation in the stable isotopes of nitrate. To display these specific characteristics related to the predominant water mass, Fig. 12.6 presents the distribution of NO_3^-, PO_4^- $\delta^{15}N$, $\delta^{18}O$, and their resulting parameters (NO_3^-/PO_4^{3-}, N*, Δ(15–18)) as latitudinal sections with an overlay of the water mass sketch (Fig. 12.5c). Nitrate and phosphate concentrations and in their ratio mirror the typical architecture of subtropical gyres expressed by sloping isoclines and illustrates the thick layer (~300 m) of nutrient-depleted surface waters in the range of the IUW and SSW (Harms et al., 2019; Fig. 12.6a, b). Nutrient concentrations remain low in the underlying SAMW, and nitrate (phosphate) concentrations of 20 μmol L^{-1} (1.5 μmol L^{-1}) are only reached within intermediate water masses. Consequently, the supply of nutrients to the euphotic layer by vertical mixing is reduced or absent in the gyre (Williams & Follows, 1998) and the external input of nutrients via lateral transfer across the gyre boundaries is a major nutrient source.

The isoclines shoal up towards the northern limp of the IOSGand nitrate and phosphate concentrations increase in the upper 500 m (Fig. 12.2). This shoaling marks the transition from subsurface waters of the IOSG to the IEW of the southern equatorial Indian Ocean and restricts the northward flow of the IUW, SSW and the SAMW (Fig. 12.6a, b). Surface waters in the gyre region have low NO_3^-/PO_4^{3-} ratios because nitrate commonly becomes depleted prior to phosphate in oligotrophic regions (Deutsch et al., 2007; Harms et al., 2019; Sarmiento & Gruber, 2006). The NO_3^-/PO_4^{3-}-ratios track the outcropping nutriclines with low ratios within the SSW towards high ratios at the northern gyre boundary and further north within the IEW (Fig. 12.6c). An average NO_3^-/PO_4^{3-}-ratio of 14.7 is calculated for the entire water column (Fig. 12.7a). This value is lower than the global ocean mean of ~16 (Redfield, 1934, 1963). As typical NO_3^-/PO_4^{3-}-ratios in the Arabian Sea are 12.81 (Codispoti et al., 2001), the influx of water masses from the Arabian Sea significantly affects the nutrient distribution in the IOSG, which is a mixed signal of water masses of South and North Indian Ocean origin. Thus, NO_3^-/PO_4^{3-}-ratios in the South Indian Ocean fall between the global ocean mean and the ratio in the Arabian Sea (Harms et al., 2019). A further indicator for changes in the nutrient ratio is N*, which marks NO_3^- deficits and excesses relative to phosphate. The Arabian Sea is characterised by an extensive oxygen deficit zone (ODZ), which induces denitrification in mid-water depths there (150–400 m; Gaye et al., 2013), causing negative N* values. This Arabian Sea N deficit is imported within the observed tongue of oxygen-depleted waters by the RSPGIW and IDW into the IOSG (Figs. 12.3i–l, 12.6d, and 12.7b).

As denitrification is biologically induced, it is associated with isotope fractionation and thus affects nitrogen isotope ratios. Because lighter isotopes are

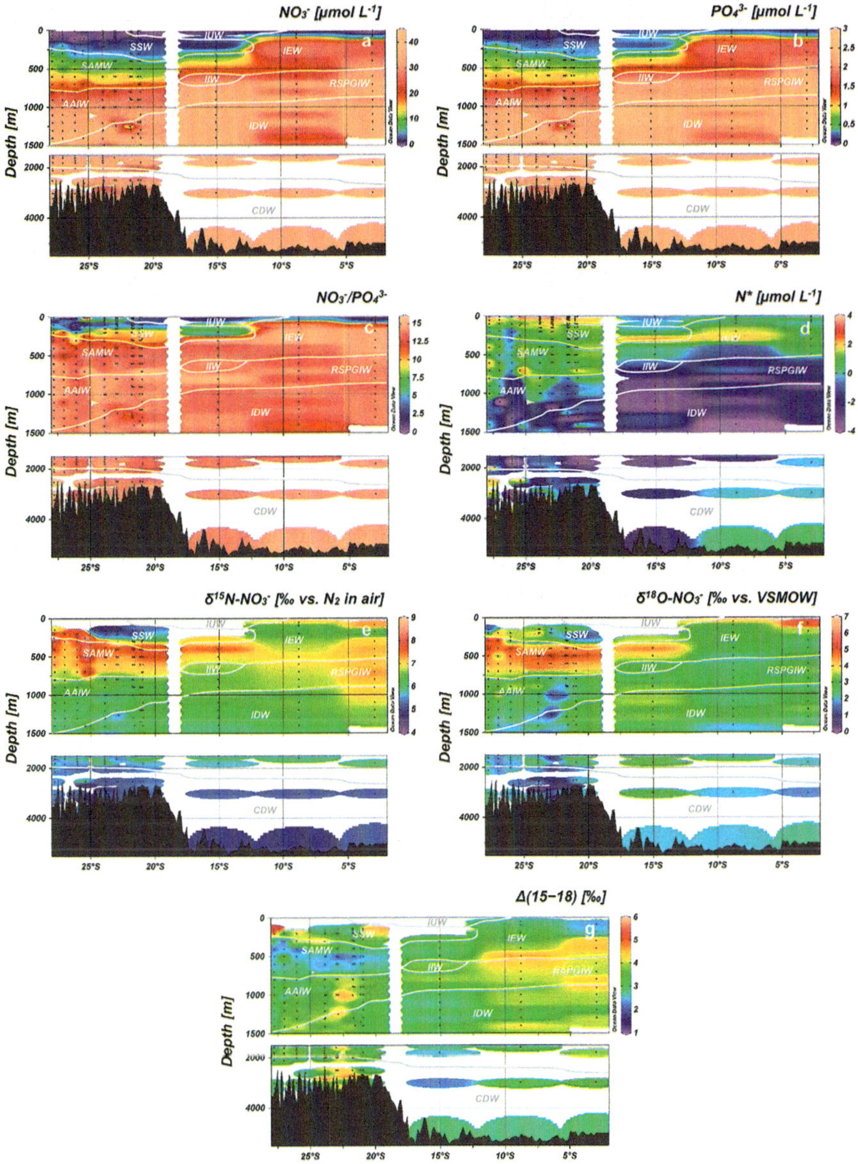

Fig. 12.6 Latitudinal profiles from south to north along the INDEX license area with an overlay of the water mass distribution (white lines; see Fig. 12.5c) in the South Indian Ocean for nitrate (**a**) phosphate (**b**), NO_3^-/PO_4^{3-} ratio (**c**), N* (**d**), stable isotopes $\delta^{15}N\text{-}NO_3^-$ (**e**) and $\delta^{18}O\text{-}NO_3^-$ (**f**), and Δ(15–18) (difference between $\delta^{15}N\text{-}NO_3^-$ and $\delta^{18}O\text{-}NO_3^-$) (**g**). Data are available at the PANGAEA DataBase at https://doi.pangaea.de/10.1594/PANGAEA.897503 and https://doi.pangaea.de/10.1594/PANGAEA.897504 (see Harms et al., 2019). For water mass abbreviations, see Fig. 12.4

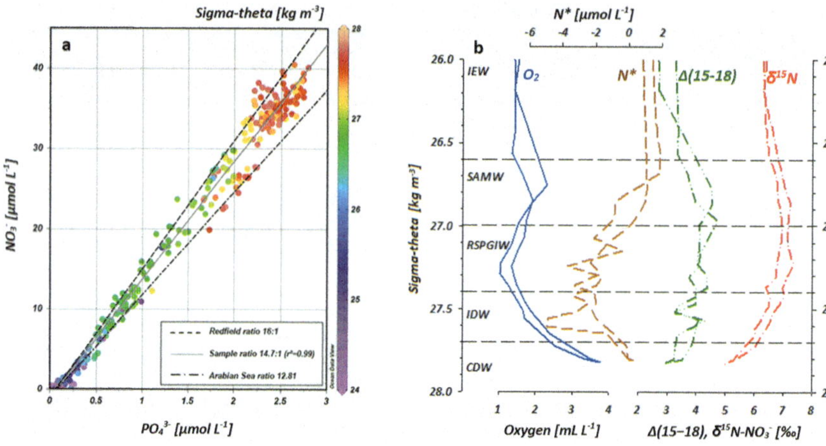

Fig. 12.7 Nitrate vs. phosphate concentrations (**a**); the regression line of the sample data (solid, grey line) indicates a ratio of ~14.7 ($r^2 = 0.99$), which is intermediate between the Redfield ratio (16:1; black, dashes line) and the mean ratio in the Arabian Sea with a slope of ~12.8 (grey, dotted-dashed line from Codispoti et al., 2001); colour-coding of dots indicates the potential density (sigma-theta; colour bar). Panel **b** shows sigma-theta profiles of oxygen concentration, N*, $\delta^{15}N$-NO_3^- and Δ(15–18), which depict the denitrification signal from the Arabian Sea in the water column profiles of the south equatorial Indian Ocean (stations at ~3°S and ~ 9°S; see Fig. 12.1). Water mass boundaries are given for the IEW, SAMW, RSPGIW, IDW and CDW (for water mass abbreviations, see Fig. 12.4)

preferentially removed during denitrification, the residual nitrate is enriched in ^{15}N (higher $\delta^{15}N$-NO_3^-) within the RSPGIW (Fig. 12.6e). This coincides with reduced N* and low oxygen concentrations (Figs. 12.5b and 12.6d). The decoupling of N and O isotopes in the nitrate pool can be highlighted by the indicator Δ(15–18). High values in the RSPGIW and maximum values in the SAMW indicate the additional input of nitrate via remineralisation (Fig. 12.6e–g). The nitrate added from the Arabian Sea is progressively replenished by ongoing organic matter respiration that brings the $\delta^{18}O$-NO_3^- value close to the oceanic average of ~2‰, while retaining the original elevated $\delta^{15}N$-NO_3^- values (Harms et al., 2019) and thus results in elevated Δ(15–18) values (Fig. 12.6g). Elevated N and O isotope ratios within the SAMW (Figs. 12.6e, f and 12.8a) are the consequence of incomplete (partial) N assimilation in the Subantarctic thermocline of the Southern Ocean (Harms et al., 2019). In general, N assimilation can induce an isotope effect of about 5–10‰ (Montoya & McCarthy, 1995; Sigman et al., 2005; Waser et al., 1998) and produces biomass that is depleted in ^{15}N. However, in oligotrophic waters, such as the IOSG, this isotopic fractionation is not expressed (Montoya et al., 2002) due to complete N assimilation that also depresses nitrate concentrations to the limit of detection. Thus, elevated isotope ratios must be produced in an ocean region of partial N assimilation: The Southern Ocean. Here, light limitation and low iron availability limit primary productivity (Boyd et al., 2000; DiFiore et al., 2006, 2010; Hutchins et al., 2001; Sigman et al., 1999), and $\delta^{15}N$-NO_3^- values of 5–9‰ were observed within Subantarctic thermocline waters of the Southern Ocean (McCartney, 1977; Sigman et al., 1999). These thermocline waters are then injected as SAMW into the IOSG, where we observe elevated $\delta^{15}N$-NO_3^- and $\delta^{18}O$-NO_3^- (Fig. 12.6e, f).

For further evidence of water mass transport from the Southern Ocean into the IOSG thermocline, the NO_3^- and $\delta^{15}N$-NO_3^- relationship in Subantarctic thermocline waters can be compared to IOSG thermocline waters (for details, see Harms et al., 2019). Using the correlation between $\delta^{15}N$-NO_3^- and the fraction that remains from N assimilation ($\ln(NO_3^-)$), the isotope fractionation effect during N assimilation can be quantified (DiFiore et al., 2006; Sigman et al., 1999, 2000). The isotope effect can then be displayed as a slope angle of a linear line in $\ln(NO_3^-)/\delta^{15}N$-NO_3^- space. We find a similar trend between the isotope effect (slope angle) of the IOSG thermocline (0.93‰; Harms et al., 2019) and the isotope effect of the Subantarctic thermocline (1.3‰ Sigman et al., 1999, 2000), but with a slightly less steep slope in the IOSG thermocline. This can be explained by vertical mixing with the overlying SSW (Harms et al., 2019), which has a fundamentally different source of water and is additionally affected by N_2 fixation.

The influx of additional nitrate via N_2 fixation into surface waters of the IOSG is identified by the parameters N*, $\delta^{15}N$-NO_3^- and Δ(15–18) (Harms et al., 2019). Increasingly positive N* values in the upper 300 m at ~20–24°S indicate a positive deviation of the NO_3^-/PO_4^{3-} ratio from the Redfield stoichiometry (Figs. 12.6d and 12.9) and thus an additional input of N to the system decoupled from P (Redfield, 1934, 1963; Gruber & Sarmiento, 1997). Due to the fact that the South Indian Ocean is little affected by the influx of nutrient-enriched mineral aerosols from atmospheric deposition (Duce et al., 2008; Duce & Tindale, 1991), airborne input is an

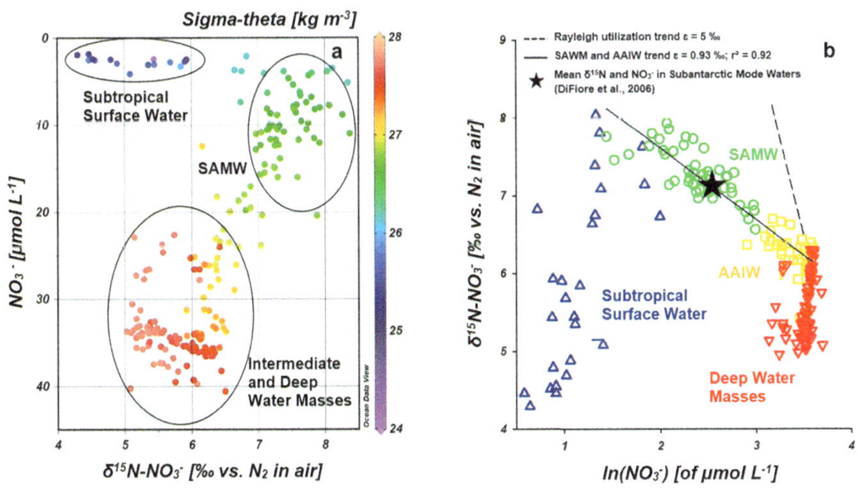

Fig. 12.8 NO_3^- concentrations vs. $\delta^{15}N$-NO_3^- (**a**) and $\delta^{15}N$-NO_3^- vs. $\ln(NO_3^-)$ (**b**) for water sampling stations within the IOSG (20.96–27.80°S); colour code of dots indicates the potential density sigma-theta (kg m^{-3}). In panel b, data are grouped for the SSW (blue), SAMW (green), AAIW (yellow) and deep water masses (red). (Reproduced from Harms et al., 2019)

unlikely source for external N. Instead, diazotrophic N_2 fixation is indicated as the origin for the new N, similar to the situation in the southwestern Pacific Ocean, where positive N* anomalies of 2 μmol kg^{-1} in surface waters reflect diazotrophic N_2 fixation (Fumenia et al., 2018). A reliable indicator for atmospheric N_2 fixation is the $\delta^{15}N\text{-}NO_3^-$, as atmospheric N assimilated by fixation produces biomass with a $\delta^{15}N$ of around ~0‰. Nitrification of this isotopically light N in diazotrophic biomass lowers the $\delta^{15}N$ NO_3^- in surface waters relative to the average oceanic NO_3^-. Consequently, the upward decrease of $\delta^{15}N\text{-}NO_3^-$values from >7‰(SAWM) to 5‰ in the upper 300 m is a clear evidence for N_2 fixation (Figs. 12.8a and 12.9). Similar upward declines were found in the North Atlantic, where N_2 fixation was verified (Bourbonnais et al., 2009; Knapp et al., 2008). Compared to $\delta^{15}N\text{-}NO_3^-$, the $\delta^{18}O\text{-}NO_3^-$ is less affected by N_2 fixation and results in a decoupling of N and O isotopes (Carpenter et al., 1997; Minagawa & Wada, 1986; Montoya et al., 2002; Wada & Hattori, 1976). Therefore, the slight upward decrease in Δ(15–18) (Fig. 12.9) is a further indication of N_2 fixation in the surface waters of the IOSG at ~20–24°S. For this region, newly fixed N via N_2 fixation provides 32–34% of the assimilated NO_3^- in the upper 200 m (Harms et al., 2019). For the subtropical North Atlantic, Bourbonnais et al. (2009) stated that N_2 fixation accounts for ~40% of newly supplied nitrate. This is slightly higher than the calculated fixation input to

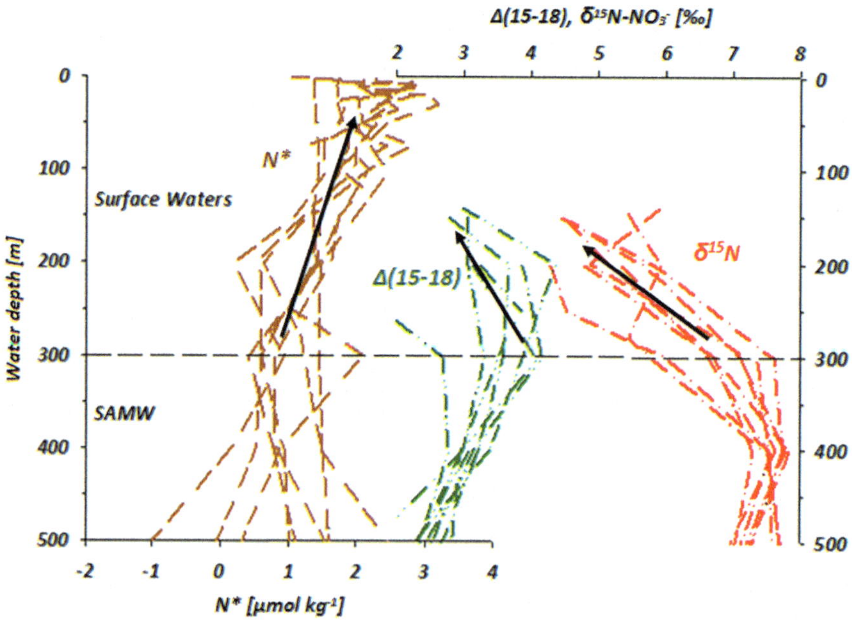

Fig. 12.9 Depth profiles of N*, $\delta^{15}N\text{-}NO_3^-$, and Δ(15–18) of water sampling stations located at ~20–24°S indicate an external input of new nitrogen via N_2 fixation to the upper 300 m. Black arrows illustrate the upward increase of N* and the decrease of $\delta^{15}N\text{-}NO_3^-$ and Δ(15–18) in the upper 300 m water depth. SAMW is the Subantarctic Mode Water

the ISOG, but they also found higher N* values, a slightly stronger upward decrease of $\delta^{15}N\text{-}NO_3^-$, and a stronger decoupling of N and O isotopes, which argues for even higher fixation rates.

4 Particulate Matter Fluxes to the Deep Ocean: Controlling Factors and Seasonal Implications

Total mass fluxes of sinking particulate matter, collected during 2014–2019 by sediment traps deployed at ~500–600 m.a.b., range between 2.3 and 22.2 mg m^{-2} d^{-1} (Ø 10.6 mg m^{-2} d^{-1}; Fig. 12.10a; Harms, 2020; Harms et al., 2021). In general, sinking particulate matter in the open ocean and the resultant pelagic sediments consist mainly of the products generated by planktic and nektic organisms in surface waters (Hüneke & Henrich, 2011). Primary producers (e.g., coccolithophores and diatoms) are the basis of the marine food web and control the biological productivity of the entire marine community by converting inorganic substances into organic matter by photosynthesis (Hüneke & Henrich, 2011; Turner, 2015). Major consumers are zooplankton such as plankticforaminifera, crustacean (e.g., amphipods, copepods) and pteropods (Turner, 2015). POM sinking out of the productive zone is generally composed of combinations of fecal pellets from zooplankton and fish, organic aggregates known as "marine snow" and detritus from dead plankton (Turner, 2015). Several phyto- and zooplanktonic organisms generate mineralised tests consisting of carbonate (planktic foraminifera, coccolithophores) or silicate (diatoms, radiolarians) (Lyle, 2014). This agrees with our observations in the IOSG, where the dominant components of the total mass flux are $CaCO_3$(~66 wt.-%) and bOpal (~14 wt.-%), followed by minor percentages of POM (~9.9 wt.-%) and LM (~9.6 wt.-%; Fig. 12.10b). Overall, the distribution of the individual components $CaCO_3$, bOpal, POM and LM is relatively constant in all sediment trap moorings (Fig. 12.10a). The high percentages of $CaCO_3$ in all traps are typical for the tropical and subtropical regions of the open Indian Ocean (Kolla et al., 1976; Lyle, 2014). The carbonate compensation depth (CCD) is at 4600–4800 m depth (Kolla et al., 1976) between 20°S and 30°S in the Indian Ocean so that an extensive solution of $CaCO_3$ components in sinking particulate matter and sediments can be excluded.

At sediment trap moorings 04-03 and 04-04, three traps were deployed to collect particles at different depths. Traps *Shallow* intercepted the particle flux that was directly exported out of the photic (low nutrient) zone. Active daily zooplankton migration extends to this depth and fluxes where higher compared to traps *Middle* (deployment depth is equal to single-sediment trap moorings; Fig. 12.10a). Traps *Deep* reveal increasing total fluxes compared to traps *Middle* due to the influx of particles from resuspension and rebound processes (Fig. 12.10a). These benthic nepheloid layer processes are crucial for the understanding of potential deep-sea mining activities. Additionally, the mooring systems of three sediment traps in a row allow us to decipher the degradation processes of particles on their way through

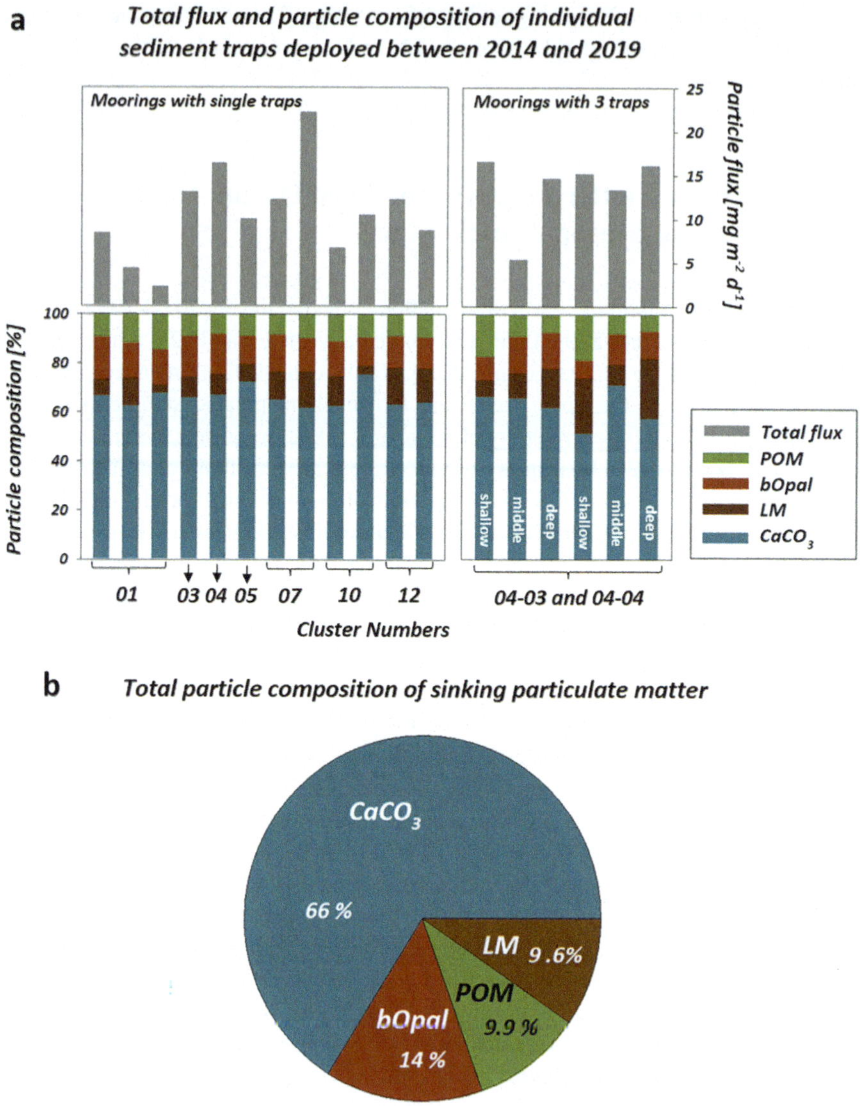

Fig. 12.10 Total mass fluxes and particle composition of the main components calcium carbonate ($CaCO_3$), particulate organic matter (POM), biogenic opal (bOpal) and lithogenic matter (LM) in wt.-% of individual sediment traps deployed during INDEX expeditions between 2014 and 2019 (a). Panel a is divided into sediment trap moorings, deployed with a single sediment trap and one with a system of three sediment traps (moorings 04-03 and 04-04). The overall averaged composition in sinking particulate matter for traps deployed at ~500–600 m.a.b. is presented in panel b

the water column. The $CaCO_3$ content is similar to the amount in the other sediment traps and there are minor differences between Traps *Shallow*, *Middle* and *Deep* (Fig. 12.10a). POC and TN gradually decrease with depth (Fig. 12.11a, b),

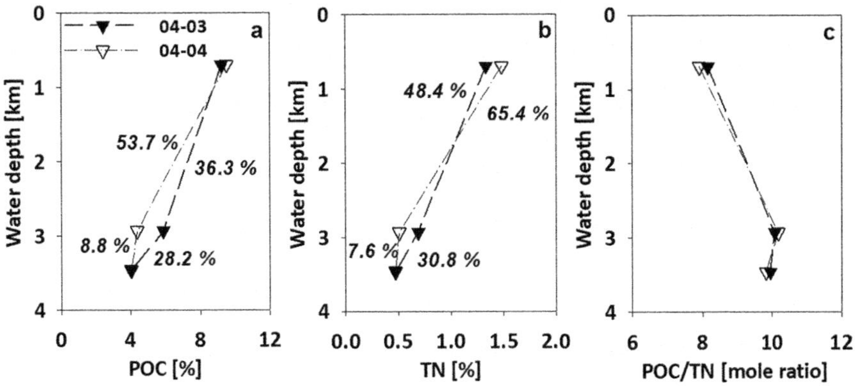

Fig. 12.11 Averaged total particulate matter flux (**a**), particulate organic carbon (POC) (**b**), total nitrogen (TN) (**c**) given in percentages (wt.-%) and the elemental ratio POC/TN (**d**) are presented as depth profiles for sediment traps *Shallow* (~695 m), *Middle* (~2930 m) and *Deep* (~3465 m and ~ 30 m.a.b.) of moorings 04-03 (black triangles) and 04-04 (open triangles) deployed between 2017 and 2019 in the INDEX license area. Percentages next to line plots in panels b and c show the relative reduction in % of POC and TN percentages (wt.-%)

reflecting the typical progressive degradation of organic components on their way down towards the ocean's interior as predicted by flux modes (e.g., Betzer et al., 1984; Martin et al., 1987; Suess, 1980) and shown in many sediment trap experiments. The gradual decrease of POC and TN between traps *Shallow* and *Middle* is more pronounced than between traps *Middle* and *Deep* (Fig. 12.11a, b). We can calculate that less than half of the POC and TN trapped in the uppermost traps reached the near-bottom traps. Moreover, TN indicates a stronger downward decrease compared to POC, leading to a downward increase in POC/TN ratios (Fig. 12.11c). This reflects the decline of effective organic matter degradation with water depth that is often expressed by an exponential function (Armstrong et al., 2001; Martin et al., 1987; Suess, 1980).

Overall, POC fluxes are 0.52 ± 0.18 mg m^{-2} d^{-1} on average and range between 0.2 to 0.9 mg m^{-2} d^{-1} for sediment traps deployed at ~500–600 m.a.b., which is roughly 5% of the total mass flux (Harms, 2020; Harms et al., 2021). Organic carbon export out of the biologically active zone, across the thermocline into the deep ocean, and its final storage in sediments drives the organic carbon pump and is thus an essential part of the global carbon cycle (Volk & Hoffert, 1985). To evaluate the efficiency of the organic carbon pump and the role of the IOSG in carbon sequestration on an ocean-wide scale, we compare our POC data with global data previously collated from the literature (e.g., Harms, 2020; Harms et al., 2021; Honjo et al., 2008; Lampitt & Antia, 1997; Lutz et al., 2007; Rixen et al., 2019a). For global comparison, we normalize POC flux data to a water depth of 2000 m. We use the often applied equation of Martin et al. (1987), $POC_{2000} = POC_{TD} \times [2000/TD]^{-0.858}$, where POC_{2000} (mg m^{-2} d^{-1}) is the POC flux normalised to a water depth of 2000 m the POC_{TD} measured at trap depth (TD; see Lampitt & Antia, 1997; Sarmiento et al.,

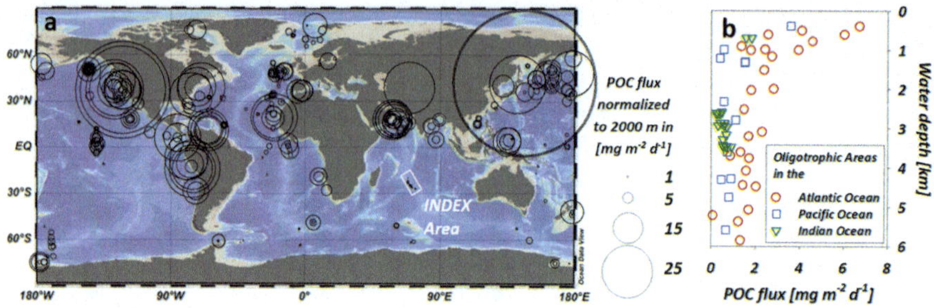

Fig. 12.12 Global map of particulate organic carbon (POC) fluxes in mg m^{-2} d^{-1} for sediment trap data collated from the literature (Harms et al., 2021) is extended by sediment trap samples in the IOSG collected during INDEX expeditions between 2014 and 2019 (a). All POC flux data were normalised to 2000 m. POC fluxes in the INDEX area (white rectangle) are comparatively low and the resulting small circles are weakly visible. Panel b shows POC flux data for the oligotrophic gyres (15–35°N, 15–35°S) against trap depths, grouped for the Pacific Ocean (grey), Atlantic Ocean (red) and INDEX Area in the Indian Ocean (green)

1998; Hedges et al., 1999; Buesseler et al., 2000; Fischer et al., 2000; Lutz et al., 2007; Harms et al., 2021). The resulting global map of normalized POC fluxes shows that fluxes in the IOSG are extremely low compared to global data (Fig. 12.12a) and even compared to other subtropical gyres in the Atlantic and Pacific Oceans (Fig. 12.12b).

The formation of sinking particulate matter is mainly determined by primary production (Betzer et al., 1984; Eppley & Peterson, 1979; Lampitt & Antia, 1997; Suess, 1980). The rate of primary production (rate of CO_2 uptake via phytoplankton photosynthesis) and the transfer on sinking POC through the thermocline and further into the deep ocean is controlled by various factors (Cram et al., 2018;Lampitt & Antia, 1997 ; Rixen et al., 2019b). The major factors controlling primary production are light, available nutrients and water column stability (Lampitt & Antia, 1997; Rixen et al., 2019b). The IOSG is a strongly stratified system with highly nutrient-depleted surface waters down to ~500 m (Harms et al., 2019). This is reflected by elevated POC/TN ratios of ~10 in particulate matter (Harms et al., 2021). This is higher than the Redfield ratio (~6.63; Redfield, 1934, 1963) and thus indicates the pronounced phytoplankton nutrient limitation (e.g., Donaghay et al., 1978; Goldman, 1980). In subtropical gyres like the IOSG, the ocean mixed layer depth (OMLD) follows the depression of the pycnoclines and thus increases towards the gyre centre (Baer et al., 2019; Sarmiento & Gruber, 2006). However, the zone of significant nutrient increase is well below the OMLD in the gyre centre and does not favour nutrient input by vertical mixing. Consequently, primary production in the IOSG is nutrient-limited and results in extremely low primary production rates of 0.12–12.4 nmol C L^{-1} h^{-1} (Baer et al., 2019). These rates are lower than those measured in other oligotrophic gyres (>20 nmol C L^{-1} h^{-1}; e.g., Williams et al., 2013) and produce the lowest oceanic POC fluxes worldwide to date (Harms, 2020; Harms et al., 2021).

The amount of the particle fraction that is exported just below the productive layer can be expressed by the export efficiency ε ($\varepsilon = POC_{export\ flux}/NPP$; Murray et al., 1989; Laws et al., 2000; Ducklow et al., 2001; Francois et al., 2002; Sarmiento & Gruber, 2006; Henson et al., 2019), while the transport of particle from the base of the productive zone towards the deep ocean is referred to the transfer efficiency (T_{eff}; Cram et al., 2018). Various factors, such as water temperature, oxygen concentrations, particle size (particle aggregation) and porosity, the sinking speed (ballast effect) and the plankton community, play a significant role in the final burial of POM/POC at the sea floor. Temperature and particle size appear to have the strongest influence on the export/transfer efficiency and positive correlations between particle export efficiency, particle size and the ballast effect have been observed (Cram et al., 2018; Weber et al., 2016). Ballast material enhances the vertical flux by promoting the aggregation of small particles that individually have an insignificant role in the vertical mass flux and transform them into fast sinking aggregates (>100 m day^{-1}) by increasing their specific density (McCave, 1975; Rixen, Gaye, Emeis, et al., 2019b; Schrader, 1971; Shanks & Trent, 1980; Suess, 1980). The increase of the sinking speeds shortens the residence time of particulate matter in the biologically active zone, lowers organic matter regeneration within the mixed layer and sub-thermocline and, thus, enhances the downward flux (Armstrong et al., 2001; Haake & Ittekkot, 1990; Hamm, 2002; Klaas & Archer, 2002). In the IOSG, $CaCO_3$ acts as the relevant ballast mineral and favours sinking speeds of 153–179 m day^{-1}, which were calculated by using Stokes' law, previous estimates on particle size, fluid viscosity (η), particle porosity (Rixen et al., 2019b) and particle composition and density of particulate matter in the IOSG (Harms et al., 2021). This is lower than estimates for the North Indian Ocean (224–230 m day^{-1}), where high LM contents (14–60 wt.-%) accelerate the sinking speed (Berelson, 2001; Rixen et al., 2019b). Furthermore, in warm and carbonate-dominated regions, like the IOSG, complex food webs lead to an enhanced regeneration of organic components in the upper ocean (Francois et al., 2002; Rixen et al., 2019b; Sarmiento & Gruber, 2006). Consequently, POC undergoes intense degradation before being exported into the deeper ocean. Thus, we do not only observe low POC fluxes but also a comparatively low POC export efficiency. This agrees with previous work on subtropical gyres (Cram et al., 2018; Henson et al., 2019; Sarmiento & Gruber, 2006). Using the well-established concept of Eppley and Peterson (1979), an export efficiency of $\varepsilon = 0.023$ can be calculated (Harms et al., 2021). Therefore, POC flux data are normalised to a depth of 130 m (equation of Martin et al., 1987), which is roughly the lower limit of the euphotic zone in our study area (NASA DataBase Giovanni, MODIS-A, Lee algorithm; Lee et al., 2007). The calculated averaged $POC_{export\ flux}$ of 7.63 mg m^{-2} d^{-1} (0.23 mol C m^{-2} $year^{-1}$) is then divided by the primary production rate of ~10 mol C m^{-2} $year^{-1}$, which was calculated for the IOSG region (see Harms et al., 2021). Consequently, only 2.3% of the produced organic carbon is exported below the biologically active zone (Fig. 12.13).

While the ε-ratio quantifies the POC fraction that is exported out of the productive layer, the transfer efficiency (T_{eff}) denotes the fraction of POC transferred from the base of the productive zone into the deep waters, bypassing remineralisation in

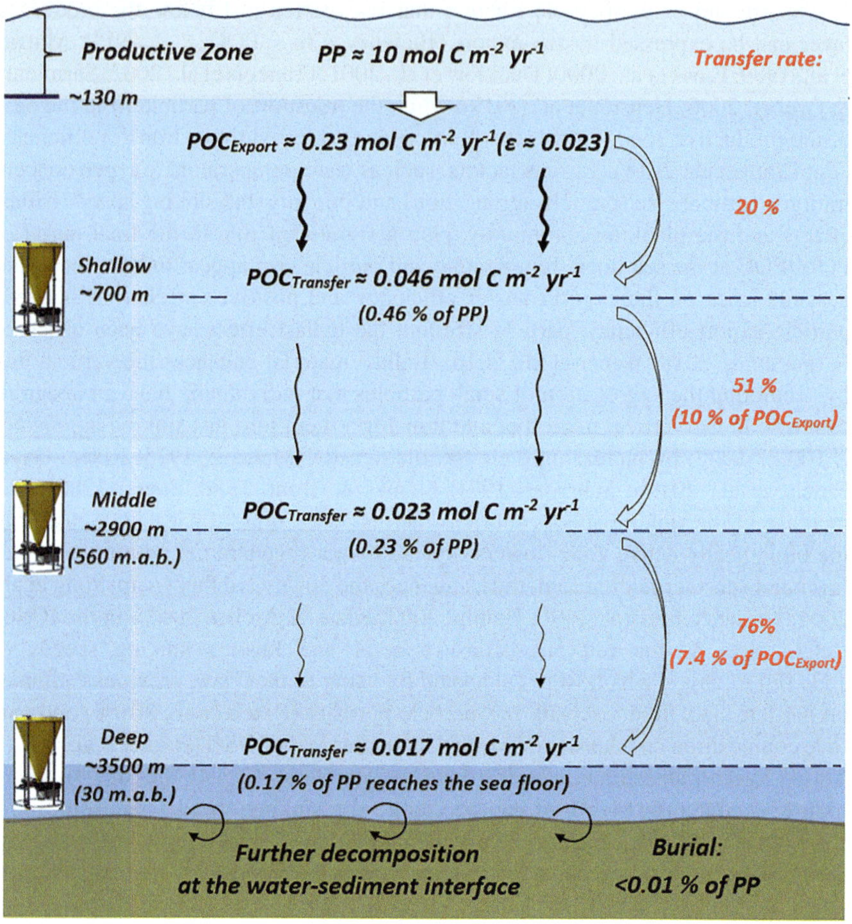

Fig. 12.13 Schematic of the particulate organic carbon export (POC_{Export}) below the productive zone, its transfer ($POC_{Transfer}$) across the sediment trap depths of traps *Shallow* (~700 m), *Middle* (~2900 m), and *Deep* (~3500 m) of moorings 04-03 and 04-04, and its final burial (POC_{Burial}) into the deep-sea sediments (first centimetre). The abbreviation m.a.b. stands for metres above bottom. Percentages in red represent the transfer rate of POC between the different depth intervals

the mesopelagic zone (150–1500 m; Martin et al., 1987; Lampitt & Antia, 1997; Passow & Carlson, 2012; Giering et al., 2014; Buesseler et al., 2020). Several studies suggested that T_{eff} is inversely related to the ε-ratio, resulting in the highest T_{eff} in low-latitude oligotrophic regions (Francois et al., 2002; Guidi et al., 2015; Lam et al., 2011; Mouw et al., 2016). However, other recent studies indicate the opposite, where T_{eff} is highest in high latitudes, intermediate in the tropics and lowest in the subtropical gyre regions (Buesseler et al., 2007; DeVries & Deutsch, 2014; Marsay et al., 2015; Weber et al., 2016). Sediment trap moorings 04-03 and 04-04 allow a first estimate of the POC fraction that is transferred to the deep ocean. Our calculations show that approximately 20% of the POC exported out of the productive zone (~130 m) reaches a water depth of ~700 m (trap *Shallow*). The percentages further decrease towards ~2900 m (trap *Middle*; 10%) and ~ 3500 m water depth (trap *Deep*; 7.4%; Fig. 12.13). The strongest decline in the POC content takes place within the surface layer (2.3%, $\varepsilon = 0.023$), while the reduction is less pronounced below 700 m water depth. This matches previous observations that POC fluxes typically decrease exponentially in the upper water column and decline almost linearly towards the deep sea (Armstrong et al., 2001; Suess, 1980). Most water-column remineralisation thus occurs at depths of less than 1000 m and most of the POC passing 1000 m water depth in fact reach the ocean floor (Armstrong et al., 2001). Overall, only 0.17% of the POC produced in the biologically active zone reach the sea floor in the IOSG (Fig. 12.13). This is in accordance with combined sediment trap studies in the open ocean where <1% of surficial productivity reach depths of 4000 m (Martin et al., 1987; Suess, 1980). Assuming that primary production generates 10 mol C m^{-2} $year^{-1}$ in the surface layer, we estimate that approximately 0.017 mol C m^{-2} $year^{-1}$is transferred to the seabed. At the sediment–water interface, further remineralisation/degradation or transport processes strongly diminish the POC that is finally accumulated in pelagic sediments. In general, bulk sedimentation rates in open ocean regions are between 0.5 to 2 cm per 1000 years (Hüneke & Henrich, 2011), while low productive areas, like the Central Pacific, have even lower sedimentation rates of 0.2–0.6 cm per 1000 years (Müller & Suess, 1979). Accordingly, less than 0.01% of the primary production becomes fossilised in pelagic sediments of the IOSG (Müller & Suess, 1979).

The seasonality in oligotrophic open ocean realms and thus in the IOSG is less pronounced compared to other open ocean regions (Baer et al., 2019; Henson et al., 2019; Williams et al., 2013). Nevertheless, fluctuations in the NPP are apparent, and maximum expansion of oligotrophic areas and lowest NPP are observed during austral summer (January), while elevated NPP occur in austral winter (July; McClain et al., 2004; Fig. 12.3). As NPP is the basis for particulate matter production and the formation of sinking particles, sediment trap data are often positively related to surface primary production (Betzer et al., 1984). Therefore, we initially assumed a coherence between total/POC fluxes and NPP data. However, our sediment trap data (2014–2019) does not reveal a positive correlation but rather a slightly negative relationship (Fig. 12.14a). Potential factors that interfere the effective transmission of seasonally induced surface production signals are variations in the organic matter remineralisation/degradation, ballast effects, particle structure (size and porosity; Francois et al., 2002; Bach et al., 2019), particle fragmentation (by zooplankton, e.g., particle-associated copepods; Cavan et al., 2019, Mayor et al., 2020), ocean circulation and physical mixing due to enhanced wind stress and the variability of the OMLD (Asper et al., 1992; Rixen et al., 2019a; Siegel et al., 1990). Especially in regions of low primary production and low seasonal variability, these mechanisms possibly alter surface production signals and temporal variations of deep-sea fluxes (Michaels et al., 1994). To determine a potential interplay of processes that modify the export of particulate matter in the IOSG, we compare the temporal

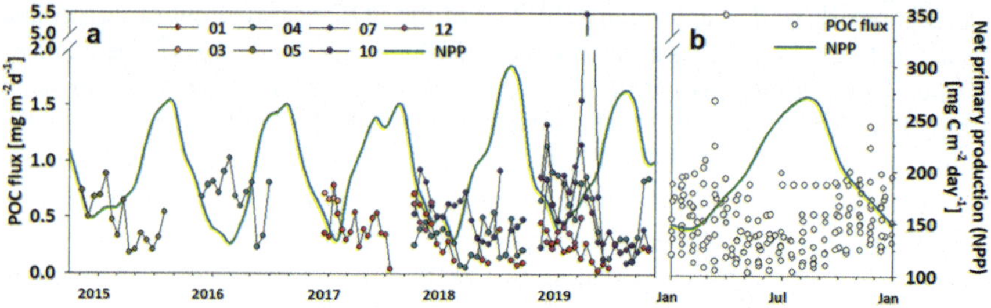

Fig. 12.14 Time series data of particulate organic carbon (POC) fluxes and net primary production (NPP) during sediment trap investigation (**a**) and combined POC fluxes for a one-year period accompanied by averaged NPP data for the period 2014–2019 in the INDEX area (**b**). Legend of panel a presents the cluster numbers of the INDEX area. NPP data were taken from the Ocean Productivity Data Base and bases on the standard VGPM algorithm (http://sites.science.oregonstate.edu/ocean.productivity/index.php, Behrenfeld & Falkowski, 1997) using data of MODIS chlorophyll-*a* concentrations, sea surface temperature (SST), SeaWiFS PAR, and estimates of euphotic zone depth from a model developed by Morel and Berthon (1989). Data were extracted from global-monthly averaged grids (original resolution: 1080 × 2160) over an area of 20–27°S and 68–74°E

variations of NPP, sea surface temperature (SST), OMLD and sea surface wind speed (Fig. 12.15a).

Elevated SSTs in austral summer reflect strong surface stratification (reduced surface ocean mixing) in the IOSG impeding upward nutrient supply to surface waters. This is reflected by a linear anti-correlation between SSTs and NPP data (Fig. 12.15b). The deeper OMLD in austral winter has the potential to increase the vertical nutrient supply and in fact leads to a positive correlation with NPP data (Fig. 12.15c). Elevated NPP in winter time may suggest accompanied increases in particulate matter fluxes. However, NPP and total mass fluxes reveal an anti-correlation (Fig. 12.15d) that corresponds to our first observation of POC fluxes, depicted in Fig. 12.14b, c. Reasons for this are combined physical and biological induced processes. First, a deep OMLD during winter (high NPP) is assumed to intensify degradation and regeneration of organic matter, thus attenuating particulate matter export (Baer et al., 2019). Furthermore, physical forcing such as changes in surface wind velocities may influence particle export (Delman et al., 2018; Rixen et al., 2019a). At several trap stations, elevated wind speeds correspond to elevated POC fluxes (Harms et al., 2021). Thus, higher-than-average deep vertical mixing may lead to increases in surface nutrient concentrations and trigger regional particulate matter export. Until now, little is known about the link between surface wind speed and particulate matter fluxes in the IOSG. Even less is known about the overall seasonal fluctuations in the gyre strength and the volume transport of in- and outflows (McMonigal et al., 2018). Studies on variations in the dynamics of eddies in the IOSG and their effects on biogeochemical processes and particulate matter fluxes are in very early stages. Enhanced cyclonic eddy-activities are observed between Australia and Madagascar in a wide band from ~20°S to 30°S, propagating westwards across our study area (Delman et al., 2018) and might well influence

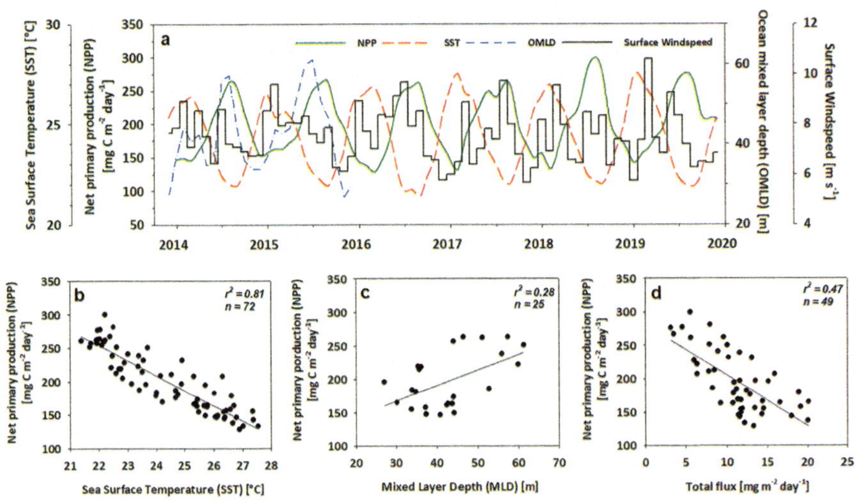

Fig. 12.15 Fluctuations of net primary production (NPP in mg C m^{-2} day^{-1}), sea surface temperature (SST), ocean mixed layer depth (OMLD), and surface wind speed (in m s^{-1}) for the time interval of sediment trap deployments between 2014 and 2019 in the INDEX area (**a**). NPP data are negatively correlated with the SST (**b**) and total particulate matter fluxes (**d**) and positively correlated with the OMLD (**c**). NPP data were taken from the Ocean Productivity Data Base (see Fig. 12.14). The SSTs were taken from MODIS-Aqua (MODISA-L3m_SSTv2014; NASA Earth DataBase Giovanni) as a monthly time series, area averaged over 9 km during 01.12.2013–01.01.2020 within a region of 21–28°S, 68–74°E. Data on the sea surface wind speeds were retrieved from MERRa-2 Model (M2TMNXFLX_5_12_4_SPEED; NASA Earth DataBase Giovanni) as monthly time series, area-averaged over 0.5 × 0.625 deg. during 01.12.2013–01.01.2020 within a region of 21–28°S, 68–74°E

particulate matter export. In the subtropical western North Pacific Ocean, cyclonic eddies are assumed to be an important mechanism for the nutrient supply to the euphotic zone and, thus, might lead to enhanced particulate matter export production (Honda et al., 2018).

On the other hand, the community structure of organisms at the sea surface and in the mesopelagic layer varies with season, even in oligotrophic regions; acts on particle size, porosity, sinking velocities and degradation/remineralisation efficiency; and hampers the transfer of primary production signals. During stages of higher nutrient supply and subsequent higher NPP (austral winter), a more complex zooplankton-dominated community can develop (Bach et al., 2019) and have the potential to decrease the export flux below the euphotic zone and the transfer efficiency through the mesopelagic via particle fragmentation (Cavan et al., 2019; Giering et al., 2014; Mayor et al., 2014). These small particles, containing free (non-associated) organic matter, can be rapidly colonised by microbes that promote remineralisation of POC to dissolved organic carbon (DOC) and further to dissolved inorganic carbon (DIC; Mayor et al., 2014, 2020; Armstrong et al., 2001). Recent data from multiple ocean regions state that ~49% of the observed decline of particle fluxes with depth in the mesopelagic can be explained by the transformation of large, fast-sinking particles into small, slow-sinking fragments (Briggs et al., 2020). Thus, during the winter season, higher NPP may lead to a more complex food web dominated by zooplankton that reduces the final export of particulate matter to the deep ocean via particle fragmentation and subsequent rise of organic matter degradation. In contrast, the predominant species during periods of reduced nutrient supply and low NPP are foraminifera and thus exhibit an anti-correlated tendency to zooplankton abundances in the oligotrophic subtropical NE Atlantic (Bach et al., 2019). As $CaCO_3$ is the dominant ballast material in the SIO gyre (Harms et al., 2021), it increases the sinking velocities of particulate matter and thus reduces the residence time of organic matter in the water column exposed by degradation. Furthermore, organic matter associated with ballast minerals is more protected than non-associated organic matter (Armstrong et al., 2001). Consequently, the export of sinking particulate matter is more efficient during summer and leads to enhanced total fluxes, although the initial particulate matter production is less. To conclude, the anti-correlated NPP and export fluxes to the deep IOSG are most likely a product of various physical and biological processes that need to be studied to obtain reliable predictions of changes in biogeochemical cycles, BCP, carbon sequestration/storage and global N- and C-budgets.

5 Conclusion

The Indian Ocean subtropical gyre (IOSG) is among the extensive oligotrophic areas that cover about 75% of the global ocean's surface. These ocean regions are expected to expand in the future due to global warming (Behrenfeld et al., 2006) and thus become more important on a global scale over the coming decades. To better understand biogeochemical processes in the IOSG and to better predict the potential consequences of warming on global marine N- and C cycles, we studied water masses using the physical properties of the water column in combination with nutrient and stable isotope distributions and the mechanisms of particulate matter fluxes from surface waters down to the deep-sea.

Nutrient distribution in the IOSG is predominantly influenced by the typical circulation pattern of subtropical gyres leading to specific physical conditions. In the centre, intense downwelling is induced by the convergence of horizontal Ekman volume flux (Williams & Follows, 2003), leading to deep thermo-, pycno- and nutriclines, which shoal towards the gyre margins and thus form strongly stratified and nutrient-depleted surface waters. Consequently, the supply of nutrients to the euphotic zone by vertical mixing is reduced or absent in the gyre and accordingly, the external input via the lateral transfer across the gyre boundaries acts as the major nutrient source. Indeed, we can illustrate that nutrient distribution and isotopic composition of nitrate are affected by the influx of different water masses of different origins, based on our water mass analysis. The nutrient and isotopic data reported in Harms et al. (2019) are among the first for the subtropical South Indian Ocean. We demonstrate the lateral influx of water masses from the Arabian Sea, imprinted by strong denitrification at mid-water depths, leading to nitrate deficits and elevated N and O isotope ratios in intermediate and deep waters of the IOSG. Furthermore, oxygen-saturated mode waters characterised by preformed nutrients and distinctly elevated nitrate isotope composition formed by partial N assimilation at high southern latitude are injected from the Southern Ocean. As an additional nitrate source, our data mirror an external input of atmospheric N via N_2 fixation. First estimates in the upper 200 m at ~20–24°S revealed that approximately 32–34% of the nitrate consumed by N assimilation is provided from newly fixed N. The IOSG has been sparsely studied in terms of N-budgets and N-transformation processes and is an area that represents those oceanic oligotrophic regions that are likely to adapt to continued warming by deepening stratification, reduced upward nutrient supply across the thermocline and decreasing biological production. Whether this is offset by enhanced N_2 fixation in warming layers remains an open question that requires dedicated follow-up studies, that is, experimental approaches, time-series observations, remote sensing and biogeochemical modelling.

The IOSG is one of the least explored ocean regions in terms of sinking particulate matter, although global oligotrophic subtropical gyres, due to their vast extent, are thought to account for half of the global organic carbon export to the ocean's interior (Emerson et al., 1997). Overall, low primary production rates induced by the strongly stratified and nutrient-depleted surface ocean lead to worldwide lowest POC fluxes (0.52 ± 0.18 mg m^{-2} d^{-1} at ~500–600 m.a.b.) and very low POC export efficiencies ($\varepsilon = 0.023$). This results in a fraction of only 0.17% of the primarily produced POC in surface waters that reaches the sea floor. Since further decomposition/remineralisation occurs at the sediment–water interface, we can calculate that less than 0.01% of the primary production in the photic zone finally accumulates in surface sediments. Although total mass fluxes are comparatively low, seasonal variations are also apparent and indicate a decoupling between NPP and the amount of sinking particulate matter to the deep ocean. A deeper ocean mixed layer in winter and enhanced particle fragmentation via a more complex and zooplankton-dominated biological community favours organic matter decomposition and degradation and leads to reduced total particulate matter fluxes during higher primary productivity. In contrast, a foraminifera-dominated plankton community increases the ballast effect and triggers the more efficient transfer of particulate matter to the deep ocean during the summer season. Nevertheless, total mass fluxes are comparatively low throughout the year. Assuming that all subtropical gyres are characterised by similarly low export rates (as postulated by Henson et al. (2019) for the oligotrophic North Atlantic) and considering that these enormous ocean areas are likely to continue to expand under climate warming, this might weaken the organic carbon pump and reduce the transfer of organic carbon to the deep oceans.

However, the magnitude and even the direction of climate change on the biogeochemistry of oligotrophic ocean areas, including their biological carbon pump, are still debated, and food web adaptions, community structure or potential changes in global or regional circulation patterns to a warmer climate have not been adequately constrained (Laufkötter et al., 2017; Laufkötter & Gruber, 2018; Passow & Carlson, 2012). Clearly, more interdisciplinary research is needed to understand how these extensive oligotrophic regions will change under global warming conditions and how they contribute to global carbon budgets. In addition, our findings on vertical particle fluxes, lateral particle transport (data not shown here) and sediment accumulation rates are important aspects of understanding the impacts of potential deep-sea mining. Our research contributes to answering pending questions about whether, when and how the overall seafloor ecosystem can recover after mining. Before intervening in the deep-sea ecosystem, it is important to clarify how and how far resuspended particles are transported from the seafloor into the water column. How long it takes for particles to redeposit after mining, and whether the ecosystem can recover in the long term. Before deep-sea mining can begin, holistic scientific studies need to be conducted to answer all of these open questions (Boetius & Haeckel, 2018).

References

Armstrong, R. A., Lee, C., Hedges, J. I., et al. (2001). A new, mechanistic model for organic carbon fluxes in the ocean based on the quantitative association of POC with ballast minerals. *Deep-Sea Research Part II, 49*(1–3), 219–236. https://doi.org/10.1016/S0967-0645(01)00101-1

Asper, V. L., Deuser, W., Knauer, G., et al. (1992). Rapid coupling of sinking particle fluxes between surface and deep ocean waters. *Nature, 357*(6380), 670–672. https://doi.org/10.1038/357670a0

Bach, L. T., Stange, P., Taucher, J., et al. (2019). The influence of plankton community structure on sinking velocity and remineralization rate of marine aggregates. *Global Biogeochemical Cycles, 33*(8), 971–994. https://doi.org/10.1029/2019GB006256

Baer, S. E., Rauschenberg, S., Garcia, C. A., et al. (2019). Carbon and nitrogen productivity during spring in the oligotrophic Indian Ocean along the GO-SHIP IO9N transect. *Deep-Sea Research Part II, 161*, 81–91. https://doi.org/10.1016/j.dsr2.2018.11.008

Baumgartner, A., & Reichel, E. (1975). *The world water balance: Mean annual global, continental and maritime precipitation, evaporation and run-off.* Elsevier Science Inc.

Behrenfeld, M. J., & Falkowski, P. G. (1997). Photosynthetic rates derived from satellite-based chlorophyll concentration. *Limnology and Oceanography, 42*, 1–20. https://doi.org/10.4319/lo.1997.42.1.0001

Behrenfeld, M. J., O'Malley, R. T., Siegel, D. A., et al. (2006). Climate-driven trends in contemporary ocean productivity. *Nature, 444*, 752–755.

Berelson, W. M. (2001). Particle settling rates increase with depth in the ocean. *Deep-Sea Res. II: Topical Studies in Oceanography, 49*(1–3), 237–251. https://doi.org/10.1016/S0967-0645(01)00102-3

Berger, W. H., Smetacek, V. S., & Wefer, G. (Eds.). (1989). *Productivity of the ocean: Present and past. Ocean productivity and paleoproductvity—An overview* (Vol. 44, pp. 1–34). John Wiley & Sons.

Betzer, P. R., Showers, W. J., Laws, E. A., et al. (1984). Primary productivity and particle fluxes on a transect of the equator at 153°W in the Pacific Ocean. *Deep-Sea Research Part I, 31*(1), 1–11. https://doi.org/10.1016/0198-0149(84)90068-2

Bindoff, N. L., & McDougall, T. J. (2000). Decadal changes along an Indian Ocean section at 32°S and their interpretation. *Journal of Physical Oceanography, 30*, 1207–1222.

Boetius, A., & Haeckel, M. (2018). Mind the seafloor. *Science, 359*(6371), 34–36. https://doi.org/10.1126/science.aap7301

Bourbonnais, A., Lehmann, M. F., Waniek, J. J., et al. (2009). Nitrate isotope anomalies reflect N2 fixation in the Azores Front region (subtropical NE Atlantic). *Journal of Geophysical Research: Oceans, 114*, 1–16. https://doi.org/10.1029/2007JC004617

Boyd, P. W., Watson, A. J., Law, C. S., et al. (2000). A mesoscale phytoplankton bloom in the polar Southern Ocean stimulated by iron fertilization. *Nature, 407*, 695–702. https://doi.org/10.1038/35037500

Briggs, N., Dall'Olmo, G., & Claustre, H. (2020). Major role of particle fragmentation in regulating biological sequestration of CO_2 by the oceans. *Science, 367*(6479), 791–793. https://doi.org/10.1126/science.aay1790

Buesseler, K. O., Steinberg, D. K., Michaels, A. F., et al. (2000). A comparison of the quantity and com-position of material caught in a neutrally buoyant versus surface-tethered sediment trap. *Deep-Sea Research Part I, 47*(2), 277–294. https://doi.org/10.1016/S0967-0637(99)00056-4

Buesseler, K. O., Lamborg, C. H., Boyd, P. W., et al. (2007). Revisiting carbon flux through the ocean's twilight zone. *Science, 316*(5824), 567–570. https://doi.org/10.1126/science.1137959

Buesseler, K. O., Boyd, P. W., Black, E. E., et al. (2020). Metrics that matter for assessing the ocean biological carbon pump. *Proceedings of the National Academy of Sciences, 117*(18), 9679–9687. https://doi.org/10.1073/pnas.1918114117

Capone, D. G., Zehr, J. P., Paerl, H. W., et al. (1997). Trichodesmium, a globally significant marine cyanobacterium. *Science, 276*, 1221–1229. https://doi.org/10.1126/science.276.5316.1221

Carpenter, E. J., Harvey, H. R., Fry, B., et al. (1997). Biogeochemical tracers of the marine cyanobacterium Trichodesmium. *Deep-Sea Research Part I, 44*, 27–38. https://doi.org/10.1016/S0967-0637(96)00091-X

Casciotti, K. L., Sigman, D. M., Hastings, M. G., et al. (2002). Measurement of the oxygen isotopic composition of nitrate in seawater and freshwater using the denitrifier method. *Analytical Chemistry, 74*, 4905–4912. https://doi.org/10.1021/ac020113w

Casciotti, K. L., McIlvin, M., & Buchwald, C. (2010). Oxygen isotopic exchange and fractionation during bacterial ammonia oxidation. *Limnology and Oceanography, 55*, 753–762.

Cavan, E. L., Laurenceau-Cornec, E. C., Bressac, M., et al. (2019). Exploring the ecology of the mesopelagic biological pump. *Progress in Oceanography, 176*, 102125. https://doi.org/10.1016/j.pocean.2019.102125

Codispoti, L., Brandes, J. A., Christensen, J., et al. (2001). The oceanic fixed nitrogen and nitrous oxide budgets: Moving targets as we enter the Anthropocene? *Scientia Marina, 65*(S2), 85–105.

Cram, J. A., Weber, T., Leung, S. W., et al. (2018). The role of particle size, ballast, temperature, and oxygen in the sinking flux to the deep sea. *Global Biogeochemical Cycles, 32*(5), 858–876. https://doi.org/10.1029/2017GB005710

Deacon, G. E. (1933). A general account of the hydrology of the South Atlantic Ocean. *Discovery Reports, 7*, 171–238.

Delman, A. S., Lee, T., & Qiu, B. (2018). Interannual to multidecadal forcing of mesoscale eddy kinetic energy in the subtropical southern Indian Ocean. *Journal of Geophysical Research, Oceans, 123*(11), 8180–8202. https://doi.org/10.1029/2018JC013945

Deuser, W. G., & Ross, E. H. (1980). Seasonal change in the flux of organic carbon to the deep Sargasso Sea. *Nature, 283*, 364–365.

Deuser, W. G., Ross, E. H., & Anderson, R. F. (1981). Seasonality in the supply of sediment to the deep Sargasso Sea and implications for the rapid transfer of matter to the deep ocean. *Deep-Sea Res. Part A. Oceanographic Research Papers, 28*(5), 495–505. https://doi.org/10.1016/0198-0149(81)90140-0

Deutsch, C., Gruber, N., Key, R. M., et al. (2001). Denitrification and N2 fixation in the Pacific Ocean. *Global Biogeochemical Cycles, 15*, 483–506. https://doi.org/10.1029/2000GB001291

Deutsch, C., Sarmiento, J. L., Sigman, D. M., et al. (2007). Spatial coupling of nitrogen inputs and losses in the ocean. *Nature, 445*, 163–167. https://doi.org/10.1038/nature05392

DeVries, T., & Deutsch, C. (2014). Large-scale variations in the stoichiometry of marine organic matter respiration. *Nature Geoscience, 7*(12), 890–894. https://doi.org/10.1038/ngeo2300

DiFiore, P. J., Sigman, D. M., Trull, T. W., et al. (2006). Nitrogen isotope constraints on subantarctic biogeochemistry. *Journal of Geophysical Research: Oceans, 111*, 1–19. https://doi.org/10.1029/2005JC003216

DiFiore, P. J., Sigman, D. M., Karsh Kristen, L., et al. (2010). Poleward decrease in the isotope effect of nitrate assimilation across the Southern Ocean. *Geophysical Research Letters, 37*, 1–5. https://doi.org/10.1029/2010GL044090

Donaghay, P. L., DeManche, J. M., & Small, L. F. (1978). On predicting phytoplankton growth rates from carbon: Nitrogen ratios. *Limnology and Oceanography, 23*(2), 359–362. https://doi.org/10.4319/lo.1978.23.2.0359

Duce, R. A., & Tindale, N. W. (1991). Atmospheric transport of iron and its deposition in the ocean. *Limnology and Oceanography, 36*, 1715–1726. https://doi.org/10.4319/lo.1991.36.8.1715

Duce, R. A., LaRoche, J., Altieri, K., et al. (2008). Impacts of atmospheric anthropogenic nitrogen on the open ocean. *Science, 320*(5878), 893–897. https://doi.org/10.1126/science.1150369

Ducklow, H. W., Steinberg, D. K., & Buesseler, K. O. (2001). Upper Ocean carbon export and the biological pump. *Oceanography Washington D. C., Oceanography Society, 14*(4), 50–58.

Duing, W. (1970). *The monsoon regime of the currents in the Indian Ocean*. Hawaii Institute of Geophysics.

Emerson, S., Quay, P., Karl, D., et al. (1997). Experimental determination of the organic carbon flux from open-ocean surface waters. *Nature, 389*(6654), 951–954. https://doi.org/10.1038/40111

Emerson, S., Mecking, S., & Abell, J. (2001). The biological pump in the subtropical North Pacific Ocean: Nutrient sources, Redfield ratios, and recent changes. *Global Biogeochemical Cycles, 15*, 535–554. https://doi.org/10.1029/2000GB001320

Emery, W. J. (2001). Water types and water masses. *Encyclopedia of Ocean Sciences, 6*, 3179–3187. https://doi.org/10.1006/rwos.2001.0108

Emery, W. J., & Meincke, J. (1986). Global water masses: Summary and review. *Oceanologica Acta, 9*, 383–391.

Eppley, R. W., & Peterson, B. J. (1979). Particulate organic matter flux and planktonic new production in the deep ocean. *Nature, 282*(5740), 677–680. https://doi.org/10.1038/282677a0

Fine, R. A. (1993). Circulation of Antarctic intermediate water in the South Indian Ocean. *Deep-Sea Research Part I, 40*, 2021–2042. https://doi.org/10.1016/0967-0637(93)90043-3

Fischer, G., Ratmeyer, V., & Wefer, G. (2000). Organic carbon fluxes in the Atlantic and the Southern Ocean: Relationship to primary production compiled from satellite radiometer data. *Deep-Sea Research Part II, 47*(9–11), 1961–1997. https://doi.org/10.1016/S0967-0645(00)00013-8

Francois, R., Honjo, S., Krishfield, R., et al. (2002). Factors controlling the flux of organic carbon to the bathypelagic zone of the ocean. *Global Biogeochemical Cycles, 16*(4), 1–20. https://doi.org/10.1029/2001GB001722

Fumenia, A., Moutin, T., Bonnet, S., et al. (2018). *Excess nitrogen as a marker of intense dinitrogen fixation in the Western Tropical South Pacific Ocean: Impact on the thermocline waters of the South Pacific*. Biogeosci. Discuss. https://doi.org/10.5194/bg-2017-557

Gaye, B., Nagel, B., Dähnke, K., et al. (2013). Evidence of parallel denitrification and nitrite oxidation in the ODZ of the Arabian Sea from paired stable isotopes of nitrate and nitrite. *Global Biogeochemical Cycles, 27*, 1059–1071. https://doi.org/10.1002/2011GB004115

Giering, S. L., Sanders, R., Lampitt, R. S., et al. (2014). Reconciliation of the carbon budget in the ocean's twilight zone. *Nature, 507*(7493), 480–483. https://doi.org/10.1038/nature13123

Godfrey, J. S., & Golding, T. J. (1981). The Sverdrup relation in the Indian Ocean, and the effect of Pacific-Indian Ocean Throughflow on Indian Ocean circulation and on the East Australian current. *Journal of Physical Oceanography, 11*(6), 771–779. https://doi.org/10.1175/1520-0485(1981)011<0771:TSRITI>2.0.CO;2

Goldman, J. C. (1980). Physiological processes, nutrient availability, and the concept of relative growth rate in marine phytoplankton ecology. In P. G. Falkowski (Ed.), *Primary productivity in the sea. Environm. Sci. Res* (Vol. 19, pp. 179–194). Springer. https://doi.org/10.1007/978-1-4684-3890-1_10

Granger, J., Sigman, D. M., Needoba Joseph, A., et al. (2004). Coupled nitrogen and oxygen isotope fractionation of nitrate during assimilation by cultures of marine phytoplankton. *Limnology and Oceanography, 49*, 1763–1773. https://doi.org/10.4319/lo.2004.49.5.1763

Gruber, N., & Sarmiento, J. L. (1997). Global patterns of marine nitrogen fixation and denitrification. *Global Biogeochemical Cycles, 11*, 235–266. https://doi.org/10.1029/97GB00077

Guidi, L., Legendre, L., Reygondeau, G., et al. (2015). A new look at ocean carbon remineralization for estimating Deepwater sequestration. *Global Biogeochemical Cycles, 29*, 1044–1059. https://doi.org/10.1002/2014GB005063

Haake, B., & Ittekkot, V. (1990). Die Wind-getriebene "biologische Pumpe" und der Kohlenstoffentzug im Ozean. *Naturwissenschaften, 77*(2), 75–79. https://doi.org/10.1007/BF01131777

Hamm, C. E. (2002). Interactive aggregation and sedimentation of diatoms and clay-sized lithogenic material. *Limnology and Oceanography, 47*(6), 1790–1795. https://doi.org/10.4319/lo.2002.47.6.1790

Harms, N. C. (2020). *The South Indian Ocean—Biogeochemical studies on water masses, nutrient and stable isotope distribution, and particulate matter in an oligotrophic ocean region.* E-Dissertation. 217 pages. https://ediss.sub.uni-hamburg.de/handle/ediss/8845

Harms, N. C., Lahajnar, N., Gaye, B., et al. (2019). Nutrient distribution and nitrogen and oxygen isotopic composition of nitrate in water masses of the subtropical southern Indian Ocean. *Biogeosciences, 16*(13), 2715–2732. https://doi.org/10.5194/bg-16-2715-2019

Harms, N. C., Lahajnar, N., Gaye, B., et al. (2021). Sediment trap-derived particulate matter fluxes in the oligotrophic subtropical gyre of the South Indian Ocean. *Deep-Sea Res. II: Topical Studies in Oceanography, 183, 104924*, 1–14. https://doi.org/10.1016/j.dsr2.2020.104924

Hedges, J. I., Hu, F. S., Devol, A. H., et al. (1999). Sedimentary organic matter preservation; a test for selective degradation under oxic conditions. *American Journal of Science, 299*(7–9), 529–555.

Henson, S., Le Moigne, F., & Giering, S. (2019). Drivers of carbon export efficiency in the global ocean. *Global Biogeochemical Cycles, 33*(7), 1–13. https://doi.org/10.1029/2018GB006158

Honda, M. C., Sasai, Y., Siswanto, E., et al. (2018). Impact of cyclonic eddies and typhoons on biogeochemistry in the oligotrophic ocean based on biogeochemical/physical/meteorological time-series at station KEO. *Progress in Earth and Planetary Science, 5*, 1–16. https://doi.org/10.1186/s40645-018-0196-3

Honjo, S., Dymond, J., Collier, R., et al. (1995). Export production of particles to the interior of the equatorial Pacific Ocean during the 1992 EqPac experiment. *Deep-Sea Research Part II, 42*(2–3), 831–870. https://doi.org/10.1016/0967-0645(95)00034-N

Honjo, S., Manganini, S. J., Krishfield, R. A., et al. (2008). Particulate organic carbon fluxes to the ocean interior and factors controlling the biological pump: A synthesis of global sediment trap programs since 1983. *Progress in Oceanography, 76*(3), 217–285. https://doi.org/10.1016/j.pocean.2007.11.003

Hüneke, H., & Henrich, R. (2011). Pelagic sedimentation in modern and ancient oceans. In H. Hüneke & T. Mulder (Eds.), *Deep sea sediments by deployments in sedimentology* (Vol. 63, pp. 215–351). Chap. 4, Elsevier. https://doi.org/10.1016/B978-0-444-53000-4.00004-4

Hutchins, D., Sedwick, P., DiTullio, G., et al. (2001). Control of phytoplankton growth by iron and silicic acid availability in the subantarctic Southern Ocean: Experimental results from the SAZ project. *Journal of Geophysical Research: Oceans, 106*, 31559–31572. https://doi.org/10.1029/2000JC000333

Ito, T., & Follows, M. J. (2005). Preformed phosphate, soft tissue pump and atmospheric CO_2. *Journal of Marine Research, 63*(4), 813–839.

Karl, D. M., Letelier, R., & Hebel, D. (1995). Ecosystem changes in the North Pacific subtropical gyre attributed to the 1991-92 El Niño. *Nature, 373*, 230–234. https://doi.org/10.1038/373230a0

Klaas, C., & Archer, D. E. (2002). Association of sinking organic matter with various types of mineral ballast in the deep sea: Implications for the rain ratio. *Global Biogeochemical Cycles, 16*(4), 1–14. https://doi.org/10.1029/2001GB001765

Knapp, A. N., DiFiore, P. J., Deutsch, C., et al. (2008). Nitrate isotopic composition between Bermuda and Puerto Rico: Implications for N_2 fixation in the Atlantic Ocean. *Global Biogeochemical Cycles, 22*, 1–14. https://doi.org/10.1029/2007GB003107

Knox, F., & McElroy, M. B. (1984). Changes in atmospheric CO_2: Influence of the marine biota at high latitude. *Journal of Geophysical Research: Atmospheres, 89*(D3), 4629–4637. https://doi.org/10.1029/JD089iD03p04629

Kolla, V., Henderson, L., & Biscaye, P. E. (1976). Clay mineralogy and sedimentation in the western Indian Ocean. Paper presented at the Deep-Sea Res. and Oceanogr. *Abstracts, 23*(10), 949–961. https://doi.org/10.1016/0011-7471(76)90825-1

Lam, P. J., Doney, S. C., & Bishop, J. K. B. (2011). The dynamic ocean biological pump: Insights from a global compilation of particulate organic carbon, CaCO3, and opal concentration profiles from the mesopelagic. *Global Biogeochemical Cycles, 25*, 1–14. https://doi.org/10.1029/2010GB003868

Lampitt, R. S., & Antia, A. N. (1997). Particle flux in deep seas: Regional characteristics and temporal variability. *Deep-Sea Research Part I, 44*(8), 1377–1403. https://doi.org/10.1016/S0967-0637(97)00020-4

Laufkötter, C., & Gruber, N. (2018). Will marine productivity wane? *Science, 359*(6380), 1103–1104. https://doi.org/10.1126/science.aat0795

Laufkötter, C., John, J. G., Stock, C. A., et al. (2017). Temperature and oxygen dependence of the remineralization of organic matter. *Global Biogeochemical Cycles, 31*(7), 1038–1050. https://doi.org/10.1002/2017GB005643

Laws, E. A., Falkowski, P. G., Smith, W. O., et al. (2000). Temperature effects on export production in the open ocean. *Global Biogeochemical Cycles, 14*(4), 1231–1246. https://doi.org/10.1029/1999GB001229

Lee, Z., Weidemann, A., Kindle, J., et al. (2007). Euphotic zone depth: Its derivation and implication to ocean-color remote sensing. *Journal of Geophysical Research, Oceans, 112*(C3), 1–11. https://doi.org/10.1029/2006JC003802

Lutz, M. J., Caldeira, K., Dunbar, R. B., et al. (2007). Seasonal rhythms of net primary production and particulate organic carbon flux to depth describe the efficiency of biological pump in the global ocean. *Journal of Geophysical Research, Oceans, 112*(C10), 1–26. https://doi.org/10.1029/2006JC003706

Lyle, M. (2014). Deep-Sea sediments. *Encyclopedia of Marine Geosciences*, 1–21. https://doi.org/10.1007/978-94-007-6644-0_53-1

Marsay, C. M., Sanders, R. J., Henson, S. A., et al. (2015). Attenuation of sinking particulate organic carbon flux through the mesopelagic ocean. *Proceedings of the National Academy of Sciences, 112*(4), 1089–1094. https://doi.org/10.1073/pnas.1415311112

Martin, J. H., Knauer, G. A., Karl, D. M., et al. (1987). VERTEX: Carbon cycling in the Northeast Pacific. *Deep Sea Research, 34*(2), 267–285.

Mayor, D. J., Sanders, R., Giering, S. L., et al. (2014). Microbial gardening in the ocean's twilight zone: Detritivorous metazoans benefit from fragmenting, rather than ingesting, sinking detritus: Fragmentation of refractory detritus by zooplankton beneath the euphotic zone stimulates the harvestable production of labile and nutritious microbial biomass. *BioEssays, 36*(12), 1132–1137. https://doi.org/10.1002/bies.201400100

Mayor, D. J., Gentleman, W. C., & Anderson, T. R. (2020). Ocean carbon sequestration: Particle fragmentation by copepods as a significant unrecognised factor? Explicitly representing the role of copepods in biogeochemical models may fundamentally improve understanding of future ocean carbon storage. *BioEssays, 42*(12), 2000149. https://doi.org/10.1002/bies.202000149

McCartney, M. S. (1977). Subantarctic mode water. In M. Angel (Ed.), *A voyage of discovery, George Deacon 70th Anniversary* (pp. 103–119). Pergamon.

McCartney, M. S. (1982). The subtropical recirculation of mode waters. *Journal of Marine Research, 40*(436), 427–464.

McCave, I. N. (1975). Vertical flux of particles in the ocean. *Deep Sea Research, 22*, 491–502.

McClain, C. R., Signorini, S. R., & Christian, J. R. (2004). Subtropical gyre variability observed by ocean-color satellites. *Deep-Sea Research Part II, 51*(1–3), 281–301. https://doi.org/10.1016/j.dsr2.2003.08.002

McDonagh, E. L., Bryden, H. L., King, B. A., et al. (2005). Decadal changes in the South Indian Ocean thermocline. *Journal of Climate, 18*(10), 1575–1590. https://doi.org/10.1175/JCLI3350.1

McMonigal, K., Beal, L. M., & Willis, J. K. (2018). The seasonal cycle of the South Indian Ocean sub-tropical gyre circulation as revealed by Argo and Satellite data. *Geophysical Research Letters, 45*(17), 9034–9041. https://doi.org/10.1029/2018GL078420

Michaels, A. F., Bates, N. R., Buesseler, K. O., et al. (1994). Carbon-cycle imbalances in the Sargasso Sea. *Nature, 372*(6506), 537–540. https://doi.org/10.1038/372537a0

Michaels, A. F., Olson, D., Sarmiento, J., et al. (1996). Inputs, losses and transformations of nitrogen and phosphorus in the pelagic North Atlantic Ocean. *Biogeochemistry, 35*, 181–226. https://doi.org/10.1007/BF02179827

Minagawa, M., & Wada, E. (1986). Nitrogen isotope ratios of red tide organisms in the East China Sea: A characterization of biological nitrogen fixation. *Marine Chemistry, 19*, 245–259. https://doi.org/10.1016/0304-4203(86)90026-5

Montoya, J. P., & McCarthy, J. J. (1995). Isotopic fractionation during nitrate uptake by phytoplankton grown in continuous culture. *Journal of Plankton Research, 17*, 439–464. https://doi.org/10.1093/plankt/17.3.439

Montoya, J. P., Carpenter, E. J., & Capone, D. G. (2002). Nitrogen fixation and nitrogen isotope abundances in zooplankton of the oligotrophic North Atlantic. *Limnology and Oceanography, 47*, 1617–1628. https://doi.org/10.4319/lo.2002.47.6.1617

Morel, A., & Berthon, J. F. (1989). Surface pigments, algal biomass profiles, and potential production of the euphotic layer: Relationships reinvestigated in view of remote-sensing applications. *Limnology and Oceanography, 34*(8), 1545–1562. https://doi.org/10.4319/lo.1989.34.8.1545

Mouw, C. B., Barnett, A., McKinley, G. A., et al. (2016). Phytoplankton size impact on export flux in the global ocean: Phytoplankton size impact on export flux. *Global Biogeochemical Cycles, 30*, 1542–1562. https://doi.org/10.1002/2015GB005355

Müller, P. J., & Suess, E. (1979). Productivity, sedimentation rate, and sedimentary organic matter in the oceans I. *Organic carbon preservation. Deep-Sea Res. A., 26*(12), 1347–1362. https://doi.org/10.1016/0198-0149(79)90003-7

Muromtsev, A. (1959). *Osnovnye cherty gidrologiiIndiiskogookeana (the main features of Indian Ocean hydrology)*. Gidrometeoizdat, Leningrad, 160 charts.

Murray, J. W., Downs, J. N., Strom, S., et al. (1989). Nutrient assimilation, export production and 234Th scavenging in the eastern equatorial Pacific. *Deep-Sea Research Part I, 36*(10), 1471–1489. https://doi.org/10.1016/0198-0149(89)90052-6

NASA Earth Data Base. (2020). *Giovanni*. https://giovanni.gsfc.nasa.gov/giovanni/. Accessed 19 May 2020.

NASA Earth Sciences. (2020). *Ocean Productivity Data Server*. http://sites.science.oregonstate.edu/ocean.productivity/index.php. Accessed 19 May 2020.

Pabortsava, K., Lampitt, R. S., Benson, J., et al. (2017). Carbon sequestration in the deep Atlantic enhanced by Saharan dust. *Nature Geoscience, 10*(3), 189–194. https://doi.org/10.1038/ngeo2899

Passow, U., & Carlson, C. A. (2012). The biological pump in a high CO_2 world. *Marine Ecology Progress Series, 470*, 249–271. https://doi.org/10.3354/meps09985

Pickard, G., & Emery, W. (1982). *Descriptive physical oceanography*. Pergamon.

Piola, A. R., & Gordon, A. L. (1989). Intermediate waters in the southwest South Atlantic. *Deep Sea Research, 36*, 1–16. https://doi.org/10.1016/0198-0149(89)90015-0

Quadfasel, D. R., & Schott, F. (1982). Water-mass distributions at intermediate layers off the Somali coast during the onset of the southwest monsoon, 1979. *Journal of Physical Oceanography, 12*, 1358–1372.

Rafter, P. A., DiFiore, P. J., & Sigman, D. M. (2013). Coupled nitrate nitrogen and oxygen isotopes and organic matter remineralization in the Southern and Pacific Oceans. *Journal of Geophysical Research: Oceans, 118*, 4781–4794. https://doi.org/10.1002/jgrc.20316

Redfield, A. C. (1934). On the proportions of organic derivatives in sea water and their relation to the composition of plankton. *James Johnstone memorial volume*, 176–192.

Redfield, A. C., Ketchum, B. H., & Richards, F. A. (Eds.). (1963). *The sea: Idea and observations. The influence of organisms on the composition of sea water* (pp. 26–77). Wiley-Interscience Publishers.

Reid, J. L. (1986). On the total geostrophic circulation of the South Pacific Ocean: Flow patterns, tracers and transports. *Progress in Oceanography, 16*, 1–61. https://doi.org/10.1016/S00796611(97)00012-8

Reid, J. L. (1989). On the total geostrophic circulation of the South Atlantic Ocean: Flow patterns, tracers, and transports. *Progress in Oceanography, 23*, 149–244. https://doi.org/10.1016/0079-6611(89)90001-3

Rixen, T., Gaye, B., & Emeis, K.-C. (2019a). The monsoon, carbon fluxes, and the organic carbon pump in the northern Indian Ocean. *Progress in Oceanography, 175*, 24–39. https://doi.org/10.1016/j.pocean.2019.03.001

Rixen, T., Gaye, B., Emeis, K.-C., et al. (2019b). The ballast effect of lithogenic matter and its influences on the carbon fluxes in the Indian Ocean. *Biogeosciences, 16*(2), 485–503. https://doi.org/10.5194/bg-16-485-2019

Sarmiento, J. L., & Gruber, N. (2006). *Ocean biogeochemical dynamics*. Princeton University Press.

Sarmiento, J. L., & Toggweiler, J. (1984). A new model for the role of the oceans in determining atmospheric pCO_2. *Nature, 308*, 621–624. https://doi.org/10.1038/308621a0

Sarmiento, J. L., Hughes, T. M. C., Stouffer, R. J., et al. (1998). Simulated response of the ocean carbon cycle to anthropogenic climate warming. *Nature, 393*(6682), 245–249. https://doi.org/10.1038/30455

Schlitzer, R. (2019). *Ocean data view*. https://odv.awi.de. Accessed 22 Feb 2020.

Schott, F. A., & McCreary, J. P. (2001). The monsoon circulation of the Indian Ocean. *Progress in Oceanography, 51*, 1–123. https://doi.org/10.1016/S0079-6611(01)00083-0

Schott, F. A., Xie, S.-P., & McCreary, J. P., Jr. (2009). Indian Ocean circulation and climate variability. *Reviews of Geophysics, 47*(1002), 1–123. https://doi.org/10.1029/2007RG000245

Schrader, H.-J. (1971). Fecal pellets: Role in sedimentation of pelagic diatoms. *Science, 174*, 55–57. https://doi.org/10.1126/science.174.4004.55

Shanks, A. L., & Trent, J. D. (1980). Marine snow: Sinking rates and potential role in vertical flux. *Deep-Sea Research Part I, 27*(2), 137–143. https://doi.org/10.1016/0198-0149(80)90092-8

Sharma, G. (1976). Transequatorial movement of water masses in the Indian Ocean. *Journal of Marine Research, 34*, 143–154.

Siegel, D. A., Granata, T. C., Michaels, A. F., et al. (1990). Mesoscale eddy diffusion, particle sinking, and the interpretation of sediment trap data. *Journal of Geophysical Research, Oceans, 95*(C4), 5305–5311. https://doi.org/10.1029/JC095iC04p05305

Sigman, D. M., Altabet, M. A., McCorkle, D. C., et al. (1999). The δ15N of nitrate in the Southern Ocean: Consumption of nitrate in surface waters. *Global Biogeochemical Cycles, 13*, 1149–1166. https://doi.org/10.1029/1999GB900038

Sigman, D. M., Altabet, M. A., McCorkle, D. C., et al. (2000). The $\delta^{15}N$ of nitrate in the Southern Ocean: Nitrogen cycling and circulation in the ocean interior. *Journal of Geophysical Research: Oceans, 105*, 19599–19614. https://doi.org/10.1029/2000JC000265

Sigman, D. M., Robinson, R., Knapp, A., et al. (2003). Distinguishing between water column and sedimentary denitrification in the Santa Barbara Basin using the stable isotopes of nitrate. *Geochemistry, Geophysics, Geosystems, 4*(1040), 1–20. https://doi.org/10.1029/2002GC000384

Sigman, D. M., Granger, J., DiFiore, P. J., et al. (2005). Coupled nitrogen and oxygen isotope measurements of nitrate along the eastern North Pacific margin. *Global Biogeochemical Cycles, 19*, 1–14. https://doi.org/10.1029/2005GB002458

Sigman, D. M., Karsh, K. L., & Casciotti, K. L. (Eds.). (2009). *Ocean process tracers: Nitrogen isotopes in the ocean. Encyclopedia of Ocean sciences* (pp. 40–54). Elsevier. https://doi.org/10.1016/B978-012374473-9.00632-9

Signorini, S. R., Franz, B. A., & McClain, C. R. (2015). Chlorophyll variability in the oligotrophic gyres: Mechanisms, seasonality and trends. *Frontiers in Marine Science, 2*(1), 1–11. https://doi.org/10.3389/fmars.2015.00001

Stramma, L. (1991). The South Indian Ocean current. *Journal of Physical Oceanography, 22*(4), 421–430. https://doi.org/10.1175/1520-0485(1992)022<0421:TSIOC>2.0.CO;2

Stramma, L. (1992). The South Indian Ocean current. *Journal of Physical Oceanography, 22*, 421–430. https://doi.org/10.1175/1520-0485(1992)022<0421:TSIOC>2.0.CO;2

Stramma, L., & Lutjeharms, J. R. (1997). The flow field of the subtropical gyre of the South Indian Ocean. *Journal of Geophysical Research: Oceans, 102*, 5513–5530. https://doi.org/10.1029/96JC03455

Suess, E. (1980). Particulate organic carbon flux in the oceans–surface productivity and oxygen utilization. *Nature, 288*(5788), 260–263. https://doi.org/10.1038/288260a0

Sverdrup, H. U., Johnson, M. W., & Fleming, R. H. (1942). The oceans: Their physics, chemistry, and general biology. *Prentice Hall New York, 7.*

Talley, L. D. (1996). Antarctic intermediate water. In G. Wefer, W. H. Berger, G. Siedler, & D. J. Webb (Eds.), *The south Atlantic—Present and past circulation* (pp. 219–238). Springer. https://doi.org/10.1007/978-3-642-80353-6_11

Talley, L. D. (2013). Closure of the global overturning circulation through the Indian, Pacific, and Southern Oceans: Schematics and transports. *Oceanography, 26*(1), 80–97. https://doi.org/10.5670/oceanog.2013.07

Toole, J. M., & Warren, B. A. (1993). A hydrographic section across the subtropical South Indian Ocean. *Deep-Sea Research Part I, 40*, 1973–2019. https://doi.org/10.1016/0967-0637(93)90042-2

Turner, J. T. (2015). Zooplankton fecal pellets, marine snow, phytodetritus and the ocean's biological pump. *Progress in Oceanography, 130*, 205–248. https://doi.org/10.1016/j.pocean.2014.08.005

Volk, T., & Hoffert, M. I. (1985). Ocean carbon pumps: Analysis of relative strengths and efficiencies in ocean-driven atmospheric CO_2 changes. In E. T. Sundquist & W. S. Broecker (Eds.), *The carbon cycle and atmospheric CO_2: Natural variations Archean to present, geophysical monograph series* (Vol. 32, pp. 99–110). AGU. https://doi.org/10.1029/GM032p0099

Wada, E., & Hattori, A. (1976). Natural abundance of 15N in particulate organic matter in the North Pacific Ocean. *Geochimica et Cosmochimica Acta, 40*, 249–251. https://doi.org/10.1016/0016-7037(76)90183-6

Wankel, S. D., Kendall, C., Pennington, J. T., et al. (2007). Nitrification in the euphotic zone as evidenced by nitrate dual isotopic composition: Observations from Monterey Bay. *Global Biogeochemical Cycles, 21*(2), 1–13. https://doi.org/10.1029/2006GB002723

Warren, B. A. (1981). Transindian hydrographic section at Lat. 18°S: Property distributions and circulation in the South Indian Ocean. *Deep Sea Research, 28*, 759–788. https://doi.org/10.1016/S0198-0149(81)80001-5

Waser, N., Harrison, P., Nielsen, B., et al. (1998). Nitrogen isotope fractionation during the uptake and assimilation of nitrate, nitrite, ammonium, and urea by a marine diatom. *Limnology and Oceanography, 43*, 215–224. https://doi.org/10.4319/lo.1998.43.2.0215

Weber, T., Cram, J. A., Leung, S. W., et al. (2016). Deep ocean nutrients imply large latitudinal variation in particle transfer efficiency. *PNAS, 113*(31), 8606–8611. https://doi.org/10.1073/pnas.1604414113

Wefer, G., & Fischer, G. (1991). Annual primary production and export flux in the Southern Ocean from sediment trap data. *Marine Chemistry, 35*(1–4), 597–613. https://doi.org/10.1016/S0304-4203(09)90045-7

Williams, R. G., & Follows, M. J. (1998). The Ekman transfer of nutrients and maintenance of new production over the North Atlantic. *Deep-Sea Research Part I, 45*, 461–489. https://doi.org/10.1016/S0967-0637(97)00094-0

Williams, R. G., & Follows, M. J. (2003). Physical transport of nutrients and the maintenance of biological production. In M. J. R. Fasham (Ed.), *Ocean biogeochemistry, Global Change—The IGBP series* (pp. 19–51). Springer. https://doi.org/10.1007/978-3-642-55844-3_3

Williams, P. J. l. B., Quay, P. D., Westberry, T. K., et al. (2013). The oligotrophic ocean is autotrophic. *Annual Review of Marine Science, 5*, 535–549. https://doi.org/10.1146/annurev-marine-121211-172335

Wong, C. S., Whitney, F. A., Matear, R. J., et al. (1998). Enhancement of new production in the northeast subarctic Pacific Ocean during negative North Pacific index events. *Limnology and Oceanography, 43*(7), 1418–1426. https://doi.org/10.4319/lo.1998.43.7.1418

Woodberry, K. E., Luther, M. E., & O'Brien, J. J. (1989). The wind-driven seasonal circulation in the southern tropical Indian Ocean. *Journal of Geophysical Research: Oceans, 94*, 17985–18002. https://doi.org/10.1029/JC094iC12p17985

Wüst, G. (1935). Die Stratosphäre. *Wissenschaftliche Ergebnisse der Deutschen Atlantischen Expedition auf dem Vermessungs- und Forschungsschiff "Meteor", 1925–1927, 6*, 109–288.

Wyrtki, K. (1971). *Oceanographic atlas of the international Indian Ocean expedition*. National Science Foundation.

Wyrtki, K. (1973). *Physical oceanography of the Indian Ocean. The biology of the Indian Ocean* (pp. 18–36). Springer. https://doi.org/10.1007/978-3-642-65468-8_3

You, Y. (1998). Intermediate water circulation and ventilation of the Indian Ocean derived from water-mass contributions. *Journal of Marine Research, 56*, 1029–1067. https://doi.org/10.1357/002224098765173455

You, Y., & Tomczak, M. (1993). Thermocline circulation and ventilation in the Indian Ocean derived from water mass analysis. *Deep-Sea Research Part I, 40*, 13–56. https://doi.org/10.1016/0967-0637(93)90052-5

Natalie C. Harms is a geoscientist and works as a research associate in the Biogeochemistry Group at the Universität Hamburg. Her field of interest is the marine cycling of elements, especially carbon and nitrogen, isotope geochemistry and biogeochemistry of amino acids in various particulate matter pools. She is an expert in sediment trap moorings and has participated in six research expeditions in the Indian Ocean.

Niko Lahajnar is a lecturer in geology and biogeochemistry at the Universität Hamburg. He heads the Teaching and Research Laboratory for Biogeochemistry and Elemental Cycles there. His research interests are in marine matter cycling with a focus on carbon and nitrogen, isotope geochemistry and amino acid biogeochemistry. He is an expert in sediment trap moorings and has conducted over 75 research expeditions in all seas and oceans from the poles to the equator.

Birgit Gaye is a biogeochemist at the Institute for Geology of the University of Hamburg. Focusses of her research are oceanic nitrogen cycling in the past and present, proxies of biogeochemical cycling preserved in sediments and paleoclimate reconstructions from marine and terrestrial records.

Tim Rixen is a biogeochemist working on carbon and nutrient cycling in the ocean and in tropical coastal ecosystems. Along the land–ocean continuum, he studies the effects of river discharges on marine biogeochemistry, while in the ocean, his work focuses mainly on upwelling systems. Upwelling systems are key elements in the marine carbon cycle, as they act as valves connecting the deep ocean to the atmosphere. In this context, Tim Rixen is particularly interested in the functioning of marine carbon pumps, as they control CO_2 fluxes at the air–sea interface and the export of carbon to the deep ocean.

Ralf Freitag is a marine geologist at the German Federal Institute for Geosciences and Natural Resources (BGR). Coming from onshore structural geology and tectonics, he focused on divergent margins and has specialised in marine hydro-acoustics and seafloor mapping. This combined knowledge of tectonics and hydro acoustic expertise led to numerous discoveries of new hydrothermal vent sites. He has more than 20 years of experience in marine research cruises and has worked for 12 years in the Indian Ocean.

Chapter 13
Multilevel Assessment and Options for the Management of Cumulative Impacts on Pelagic Ecosystems in the North-Eastern Tropical Pacific Ocean

Virginie Tilot, Léa Fourchault, Alain Jeudy de Grissac, Jérôme Mallefet, and Juan Moreno Navas

Abstract The multilevel assessment of the water column located above the seabed of the Clarion-Clipperton Zone, an area rich in mineral resources targeted for the mining industry, first reveals that interconnected ecological and oceanographic processes support numerous ecosystem services. As a result, the cumulative impacts of fishing, deep-sea mining, shipping, and climate change on organisms across the epi-, meso-, and bathy-pelagic layers of this region are highlighted, as well as their impact on ecosystem services. Finally, management measures and conservation strategies are proposed, highlighting the socio-ecological interconnectivity between marine life and humans, to foster sustainable ecosystem services for global human well-being and to ensure sustainable livelihoods for the Pacific communities.

V. Tilot (✉)
Muséum National d'Histoire Naturelle, Paris, France

Académie Royale des Sciences d'Outre-Mer, Brussels, Belgium
e-mail: v.tilot@wanadoo.fr

L. Fourchault
Institute of Natural Sciences, Brussels, Belgium

A. J. de Grissac
Académie des Sciences d'Outremer, Paris, France

J. Mallefet
Université Catholique de Louvain, Laboratoire de Biologie Marine, Ottignies-Louvain-la-Neuve, Belgium

J. M. Navas
Centro Nacional Instituto Español de Oceanografía, CSIC. Centro Oceanográfico de Málaga, Málaga, Spain

R. Sharma (ed.), *Deep-Sea Mining and the Water Column*,
https://doi.org/10.1007/978-3-031-59060-3_13

Keywords Deep-sea mining · Water column · Pelagic · Ecosystem services · Fisheries · Anthropogenic and natural impacts · Sediment plumes · Conservation strategies · Socio-ecological interconnectivity · Management measures

1 Introduction

The Clarion-Clipperton Zone (CCZ), located in the north-eastern tropical Pacific (118°–157° W/9°–16° N), holds the most commercially interesting marine polymetallic nodule deposits globally (Morgan, 2000; Petersen et al., 2016). Following prospects of deep-sea mining (DSM) activities, a comprehensive study of the biodiversity and distribution of epibenthic megafauna of the CCZ, serving as a baseline study (Tilot, 2006a, b), led to the development of several propositions of conservation strategy and management tools in view of the assessed impacts (ESCO CNRS IFREMER, 2014; Tilot, 2010, 2013, 2014, 2019; Tilot et al., 2018).

This chapter furthers this approach to highlight potential cumulative impacts resulting from DSM and other anthropogenic activities as well as global change on ecosystems found throughout the eastern Pacific's water column (Willett et al., 2006; Tilot, 2010, 2019; Jones et al., 2017; Levin et al., 2018; Tilot et al., 2018; Sharma, 2019; Christiansen et al., 2020; Blanluet et al., 2023). Strategies for the conservation of pelagic ecosystems in this area are suggested, highlighting the socio-ecological interconnectivity between marine life and humans to ensure sustained ecosystems services for human well-being and sustainable livelihoods for Pacific communities (Tilot et al., 2021a, b, c).

2 Physical and Chemical Properties of the Water Column in the CCZ

Marine ecosystems in open systems are in constant flux and can be affected by marine and climatic parameters generated locally and elsewhere. Such ecosystems are characterized by short-term perturbations and disruptions as well as by more regular, seasonal patterns according to which marine species adapt their life cycle patterns. The physical and chemical characteristics shaping pelagic ecosystems in the tropical north-eastern Pacific display a considerable regional-scale variation, as outlined in the following paragraphs.

2.1 Surface Currents and Eddies

Surface currents influence many of the ocean patterns, dynamics, and processes observed in the CCZ. The eastern Pacific is characterized by two cold-water boundary currents: the southward flowing California Current in the north and the northward flowing Peru Current in the south (Fiedler & Talley, 2006).

The CCZ itself is dominated by the North Equatorial Current (NEC, 8 °N – 20°N). This broad westward flow of the subtropical gyre brings comparatively richer waters from the easternmost Pacific Ocean to the CCZ (Fiedler & Talley, 2006; Kessler, 2006). Another important current in the southern CCZ is the strong eastward flowing North Equatorial Counter-current (NECC, 5°N and 10°N). Both the NEC and NECC have mean speeds of 0.4 knots at a depth of 20 m, decreasing to 0.2 knots at 300 m to become relatively weak (0.08–0.15 knots) at 300–4800 m depths (Worthington, 1981).

These equatorial currents are complemented by eddies, i.e., circulations of about 50–200 km diameter in surface, often variable over a period of weeks to months (Talley et al., 1998). Their presence results in sea-surface elevation variability of 4 cm or more over the CCZ, caused by excess kinetic energy (Hollister & Nowell, 1991; Kontar & Sokov, 1994). The occurrence of eddies is higher over the eastern part of the CCZ, where it produces greater mixing (Chang et al., 2012; Purkiani et al., 2020).

2.2 *Midwater Circulation*

Maps of net transport and circulation from 500 to 1500 m in the equatorial Pacific exhibit two almost symmetrical cyclonic gyres between 15°S and 15°N for the Equatorial Pacific Intermediate Water (EqIW) (Bostock et al., 2010; Reid, 1997). Intermediate waters in the CCZ are formed through influx of North Pacific Intermediate Waters into the North EqIW, and circulation patterns broadly mirror surface circulation (Fig. 13.1, Bostock et al., 2010).

2.3 *Deep Circulation*

Bottom currents in the north-eastern tropical Pacific also broadly mirror surface currents, and are predominantly weak (Shor, 1959), between 2 and 25 cm/s (Amos et al., 1977). The flow is characterized by alternating periods of stronger, quasi-unidirectional currents and periods of sluggish water movement (McCartney, 1982; Kontar & Sokov, 1994, 1997). Specifically, sediment distribution in the deep Pacific basin is controlled by the western, northward flowing Lower Circum Polar Water (LCPW), which is characterized by low bottom current velocities and high oxygen concentrations. The LCPW becomes the North Pacific Deep Water (NPDW) in the central-north Pacific, as the LCPW warms and gains silica. The southward flow of NPDW bounds the CCZ in the east, bringing these deep polar waters (Hayes, 1979; Mewes et al., 2014).

The surface and deeper currents are strongly coupled to each other, as evidenced in ocean modeling, where subsurface flows dive down the water column to merge into the deep circulation. There are local accelerations of flow caused by the morphology of the seabed, where rises and seamounts strongly steer bottom flows. Several hydrographic disturbances, such as periodic benthic storms, have been recorded during long-term in situ monitoring (Gardner et al., 1984; Richardson

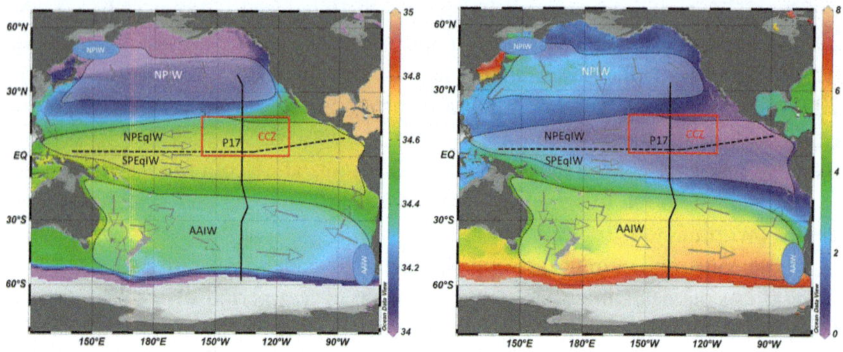

Fig. 13.1 Map view of the (**a**) salinity and (**b**) oxygen concentration (mmol l^{-1}) at the Pacific basin scale (along the potential density (sigma theta) of 27.1 kg m^{-3}), highlighting currents and water masses. The possible midwater circulation is depicted by the gray arrows. Formation regions of North Pacific Intermediate Water (NPIW) and Antarctic Intermediate Water (AAIW) are represented in blue circles. Intermediate waters are shaded in light gray (NEqPIW—North Equatorial Pacific Intermediate Water; SEqPIW—South Equatorial Pacific Intermediate Water), separated by regions of mixing. The CCZ is framed by the red box. (Adapted from Bostock et al. 2010)

et al., 1993; Kontar & Sokov, 1994; Aller, 1997). During benthic storms, cold currents flowing toward the equator in the deep ocean are accelerated into powerful sediment-transporting events, associated with a deep-reaching anticyclonic eddy (Hollister & Nowell 1991).

2.4 *Water Masses*

The water body in the CCZ is divided into three layers: an upper mixed layer that overlies a strong permanent thermocline and a deep layer.

First, the *upper* 500 m of the north-eastern tropical Pacific can be divided into three layers: (1) a top layer of weak temperature and salinity gradients extending from the surface up to 100 m depth, (2) a sharp thermocline with temperature gradients from 28 to 12 °C, over a water depth of approximately 100 m, and (3) a region of lower stratification below the thermocline, with a salinity maximum (Wijffels et al., 1996). These upper layers include the Pacific Equatorial Water, strongly linked to the El-Nino cycle, and the Eastern North Pacific Central Water (Emery, 2003).

The upper layer varies in thickness as it is related to seasonal variations in wind and wave conditions. Overall, temperatures are several degrees cooler in the north-eastern tropical Pacific due to the thermal structure of the upper ocean where the surface water mass layer is shallower (<100 m thick) and the thermocline closer to the surface caused by the upwelling along the coasts of the Americas and the eastern boundary currents which flow toward the equator (Wyrtki & Kilonsky, 1984).

Second, *intermediate waters* in oceanic basins comprise a range of depths between 550 and 1500 m, as the strong permanent thermocline separating the surface and intermediate waters exhibits seasonal variability (Emery, 2003). Despite this variability, three major intermediate water masses have been identified in the Pacific Ocean, based on temperature and salinity diagrams: North Pacific Intermediate Waters, Antarctic Intermediate Waters (Bostock et al., 2010; Dickson et al., 2001; Talley, 1993), and an Equatorial Pacific Intermediate Water, which dominates the CCZ (Figs. 13.1 and 13.2, Bingham & Lukas, 1995; Bostock et al., 2010; Wijffels, 1996; Bostock et al., 2010).

The North Pacific Intermediate Water appears from 400 to 800 m and is characterized by a salinity minimum (reaching 33.8 ppm/ppt), low oxygen concentrations (50–150 mmol kg-1), and low density (26.4–27.2 g/cm^3) (Dickson et al., 2001, Talley et al., 2011). Antarctic Intermediate Waters presents a high oxygen content (200–250 mmol kg^{-1}), a minimum salinity (34.3–34.533.8 ppm/ppt), and a temperature range from 3.5 to 10 °C, resulting in an average density of 27.1 g/cm^3 (Bostock et al., 2010; Tsuchiya & Talley, 1996). The coldest branch of the Antarctic Intermediate Water at 3.1 °C (McCartney, 1982; Talley, 1993) is slightly present in the CCZ (0.9%) (Catala et al., 2015a, b; Tilot et al., 2018).

From the analysis of water samples collected between 7 °N – 20 °N and 99 °W – 167 °W during the Malaspina 2010 circumnavigation, the thermohaline and intermediate waters in the CCZ present unique features. The water masses associated to this region undergo excessive aging, i.e., the oxygen consumption is higher than expected (Catalá et al., 2015a, b; Feely et al., 2004). This anomaly is associated with (1) the comparatively high biological productivity of the south-eastern part of the area, which exports organic material from the surface to the deep ocean (Feely et al., 2004) and (2) the sluggish deep circulation of the North Pacific, which facilitates organic matter sinking.

Last, the *deep-water* stratification near the seabed is weak in the CCZ (Hayes, 1979), in a very narrow range of temperature and salinity (1.2 °C and 34.70 PSU). These oldest water masses, located in the north-eastern Pacific between 1800 and 3500 m (Talley et al., 2011), are characterized by a distinct biogeochemistry, and greater molecular diversity, heterogeneity, and complexity of dissolved organic matter (Dittmar, 2015). Local microbial communities are expected to be adapted to hypoxic conditions, resulting in a different deep-sea microbial biodiversity. Both biochemistry and microbial communities are likely to influence the nodule ecosystem via pelagic-benthic transfers to the sediments.

2.5 *Topography, Habitats, and Eco-hydrodynamics*

The bottom depth in CCZ varies between 3800 and 4000 m in the eastern region and between 5400 and 5600 m in the western region. The seabed topography is mainly composed of abyssal hills, elongated in a north to south direction and running parallel to each other (Fig. 13.3). These hills have irregular, elliptically domed summits,

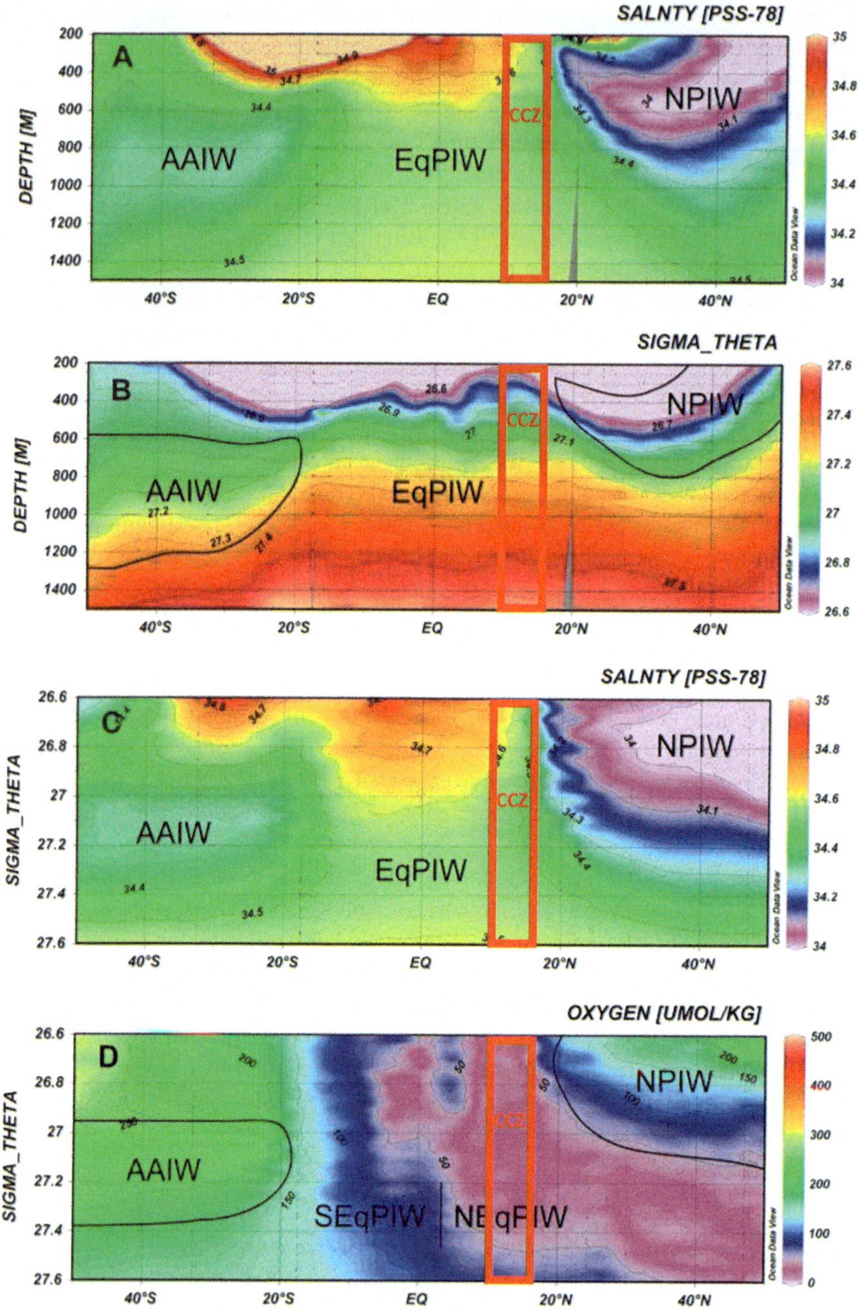

Fig. 13.2 North to south transect along the 140°W Meridian showing the intermediate waters characterizing the Pacific. The CCZ is located between 9 and 16°N (red box). Distribution of different parameters and derived variables with depth (200–1500 m): (**a**) Salinity and (**b**) potential density (sigma theta), with the salinity contour overlaid from A. Distribution of different parameters with potential density: (**c**) Salinity and (**d**) oxygen, with the contour overlaid from **c**. (Adapted from Bostock et al. 2010)

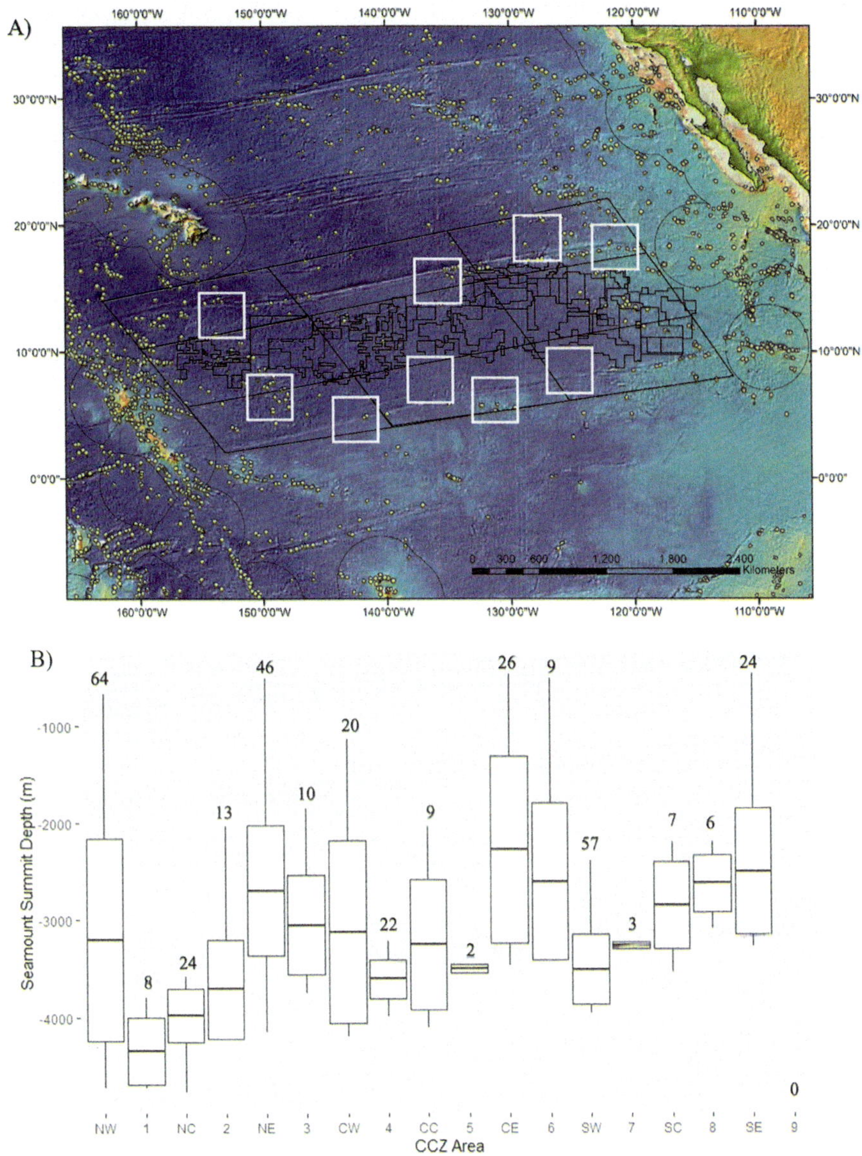

Fig. 13.3 Topographic characteristics of the CCZ: knolls, seamounts, and depth (Washburn et al., 2021). Depth (in meters) at a 1-minute resolution downloaded from ETOPO1. There is a bar for each CCZ subregion (NW northwest, NC northcentral, NE northeast, CW central west, CC central central, CE central east, SW southwest, SC southcentral, SE southeast) and Areas of Particular Environmental Interest (APEI) designated by ISA (APEIs 1–9). Each APEI follows the subregion it is meant to represent. Seamounts (**a**) and knolls (**c**) (in yellow) were identified using global bathymetric data at a 30 arc second resolution. Bar graphs for depth at a 0.5-degree resolution (in meters). Thick, black bars represent means, the lower and upper edges of the boxes represent 1 standard deviation, and the lower and upper edges of the lines represent minimums and maximums

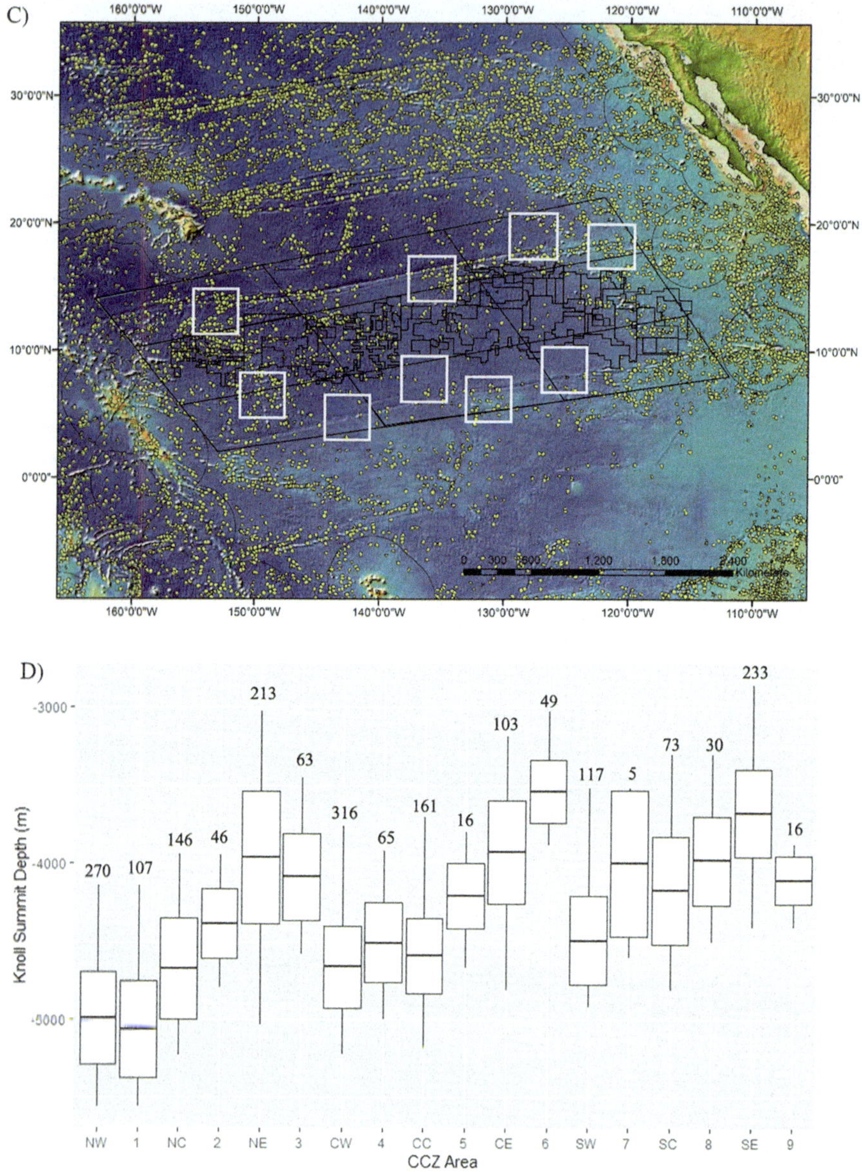

Fig. 13.3 (continued)

and steep, sometimes escarpment-like sides. Vertical cliffs of calcareous clay up to 40 m high occur on the flanks of the hills, and large potholes have been found on the crests that rise 100–300 m above the lowest areas (Menard, 1964; Halbach et al., 1988; Juan et al., 2018; Kotlinski, 1998; Washburn et al., 2021). Sediments are covered by nodule fields interspersed with some nodule-free sediment areas (Hoffert,

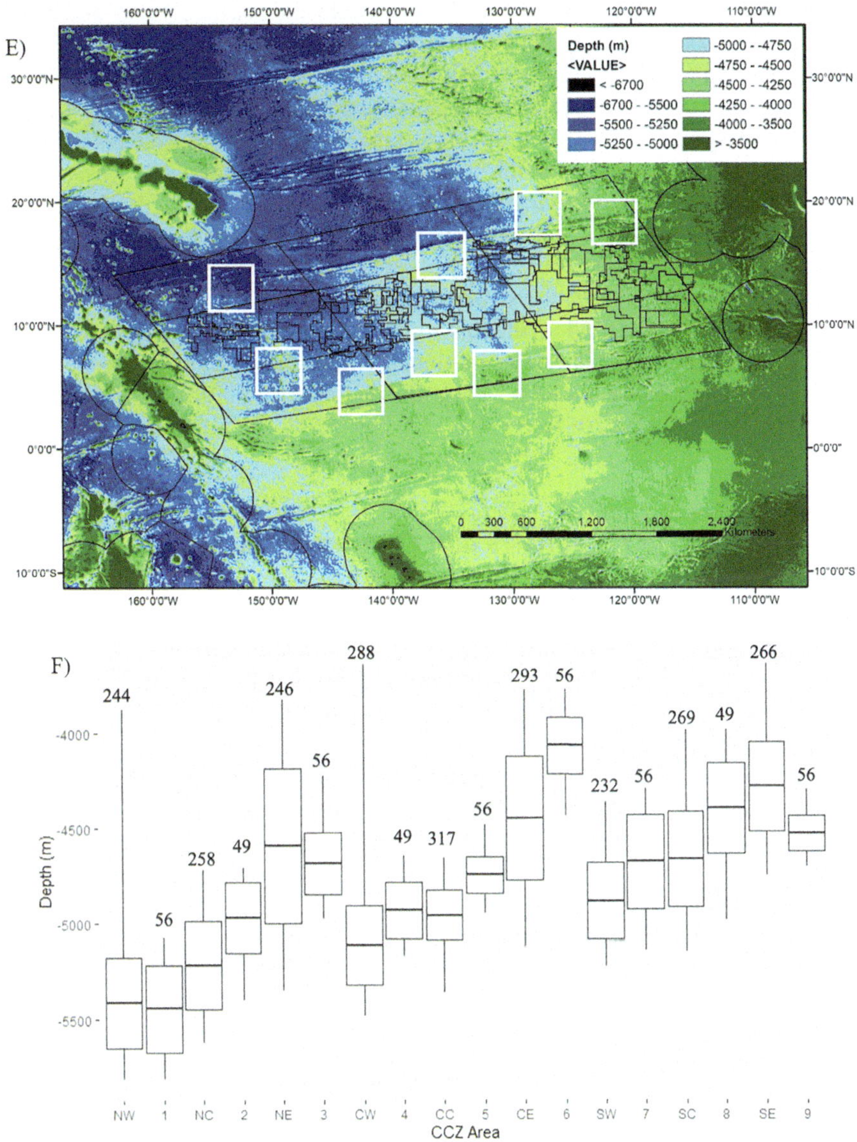

Fig. 13.3 (continued)

2008). In the eastern tropical Pacific, sediment coverage is about 20% thicker within morphological depressions than on the ridges. Sediment thickness increases westward from 50 to 150 m between 120° and 155°W longitude. Locally at some sites, seabed morphology comprises horsts and grabens that tend to generate stronger currents on slopes. Under the influence of sedimentary processes, the contours tend to soften (Shor, 1959).

The CCZ hosts 387 seamounts (>1000 m from the sea floor) and numerous knolls (200–1000 m) and highs (<200 m), especially in its eastern and western peripheries (Macdonald et al., 1996; Washburn et al., 2021) (Fig. 13.3a–f). Cobaltiferous crusts have been found on certain escarpments, rocky plateaus, ridges, and seamounts in the CCZ. Their thickness ranges from a few decimeters to a few meters, depending on the age of the seamounts (Macdonald et al., 1996; Washburn et al., 2021). This habitat heterogeneity locally increases the strength of currents, which influences biodiversity, especially with regard to suspension feeders (Tilot, 2006a, b, Tilot et al., 2018).

Larger seamounts, particularly when closer to the surface, can create local upwellings, eddies, and Taylor columns—a fluid dynamics phenomenon enhanced by the Coriolis effect. This results in higher surface productivity and supports a wide range of pelagic species, including sharks, tunas, and billfish (Morato et al., 2010; Watling & Auster, 2017).

2.6 Geophysical Parameters and Primary Productivity

The CCZ is in the pelagic bioregion of the Eastern Tropical Pacific (ETP) (UNESCO, 2009). The entire ETP is defined by a strong, shallow thermocline, a pronounced oxygen minimum zone, ranging from 250 to 2250 m with the lowest values at 500–750 m, and decadal El Niño Southern Oscillation (ENSO) variability (Willett et al., 2006; Longhurst, 2007).

In the eastern tropical Pacific, nitrate and phosphate concentrations are low in the upper layer due to uptake by photosynthesizing organisms. A nutrient maximum appears in the intermediate layer—centered at 1000 m depth, where rates of decomposition are high and rates of photosynthesis are low—followed by a decrease in nitrate concentration with increasing depth. Dissolved silica concentrations are low in the upper mixed layer due to uptake by phytoplankton, with dissolved silica concentrations increasing with increasing water column depth (Pennington et al., 2006; Vasiliu et al., 2022).

The overall primary productivity, particulate organic carbon fluxes, and trophic input primary productivity in the CCZ are comparatively low (Smith & Demopoulos, 2003; Pennington et al., 2006). For instance, particulate organic carbon fluxes range from 0.5 to 1.6 g C $m^{-2}year^{-1}$, placing the CCZ within the broadly defined mesotrophic abyss, as defined by Hannides and Smith (2003). However, considerable variation in the vertical flow of particles has been described (Karl et al., 1996; Baldwin et al., 1998; Karl, 2002; Smith & Rabouille, 2002), in line with seasonal patterns, upwelling, and El Niño events (Dymond & Collier, 1988; Ruhl & Smith, 2004). The openness of the ocean system leads to a dynamic and highly variable benthic-pelagic coupling (Rowe, 1971; Pfannkuche & Lochte, 1993) which results in variable sinking patterns of marine snow and consequently patches of deposition of phytodetritus on the seafloor (Smith et al., 1997, 2008a; Beaulieu & Smith, 1998). This causes significant changes in the structure of both benthic and pelagic faunal

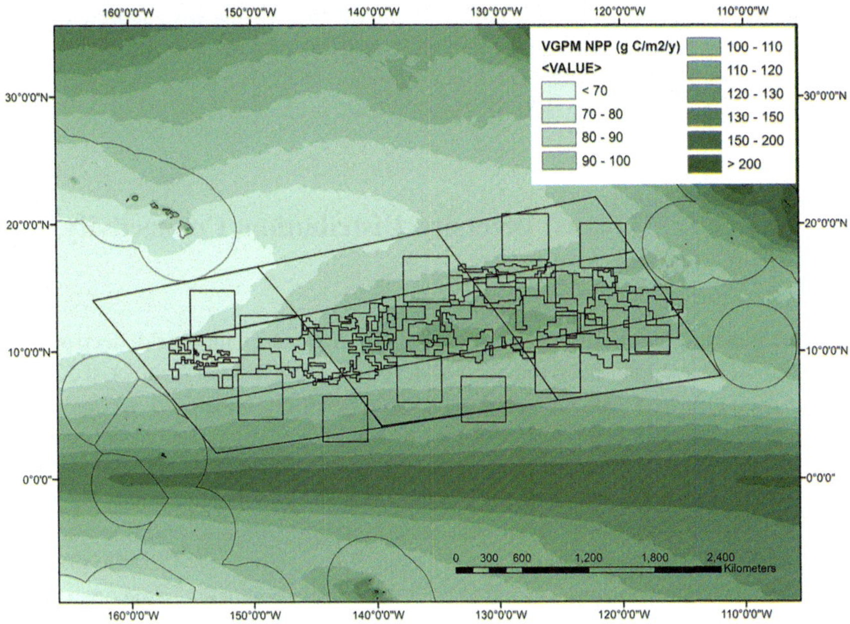

Fig. 13.4 Net primary productivity over the CCZ (Washburn et al., 2021). VGPM: vertically generalized production model. NPP: net primary productivity. Dissolved inorganic nitrogen (DIN) and dissolved inorganic phosphorus (DIP) display spatial variability in the upper ocean, with higher nutrient concentrations at low latitudes and a marked nutricline in the eastern region at ~150 m (DIN, ~32.0 μmol/L; DIP, ~2.4 μmol/L). Conversely, they are lower (DIN, ~1.2 μmol/L; DIP, ~0.1 μmol/L; at ~150 m) at high latitudes and in the western region, where there is no nutricline. Nutrient concentrations are more homogeneous spatially at greater depths. (Han et al., 2022)

assemblages and measurable effects on ecosystem functions (Smith et al., 1993; Lauerman et al., 1997; Drazen et al., 1998; Thurston et al., 1998; Danovaro, 2008; Loreau, 2008).

Overall, a marked spatial variation in net primary productivity can be observed in the water column above the CCZ, with east-west and north-south gradients in nutrient concentrations for upper layers (Fig. 13.4, Washburn et al., 2021). In the eastern CCZ, a moderate temporal variation has recently been observed, with higher temperatures, chlorophyll a, and particulate organic carbon concentrations from June to November (Gillard et al., 2022). This seasonality is likely caused by the meridional migration of the intertropical convergence zone (Amador et al., 2006). Nonetheless, productivity levels remain comparatively stable throughout the year (Fiedler & Talley, 2006).

Other than the varying vertical deposition of phyto-detritus, three major factors contribute to trophic input and distribution, hence influencing the structure of faunal assemblages. First, currents originating in zones of higher primary production can horizontally contribute suspended nutritive sediment particles. Second, certain edaphic and hydrological conditions above the bottom layer, likely related to the topography of the ocean floor, can favor the collection of nutritive sediment particles. Last, the sinking carcasses of pelagic organisms provide a significant input of carbon to benthic food webs (Dayton & Hessler, 1972; Haedrich & Rowe, 1977; Smith & Baco, 2003; Tyler, 2003).

3 Pelagic Life Forms: Diversity, Distribution, Connectivity, and Ecosystem Services

Despite their importance as the greatest source of biomass in the world, deep pelagic communities have long been the most understudied organisms (Webb et al., 2010). In the mesopelagic (200–1000 m depth), bathypelagic (1000–4000 m), and abyssopelagic realms (4000–6000 m), taxonomy, biodiversity, and functional relationships within ecosystems are still poorly described (Robison, 1992; Christiansen et al., 2020). However, deep pelagics are likely to play a significant role in the provision of ecosystem services on a global scale, nutrient cycling supporting fisheries, and carbon sequestration regulating the global climate (Jones et al., 1994; O'Leary & Roberts, 2017). Well-known examples of ecosystem engineers in the open ocean include phytoplankton and zooplankton (Jones et al., 1994; Breitburg et al., 2010), and baleen and sperm whales (Roman et al., 2014).

Numerous pelagic species use the CCZ for feeding, migration, and reproduction near the surface and extend their range to greater depths into the meso- and even bathypelagic (1000–4000 m) layers (Thorold et al., 2014; Abascal et al., 2010; Fuller et al., 2015; Howey et al., 2016).

3.1 Phytoplankton

Primary production of phytoplankton is dominant in the open ocean and comprises nearly half of the annual global net primary production (Field et al., 1998; Falkowski, 2012; Siegel et al., 2013). Through photosynthesis, organic matter fuels the metabolic demands of marine ecosystems and has an important regulatory function in the global carbon cycle, acting as biological pump sending carbon dioxide to the deep ocean (Siegel et al., 2013; Falkowski, 2012). Evidence has shown that changes in temperature, irradiance, nutrient, and physical circulation of the atmosphere and ocean affect phytoplankton production, as seen with algal blooms associated with El Niño (Falkowski, 2012; Not et al., 2012; Siegel et al., 2013; Chen et al., 2017). Notably, long-term data from the north-eastern Pacific show that variations in climate and surface primary productivity, as indicated for instance by the Bakun Upwelling Index, were significantly correlated with the flux of particulate organic carbon at 3500 m depth when lagged by 2–3 months (Smith et al., 2009).

Primary productivity in the CCZ is spatially patchy with significant temporal variability preventing homogenous mixing of deeper nutrient-rich waters with the surface waters (Peña et al., 1990; Pennington et al., 2006). The vertical distribution of phytoplankton is characterized by a peak in abundance within the mixed layer (first 50 m) of the water column, where the flux of organic matter is maximal (Kirchman, 2008). Because of the elevated primary productivity in the south-eastern part of the epipelagic layer of the CCZ, the water column below this layer is considered a hotspot for abyssal biodiversity in the region (Smith et al., 2008, Hernández-León et al., 2020).

3.2 *Zooplankton*

The large deep-sea zooplankton biomass mirrors primary production in the global ocean (Hernández-León et al., 2020). Zooplankton also participate in the fixation of carbon from the surface to the seabed. Through their vertical migrations (up to 500 m), zooplanktonic organisms ingest food near the surface, which becomes available to deep-sea organisms through excretion, respiration, or predation in deeper layers (Jónasdóttir et al., 2015; Sutton et al., 2017; Turner, 2015). This active flux likely represents 10–50% of the total vertical flux of carbon downward from surface waters (Turner, 2015). Global annual respiration of meso-zooplankton (e.g., copepods) across all depth zones of the ocean has been estimated between 8.8 and 17.2 billion tonnes of carbon, representing 17–32% of global open ocean primary productivity (Hernández-León & Ikeda, 2005).

Zooplankton in the CCZ includes both primary and secondary consumers of a broad taxonomic composition, such as fish larvae (ichthyoplankton), coelenterates (e.g., jellyfish), copepods, and chaetognaths (arrow worms). Within the eastern tropical Pacific, copepods and krill (euphausiid crustaceans) are the most important zooplankton groups in terms of biomass (Brinton, 1979). The vertical distribution of zooplankton is forced by the strong thermocline and the oxygen minimum zone, which occurs below the thermocline and extends down to 600–800 m beneath the surface. The abundance of zooplankton in the CCZ fluctuates relatively little throughout the year, as it is the case for most of the open ocean in the ETP area (Fernández Álamo & Färber Lorda, 2006).

3.3 *Invertebrate Nekton*

Invertebrate nekton in the CCZ, though poorly known, predominantly belongs to the cephalopods (e.g., phylum Mollusca) or crustaceans (e.g., phylum Arthropoda). Pelagic cephalopods, including squids, form the intermediate step between the micronekton and top pelagic predators such as commercially important tuna and swordfish, mammals, seabirds, and sharks (Young et al., 2013). In line with its dependence on nutrients from epipelagic layers, deep-sea nekton is likely more abundant in the eastern part of the CCZ.

3.4 *Mesopelagic Fishes*

Mesopelagic fishes are likely the most abundant vertebrates on Earth, with an estimated global biomass of 10 billion tons, representing the largest structuring biomass in the open sea (Irigoien et al., 2014). Due to limited food availability, fishes in the deeper parts of the midwater environment typically have lower metabolic rates, softer skeletons, and reduced protein levels, and they respond to cyclic hydrodynamic patterns and periodic fluctuations of variables such as temperature, salinity, phytopigments, and oxygen concentration as well as ecological needs, including food acquisition, reproduction, or avoiding predators (Dietz, 1962) aggregating in moving deep scattering layers. Like their zooplanktonic prey, deep pelagic organisms are also known to vertically migrate, displacements known as diel vertical migrations (DVMs) when responding to a solar-based synchronization between the photic and aphotic realms (Brierly, 2014; Robinson et al., 2010), hence contributing to connectivity throughout the water column and supplying food to deep seafloor communities through the active downward transport of carbon and nutrients.

In the eastern Pacific, mesopelagic fish export about 15% of epipelagic carbon to deeper layers through their vertical migrations (Davison et al., 2013). Their daily mass migration also modifies fluxes of nutrients and oxygen (Robinson et al., 2010; Bianchi et al., 2013a, b) and helps support higher trophic levels across depths, including commercially important tunas and billfish, sharks, marine mammals, and small pelagic organisms (Potier et al., 2007; Johnson et al., 2011; Duffy et al., 2017), as well as benthic dwellers (Smith et al., 2008a, 2011). The eastern Pacific stands out in its proportion of vertically migrating fishes: strong differences have been observed, for instance, in the proportion of deep scattering layers migrating out of the mesopelagic at night, from 20% in the Indian Ocean to 90% in the eastern Pacific Ocean (Klevjer et al., 2016). The fish families Myctophidae and Gonostomatidae (Actinopterygii) dominate marine vertebrates caught in midwater trawls around the globe (Sutton et al., 2010) and account for a large portion of mesopelagic scattering layers (Perelman et al., 2021), including in the eastern Pacific.

Overall, two mesopelagic ecoregions have been delimited close to the CCZ with spatially distinct faunal assemblages (Sutton et al., 2017): the Northern Central Pacific (oligotrophic gyre, with different taxa compared to the equatorial Pacific) and the Eastern Tropical Pacific (eutrophic, with endemic taxa adapted to low oxygen).

3.5 *Bioluminescence*

Bioluminescence is considered to play an important role in deep-sea community structuration (Widder 2010; Haddock et al. 2017). Bioluminescence is used both for interspecific relations such as predation (Partridge & Douglas; 1995) and predation avoidance (Mensinger & Case, 1990; Robison, 1992; Pietsch, 2009). It also plays an essential role in intraspecific relations such as school formation, specific recognition, and matching for reproduction (Herring, 2007). Deep pelagic waters are characterized by a high level of transparency in most cases, wherefore bioluminescence is an effective means of communication that spreads over long distances, particularly in mesopelagic organisms (Johnsen et al., 2004; Claes et al., 2014a, b), given the highly developed visual systems observed in these organisms (Denton 1990; Warrant et al., 2003; Nilsson et al., 2012; de Busserolles et al., 2017).

In pelagic zones, it has been estimated that more than 75% of organisms are capable of emitting light (Martini & Haddock, 2017), while in the benthic zone, the number of bioluminescent species is estimated to be at least 50% (Martini et al., 2019). Diel vertical migrations (DVMs) of bioluminescent organisms also supply food to deep seafloor communities through the active downward transport of carbon and nutrients, thus playing a role in the benthopelagic coupling, a key factor to the ocean biological pump (Chatzievangelou et al., 2021).

3.6 *Apex Predators and Other Charismatic Megafauna*

The eastern Pacific is rich in Apex predators and charismatic megafauna, including blue and mako sharks or leatherback and logger head turtles, which have a widespread distribution and, as generalist meso-predators in oceanic food webs, are key components of pelagic communities and high sea ecosystems (IATTC, 2014) (Fig. 13.5). Several commercially important marine species such as billfish, tuna, sharks, larger squids, and rays regularly deep-dive into the meso- and even bathypelagic (1000–4000 m) layers (Thorold et al., 2014; Abascal et al., 2010; Fuller et al., 2015; Howey et al., 2016). In particular, key commercial species, such as yellowfin tuna, bigeye tuna, and swordfish, range to a depth of 1000 m, while species of cultural importance to Pacific Islanders, such as bottlenose whales or sperm whales, can reach 1500 and 3000 m depth, respectively (pilot whales and Cuvier's whales (*Ziphius cavirostris*) can also dive to 3000 m depth) (Fisheries and Aquaculture Department, 2000; Block et al., 2011; Schorr et al., 2014).

By the structure of their communities and the food webs they form, these Apex predators provide and sustain many of high-sea ecosystem services including provisioning, regulating, and cultural services (Juan-Jorda et al., 2015; Enquist et al., 2020). Specifically, tunas are known to move within the mixed layer during daytime but can descend below the mixed layer at night and will thus be directly impacted

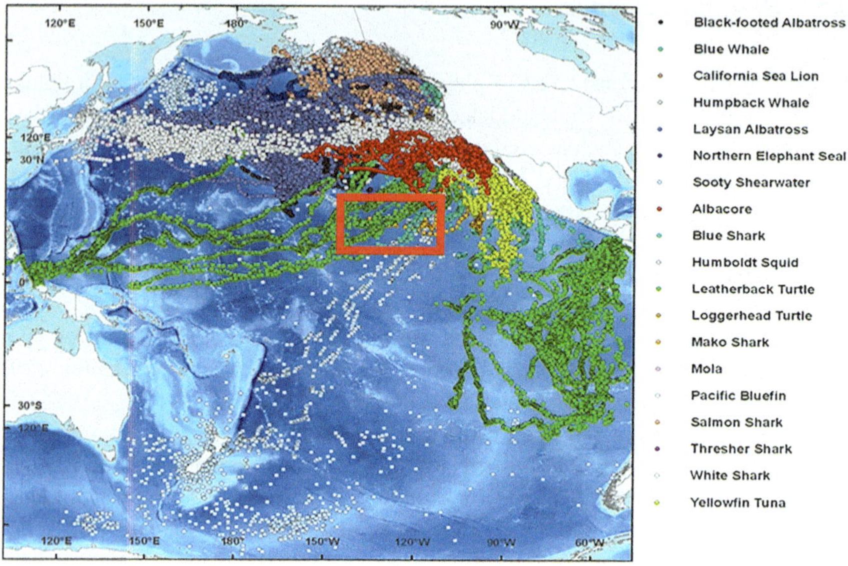

Fig. 13.5 Apex predators and charismatic megafauna tracks recorded monthly each year throughout the "Tagging of pelagic Predators' Program" (TOPP) in the Pacific (Costa et al., 2010). The CCZ is framed in red

by deep-sea mining activities if return waters are released in midwater layers (Schaefer et al., 2007).

The CCZ is also a major open ocean habitat for migratory marine species. Several whale species follow migratory routes through the CCZ to feed, driven by surface water temperature and primary productivity (Block et al., 2011; Tilot et al., 2023). Further, five sea turtle species are found in the CCZ, the olive ridley turtle (*Lepidochelys olivacea*) being the most abundant (Pitman, 1993) and the loggerhead (*Caretta caretta*), leatherback (*Dermochelys coriacea*), and green turtle (*Chelonia mydas*) classified as threatened (IUCN Red List—Polidoro et al., 2012).

Importantly, many of these Apex predators and marine megafauna can be considered ecosystem engineers, similar to zoo- and phytoplankton: they have lasting impacts on the trophic structure—keeping in check mid-size predators (Heithaus et al., 2008; Ritchie & Johnson, 2009; Ferretti et al., 2010; Ortuño Crespo & Dunn, 2017), which impacts phytoplankton growth and consequently geophysical parameters within the water column. For instance, sharks' and tunas' deep-dives for foraging and excretions in shallow waters contribute to upward nutrient shuttling (Thorold et al., 2014; Kiszka et al., 2015, 2022). Similarly, baleen and toothed whales are also considered ecosystem engineers or "whale pumps" because of their role in the upward nutrient pump and the deep sea ecosystems they sustain with their carcasses (Smith, 2006; Smith & Baco, 2003; Roman et al., 2014; Smith et al., 2014; Wang et al., 2021; Tilot et al., 2023).For example, annual Southern Ocean sperm whale faeces pump about 50 tonnes of iron ingested in deeper layers to the ocean surface

(Lavery et al., 2010). The iron contained in these faeces fosters phytoplankton growth, resulting in new primary production that fixes 200 000 tonnes of atmospheric carbon to the deep-sea annually (Lavery et al., 2010) – a necessary step in the absorption of greenhouse gasses, albeit insufficient to counter excess anthropogenic emissions globally (Meynecke et al., 2023).

Declining megafauna populations is linked to lower primary productivity because of lower upward nutrient cycling (Nicol et al., 2010; Roman & McCarthy, 2010) and changes in food-web structures and biogeochemical parameters (Lavery et al., 2010; Roman & McCarthy, 2010), resulting in overall decreased potential for organic carbon sequestration (Lavery et al., 2010; Pershing et al., 2010). The consequences of local declines in marine megafauna therefore extend much beyond the loss of culturally important species.

3.7 Interspecific Interactions

Interspecific interactions, mutualistic, antagonistic, or competitive, are important drivers of distribution patterns, local community assemblies, and evolutionary changes (Pantel et al., 2017). Several interspecific interactions occur in the CCZ. A tuna-dolphin-seabird interaction has been described in the eastern tropical Pacific region between 5°N and 30°N (Au & Perryman, 1985; Au & Pitman, 1986). Yellowfin tunas spotted (*Stenella attenuata*) and spinner (*Stenella longirostris*) dolphins drive small prey toward the surface, where boobies (*Sula* spp.), wedge-tailed shearwaters (*Puffinus pacificus*), jaegers (*Stercorarius* spp.), sooty terns (*Sterna fuscata*), and frigatebirds (*Fregata* spp.) forage on these preys. While this led to dolphins being recorded as bycatch in tuna fisheries, current changes in fishing practices are supporting an increase in dolphin populations (Ballance et al., 2021). Other interspecies interactions involve seabirds resting on sea turtles (e.g., boobies, black terns, and frigatebirds on olive ridley turtles) to reduce predation, to rest, and to feed on fish associated with the turtles (Pitman, 1993).

4 Deep-Sea Mining in the CCZ: Predicted Impacts on Marine Life and Human Activities

Areas of interest for the deep-sea mining industry can overlap with areas of interest for several other human activities (Fourchault et al., 2024). Related stressors, such as fishing, shipping, plastic pollution, and the suite of varying geochemical factors linked to climate change, are thus to be considered when evaluating synergies in negative anthropogenic impacts (Levin et al., 2018; Sharma, 2019; Ramirez-Llodra, 2020; Bell et al., 2021; van der Grient & Drazen, 2021) and incorporating ecosystem services into environmental management of deep-seabed mining (Le et al., 2017). Marine organisms in the water column of the eastern Pacific face cumulative stressors from diverse human activities, and especially fishing, in addition to the stressors inherent to deep-sea mining (Willett et al., 2006; Chang et al., 2012; Blanluet et al., 2023). The following paragraphs will highlight the foreseen impacts of deep-sea mining on pelagic organisms in addition to those reported in Drazen et al. (2020), followed by a brief description of cumulative stressors that may reinforce the impact of mining activities.

4.1 Predicted Impacts of DSM

DSM, in particular for polymetallic nodules, involves the exploration, extraction, transportation, and processing of minerals retrieved from the ocean floor and transiting through the water column to the surface before being shipped landwards. DSM is therefore tri-dimensional: ore is collected on the seabed, before being pumped upward through the water column as concentrated slurry (Fig. 13.6, adapted from Christiansen et al., 2020). This implies the release of a return plume, composed of seawater and low concentration seabed sediments, either at midwater or near the seabed before the ore is shipped to a terrestrial processing center (Miller et al., 2018).

Disturbances at the point of collection and return water outlet include sediment plumes that cause oxygen depletion, increased turbidity, possible dispersion of pathogenic material, and increased proportion of mineral particles, leading to nutritional deficiencies and repercussions at all trophic levels of the food web (for a full review, see Christiansen et al., 2020; Drazen et al., 2020). For example, suspension feeders, such as copepods, polychaetes, salps, and appendicularians forming an important basal group of the food web, may suffer from clogging of their mucous filter nets, resulting in death by starvation. Further, small particles may settle on gelatinous plankton, causing buoyancy issues. Bioluminescence processes by living organisms may also be hindered affecting communication and prey capture. Metals from crushed ore materials, including toxic elements, could enter the food web, bioaccumulate, and be released in the midwater and deep-sea ecosystems through carcasses.

Physical harm to living organisms through collisions, anthropogenic interference, and pollution will be caused by the increase in the number of vessels and operation platforms (Ritter & Panigada, 2019).

Consequently, DSM could result in increased mortality, altered migration patterns (both horizontal and vertical), decreased fitness, and shifts in community composition (Drazen et al., 2019).

Seabed collection and return water plumes will release sediments, the granulometry of which will determine their residence time in the water column. Discharge

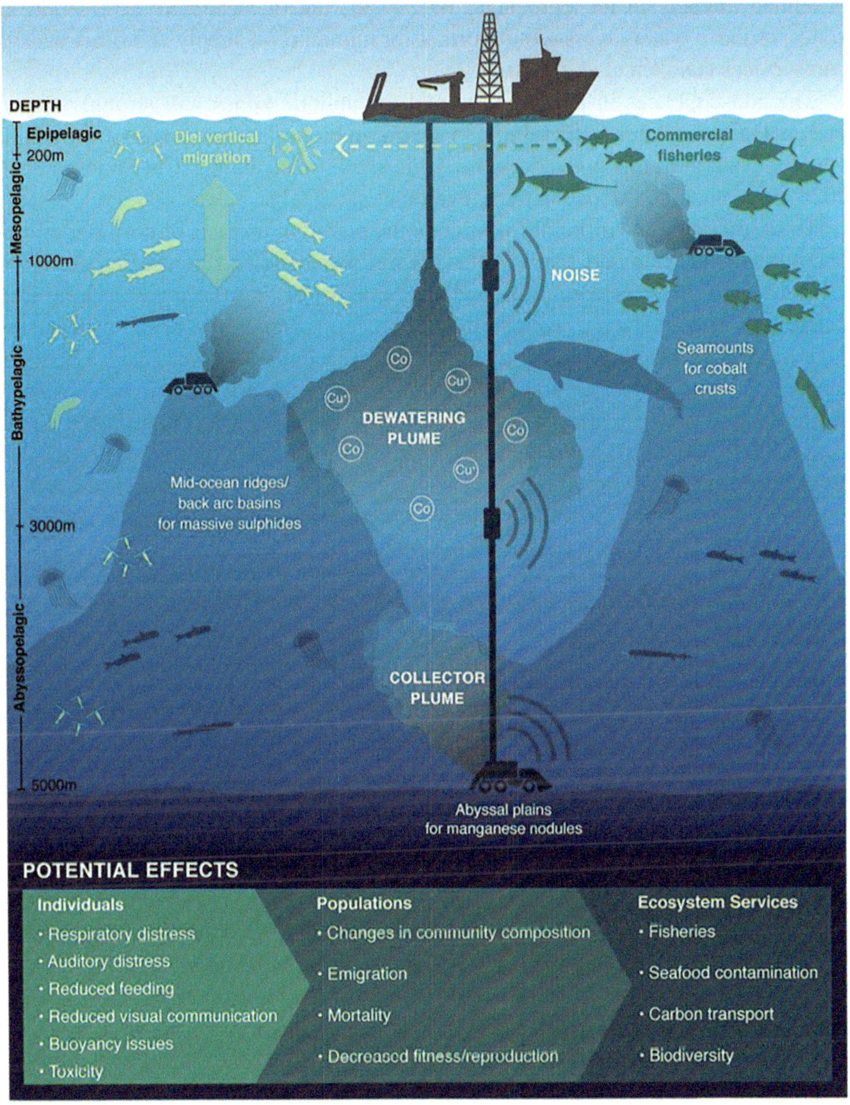

Fig. 13.6 Graphic representation of potential impacts from deep-sea mining (Reproduced from Drazen et al. 2020). Mining-generated sediment plumes and noise have a variety of possible effects on pelagic taxa. (Organisms and plume impacts are not to scale.) (Image credit: Amanda Dillon (graphic artist))

plumes in midwaters or at the seabed are estimated at 400 t day^{-1} dry solids suspended in about 50,000 t of water per collector (Christiansen et al., 2020). DSM dewatering discharge depth is a first key mitigation measure and should be below the mesopelagic/bathypelagic transition (i.e., below a depth of 1500–2000 m) to minimize impacts in the epipelagic (0–200 m) and mesopelagic (200–1000 m) zones, as these water layers are important for human food supply and other ecosystem services (Drazen et al., 2020).

The consequences in the euphotic and the aphotic zones will strongly impact mesopelagics that play a crucial role in the ocean biological pump with the transport of carbon and nutrients to the deeper pelagic and benthic realms (Christiansen et al., 2020; Drazen et al., 2020, Chatzievangelou et al., 2021).

To summarize, it is difficult to estimate the potential impacts of multiple mining operations over time and space with varying extracting and processing technologies, as there are considerable knowledge gaps regarding the possible long-term deleterious effects of deep seabed mining on deep-sea and open ocean ecosystems. Therefore, it is essential to better understand the unique habitats and associated faunal communities that are found in the CCZ and to quantify the crucial ecosystem functions and services they deliver (Glover & Smith, 2003; Thiel, 2003; Smith et al., 2008a; Tilot, 2010; Woodall et al., 2014; Levin et al., 2016, 2018; Le et al., 2017; Miller et al., 2018, 2021).

4.2 *Deep-Sea Mining, Fishing, and Climate Change*

As connections between the different layers of the ocean are better studied, there is growing evidence that both the water column and the seabed are affected by natural variations (seasonal climatic variation, El Niño events, benthic storms) and human activities (fishing, shipping, greenhouse gas emissions resulting in ocean acidification and changes in temperatures).

The fishing industry has become a major threat to pelagic organisms, which are increasingly exploited in response to depleted coastal stocks, growing fish-protein demand, the use of governmental subsidies, and technological innovation (Sumaila et al., 2015 in Van der Grient & Drazen, 2021; Ortuno Crespo & Dunn, 2017). Over the last seven decades, there has been a transition from large, long-lived, high trophic-level, piscivorous fish toward smaller, short-lived, low trophic-level invertebrates and planktivorous pelagic fish. Tunas and billfish stocks have declined by 60% on average in the last five decades, globally (Juan-Jorda et al., 2011).

Pacific Islanders are directly impacted by the diminishing trophic support of fisheries as they heavily rely on marine fisheries for protein and income (Johannes, 1987; Lieber, 1994; Rabbitt et al., 2019). The main tuna fishing activities globally occur within Pacific Island national waters, including waters of State Parties to the 1982 Nauru Agreement*, a regional cooperation with strong conservation benefits (Thakur, 1991; South et al., 2004). These provide 60% of the west and central Pacific Ocean tuna catch, and 25–30% of canned tuna worldwide, with a value of approx. USD 4.5 billion annually (Brouwer et al., 2018; The Western and Central Pacific Fisheries Commission, 2019).

(*The subregional Pacific Nauru Agreement concerning Cooperation in the Management of Fisheries of Common Interest was signed in 1982 between the

Federated States of Micronesia, Kiribati, the Marshall Islands, Nauru, Palau, Papua New Guinea, Solomon Islands, and Tuvalu. This agreement subsequently endorsed by the Western and Central Pacific Fisheries Commission includes presently all Distant-Water Fishing Nation fleets and coastal states that participate in the western and central Pacific tuna fisheries, the major commercial fisheries resource harvested in the Pacific (Thakur, 1991; South et al., 2004).)

Long-term datasets from the Pacific Ocean show that deep-sea communities are strongly affected by climate variation (Smith et al., 2009; Levin et al., 2018; Sharma, 2019), as are those of the upper ocean (Levitus et al., 2000; Barnett et al., 2005). For example, significant shifts in tuna biomass are likely to occur from west to east and from national jurisdictions to high seas (Bell et al., 2021), with a significant overlap of predicted tuna biomass distribution (both for skipjack, bigeye, and yellowfin species) and polymetallic nodule areas in the CCZ (Fig. 13.7). These stocks would therefore be under simultaneous pressure from the fishing and the deep-sea mining industries (van der Grient & Drazen, 2021) that would cause a reduction in the provisioning of commercial fish species and biomass and an increase in toxic metals throughout pelagic food webs and hence seafood supply (Drazen et al., 2019).

This would affect the eastern Pacific where at least ten countries derive 10% or more of their regional fisheries management organization from areas overlapping with mining areas (WCPFC, 2019; Van der Grient & Drazen, 2021).

More specifically in the CCZ, anthropogenic and climate change disturbances have been evidenced with two areas of high bycatch estimates in international waters near Hawai'i and Mexico (Halpern et al., 2013, 2015).

Mitigating the cumulative impacts of industrial fishing and deep-sea mining on pelagic fisheries will thus be essential to safeguard economically important fishing activities and to ensure food safety. In this perspective, the ISA suggests that several priority areas for research and monitoring should be designated in the CCZ (ISA, 2023). Currently, several global ocean models are available at the spatial resolution needed to better understand the deep-sea mining interactions with fisheries (Fagundes et al. 2020; Ringler et al. 2013; Xiao et al., 2022).

4.3 Deep-Sea Mining and Shipping

The mining of polymetallic nodules in the CCZ will result in higher shipping densities above the relatively unvisited surface waters of the CCZ and across the Pacific to transport the ore for on-land processing. Consequently, the risk of collision, anthropogenic interference, and pollution with megafauna will increase (Ritter & Panigada, 2019). Further, the noise and light emitted by these ships will further influence the behavior of marine fauna relying on physical cues for orientation and communication. Auditory masking will likely occur in the epipelagic layer, with surface support vessels producing low frequency and broadband sound similar to frequencies emitted by commercial shipping vessels (Thompson et al., 2023). Throughout the water column, further auditory masking is linked with acoustic

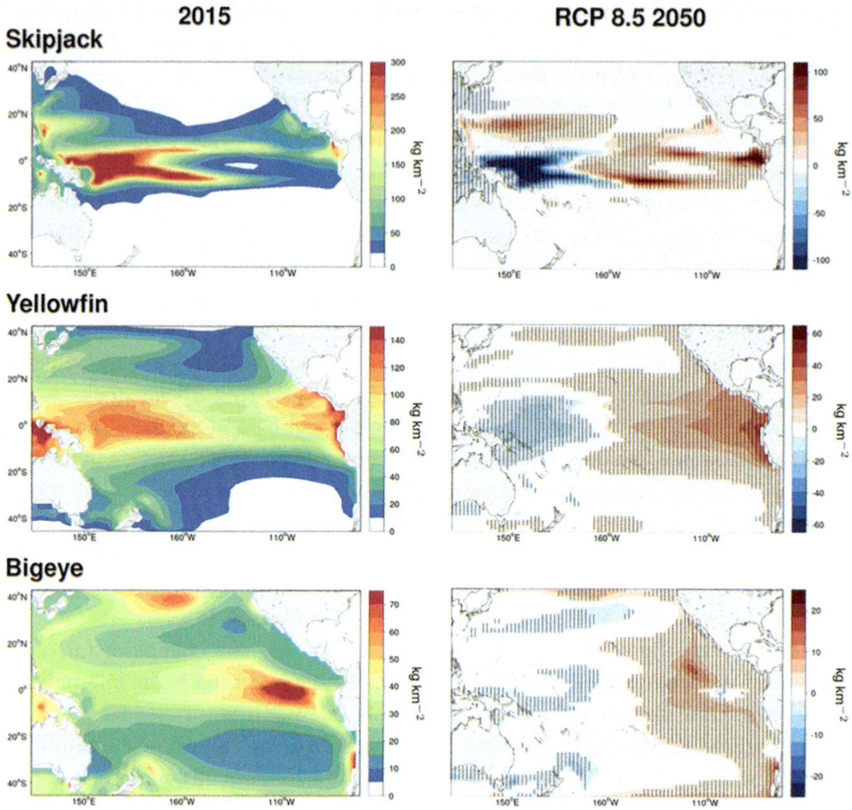

Fig. 13.7 Predicted shifts in tuna biomass distributions (Bell et al., 2021). Average biomass distributions (kg km^{-2}) of skipjack, yellowfin, and bigeye tuna in the Pacific Ocean basin for 2015 (2011–2020) (left) and mean anomalies (kg km^{-2}) from the average 2015 biomass distribution of each tuna species projected to occur by 2050 (2044–2053) under the RCP 8.5 greenhouse gas emissions scenario (right). Shading indicates areas where projections from four Earth System Models agree in the sign of change

telemetry for remotely operated vehicles, which overlaps with Delphinidae and Odontoceti clicks, and dredging noise, which overlaps with the wavelengths of both small (e.g., humpback whales) and larger (e.g., blue whales) Mysticeti (Thompson et al., 2023, and references therein).

Finally, cumulative impacts would occur by the fact that navigational routes cross the CCZClarion Clipperton zone (CCZ) area. In particular in the southern part of the CCZ, busy shipping lanes link the Panama Canal with east Asia and in the North and East (Schwer & McGillivary, 2007; Halpern et al., 2013, 2015; Avila et al., 2018; Tilot et al., 2023).

5 DSM and Ecosystem Services of the Pelagic Ecosystems

Ecosystem services are the products of healthy, diverse, and functioning ecosystems and their.

associated living organisms, which contribute to human well-being (Rogers et al. 2014). The high seas, in particular, are known to provide several key ecosystem services (Fig. 13.8) that benefit human societies (Armstrong et al., 2012; Santora et al., 2021; Xuan et al., 2021; La Bianca et al., 2023):

- Existence and stewardship values of all organisms.
- Supporting services (e.g., biomass production, geochemical cycling, habitat provision).
- Provisioning services (e.g., halieutic resources, raw materials and genetic resources for pharmaceutical or other industrial applications). Examples of raw materials include fisheries (Victorero et al., 2018) and deep-sea minerals (Thurber et al., 2014). Examples of genetic resources include bamboo corals that are used as a model for synthetic human bone replacements (Ehrlich et al., 2006) and sponge spicules that are superconductors for light (Brummer et al., 2008). Deep-sea sponges and associated microorganisms are also a source of bioactive metabolites for pharmaceutical applications such as anti-tumor, antibacterial, antiviral, toxin inhibitors, and anti-inflammatory metabolites (Rateb & Ebel, 2011; Batista-Garcıá et al., 2017; Koch et al., 2021).
- Regulating services (e.g., climate regulation, nutrient cycling, waste treatment). In the context of climate change, the finding of the role of whale-fall ecosystems in carbon sequestration is of importance (Smith et al., 2014). In the same perspective, deep demosponges have a role in filtering out bacteria and dissolved

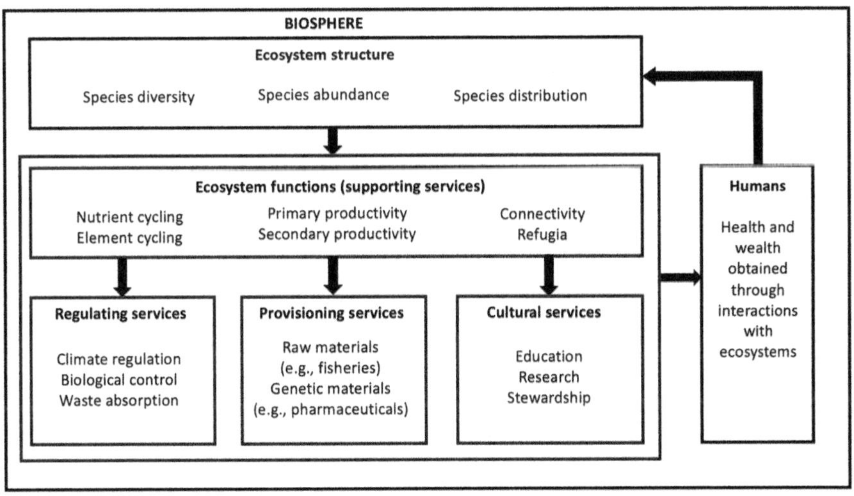

Fig. 13.8 Key services provided by healthy high seas ecosystems for human well-being

organic carbon from seawater (Kutti et al., 2013; Cathalot et al., 2015; Leys et al., 2018).
- Cultural, artistic, educational, and scientific services (e.g., information for research, education, leisure, inspiration for culture, art, and design) (Turner et al., 2020).

So far, only a few attempts have been made to quantify marine ecosystem services (Sumaila et al., 2007, 2015; White & Costello, 2014). La Bianca et al. (2023) highlight the importance of valuing supporting services, which have been neglected in recent studies. Indeed, studies more frequently quantify provisioning services, as these directly benefit commercial activities. Nonetheless, to identify the importance of an ecosystem for fish and invertebrate catch, the number and volume of breeding or spawning grounds, the nursery habitat(s), and the primary production for trophic support—which are all supporting services—need to be evaluated (Le et al., 2017; La Bianca et al., 2023). Other ecosystem functions, such as survival and recruitment rates and larvae and juvenile populations' linkage to adult populations, are also key to apply economic value to an ecosystem (Le et al., 2017). Similarly, among regulating services, carbon sequestration is relatively better studied and is known to depend on functions such as primary productivity at the surface, carbon flux to the bottom, and degradation and burial rates of organic carbon (Smith et al., 2015a, b).

Generating and merging the economic values of all services provided by ecosystems found throughout the water column above mining sites is therefore a complex but necessary task, which will allow to balance the costs and benefits of conserving habitats and wildlife (Jobstvogt et al., 2014; Fourchault et al., 2024).

Deep-sea mining activities will impact ecosystem services provided by benthic, demersal, and pelagic organisms living near or migrating close to commercially interesting mineral deposits (Glover & Smith, 2003; Drazen et al., 2020; La Bianca et al., 2023). Industrial deep-sea mining processes will have various ecological consequences, which will impact the health of marine organisms and the quality of the ecosystem services they provide (Table 13.1). Even the earliest mining activity would have an environmental impact at a regional scale, disturbing ecosystems both on collected sites of the benthos and in the water column, surface, and above mined sites. Impacts in larger areas would depend on oceanographic and atmospheric processes occurring in the area (Glover & Smith, 2003; Drazen et al., 2020).

We thus need standardized procedures both to understand the mining-induced consequences on ecosystem structures (Van Dover et al., 2017; Niner et al., 2018) and to gain knowledge on how changes in ecosystem structures will impact ecosystem functions and services (Ginzky et al., 2020).

Table 13.1 Potential impacts of deep-sea mining on ecosystem services throughout the water column

Industrial process	Ecological consequences	Impacted organisms	Impacted ecosystem services
Ore collection	Sediment compaction Habitat alteration Increased turbidity Light pollution Sound pollution Accidental release of chemicals	Benthic and demersal species: negatively impacted respiratory systems, feeding efficiency, behaviour, growth, reproduction, mortality (e.g., Christiansen et al., 2020)	***Supporting services*** Habitat provision (e.g., diminished breeding grounds and refugia) Biomass production (through disrupted trophic chains) ***Regulating services*** Climate regulation (e.g., lower carbon sequestration because of lower biomass levels; lower carbon capture from lower phytoplankton growth because of reduced upward nutrient shuttling from deep-diving predators; increased turbidity negatively impacting carbon fixation by mesopelagics) (but see Orcutt et al., 2020) Biological control and waste treatment (e.g., reduced filtration of bacteria and dissolved organic carbon through deep sponges) ***Provisioning services*** Genetic resources (e.g., fewer biomaterials, pharmaceuticals, industrial agents) Trophic support for fisheries (e.g., reduced feeding efficiency of tuna, swordfish, squid, and depletion of slow-growth deep-sea fisheries) Geochemical cycling (e.g., lower upward nutrient shuttling through dial vertical migration, deep-diving predators, or marine mammals) ***Cultural services*** Historical archives (e.g., destruction of fossils records in sediments) Research and education Stewardship and bequest value Outreach value attached to marine mammals and deep-sea octopus (Tilot et al., 2021b) ***Existence and stewardship value services***
Pumping of ore through riser systems	Removal of ambient water Noise pollution Potential accidental leaks leading to sediment plumes	Benthic and demersal species: negatively impacted larval dispersal (Christiansen et al., 2020) Marine mammals and species relying on sound: negatively impacted communication, behaviour, reproduction, growth (Thompson et al., 2023)	
Release of return water through discharge systems	Sediment plumes leading to increased turbidity, release of chemicals and biohazards Cold, slightly saltier water release	Mid-water organisms: negatively impacted feeding efficiency, respiratory processes, behaviour, and reproduction (e.g., Haddock et al., 2017)	
Production support vessels	Noise pollution Light pollution Collision risk	Epipelagic organisms, including marine mammals (e.g., Thompson et al., 2023) Mid-water and deep pelagics relying on trophic input from upper layers (Drazen et al., 2020)	

6 Conservation and Sustainable Management Strategies

6.1 Different Options for Managing Conservation and Sustainable Development in the High Seas and in the CCZ

Customary marine resource management incorporating traditional knowledge and beliefs have long been practiced in the Pacific to conserve and manage sustainably marine areas and species (https://www.sprep.org). These managed areas are now called "Locally Managed Marine Areas" (LMMAs) and are part of a network shared by Fiji, Papua New Guinea, and other islands (https://www.lmmanetwork.org). A portion of the fishing grounds is usually set aside as a no-fishing zone to safeguard the future sustainability of fisheries resources (Tawake 2004; Veitayaki et al., 2003). The success of each area in the network is measured by the health of the ecosystem, habitats, and species, along with the reduction of threats and the well-being of people who use the resources (Tawake & Tuivanuavou, 2004).

It is now widely accepted that as a science-based tool of conservation, Marine Protected Areas (MPAs) are particularly effective and serve as enforceable tools to safeguard marine biodiversity from trophic downgrading on the oceans (Estes et al., 2011; Jeudy de Grissac, 2015) and from deleterious impacts from anthropogenic activities (Agardy, 1994; Margules & Pressey, 2000; Roberts et al., 2001; Edgar et al., 2007; Gaines et al., 2010; Halpern et al., 2010; Duarte et al., 2020).

The numbers and extent of MPAs have increased steadily and more substantially in recent years with the adoption by the United Nations Convention on Biological Diversity (CBD) of a new target at the horizon 2030, the 30 × 30 proposal, increasing the effective conservation target to at least 30% of the world's lands, inland waters, coastal areas, and oceans, with emphasis on areas of particular importance for biodiversity and ecosystem functioning and services. In particular, the CBD recommends prioritizing ecologically representative, well-connected, and equitably-governed systems of protected areas and other effective area-based conservation, recognizing indigenous and traditional territories and practices, which directly addresses the situation of managing human activities in the high seas (https://www.cbd.int/article/draft-1-global-biodiversity-framework).

Therefore, there is an increased interest toward finding strategies that synergize conservation and exploitation of marine ecosystems and resources (Kriegl et al., 2021). An ongoing multidisciplinary consensus stresses the importance of MPAs promoting ecosystem resilience (Peterson et al., 1998; Folke et al., 2004) and rebuilding life in the ocean (Duarte et al., 2020) and as "key components of global marine conservation strategies" (Lubchenco et al., 2003; Gaines et al., 2010; McCrea-Stub et al., 2011).

Recognizing the importance and value of deep-sea ecosystems and the biodiversity they contain, the United Nations General Assembly (UNGA) resolution 61/105 in 2006 called "upon States to take action immediately, individually, and through regional fisheries management organizations and arrangements, and consistent with the precautionary approach and ecosystem approaches, to sustainably manage fish stocks and protect Vulnerable Marine Ecosystems (VMEs), including seamounts (Clark et al., 2011, 2016), hydrothermal vents and cold water corals, from destructive fishing practices" (https://www.fao.org/in-action/vulnerable-marine-ecosystems).

Presently, VMEs target areas could be extended to areas that would be impacted from other human activities, such as deep-sea mining.

In 2008, a process was put in place by the CBD, to recognize « Ecologically and Biologically Significant Areas (EBSAs) » on seven scientific criteria: uniqueness or rarity, special importance for life stages of species, importance for threatened, endangered, or declining species and/or habitats, vulnerability, fragility, sensitivity or slow recover, biological productivity, biological diversity, and naturalness (https://www.cbd.int/ebsa).

The International Maritime Organization (IMO) has adopted the Resolution A.720 (17) concerning guidelines for the identification and designation of Particularly Sensitive Sea Areas (PSSAs) for species, habitats, and seascapes and has established special protective measures that ships must follow. IMO's PSSA program supports the UN Sustainable Development Goal 14 to conserve and sustainably use the oceans, seas, and marine resources for sustainable development. It also supports other protective measures and mechanisms, such as the UNESCO World Heritage Marine Sites, and Ecologically or Biologically Sensitive Areas under the Convention on Biological Diversity. So far, 15 PSSAs in the world have been identified and designated. For the Pacific, they include the Great Barrier Reef.

Other Effective Area-Based Conservation Measures (OECMs) allow to achieve conservation objectives mainly as a by-product of other management strategies. These are geographically defined and governed and managed in ways that achieve positive and sustained long-term outcomes for in situ conservation of biodiversity, with associated ecosystem functions and services including cultural, spiritual, socio-economic, and other locally relevant values (https://biodiversity.europa.eu/europes-biodiversity/protected areas/other-effective-area-based-conservation-measures).

The emerging internationally legally binding instrument under the auspices of the United Nations for the conservation and sustainable use of marine biodiversity from the seafloor to the surface is the new UN agreement on Biodiversity Beyond National Jurisdiction (BBNJ), or the BBNJ Treaty, adopted by the UN, June 19, 2023. Its role is « to ensure the long-term conservation and sustainable use of marine biological diversity of areas beyond national jurisdiction through effective implementation of the relevant provisions of the Convention and further international cooperation and coordination ». "The common set of overarching goals and objectives of the BBNJ to ensure the ecological integrity of ocean ecosystems (Kim & Bosselmann, 2015; Kim & van Asselt, 2016) are the following:

- Marine genetic resources, including the sharing of benefits.
- Area-based management tools, including MPAs.
- Environmental impact assessments.

– Capacity-building and transfer of marine technology" (https://www.un.org/bbnj).

In the Areas Beyond National Jurisdiction (ABNJ), the importance of protecting and conserving natural resources and ecosystems is outlined as ecosystems are ecologically closely interconnected. However, it lacks a coherent and holistic governance based on precaution and knowledge-based decisions (Gjerde & Yadav, 2017; Gjerde et al., 2019).

The concept of ecosystem-based management (EBM) provides the basis for more coordinated, holistic, and transboundary governance of interrelated areas of the ocean that are politically divided, such as the High Seas and the Area in ABNJ (De Lucia, 2019; Gjerde & Wright, 2019; Tunnicliffe et al., 2020; Warner, 2016; Tilot, 2023).

– "Oceanic Marine Protected Areas" (OMPAs) (defined here as waters >200 m depth) have emerged from the ratification of the BBNJ Treaty (Blanluet et al., 2023).
– "Large Marine Protected Areas" (LMPAs) (defined for areas >100,000 km^2) have been created in the current decade, often in networks (Corrigan & Kershaw, 2008; Game et al., 2009; O'Leary et al., 2012; Gianni et al., 2016). With LMPAs, all natural processes and habitats, including deep sea and deep-sea habitats, are protected. Evidence has shown that LMPAs are much more cost-effective to implement and manage as smaller MPAs, especially when including corridors for migrating species (McCrea-Stroub et al., 2011), particularly with the ratification of the BBNJ Treaty.

Concerning the CCZ, two large EBSAs are bordering the CCZ to the north and the south: (1) the North Pacific Transition Zone, bounded to thermohaline fronts and eddies, a highly productive habitat and migratory corridor for many pelagic predators including endangered and commercially important species (2) and the Equatorial High-Productivity Zone, a large-scale oceanic feature supporting high primary production over a large area and a strong benthic-pelagic coupling strongly related to surface primary productivity. Both EBSAs are rated "high" for uniqueness, due to their large scale as well as being major components of the oceanographic processes within the Pacific Ocean; they are also rated as "high" for primary productivity (https://www.cbd.int/ebsa).

The CCZ also includes several Important Bird Areas (IBAs), in particular for petrels and shearwaters, and Important Marine Mammal Areas (IMMAs). The area is also under the influences of the Clipperton Fracture Zone Petrel foraging area, the Clipperton Atoll, a highly endemic habitat and an outpost for the migratory flux coming from the west, and the North-East Pacific White Shark Offshore Aggregation Area. Evidence has shown as well that, unrelated to geographic distance, coastal regions are connected to ABNJ, such as the CCZ, both through larval dispersal and through the potential dispersal of pollutants (Popova et al., 2019).

In the CCZ, the recognition of marine protected areas "where sea-floor mining and other deleterious activities are not allowed" is considered impracticable. Instead, the International Seabed Authority (ISA) has designated thirteen "Areas of Particular

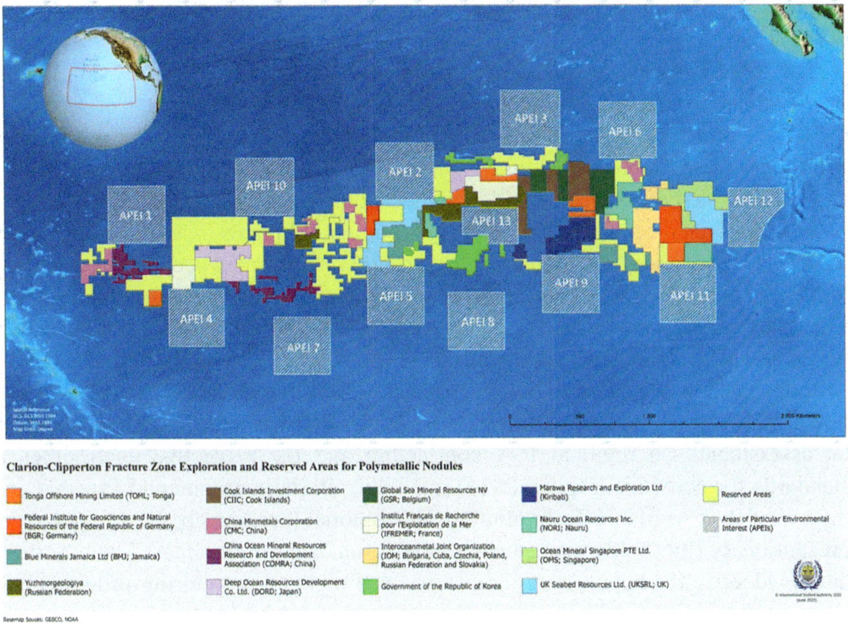

Fig. 13.9 Exploration Area and the thirteen Reserved Areas (APEIs) for polymetallic nodules in the Clarion-Clipperton Fracture Zone. (https://www.isa.org.jm/maps/clarion-clipperton-fracture-zone/)

Environmental Interest (APEI)," or "Impact Reference Zones," or "Preservation Reference Zones" (Smith et al., 2008b; Wedding et al., 2015), the purposes of which would be to assist the ISA in taking decisions concerning the modalities of nodule exploitation and its impact both on the ecosystems concerned and on the future of mining activities (Fig. 13.9).

In order to serve as valid reference sites, these APEIs should be fully representative of the areas to be exploited in terms of depth, physical and chemical nature of the substratum, associated faunal communities and ecosystems, overlying chemistry and oceanography of the water column, and broader climatology, all of which ideally require monitoring at different temporal and spatial scales (Freiwald & Roberts, 2006; Tilot, 2010; Oesterwind et al., 2016; Tilot et al., 2018). They would also need to remain effectively unimpacted and to have extensive buffer zones, as generally have other MPAs, especially in the high seas.

However, these conditions are not entirely respected as some APEIs are not fully representative of the mining sites, with some containing other habitats, such as seamounts, than those targeted by nodule mining (Tilot, 2010; Tilot et al., 2018; Jones et al., 2021). APEIs may nonetheless be impacted by the plumes deriving from the mining technology that would occur throughout the whole water column. As well according to the topography, the oceanographic processes and associated life cycles, plumes may adopt several patterns and travel over great distances (Weaver et al., 2022).

The APEI network, presently designed only for reference purposes, should thus be associated with management activities to help mitigate the impacts of mining in the CCZ (Tilot, 2010; Tilot et al., 2018; Kriegl et al., 2021). A network of flexible MPAs, preferably linked to tridimensional LMPAs, with corridors and buffer zones and expandable boundaries, would be the best option (Tilot et al., 2023).

ISA has been developing Regional Environmental Management Plans (REMPs): *a proactive area-based management tool to support informed decision-making that balances resource development with conservation* (ISA, 2018, 2020). ISA also considers the potential interactions between fishing and mineral resource-related activities in areas ABNJ with a spatial analysis (ISA, 2023).

If designed as a legally binding instrument, REMPs would allow the ISA to function as a proper regulator, i.e., through feedback of REMP cumulative environmental assessments on regional ISA contracting and the respective environmental standards for activities (Jaekel, 2016), including Best Environmental Practice and the use of Best Available Technologies. The regional level may be best suited to set precautionary thresholds for "effective protection," "harmful effects," and "serious harm," identify the appropriate indicators for a regional monitoring program, and carry out the necessary environmental assessments that leads to an integrated management and embrace reviews in an adaptive review cycle.

The newly adopted High Seas Treaty could provide a critical impetus to secure the necessary strategic perspective of marine spatial planning for integrating sectoral management tools such as the ISA REMPs into the global biodiversity conservation agenda in ABNJ with EBSAs, VMEs, IMMAs, LMPAs, PSSAs, and OECMs in relation to risks from mining-related activities as all ecosystems in the high seas are closely interconnected (Ardron et al., 2008; Maes, 2008; Tilot, 2010).

6.2 Deep-Sea Mining and Traditional and Cultural Dimension in Managing Marine Resources

Pacific Islanders, in particular Indigenous Peoples and Local Communities (IPLCs), embrace their role as custodians of ecosystems and of species, which they perceive as interconnected. IPLCs are particularly attached to migratory species and traditional navigation routes in the open Ocean (Johannes, 1981; D'Arcy, 2006; Ey & Sherval, 2016; Eckstein & Schwarz, 2019). Similar to the sensory navigation used by sea turtles, sharks, and whales, Pacific Islanders practice the traditional art of navigating, using the motion of stars, the shape of clouds, the color of the sea, the movements of marine species, the shape of waves, currents, and water temperature (Lohmann et al., 2008), envisioning the Pacific as a "Sea of Islands" (Hau'ofa, 1994, 2008). Deep-sea mining activities therefore cannot be conceptualized as being far off, or beyond Pacific Islanders' territories, despite their location outside of official national boundaries.

Traditional knowledge and customary practices have long been practiced in the Pacific to conserve and manage sustainably marine areas and species (https://www.sprep.org). Presently called LMMAs, as described previously, they form a network (https://www.lmmanetwork.org).

There is a long-standing call of civil society of the Pacific Island Countries to preserve the marine environment and the resources and to prioritize the health and values of Pacific communities, in particular traditional knowledge, e.g.,10 year moratorium on DSM from 2020 to 2030 to allow for prudent research in marine areas under national jurisdiction of the Pacific nations (Fiji, PNG, Vanuatu, Cook). The Cook islands has incorporated traditional practices such as the "rahui" in its strategy of ocean governance over its extended EEZ (Govan et al., 2009). The "rahui" is a complex system of rules and prohibitions regulating interactions between specific peoples and organisms (Bambridge, 2016). The subregional Pacific Nauru Agreement (PNA) concerning Cooperation in the Management of Fisheries of Common Interest signed in 1982 outlines strong conservation benefits in the region (Tilot et al., 2021a). The traditional and cultural dimension of Pacific Island Countries is also taken in account in the recent agreement on Biodiversity Beyond National Jurisdiction (https://www.un.org.bbnj).

Most Pacific Islands countries have ratified or acceded to global, regional, or sectoral instruments relevant to ensuring the protection of the marine environment and biodiversity or more exactly "the protection of living things, their physical environment and/or the ways in which holders of traditional knowledge utilize them for maintaining, facilitating, or improving human life" that would be impacted by DSM activities, such as with the CBD in 1992, the Madang Guidelines on Principles for the Development of National Offshore Mineral Policies in 1999 (https://pacific--data.sprep.org/dataset/madang-guidelines-principles-development-national-offshore-mineral-policies), and the Noumea Convention for the Protection of Natural Resources and the Environment of the South Pacific Region in 1986 which contains an indirect reference to the cultural value of areas and the exercise of traditional customary rights in one of its protocols (https://www.sprep.org/convention-secretariat/noumea-convention) (Tilot et al., 2021a). There is also "the Pacific Plan for strengthening Regional Cooperation and Integration" with a strategic framework of regional policies, approved by the Pacific Islands Forum leaders in 2005 (https://www.pacificplan.org).

Traditional dimensions of seabed resource management are clearly integrated in the national laws in some countries, such as for the Cook Islands with its LMPA, vast multipurpose MPA, covering their entire territorial sea and exclusive economic zone which encompasses a huge share of the world's currently known cobalt (Cook Islands Marae Moana (=sacred ocean) Act 2017). In this perspective, the Cook Islands, using the traditional "rahui" concept, balance the economic interests like tourism, fishing, and deep-sea mining with conserving marine biodiversity and ecology, in order to maximize the benefits for the current and future generations of the Islanders (Willaert, 2020; Tilot et al., 2021a).

LMPAs offer unique opportunities to protect also cultural seascapes and to enable indigenous communities to maintain their traditional practices while participating in management (Bambridge & d'Arcy, 2014; Lewis et al., 2017). North of the CCZ, the Hawaiian marine national monument, Papahanaumokuakea, a very large marine protected area of 1.508870 km^2 and World Heritage site, was created in particular for biocultural reasons.

6.3 *Proposed Marine Spatial Planning and Management Tools in the CCZ*

Marine spatial planning in the high seas and deep sea needs to pursue an integrated approach that considers transparency and all stakeholders and sectors of activity and fully appreciates the tridimensional status of the oceans, in conformity with the principle of the ecosystem approach to management (EAM), and the commitments enshrined in the Convention on Biological Diversity health (https://cbd.int/ecosystem), (Tilot, 2004; Ardron et al., 2008; Cormier et al., 2017, 2019; Brito-Morales et al., 2022; Fourchault et al., 2024).

An example of an application of a strategic approach to marine spatial planning based on EAM is the marine mammal sanctuary "Pelagos" shared by three countries in an upwelling area of importance for cetacean in the Mediterranean Sea (Tilot, 2004; Ardron et al., 2008). The "Pelagos" agreement is looking at the reduction of all potential impacts on cetaceans, including pollution, navigation, nautical tourism, whale watching, and noise, and adopts necessary regulations related to the surface of the sea, the water column, and the sea bottom.

Concerning the management of a vast area in the high seas targeted by DSM, such as the CCZ, the marine spatial planning strategy would require a comprehensive assessment of baseline environmental and oceanographic parameters and of ecosystems as well as of natural, anthropic and cumulative impacts. This comprehensive assessment should be performed at temporal and spatial scales (Ardron et al., 2008; Foley et al., 2010). Marine spatial planning is generally executed through a Transboundary Diagnostic Analysis (TDA) and then a Strategic Environmental Assessment (SEA) in ocean governance (Gjerde & Yadav, 2017).

Because pelagic ecosystems are dynamic with vertical and horizontal migrations, and because mining effects are probably drifting as well, a Lagrangian (drifting) perspective is needed in monitoring and management. Consequently, potential broad-scale impacts to pelagic ecosystems should be considered with the choice of stations for environmental baseline and monitoring studies and other area-based management and conservation measures, in particular the choice of APEIs (Drazen et al., 2019) and possibly oMPAs, as highlighted above. Also, management systems should incorporate elements of adaptive or reactive management, so that preventive action can be taken as soon as new information becomes available (Cormier et al., 2017).

6.3.1 Metrics for Biodiversity Assessments, Ecosystem Dynamics, and State of Health

Key water column parameters further include turbidity, current speed, dissolved oxygen concentration, primary production, and alkalinity. Most of these variables can be measured with sensors, such as lowered acoustic Doppler current profilers (LADCPs) for current velocity.

Concerning taxonomic diversity measures of biodiversity, at both within-habitat and within-region scales, it is a priority to develop metrics, adapted to assess and monitor ecosystems where full species and habitats lists are unavailable as in the high seas, in particular in the CCZ (Warwick & Clarke, 2001; Gray, 1997; Price, 1999; Izsak & Price, 2001), as well as process-oriented metrics that account for ecosystem dynamics across temporal and spatial scales (Steneck, 2001; Price et al., 2007). Key functional groups, ecological roles, and species interactions (Hughes et al., 2005) can be assessed by measures that are not too sensitive to sample size such as average taxonomic distinctiveness (i.e., a measure of the mean taxonomic distance between species within a sample or study area) and variation in taxonomic distinctiveness (i.e., a measure of the variance in taxonomic distance between species) (Warwick & Clarke, 2001). Complementing taxonomic distinctiveness, the concept of taxonomic similarity (i.e., the mean taxonomic distance between species from different samples or sites), would enable comparison across areas at different scales being not too sensitive to sampling intensity (Izsak & Price, 2001; Price, 2002; Price & Izsak, 2005).

Further, diversity indices, such as Environmental Sensitivity or Vulnerability Indices, can be used to prioritize different conservation areas. These indices are often coupled to Geographic Information System (GIS) to indicate biological sensitivity to different classes of impact and the likelihood of occurrence of these impacts (Cogan et al., 2009). ESI maps have been widely applied to the marine environment, first for risk management of spills in the oil industry (Buckley, 1982; Tortell, 1992; Jensen et al., 1993). ESI maps, when complemented by maps representing socio-economic data, can sharpen the collaboration between managers and stakeholders to identify vulnerable locations, establish protection priorities, and identify strategies to minimize undesirable consequences. The value of habitats as surrogates for ecosystem services needs further consideration, as there can be important co-benefits or opportunity costs associated with either their conservation or their loss (Fraschetti, 2012; Fourchault et al., 2024).

6.3.2 Reference Levels and Alert Systems for Spatial Marine Management

The establishment of reference levels on which management decisions can be made and clear protocols to evaluate trade-offs (Link & Browman, 2017) are needed for the spatial planning and management of the CCZ. Alert systems, in contrast, are still novel in their application to marine areas, but have, for example, been proposed in relation to coastal areas and MPAs in the Red Sea, Gulf of Oman, and Arabian Gulf

(Tilot, 2016; IUCN/ROWA, 2016). Levels of key biological (e.g., live coral cover) or physical (e.g., sediment load) or chemical (e.g., oil hydrocarbon) parameters are set to trigger management action to halt the damaging activity, such as an oil leak and implement restitution activities (e.g., oil spill clear up) as previously determined. Often, alert systems discriminate more than one threshold, using a traffic lights system (Halliday et al., 2001; Ceriola et al., 2007; Barange et al., 2010).

6.3.3 Rapid Environmental Assessment and Other Monitoring Techniques

The need for marine biological observations and monitoring strategies to inform science, early warnings, climate projections, protect ocean health, and support the blue economy leads to the identification "Essential Ocean Variables (OEVs)" by the Global Ocean Observing System (GOOS) expert panels (https://goosocean.org/what-we-do/framework/essential-ocean-variables/) (Estes et al., 2021).

Ecosystem surveys or holistic monitoring, in particular when combined with numerical modeling, have proven to be more efficient to resolve ecosystem status and changes in the full temporal and spatial domains, but to obtain higher precision for some stock estimations, there will always be a need for single species/group surveys (Eriksen et al., 2018).

Concerning the design and location of newly declared oceanic MPAs (oMPAs), as described previously, new sampling platforms and methods will serve to best protect biodiversity across the water column (Blanluet et al., 2023). Indeed, clear and measurable conservation objectives and management targets have to be set and evaluated within solid monitoring schemes. The design of a clearly structured evaluation framework, similar to the drivers-pressures-state-impact-response framework (DPSIR), may support this process (e.g., Gabrielsen & Bosch, 2003; Oesterwind et al., 2016).

Rapid environmental (or ecological) assessment (REA) has been developed, typically for the assessment of large areas (Price, 1999; Tilot, 2014). REAs have in common that they accept semi-quantitative abundance estimates of biological taxa and of environmental factors, often assessed on a predefined (often 5- or 6-, or 10-point) scale (Tait & Dipper, 1998). Further, in order to assess the diversity or monitor the presence of keystone trophic groups, REAs restrict species identification to preselected orders or families, for example, echinoderms or fish (Hourigan et al., 1988; Roberts et al., 1988; Montevecchi, 1993; Whitfield & Elliott, 2002; Price, 2004). While individual estimates may lack precision, statistical methods can be applied to the data, enabling quantitative conclusions. Subsequent refinement for specific areas or taxa can then be obtained through archived imagery or additional genetic or tag-based studies.

The moving vessel profiler (MVP) is a rapid assessment tool for the collection of water column profiles and sediment classification used in the water column (Paka et al., 2020). It has been effectively used to collect mid-water oceanographic information to support science and hydrography survey operations.

The use of productivity and susceptibility analysis (PSA) and a rapid management-risk assessment could be used to evaluate the vulnerability of some pelagic species when traditional quantitative methods for population assessment are difficult to use for pelagic predator species that require adequate protection measures (Furlong-Estrada et al., 2017).

Concerning the CCZ, it is essential to establish a permanent or semi-permanent tridimensional monitoring system for the water column, in addition to monitoring the seabed, to record key environmental variables related to the different mining activities that will take place in different layers.

However, to better understand pelagic habitat use, the "Tagging of pelagic Predators' Programme" (TOPP) has been effective. TOPP integrates biological and physical oceanographic data, in collaboration with other ocean monitoring initiatives such as GLOBEC, and the IGBP's core project, initiated by the IOC of UNESCO in 1991, to understand how global change affects the abundance, diversity, and productivity of marine populations (Hofman et al., 2002). This approach is based on synergism between scientists, industry, and governments, using animals as ocean sensors (mirroring Pacific Islanders' sensorial navigation, as previously described) (Bambridge et al., 2019; Tilot et al., 2021a, b, c).

The TOPP collects time-series data from marine migratory species supplemented by pelagic and aerial ecosystems of the eastern Pacific (Block et al., 2003). It targets key species such as air-breathing vertebrates, birds and marine mammals (pink and sooty shearwaters, laysan albatross, black footed albatross, leatherback turtles, loggerhead turtles, blue whales, fin whales and humpback whales, sperm whales), and other marine vertebrates (bluefin tuna, yellow fin tuna, white shark, salmon shark, mako shark, blue shark, *Dosidicus gigas* squid, *Mola mola*). These key species carry a variety of archival electronic tags (pressure time at depth, temperature, external environment, internal body, light level, and geolocation) that allow to record physical and chemical parameters and deduce biochemical and productivity parameters.

In this perspective, the TOPP framework, with its various tools, logistics, and oceanographic integration, is a key asset for monitoring biodiversity, critical habitats, distribution, and abundance of key species in the high seas worldwide, in particular in the CCZ. It could be used to assess the uniqueness of species assemblages, critical habitats, and the range of the movements into and out of the ecosystem (Block et al., 2003; Harrison et al., 2018). Further, standardized methods could enable comparisons between oceanic regions of the world thus facilitating the assessment and management of marine resources.

6.3.4 Indicator Species

Indicators are key tools used to assess the ecological status of the environment for ecosystem-based management (Whitfield & Elliott, 2002; Link & Browman, 2017). Monitoring a group of sentinel species (from a taxonomic and functional point of view) provides useful insights into an ecosystem or a habitat condition under the pressure of natural and anthropogenic disturbances. The monitoring of sentinel species or communities should be adapted to the pelagic domain and its different water layers including trophic processes and migrations. It would also enable cost-effective monitoring of the state of the water column throughout mining operation timeframes.

The TOPP program as described previously targets key species in the north-eastern Pacific, some of which can serve as sentinel species. Some of these species are known to dive far into the mesopelagic and sometimes abyssopelagic layers (e.g., tuna or sperm whales), hence serving as sentinel species for layers in which human monitoring is restricted.

Bioluminescent species, e.g., Myctophidae fishes, gelatinous zooplankton, crustacean, jellyfishes, cephalopods (Irigoien et al., 2014), could also serve to monitor the mesopelagic layer. Their Diel vertical migrations (DVMs) between photic and aphotic realms play a major role in benthopelagic coupling, trophically and behaviorally, by acting as carriers in the transport of carbon and nutrient to deep epibenthic communities, a key factor of the ocean biological pump. There is thus a need to integrate the efficient detection and quantification of bioluminescence, coupled to acoustic backscatter and video imagery from ultra-low-light cameras, into large-scale monitoring programs. This would enable to monitor proxies for the mass and energy transfer from the upper ocean into the deep seabed and consequently the effects of climate change (Levin et al., 2018). It could also enable to monitor the role of deep scattering DVMs in the behavioral responses and abundance and structure of deep-sea benthic communities (Chatzievangelou et al., 2021).

Further into the abyssopelagic layer, sometimes several hundreds of meters above the seabed, occur benthopelagic holothurians such as bioluminescent *Enypniastes eximia* and *Peniagone leander* (Tilot, 2006a, b; Rogacheva et al., 2012), which can serve as indicator species for near-bottom regions.

There are also Sentinels of Seabed (SOS) indicators, typical and sensitive species that enable to assess the state of benthic habitats (Serrano et al., 2022). Their use is operated in two stages with first a "typical species set" using intra-habitat similarity and frequency under reference conditions, then the "sentinel species set" generated by selecting the most sensitive species from the typical species set. This monitoring strategy using "sentinel" indicators recalls the series of monitoring program used to evaluate coral reef ecosystems such as with Reef Check (Hodgson et al., 2006).

6.3.5 Eco-hydrodynamics and Predictive Modeling

The use of mathematical and geospatial tools, such as 3D hydrodynamic models and particle dispersion models, species distribution models, combined with Geographic Information Systems (GIS), is a novel approach to understand how topography, hydrography, and species ecology influence the structure of faunal assemblages (Nihoul, 1981; Nihoul & Djenidi, 1991; Harkema & Weatherly, 1996; Moreno Navas et al., 2014). Eco-hydrodynamic modeling can also provide a better understanding of the interactions between variables such as food supply, larval transport, community composition, sediment dynamics, and pollutant dispersion, in response to hydrological factors such as currents, internal waves, upwelling, downwelling, and bottom topography (Henry et al., 2013).

Such an approach may also be used to predict the effects of anthropogenic activities. Eco-hydrodynamics have for example been applied to a cold-water coral reef ecosystem to predict the effects of bottom fishing and renewable energy installations on reef connectivity and ecosystem function (Watson & Morato, 2004; Henry et al., 2013; Moreno Navas et al., 2014; Roberts & Cairns, 2014).

The effectiveness of both sensitivity maps and alert systems will be greatly enhanced if eco-hydrodynamic models can be incorporated in them, enabling managers to visualize more directly the way in which short- or long-term changes in hydrodynamic environment will impact the biota (Moreno Navas et al., 2014; Roberts & Cairns, 2014).

7 Conclusion

Based on our analysis, we conclude that for deep-sea mining in the CCZ, it is essential to adopt a marine spatial planning and management strategy, based on an ecosystem approach, prior to decision-making with precautionary management measures, adapted preserved areas, and monitoring programs to avoid harm to marine ecosystems from the surface to the seabed. Research is highly recommended to better understand midwater ecosystems, the structure and functions of their components and respective ecosystem services in the ocean biological pump. It is also necessary to harness technological developments to better understand the impacts of deep-sea mining on faunal community dynamics, in particular local and migratory species, to assess possible resilience and adaptations over extended spatial and temporal scales to anthropogenic and natural impacts incrementally to global change.

Scientific evidence shows that for mining operations, in particular midwater discharges, to avoid epi- and mesopelagic zones because of the essential ecosystem services contained in these water layers where primary productivity is essential for all trophic foodwebs and where major fisheries operate. Consequently, discharges would have to be performed in deeper water layers and in the bathyal and abyssal domains, although faunal assemblages and life processes are poorly known at those depths. When considering vertical migrations and life cycles, would these depths be less vulnerable? Recent data have evidenced that climate change could also affect the deep seabed and that connectivity is an important factor for life cycles in the high seas and deep sea.

There is an urgent need to develop adapted management tools to monitor mining operations and identify suitable standards, thresholds, and indicators of impacts on pelagic communities. Pelagic research would need to become a key component of environmental baseline studies and monitoring efforts to ensure that the risks linked to ore collection, transport, and midwater dewatering plumes are fully evaluated. Thus, understanding the biodiversity and dynamics of midwater ecosystems and their value in terms of ecosystem services is an imperative for the scientific community, contractors, managers, and the full suite of stakeholders.

In the selection of conservation measures and area-based management strategies, predicted broad-scale and cumulative impacts to pelagic ecosystems should be considered. When considering conservation areas that would be representative of the whole water column connectivity of ecosystems from seabed to the surface, fixed preserved areas in the high seas, such as the APEIs designed by ISA in the CCZ, are not adapted. The design of moving tridimensional conservation areas with buffer zones would be more adapted to changes in space and time of interconnected ecosystems, playing key roles in the ocean biological pump. These types of large dynamic conservation areas would be essential for Apex predators and highly culturally esteemed migratory species. The best choice would be therefore to set a dynamic network of preserved areas with corridors and flexible boundaries linked to the LMPAs network in the region and the whole Pacific (as well as the network of OMPAs defined by the recently UN adopted High Seas Treaty). Also, it should be encouraged to link actions in the CCZ with other conservation and sustainable development initiatives in the region and the whole Pacific, in particular with the Pacific Island States.

The use of ecohydrodynamics modeling over bottom habitats has proven to enable better management strategies for sustainable fisheries in relatively shallow waters. If adapted to deep sea habitats, it could enable to better understand the dynamics that topography and substrate heterogeneity trigger and thus identify preferential habitats of faunal communities, as on seamounts, and thus to enable to better preserve habitats and associated faunal biodiversity.

As bioluminescence is produced by marine organisms in the whole water column as on the benthos, this phenomenon could be used as a natural indicator of good health of an ecosystem, and more specifically to identify sentinel species for each functional and trophic guild at different depths and to monitor their behavioral responses to gradual impacts in order to identify thresholds for decision-making. The TOPP program targeting key species among Apex predators and associated food web performing horizontal and vertical migrations would also be of good use to monitor naturally the water column in the CCZ using living organisms as sensors and to compare results with other regions of the Pacific as well as globally.

Acknowledgments Special thanks to Ulysses Shen, first grader at Sligo Creek Elementary school, Silver Spring, Maryland, USA, for the first input to the paper drafting fact sheets on basis of Sophie Webb's book « Far from Shore, Chronicles of an Open Voyage » and the program for schools from NOAA « Ocean for Life/National Marine Sanctuaries (https://sanctuaries.noaa.gov). Many thanks to Ganiyat Saliu (Vrije Universiteit Brussels) for proofreading and formatting the bibliography.

Conflict of Interest Statement The authors declare that the research was conducted in the absence of any commercial or financial relationships that could be construed as a potential conflict of interest.

References

Abascal, F. J., Mejuto, J., Quintans, M., & Ramos-Cartelle, A. (2010). Horizontal and vertical movements of swordfish in the Southeast Pacific. *ICES Journal of Marine Science, 67*(3), 466–474. https://doi.org/10.1093/icesjms/fsp252

Agardy, T. M. (1994). Advances in marine conservation: the role of marine protected areas. *Trends in Ecology & Evolution, 9*, 267–270. https://doi.org/10.1016/0169-5347(94)90297-6

Aller, R. (1997). Benthic community response to temporal and spatial gradients in physical disturbance within a deep-sea western boundary region. *Deep Sea Research Part I, 44*, 39–69. https://doi.org/10.1016/S0967-0637(96)00092-1

Amador, J. A., Alfaro, E. J., Lizano, O. G., & Magaña, V. O. (2006). Atmospheric forcing of the eastern tropical Pacific: A review. *Progress in Oceanography, 69*, 101–142. https://doi.org/10.1016/j.pocean.2006.03.007

Amos, A. F., Mos, A. F., Roels, O. A., Garside, C., & Malone, T. C. (1977). Environmental aspects of nodule mining. In G. P. Glasby (Ed.), *Marine manganese deposits* (pp. 391–437). Elsevier Publication Company.

Ardron, J., Gjerde, K., Pullen, S., & Tilot, V. (2008). Marine spatial planning in the high seas. *Marine Policy, 32*, 832–839. https://doi.org/10.1016/j.marpol.2008.03.018

Armstrong, C. W., Foley, N. S., Tinch, R., & van den Hove, S. (2012). Services from the deep: Steps towards valuation of deep sea goods and services. *Ecosystem Services, 2*, 2–13. https://doi.org/10.1016/j.ecoser.2012.07.001

Au, D. W., & Perryman, W. L. (1985). Dolphin habitats in the eastern tropical Pacific. *Fisheries Bulletin, 83*, 623–644.

Au, D. W., & Pitman, R. L. (1986). Seabird interactions with dolphins and tuna in the eastern tropical Pacific. *Condor*, 304–317. https://doi.org/10.2307/1368877

Avila, I., Kaschner, K., & Dormann, C. F. (2018). Current global risks to marine mammals : Taking stock of threats. *Biological Conservation, 221*, 44–58. https://doi.org/10.1016/j.biocon.2018.02.021

Baldwin, R. J., Glatts, R. C., & Smith, K. L. (1998). Particulate matter fluxes into the benthic boundary layer at a long time-series station in the abyssal NE Pacific: Composition and fluxes. *Deep Sea Research Part II, 45*, 643–665. https://doi.org/10.1016/S0967-0645(97)00097-0

Ballance, L. T., Gerrodette, T., Lennert-Cody, C. E., Pitman, R. L., & Squires, D. (2021, 2021). A history of the tuna-dolphin problem: Successes, failures and lessons learned. Review article. *Frontiers in Marine Science, Section Marine Conservation and Sustainability*. https://doi.org/10.3389/fmars.2021.754755

Bambridge, T. (2016). *Rahui: Legal pluralism in Polynesian traditional management of resources and territories* (269 pp). Pacific Series. Australian National University Press.

Bambridge, T., & D'Arcy, P. (2014). Large-scale marine protected areas in the Pacific: Cultural and social perspectives. In F. Feral & B. Salvat (Eds.), *Gouvernance, enjeux et mondialisation des grandes aires marines protegées: recherche sur les politiques environnementales de zonage maritime, le challenge maritime de la France de Mediterranée et d'Outre-mer* (pp. 113–132). l'Harmattan.

Bambridge, T., Gaulme, F., Montet, C., & Paulais, T. (2019). *Communs et oceans. Le « Rahui » en Polynésie* (175 pp). Au vent des îles.

Barange, M., Field, J. G., Harris, R. P., Hofmann, E. H., Perry, R. I., Werner, F. E., & (eds.). (2010). *Marine ecosystems and global change*. Oxford University Press. https://doi.org/10.1093/acprof:oso/9780199558025.001.0001

Barnett, T. P., Adam, J. C., & Lettenmaier, D. P. (2005). Potential impacts of a warming climate on water availability in snow-dominated regions. *Nature, 438*, 303–309. https://doi.org/10.1038/nature04141

Batista-Garcia, R. A., Kumar, V. V., Ariste, A., Tovar-Herrera, O. E., Savary, O., Peidro-Guzman, H. (2017). Simple screening protocol for identification of potential mycoremediation tools for the elimination of polycyclic aromatic hydrocarbons and phenols from hyperalkalophile industrial effluents. *Journal of Environmental Management, 198(Pt 2), 1–11.* https://doi.org/10.1016/j.jenvman.2017.05.010

Beaulieu, S. E., & Smith, K. L., Jr. (1998). Phytodetritus entering the benthic boundary layer and aggregated on the sea floor in the abyssal NE Pacific: Macro- and microscopic composition. *Deep Sea Research Part II, 45*, 781–815. https://doi.org/10.1016/S0967-0645(98)00003-4

Bell, J. D., Senina, I., Adams, T., Aumont, O., Calmettes, B., Clark, S., et al. (2021). Pathways to sustaining tuna-dependent Pacific Island economies during climate change. *Nature Sustainability, 4*, 900–910. https://doi.org/10.1038/s41893-021-00745-z

Bianchi, D., Galbraith, E. D., Carozza, D. A., Mislan, K. A. S., & Stock, C. A. (2013a). Intensification of open ocean oxygen depletion by vertically migrating animals. *Nature Geoscience, 6*, 545–548. https://doi.org/10.1038/ngeo1837

Bianchi, D., Stock, C. A., Galbraith, E. D., & Sarmiento, J. L. (2013b). Diel vertical migration: Ecological controls and impacts on the biological pump in a one-dimensional ocean model. *Global Biogeochemical Cycles, 27*(2), 478–491. https://doi.org/10.1002/gbc.20031

Bingham, F. M., & Lukas, R. (1995). The distribution of intermediate water in the western equatorial Pacific during January–February 1986. *Deep Sea Research Part I: Oceanographic Research Papers, 42*, 1545–1573. https://doi.org/10.1016/0967-0637(95)00064-D

Blanluet, A., Game, E. T., Dunn, D. C., Everett, J. D., Lombard, A. T., & Richardson, A. J. (2023). Evaluating ecological benefits of oceanic protected areas. *Trends in Ecology & Evolution, 39*, 175. https://doi.org/10.1016/j.tree.2023.09.003

Block, B. A., Costa, D. P., Boehlert, G. W., & Kochevar, R. E. (2003). Revealing pelagic habitat use: The tagging of Pacific pelagics program. *Oceanologica Acta, 25*, 255–266. https://doi.org/10.1016/S0399-1784(02)01212-4

Block, B. A., Jonsen, I., Jorgensen, S., Winship, A. J., Shaffer, S. A., Bograd, S. J., et al. (2011). Tracking apex marine predator movements in a dynamic ocean. *Nature, 475*, 86–90. https://doi.org/10.1038/nature10082

Bostock, H. C., Opdyke, B. N., & Williams, M. J. M. (2010). Characterising the intermediate depth waters of the Pacific Ocean using δ13C and other geochemical tracers. *Deep Sea Research Part I: Oceanographic Research Papers, 57*(7), 847–859. https://doi.org/10.1016/j.dsr.2010.04.005

Breitburg, D. L., Crump, B. C., Dabiri, J. O., & Gallegos, C. L. (2010). Ecosystem engineers in the pelagic realm: Alteration of habitat by species ranging from microbes to jellyfish. *Integrative and Comparative Biology, 50*, 188–200. https://doi.org/10.1093/icb/icq051

Brierley, A. S. (2014). Diel vertical migration. *Current Biology, 24, R1074–R1076.* https://doi.org/10.1016/j.cub.2014.08.054

Brinton, E. (1979). Parameters relating to the distributions of planktonic organisms, especially euphausiids in the eastern tropical Pacific. *Progress in Oceanography, 8*, 125–189. https://doi.org/10.1016/0079-6611(79)90001-6

Brito-Morales, I., Schoeman, D. S., Everett, J. D., Klein, C. J., Dunn, D. C., Garcia, J., et al. (2022). Towards climate-smart, three-dimensional protected areas for biodiversity conservation in the high seas. *Nature Climate Change, 12*, 402–407. https://doi.org/10.1038/s41558-022-01323-7

Brouwer, S., Piling, G., Hampton, J., Williams, P., Vincent, M., & Peatman, T. (2018). *Oceanic fisheries programme. Tuna fisheries assessment report no 19*. Pacific Community.

Brummer, F., Pfannkuchen, M., Baltz, A., Hauser, T., & Thiel, V. (2008). Light inside sponges. *Journal of Experimental Marine Biology and Ecology, 367*, 61–64. https://doi.org/10.1016/j.jembe.2008.06.036

Buckley, R. C. (1982). Environmental sensitivity mapping – What, why and how. *Minerals and the Environment, 4*, 151–155. https://doi.org/10.1007/BF02085976

Catalá, T. S., Reche, I., Fuentes-Lema, A., Romera-Castillo, C., Nieto-Cid, M., Ortega-Retuerta, E., et al. (2015a). Turnover time of fluorescent dissolved organic matter in the dark global ocean. *Nature Communications, 6*, 5986. https://doi.org/10.1038/ncomms6986

Catalá, T. S., Reche, I., Álvarez, M., Khatiwala, S., Guallart, E. F., Benítez-Barrios, V. M., et al. (2015b). Water mass age and ageing driving chromophoric dissolved organic matter in the dark global ocean. *Global Biogeochemical Cycles, 29*, 1–18. https://doi.org/10.1002/2014GB005048

Cathalot, C., Van Oevelen, D., Cox, T. J. S., Kutti, T., Lavaleye, M., Duineveld, G., et al. (2015). Cold-water coral reef sand adjacent sponge grounds : Hotspots of benthic respiration and organic carbon cycling in the deep sea. *Frontiers in Marine Science, 2*, 1–12. https://doi.org/10.3389/fmars.2015.00037

Ceriola, L., Ungaro, N., & Toteda, F. (2007). A "Traffic Light" approach for the assessment of the Broadtail shortfin squid *Illex coindetii* (Verany, 1839) in the Southern Adriatic Sea (Central Mediterranean). *Reviews in Fish Biology and Fisheries, 17*, 145. https://doi.org/10.1007/s11160-006-9019-5

Chang, C.-H., Xie, S.-P., Schneider, N., Qiu, B., Small, J., Zhuang, W., et al. (2012). East Pacific Ocean eddies and their relationship to subseasonal variability in Central American wind jets. *Journal of Geophysics Research, 117*, C10001. https://doi.org/10.1029/2011JC007315

Chatzievangelou, D., Bahamon, N., Martini, S., Del Rio, J., Riccobene, G., Tangherlini, M., et al. (2021). Integrating diel vertical migrations of bioluminescent deep scattering layers into monitoring programs. *Frontiers in Marine Science, 8*. https://doi.org/10.3389/fmars.2021.661809

Chen, W., Zhang, T., & Guan, L. (2017). Radiation transfer in the ocean and ocean color. In *Comprehensive remote sensing* (pp. 43–78). Elsevier. https://doi.org/10.1016/B978-0-12-409548-9.10397-5

Christiansen, B., Denda, A., & Christiansen, S. (2020). Potential effects of deep seabed mining on pelagic and benthopelagic biota. *Marine Policy, 2020*, 103442. https://doi.org/10.1016/j.marpol.2019.02.014

Claes, J. M., Nilsson, D. E., Straube, N., et al. (2014a). Iso-luminance counter-illumination drove bioluminescent shark radiation. *Scientific Reports, 4*, 4328. https://doi.org/10.1038/srep04328

Claes, J. M., Partridge, J. C., Hart, N. S., Garza-Gisholt, E., Ho, H.-C., Mallefet, J., et al. (2014b). Photon hunting in the twilight zone: Visual features of mesopelagic bioluminescent sharks. *PLoS One, 9*(8), e104213. https://doi.org/10.1371/journal.pone.0104213

Clark, M. R., Watling, L., Rowden, A., Guinotte, J. M., & Smith, C. R. (2011). A global seamount classification to aid the scientific design of marine protected area networks. *Ocean and Coastal Management, 54*, 19–36. https://doi.org/10.1016/j.ocecoaman.2010.10.006

Clark, M. R., Althaus, F., Schlacher, T. A., Williams, A., Bowden, D. A., & Rowden, A. A. (2016). The impacts of deep-sea fisheries on benthic communities: A review. *ICES Journal of Marine Science, 73*, 151–169. https://doi.org/10.1093/icesjms/fsv123

Cogan, C. B., Todd, B. J., Lawton, P., & Noji, T. T. (2009). The role of marine habitat mapping in ecosystem-based management. *ICES Journal of Marine Science, 66*, 2033–2042. https://doi.org/10.1093/icesjms/fsp214

Cormier, R., Elliott, M., & Rice, J. (2019). Putting on a bow-tie to sort out who does what and why in the complex arena of marine policy and management. *Science of the Total Environment, 648*, 293–305. Elsevier.

Cormier, R., Kelble, C. R., Anderson, M. R., Allen, J. I., Grehan, A., & Gregersen, O. (2017). Moving from ecosystem-based policy objectives to operational implementation of ecosystem-based management measures. *ICES Journal of Marine Science, 74*(1), 406–413. https://doi.org/10.1093/icesjms/fsw181

Corrigan, C., & Kershaw, F. (2008). *Working toward high seas marine protected areas: An assessment of progress made and recommendations for collaboration*. UNEP/WCMC.

Costa, D., Block, B., Bograd, S., Fedak, M., & Gunn, J. (2010). TOPP as a marine life observatory: Using electronic tags to monitor the movements, behaviour and habitats of marine invertebrates. *Proceedings of the OceanObs, 9*, 21–25. https://doi.org/10.5270/OceanObs09.cwp.19

Danovaro, R. (2008). Exponential decline of deep-sea ecosystem functioning linked to benthic biodiversity loss. *Current Biology, 18*, 1–8. https://doi.org/10.1016/j.cub.2007.11.056

D'Arcy, P. (2006). *The people of the sea: Environment, identity and history in Oceania* (312 pp). University of Hawai'i Press.

Davison, P. C., Checkley, D. M., Jr., Koslow, J. A., & Barlow, J. (2013). Carbon export mediated by mesopelagic fishes in the Northeast Pacific Ocean. *Progress in Oceanography, 116*, 14–30. https://doi.org/10.1016/j.pocean.2013.05.013

Dayton, P. K., & Hessler, R. R. (1972). Role of biological disturbance in maintaining diversity in the deep-sea. *Deep Sea Research, 19*, 199–208. https://doi.org/10.1016/0011-7471(72)90031-9
de Busserolles, F., Cortesi, F., Helvik, J. V., Davies, W. I. L., Templin, R. M., Sullivan, R. K. P., et al. (2017). Pushing the limits of photoreception in twilight conditions: The rod-like cone retina of the deep-sea pearlsides. *Science Advances, 3*(11). https://doi.org/10.1126/sciadv.aao4709
De Lucia, V. (2019). The BBNJ negotiations and ecosystem governance in the Arctic. *Marine Policy, 142*, 103756. https://doi.org/10.1016/j.marpol.2019.103756
Denton, E. J. (1990). Light and vision at depths greater than 200 metres. In P. J. Herring, A. K. Campbell, M. Whitfield, & L. Maddock (Eds.), *Light and life in the sea* (pp. 127–148). Cambridge University Press.
Dickson, B., Hurrell, J., Bindoff, N., Wong, A., Arbic, B., Owens, B., et al. (2001). Chapter 7.3 The world during WOCE. *International Geophysics, 77*, 557–583. https://doi.org/10.1016/S0074-6142(01)80139-X
Dietz, R. S. (1962). The sea's deep scattering layers. *Scientific American, 207*, 44–51. https://doi.org/10.1038/scientificamerican0862-44
Dittmar, T. (2015). Reasons behind the long-term stability of dissolved organic matter. In A. D. Hansell & C. A. Carlson (Eds.), *Biogeochemistry of marine dissolved organic matter* (2nd ed., pp. 369–388). Elsevier Academic Press.
Drazen, J. C., Baldwin, R. J., & Smith, K. L., Jr. (1998). Sediment community response to a temporally varying food supply at an abyssal station in the NE Pacific. *Deep Sea Research II, 45*, 893–913. https://doi.org/10.1016/S0967-0645(98)00007-1
Drazen, J. C., Smith, C. R., Gjerde, K., Whitlow, A., Black, J., Carter, G., et al. (2019). Report of the workshop: "Evaluating the nature of midwater mining plumes and their potential effects on midwater ecosystems". *RIO Research Ideas and Outcomes, 5*, e33527. https://doi.org/10.3897/rio.5.e33527
Drazen, J. C., Smith, C. R., Gjerde, K. M., Haddock, S. H. D., Carter, G. S., & Choy, C. A. (2020). Midwater ecosystems must be considered when evaluating environmental risks of deep-sea mining. *Proceedings of the National Academy of Sciences, 117*, 17455–17460.
Duarte, C. M., Agusti, S., Barbier, E., Britten, G. L., Castilla, J. C., Gattuso, J. P., et al. (2020). Rebuilding marine life. *Nature, 580*, 39–51. https://doi.org/10.1038/s41586-020-2146-7
Duffy, L. M., Kuhnert, P. M., Pethybridge, H. R., Young, J. W., Olson, R. J., Logan, J. M., et al. (2017). Global trophic ecology of yellowfin, bigeye, and albacore tunas: Understanding predation on micronekton communities at ocean-basin scales. *Deep Sea Research Part II: Topical Studies in Oceanography, 140*, 55–73. https://doi.org/10.1016/j.dsr2.2017.03.003
Dymond, J., & Collier, R. (1988). Biogenic particle fluxes in the equatorial Pacific: Evidence for both high and low productivity during the 1982–1983 El Niño. *Global Biogeochemical Cycles, 2*, 129–137. https://doi.org/10.1029/GB002i002p00129
Eckstein, L., & Schwarz, A. (2019). The making of Tupaia's map: A story of the extent and mastery of polynesian navigation, competing systems of wayfinding on James Cook's endeavour, and the invention of an ingenious cartographic system. *The Journal of Pacific History, 54*, 1–95. https://doi.org/10.1080/00223344.2018.1512369
Edgar, G. J., Russ, G. R., & Babcock, R. C. (2007). Marine protected areas. In S. D. Connell & B. M. Gillanders (Eds.), *Marine ecology* (pp. 533–555). Oxford University Press. ISBN: 0195553020.
Ehrlich, H., Etnoyer, P., Litvinov, S.D., Olennikova, M. M., Domaschke, H., Hanke,T., et al., (2006). Biomaterial structure in deep-sea bamboo coral *(Anthozoa:Gorgonacea:Isididae):* Perspectives for the development of bone implants and templates for tissue engineering. Materials Science & Engineering Technology, 37, 552–557. https://doi.org/10.1002/mawe.200600036.
Emery, W. J. (2003). Ocean circulation, water types and water masses. In *Encyclopedia of atmospheric sciences* (pp. 1556–1567). https://doi.org/10.1006/rwos.2001.0108
Enquist, B. J., Abraham, A. J., Harfoot, M. B. J., Malhi, Y., & Doughty, C. E. (2020). The megabiota are disproportionately important for biosphere functioning. *Nature Communications, 11*, 699. https://doi.org/10.1038/s41467-020-14369-y

Eriksen, E., Gjostaeter, H., Prozorkevich, D., Shamray, E., Dolgov, A., Skern-Mauritzen, M., et al. (2018). From single species surveys towards monitoring of the Barents Sea ecosystem. *Progress in Oceanography, 166*, 4–14. https://doi.org/10.1016/j.pocean.2017.09.007

ESCO CNRS IFREMER. (2014). Rapport d'expertise collective, Juin (2014). *Impacts environnementaux de l'exploitation des ressources minérales marines profondes*. 939pp.

Estes, J. A., Terborgh, J., Brashares, J. S., Power, M. E., Berger, J., Bond, W. J., et al. (2011). Trophic downgrading of planet Earth. *Science, 333*(6040), 301–306. https://doi.org/10.1126/science.1205106

Estes, M., Anderson, C., Appeltans, W., Bax, N., Bednarsek, N., Canonico, G., et al. (2021). Enhanced monitoring of life in the sea is a critical component of conservation management and sustainable economic growth. *Marine Policy, 132*, 104699. https://doi.org/10.1016/j.marpol.2021.104699

Ey, M., & Sherval, M. (2016). Exploring the minescape: Engaging with the complexity of the extractive sector: Exploring the minescape. *Area, 48*, 176–182. https://doi.org/10.1111/area.12245

Fagundes, M., Litvin, S. Y., Micheli, F., De Leo, G., Boch, C. A., Barry, J. P., et al. (2020). Downscaling global ocean climate models improves estimates of exposure regimes in coastal environments. *Scientific Reports, 10*, 14227. https://doi.org/10.1038/s41598-020-71169-6

Falkowski, P. (2012). The power of plankton. *Nature, 483*, 17–20. https://doi.org/10.1038/483S17a

FAO. Fisheries and Aquaculture Department. (2000). Biological Characteristics of Tuna. *Text by Michel Goujon and Jacek Majkowski*. Available online at: http://www.fao.org/fishery/topic/16082/en

Feely, R., Sabine, C., Lee, K., Berelson, W., Kleypas, J., Fabry, V., et al. (2004). Impact of anthropogenic CO_2 on the $CaCO_3$ system in the oceans. *Science, 305*, 362–366. https://doi.org/10.1126/science.1097329

Fernández Álamo, M. A., & Färber Lorda, J. (2006). Zooplankton and the oceanography of the eastern tropical Pacific: A review. *Progress in Oceanography, 69*(2–4), 318–359. https://doi.org/10.1016/j.pocean.2006.03.003

Ferretti, F., Worm, B., Britten, G., Heithaus, M., & Lotze, H. (2010). Patterns and ecosystem consequences of shark declines in the ocean. *Ecology Letters, 13*(8), 1055–1071. https://doi.org/10.1111/j.1461-0248.2010.01489.x

Field, C. B., Behrenfeld, M. J., Randerson, J. T., & Falkowski, P. (1998). Primary production of the biosphere: Integrating terrestrial and oceanic components. *Science, 281*(10), 237–240. https://doi.org/10.1126/science.281.5374.237

Fiedler, P. C., & Talley, L. D. (2006). Hydrography of the eastern tropical Pacific: A review. *Progress in Oceanography, 69*, 143–180. https://doi.org/10.1016/j.pocean.2006.03.008

Foley, M. M., Halpern, B. S., Micheli, F., Armsby, M. H., Caldwell, M. R., Crain, C. M., et al. (2010). Guiding ecological principles for marine spatial planning. *Marine Policy, 34*, 955–966. https://doi.org/10.1016/j.marpol.2010.02.001

Folke, C., Carpenter, S., Walker, B., & Scheffer, M. (2004). Regime shifts, resilience and biodiversity in ecosystem management. *Annual Review of Ecology, Evolution and Systematics, 35*, 557–581. https://doi.org/10.1146/annurev.ecolsys.35.021103.105711

Fourchault, L., Dahdouh-Guebas, F., Dunn, D. C., Everett, J. D., Hanson, J. O., Buenafe, K. C. V., et al. (2024). Generating affordable protection of high seas biodiversity through cross-sectoral spatial planning. *OneEarth, 7*, 253. https://doi.org/10.1016/oneear.2023.12.006

Fraschetti, S. (2012). Mapping the marine environment: Advances, relevance and limitations in ecological research and in marine conservation and management. *Biologia Marina Mediterranea, 19*, 79–83.

Freiwald, A., & Roberts, J. M. (2006). Cold-water corals and ecosystems. In A. Freiwald & J. M. Roberts (Eds.), *Erlangen earth conference series* (p. 1243). Springer.

Fuller, D. W., Schaefer, K. M., Hampton, J., Caillot, S., Leroy, B. M., & Itano, D. G. (2015). Vertical movements, behavior, and habitat of bigeye tuna (*Thunnus obesus*) in the equatorial Central Pacific Ocean. *Fisheries Research, 172*, 57–70. https://doi.org/10.1016/j.fishres.2015.06.024

Furlong-Estrada, E., Galvan-Magana, F., & Tovar-Avila, J. (2017). Use of the productivity and susceptibility analysis and a rapid management-risk assessment to evaluate the vulnerability of sharks to evaluate the vulnerability of sharks caught off the west coast of Baja California Sur, Mexico. *Fisheries Research, 194*, 197–208. https://doi.org/10.1016/j.fishres.2017.06.008

Gabrielsen, P., & Bosch, P. (2003). *Environmental indicators: Typology and use in reporting*. Internal working paper, Copenhagen.

Gaines, S. D., White, C., Carr, M., & Palumbi, S. R. (2010). Designing marine reserve networks for both conservation and fisheries management. *Proceedings of the National Academy of Sciences of the United States of America, 107*(43), 18286–18293. https://doi.org/10.1073/pnas.0906473107

Game, E. T., Grantham, H. S., Hobday, A. J., Pressey, R. L., Lombard, A. T., Beckley, L., et al. (2009). Pelagic protected areas: The missing dimension in ocean conservation. *Trends in Ecology & Evolution, 24*, 360–369. https://doi.org/10.1016/j.tree.2009.01.011

Gardner, W. D., Sullivan, L. G., & Thorndike, E. M. (1984). Long term photographic, current, and nephelometer observations of manganese nodule environments in the Pacific. *Earth and Planetary Science Letters, 70*, 95–109. https://doi.org/10.1016/0012-821X(84)90212-7

Gianni, M., Fuller, S. D., Currie, D. E. J., Schleit, K., Goldsworthy, L., Pike, B., et al. (2016). *How much longer will it take? A ten-year review of the implementation of United Nations General Assembly resolutions 61/105, 64/72 and 66/68 on the Management of Bottom Fisheries in areas beyond National Jurisdiction*. Deep Sea Conservation Coalition. Available online at: http://www.savethehighseas.org/publicdocs/DSCC-Review-2016_Launch-29-July.pdf

Gillard, B., Harbour, R. P., Nowald, N., Thomsen, L., & Iversen, M. H. (2022). Vertical distribution of particulate matter in the clarion clipperton zone (German Sector)—Potential impacts from deep-sea mining discharge in the water column. *Frontiers in Marine Science, 9, 820947.* https://doi.org/10.3389/fmars.2022.820947

Ginzky, H., Singh, P., & Markus, T. (2020). Strengthening the international seabed authority's knowledge-base: Addressing uncertainties to enhance decision-making. *Marine Policy, 114, 103823. ISSN 0308-597X.* https://doi.org/10.1016/j.marpol.2020.103823

Gjerde, K. M., & Yadav, S. S. (2017). Polycentricity and regional Ocean governance: Implications for the emerging UN Agreement on Marine Biodiversity Beyond National Jurisdiction. *Frontiers in Marine Science, Section of Marine Affairs and Policy, 8*. https://doi.org/10.3389/fmars.2021.704748

Gjerde, K., & Wright, G. (2019). *Towards ecosystem-based management of the Global Ocean: Strengthening regional cooperation through a new agreement for the conservation and sustainable use of marine biodiversity in areas beyond National Jurisdiction* (Vol. 132, p. 28). STRONG High Seas Project.

Gjerde, K., Clark, N., & Harden-Davies, H. (2019). *Building a platform for the future: The relationship of the expected new agreement for marine biodiversity in areas beyond National Jurisdiction and the UN convention on the law of the sea* (Vol. 33, p. 1). Koninklijke Brill NV.

Glover, A. G., & Smith, C. R. (2003). The deep-sea floor ecosystem: Current status and prospects of anthropogenic change by the year 2025. *Environmental Conservation, 30*, 219–241. https://doi.org/10.1017/S0376892903000225

Govan, H., Tawake, A., Tabunakawai, K., Jenkins, A., Lasgorceix, A., Schwarz, A.-M., et al. (2009). *Status and potential of locally-managed marine areas in the South Pacific: Meeting nature conservation and sustainable livelihood targets through wide-spread implementation of LMMAs* (p. 134). SPREP/WWF/WorldFish-Reefbase/CRISP.

Gray, J. S. (1997). Marine biodiversity: Patterns, threats and conservation needs. *Biodiversity and Conservation, 6*, 153–175. https://doi.org/10.1023/A:1018335901847

Haddock, S. H. D., Christianson, L. M., Francis, W. R., Martini, S., Dunn, C. W., Pugh, P. R., et al. (2017). Insights into the biodiversity, behavior, and bioluminescence of deep-sea organisms using molecular and maritime technology. *Oceanography, 30*(4), 38–47. https://doi.org/10.5670/oceanog.2017.422

Haedrich, R. L., & Rowe, G. T. (1977). Megafaunal biomass in the deep-sea. *Nature, 269*, 141–142. https://doi.org/10.1038/269121a0

Halbach, P., Friedrich, G., & von Stackelberg, U. (1988). *The manganese Nodule Belt of the Pacific Ocean: Geological environment, nodule formation, and mining aspects*. Enke Verlag.

Halliday, R. G., Fanning, L. P., & Mohn, R. K., (2001). Use of traffic light method in fishery management planning. *Canadian Science Advisory Secretariat, Res. Doc.* No 2001/108, 41. Available online at: http://www.dfo-mpo.gc.ca/csas/Csas/DocREC/2001/RES2001_108e.pdf

Halpern, B. S., Lester, S. E., & McLeod, K. L. (2010). Placing marine protected areas onto the ecosystem-based management seascape. *Proceedings of the National Academy of Sciences of the United States of America, 107*(43), 18312–18317. https://doi.org/10.1073/pnas.0908503107

Halpern, B., Frazier, M., Afflerbach, J., Stewart, J. S., Micheli, F., O'Hara, C., et al. (2013). Recent pace of change in human impact on the world's ocean. *Scientific Reports, 9*(1), 11609. https://doi.org/10.1038/s41598-019-47201-9

Halpern, B. S., Frazier, M., Potapenko, J., Casey, K. S., Koenig, K., Longo, C., et al. (2015). Spatial and temporal changes in cumulative human impacts on the world's ocean. *Nature Communications, 6*, 7615. https://doi.org/10.1038/ncomms8615

Han, A., Wang, Y., Huo, Y., et al. (2022). Nutrient distributions and nitrogen-anomaly (N*) in the tropical North Pacific Ocean. *Acta Oceanologica Sinica, 41*, 23–33. https://doi.org/10.1007/s13131-021-1918-8

Hannides, A. K., & Smith, C. R. (2003). The northeastern Pacific abyssal plain. In K. D. Black & G. B. Shimmield (Eds.), *Biogeochemistry of marine systems, Chapter 10* (pp. 208–237). Blackwell Science.

Harkema, R., & Weatherly, G. L., (1996). *A compilation of moored currentmeter data from the southern boundary of the Brazil basin for the deep basin experiment September 1993–March 1995*. Technical Report CMF 96 01.

Harrison, A.-L., Costa, D. P., Winship, A. J., Benson, S. R., Bograd, S. J., Antolos, et al. (2018). The political biogeography of migratory marine predators. *Nature Ecology and Evolution, 2*, 1571–1578. https://doi.org/10.1038/s41559-018-0646-8

Hau'ofa, E. (1994). Our sea of islands. *The Contemporary Pacific: A Journal of Island Affairs, 6*, 148–161.

Hau'ofa, E. (2008). *We are the ocean: Selected works* (188 pp). University of Hawaii Press.

Hayes, S. (1979). Benthic current observations at Domes sites A, B, C in the tropical central North Pacific Ocean. In J. L. Bischoff & D. Z. Piper (Eds.), *Marine geology and oceanography of the Pacific Manganese Nodule Province* (pp. 83–112). Plenum Press.

Heithaus, M. R., Frid, A., Wirsing, A. J., & Worm, B. (2008). Predicting ecological consequences of marine top predator declines. *Trends in Ecology & Evolution, 4*, 202–210. https://doi.org/10.1016/j.tree.2008.01.003

Henry, L.-A., Moreno Navas, J., & Roberts, J. M. (2013). Multi-scale interactions between local hydrography, seabed topography and community assembly on cold-water coral reefs. *Biogeosciences, 10*, 2737–2746. https://doi.org/10.5194/bg-10-2737-2013

Hernández-León, S., & Ikeda, T. (2005). A global assessment of mesozooplankton respiration in the ocean. *Journal of Plankton Research, 27*(2), 153–158. https://doi.org/10.1093/plankt/fbh166

Hernández-León, S., Kooplemann, R., Fraile-Nuez, E., Bode, A., Mompeán, C., Irigoien, et al. (2020). Large deep-sea zooplankton biomass mirrors primary production in the global ocean. *Nature Communications, 11*(1), 1–8. https://doi.org/10.1038/s41467-020-19875-7

Herring, P. J. (2007). Sex with the lights on? A review of bioluminescent sexual dimorphism in the sea. *Journal of the Marine Biological Association of the United Kingdom, 87*(4), 829–842. https://doi.org/10.1017/S0025315407056433

Hodgson, G., Hill, J., Kiene, W., Maun, L., Lihaly, J., Liebeler, J., et al. (2006). *Reef check instruction manual: A guide to reef check coral reef monitoring* (95 pp). Reef Check Foundation.

Hoffert, M. (2008). *Les Nodules Polymétalliques Dans Les grands Fonds Océaniques, une Extraordinaire Aventure Minière et Scientifique Sous-Marine, Editions VUIBERT* (430 pp). Société géologique de France.

Hofman, E. E., Klinck, J. M., Costa, D. P., Daly, K. L., Torres, J. J., & Fraser, W. R. (2002). U.S. Southern Ocean Global Ocean ecosystems dynamics program. *Oceanography, 14*, 67–74. https://doi.org/10.5670/oceanog.2002.22

Hollister, C. D., & Nowell, A. R. M. (1991). HEBBLE epilogue. *Marine Geology, 99*, 445–460. https://doi.org/10.1016/0025-3227(91)90055-9

Hourigan, T. F., Timothy, C. T., & Reese, E. S. (1988). Coral reef fishes as indicators of environmental stress in coral reefs. In D. F. Soule & G. S. Kleppel (Eds.), *Marine organisms as indicators* (pp. 107–135). Springer.

Howey, L., Tolentino, E., Papastamatiou, Y., Brooks, E., Abercrombie, D., Watanabe, Y., et al. (2016). *Ecology and Evolution, 6*(15), 5290–5304. https://doi.org/10.1002/ece3.2260

Hughes, T. P., Bellwood, D. R., Folke, C., Steneck, R., & Wilson, J. (2005). New paradigms for supporting the resilience of marine ecosystems. *Trends in Ecology & Evolution, 20*, 380–385. https://doi.org/10.1016/j.tree.2005.03.022

Inter-American Tropical Tuna Commission (IATTC). (2014). *The quarterly report of the Inter-American Tropical Tuna Commission*. Compean, G. Headquarters and main laboratory La Jolla, California, USA. www.iattc.org

International Seabed Authority (ISA). (2018). *Summary report, 16–26 July 2018. 2nd Part of the 24th Session of the International Seabed Authority*. Available online at: https://enb.iisd.org/events/2nd-part-24th-session-international-seabed-authority/summary-report-16-26-july-2018

International Seabed Authority (ISA). (2020). *Proposal for a template with minimum requirements for regional environmental management plans: A proposal for a standardized approach*. International Seabed Authority. Available online at: https://www.isa.org.jm/protection-of-the-marine-environment/regional-environmental-management-plans/

International Seabed Authority (ISA). (2023). Potential interactions between fishing and mineral resource-related activities in areas beyond national jurisdiction: A spatial analysis. *ISA Technical Study No. 33.*, 61pp. International Seabed Authority (ISA). Available online at: https://www.isa.org.jm/publications/technical-study-33-potential-interactions-between-fishing-and-mineral-resource-related-activities-in-areas-beyond-national-jurisdiction-a-spatial-analysis

Irigoien, X., Klevjer, T. A., Rostad, A., Martinez, U., Boyra, G., Acuna, J. L., et al. (2014). Large mesopelagic fishes biomass and trophic efficiency in the open ocean. *Nature Communications, 5*, 3271. https://doi.org/10.1038/ncomms4271

IUCN/ROWA. (2016). *Ecological baseline assessment using environmental sensitivity index mapping for Al Daymaniyat Islands, Oman Sultanate*. IUCN/ROWA Report, 170pp.

Izsak, C., & Price, A. R. G. (2001). Measuring S-diversity using a taxonomic similarity index, and its relation to spatial scale. *Marine Ecology Progress Series, 215*, 69–77. https://doi.org/10.3354/meps215069

Jensen, J. R., Narumalani, S., Weatherbee, O., Murday, M., Sexton, W. J., & Green, C. (1993). Coastal environmental sensitivity mapping for oil spills in The United Arab Emirates using remote sensing and GIS technlogy. *Geocarto International, 8*, 5–13. https://doi.org/10.1080/10106049309354404

Jeudy de Grissac, A., (2015). Reviewing existing Mediterranean models of governance of MPAs with artisanal fisheries. In *Fishing Governance in MPAs : Potentialities for blue economy (FISHMPABLUE)*. 54pp. Available online at: https://www.iucn.org/sites/dev/files/content/documents/fishmpablue_governance_models_review_existing_final_oct.pdf

Jaekel, A. (2016). Deep seabed mining and adaptive management: The procedural challenges for the international seabed authority. Kingston, Jamaica, pp.1–15.

Johnson, C., Runge, J., Curtis, K. A., Durbin, E. G., Hare, J. A., Incze, L. S. et al. (2011). Biodiversity and ecosystem function in the Gulf of Maine: Pattern and role of Zooplankton and Pelagic Nekton. Published January 2011. *PLOS One*. https://doi.org/10.1371/journal.pone.0016491

Jobstvogt, N., Hanley, N., Hynes, S., Kenter, J., & Witte, U. (2014). Twenty thousand sterling under the sea: Estimating the value of protecting deep-sea biodiversity. *Ecological Economics, 97*, 10–19. https://doi.org/10.1016/j.ecolecon.2013.10.019

Johannes, R. (1981). *Words of the lagoon: Fishing and marine Lore in the Palau District of Micronesia*. University of California Press. https://doi.org/10.1525/9780520321397

Johannes, R. (1987). Traditional beliefs and fisheries management in Oceania. *Naga, 10*, 5.

Johnsen, S., Widder, E. A., & Mobley, C. D. (2004). Propagation and perception of bioluminescence: Factors affecting counterillumination as a cryptic strategy. *The Biological Bulletin, 207*, 1–16. https://doi.org/10.2307/1543624

Jónasdóttir, S. H., Visser, A. W., Richardson, K., & Heath, M. R. (2015). Seasonal copepod lipid pump promotes carbon sequestration in the deep North Atlantic. *Proceedings of the National Academy of Sciences, 112*(39), 12122–12126. https://doi.org/10.1073/pnas.1512110112

Jones, C. G., Lawton, J. H., & Shachak, M. (1994). Organisms as ecosystem engineers. *Oikos, 69*, 373–386. https://doi.org/10.2307/3545850

Jones, D. O. B., Kaiser, S., Sweetman, A. K., Smith, C. R. R., Menot, L., Vink, A., et al. (2017). Biological responses to disturbance from simulated deep-sea polumetallic nodule mining. *Plos One, 12*. https://doi.org/10.1371/journal.pone.0171750

Jones, D. O. B., Simon-Lledo, E., Amon, D., Bett, B., Caulle, C., Clément, L., et al. (2021). Environment, ecology and potential effectiveness of an area protected from deep-sea mining (Clarion Clipperton Zone, abyssal Pacific). *Progress in Oceanography, 197*, 102653. https://doi.org/10.1016/j.pocean.2021.102653

Juan, C., Van Rooij, D., & De Bruycker, W. (2018). An assessment of bottom current controlled sedimentation in Pacific Ocean abyssal environments. *Marine Geology, 403*(2018), 20–33. https://doi.org/10.1073/pnas.1107743108

Juan-Jordá, M. J., Mosquira, I., Cooper, A. B., Freire, J., & Dulvy, N. K. (2011). Global population trajectories of tunas and their relatives. *Proceedings of the National Academy of Sciences of the United States of America, 108*(51), 20650–20655. https://doi.org/10.1073/pnas.1107743108

Juan-Jordá, M. J., Arrizabalga, H., Dulvy, N. K., Cooper, A. B., & Murua, H. (2015). Preliminary review of ICCAR and IATTC progress in applying an ecosystem approach to fisheries management. *Collective Volume of Scientific Papers ICCAT, 71*(6), 2958–2976.

Karl, D. M. (2002). Nutrient dynamics in the deep blue sea. *Trends in Microbiology, 10*, 410–418. https://doi.org/10.1016/S0966-842X(02)02430-7

Karl, D. M., Christian, J. R., Dore, J. E., Hebel, D. V., Letelier, R. M., Tupas, L. M., et al. (1996). Seasonal and interannual variability in primary production and particle flux at Station ALOHA. *Deep Sea Research II, 43*, 539–568. https://doi.org/10.1016/0967-0645(96)00002-1

Kessler, W. (2006). The circulation of the eastern tropical Pacific: A review. *Progress in Oceanography, 69*, 181–217. https://doi.org/10.1016/j.pocean.2006.03.009

Kim, R. E., & Bosselmann, K. (2015). Operationalizing sustainable development: Ecological integrity as a Grundnorm of international law. *Review of European, Comparative & International Environmental Law, 24*, 194. https://doi.org/10.1111/reel.12109

Kim, R. E., & van Asselt, H. (2016). Global governance: Problem shifting in the anthropocene and the limits of international law. In E. Morgera & K. Kulovesi (Eds.), *Research handbook on international law and natural resources* (pp. 473–495). Elgar Publishing.

Kirchman, D. L. (2008). *Microbial ecology of the oceans* (2nd ed.). Wiley. https://doi.org/10.1002/9780470281840

Kiszka, J., Heithaus, M., & Wirsing, A. (2015). Behavioural drivers of the ecological roles and importance of marine mammals. *Marine Ecology Progress Series, 523*, 267. https://doi.org/10.3354/meps11180

Kiszka, J., Heithaus, M., & Woodstock, M. (2022). Functional roles and ecological importance of small cetaceans in aquatic ecosystems. *Frontiers in Marine Science Section of Marine Megafauna, 9*. https://doi.org/10.3389/fmars.2022.803173

Klevjer, T. A., Irigoien, X., Røstad, A., Fraile-Nuez, E., Benítez-Barrios, V. M., & Kaartvedt, S. (2016). Large scale patterns in vertical distribution and behaviour of mesopelagic scattering layers. *Scientific Reports, 6*, 19873. https://doi.org/10.1038/srep19873

Koch, M. J., Hesketh-Best, P. J., Smerdon, G., Warburton, P. J., Howell, K., & Upton, M. (2021). Impact of growth media and pressure on the diversity and antimicrobial activity of isolates from two species of hexactinellid sponge. *Microbiology (United Kingdom), 167*(12). https://doi.org/10.1099/MIC.0.001123

Kontar, E. A., & Sokov, A. V. (1994). A benthic storm in the northeastern tropical Pacific over the fields of manganese nodules. *Deep Sea Research, 41*(7), 1069–1089. https://doi.org/10.1016/0967-0637(94)90019-1

Kontar, E. A., & Sokov, A. V. (1997). On the benthic boundary layer's dynamics. *Journal of Marine Systems, 11*(3–4), 369–385. https://doi.org/10.1016/S0924-7963(96)00131-5

Kotlinski, R. (1998). The present state of knowledge on oceanic polymetallic ores as exemplified by Interoceanmetal Joint Organization's activity. *Mineralogia Polonica, 29*, 77–89.

Kriegl, M., LLosvay, X.E., von Dorrien,C., Oesterwind, D., (2021). Marine protected areas; at the crossroads of nature conservation and fisheries management. Fronties in Marine Science Section Marine Conservation and Sustainability.8-2021. https://doi.org/10.3389/fmars.2021.676264.

Kutti, T., Bannister, R. J., & Fosså, J. H. (2013). Community structure and ecological function of deep-water sponge grounds in the traenadypet MPA–northern Norwegian continental shelf. *Continental Shelf Research, 69*, 21–30. https://doi.org/10.1016/j.csr.2013.09.011

La Bianca, G., Rees, S., Attrill, M., Lombard, A. T., McQuaid, K. A., Niner, H. J., et al. (2023). A standardised ecosystem services framework for the deep sea. *Frontiers in Marine Science, 10*, 1176230. https://doi.org/10.3389/fmars.2023.1176230

Lauerman, L. M., Smoak, J. M., Shaw, T. J., Moore, W. S., & Smith, K. L. (1997). 234 Th and 210 Pb evidence from rapid ingestion of settling particles by mobile epibenthic megafauna in the abyssal NE Pacific. *Limnology and Oceanography, 42*, 589–595.

Lavery, T., Roudnew, B., & Gill, P. C. (2010). Iron defecation by sperm whales stimulates carbon export in the Southern Ocean. *Proceedings of the Royal Society, B277*(1699), 3527–3531. https://doi.org/10.1098/rspb.2010.0863

Le, J. T., Levin, L. A., & Carson, R. T. (2017). Incorporating ecosystem services into environmental management of deep-seabed mining. *Deep Sea Research Part II Topical Studies in Oceanography, 137*, 486–503. https://doi.org/10.1016/j.dsr2.2016.08.007

Levin, L., Mengerink, K., Gjerde, K., Rowden, A., Van Dover, C. L., Clark, M. R., et al. (2016). Defining « serious harm » to the marine environment in the context of deep-seabed mining. *Marine Policy, 74*, 245–259. https://doi.org/10.1016/j.marpol.2016.09.032

Levin, L., Baker, M., & Thompson, A. (2018). *Deep-ocean climate change impacts on habitat, fish and fisheries* (166 pp). FAO.

Levitus, S., Antonov, J. I., Boyer, T. P., & Stephens, C. (2000). Warming of the world ocean. *Science, 287*, 2225–2229. https://doi.org/10.1126/science.287.5461.2225

Lewis, N., Day, J. C., Wilhem, A., Wagner, D., Gaymer, C., Parks, J., et al. (2017). *Large-scale marine protected areas: Guidelines for design and management*. Best practice protected area guidelines series, no. 26, Gland Switzerland, IUCN, XXVII+120pp. Available online at: https://www.iucn.org/pa_Guidelines

Leys, S. P., Kahn, A. S., Fang, J. K. H., Kutti, T., & Bannister, R. J. (2018). Phagocytosis of microbial symbionts balances the carbon and nitrogen budget for the deep-water boreal sponge *Geodia barretti*. *Limnology and Oceanography, 63*(1), 187–202. https://doi.org/10.1002/lno.10623

Lieber, M. D. (1994). Traditional fishing on a Polynesian atoll. In: *SPC marine resource management and knowledge information bulletin* #4 – September 1994 (Noumea: SPC), 12–17.

Link, J. S., & Browman, H. I. (2017). Operationalizing and implementing ecosystem-based management. *ICES Journal of Marine Science, 74*, 379–381. https://doi.org/10.1093/icesjms/fsw247

Lohmann, K., Lohmann, C., & Endres, C. (2008). The sensory ecology of ocean navigation. *The Journal of Experimental Biology, 211*, 1719–1728. https://doi.org/10.1242/jeb.015792

Longhurst, A. R. (2007). *Ecological geography of the sea* (2nd ed.). Burlington Academic Press.

Loreau, M. (2008). Biodiversity and ecosystem functioning: The mystery of the deep sea. *Current Biology, 18*, R126–R128. https://doi.org/10.1016/j.cub.2007.11.060

Lubchenco, J., Palumbi, S. R., Gaines, S. D., & Andelman, S. (2003). Plugging a hole in the ocean: The emerging science of marine reserves. *Ecological Applications, 13*, 3–7. https://doi.org/10.1890/1051-07612003013[0003:PAHITO]2.0.CO;2

Macdonald, K. C., Fox, P. J., Alexander, R. T., Pockalny, R., & Gente, P. (1996). Volcanic growth faults and the origin of Pacific abyssal hills. *Nature, 380*(1996), 125–129. https://doi.org/10.1038/380125a0

Maes, F. (2008). The international legal framework for marine spatial planning. *Marine Policy, 32*, 797–810. https://doi.org/10.1016/j.marpol.2008.03.013

Margules, C. R., & Pressey, R. L. (2000). Systematic conservation planning. *Nature, 405*, 243–253.

Martini, S., & Haddock, S. H. (2017). Quantification of bioluminescence from the surface to the deep sea demonstrates its predominance as an ecological trait. *Scientific Reports, 7*(1), 45750. https://doi.org/10.1038/srep45750

Martini, S., Kuhnz, L., Mallefet, J., & Haddock, S. H. (2019). Distribution and quantification of bioluminescence as an ecological trait in the deep sea benthos. *Scientific Reports, 9*(1), 14654. https://doi.org/10.1016/0077-7579(86)90040-2

McCartney, M. S. (1982). The subtropical recirculation of mode waters. *Journal of Marine Research, 40*, 427–464.

McCrea-Stub, A., Zeller, D., Sumaila, U. R., Nelson, J., Balmford, A., & Pauly, D. (2011). Understanding the cost of establishing marine protected areas. *Marine Policy, 35*(1), 1–9. https://doi.org/10.1016/j.marpol.2010.07.001

Menard, H. W. (1964). Chemical analyses of manganese nodules and crusts from the Pacific Ocean using X-ray fluorescence. In H. W. Menard, E. Goldberg, & H. E. Hawkes (Eds.), *Composition of Pacific sea-floor manganese nodules*. PANGEA. https://doi.org/10.1594/PANGAEA.859807

Mensinger, A. F., & Case, J. F. (1990). Luminescent properties of deep sea fish. *Journal of Experimental Marine Biology and Ecology, 144*, 1–15. https://doi.org/10.1016/0022-0981(90)90015-5

Mewes, K., Mogollón, J. M., Picard, A., Rühlemann, C., Kuhn, T., Nöthen, K., et al. (2014). Impact of depositional and biogeochemical processes on small scale variation in nodule abundance in the Clarion-Clipperton Fracture Zone. *Deep-Sea Research Part I: Oceanographic Research Papers, 91*, 125–141. https://doi.org/10.1016/j.dsr.2014.06.001

Meynecke, J.-O., Samantha, S., de Bie, J., Seyboth, E., Prakash Dey, S., Fearon, G., et al. (2023). Do whales really increase the oceanic removal of atmospheric carbon? *Fronteirs in Marine Science Hypothesis and Theory Article. Section Marine Megafauna, 10*. https://doi.org/10.3389/fmars.2023.1117409

Miller, K. A., Thompson, K., Johnston, P., & Santillo, D. (2018). An overview of seabed mining including the current state of development, environmental impacts, and knowledge gaps. *Frontiers in Marine Science, 4*, 418. https://doi.org/10.3389/fmars.2017.00418

Miller, K. A., Brigden, K., Santillo, D., Currie, D., Johnston, P., & Thompson, K. F. (2021). Challenging the need for deep seabed mining from the perspective of metal demand, biodiversity, ecosystems services, and benefit sharing. *Frontiers in Marine Science, 8*, 706161. https://doi.org/10.3389/fmars.2021.706161

Montevecchi, W. A. (1993). Birds as indicators of change in marine prey stocks. In R. W. Furness & J. J. D. Greenwood (Eds.), *Birds as monitors of environmental change* (pp. 217–266). Chapman and Hall.

Morato, T., Hoyle, S. D., Allain, V., & Nicol, S. J. (2010). Seamounts are hotspots of pelagic biodiversity in the open ocean. *Proceedings of the National Academy of Sciences of the United States of America, 107*, 9707–9711. https://doi.org/10.1073/pnas.0910290107

Moreno Navas, J., Miller, P. I., Henry, L. A., Hennige, S. J., & Roberts, J. M. (2014). Ecohydrodynamics of cold-water coral reefs: A case study of the Mingulay Reef Complex (Western Scotland). *PLoS One, 9*, e98218. https://doi.org/10.1371/journal.pone.0098218

Morgan, C. L. (2000). Resource estimates of the Clarion–Clipperton manganese nodule deposits. In D. S. Cronan (Ed.), *Handbook of marine mineral deposits* (pp. 145–170). CRC Press.

Nicol, S., Bowie, A., Jarmon, S., Lannuzel, D., Meiners, K. M., et al. (2010). Southern Ocean iron fertilization by baleen whales and Antarctic krill. *Fish and Fisheries, 11*, 203–209. https://doi.org/10.1111/j.1467-2979.2010.00356.x

Nihoul, J. C. J. (1981). Marine hydrodynamics at ecological scales. In J. C. J. Nihoul (Ed.), *Ecohydrodynamics* (pp. 1–12). Elsevier Press.

Nihoul, J. C. J., & Djenidi, S. (1991). Hierarchy and scales in marine ecohydrodynamics. *Earth-Science Reviews, 31*, 255–277.

Nilsson, D. E., Warrant, E. J., Johnsen, S., Hanlon, R., & Shashar, N. (2012). A unique advantage for giant eyes of giant squid. *Current Biology, 22*, 683–688. https://doi.org/10.1016/j.cub.2012.02.031

Niner, H. J., Ardron, J. A., Escobar, E. G., Gianni, M., Jaeckel, A., Jones, D. O. B., et al. (2018). Deep-Sea mining with no net loss of biodiversity-an impossible aim. *Frontiers in Marine Science, 5*, 53.

Not, F., Siano, R., Kooistra, W. H. C. F., Simon, N., Vaulot, D., & Probert, I. (2012). Diversity and ecology of eukaryotic marine phytoplankton. *Advances in Botanical Research, 64*. Elsevier. https://doi.org/10.1016/B978-0-12-391499-6.00001-3

Oesterwind, D., Rau, A., & Zaiko, A. (2016). Drivers and pressures—Untangling the terms commonly used in marine science and policy. *Journal of Environmental Management, 181*, 8–15. https://doi.org/10.1016/j.jenvman.2016.05.058

O'Leary, B. C., & Roberts, C. M. (2017). The structuring role of marine life in Open Ocean habitat: Importance to international policy. *Frontiers in Marine Science, 4*, 268. https://doi.org/10.3389/fmars.2017.00268

O'Leary, B. C., Brown, R. L., Johnson, D. E., Von Nordheim, H., Ardron, J., Packeiser, T., et al. (2012). The first network of marine protected areas (MPAs) in the high seas: The process, the challenges and where next. *Marine Policy, 36*, 598–605. https://doi.org/10.1016/j.marpol.2011.11.003

Orcutt, B. N., Bradley, J. A., Brazlton, W. J., Estes, E. R., Goordial, J. M., & Huber, J. A. (2020). Impacts of deep-sea mining on microbial ecosystem services. *Limnology and Oceanography, 65*(7), 1489–1510. https://doi.org/10.1002/lno.11403

Ortuno Crespo, G., & Dunn, D. (2017). A review of the impacts of fisheries on open-ocean ecosystems. *ICES Journal of Marine Science, 74*(9), 2283–2297. https://doi.org/10.1093/icesjms/fsx084

Paka, V., Golenko, M., Kondrashov, A., Korzh, A., Lander, M. R., Obleukhov, S. D., et al. (2020). On improvement and further development of the moving vessel vertical profiler. *Journal of Oceanological Research, 48*(4). https://doi.org/10.29006/1564-2291.jor-2020.48(4).6

Pantel, J. H., Bohan, D. A., Calcagno, V., David, P., Duyck, P.-F., & Kamenova, S. (2017). Chapter six – 14 questions for invasion in ecological networks. *Advances in Ecological Research, 56*, 293–340. https://doi.org/10.1016/bs.aecr.2016.10.008

Partridge, J. C., & Douglas, R. H. (1995). Far-red sensitivity of dragon fish. *Nature, 375*, 21–22. https://doi.org/10.1038/375021a0

Peña, M. A., Lewis, M. R., & Harrison, W. G. (1990). Primary productivity and size structure of phytoplankton biomass on a transect of the equator at 135°W in the Pacific Ocean. *Deep Sea Research Part A: Oceanographic Research Papers, 37*(2), 295–315. https://doi.org/10.1016/0198-0149(90)90129-J

Pennington, T. J., Mahoney, K. L., Kuwahara, V. S., Kolber, D. D., Calienes, R., & Chavez, F. P. (2006). Primary production in the eastern tropical Pacific: A review. *Progress in Oceanography, 69*, 285–317. https://doi.org/10.1016/j.pocean.2006.03.012

Perelman, J., Firing, E., Van der Grient, J., Jones, B., & Drazen, J. (2021). Mesopelagic scattering layer behaviors across the clarion-Clipperton zone : Implications for deep-sea mining. *Frontiers in Marine Science, Section Deep-Sea Environments and Ecology, 8*. https://doi.org/10.3389/fmars.2021.632764

Pershing, A. J., Christensen, L. B., Record, N. R., Sherwood, G., & Stetson, P. B. (2010). The impact of whaling on the ocean carbon cycle : Why bigger was better. *PLoS One, 5*, e12444. https://doi.org/10.1371/journal.pone.0012444

Peterson, G., Allen, C. R., & Holling, C. S. (1998). Ecological resilience, biodiversity, and scale. *Ecosystems, 1*, 6–18. https://doi.org/10.1007/s100219900002

Petersen, S., Kratschell, A., Augustin, N., Jamieson, J., Hein, J. R., & Hannington, M. D. (2016). News from the seabed: Geological characteristics and resource potential of deep-sea mineral resources. *Marine Policy, 70*, 175–187. https://doi.org/10.1016/j.marpol.2016.03.012

Pfannkuche, O., & Lochte, K. (1993). Open ocean pelago-benthic coupling: Cyanobacteria as tracers of sedimenting salp feces. *Deep Sea Research Part I, 40*(4), 727–737. https://doi.org/10.1016/0967-0637(93)90068-E

Pietsch, T. W. (2009). *Oceanic anglerfishes: Extraordinary diversity in the Deep Sea* (576 pp). University of California Press. ISBN: 9780520942554.

Pitman, R. L. (1993). Seabird associations with marine turtles in the Eastern Pacific Ocean. *Colonial Waterbirds, 16*(2), 194–201. https://doi.org/10.2307/1521438

Polidoro, B. A., Brooks, T., Carpenter, K. E., Edgar, G. J., Henderson, S., Sanciangco, et al. (2012). Patterns of extinction risk and threat for marine vertebrates and habitat forming species in the Tropical Eastern Pacific. *Marine Ecology Progress Series, 448*, 93–104. https://doi.org/10.3354/meps09545

Popova, E., Vousden, D., Sauer, W. H. H., Mohammed, E. Y., Allain, V., Downey-Breedt, N., et al. (2019). Ecological connectivity between the areas beyond National Jurisdiction and coastal waters: Safeguarding interests of coastal communities in developing countries. *Marine Policy, 104*, 90–102. https://doi.org/10.1016/j.marpol.2019.02.050

Potier, M., Marsac, F., Cherel, Y., Lucas, V., Sabatié, R., Maury, O., et al. (2007). Forage fauna in the diet of three large pelagic fishes (lancetfish, swordfish and yellowfin tuna) in the western equatorial Indian Ocean. *Fisheries Research, 83*(1), 60–72. https://doi.org/10.1016/j.fishres.2006.08.020

Price, A. R. G. (1999). Broadscale coastal environmental assessment of the Chagos archipelago. In C. R. C. Sheppard & M. R. D. Seaward (Eds.), *Ecology of the Chagos archipelago* (pp. 285–296). Westbury Publishing/Linnean Society Occasional Papers.

Price, A. R. G. (2002). Simultaneous 'hotspots' and 'coldspots' of marine biodiversity and implications for global conservation. *Marine Ecology Progress Series, 241*, 23–27. https://doi.org/10.3354/meps241023

Price, A. R. G. (2004). Rapid coastal environmental assessment. In *Standard survey methods for the Red Sea and Gulf of Aden (PERSGA/GEF). PERSGA technical series 10* (pp. 1–2). PERSGA.

Price, A. R. G., & Izsak, C. (2005). Is the Arabian Gulf really such a low spot of biodiversity?: Scaling effects and management implications. *Aquatic Ecosystem Health & Management, 8*, 363–366. https://doi.org/10.1080/14634980500457757

Price, A. R. D., Keeling, M., & Stewart, I. (2007). A robustness metric integrating spatial and temporal information: Application to coral reefs exposed to local and regional disturbances. *Marine Ecology Progress Series, 331*, 101–108. https://doi.org/10.3354/meps331101

Purkiani, K., Paul, A., Vink, A., Walter, M., Schulz, M., & Haeckel, M. (2020). Evidence of eddy-related deep ocean current variability in the North-East Tropical Pacific Ocean induced by remote gap winds. *Biogeosciences, 17*, 6527–6544. https://doi.org/10.5194/bg-17-6527-2020

Rabbitt, S., Lilley, I., Albert, S., & Tibbetts, I. (2019). What's the catch in who fishes? Fisherwomen's contributions to fisheries and food security in Marovo Lagoon, Solomon Islands. *Marine Policy, 108*, 103667. https://doi.org/10.1016/j.marpol.2019.103667

Ramirez-Llodra, E. (2020). Deep-sea ecosystems: Biodiversity and anthropogenic impacts. In *The law of the seabed (Chap. 2, pp. 36–60).* https://doi.org/10.1163/9789004391567_004

Rateb, M. E., & Ebel, R. (2011). Secondary metabolites of fungi from marine habitats. *Natural Product Reports, 28*(2), 290–344. https://doi.org/10.1039/c0np00061b

Reid, J. L. (1997). On the total geostrophic circulation of the Pacific Ocean: Flow patterns, tracers and transports. *Progess in Oceanography, 39*(4), 263–352. https://doi.org/10.1016/S0079-6611(97)00012-8

Richardson, M. J., Weatherly, G. L., & Gardner, W. D. (1993). Benthic storms in the Argentine basin. *Deep Sea Research Part II, 40*, 975–987. https://doi.org/10.1016/0967-0645(93)90044-N

Ringler, T., Petersen, M., Higdon, R. L., Jacobsen, D., Jones, P. W., & Maltrud, M. (2013). A multi-resolution approach to global ocean modelling. *Ocean Modelling, 69*, 211–232. ISSN 1463-5003.

Ritchie, E. G., & Johnson, C. N. (2009). Predator interactions, meso-predator release and biodiversity conservation. *Ecological Letters, 12*, 982–998.

Roberts, M., & Cairns, S. (2014). Cold-water corals in a changing ocean. *Current Opinion in Environment Sustainability, 7*, 118–126. https://doi.org/10.1016/j.cosust.2014.01.004

Ritter, F., & Panigada, S. (2019). Chapter 28-Collisions of vessels with cetacean-the underestimated threat. In *World seas : An environmental evaluation (second edition). Volume III: Ecological Issues and Environmental Impacts* (pp. 531–547). https://doi.org/10.1016/B978-0-12-805052-1.00026-7

Roberts, C. M., Ormond, R. F. G., & Shepherd, A. D. (1988). The usefulness of butterflyfishes as environmental indicators on coral reefs. In J. H. Choat, D. Barnes, M. A. Borowitzka, J. C. Coll, P. J. Davies, P. Flood, & B. G. Hopley (Eds.), *Proceedings of the 6th interntional coral reef symposium: Vol. 2: Contributed papers*.

Roberts, C. M., Bohnsack, J. A., Gell, F., Hawkins, J. P., & Goodridge, R. (2001). Effects of marine reserves on adjacent fisheries. *Science, 294*, 1920–1923. https://doi.org/10.1126/science.294.5548.1920

Robinson, C., Steinberg, D. K., Anderson, T. R., Arístegui, J., Carlson, C. A., Frost, J. R., et al. (2010). Mesopelagic zone ecology and biogeochemistry—A synthesis. *Deep Sea Research Part II, 57*, 1504–1518. https://doi.org/10.1016/j.dsr2.2010.02.018

Robison, B. H. (1992). Bioluminescence in the benthopelagic holothurian *Enypniastes eximia*. *Journal of the Marine Biological Association of the United Kingdom, 72*, 463–472. https://doi.org/10.1017/S0025315400037826

Rogacheva, A., Gebruk, A., & Alt, C. (2012). Swimming deep-sea holothurians (*Echinodermata: Holothuroidea*) on the northern Mid-Atlantic Ridge. *Zoosymposia, 7*, 213–224. https://doi.org/10.11646/zoosymposia.7.1.19

Rogers, A. D., Sumalia, U. R., & Hussain, S. S., Baulcomb, C. (2014). The high sea and us. Understanding the value of high-seas ecosystems. *Global Ocean Commission*. www.globaloceancommission.org

Roman, J., & McCarthy, J. (2010). The whale pump : Marine mammals enhance primary productivity in a coastal basin. *PLoS One, 5*, e13255. https://doi.org/10.1371/jounal.pone.0013255

Roman, J., Estes, J. A., Morissette, L., Smith, C., Costa, D. P., McCarthy, J., et al. (2014). Whales as marine ecosystem engineers. *Frontiers in Ecology and the Environment, 12*, 377–385. https://doi.org/10.1890/130220

Rowe, G. T. (1971). Benthic biomass and surface productivity. In J. D. Costlow (Ed.), *Fertility of the sea* (pp. 97–122). Gordon and Breach.

Ruhl, H. A., & Smith, K. L., Jr. (2004). Shifts in deep-sea community structure linked to climate and food supply. *Science, 305*, 513–515. https://doi.org/10.1126/science.1099759

Santora, J. A., Schroeder, I. D., Bograd, S. J., Chavez, F. P., Cimino, M. A., Fiechter, J., et al. (2021). Pelagic biodiversity, ecosystem function, and services. *Oceanography, 34*(2), 16–37. https://www.jstor.org/stable/27085036

Schaefer, K. M., Fuller, D. W., & Block, B. A. (2007). Movements, behavior, and habitat utilization of yellowfin tuna (*Thunnus albacares*) in the northeastern Pacific Ocean, ascertained through archival tag data. *Marine Biology, 152*(3), 503–525. https://doi.org/10.1007/s00227-007-0689-x

Schorr, G. S., Falcone, E. A., Moretti, D. J., & Andrews, R. D. (2014). First long-term behavioral records from Cuvier's beaked whales (*Ziphius cavirostris*) reveal record-breaking dives. *PLoS One, 9*, e92633. https://doi.org/10.1371/journal.pone.0092633

Schwer, K., & Mcgillivary, P. (2007). Marine ship automatic identification system (AIS) for enhanced coastal security capabilities: An oil spill tracking application. *Conference OCEANS, 2007*. https://doi.org/10.1109/OCEANS.2007.4449285

Serrano, A., de la Torriente, A., Punzon, A., Blanco, M., Bellas, J., Duran-Munoz, P., et al. (2022). Sentinels of seabed (SoS) indicator: Assessing benthic habitats condition using typical and sensitive species. *Ecological Indicators, 140*, 108979. https://doi.org/10.1016/j.ecolind.2022.108979

Sharma, R. (2019). Environmental issues of deep sea mining. In *Impacts, consequences and perspectives in deep sea mining* (p. 577). Springer Nature.

Shor, G. (1959). Reflexion studies in the eastern equatorial Pacific. *Deep Sea Research, 5*, 283–285.

Siegel, D. A., Behrenfeld, M. J., Maritorena, S., McClain, C. R., Antoine, D., Bailey, S. W., et al. (2013). Regional to global assessments of phytoplankton dynamics from the SeaWiFS mission. *Remote Sensing of the Environment, 135*, 77–91. https://doi.org/10.1016/j.rse.2013.03.025

Smith, A. D., Brown, C. J., Bulman, C. M., Fulton, E. A., Johnson, P., Kaplan, I. C., et al. (2011). Impacts of fishing low–trophic level species on marine ecosystems. *Science, 333*(6046), 1147–1150. https://doi.org/10.1126/science.1209395

Smith, C. R. (2006). Bigger is better : The role of whales as detritus in marine ecosystems. In J. A. Estes, P. De-Master, D. F. Doak, T. M. Williams, & R. L. Brownell Jr. (Eds.), *Whales, whaling and ocean ecosystems* (pp. 286–300). University of California Press.

Smith, C. R., & Rabouille, C. (2002). What controls the mixed-layer depth in deep-sea sediments? The importance of POC flux. *Limnology and Oceanography, 49*, 620–622. https://doi.org/10.4319/lo.2004.49.2.0620

Smith, C. R., & Baco, A. R. (2003). Ecology of whale falls at the deep-sea floor. *Oceanography and Marine Biology: An Annual Review, 41*, 311–354.

Smith, C. R., & Demopoulos, A. (2003). Ecology of the deep. Pacific Ocean floor. In P. A. Tyler (Ed.), *Ecosystems of the world, 28: Ecosystems of the deep ocean* (pp. 179–218). Elsevier.

Smith, C. R., Berelson, W., Demaster, D. J., Dobbs, F. C., Hammond, D., Hoover, D. J., et al. (1997). Latitudinal variations in benthic processes in the abyssal equatorial Pacific: Control by biogenic particle flux. *Deep Sea Research Part II, 44*, 2295–2317.

Smith, C. R., De Leo, F. C., Bernardino, A. F., Sweetman, A. K., & Arbizu, P. M. (2008a). Abyssal food limitation, ecosystem structure and climate change. *Trends in Ecology & Evolution, 23*(9), 518–528. https://doi.org/10.1016/j.tree.2008.05.002

Smith, C.R., Gaines, S., Watling, L., Friedlander, A., Morgan, C., Thurnherr, A., et al. (2008b). Rationale and recommendations for the establishment of preservation reference areas for nodulemining in the Clarion-Clipperton Zone. *Fourteenth session. Kingston, 26 May–6 June 2008, sessions/2008/documents: Legal and Technical Commission, International Seabed Authority. Technical document no. ISBA/14/LTC/2*. Available online at: https://www.isa.org.jm/(click/contractors/exploration-areas/ClarionClippertonFractureZone)

Smith, C. R., Bernardino, A. F., Baco, A., Hannides, A. K., & Altamira, I. (2014). The seven-year enrichment : Macrofaunal succession in deep-sea sediments around a 30-tonne whalefall in the Northeast Pacific. *Marine Ecology Progress Series, 515*, 133. https://doi.org/10.3354/meps10955

Smith, C. R., Glover, A., Treude, T., Higgs, N., & Amon, D. (2015a). Whale-fall ecosystems : Recent insights into ecology, paleoecology and evolution. *Annual Review of Marine Science, 2015*(7), 571–596. https://doi.org/10.1146/annurev-marine-010213-135144

Smith, K. L. Jr., Kaufmann, R. S., & Wakefield, W. (1993). Mobile megafaunal activity monitored with a time-lapse camera in the abyssal North Pacific. *Deep Sea Research Part I: Oceaonographic Research Papers, 40*(11–12), 2307–2324.

Smith, K. L. Jr., Ruhl, H. A., Kaufmann, R. S., & Kahru, M. (2008). Tracing abyssal food supply back to upper-ocean processes over a 17-year time series in the NE Pacific. *Limnology and Oceanography, 53*, 2655–2667.

Smith, K. L. Jr., Ruhl, H., Bett, B., Billett, D., Lampitt, R., & Kaufmann, R. (2009). Climate, carbon cycling and deep-ocean ecosystems. *Proceedings. National Academy of Sciences. United States of America, 106*, 19211–19218. https://doi.org/10.1073/pnas.0908322106

Smith, R. W., Bianchi, T. S., Allison, M., Savage, C., & Galy, V. (2015b). High rates of organic carbon burial in fjord sediments globally. *Nature Geoscience, 8*, 450–453. https://doi.org/10.1038/NGEO02421

South, G. R., Skelton, P. A., Veitayaki, J., Resture, A., & Carpenter, C. (2004). The global international waters assessment for the Pacific Islands: Aspects of transboundary water and coastal fisheries issues. *Ambio, 33*, 98–106. https://doi.org/10.1579/0044-7447-33.1.98

Steneck, R. S. (2001). Functional groups. In S. A. Levin (Ed.), *Encyclopedia of biodiversity* (Vol. 3, pp. 121–139). Academic.

Sumaila, U. R., Zeller, D., Watson, R., Alder, J., & Pauly, D. (2007). Potential costs and benefits of marine reserves in the high seas. *Marine Ecology Progress Series, 345*, 305–310. https://doi.org/10.3354/meps07065

Sumaila, U. R., Lam, V. W. Y., Miller, D. D., Teh, L., Watson, R. A., Zeller, D., et al. (2015). Winners and losers in a world where the high seas is closed to fishing. *Scientific Reports, 5*, 8481. https://doi.org/10.1038/srep08481

Sutton, T. T., Wiebe, P. H., Madin, L., & Bucklin, A. (2010). Diversity and community structure of pelagic fishes to 5000 m depth in the Sargasso Sea. *Deep Sea Research Part II Topical Studies in Oceanography, 57*, 2220–2233. https://doi.org/10.1016/j.dsr2.2010.09.024

Sutton, T. T., Clark, M. R., Dunn, D. C., Halpin, P. N., Rogers, A. D., Guinotte, J., et al. (2017). A global biogeographic classification of the mesopelagic zone. *Deep Sea Research Part I Oceanography Research Papers, 126*, 85–102. https://doi.org/10.1016/j.dsr.2017.05.006

Tait, R. V., & Dipper, F. A. (1998). *Elements of marine ecology* (4th ed.). Butterworth Heinemann.

Talley, L. D. (1993). Distribution and formation of North Pacific intermediate water. *Journal of Physical Oceanography, 23*, 517–537. https://doi.org/10.1175/1520-0485(1993)023

Talley, L. D., Fryer, G., & Lumpkin, R. (1998). Physical oceanography of the tropical Pacific. Chapter 2. In M. Rapaport (Ed.), *Geography of the Pacific Islands* (pp. 19–32). Bess Press.

Talley, L. D., Pickard, G., Emery, W., & Swift, J. (2011). *Descriptive physical oceanography: An introduction* (6th ed.). Elsevier Academic Press.

Tawake, A. (2004). *Human impacts on coastal fisheries in rural communities and their conservation approach: A case study on kaikoso (Anadara spp.) fishery in the coastal areas of Verata district, Fiji. MSc thesis, University of the South Pacific, Suva, Fiji.*

Tawake, A., & Tuivanuavou, S. (2004). Community involvement in the implementation of ocean policies: The Fiji locally managed marine areas (FLMMAs) network. In *Tabus or not taboos: How to use traditional environmental knowledge to support sustainable development of marine resources in Melanesia, pp. 26–27. SPC Traditional Marine Resource Management and Knowledge Information Bulletin 17:14–29.*

Thakur, R. (1991). *The South Pacific: Problems, issues and prospects* (248 pp). St.Martin's Press.

The Western and Central Pacific Fisheries Commission (WCPFC). (2019). The Western and Central Pacific tuna fishery: 2018. *Overview and status of stocks.* Available online at: www.wcpfc.int/statistical-bulletins

Thiel, H. (2003). Anthropogenic impacts on the deep-sea. Chapter 13. In P. A. Tyler (Ed.), *Ecosystems of the world: The deep sea.* Elsevier.

Thompson, K. F., Miller, K. A., Wacker, J., Derville, S., Laing, C., Santillo, D., et al. (2023). Urgent assessment needed to evaluate potential impacts on cetaceans from seabed mining. *Frontiers in Marine Science Section Marine Magafauna, 10-2023.* https://doi.org/10.3389/fmars.2023.1095930

Thorold, S., Afonso, P., Fontes, J., Braun, C., Santos, R., Skomal, G., et al. (2014). Extreme diving behaviour in devil rays links surface waters and the deep ocean. *Nature Communications* 5, Article number: 4274. https://doi.org/10.1038/ncomms5274.

Thurber, A. R., Sweetman, A. K., Narayanaswamy, B. E., Jones, D. O. B., Ingels, J., & Hansman, R. L. (2014). Ecosystem function and services provided by the deep sea. *Biogeosciences, 11*(14), 3941–3963. https://doi.org/10.5194/bg-11-3941-2014

Thurston, M. H., Rice, A. L., & Bett, B. J. (1998). Latitudinal variation in invertebrate megafaunal abundance and biomass in the North Atlantic Ocean abyss. *Deep Sea Research Part II, 45*, 203–224. https://doi.org/10.1016/S0967-0645(97)00077-5

Tilot, V. (2004). *Management plan for the marine mammals sanctuary in the Mediterranean Sea, "Pelagos"*. Adopted by the governments of France, Italy and Monaco in October 2004, Signatories to the Agreement on the Creation of a Sanctuary for Marine Mammals, pp. 107.

Tilot, V. (2006a). Biodiversity and distribution of the megafauna Vol 1. The polymetallic nodule ecosystem of the eastern equatorial Pacific Ocean. *Unesco/Intergovernmental Oceanographic Commission. Technical Series 69,* Project Unesco COI/Min, Vlanderen. Available online at: http://unesdoc.unesco.org/images/0014/001495/149556f.pdf

Tilot, V., (2006b). Biodiversity and distribution of the megafauna Vol 2. Annotated photographic atlas of the echinoderms of the clarion-Clipperton fracture zone. *Unesco/Intergovernmental Oceanographic Commission. Technical Series 69,* Project Unesco/COI/Min, Vlanderen. Available online at: http://unesdoc.unesco.org/images/0014/001495/149556f.pdf

Tilot, V., (2010). Biodiversity and distribution of faunal assemblages. Vol 3. Options for the management and conservation of the nodule ecosystem in the Clarion Clipperton Fracture Zone (NE Pacific Ocean): Scientific, legal and institutional aspects. *Unesco/Intergovernmental Oceanographic Commission. Technical Series* 69, Project Unesco COI/Min Vlanderen. Available online at: https://unesdoc.unesco.org/images/0014/001495/149556f.pdf

Tilot, V. (2013). Tools for marine spatial management in deep seas (A guide for research, conservation and development practitioners). In *Proceedings of the international marine protected areas congress (IMPAC3).* IUCN/Agence des Aires Marines Protégées (Fr) (Marseille).

Tilot, V. (2014). An innovative multilayer rapid ecological assessment adapted to monitoring and management of Deep Sea ecosystems. In *Proceedings deep sea mining summit 2014 (London).*

Tilot, V. (2016). *General features of the coral reef communities of the Daymaniyat Islands, Oman Sultanate and designation of management indexes for environmental sensitivity mapping.* IUCN/ROWA Report.

Tilot, V. (2019). Assessment of Deep-Sea faunal communities-indicators of environmental impact. Chapter 4, pp. 147–189. *Environmental issues of deep-sea mining. impacts, consequences and policy perspectives* (577 pp). https://www.springer.com/us/book/9783030126957, Springer International Publishing. https://doi.org/10.1007/978-3-030-12696-4

Tilot, V. (2023). Integrated ecosystem-based for regional projects in marine environment; from coasts, islands, high seas and the deep sea. *Proceedings in a RAOS special issue of the International meeting of the "Académie des Sciences d'Outre-Mer", France and l'Académie royale des Sciences d'Outremer, Belgique"* on: "Governance. Marine resource management, biodiversity conservation, food and maritime security and sustainable development". In press.

Tilot, V., Ormond, R., Moreno-Navas, J., & Catala, T. (2018). The benthic megafaunal assemblages of the CCZ (Eastern Pacific) and an approach to their management in the face of threatened anthropogenic impacts. *Frontiers in Marine Science, 5*, 7. https://doi.org/10.3389/fmars.2018.00007

Tilot, V., Willaert, K., Guilloux, B., Chen, W., Mulalap, C. Y., Gaulme, F., Bambridge, T., Peters, K., & Dahl, A. (2021a). Traditional dimensions of seabed resource management in the context of Deep Sea Mining in the Pacific: Learning from the socio-ecological interconnectivity between Island communities and the ocean realm. *Frontiers in Marine Science, special issue: Integrated Marine Biosphere Research: Ocean Sustainability, Under Global Change, for the Benefit of Society, 8*, 637938. https://doi.org/10.3389/fmars.2021.637938

Tilot, V., Willaert, K., Guilloux, B., Mulalap, C. Y., Gaulme, F., Bambridge, T., Kacenelenbogen, E., Peters, K., Navas, M., & Dahl, A. (2021b). Traditional and cultural/societal dimensions of seabed resource management and present applicable legal frameworks in the Pacific. In R. Sharma (Ed.), *Perspectives in deep-sea mining*. Springer. ISBN103030879 81X.

Tilot, V., Guilloux, B., Willaert, K., Mulalap, C. Y., Gaulme, F., Bambridge, T., Kacenelenbogen, E., Jeudy de Grissac, A., Peters, K., Moreno Navas, J., & Dahl, A. (2021c). The concept of Oceanian Sovereignity in the context of deep sea mining in the Pacific. *Frontiers in Marine*

Science, Special issue: Integrated Marine Biosphere Research: Ocean Sustainability, Under Global Change, for the Benefit of Society, 8. https://doi.org/10.3389/fmars.2021.756072

Tilot, V., Fourchault, L., & Jeudy de Grissac, A. (2023). Marine mammal communities and human activities in the north-eastern tropical Pacific: Conservation and management strategies. In R. Sharma (Ed.), *Deep-sea mining and the water column*. Springer. In Press.

Tortell, P. (1992). Coastal zone sensitivity mapping and its role in marine environmental management. *Marine Pollution Bulletin, 25*(1–4), 88–93. https://doi.org/10.1016/0025-326X(92)90193-A

Tsuchiya, M., & Talley, L. D. (1996). Water property distributions along an eastern Pacific hydrographic section at 135W. *Journal of Marine Research, 54*, 541–564. https://doi.org/10.1029/97JC03415

Tunnicliffe, V., Metaxas, A., Le, J., Ramirez-Llodra, E., & Levin, L. A. (2020). Strategic environmental goals and objectives: Setting the basis for environmental regulation of deep seabed mining. *Marine Policy, 114*(1–10), 103347. https://doi.org/10.1016/j.marpol.2018.11.010

Turner, J. T. (2015). Zooplankton fecal pellets, marine snow, phyto-detritus and the ocean's biological pump. *Progress in Oceanography, 130*, 205–248. https://doi.org/10.1016/j.pocean.2014.08.005

Turner, P. J., Cannon, S., De Land, S., Delgado, J. P., Eltis, D., Halpin, P. N., et al. (2020). Memorializing the middle passage on the Atlantic seabed in areas beyond national jurisdiction. *Marine Policy, 122*, 104254. https://doi.org/10.1016/j.marpol.2020.104254

Tyler, P. A. (2003). Ecosystems of the world. 28. In *Ecosystems of the deep oceans*. Elsevier Press. ISBN: 9780080494654.

UNESCO. (2009). Global open oceans and deep seabed (GOODS) biogeographic classification. *IOC Technical Series No. 84*. UNESCO.

Van Dover, C. L., Ardron, J. A., Escobar, E., Gianni, M., Gjerde, K. M., Jaeckel, A., et al. (2017). Biodiversity loss from dee-sea mining. *Nature Geoscience, 10*, 464–465. https://doi.org/10.1038/ngeo2983

Van der Grient, J. M. A., & Drazen, J. C. (2021). Potential spatial intersection between high-seas fisheries and deep-sea mining in international waters. *Marine Policy, 129*(2021), 104564. https://doi.org/10.1016/j.marpol.2021.104564

Vasiliu, D., Preuss, I., & Muresan, M. (2022). *Water column characteristics in the Clarion Clipperton Fracture Zone, north-eastern tropical Pacific Ocean*. https://doi.org/10.5281/zenodo.7491412

Veitayaki, J., Aalbersberg, B., & Tawake, A. (2003). Empowering LocalCommunities: Case Study of Votua, Ba, Fiji. *Ocean Yearbook Online, 17(1), 449–463*. https://doi.org/10.1163/221160003X00177

Victorero, L., Watling, L., Palomares, M. L. D., & Nouvian, C. (2018). Out of sight, but within reach: A global history of bottom-trawled deep-sea fisheries from > 400 m depth. *Frontiers in Marine Science, 5*, 98. https://doi.org/10.3389/fmars.2018.00098

Wang, Z.-T., Supin, A. Y., Akamatsu, T., Duan, P.-X., Yang, Y.-N., Wang, K.-X., et al. (2021). Auditory evoked potential in stranded melon-headed whales (Peponocephala electra): With sever hearing loss and possibly caused by anthropogenic noise pollution. *Ecotoxicology and Environmental Safety, 228*, 113047. https://doi.org/10.1016/j.ecoenv.2021.113047

Warner, R. (2016). *Strategic environmental assessment (SEA) and its application to marine areas beyond National Jurisdiction (ABNJ)*. University of Wollongong.

Warrant, E. J., Collin, S. P., & Locket, N. A. (2003). Eye design and vision in deep-sea fishes. In *Sensory processing in aquatic environments* (pp. 303–322). Springer. https://doi.org/10.1007/978-0-387-22628-6_16

Warwick, R. M., & Clarke, K. R. (2001). Practical measures of marine biodiversity based on relatedness of species. *Oceanography and Marine Biology: An Annual Review, 39*, 207–231.

Washburn, T. W., Menot, L., Bonifácio, P., Pape, E., Błażewicz, M., Bribiesca Contreras, G., et al. (2021). Patterns of macrofaunal biodiversity across the Clarion-Clipperton Zone: An area targeted for seabed mining. *Frontiers in Marine Science, 8*, 626571. https://doi.org/10.3389/fmars.2021.626571

Watling, L., & Auster, P., (2017). Seamounts on the high seas should be managed as vulnerable marine ecosystems. *Opinion Article Frontiers in Marine Science. Section Deep-Sea Environments and Ecology, 4-2017*. https://doi.org/10.3389/fmars.2017.00014

Watson, R., & Morato, T. (2004). Exploitation patterns in seamount fisheries: A preliminary analysis. In T. Morato & D. Pauly (Eds.), *Seamounts: Biodiversity and fisheries, Fisheries Centre research report, 12* (pp. 61–66). University of British Columbia.

Weaver, P. P. E., Aguzzi, J., Boschen-Rose, R. E., Colaço, A., de Stigter, H., Gollner, S., et al. (2022). Assessing plume impacts caused by polymetallic nodule mining vehicles. *Marine Policy, 139*, 105011. https://doi.org/10.1016/j.marpol.2022.105011

Webb, T., Vanden Berghe, E., & O'Dor, R. (2010). Biodiversity's big wet secret: The global distribution of marine biological records reveals chronic under-exploration of the deep Pelagic Ocean. *PLoS One, 5*, e10223. https://doi.org/10.1371/journal.pone.0010223

Wedding, L. M., Reiter, S. M., Smith, C. R., Gjerde, K. M., Kittinger, J. N., et al. (2015). Managing mining of the deep seabed. *Science, 349*(6244), 144–145. https://doi.org/10.1126/science.aac6647

White, C., & Costello, C. (2014). Close the high seas to fishing? *PLoS Biology, 12*(3), e1001826. https://doi.org/10.1371/journal.pbio.1001826

Whitfield, A. K., & Elliott, M. (2002). Fishes as indicators of environmental and ecological changes within estuaries: A review of progress and some suggestions for the future. *Journal of Fish Biology, 61*, 229–250. https://doi.org/10.1111/j.1095-8649.2002.tb01773.x

Widder, E. A. (2010). Bioluminescence in the ocean: Origins of biological, chemical, and ecological diversity. *Science, 328*(5979), 704–708. https://doi.org/10.1126/science.1174269

Wijffels, S. E., Toole, J. M., Bryden, H. L., Fine, R. A., Jenkins, W. J., & Bullister, J. L. (1996). The water masses and circulation at 10°N in the Pacific. *Deep-Sea Research Part I: Oceanographic Research Papers, 43*, 501–544. https://doi.org/10.1016/0967-0637(96)00006-4

Willaert, K. (2020). Crafting the perfect deep sea mining legislation: A patchwork of national laws. *Marine Policy, 119*, 104055. https://doi.org/10.1016/j.marpol.2020.104055

Willett, C. S., Leben, R., & Lavin, M. (2006). Eddies and tropical instability waves in the eastern tropical Pacific : A review. *Progress in Oceanography, 69*(2), 218–238. https://doi.org/10.1016/j.pocean.2006.03.010

Woodall, L., Sanchez-Vidal, A., Canals, M., Paterson, G. L. J., Coppock, R., Sleight, R., et al. (2014). The deep sea is a major sink for microplastic debris. *Royal Society Open Science, 1*, 140317. https://doi.org/10.1098/rsos.140317

Worthington, L. V. (1981). The water masses of the World Ocean: Some results of a fine-scale census. CHAP 2. In Warren (Ed.), *Evolution of physical oceanography: Scientific surveys in honor of Henry Stommel* (pp. 42–69).

Wyrtki, K., & Kilonsky, B. (1984). Mean water and current structure during the Hawaii-to Tahiti shuttle experiment. *Journal of Physical oceanography, 14(2), 242–254.* https://doi.org/10.1175/1520-0485(1984)014<0242:MWACSD>2.0.CO;2

Xiao, B., Qiao, F., Shu, Q., Yin, X., Wang, G., & Wang, S. (2022). The development and validation of a global 1/32° surface-wave–tide–circulation coupled ocean model: FIO-COM32. *Geoscientific Model Development, 16*, 1755–1777. https://doi.org/10.5194/gmd-2022-52

Xuan, B. B., Armstrong, C. W., Ankamah-Yeboah, I., Hynes, S., & Needham, K. (2021). Valuing high seas ecosystem conservation. *Conservation Biology, 35*(5), 1405–1416. https://doi.org/10.1111/cobi.13720

Young, J. W., Olson, R. J., & Rodhouse, P. G. K. (2013). The role of squids in pelagic ecosystems: An overview. *Deep Sea Research Part II: Topical Studies in Oceanography, 95*, 3–6. https://doi.org/10.1016/j.dsr2.2013.05.008

Dr. Virginie Tilot holds a PhD in biological oceanography, an ecological engineer degree (conservation and fisheries), and an MSc. She is honorary Attaché of the National Museum of Natural History, Paris (France), and is corresponding member of the Académie des Sciences d'Outre-Mer of France and a titulary member of the Académie Royales des Sciences d'Outre-Mer of Belgium. Presently she is a research scholar and senior international expert on marine environment, fisheries, deep-sea ecology, coral reefs, biological oceanography, and conservation and sustainable development for diverse national and international organizations, institutions, universities, in particular EU and agencies of the United Nations (UNEP, UNDP, FAO, UNESCO/IOC). She has more than 35 years of experience, and she produced a total of 115 publications. She participated in 37 regional projects in the world, mainly in the Indian and Pacific Oceans and the Red Sea .

Concerning the domain of deep-sea mining, Dr. Tilot has issued recently four peer-reviewed publications, several communications in international meetings, a series of three volumes in the UNESCO/IOC Technical Series on *the Biodiversity and distribution of faunal assemblages associated to polymetallic nodule ecosystem and options for the conservation and the management facing deep sea mining*, and two chapters in two open access books edited by Springer on "*Environmental Issues of Deep-Sea Mining. Impacts, Consequences and Policy perspectives*" and on "Perspectives in Deep-sea mining". This is the third commitment with two chapters for the Springer book on "Deep-sea mining and water column."

Léa Fourchault holds an Erasmus Mundus MSc in tropical ecosystems and biodiversity. She is a young researcher interested in socio-ecological interactions, now working at the Royal Belgian Institute of Natural Sciences. Her current focus is on the resolution of taxonomic and phylogenetic uncertainties in under-researched vertebrates, using molecular techniques. Key research outcomes include a better understanding of the interrelation between humans, biodiversity, and ecosystem health .

She gained experience in marine spatial planning and spatial prioritization techniques under the mentorship of Professor A. J. Richardson (Queensland University) and has recently published a paper on generating affordable protection to high seas biodiversity. Economic and political feasibility is a key part of her research, wherefore she developed a cross-sectoral prioritization approach that reduces the projected opportunity cost of MPA implementation to the fishing, shipping, and deep-sea mining sectors simultaneously.

Dr. Alain Jeudy de Grissac holds a PhD in marine geology, sedimentology, and oceanology from the University of Marseilles (France). With more than 40 years of experience with different international organizations or institutions, he is presently corresponding member of the Académie des Sciences d'Outremer (France) and still active as an independent expert specialized in projects related to marine and coastal conservation, natural and cultural heritage, local community development and planning for sustainable development. His main focus is on marine and coastal protected areas, from the identification, creation, management, and evaluation of individual sites or networks of sites to the development of national policies or action plans, improvement of national legislation, and training of national administrations. He participated in the creation of more than 70 marine protected areas in the world and was managing some of them. Concerning the domain of deep-sea mining, Dr. Jeudy de Grissac was mandated by IFREMER to participate in research cruises in the Pacific to assess polymetallic nodules deposits in the CCZ. He has also participated in different technical meetings and the drafting of guidelines on deep-sea conservation and management in the world oceans.

Dr. Jérôme Mallefet holds a PhD degree focused on marine bioluminescence in fishes from Université catholique de Louvain (Belgium). He was a postdoctoral research scientist, working mainly on immunohistochemistry research of the nervous control of a luminous fish in Montreal University (Canada). He gets a position as research associate of National Sciences Foundation of Belgium (FNRS) at UCLouvain. He leads the marine biology laboratory and is currently working on bioluminescence biodiversity using different model species within cnidarians, worms, echinoderms, fishes. He has published more than 116 papers in peer review international journals and conference proceedings. He is a member of the scientific board of the International Bioluminescence and Chemiluminesnce Society. As such, he is concerned by the disturbance deep-sea mining will induce in the CCZ (Clarion Clipperton Zone, Eastern Pacific); sentinel glowing species should be developed before they disappear.

Dr. Juan Moreno Navas has an honors degree in marine science from Cadiz University (Spain) and a PhD in marine aquaculture spatial planning from Stirling University (UK). His PhD degree focused on marine vulnerability, spatial analysis, and hydrodynamic and ecological models. He was a postdoctoral research scientist, working mainly on GIS, spatial analysis, and biodiversity in cold water coral habitats at Heriot-Watt University (UK). He joined the Physical Oceanography Research Group at Málaga University (Spain) during 2 years as research associate working on hydrodynamic modeling. He is currently working as a research technician in artisanal fisheries management and modeling at the Spanish Institute of Oceanography (CNIEO-CSIC). Concerning the domain of deep-sea mining, he is coauthor of a publication on "The benthic megafaunal assemblages of the CCZ (Clarion Clipperton Zone, Eastern Pacific) and an approach to their management in the face of threatened anthropogenic impacts."

Chapter 14
Marine Mammal Communities and Human Activities in the North-Eastern Tropical Pacific: Conservation and Management Strategies

Virginie Tilot, Lea Fourchault, and Alain Jeudy de Grissac

Abstract Marine mammals could be impacted by deep-sea mining activities because of their physiological and behavioural characteristics, their migration patterns and their ecology, although there are knowledge gaps concerning them for the Clarion-Clipperton Zone (CCZ). This chapter aims at reviewing the current state of marine mammal populations and their associated ecosystems in the water column of the tropical north-eastern Pacific. Specifically, we assess their vulnerability to natural and anthropogenic impacts, in particular to deep-sea mining in the mineral-rich CCZ. As there is growing evidence that marine mammal communities and other apex predators play a critical role in ecosystem structures and functions, we outline their vulnerability and the existing conservation measures for marine mammals in the Pacific. We then propose to enhance knowledge in different domains of research linked to marine mammals and to adapt conservation strategies to ensure their well-being and the continuity of the ecosystem services they provide to the oceans and human societies in integration with other fields of ocean management.

Keywords Cetaceans · Interconnectivity · Nutrient cycling · Climate change · Deep-sea mining · Collision · Bioaccumulation · By-catch · Plastic pollution · Ocean stewardship · BBNJ · Marine protected areas · Ocean governance

---

V. Tilot (✉)
Muséum National d'Histoire Naturelle, Paris, France

Académie Royale des Sciences d'Outre-Mer, Brussels, Belgium
e-mail: v.tilot@wanadoo.fr

L. Fourchault
Institute of Natural Sciences, Brussels, Belgium

A. Jeudy de Grissac
Académie des Sciences d'Outre-mer, Paris, France

R. Sharma (ed.), *Deep-Sea Mining and the Water Column*,
https://doi.org/10.1007/978-3-031-59060-3_14

1 Introduction

Following a comprehensive study of the biodiversity and distribution of epibenthic megafauna and polymetallic nodule deposits in the Clarion-Clipperton Zone (CCZ) (ESCO CNRS IFREMER, 2014; Tilot, 2006a, b) and the development of a number of management and conservation strategies and tools (Ardron et al., 2008; Académie des Sciences d'Outremer, 2010; Tilot, 2010, 2013, 2014, 2019; Tilot et al., 2018), in particular for marine mammals (Tilot, 2004; Tilot et al., 2016), this chapter aims at assessing the presence, distribution and state of marine mammal populations and their associated ecosystems in the water column of the tropical north-eastern Pacific.

Marine mammal communities and their associated ecosystems play a critical role in the provision of numerous ecosystem services, ranging from climate regulation and nutrient cycling to aesthetic and recreational values. To best conserve and manage these populations, their vulnerability to natural and anthropogenic impacts, in particular to deep-sea mining, plastic pollution and climate change, needs to be assessed and mitigation measures proposed.

Measures to safeguard marine mammals, as they migrate on varying patterns over great distances, should *be flexible in space and time and actions and initiatives should be coordinated* on a regional basis with an ecosystem-based approach (Hoyt, 2011).

2 Topographical, Geo-physical and Hydrological Characteristics Associated with Marine Mammal Communities

Marine mammals can be associated with several topographic and geo-physical features, in particular with the seamounts, ridges and underwater volcanoes generally altering the flow of ocean currents (Kaschner, 2007). In the CCZ, seamounts rise from the surrounding abyssal plains and hills to such large heights that they increase hydrological flow rates. On such topographical features, especially with higher substrate heterogeneity, food particles are trapped and concentrated on summits and flanks. Suspension-feeding animals thrive there, such as Porifera, Scleractinia, Antipatharia, Actiniaria, Octocorallia, Stylasterida, Brachiopoda, Crinoidea and Tunicata (Tilot, 2006b, 2019; Rogers et al., 2007; Rogers, 2018; Clark et al., 2012). There is evidence of the interactions of zooplankton, micronekton and macronekton communities with benthic ecosystems on these elevated features. Seamounts with their concentration of preys attract foraging large pelagic predators, such as Odontocetes and Mysticetes, tuna and other ocean predators (Letessier et al., 2019), the reason for which pelagic fisheries, such as tuna fisheries, are often concentrated around seamount ecosystems (Morato et al., 2010). Seamounts may have several functions for large predators, as hypothesized by Genin (2004) as "topographic blockages", "feed-rest" sites for predators following vertical migrations of the nekton and current patterns, "concentration/retention of prey", "availability of prey from alternative habitats", "spawning grounds of fish" and "navigational waypoints".

As well, persistent hydrological features, such as oceanic fronts, have proven to enhance biological activity and drive marine mammal distribution and migration patterns. Seamounts with a higher frequency of oceanic fronts are more likely to aggregate large predators, for example tunas and cetacean, than average seamounts (Mate et al., 2020).

The northeastern Pacific blue whales, an endangered species, may be the largest remnant population in the world. These whales are stenophagic predators, feeding almost exclusively on euphausiids (Sears & Perrin, 2009). They are common in areas of upwelling and oceanic fronts, where phytoplankton and, consequently, euphausiid concentrations are high (Fiedler et al. 1998; Etnoyer et al. 2006, Branch et al. 2007). Diel patterns in blue whale match the diel patterns in the vertical distribution of their prey (Stafford et al. 2005; Bailey et al., 2010).

3 Characteristics, Main Threats and Conservation Status of Whales and Oceanic Dolphins in the North Eastern Pacific Ocean

On the basis of the literature (Hoyt, 2004, 2011; Würsig & Kovacs, 2018; Palacios et al., 2023), over 20 species of toothed whales and seven species of baleen whales can be found in the eastern Pacific Ocean, in particular in the CCZ. Most baleen whales and the large-toothed sperm whales undertake large yearly migrations between their high-latitude feeding grounds and their low-latitude breeding grounds, although variations in distances and distribution patterns may occur due to environmental change (Bailey et al., 2010). Their characteristics, main threats and conservation status are detailed in Table 14.1.

Whales have varied diets with many populations feeding on small schooling fish ranging from swarming mysids and amphipods, crabs, cephalopods, anchovies, sardines, mackerels, herrings, capelins and cods. They often have fishing techniques in groups, such as bubble nets, to concentrate preys (Wiley et al., 2011) and diel patterns to follow the distribution pattern of their preys and specific oceanographic conditions, such as peaks in upwelling of highly productive waters (Bailey et al., 2010). They can dive at deep depths, 3000 m, yet recorded, in search of food (Irvine et al., 2017).

Formerly heavily hunted for some species, whales have recovered since the International Whaling Commission (IWC) protection in 1946 (IWC was established in 1946 as the global body responsible for the management of whaling and conservation of whales). Presently, whales face risks from entanglement in fishing gear, vessel strikes, vessel-based harassment, pollution (marine litter, chemical pollution, underwater noise), whale watching, scientific whaling and subsistence whaling in some countries (Lagerquist et al., 2008; Martinez-Loustalot et al., 2022; Pelayo-Gonzalez et al., 2022).

Table 14.1 Characteristics, main threats and conservation status of Baleen, toothed whale and oceanic dolphin populations in the north-eastern Pacific Ocean

Baleen whales	Distribution/migration	NE Pacific	S Pacific	SE Pacific	Threats/impacts	Global change	Conservation status
Gray whales Eastern N Pacific Gray whale, *Eschrichtius robustus*	Gray whales migrate in pods upto 16 for more than 10,000 miles yearly. Larger populations occur on feeding and breeding grounds. Their feeding is mainly on swarming mysids and amphipods in the north, otherwise opportunistic elsewhere. In the eastern Pacific, they usually travel 5000–7000 miles along the coastline from Bering, Chukchi and Beaufort Seas in the Arctic, where they forage in the summer, to their winter mating and birthing grounds in Mexico's Baja peninsula area. However, some populations of hundreds, the "Pacific Coast Feeding Group", seem to have settled between southeast Alaska and northern California.	x			Gray whales were hunted close to extinction for commercial purposes in the past; their populations in the Eastern North Pacific have now recovered pre-exploitation numbers due to international conservation measures (IWC protection in 1946).	A continuing series of die-offs of eastern Pacific gray whales occurring in foraging of the Arctic have reduced the populations to 25% over a few years according to changing Arctic conditions impacting the ecological system (Stewart et al., 2023). A total of 606 gray whales were stranded (Unusual Mortality Event, UME), between 2019 and 2022, along the west coast of North America and Central America, likely due to climate change (Baird et al., 2022).	Endangered (NOAA fisheries), least concern (IUCN) Population: ~27,000

Humpback whales *Megaptera novaeangliae*	From their feeding grounds along the Bering Strait to their calving grounds in Hawaiian waters, humpback whales cover over 5000 miles in their yearly migration (Palacios et al., 2019a). They are the most acrobatic whales, they use bubble nets, bubble clouds, tail flicks and other techniques to help concentrate krill and small schooling fish for easier feeding, often gathering in pods of 20 or more whales. Complex songs are part of the breeding display.	x	x	x	Since IWC protection in 1946, populations have recovered. During their migration, they encounter "the great Pacific garbage patch", which contains numerous ghost nets (Palacios et al., 2023). Although humpback whales are no longer heavily hunted, they presently face threats from entanglement in fishing gear, vessel strikes, vessel-based harassment and pollution (especially noise), in particular for populations off the US west coast and off Central America (Lagerquist et al., 2008; Martinez-Loustalot et al., 2022; Pelayo-Gonzalez et al., 2022).	Global change impacts the distribution and abundance of their prey, especially krill, leading to changes in foraging behaviour, migration patterns and diminished reproduction and population abundance (Clapham, 2018; Pallin et al., 2018; Tulloch et al., 2019).	Vulnerable, least concern (IUCN) Population: ~84,000

(continued)

Table 14.1 (continued)

Baleen whales	Distribution/migration	NE Pacific	S Pacific	SE Pacific	Threats/impacts	Global change	Conservation status
Common Minke whales *Balaenoptera acutorostrata*	These whales are widely distributed and have varied diets with many populations feeding on small schooling fish ranging from anchovies to mackerels, herrings, capelins and cods. Minke whales' seasonal migrations are less predictable and less well-defined than for other species of baleen whales. In the north-eastern Pacific, the common Minke whales are widespread, with a few known areas of aggregation (Towers et al., 2013).	x	x	X	Common Minke whales encounter the same threats than the other whales (underwater noise, ship strikes, marine litter and chemical pollution, entanglement in fishing gear), A certain amount of "scientific whaling" (Japan), commercial whaling (Norway, Iceland) and subsistence whaling (Greenland) occur although IWC great whales protection (Risch et al., 2019).	These species are also at risk from multiple environmental pressures caused by climate change. Thus, competition with other predators for a limited prey resource could cause displacement to areas with higher predation risk from from killer whales (Higdon & Ferguson, 2009)	Near threatened (NOAA), least concern (IUCN) Population: ~200,000

Bryde's whales *Balaenoptera edeni*	These whales have a wide distribution in tropical, subtropical and warm temperate waters around the world. Some populations migrate seasonally, others are residents closer to the equator. Bryde's whales are usually seen alone; however, groups can be seen in feeding areas, where they display a wide array of novel behaviors than other species of baleens. Their preys are krill, copepods, red crabs and shrimp, as well as a variety of schooling fish, such as herring, mackerel, pilchards and sardines. They use different methods to feed in the water column, including skimming the surface, lunging and creating bubble nets.	x	x	x	They are the most commonly reported species struck by vessels in the eastern N Pacific. Underwater noise can impact their behaviour by hindering their vocal communication and thus their ability to find food and mate, avoid predators and navigate. Bryde's whales are presently used for "scientific whaling" in Japan and commercial whaling in Indonesia and the Philippines.	With their flexible and diverse foraging strategies, contrasting movement patterns and broad prey preferences, Bryde's whales may adapt well to global change (Constantine et al., 2018).	Least concern (IUCN) Population: ~90,000–100,000

(continued)

Table 14.1 (continued)

Baleen whales	Distribution/migration	NE Pacific	S Pacific	SE Pacific	Threats/impacts	Global change	Conservation status
Sei whales *Balaenoptera musculus*	They live in subtropical, temperate and subpolar waters around the world. They make seasonal migrations from low-latitude wintering areas to high-latitude summer feeding grounds. Migrations in both hemispheres are segregated based on age, sex and reproductive condition. Sei whales use primarily pelagic habitats and are most often found in deeper waters. The main characteristic of the Sei whale feeding habitat is a high pelagic concentration of zooplankton, especially copepods, but they also feed on fish.	x	x	x	Sei whales are threatened by habitat loss due to climate change, pollution, vessel strikes and entanglement in fishing gear, as well as inter-specific competition, disease and human-induced stresses. The level of competition between Sei whales and other whale species is not known, but the most likely competitors would be Fin and Right whales, given their sympatric distribution and similarities in diet. Similar to Bryde's whales, Sei whales are used for "scientific whaling" in Japan.	Climate change has been evidenced to impact Sei whales, with the biggest single baleen whale mass mortality event caused by a harmful toxic algal bloom caused by the intense El Niño event of 2014–2016 (Häussermann et al., 2017).	Endangered Population: ~80,000

Fin whales *Balaenoptera physalus*	Fin whales are primarily oceanic; they migrate from Baja California, where they calve and mate, to the Bering Sea, where they feed (Clapham, 2001). However, there are some sedentary populations more coastal. These baleen whales prey upon euphausiids and small schooling fish that regularly aggregate into large schools in the upper water layers, in fewer than 280 m depth (Mizroch et al., 2009). These whales are the fastest, attaining high speeds, close to 40 km/h.	x	x	x	Vessel strikes are the main impact for fin whales. A sedentary population located inside the Gulf of California is frequently impacted by ship strikes (Jimenez et al., 2019). Environmental change including habitat loss, toxics and climate change put fin whales at risk. Commercial whaling remains a threat for fin whales.	Extreme climate variability already appears to have had an impact fin whales. It was implicated in the Unusual Mortality Event of Fin and Humpback Whales along the coast of British Columbia and Alaska from 2015 to 2016 (COSEWIC, 2021).	Endangered (NOAA), vulnerable (IUCN) Population: ~100,000

(continued)

Table 14.1 (continued)

Baleen whales	Distribution/migration	NE Pacific	S Pacific	SE Pacific	Threats/impacts	Global change	Conservation status
Blue whales *Balaenoptera musculus*	The north-eastern Pacific population of blue whales is the largest remnant population in the world. These endangered whales are stenophagic predators, feeding almost exclusively on euphausiids (Sears & Perrin, 2009). They are common in areas of upwelling and oceanic fronts, where phytoplankton and, consequently, euphausiid concentrations are high (Fiedler et al. 1998; Etnoyer et al. 2006; Branch et al. 2007). They generally follow the diel patterns in the distribution of their preys (Stafford et al. 2005; Bailey et al., 2010). The California population migrates between feeding grounds off the USA's west coast and breeding grounds in the Gulf of California and the Costa Rica Dome (large, stationary eddy centred near 9°N, 89°W). This migration coincides with the spring peak in the upwelling of highly productive waters hosting large standing stocks of euphausiids (Bailey et al., 2010; Irvine et al., 2014; Mate et al., 1999; Blevins et al., 2022; Palacios et al., 2019b).	x	x	x	As the largest whales, they were hunted as much as rorquals. There are signs of recovery since protection by the IWC in 1965, especially stocks feeding off California. Present threats to blue whales include habitat loss, entanglement in fishing gear, chemical and sound pollution and overfishing of krill.	Negative impacts of climate change and ocean acidification on krill are deeply concerning as krill play such a pivotal role in the survival of the Blue Whale. In response to environmental changes affecting their prey, patterns of northward migration may vary significantly (Bailey et al., 2010).	Endangered Population: ~5000–15,000

Sperm whales *Physeter macrocephalus*	The largest toothed whales are widely distributed. They live in social matriarchal family groups of up to 50 whales, migrating over 4000 km in tropical and temperate oceanic waters. Movements are integral to the complex population ecology of sperm whales. Social units adapt their movements to the temporal and spatial scales of the distribution of resources, leading to density-dependent habitat selection (Whitehead et al., 2008). Sperm whales usually dive between 300 and 800 m, with a maximum of 3000 m, yet recorded, in search of food (Irvine et al., 2017). They prey on a large variety of fish and cephalopods, their major prey.	x	x	x	Sperm whales were a major target of the whale industry for 3 centuries. Presently, the major threats are vessel strikes, entanglement in fishing gear (trap lines, pots, gillnets..), ocean noise, marine debris, oil spills and contaminants	Sperm whales face a range of challenges driven by climate change such as rising water temperatures, disappearing habitats, changing prey distributions and availability, higher inter and intra species competition occupying with shifting ecological niches. The effect on whales of ocean acidification on organisms that constitute the base of food chains and provide important habitats for their preys would trigger behavioral changes solved through their mobility. In addition, disease outbreaks in marine ecosystems could trigger increased stress and expansion of vectors of disease. The incremental effect from the increased encroachment of anthropogenic activity will exacerbate the effects of climate change	Vulnerable (IUCN) Population: ~350,000

(continued)

Table 14.1 (continued)

Toothed whales	Distribution/Migration	NE Pacific	S Pacific	SE Pacific	Threats/impacts	Global change	Conservation status
Pygmy sperm whales *Kogia breviceps* and **Dwarf sperm whales** *Kogia simus*	Pygmy sperm and dwarf sperm whales live in oceanic warm tropical waters and feed on deep-sea pelagic squids, fish and shrimps. They are often seen in small groups of 5–6 individuals.	x	x	X	They are very difficult to detect if seas are not calm. They can be found in gillnet fisheries.		Lower risk
Beaked whales Cuvier's beaked whale *Ziphius Cavirostris,* Perrin's beaked whale *Mesoplodon perrini,* Pygmy beaked whale *M. peruvianus,* Hubb's beaked whale *M. carlhubbsi,* Ginko-toothed beaked whale M. *ginkogodens,* Stejneger's beaked whale *M. stejnegeri,* Blainville's beaked whale *M. densirostris*	Generally beaked whales Ziphiids are oceanic, mainly found in tropical and warm temperate waters and generally dive in depths over 1000-2000m (Dalebout et al., 2002). They swim in groups segregated on the basis of sex, staying close together. They feed on deep-sea cephalopods and fish, as well as benthic crustacean. Cuvier's beaked whales are fairly common in the eastern tropical Pacific.	x	x	x	Often caught as bycatch in fisheries with certain types of fishing gear such as deepwater gill nets used for capturing large pelagic fish species. Like other marine mammals, it may also be negatively impacted by navy sonar and loud noises generated by seismic exploration.	Climate change may also cause their range to change or grow smaller	Data deficient

Toothed whales							
Ocean dolphins	**Distribution/migration**	**NE Pacific**	**S Pacific**	**SE Pacific**	**Threats/impacts**	**Global change**	**Conservation status**
Rough-toothed dolphin *Steno bredanensis*,	They are tropical to subtropical species inhabiting deep oceanic waters. They feed on cephalopods and fish (in particular of large *Coryphaena hippurus* (Pitman & Stinchcomb, 2002) and live in relatively open societies, from small groups to herds of hundreds, often associated with other cetaceans in the eastern tropical Pacific and floating objects.	x	x	X	In the eastern tropical Pacific, they have been reported as bycatch of fisheries, e.g. purse seines, longlines and drift gillnets.	With climate change, harmful algal blooms would occur more frequently causing impacts on ecosystems by associated biotoxicosis. Such events would be at the origin of mass mortalities of rough-toothed dolphins (Fernandez et al., 2022)	Data deficient
Bottle-nose dolphin *Tursiops truncatus*,	These dolphins are found in temperate and tropical waters in the world, mostly in coastal areas. In the eastern tropical Pacific, they are also pelagic where they can be seen in big herds of several hundreds. They are often associated with other cetacean. Some are migratory. They are opportunistic feeders.	x	x	X	Bottle-nose dolphins are bycatch of fisheries with shrimp trawls, shark nets, gillnets and purse seines (those from multi-national tuna purse seine fisheries in the eastern tropical Pacific). They are also killed in harpoon and drive fisheries. Coastal populations have been affected by live capture removals and habitat change	Evidence that extreme climate change events is the biggest threat with high intensity epizooties to populations of bottlenose dolphins. Climate change will affect the abundance and distribution of prey, major loss of habitats, the metabolism of dolphins, the reduction of reproductive success, more competition (Reed et al., 2020).	Data deficient

(continued)

Table 14.1 (continued)

Toothed whales	Distribution/Migration	NE Pacific	S Pacific	SE Pacific	Threats/impacts	Global change	Conservation status
East Pacific offshore spotted dolphin *Stenella attenuata*, Gray's spinner dolphin *Stenella longirostris*, Striped dolphin *Stenella Coeruleoalba*, Short-beaked common dolphin *Delphinus delphis*, Fraser's dolphin *Lagenodelphis hosei*	The most abundant dolphin species in the eastern tropical Pacific. They often live in association with other species, as yellowfin and skipjack tunas, other dolphins and pelagic predators. Offshore populations may occur in the thousands. Spotted dolphins feed mainly the day on epipelagic fish and squid while spinner dolphins feed at night on midwater fish and squid. Striped dolphins feed on small mid-water squid and fish, especially lanternfish. Fraser's dolphins feed on mid-water squids, fish and crustaceans.	x	x	x	Often bycatch of tuna fisheries (entanglement in tuna purse seines), causing significant reduction of the populations in the eastern tropical Pacific.		Data deficient
Risso's dolphin *Grampus griseus*	They live in deep oceanic waters widely distributed. Herds of thousands may occur associated to other dolphins. They seem to prefer squids.	x	x	X	They are caught as bycatch in drive, gillnet, seine and harpoon fisheries but can be hunted for human consumption and fish bait.		Data deficient

Melon-headed whale *Peponocephala electra*	Highly social, they occur generally in pods of 100 to thousands and associate with other species.	x	x	X	Often taken in pursue and driftnet fisheries, some for display in oceanaria.		Lower risk
Pygmy killer whale *Feresa attenuate,* False killer whale *Pseudorca crassidens*	Small herds that sometimes attack other dolphins in tuna fishery interactions, generally eat squids and fish.	x	x	x	Can be caught in drive, lines and driftnets. False killer whales are the most common species in mass strandings.		Data deficient
Orca or killer whale *Orcinus orca*	Residents mainly eat fish while transients eat marine mammals, fish, cephalopods and occasionally seabirds and marine turtles.	x	x	X	Often killer whales are seen as competitors for fishermen and may be killed. Small numbers are also taken incidentally in fisheries.		Lower risk (conservation dependent)
Short-finned pilot whale *Globicephale macrorhynchus*	Commonly associated with other species in the eastern Pacific. Pods of several to hundreds occur. They could be nomadic in association to seasonal abundance of squid abundance and to major change in current structure or relief topography. There is also a seasonal distribution with depth.	x	x	X	Among the most frequently mass-strand. They can be captured live for display and research. Incidental catches occur in fishing gear. A wide variation in contaminant loads occurs in short-finned pilot whales. Whale watching vessels can potentially cause short-term disturbance (Glen, 2003).		Lower risk

Adapted from Hoyt (2004, 2011), Würsig and Kovacs (2018), and Palacios et al. (2023)

Pacific Ocean dolphins feed on a variety of prey, as mostly opportunistic feeders, such as cephalopods and fish, from small schooling fish (capelin, sardines, and herring) to tunas and coryphaenas. There is a seasonal distribution with depth in association with seasonal prey abundance and oceanographic conditions. They form societies from small groups to herds of hundreds and often associate with other cetaceans. Some populations are residents, others nomadic.

A primary threat to ocean dolphins is entanglement in fishing gear, such as gillnets, trawls and longlines. Often caught as bycatch in high seas fisheries, especially with tuna fisheries, where they interact. Noise interference hinders their ability to use sound, which, in turn, disturbs their ability to communicate, choose mates, find food, avoid predators and navigate.

As detailed in Table 14.1, baleen whale migrations typically consist of annual movements between productive, high-latitude feeding grounds and unproductive, low-latitude breeding grounds. The migrations of some baleen whales encompass the longest movement (approximately 8300 km) ever recorded for mammals (Corkeron & Connor, 1999; Rasmussen et al. 2007). Some whales are known to frequent multiple breeding grounds within one breeding season, migrating between high-latitude feeding grounds, and Mexican and Hawaiian breeding grounds within the same year with some temporal and spatial changes due to climate change (Fig. 14.1) (Darling et al., 2022).

A diversity of whale populations use « blue corridors », interconnecting areas over thousands of kilometres of productive oceanographic conditions. These can be found along the Bering Strait and Antarctic peninsula to the temperate and tropical Pacific (Palacios et al., 2019a). While the CCZ is not part of a defined "blue corridor", it is part of an oceanic habitat where migratory and resident marine species live such as sharks, turtles, tuna, dolphins and whales.

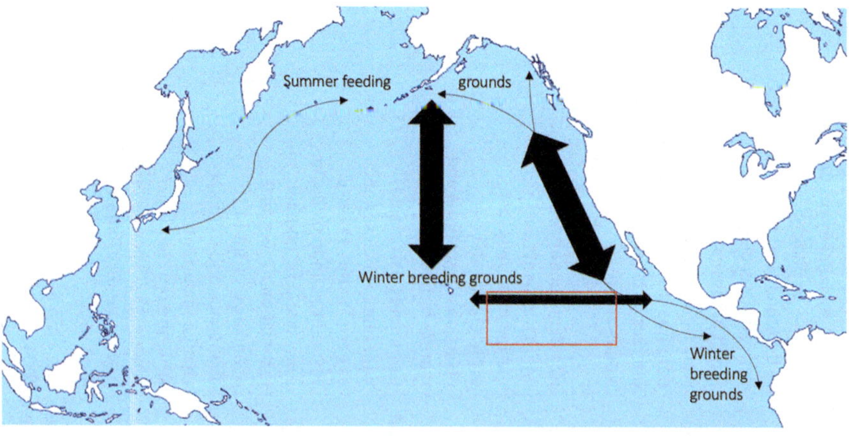

Fig. 14.1 Humpback whale migration patterns across the northern Pacific Ocean (adapted from NOAA and Darling et al., 2022). Black arrows represent migration patterns. The CCZ is located in the red box

Presently, whale and dolphin populations and their preys are impacted by entanglement in fishing gear (bycatch, ghost gear), ship strikes, chemical pollution and underwater noise pollution and loss of habitat.

Cetaceans are also vulnerable to climate change (Albouy et al., 2020) and react differently, according to a literature review (van Weelden et al., 2021), cetacean populations have different responses. Some would make spatial or temporal changes, by shifting towards preferred Sea surface temperatures (SSTs), oceanographic conditions and habitats, as well as prey distribution and abundance or altering the timing of their migrations; some would be under extreme pressure, reducing reproductive success and increasing the risk of extinction while other species would adapt and even benefit on climate change effects on ecosystems. Due to intra-and inter-specific competition, genetic alterations, algal blooms and enhanced mercury bioaccumulation, ecosystem-level changes among cetacean populations would occur as evidenced in some case studies (van Weelden et al., 2021). Therefore, research is needed to assess and predict future impacts based on modelling, at a species and population level, locally and regionally, as most species are mobile, in order to design adapted mitigation and conservation measures.

There is an increasing chronic background anthropogenic noise in the oceans mostly coming from the growing number of container ships, seismic surveys to find oil and gas deposits, naval sonars, drilling, pile-driving and other construction, icebreakers, oceanographic experiments, wind farms, acoustic, harassment devices, recreational boating, commercial sonar, etc. (Hildebrand, 2005). This will affect whales and dolphins, which use the special acoustical properties of water, matched to their finely tuned sensory systems, for communication, echolocation and other social behaviours.

More specifically in the CCZ, anthropogenic and climate change disturbances have been evidenced by Halpern et al. (2013, 2015) in the East and West by two areas of high bycatch estimates—about 0.005–0.01 metric tonnes of yearly bycatch per square meter for every gram of NPP in international waters near Hawai'i and Mexico on the 20°N parallel (Fig. 14.2a). In the South, the CCZ is bound by a busy shipping lane linking the Panama Canal with East Asia (Fig. 14.2b). This implies the presence of fishing gear including ghost nets, vessel strikes, vessel-based harassment and underwater noise (Lagerquist et al., 2008; Martinez-Loustalot et al., 2022; Pelayo-Gonzalez et al., 2022; Thompson et al., 2023). Also about 66% of marine mammal species are directly affected by plastic waste: 41% by entanglement and about half by ingestion (Kühn et al., 2015). Concerning evidence of climate change, in the North and East, the CCZ is bound by areas of changing sea-surface temperatures, with an observed decrease in temperatures in the North and increase in the East (Fig. 14.2c).

Avila et al. (2018) recently evaluated the number of whale species at risk of cumulative anthropogenic impacts in the Pacific Ocean. The greatest number of toothed whales at risk are located off the Californian and Mexican coasts, as well as in the eastern Pacific between 5°N and 10°N (Fig. 14.3). Except for the easternmost part, the CCZ is an area of low risk of cumulative anthropogenic impacts for toothed

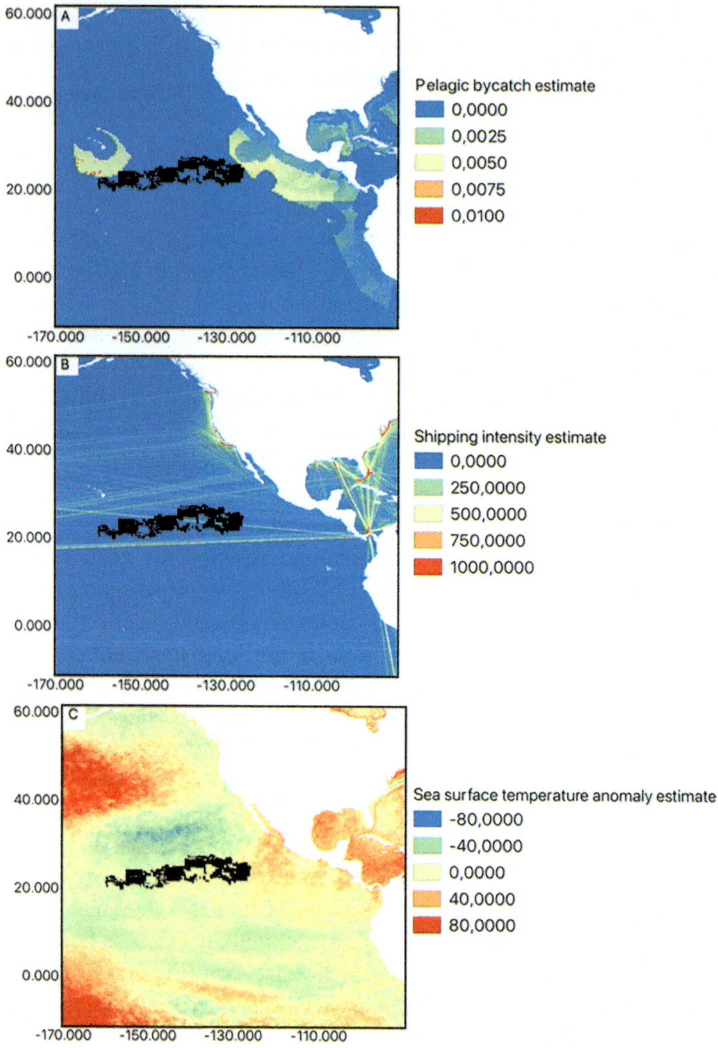

Fig. 14.2 Anthropogenic disturbances (fishing, shipping) and climate change in the tropical north-eastern Pacific. The CCZ is displayed in black in all maps. Estimates obtained by Halpern et al. (2013, 2015) by (**a**) dividing yearly metric tonnes of catch for industrial pelagic fishing using high bycatch gear types by yearly Net Primary Productivity (monthly data averaged for yearly NPP estimate) to standardize by the region's productivity; (**b**) creating a 1 km buffer around voluntary ship track data to account for shipping lane size; (C) subtracting the number of extreme sea-surface temperature (SST) weeks during a five year period (2008–2013) from the number of extreme SST weeks during a baseline 5-year period (1985–1989). An extreme week is defined as having an SST anomaly (weekly SST—weekly climatological SST) exceeding 1 SD of the anomalies calculated across all years (1982–2017) for that week

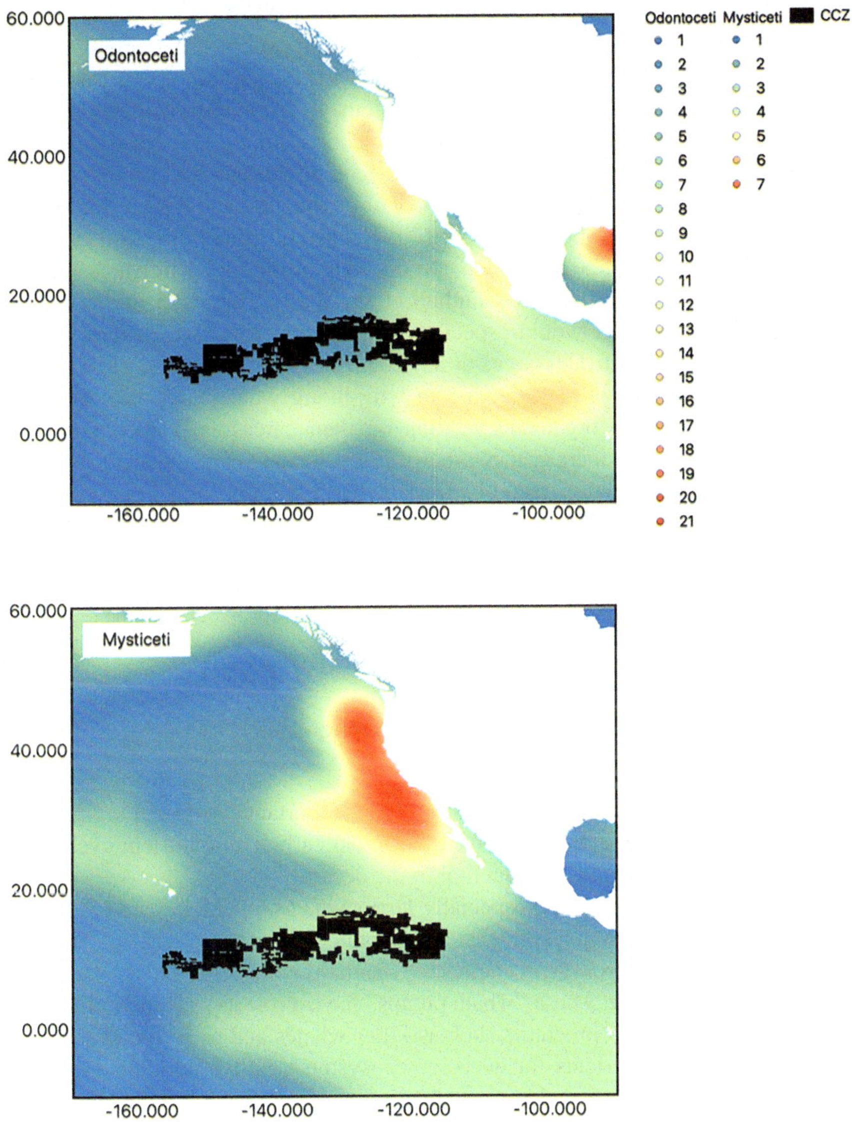

Fig. 14.3 Number of species threatened by cumulative anthropogenic impacts in the eastern Pacific. Each cell represents the number of marine mammal species (Odontoceti or Mysticeti) that are disturbed by anthropogenic activities. (Based on data from Avila et al., 2018)

whales. For baleen whales, waters off the Californian coast represent a risk for the greatest number of species while half of the species are at risk in the north-eastern part of the CCZ (Fig. 14.3).

Deep-sea mining operations will further increase these stresses on marine mammal populations through several mechanisms. First, noise will be emitted on the seabed during ore extraction, during the upward pumping of sludge and in the epipelagic layer both during offshore processing and transport of extracted minerals towards onshore facilities (Miller et al., 2018; Thompson et al., 2023). As mining activities are planned to be performed continuously, the constant auditory masking will likely disrupt echolocation and communication among and between marine mammal populations, resulting in decreased feeding, mating and social cohesion (Thompson et al., 2023). The possible impact of extraction and dewatering plumes on organisms at lower trophic levels will likely have repercussions throughout the food web (Drazen et al., 2020), which could lead to nutritional deficiencies in cetaceans, especially in odontocetes (Southall et al., 2019). The accidental release of chemicals into the water would especially impact cetaceans, which tend to have high relative fat content (Desforges et al., 2018; de Oliveira-Ferrera et al., 2023). Toxic elements would bio-accumulate in long-living whales and then would be transferred to deep-sea organisms feeding on their carcasses (Debenham et al., 2004). In addition, the increased number of vessels and surface-water operations platforms will increase the risk of physical harm through collisions (Ritter & Panigada, 2019). In this perspective, it is essential to optimise conservation and management measures, as further proposed, to ensure that the additional stressors brought out by deep-sea mining activities would minimally impact marine mammal populations.

4 Marine Mammals and Ecosystem Services

Pelagic ecosystems in the Pacific region provide important ecosystem services (De Groot et al., 2012) through their impact on global carbon and nutrient cycles, and their contribution to biomass production in the ocean.

In particular, cetaceans, and especially larger whales, play a key role in marine ecosystems due to their physiological and behavioural characteristics. They are referred to as "marine ecosystem engineers" (Roman et al., 2014) and are crucial for oceanic nutrient cycling. These "whale pumps" have a role as predators of fish and invertebrates, as prey to predators such as killer whales and as a crucial source of organic carbon and nutrients for deep-sea ecosystems with whale falls (Riisager-Simonsen et al., 2020). They also create feeding opportunities for marine birds (Dill et al., 2003). For example, annual Southern Ocean sperm whale faeces pump about 50 tonnes of iron ingested in deeper layers to the ocean surface, which, in turn, fosters phytoplankton growth when excreted at the surface, resulting in new primary production that fixes 200,000 tonnes of atmospheric carbon to the deep-sea annually (Lavery et al., 2010). Dead whale falls can sustain up to 407 species of decomposer-dominated benthic ecosystems for over a century (Smith & Baco, 2003; Smith, 2006; Smith et al., 2008, 2014, 2015; Wang et al., 2021).

Moreover, cetaceans are particularly relevant from an ecosystem-service perspective because of their large distribution ranges, migratory behaviours, past histories of exploitation, present iconic status across much of the world, cultural and economic importance for indigenous people, regulating effect on ecosystem structure and monetary value for tourism (O'Connor, 2009; Pompa et al., 2011; De Groot et al., 2012). Indeed, large body size, carnivory and endothermic costs lead to high caloric demands in mammalian predators, which may make them particularly vulnerable to changes in prey abundance (Acevedo-Gutiérrez et al., 2002; Benson et al., 2002; Croll et al. 2005). The disappearance of such apex predators is linked to the effects of trophic downgrading on process, function and resilience in global systems (Estes et al., 2011). As a consequence, ecosystems with fewer marine mammals will likely provide fewer ecosystem services and would eventually collapse, including deep-sea ecosystems relying on whale falls (Smith et al., 2014, 2015; Wang et al., 2021).

Baleen whales, in particular blue, fin, humpback and minke whales, play a key role in geochemical cycling. Indeed, by releasing their P-, N- and Fe-rich faeces in the epipelagic layer, Odontocetes and Mystecetes contribute to the growth of phytoplankton, which fixates atmospheric carbon (Fig. 14.4). This abundance of phytoplankton contributes to climate regulation and serves as a trophic basis for several species of commercial importance, including swordfish and tuna species. The phytoplankton is then preyed upon by zooplankton, which is preyed upon both by Mysticetes and by mesopelagic organisms, which in turn are preyed by Odontocetes.

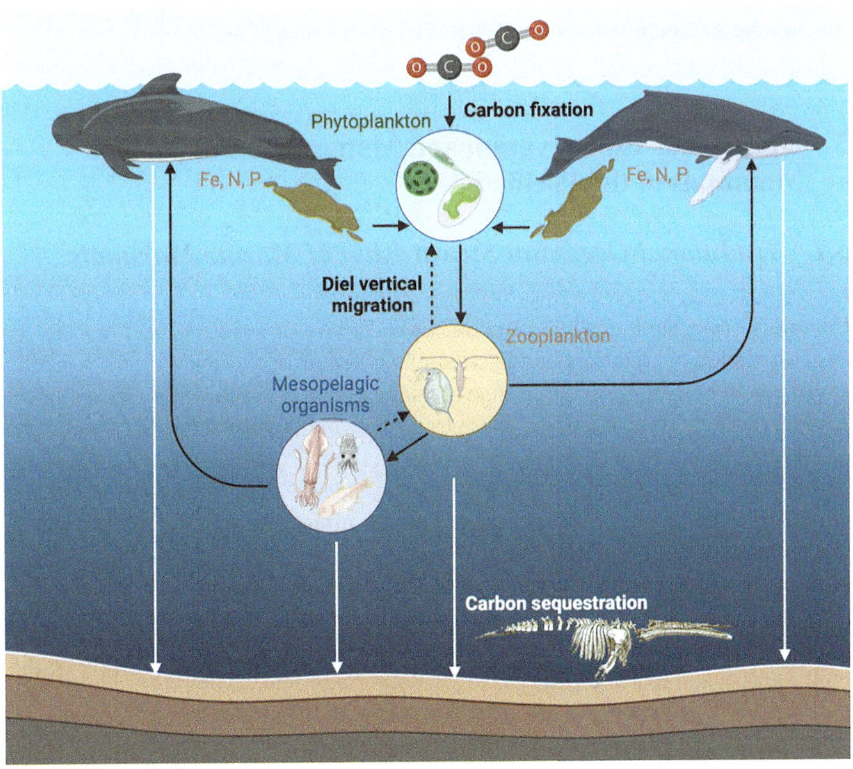

Fig. 14.4 The role of marine mammals in atmospheric carbon sequestration. (Adapted from Palacios et al., 2019a)

Marine falls provide carbon sequestration, a key source of nutrients for deep-sea organisms, where carcasses serve as the basis of the trophic chain in numerous deep-sea ecosystems (Smith & Baco, 2003; Debenham et al., 2004). Their importance for nutrient recycling (Doughty et al., 2016; Enquist et al., 2020) has been emphasized by recent results using high-resolution foraging measurements (Savoca et al., 2021). Protecting whale populations would thus augment primary productivity and restore the structure and function of ocean ecosystems to better sustain climate regulation and productive fisheries (Estes et al., 2016).

The movement patterns of small cetaceans, often high local and regional abundances across a range of ecosystems, and high metabolic rates suggest that these marine mammal communities could have large effects on ecosystem structure, dynamics and function through a variety of mechanisms. These include top-down (e.g., direct predation and risk effects) and bottom-up effects (e.g., translocation of nutrients within and across ecosystems), but also behaviour-mediated processes where these predators can facilitate access to resources to other predators or modify the physical properties of habitat (e.g., bioturbation). Most small cetaceans can be consumed by other marine predators, particularly killer whales (*Orcinus orca*) and large sharks. Mass-balance ecosystem models suggest that small cetaceans may impact the populations of short-lived prey species (particularly fish and cephalopods), but other factors (e.g. eutrophication and fisheries) also affect ecosystem functioning and population trends (Kiszka et al., 2015, 2022).

5 Options for Conservation and Management of Marine Mammals in the Pacific

5.1 *Traditional Oceanian Stewardship of Marine Mammals*

Centuries of acquaintance with the interconnectivity of terrestrial and marine ecosystems underlay traditional indigenous knowledge in the Pacific and constitute keystones to holistic forms of marine resource management adapting to ecological, societal and economic changes. These traditional management practices and forms, originating from complex community rules in many Pacific Islands, generally preserve or protect areas in an equitable manner and vastly precede their earliest formulations in Western models for marine conservation (Govan et al., 2009; Veitayaki, 2011; Bambridge, 2016; Bambridge et al., 2019; Tilot et al., 2021a).

Temporal closures and "tabu"(Fiji) or "rahui" (Cook islands) restrictions on certain areas, species and persons in traditional customary pratices are widely used to reduce intensive harvest of spawning fishes or other predictable aggregations (e.g., migration routes) (Johannes, 1978, 1981; Bambridge et al., 2019). In order to ensure the well being of the people and their environment, these concepts cover a complex system of community rules and prohibitions in many Pacific Island traditions and uses such as in Fiji, Samoa, Kiribati, Rapanui, Tahiti, hawaii, Vanuatu and Tonga (Grey, 2019; Juster, 2016; Mead, 2003; Love, 2021).

Whales, apex predators and large megafauna are highly esteemed by many peoples from the Pacific Islands region as they feature in legends and myths, serving in many roles, such as guardians of ocean voyagers or as reincarnated ancestors (Johannes, 1981; D'Arcy, 2006). Also, indigenous knowledge on fishing and megafauna, including whale populations, has been fundamental to environmental, cultural and livelihood sustainability of Pacific peoples, in particular in the off-shore (Nordhoff, 1930; Ruddle et al., 1992; Flood et al., 1999; Firestone & Lilley, 2007; Hind, 2015; Kitolelei et al., 2021).

Indigenous Peoples and Local Communities (IPLCs) have a role as custodians of significant ecosystems and of species generally traveling between coasts and high seas (Ey & Sherval, 2016; Eckstein & Schwarz, 2019). Their rights are respected in different instruments in the Pacific Nations, in particular in the recent agreement on Biodiversity Beyond National Jurisdiction (https://www.un.org›bbnj) (Vierros et al., 2020; Tilot et al., 2021a, b, c).

Presently, there is a network of "Locally Managed Marine Areas" (LMMAs) originated by community- and village-based marine management initiatives involving Fiji, Papua New Guinea and other islands (see www.lmmanetwork.org). Despite the overall emphasis on cetaceans in marine mammal sanctuaries, there is a recognition and understanding among Pacific islanders that Marine Protected Areas (MPAs) must serve a variety of purposes, from broad ecosystem and ecosystem services protection to ensuring future livelihoods for local people through tourism, sustainable fishing and other initiatives. As an example, the network of 400 LMMAs in Fiji, which are linked to a large whale sanctuary, is a successful example of traditional management integrated to science-based management (Thaman et al., 2017). A portion of the fishing grounds is usually set aside as a no-fishing zone to safeguard the future sustainability of fisheries resources (Tawake et al., 2001; Tawake & Tuivanuavou, 2004; Veitayaki et al., 2003). This LMMA network not only focuses on management but also on raising awareness, informing policies and sharing information at the national and international levels (Jupiter et al., 2014). The success of each LLMA is measured by the state of health at different levels, ecosystem, habitats and species along with the reduction of threats and the well-being of people who use the resources (Tawake & Tuivanuavou, 2004).

The national monument Papahânaumokuâkea, built for bio-cultural reasons, is cooperatively managed to ensure ecological integrity and achieve strong, long-term protection and perpetuation of Northwestern Hawaiian Island ecosystems, Native Hawaiian culture, and heritage resources for current and future generations (https://www.papahanaumokuakea.gov).

The Cook island's vast multipurpose marine park, "Marae Moana" ("ocean sanctuary"), covers their entire territorial sea and EEZ and includes a large whale sanctuary (Cook Islands Marae Moana Act 2017). The traditional "rahui concept" is integrated into the national law (Willaert, 2020; Tilot et al., 2021a), signifying the importance of the ocean space to Cook Islanders (Flood et al., 1999; International Union for Conservation of Nature, 2018). This holistic approach of Cook's national law in managing marine resources recognizes the connectivity between marine ecosystems and aims to promote shared use and sustainable development. National ownership and widespread support are ensured by promoting transparency and by involving the island communities in the decision-making processes (Tilot et al., 2021a).

5.2 Science-Based Marine Protected Areas and Other Measures of Protection and Management for Cetacean

Within the perspective of Western models for marine conservation (in contrast to traditional practices as previously described). Marine Protected Areas (MPAs) are particularly effective if these are large enough for marine mammals and serve as enforceable tools to safeguard pelagic ecosystems from anthropogenic activities, (Agardy, 1994; Jeudy de Grissac, 2003; Halpern et al., 2010; Duarte et al., 2020). They often are designed with a scientifically based ecosystem-based management approach, which appears best adapted to cetaceans, as it enables to maintain ecosystem integrity and minimize adverse natural and anthropogenic impacts.

The different area-based conservation measures for cetaceans, adopted by a country, a group of countries at a regional level or by an inter-governmental organisation, are defined as follows:

- Whale sanctuaries, which are designed to promote non-lethal scientific research; collaboration for information exchange, education and awareness initiatives; monitoring of whales; identification of future threats to whales and appropriate action to counter those threats; and development of mechanisms for the regular evaluation of management programmes in terms of appropriate objectives.
- Particularly Sensitive Sea Areas (PSSAs) for species, habitats and seascapes that are identified and established by the International Maritime Organisation (IMO). IMO's PSSA programme supports the UN Sustainable Development Goal 14. IMO also supports UNESCO World Heritage Marine Sites and Ecologically or Biologically Sensitive Areas (EBSAs) under the Convention on Biological Diversity.
- *Other Effective area-based Conservation Measures (OECMs) allow to achieve conservation objectives mainly as a by-product of other management strategies.* (https://biodiversity.europa.eu/europes-biodiversity/protected-areas/other-effective-area-based-conservation-measures).

Several organisations have also developed methodologies to assist the countries in identifying important sites for biodiversity conservation, which include the following:

- "Important Marine Mammal Areas" (IMMAs), developed by the IUCN Joint SSC/WCPA Marine Mammal Protected Areas Task Force. IMMAs are developed on the basis of critical aspects of marine mammal biology, ecology and population structure such as vulnerability, distribution, abundance, special attributes and key life cycle activities. The creation of a network of IMMAs represents a cost-effective approach to conservation. (https://www.marinemammalhabitat.org/immas

- "Ecologically and Biologically Significant Areas" (EBSAs) are recognized by the United Nations Convention on Biological Diversity (CBD). EBSAs are identified on the basis of the following scientific criteria: uniqueness or rarity, special importance for life stages of species, importance for threatened, endangered or declining species and/or habitats, vulnerability, fragility, sensitivity or slow recover, biological productivity, biological diversity and naturalness (https://www.cbd.int/ebsa).
- "Key Biodiversity Areas" (KBAs) have been identified through the IUCN Standard (Johnson & Weaver, 2014; IUCN, 2016). KBAs are the most diverse places in the world for species and their habitats (https://www.iucn.org/resources/conservation-tool/key-biodiversity-areas). As marine mammals are indicators of ocean ecosystem health, they will thus support existing EBSAs and KBAs designations as a basis for promoting environmental protection and developing management plans for specific areas in the world's oceans.
- "Vulnerable Marine Ecosystems" (VMEs) emerged as a concept from discussions at the United Nations General Assembly (UNGA) and gained momentum after UNGA resolution 61/105 in 2006. FAO adopted the concept for the management of deep-sea fisheries and produced guidelines (FAO DSF Guidelines) which are now used in the management of fisheries in the high seas, in Areas Beyond National Jurisdiction (ABNJ). VMEs are groups of species, communities or habitats, including seamounts, hydrothermal vents and cold water corals that may be vulnerable to impacts from destructive fishing activities. The management of their protection should be consistent with the precautionary approach and ecosystem approaches (Clark et al., 2016) (https://www.fao.org/in-action/vulnerable-marine-ecosystems). Presently, VMEs target also other destructive anthropogenic activities, such as deep-sea mining.
- "Oceanic Marine Protected Areas" (OMPAs) (defined here as waters >200 m depth) have emerged from the ratification of the United Nations High Seas Treaty (also known as the Biodiversity Beyond National Jurisdiction Treaty—BBNJ Treaty) (Blanluet et al., 2023).
- "Large Marine Protected Areas" (LMPAs) (defined for areas >100,000 km^2) have been created in the current decade and often set in networks (Corrigan & Kershaw, 2008; Game et al., 2009; O'Leary et al., 2012; Gianni et al., 2016), in particular recently with the ratification of the BBNJ Treaty. Evidence has shown that LMPAs are much more cost-effective to implement and manage than smaller MPAs (McCrea-Strub et al., 2010).

The advantage LMPAs is the interconnection of ecosystems (from the surface, and above, to pelagic and deep-sea ecosystems) favoring corridors for migratory species (whales, tunas, turtles, etc.). LMPAs also offer unique opportunities to protect cultural seascapes and to enable indigenous communities to maintain their traditional practices while participating in management (Bambridge & d'Arcy, 2014;

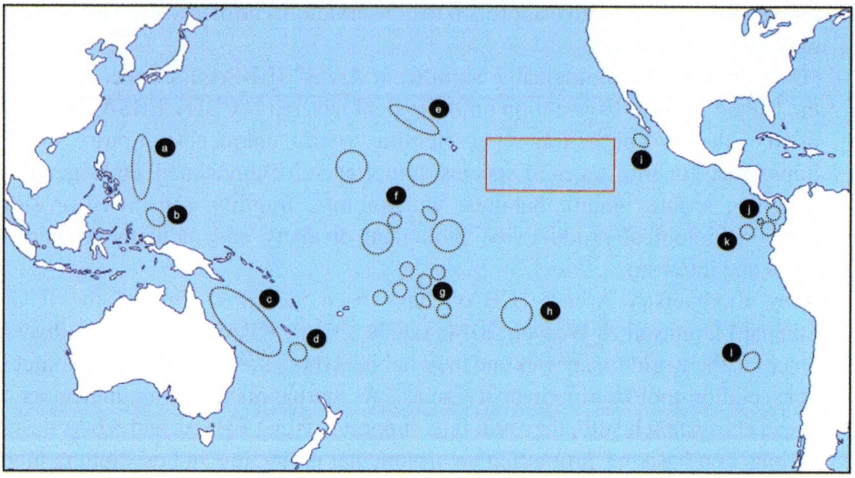

Fig. 14.5 Large marine protected areas in the inter-tropical Pacific ocean (Adapted from UNEP-WCMC and IUCN 2017). The CCZ is located in the red box. MPAs are lettered as follows: (a) Northern Mariana islands; (b) Palau marine mammal sanctuary; (c) Coral Sea conservation, Great Barrier Reef park; (d) New Caledonia whale sanctuary; (e) North-western Hawaiian islands (f) Pacific remote islands, American Samoa whale and turtle sanctuary; (g) Cook Islands whale sanctuary, Fiji whale sanctuary, Phoenix islands protected area (PIPA), Tokelau whale sanctuary, Samoa whale, turtle and shark sanctuary, Tonga whale sanctuary, Vanuatu marine mammal sanctuary; (h) French Polynesia marine mammal sanctuary; (i) Mexican whale refuge; (j) East Pacific marine conservation corridor, Costa Rica Dome, Ecuador whale natural refuge + Coasta Rica whale and dolphin sanctuary, Panama marine corridor (k) Galapagos marine resources reserve and whale sanctuary; (l) Motu Motiro Hiva marine park

Lewis et al., 2017). Concerning global warming, LMPAs with corridors and expandable boundaries and covering different oceanic temperatures and processes, are certainly better suited to marine mammals than restricted MPAs, such as those limited to a single marine region, which would probably lead them to extinction, in the absence of knowing precisely how species and ecosystems will adapt to climate change (Hoyt, 2011, 2021).

LMPAs could also be valuable for controlling most anthropogenic noise sources (high-intensity low-frequency sounds such as low-frequency active sonar or seismic surveys associated with oil and gas exploration), which are evidenced to impact whale populations (Hoyt, 2011). Special noise-threat buffer zones have proven to be efficient when large enough and set up around MPAs to provide adequate or precautionary distances between noise sources and known or suspected cetacean habitats (Engel in Agardy et al., 2007; Hoyt, 2021; Grau Tomas & Garcia Sanabria, 2022).

The main established LMPAs (>1000,00 km^2) in the tropical Pacific are displayed in Fig. 14.5 and the proposed and established high seas/EEZ and LMPAs in the three marine regions of the Pacific are displayed in Box 14.1.

Box 14.1: Established and proposed high seas/EEZ LMPAs >100,000 km^2 in the three marine regions of the Pacific (adapted from Hoyt, 2011)

1. North and South Pacific

United States:

- **American Samoa Whale and turtle sanctuary** (404,391 km^2): Humpback whales, *Megaptera novaeangliae,* sperm whale *Physeter macrocephalus;* false killerwhale, *Pseudorca crassidens.*
- **The national monument Papahânaumokuâkea** (362,075 km^2) **in the Hawaiian archipelago**: Spinner dolphin, *Stenella longirostris;* bottlenose dolphin, *Tursiops truncatus;* humpback whale, *Megaptera novaeangliae;* pantropical spotted dolphin, *Stenella attenuata;* killer whale, *Orcinus orca;* various beaked whale species.

  linked to Hawaiian islands humpback whale national marine sanctuary (3368km^2) proposed for extension: Humpback whale, *Megaptera novaeangliae,* bottlenose dolphin, *Tursiops truncatus;* spinner dolphin, *Stenella longirostris;* pantropical spotted dolphin, *Stenella attenuata;* rough-toothed dolphin, *Steno bredanensis;* false killer whale, *Pseudorca crassidens;* shortfinned pilot whale, *Globicephala macrorhynchus;* some 24 cetacean species.
- **Northern Mariana islands**: Mariana trench marine national monument (246,624 km^2): Sperm whales *Physeter macrocephalus*; humpback whale, *Megaptera novaeangliae;* up to 17 other cetacean species.
- **Pacific Remote Islands:** Pacific Remote Islands Marine National Monument (224.311 km^2): Sperm whale, *Physeter macrocephalus*; other whales and apex predators.

Australia: Coral sea conservation zone proposed for protection (972,000 km^2), dwarf mink whale, *Balaenoptera acutorostrata;* humpback whale, *Megaptera novaeangliae;* blue whale, *Balaenoptera musculus;* sperm whale, *Physeter macrocephalus;* melon-headed whale, *Peponocephala electra;* 28 cetacean species.

Cook Islands: Cook Islands vast multipurpose marine park "Marae Moana" (1,960,135 km^2) including a whale sanctuary: Humpback whale, *Megaptera novaeangliae;* sperm whale, *Physeter macrocephalus;* Blainville's beaked whale, *Mesoplodon densirostris;* Cuvier's beaked whale, *Ziphius cavirostris;* spinner dolphin, *Stenella longirostris;* Fraser's dolphin, *Lagenodelphis hosei;* killer whale, *Orcinus orca* and other cetacean species.

Fiji: Fiji whale sanctuary (1281122km^2)**:** Humpback whale, *Megaptera novaeangliae;* sperm whale, *Physeter macrocephalus;* spinner dolphin, *Stenella longirostris* and other species..

(continued)

Box 14.1: (continued)

France:

- **French Polynesia: French polynesia Marine Mammal sanctuary** (4,767,242 km^2): Humpback whale, *Megaptera novaeangliae,* spinner dolphin, *Stenella longirostris;* roughtoothed dolphin, *Steno bredanensis;* melon-headed whale, *Peponocephala electra;* Fraser's dolphin, *Lagenodelphis hosei;* 10 other species.
- **New Caledonia: New Caledonia whale sanctuary** (1,426,847 km^2): humpback whale, *Megaptera novaeangliae,* blue whale, *Balaenoptera musculus;* spinner dolphin, *Stenella longirostris;*bottlenose dolphin, *Tursiops* spp.*;* 10 others including beaked whales.

Kiribati: Phoenix islands protected area (PIPA) (410500km^2): Sperm whale *Physeter macrocephalus;* spinner dolphin, *Stenella longirostris*; bottlenose dolphin, *Tursiops truncatus;* other species.

One of the world's largest MPAs. It includes underwater mountains and other deep sea habitat, virtually untouched by humans. In 2009 a sister sanctuary relationship was established with the Papahanaumokuakea Marine National Monument in the Northwestern Hawaiian Islands.

New Zealand:

- **Tokelau Whale sanctuary** (319,031 km^2): Humpback whale, *Megaptera novaeangliae*; sperm whale, *Physeter macrocephalus*; spinner dolphin, *Stenella longirostris.*
- **NIUE whale sanctuary** (316,629 km^2): Humpback whale, *Megaptera novaeangliae*, year-round: spinner dolphin, *Stenella longirostris*; false killer whale, *Pseudorca crassidens.*

Palau: Palau Marine Mammal sanctuary (EEZ): (600,000 km^2): Sperm whale *Physeter macrocephalus*; at least 10 other cetacean species.

Papua New Guinea: Papua New Guinea Whale sanctuary (2,396,214 km^2): Sperm whale *Physeter macrocephalus,* sei whale, *Balaenoptera borealis;* Indo-Pacific humpback dolphin, *Sousa chinensis;* at least 10 other species.

Samoa: Samoa Whale, turtle and shark sanctuary (131,812 km^2): Humpback whale *Megaptera novaeangliae*; sperm whale, *Physeter macrocephalus*; spinner dolphin, *Stenella longirostris;* false killer whale, *Pseudorca crassidens*; 10 other species.

Tonga: Tongan whale sanctuary (664,853 km^2): Humpback whale *Megaptera novaeangliae*; sperm whale, *Physeter macrocephalus*; spinner dolphin, *Stenella longirostris*; 9 other cetacean species.

Vanuatu: Vanuatu Marine Mammal Sanctuary (639,391km^2): Humpback whale, *Megaptera novaeangliae;* spinner dolphin, *Stenella longirostris,* 2 other species.

(continued)

Box 14.1: (continued)

Chili: East of Easter Island: Motu Motiro Hiva Marine Park, proposed for expansion to cover (411,717 km^2): Similar to easter island Rapa Nui: Sperm whale *Physeter macrocephalus*; common dolphin, *Delphinus* spp.; *Stenella* and *Globicephala* spp.; humpback whale, *Megaptera novaeangliae*; pygmy blue whale, *Balaenoptera musculus brevicauda.*

2. North East Pacific

The eastern North Pacific population has the greatest abundance (2000–3000 individuals) and the healthiest of the 10 locations where blue whales are studied in the world (Bailey et al., 2010).

There are known blue whale habitats off California in and around the four large national marine sanctuaries within the US EEZ, with regular migrations to México and the Gulf of California (Calambokidis & Barlow, 2004; Bailey et al., 2010) and summer movements as far north as offshore British Columbia and Gulf of Alaska waters, particularly during El Niño years, when krill concentrations may have been reduced off California (Bailey et al., 2010).

Mexico: Mexican whale refuge (3,269,386 km^2), extends close to Clipperton island: Gray whale *Eschrichtius robustus*; humpback whale, *Megaptera novaeangliae*; Bryde's whale, *Balaenoptera edeni*; fin whale, *Balaenoptera physalus*; blue whale, *Balaenoptera musculus*; bottlenose dolphin, *Tursiops truncatus*; 39 total cetacean species found.

3.South East Pacific

- **Costa Rica, Nicaragua, El Salvador, Guatemala and México ± High seas: The high seas MPA Costa Rica Dome (970,000 km^2): (Proposed):** Unique year-round habitat for blue whales *Balaenoptera musculus* for feeding and breeding dwelling on highly productive seas due to nutrient-rich upwellings (Fiedler & Talley, 2006). There is also evidence that there is considerable migration with other whales. The Dome varies in size and position throughout the year, which is experienced in other upwelling areas targeted by conservation, e.g. « Pelagos, the Mediterranean marine sanctuary » (Tilot, 2004). Dynamic conservation measures have to be adapted to those areas with an open tri-dimensional approach.

  Other cetaceans dwell on the Costa Rica Dome: Pantropical spotted dolphin, *Stenella attenuata,* bottlenose dolphin, *Tursiops truncatus,* common dolphin, *Delphinus* spp., spinner dolphin, *Stenella longirostris*, Bryde's whale, *Balaenoptera edeni*, killer whale, *Orcinus orca*, rough-toothed dolphin, *Steno bredanensis*, pilot whales, *Globicephala* spp., Risso's dolphin, *Grampus griseus*, striped dolphin, *Stenella coeruleoalba*; Cuvier's beaked whale, *Ziphius cavirostris* and other beaked whales, *Mesoplodon* spp.
- **The East Pacific Conservation Marine Corridor (high seas, in the waters of Colombia, Costa Rica, Ecuador and Panama)** (2.110.000 km2), formerly the Eastern Tropical Pacific Seascape, the initiative includes a cor-

(continued)

Box 14.1: (continued)

ridor for migrating cetaceans. This corridor was agreed between the United Nations Foundation, Conservation International and the UNESCO World Heritage Centre. It is not an MPA per se but a marine management area of MPAs which functions as a network initiative connecting Cocos Island Marine and Terrestrial Conservation Area and Las Baulas National Marine Park (Costa Rica), Coiba Island National Park (Panamá), Gorgona National Park and Malpelo Island Flora and Fauna Sanctuary (Colombia), and Galápagos National Park, Marine Resources Reserve and Whale Sanctuary (Ecuador). The Ecuadorian fishing industry, however, has opposed the initiative on the grounds that it challenges fishing interests in the region.

- The cetacean present are: Blue whale *Balaenoptera musculus*; humpback whale, *Megaptera novaeangliae*; sperm whale, *Physeter macrocephalus*; various beaked whales and many others.
- <u>**Panama**</u>: Panama Marine Corridor (331,465 km^2): 20–25 cetacean species.
- <u>**Costa Rica**</u>**: Coasta Rican whale and dolphin sanctuary (220,856 km^2):** humpback whale, *Megaptera novaeangliae*; bottlenose dolphin, *Tursiops truncatus*; spinner dolphin, *Stenella longirostris;* 29 species of whales and dolphins.
- <u>**Ecuador:**</u>

  - **Whale Natural Refuge** (1,072,533 km^2): Humpback whale, *Megaptera novaeangliae*; Bryde's whale, *Balaenoptera edeni*; fin whale, *Balaenoptera physalus*; blue whale, *Balaenoptera musculus*; bottlenose dolphin, *Tursiops truncatus*; pantropical spotted dolphin, *Stenella attenuata*; spinner dolphin, *Stenella longirostris*; common dolphin, *Delphinus* spp.*;* killer whale, *Orcinus orca*; sperm whale, *Physeter macrocephalus*; various *beaked whales*.
  - ***Galapagos MPA*** **(158,000 km^2) Galapagos marine resources reserve and whale sanctuary:** A natural laboratory for evolution, high biodiversity among which waters for cetacean mating and breeding and migrating station. The cetacean are: bottlenose, striped and short-beaked common dolphins, short-finned pilot and sperm whales. Humpback whale, *Megaptera novaeangliae*; Bryde's whale, *Balaenoptera edeni*; sei whale, *Balaenoptera borealis*; blue whale, *Balaenoptera musculus*; bottlenose dolphin, *Tursiops truncatus*; pantropical spotted dolphin, *Stenella attenuata*; spinner dolphin, *Stenella longirostris;* striped dolphin, *Stenella coeruleoalba*; short-beaked common dolphin, *Delphinus delphis*; Fraser's dolphin, *Lagenodelphis hosei*; rough-toothed dolphin, *Steno bredanensis*; Risso's dolphin, *Grampus griseus*; melon-headed whale, *Peponocephala electra*; pygmy killer whale, *Feresa attenuata*; false killer whale, *Pseudorca crassidens*; killer whale, *Orcinus orca*; short-finned pilot whale, *Globicephala macrorhynchus*; beaked whales; ginkgo-toothed beaked whale, *Mesoplodon ginkgodens*; Blainville's beaked whale, *Mesoplodon densirostris*; sperm whale, *Physeter macrocephalus*; pygmy sperm whale, *Kogia breviceps*; dwarf sperm whale, *Kogia sima*; Cuvier's beaked whale, *Ziphius cavirostris*.

5.3 *Existing Area-Based Important Sites for Biodiversity Conservation in the CCZ*

Two large **EBSA**s are bordering the CCZ to the north and the south (Clark et al, 2014):

- The North Pacific Transition Zone, bounded to thermohaline fronts and eddies, a highly productive habitat and migratory corridor for many pelagic predators including endangered and commercially important species.
- The Equatorial High-Productivity Zone, a large-scale oceanic feature supporting high primary production over a large area and a strong benthic-pelagic coupling strongly related to surface primary productivity. Historically, high sperm whale abundance was recorded in this large-scale oceanographic feature, highly influenced by El Niño events and therefore potentially susceptible to climate change.

Both EBSAs are rated "high" for uniqueness, due to their large scale as well as being a major component of the oceanographic processes within the Pacific Ocean; they are also rated as "high" for primary productivity (https://www.cbd.int/ebsa).

In addition, the eastern Pacific hosts several Important Marine Mammal Areas (IMMAs) to the south and northwest of the CCZ, in addition to numerous Important Bird Areas (IBAs), in particular for petrels and shearwaters, overlapping with the CCZ (Fig. 14.6).

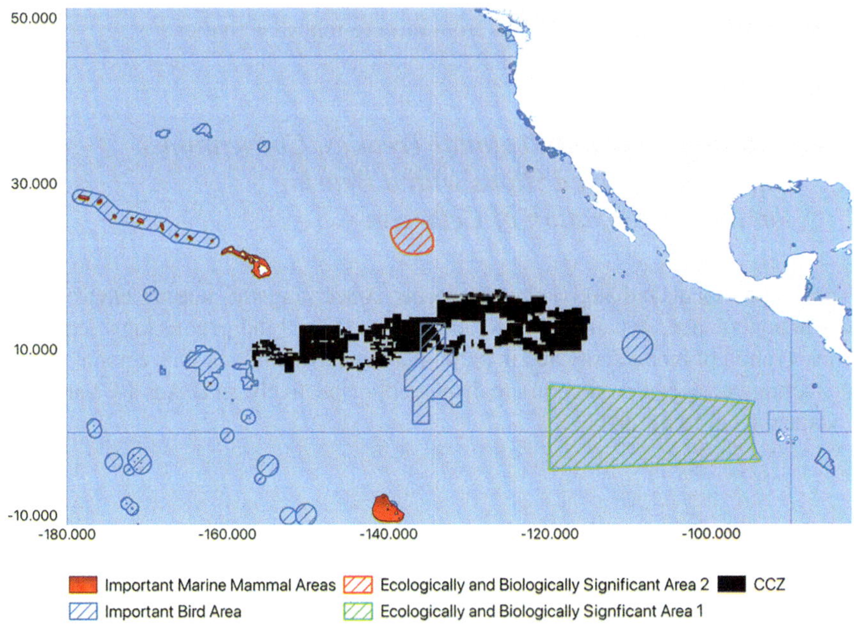

Fig. 14.6 The CCZ area and important biodiversity areas. EBSA 1: High Primary Productivity Area; EBSA 2: White Shark Offshore Aggregation Area

The International Seabed Authority (ISA) has designated thirteen "Areas of Particular Environmental Interest (APEI)" or "Impact Reference areas" (Smith et al., 2008; Wedding et al., 2015, 2017), in order to assist the ISA in making decisions concerning the modalities of nodule exploitation and its impact on the ecosystems. However, after analysis, some are not representative of the areas targeted for mining (Tilot, 2010; Tilot et al., 2018; Jones et al., 2021). They would be impacted by the plumes derived from mining operations in the water column, due to the topography and oceanographic processes and associated ecosystems (Tilot et al., 2018; Weaver et al., 2022).

The APEI network should be integrated in the management of the whole region, in particular to the network of LMPAs, to help mitigate the impacts of mining in the CCZ (Tilot, 2010; Tilot et al., 2018; Kriegl et al., 2021).

Concerning the protection of marine mammals, a network of flexible MPAs, preferably tridimensional LMPAs, with corridors and buffer zones and expandable boundaries, would be better, as discussed previously. These buffer zones should probably be expanded as "special noise-threat buffer zones" to mitigate the background sound issue with potential continuous deep-sea mining–related activities.

To enhance integrated environmental governance, the International Seabed Authority has been developing Regional Environmental Management Plans (REMPs) (Christiansen et al., 2022). Marine mammal Conservation strategies in the CCZ should be clearly integrated in these REMPs. The agreed upon High Seas Treaty could eventually provide a critical impetus for integrating sectoral management tools such as REMPS into the global biodiversity conservation agenda in Areas Beyond National Jurisdiction (ABNJ) with EBSAs, VMEs, IMMAs, LMPAs, PSSAs, OECMs in relation to risks from mining-related activities.

5.4 International and Regional Treaties, Conventions, Agreements and aAtions with a Bearing on the Conservation of Cetacean

The international and regional instruments concerning the marine environment, ocean governance, sustainable development, traditional indigenous knowledge, the conservation of biodiversity and in particular marine mammals and regional actions concerning cetacean protection are displayed, respectively, in Boxes 14.2 and 14.3 (adapted from Hoyt, 2011).

Box 14.2: The international and regional instruments concerning the marine environment, ocean governance, sustainable development, traditional indigenous knowledge, the conservation of biodiversity and in particular marine mammals (adapted from Hoyt, 2011)

International Instruments

- CITES—Convention on International Trade in Endangered Species of Wild Fauna and Flora
- CMS or Bonn Convention—Convention on the Conservation of Migratory Species of Wild Animals
- CBD—Convention on Biological Diversity and the Jakarta Mandate
- ICRW—International Convention for the Regulation of Whaling (convention responsible for establishing the International Whaling Commission—IWC)
- MARPOL Agreement—International Convention for the Prevention of Pollution from Ships put into practice by IMO (International Maritime Organization)
- UNCLOS—United Nations Convention on the Law of the Sea in particular the sections on Environment Regime and on Fisheries Regime and the new High Seas Treaty (adopted by the UN, June 19, 2023)
- UNEP (United Nations Environment Programme) Regional Seas Programme
- United Nations Agreement on Straddling Fish Stocks (1995)
- UNESCO Man and the Biosphere Programme (MAB)
- United Nations Conference on Environment and Development (UNCED): Agenda 21
- World Heritage Convention (WHC)
- World Summit on Sustainable Development (WSSD)

Regional Instruments

- Apia Agreement— he Agreement establishing the South Pacific Regional Environment Programme Apia, 1993
- Apia Convention—The Convention on Conservation of Nature in the South Pacific adopted in 1976, suspended in 2006 as replaced by the South Pacific Regional Environmental Programme SPREP.
- CCAMLR—Convention on the Conservation of Antarctic Marine Living Resources
- CNSP—Convention on Conservation of Nature in the South Pacific
- Lima Convention or CPPS—Convention for the Protection of the Marine Environment and Coastal Area of the South-East Pacific
- Noumea Convention—Convention for the Protection of the Natural Resources and Environment of the South Pacific Region and its protocols
- PCSP—Permanent Commission and Agreements of the Conference on the Use and Conservation of the Marine Resources of the South Pacific

(continued)

Box 14.2: (continued)

- SPREP—the Secretariat of the Pacific Regional Environment Programme established in 1982 and renamed in 2004 as the Secretariat of the Pacific Regional Environmental Programme
- Waigani Convention—Convention to Ban the Importation into Forum Island Countries of Hazardous and Radioactive Wastes and to Control the Transboundary Movement and Management of Hazardous Wastes within the South Pacific Region

Cetacean Instruments

- Pacific Cetaceans MoU—Memorandum of Understanding for the Conservation of Cetaceans and Their Habitats in the Pacific Islands Region

Box 14.3: Regional Actions Concerning Cetacean Protection

- **In the South East Pacific**
  - The Antigua Convention is a UNEP Convention for Cooperation in the Protection and Sustainable Development of the Marine and Coastal Environment of the North East Pacific.
  - The Lima Convention countries have agreed to create a regional network of marine and coastal protected areas (MCPAs) to strengthen management through capacity building and networking. The Southeast Pacific regional network includes 91 MCPAs covering 3 million km^2 (CPPS, 2010). The Lima Convention implementation is now one of the responsibilities of the SPREP.
- **In the North East Pacific**

The Commission for Environmental Cooperation (CEC) established the North American MPA Network (NAMPAN) and adopted the North East Pacific Marine Region as a pilot project for the Baja to Bering (B2B) Marine Conservation Initiative.

- **In the South Pacific**

An International High Seas South Pacific Whale sanctuary, which would cover most of the South Pacific Ocean and adjoin the Southern Ocean Sanctuary, has been proposed by the SPREP in 2001 to protect the heavily whaled populations recovering in the Pacific. It has been rejected by the International Whaling Commission. However, it could be now considered under the new High Seas Treaty.

- **Antarctica and its surrounding seas**

The Southern Ocean Sanctuary, which includes 7 baleens and 13 toothed whales, is part of a portfolio of treaties and other measures to protect Antarctica and its surrounding seas.

The Secretariat of the Pacific Regional Environment Programme (SPREP) is the regional organisation established by the governments and administrations of the Pacific charged with protecting and managing the environment and natural resources of the Pacific. It includes 21 Pacific island member countries and territories (American Samoa, Cook Islands, Federated States of Micronesia, Fiji, French Polynesia, Guam, Kiribati, Marshall Islands, Nauru, New Caledonia, Niue, Northern Marianas, Palau, Papua New Guinea, Samoa, Solomon Islands, Tokelau, Tonga, Tuvalu, Vanuatu and Wallis & Futuna), and 5 metropolitan countries (Australia, France, New Zealand, the United Kingdom and the United States) with direct interests in the region (https://www.sprep.org/about-us).

The Pacific Cetaceans Memorandum of Understanding (MoU) for the Conservation of Cetaceans and their Habitats in the Pacific Island Region has been concluded under the auspices of the Convention on Migratory Species of Wild Animals (CMS), and in collaboration with the Pacific Regional Environment Programme (SPREP). The MoU provides an international framework of 22 states for coordinated conservation efforts to improve the conservation status of the Pacific Islands Cetaceans and came into effect on 15 September 2006 (https://www.sprep.org/about-us).

These two organisations are the most appropriate to host and implement a Pacific marine mammals Conservation Agreement that could look like the one developed for the Mediterranean and Black Seas: the Agreement for the Conservation of Cetaceans in the Black and Mediterranean Seas (ACCOBAMS), presented in the following section.

5.5 *ACCOBAMS, a Potential Model for a Pacific Cetaceans Conservation Agreement*

The Agreement on the Conservation of Cetaceans of the Black Sea, Mediterranean Sea and contiguous Atlantic area (ACCOBAMS)* is a legal conservation tool based on cooperation between 24 state parties (Jeudy de Grissac, 2015). Its purpose is to reduce threats to cetaceans notably by improving current knowledge on these animals. This intergovernmental agreement provides the demonstration of the commitment of riparian Countries to preserve all species of cetaceans and their habitats within the geographical Agreement area by the enforcement of more stringent measures than those defined in the texts adopted previously.

* ACCOBAMS being a formal agreement between 24 countries for the conservation of cetaceans, the source is not the report of the scientific committee, but the website of this agreement. The content of the agreement being approved by Contracting Parties in formal meetings, the approved wording or structure or decisions have to be quoted as they have been written without any change. The most

recent meeting of Parties took place in 2022. The relevant information is displayed in Boxes 14.4, 14.5, 14.6, 14.7, 14.8, 14.9.

Box 14.4: ACCOBAMS Introduction (https://accobams.org/about/introduction/)

ACCOBAMS results from consultations between Secretariats of four Conventions:

- The **Barcelona Convention** for the Protection of the Marine Environment and the Coastal Region of the Mediterranean and its Protocol concerning Specially Protected Areas and Biological Diversity in the Mediterranean.
- The **Bonn Convention** on the Conservation of Migratory Species of Wild Animals.
- The **Bern Convention** on the Conservation of European Wildlife and Natural Habitats.
- The **Bucharest Convention** on the Protection of the Black Sea Against Pollution.

Finally, given the migratory characteristics of these species, the Agreement was established under the auspices of the Bonn Convention (UNEP/CMS). ACCOBAMS was signed on 24 November 1996 and entered into force on 1 June 2001.

The Agreement Area consists of all the maritime waters of the Black Sea, the Mediterranean and the contiguous Atlantic area West of the Straits of Gibraltar. The Area includes the "Pelagos Sanctuary", dedicated to marine mammals in the North-West Mediterranean and established by France, Italy and Monaco (Tilot, 2004; Nortarbartolo di Sciara et al., 2008).

ACCOBAMS is the first Agreement binding the Countries of these sub-regions to work together for cetacean conservation. One innovative aspect of this Agreement is to include non-riparian countries whose maritime activities are likely to jeopardize cetacean conservation. In 2010, Parties to ACCOBAMS adopted a Resolution to extend the geographical scope of the Agreement Area to the Exclusive Economic Zones of Spain and Portugal in the Atlantic Ocean.

In 2022, the Agreement had 24 Parties: Albania, Algeria, Bulgaria, Croatia, Cyprus, Egypt, France, Georgia, Greece, Italy, Lebanon, Libya, Malta, Monaco, Montenegro, Morocco, Portugal, Romania, Slovenia, Spain, Syria, Tunisia, Turkey and Ukraine.

In addition, the Scientific Committee is preparing draft resolutions concerning cetacean conservation to be delivered to the Secretariat, then discussed and amended by the Parties before approval. The following table presents the main topics covered by resolutions for conservation actions (https://accobams.org/documents-resolutions/resolutions/).

Box 14.5: ACCOBAMS Strategy (https://accobams.org/documents-resolutions/accobams-strategy/)
The ACCOBAMS Strategy

The strategy's general objective is to improve the conservation status of cetaceans and of their habitats in the area of competence of the Agreement by 2023. The strategy incudes 10 specific objectives, 5 for the good management of the agreement (MA) and 5 for the conservation of cetacean and their habitats (CA).

For the **good management of the Agreement**, the five specific objectives are:

A1 Improve communication across, up and down ACCOBAMS as an organization
A2 Strengthen involvement of all key stakeholders in ACCOBAMS's operations, including all riparian countries
A3 Ensure adequate funding, particularly for conservation activities
A4 Improve the level of implementation of and compliance with ACCOBAMS resolutions as well as the monitoring of its progress
A5 Ensure implementation of the ACCOBAMS's cetacean conservation standards in the adjacent areas in close cooperation with other CMS instruments

For the **cetacean's conservation actions**, the 5 specific objectives are:

B1 Improve the knowledge about the state of cetaceans
B2 Reduce human pressures on cetaceans, particularly those related to interaction with fisheries, habitat loss and degradation
B3 Enhance public awareness about cetaceans
B4 Improve capacities of national organizations and experts
B5 Enhance effective conservation of cetaceans'critical habitats

The constituency of ACCOBAMS includes eight elements and the implementation of the strategy is linked to the activities of each constituent of the agreement.

- The Parties (Ps, represented by their National Focal Points - NFPs).
- The Bureau (B) and the Extended Bureau (EB).
- The Permanent Secretariat (PS—assisting the Parties).
- The Follow-up Committee (FC—evaluating the compliance).
- The Scientific Committee (SC—defining the scientific strategy).
- The Meeting of National representatives.
- The two sub-regional coordination units (SRCUs) linked to the areas of competence of the Barcelona Convention for the Mediterranean and the Bucharest Convention for the Black Sea.

(continued)

Box 14.5: (continued)

- The partners, divided into three categories.
  - The collaborating or cooperating partners/entities (CPs),
  - The international partners (IPs).
  - The national partners (NPs, agreed upon by and reporting to each Party).

Meetings of the Parties take place every 3 years, every year for the Scientific Committee. The Scientific Committee is responsible for the preparation of guidelines, best practices and procedures related to the conservation of cetaceans in the area of competence of ACCOBAMS

Box 14.6: Guidelines, best practices and procedures related to the conservation of cetaceans in the area of competence of ACCOBAMS (https://accobams.org/documents-resolutions/guidelines/)
ACCOBAMS guidelines (G), best practices (BP) and procedures (P) related to the conservation of cetaceans

- G for technical measures to minimize cetaceanfishery conflicts
- G for BP and P for addressing cetacean mortality events related to chemical, acoustic and biological pollution
- G for comprehensive population estimates and distribution
- G for a coordinated cetacean stranding response during mortality events caused by infectious agents and harmful algal blooms
- G for implementing a label for commercial whalewatching activities
- G for commercial cetaceanwatching activities
- G for the criteria for the selection of protected areas
- G for the establishment and management of marine protected areas for cetaceans
- G for the development of national networks of cetacean strandings monitoring
- G for a coordinated cetacean stranding response
- G for the establishment of tissue banks and ethical code
- G for the purpose of nonlethal in situ research
- G for the use of acoustic deterrent devices
- G for the release of captive cetaceans into the wild
- G to address the impact of anthropogenic noise on marine mammals

(continued)

Box 14.6: (continued)

Best practices and procedures

- BP and P for addressing cetacean mortality events related to chemical, acoustic and biological pollution
- BP on population genetics
- BP to assess the impact of chemical pollution on cetaceans and to measure the chemical contamination on cetaceans
- BP for cetacean post-mortem investigation and tissue sampling and for the development of diagnostic frameworks for specific threats to be investigated during cetaceans' strandings

Presented here as a model, are the whale watching guidelines that have been prepared by ACCOBAMS for the Mediterranean and the Black Seas, with the objective to limit the interferenc, disturbance and impacts on marine mammals of the region, which are facing a strong increase in the development of tourism activity (ACCOBAMS, 2023).

Box 14.7: Main topics covered by the resolutions for conservation Actions by ACCOBAMS (https://accobams.org/conservations-action)
Conservation Actions (CA)

CA1- Concerning the improvement of knowledge for conservation

1.1- Centralize, organize and disseminate the existing knowledge on cetaceans, their habitat, the pressures and impacts, the national institutions, legislations and capacities

1.2- Identify the gaps in knowledge and propose actions or programmes to improve the knowledge on cetaceans

CA2- Concerning the development of conservation activities and management measures

2.1- Identify the pressures on cetaceans and propose measures to reduce the impacts on the species or their habitats

2.1.1 Interaction Between Fisheries and Cetaceans

Guidelines for the Use of Acoustic Deterrent Devices
Pelagic Gillnets
Assessment and Mitigation of the Adverse Impacts of Interactions between Cetaceans and Fishing Activities
Prey Depletion
Use of Driftnets

(continued)

Box 14.7: (continued)

2.1.2 Anthropogenic Noise

Assessment and Impact Assessment of Man-Made Noise
Guidelines to Address the Impact of Anthropogenic Noise on Marine Mammals

2.1.3 Ship Strikes
2.1.4 Cetacean Watching

Commercial Whale-Watching: Towards a Label
Guidelines for Commercial Cetacean-Watching Activities
Certification for Highly Qualified Marine Mammals Observers

2.1.5 Climate Change
2.1.6 Captivity and Release

Dolphin Interaction Programmes
Guidelines on the Release of Cetaceans into the Wild
Live Removals of species

2.1.7 Functional Stranding Networks and Responses to Emergency Situations

Cooperation Between National Networks of Cetacean Strandings and the Creation
of a Database
Facilitation of Exchange of Tissue Samples
Guidelines for a Coordinated Cetacean Stranding Response
Cetacean Live Strandings
Best Practices in Monitoring and Management of Cetacean Stranding
Cetacean Stranding Networks

2.1.8 Marine Litter

Marine Litter and Chemical Pollution

2.2 *Identify cetaceans' critical habitats and based on the existing pressures propose changes in the national legislation (environment, fisheries, maritime transport,* etc.*) and support the implementation of area-based conservation measures*

2.2.1 Species Conservation Plans

Conservation Plans for Specific Species or Specific Areas
Conservation Management Plans

2.2.2 Protected Areas for Cetaceans

Marine Protected Areas of Importance for Cetacean Conservation
New Areas of Conservation of Cetacean Habitats

2.2.3 Citizen Science

Box 14.8: The "ACCOBAMS Survey Initiative" (ASI) (https://accobams.org/use-of-new-technologies/)

The "ACCOBAMS Survey Initiative" (ASI) was an unprecedented regional effort on cetacean monitoring in the Mediterranean and Black Seas, developed and implemented by the Permanent Secretariat of ACCOBAMS, which also ensures the overall coordination of the project, with the support of a Project Steering Committee composed of the Regional Activity Centre for Specially Protected Areas (UNEP-MAP-SPA/RAC), the IUCN Centre for Mediterranean Cooperation, the French Agency for Biodiversity, the Italian Institute for Environmental Protection and Research and the PELAGIS Observatory of the University of La Rochelle in France.

The overall objective of the project was to ensure at the macro-regional level that public and private stakeholders in the Mediterranean/Black Sea region "have a coherent monitoring system for the species concerned in the Mediterranean/Black Sea ecoregion, based on objective, robust and comparable data, with a view to improving the conservation status of these species and their habitats through appropriate management for the good status of the marine and coastal environments in the area of the study.

The ASI-specific objectives are to contribute primarily to:

- Providing a deeper understanding of the abundance and distribution of the species in question at the Mediterranean Sea/Black Sea macro-regional level, based on undisputed data
- Strengthening international cooperation and synergy between the countries of the macro-region to optimise and increase the efficiency of their efforts to establish a transnational approach to conserving the species concerned
- Characterising the good environmental status of the study zone
- Maintaining and/or restoring this good status as well as the critical and preferred habitats of the species in question, both marine and coastal
- Progressing research/development in this field
- Providing guidance on the measures required to adapt to climate change

The ASI has seen the participation of 20 Mediterranean countries, over 100 scientific observers and over 30 organizations. Eight planes and six boats were mobilized, with respectively 70,000 km of aerial survey, and more of 22,000 km of boat survey.

5.6 Research on Cetacean Populations and Monitoring

Populations of whales, in particular blue whales, have been studied using acoustic, line-transect and photo-identification techniques in different parts of their range (e.g., Stafford et al. 1998, 1999, 2006; Calambokidis & Barlow, 2004), but little is known of their fine-scale movements, migratory routes and speeds and present changes in spatio-temporal patterns (Bailey et al., 2010; Darling et al., 2022).

Box 14.9: ACCOBAMS whale-watching guidelines (https://accobams.org/wp-content/uploads/2023/01/MOP8.Doc31_Annex13_Res8.19.pdf)
ACCOBAMS Whale-watching guidelines

Nine topics are considered to evaluate the potential impacts:

- Measures to assess target populations and the potential impacts of tourism, including the concept of "carrying capacity".
- Monitoring and adaptive management.
- Development of effective management teams (stakeholder involvement).
- Licensing or certification measures.
- Measures to regulate approaches, frequency, length and type of exposure in encounters with cetaceans.
- Monitoring and enforcement.
- Time/area closures to provide additional protection.
- Promoting education and awareness raising.
- Whale watching as a platform of opportunity for scientific data collection.

An overview of the types of studies that have been conducted to monitor and measure the potential impact of whale-watching activities on cetacean populations is displayed in Table 14.2.

Also, the development of tag design and attachment methods has enabled to record year-round movements and fine-scale movements (Bailey et al., 2010). In particular, the Tagging Of Pelagic Predators (TOPP) programme has proven to be most relevant to monitor marine mammal communities in the area as whales are part of the sentinel migratory species that are tagged and monitored at different spatio-temporal scales in the eastern Pacific (Block et al., 2003, 2011). With the TOPP programme, whales carry a variety of archival electronic tags (pressure-time at depth, temperature, external environment, internal body, light level, and geolocation), that allow to record a complete set of parameters in three dimensions, physical and chemical parameters, and deduce oceanographic, biochemical and productivity parameters in pelagic and aerial ecosystems. As marine mammal communities dive into the mesopelagic, bathyal and, sometimes, abyssopelagic layers of the water column, they are able to monitor vertical water depths where human monitoring is restricted to high technology and costs.

The TOPP programme operates in collaboration with other ocean monitoring initiatives such as GLOBEC, and the IGBP's core project, initiated by the IOC of UNESCO in 1991. It is based on synergism between scientists, industry, and governments, using animals as ocean sensors to understand how global change affects the abundance, diversity and productivity of marine populations (Hofman et al., 2002).

One could debate on the fact that this approach, using the environment and animals as sensors, somewhat mirrors Pacific Islanders' "sensorial navigation" as their world is conceived as a vast kinship network with continual interactions (Lohmann et al., 2008; Bambridge, 2016; Tilot et al., 2021b, 2023).

Table 14.2 Monitoring methods used to assess and monitor the potential impacts of whale watching activities on cetacean populations

Category of study	Research platforms	Indicator or measure of potential impact
Behavioural responses linked to the presence of whale-watching vessels	Observations of behaviour from research vessels, shore-based observation stations, tethered balloons, to observe actual whale-watching vessels and reactions or experimental approaches and reactions	Changes in diving or surfacing patterns Changes in swimming speed and direction Changes in time spent feeding Changes in time spent resting
Behavioural responses linked to noise	Passive acoustic monitoring or vessel-operated hydrophones	Changes in vocal behaviour
Modelling of how behaviour changes translate into potential long-term impacts on fitness	Studies that measure behaviour changes above include modelling of long-term and cumulative effects persistent shifts in energy budgets	Potential energy deficits accrued over time by decreased feeding or resting, or increased traveling or surface-active behaviour
Monitoring vessel behaviour around target species	Observations can be made from a research vessel, or land-based station, or by researchers (anonymously) placed on whale-watching vessels	Rates of categories of vessel behaviour that fall within, or outside of established standards in terms of distance, speed, approaches to species or group categories that are off limits, etc.
Monitoring tourist perceptions of whale watching experiences	Interview surveys are conducted with tourists following their participation in whale watching tours before and after surveys can be used to measure knowledge gain or change in awareness or attitudes	Rates of satisfaction or dissatisfaction with elements of whale watching experiences, or changes in levels of awareness, knowledge or motivation after participating in whale watching

Adapted from ACCOBAMS, MOP8, 2022, Doc 31, Annex13, Res. 8.19, Annex 1 (ACCOBAMS, 2023)

The TOPP framework, is therefore an asset for monitoring the biodiversity, critical habitats, distribution, migrations and abundance of key species, such as marine mammal communities and their associated ecosystems in the oceans (Block et al., 2011; Harrison et al., 2018). Further, this standardized method, presently applied worldwide, enables comparisons, benchmarking and threshold values determination between different oceanic regions of the world, all necessary information for the development of adapted resource management purposes.

6 Conclusions and Recommendations

Humpback whales, and potentially other unrecorded marine mammals, regularly cross the CCZ during their migrations, as they sometimes travel to multiple breeding and feeding grounds within a year, even if migratory patterns to preferential areas are modified on spatial and temporal scales as an effect of climate change.

Cetaceans play a key role in marine ecosystem services as whale pumps for geochemical and oceanic nutrient cycling supporting primary production and fisheries and for carbon cycling supporting climate regulation. Whale carcasses are also a crucial source of organic carbon and nutrients for deep-sea ecosystems. Further, Pacific Islanders assign a great spiritual value to marine mammals, and their recreational value, for instance through whale-watching, as acknowledged worldwide.

The majority of marine mammal species found in the high seas are impacted by multiple anthropogenic stressors as entanglement in fishing gear (bycatch, ghost gear..), ship strikes, chemical, plastic and waste pollution, underwater noise, competition with fisheries over preys. With the additional impact of climate change and cumulative pressures, cetaceans are more subjected to loss of preferential habitats and preys, potential nutritional deficiencies, increased intra and inter-specific competition, genetic alterations, reduced reproductive success causing increased mortality and ultimately a risk of their extinction.

Deep-sea mining would be a supplementary source of stress, through increased collision risk with mining and cargo vessels and offshore treatment platforms; through near constant auditory masking throughout the water column, hindering communication, echolocation and prey capture; and through increased risk of bioaccumulation of toxic components, such as heavy metals.

Key recommendations for the protection of marine mammal communities in the CCZ are the following:

- To support the development of high-resolution measurements to enhance knowledge on marine mammal population distribution and functioning in the CCZ and in relation with the other regions of the Pacific.
- To identify and implement innovative dynamic and seasonal ocean management measures across critical habitats.
- To prioritize research on tolerance and responses of marine mammals towards increased natural and anthropogenic pressures, in particular deep sea mining, identifying thresholds for management.
- To link actions for marine mammal populations conservation/management in the CCZ to a regional basis in NE and SE tropical Pacific.
- To promote a network of LMPAs with corridors, with flexible or expandable boundaries for the cetacean contributing to ecological connectivity in the region protecting essential habitats and migratory routes on a regional scale for marine mammals, with the aim of protecting 30% of the Pacific in line with what was agreed at the Convention on Biological Diversity at COP15.
- To link actions to existing initiatives in the 3 marine regions of the Pacific.

- To implement effective ocean management and cooperative arrangements within the Global Oceans Treaty at the United Nations (BBNJ) and into the global biodiversity conservation agenda in Areas Beyond National Jurisdiction (ABNJ) with EBSAs, VMEs, IMMAs, LMPAs, PSSAs, OECMs in relation to risks from mining-related activities.
- To support the full, effective and equitable participation of Indigenous Peoples and local communities (IPLCs) in governance in ABNJ and in the design and implementation of MPA networks and OECMs such as in the Cook Islands, the network of LMMAs in Fiji and the national monument Papahânaumokuâkea of the Hawaiian archipelago.
- To reduce cumulative threats through cooperation from governments, industry and individuals in the eastern Pacific and safeguard whale migration in collaboration with the International Whaling Commission (IWC) and the International Maritime Organization (IMO).
- To enforce noise exclusion regulations and large special noise-threat buffer zones to provide precautionary distances between noise sources from deep sea mining and cetacean habitats.
- To reduce plastic and other pollution, including supporting the new UN Plastics Treaty.
- To achieve zero whales entanglements by fisheries.
- To eliminate and clean up ghost gear including supporting the Global Ghost Gear Initiative.
- To reroute shipping lanes away from critical whale habitats including seasonal migration areas, adapting to new migratory patterns in space and time due to climate change.

Acknowledgements Special thanks to Ulysses Shen, first grader at Sligo Creek Elementary school, Siver Spring, Maryland, USA for the first input to the paper drafting fact sheets on basis of Sophie Webb's book « Far from shore, chronicles of an open voyage » and the program for schools from NOAA's « Ocean for Life/National Marine Sanctuaries (https://sanctuaries.noaa.gov) and webinars with Claire Fackler from NOAA's National Education Liason.

Many thanks to Ganiyat Saliu (Vrije Universiteit Brussels) for proofreading and formatting the bibliography.

Conflict of Interest Statement The authors declare that the research was conducted in the absence of any commercial or financial relationships that could be construed as a potential conflict of interest.

References

Albouy, C., Delattre, V., Donati, G., Frölicher, T. L., Albouy-Boyer, S., Rufino, M., et al. (2020). Global vulnerability of marine mammals to global warming. *Scientific Reports, 10*, 548. https://doi.org/10.1038/s41598-019-57280-3

Académie des Sciences d'Outre-Mer (ASOM). (2010). Recommendations on the conservation and management of biodiversity of three deep-sea ecosystems targeted by mining (polymetallic

nodules, cobalt rich ferro-manganese crusts and hydrothermal sulfurs). In *Proceedings of the international workshop organized at UNESCO, 15 December 2010.*
Acevedo-Gutiérrez, A., Croll, D. A., & Tershy, B. R. (2002). High feeding costs limit dive time in the largest whales. *The Journal of Experimental Biology, 205*, 1747–1753. https://doi.org/10.1242/jeb.205.12.1747
Agreement on the Conservation of Cetaceans Of the BlAck Sea, Mediterranean Sea and contiguous Atlantic area (ACCOBAMS). (2023). *Resolution 8.19. Commercial cetacean-watching activities in the ACCOBAMS AREA*. https://accobams.org/wp-content/uploads/2023/01/MOP8.Doc31_Annex13_Res8.19.pdf
Agardy, T. M. (1994). Advances in marine conservation: the role of marine protected areas. *Trends in Ecology & Evolution, 9*, 267–270. https://doi.org/10.1016/0169-5347(94)90297-6
Agardy, T., Aguilar, N., Cañadas, A., Engel, M., Frantzis, A., Hatch, L. et al. (2007). A global scientific workshop on spatio-temporal management of noise. *Report of the Scientific Workshop, Puerto Calero,* Lanzarote June 4–6, Dokumente des Meeres GmbH, Darmstadt. 44 pp.
Ardron, J., Gjerde, K., Pullen, S., & Tilot, V. (2008). Marine spatial planning in the high seas. *Marine Policy, 32*, 832–839. https://doi.org/10.1016/j.marpol.2008.03.018
Avila, I., Kaschner, K., & Dormann, C. F. (2018). Current global risks to marine mammals: taking stock of threats. *Biological Conservation, 221*, 44–58. https://doi.org/10.1016/j.biocon.2018.02.021
Bailey, H., Mate, B. R., Palacios, D. M., Irvine, L., Bograd, S. J., & Costa, D. P. (2010). Behavioural estimation of blue whale movements in the Northeast Pacific from state-space model analysis of satellite tracks. *Endangered Species Research, 10*, 93–106. https://doi.org/10.3354/esr00239
Baird, R. W., James, J., Mata, C., & Hughes, M. (2022). Two gray whale (*Eschrichtius robustus*) sightings off Hawai'i Island: The first records for the Central Tropical Pacific. *Aquatic Mammals, 48*, 432–435. https://doi.org/10.1578/am.48.5.2022.432
Bambridge, T. (2016). *The Rahui: Legal pluralism in Polynesian traditional management of resources and territories* (269 pp). Pacific Series. Australian National University Press. ISBN: 9781925022797.
Bambridge, T., & D'Arcy, P. (2014). Large-scale marine protected areas in the Pacific: Cultural and social perspectives. In F. Feral & B. Salvat (Eds.), *Gouvernance, enjeux et mondialisation des grandes aires marines protegées: recherche sur les politiques environnementales de zonage maritime, le challenge maritime de la France de Mediterranée et d'Outre-mer* (pp. 113–132). l'Harmattan.
Bambridge, T., Gaulme, F., Montet, C., & Paulais, T. (2019). *Communs et oceans. Le Rahui en Polynésie* (p. 175). Au vent des îles.
Benson, S. R., Croll, D. A., Marinovic, B. B., Chavez, F. P., & Harvey, J. T. (2002). Changes in the cetacean assemblage of a coastal upwelling ecosystem during El Niño 1997–98 and La Niña 1999. *Progress in Oceanography, 54*, 279–291. https://doi.org/10.1016/S0079-6611(02)00034-X
Blanluet, A., Game, E. T., Dunn, D. C., Everett, J. D., Lombard, A., & Richardson, A. (2023). Evaluating ecological benefits of oceanic protected areas. *Trends in Ecology and Evolution, 39*(2), 175–187. https://doi.org/10.1016/j.tree.2023.09.003
Blevins, C., Busquets-Vass, G., Pardo, M. A., Gendron, D., Jacobsen, J. K., Gómez-Díaz, F., et al. (2022). Sex- and age-specific migratory strategies of blue whales in the northeast Pacific Ocean. *Frontiers in Marine Science, 9*(2022). https://doi.org/10.3389/fmars.2022.944918
Block, B. A., Costa, D. P., Boehlert, G. W., & Kochevar, R. E. (2003). Revealing pelagic habitat use: the tagging of Pacific pelagics program. *Oceanologica Acta, 25*, 255–266. https://doi.org/10.1016/S0399-1784(02)01212-4
Block, B., Jonsen, I., Jorgensen, S., et al. (2011). Tracking apex marine predator movements in a dynamic ocean. *Nature, 475*, 86–90. https://doi.org/10.1038/nature10082
Branch, T. A., Stafford, K. M., Palacios, D. M., Allison, C., Bannister, J. L., Burton, C. L. K., et al. (2007). Past and present distribution, densities and movements of blue whales *Balaenoptera musculus* in the southern hemisphere and northern Indian Ocean. *Mammal Review, 37*, 116–175. https://doi.org/10.1111/j.1365-2907.2007.00106.x

Calambokidis, J., & Barlow, J. (2004). Abundance of blue and humpback whales in the eastern North Pacific estimated by capture-recapture and line-transect methods. *Marine Mammal Science, 20*, 63–85. https://doi.org/10.1111/j.1748-7692.2004.tb01141.x

Christiansen, S., Durussel, C., Guilhon, M., Singh, P., & Unger, S. (2022). Towards an Ecosystem Approach to management in Areas Beyond National Jurisdiction: REMPs for Deep Seabed Mining and the proposed BBNJ instrument. *Frontiers in Marine Science Section Marine Affairs and Policy, 9-2022*. https://doi.org/10.3389/fmars.2022.720146

Clapham, P. (2001). Why do baleen whales migrate? A response to Corkeron and Connor. *Marine Mammal Science, 17*(2), 432–436. https://doi.org/10.1111/j.1748-7692.2001.tb01289.x

Clapham, P. (2018). Humpback Whale: *Megaptera novaeangliae*. In *Encyclopedia of Marine Mammals* (3rd ed., pp. 489–492). Academic. https://doi.org/10.1016/B978-0-12-804327-1.00154-0

Clark, M. R., Schlacher, T. A., Rowden, A. A., Stocks, K. I., & Consalvey, M. (2012). Science priorities for Seamounts: Research links to conservation and management. *PLoS One*. https://doi.org/10.1371/journal.pone.0029232

Clark, M. R., Rowden, A. A., Schlacher, T. A., Guinotte, J., Dunstan, P. K., & Williams, A. (2014). Identifying Ecologically or Biologically Significant Areas (EBSA): A systematic method and its application to seamounts in the South Pacific Ocean. *Ocean and Coastal Management, 91*(2014), 65–79. https://doi.org/10.1016/j.ocecoaman.2014.01.016

Clark, M. R., Watling, L., Rowden, A., Guinotte, J. M., & Smith, C. R. (2016). A global seamount classification to aid the scientific design of marine protected area networks. *Ocean and Coastal Management, 54*(2011), 19–36. https://doi.org/10.1016/J.OCECOAMAN.2010.10.006

Constantine, R., Iwata, T., Nueukirk, S., & Penry, G. (2018). Future directions in research on Bryde's whales. *Frontiers in Marine Science Section Marine Megafauna, 5*. https://doi.org/10.3389/fmars.2018.00333

Cook Islands Marae Moana Act. (2017). Available online at: www.seabedmineralsauthority.gov.ck/s/Marae-Moana-Act-2017.pdf

Corkeron, P. J., & Connor, R. C. (1999). Why do baleen whales migrate? *Marine Mammal Science, 15*, 1228–1245. https://doi.org/10.1111/j.1748-7692.2001.tb01289.x

Corrigan, C., & Kershaw, F. (2008). *Working toward high seas marine protected areas: An assessment of progress made and recommendations for collaboration*. UNEP/WCMC.

Committee on the Status of Endangered Wildlife in Canada (COSEWIC). (2021). *Annual report*. https: https://www.canada.ca/en/environment-climate-change/services/species-risk-public-registry/publications/canadian-wildlife-species-risk-2021.html

CPPS (Convention for the Protection of the Marine Environment and Coastal Area of the South-East Pacific). (2010). Red regional de áreas costeras y marinas protegidas del Pacífico Sudeste, Comisión Permanente del Pacífico references 401 Sur, Guayaquil, Ecuador, 26pp. www.cpps--int.org/spanish/planaccion/AMCP_version2010.pdf

Croll, D. A., Marinovic, B., Benson, S., Chavez, F. P., Black, N., Ternullo, R., et al. (2005). From wind to whales: Trophic links in a coastal upwelling system. *Marine Ecology Progress Series, 289*, 117–130. https://doi.org/10.3354/meps289117

Dalebout, M. L., Mead, J. G., Baker, C. S., Baker, A. N., & Van Helden, A. L. (2002). A new species of beaked whale *Mesoplodon perrini* sp.n. (Cetacea: Ziphiidae) discovered through phylogenetic analyses of mitochondrial DNA sequences. *Marine Mammal Science, 18*(3), 577–608. https://doi.org/10.1111/j.1748-7692.2002.tb01061.x

D'Arcy, P. (2006). *The people of the sea: Environment, identity and history in oceania* (312 pp). University of Hawai'i Press. ISBN-13: 9780824832971.

Darling, J. D., Audley, K., Cheeseman, T., Goodwin, B., Lyman, E. G., & Urban, R. J. (2022). Humpback whales (*Megaptera novaeangliae*) attend both Mexico and Hawaii breeding grounds in the same winter: mixing in the northeast Pacific. *The Royal Society Biology Letters, 18*, 20210547. https://doi.org/10.1098/rsbl.2021.0547

Debenham, N. J., Lambshead, P. J., Ferrero, T. J., & Smith, C. R. (2004). The impact of whale falls on nematode abundance in the deep sea. *Deep sea Research Part I: Oceanographic Research Papers, 51*(5), 701–706. https://doi.org/10.1016/j.dsr.2004.02.004

De Groot, R., Brander, L., van der Ploeg, S., Costanza, R., Bernard, F., Braat, L., et al. (2012). Global estimates of the value of ecosystems and their services in monetary units. *Ecosystem Services, 1*, 50–61. https://doi.org/10.1016/j.ecoser.2012.07.005

De Oliveira-Ferreira, N., Santos-Neto, E. B., Manhaes, B. M. R., Carvalho, V. L., Goncalves, L., & de Castilho, P. V. (2023). The deep dive of organohalogen compounds: Bioaccumulation in the top predators of mesopelagic trophic webs, pygmy and dwarf sperm whales, from the Southwestern Atlantic ocean. *Chemosphere, 345*, 140456. https://doi.org/10.1016/j.chemosphere.2023.140456

Desforges, J. P., Hall, A., McConnell, B., Rosing-Asvid, A., Barber, J. L., & Brownlow, A. (2018). Predicting global killer whale population collapse from PCB pollution. *Science, 361*(6409), 1373–1376. https://doi.org/10.1126/science.aat1953

Dill, L. M., Heithaus, M. R., & Walters, C. J. (2003). Behaviorally mediated indirect interactions in marine communities and their conservation implications. *Ecology, 84*(2003), 1151–1157. https://doi.org/10.1890/0012-9658(2003)084[1151:BMIIIM]2.0.CO;2

Doughty, C. E., Roman, J., Faurby, S., Wolf, A., Haque, A., Bakker, E. S., et al. (2016). Global nutrient transport in a world of giants. *Proceedings of the National Academy of Sciences USA, 113*(4), 863–873. https://doi.org/10.1073/pnas.1502549112

Drazen, J. C., Smith, C. R., Gjerde, K. M., Haddock, S. H. D., Carter, G. S., & Choy, C. A. (2020). Midwater ecosystems must be considered when evaluating environmental risks of deep-sea mining. *Proceedings of the National Academy of Sciences USA, 117*, 17455–17460.

Duarte, C. M., Agusti, S., Barbier, E., Britten, G. L., Castilla, J. C., Gattuso, J. P., et al. (2020). Rebuilding marine life. *Nature, 580*, 39–51. https://doi.org/10.1038/s41586-020-2146-7

Eckstein, L., & Schwarz, A. (2019). The making of Tupaia's map: A story of the extent and mastery of polynesian navigation, competing systems of wayfinding on James Cook's endeavour, and the invention of an ingenious cartographic system. *Journal of Pacific History, 54*, 1–95. https://doi.org/10.1080/00223344.2018.1512369

Enquist, B. J., Abraham, A. J., Harfoot, M. B. J., Malhi, Y., Doughty, C. E., et al. (2020). The megabiota are disproportionately important for biosphere functioning. *Nature Communications, 11*, 699.

ESCO CNRS IFREMER. (2014). Rapport d'expertise collective, Juin (2014). Impacts environnementaux de l'exploitation des ressources minérales marines profondes. 939pp.

Estes, J. A., Terborgh, J., Brashares, J. S., Power, M. E., Berger, J., Bond, W. J., et al. (2011). Trophic downgrading of planet Earth. *Science, 333*(6040), 301–306. https://doi.org/10.1126/science.1205106

Estes, J. A., Heithaus, M., McCauley, D. J., Rasher, D. B., & Worm, B. (2016). Megafaunal impacts on structure and function of ocean ecosystems. *Annual Review of Environment and Resources, 41*, 83–116. https://doi.org/10.1146/annurev-environ-110615-085622

Etnoyer, P., Canny, D., Mate, B. R., Morgan, L. E., Ortega-Ortiz, J. G., & Nichols, W. J. (2006). Sea-surface temperature gradients across blue whale and sea turtle foraging trajectories off the Baja California Peninsula, Mexico. *Deep-Sea Research Part II, 53*, 340–358. https://doi.org/10.1016/j.dsr2.2006.01.010

Ey, M., & Sherval, M. (2016). Exploring the minescape: engaging with the complexity of the extractive sector: exploring the minescape. *Area, 48*, 176–182. https://doi.org/10.1111/area.12245

Fernandez, A., Sierra, E., Arbelo, M., Gago-Martinez, A., Leao Martins, J. M., Garcia Alvarez, N., et al. (2022). First case of Brevetoxicosis linked to rough-toothed dolphin (*Steno bredansis)* mass mortality event in eastern central Atlantic ocean: a climate change effect? *Frontiers in Marine Science Section Marine Megafauna, 9*. https://doi.org/10.3389/fmars.2022.834051

Fiedler, P. C., Reilly, S. B., Hewitt, R. P., Demer, D., et al. (1998). Blue whale habitat and prey in the California Channel Islands. *Deep-Sea Research Part II, 45*, 1781–1801. https://doi.org/10.1016/S0967-0645(98)80017-9

Fiedler, P. C., & Talley, L. D. (2006). Hydrography of the eastern tropical Pacific: a review. *Progress in Oceanography, 69*, 143–180. https://doi.org/10.1016/j.pocean.2006.03.008

Firestone, J., & Lilley, J. (2007). Aboriginal subsistence whaling and the right to practice and revitalize cultural traditions and customs. *Journal of International Wildlife Law Policy, 8*, 177–219. https://doi.org/10.1080/13880290590965339

Flood, N. B., Strong, B. E., & Flood, W. (1999). *Pacific Island legends: Tales from Micronesia, Melanasia, Polynesia and Australia* (p. 266). Bess Press.

Game, E. T., Grantham, H. S., Hobday, A. J., Pressey, R. L., Lombard, A. T., Beckley, L., et al. (2009). Pelagic protected areas: the missing dimension in ocean conservation. *Trends in Ecology & Evolution, 24*, 360–369. https://doi.org/10.1016/j.tree.2009.01.011

Genin, A. (2004). Bio-physical coupling in the formation of zooplankton and fish aggregations over abrupt topographies. *Journal of Marine Systems, 50*, 3–20. https://doi.org/10.1016/j.jmarsys.2003.10.008

Gianni, M., Fuller, S., Currie, D. E. J., Schleit, K., Pike, B., Goldsworthy, L., et al. (2016). "How much longer will it take? A ten-year review of the implementation of United Nations General Assembly resolutions 61/105, 64/72 and 66/68 on the management of bottom fisheries in areas beyond national jurisdiction". Deep Sea Conservation Coalition, 131pp. July 2016.

Glen, R. (2003). Behavioural reponses of the short-finned pilot whale, *Globicephala macrorhynchus*, in relation to the number of surrounding whale-watching vessels. Annual Meeting of the European Cetacean Society, Tenerife, Spain.

Govan, H., Tawake, A., Tabunakawai, K., Jenkins, A., Lasgorceix, A., Schwarz, A.-M., et al. (2009). *Status and potential of locally- managed marine areas in the South Pacific: Meeting nature conservation and sustainable livelihood targets through wide-spread implementation of LMMAs* (134 pp). SPREP/WWF/WorldFish-Reefbase/CRISP.

Grau Tomas, E., & Garcia Sanabria, J. (2022). Comparative analysis of marine-protected area effectiveness in the protection of marine mammals: Lessons learned and recommendations. *Fronties in Marine Science Section Marine Affairs and Policy*, 9. https://doi.org/10.3389/fmars.2022.940803

Grey, G. (2019). *Polynesian mythology. The myths, legends, songs and ancient traditional history of the New Zealanders and Pacific Islanders* (152 pp). Pantianos Classics. ISBN-10: 1789871654.

Halpern, B. S., Lester, S. E., & McLeod, K. L. (2010). Placing marine protected areas onto the ecosystem-based management seascape. *Proceedings of the National Academy of Sciences USA, 107*, 18312–18317. https://doi.org/10.1073/pnas.0908503107

Halpern, B., Frazier, M., Afflerbach, J., Stewart, J. S., Micheli, F., O'Hara, C., et al. (2013). Recent pace of change in human impact on the world's ocean. *Scientific Reports, 9*(1), 11609. https://doi.org/10.1038/s41598-019-47201-9

Halpern, B. S., Frazier, M., Potapenko, J., Casey, K. S., Koenig, K., Longo, C., et al. (2015). Spatial and temporal changes in cumulative human impacts on the world's ocean. *Nature Communications, 6*, 7615. https://doi.org/10.1038/ncomms8615

Harrison, A.L., Costa, D. P., Winship, A. J., Benson, S. R., Bograd, S. J., Antolos et al., (2018). The political biogeography of migratory marine predators. Nature Ecology & Evolution, 2, 1571–1578. https://doi.org/10.1038/s41559-018-0646-8.

Häussermann, V., Gutstein, C. S., Bedington, M., Cassis, D., Olavarria, C., Dale, A. C., et al. (2017). Largest baleen whale mass mortality during strong El Niño event is likely related to harmful toxic algal bloom. *PeerJ, 5*, e3123. https://doi.org/10.7717/peerj.3123

Higdon, J. W., & Ferguson, S. H. (2009). Loss of Arctic sea ice causing punctuated change in sightings of killer whales (*Orcinus orca*) over the past century. *Ecological Applications, 19*, 1365–1375. https://doi.org/10.1890/07-1941.1

Hildebrand, T. (2005). Shipboard acoustic doppler current profiling during cruise MW9403 (SAC ID 00108). *PANGAEA*. https://doi.org/10.1594/PANGAEA.31938

Hind, E. J. (2015). A review of the past, the present, and the future of fishers' knowledge research: a challenge to established fisheries science. *ICES Journal of Marine Science, 72*, 341–358. https://doi.org/10.1093/icesjms/fsu169

Hofman, E. E., Klinck, J. M., Costa, D. P., Daly, K. L., Torres, J. J., & Fraser, W. R. (2002). U.S. southern ocean global ocean ecosystems dynamics program. *Oceanography, 14*, 67–74. https://doi.org/10.5670/oceanog.2002.22

Hoyt, E. (2004). *Marine protected areas for whales, dolphins and porpoises. A world handbook for Cetacean habitat conservation* (492 pp). Earthscan. ISBN 1-84407-064-6.

Hoyt, E. (2011). *Marine Protected Areas for whales, dolphins and porpoises. A world handbook for Cetacean Habitat Conservation and Planning*. (Second edition, revised, 477 pp). ISBN: 978-1-84407-762-5 and 978-1-84407-763-2.

Hoyt, E. (2021). Benefits and pitfalls of MPAs as a conservation tool for cetaceans. In L. Nunny (Ed.), *Under pressure the need to protect whales and dolphins in European waters. An OceanCare report* (pp. 86–95). Wild Animal Welfare. ISBN 978-3-033-08516-9.

International Union for Conservation of Nature (IUCN). (2016). *A global standard for the identification of key biodiversity areas, version 1.0* (1st ed.). IUCN. ISBN: 978-2-8317-1835-4.

International Union for Conservation of Nature (IUCN). (2018). *The journey of Cook Islands: Marae Moana*. https://www.iucn.org/news/oceania/201806/journey-cook-islands-marae-moana. Accessed 11 Mar 2021.

Irvine, L. M., Mate, B. R., Winsor, M. H., Palacios, D. M., Bograd, S. J., Costa, D. P., et al. (2014). Spatial and temporal occurrence of blue whales off the U.S. West Coast, with implications for management. *PLoS One, 9*(7), e102959. https://doi.org/10.1371/journal.pone.0102959

Irvine, L., Palacios, D. M., Urban, J., & Mate, B. (2017). Sperm whale dive behavior characteristics derived from intermediate-duration archival tag data. *Ecology and Evolution, 7*(19), 7822–7837. https://doi.org/10.1002/ece3.3322

Jeudy de Grissac, A. (2003). *Protection de la Nature. Aires protégées*. Encyclopædia Universalis. 13. Available online at: https://www.universalis.fr/encyclopedie/protection-de-la-nature-aires-protegees/#i_0

Jeudy de Grissac, A. (2015). Reviewing existing Mediterranean models of governance of MPAs with artisanal fisheries. In Fishing Governance in MPAs: potentialities for blue economy. *FISHMPABLUE*. 54. Available online at: https://www.iucn.org/sites/dev/files/content/documents/fishmpablue_governance_models_review_existing_final_oct.pdf

Jimenez Lopez, M. E., Palacios, D. M., Legorreta, A. J., Urban, J., & Mate, B. R. (2019). Fin whale movements in the Gulf of California, Mexico, from satellite telemetry. *PLoS One, 14*(1), e0209324. https://doi.org/10.1371/journal.pone.0209324

Johannes, R. (1978). Traditional marine conservation methods in Oceania and their demise. *Annual Review of Ecology and Systematics, 9*, 349–364. https://doi.org/10.1146/annurev.es.09.110178.002025

Johannes, R. (1981). *Words of the Lagoon: Fishing and Marine Lore in the Palau District of Micronesia*. University of California Press. https://doi.org/10.1525/9780520321397

Johnson, P., & Weaver, D. (2014). Ecologically or biologically significant marine areas. In N. Dudley et al. (Eds.), *Applications of key biodiversity areas: End-user consultations* (pp. 22–24). IUCN. Online at: https://portals.iucn.org/library/node/44911

Jones, D. O., Simon-Lledo, E., Amon, D., Bett, B. J., Caulle, C., Clément, L., et al. (2021). Environment, ecology, and potential effectiveness of an area protected from deep-sea mining (Clarion Clipperton Zone, abyssal Pacific). *Progress in Oceanography, 197*, 102653. https://doi.org/10.1016/j.pocean.2021.102653

Jupiter, S., Cohen, P., Weeks, R., Tawake, A., & Govan, H. (2014). Locally managed marine areas: multiple objectives and diverse strategies. *Pacific Conservation Biology, 20*, 165–179. https://doi.org/10.1071/PC140165

Juster, A. (2016). Histoire et Culture d'Océanie : le rahui, une notion structurante de la société, en intime relation avec le pouvoir politique et le sacré. *Outremers 360°*. https://outremers360.com/culture/histoire-culture-doceanie-le-rahui-une-notion-structurante-de-la-societe-en-intime-relation-avec-le-pouvoir-politique-et-le-sacre-par-alexandre-juster

Kaschner, K., (2007). Air breathing visitors to seamounts : marine mammals. Wiley online. 10 January 2007. T. Morato, P. Hart, M. Clark, N. Haggan, R. Santos (Eds.). https://doi.org/10.1002/9780470691953.ch12a

Kiszka, J., Heithaus, M., & Wirsing, A. (2015). Behavioural drivers of the ecological roles and importance of marine mammals. *Marine Ecology Progress Series, 523*. https://doi.org/10.3354/meps11180

Kiszka, J., Heithaus, M., & Woodstock, M. (2022). Functional roles and ecological importance of small cetaceans in aquatic ecosystems. *Frontiers in Marine Science, 9*. https://doi.org/10.3389/fmars.2022.803173

Kitolelei, S., Thaman, R., Veitayaki, J., Breckwoldt, A., & Piovano, S. (2021). Na Vuku Makawa ni Qoli: Indigenous Fishing Knowledge (IFK) in Fiji and the Pacific. *Frontiers in Marine Science, 8*, 684303. https://doi.org/10.3389/fmars.2021.684303

Kriegl, M., Elias Ilosvay, X., von Dorrien, C., & Oesterwind, D. (2021). Marine Protected Areas: at the crossroads of Nature Conservation and Fisheries Management. *Frontiers in Marine Science Section Marine Conservation and Sustainability, 8*. https://doi.org/10.3389/fmars.2021.676264

Kühn, S., Bravo Rebolledo, E. L., & van Franeker, J. A. (2015). Deleterious effects of litter on marine life. In M. Bergmann, L. Gutow, & M. Klages (Eds.), *Marine anthropogenic litter* (pp. 75–116). Springer International Publishing. https://doi.org/10.1007/978-3-319-16510-3_4

Lagerquist, B. A., Mate, B. R., Ortega-Ortiz, J. G., Winsor, M., & Urbán-Ramirez, J. (2008). Migratory movements and surfacing rates of humpback whales (*Megaptera novaeangliae*) satellite tagged at Socorro Island, Mexico. *Marine Mammal Science, 24*, 815–830. https://doi.org/10.1111/j.1748-7692.2008.00217.x

Lavery, T., Roudnew, B., & Gill, P. C. (2010). Iron defecation by sperm whales stimulates carbon export in the Southern Ocean. *Proceedings of the Royal Society, B277*(1699), 3527–3531. https://doi.org/10.1098/rspb.2010.0863

Letessier, T., Mouillot, D., Bouchet, P., Vigliola, L., Fernandes, M. C., & Thompson, C. (2019). Remote reefs and seamounts are the last refuges for marine predators across the Indo-Pacific. *PLos Biology, 17*(8). https://doi.org/10.1371/journal.pbio.3000366

Lewis, N., Day, J., Wilhelm, A., Wagner, D., Gaymer, C., Parks, J. et al. (2017). Large-Scale Marine Protected Areas: guidelines for design and management. Prepared by *Big Ocean and the IUCN WCPA Large-Scale MPA Task Force*. 152pp. *Best Practice Protected Area Guidelines Series N°. 26*. Craig Groves, Series Editor. ISBN: 978-2-8317-1880-4.

Lohmann, K., Lohmann, C., & Endres, C. (2008). The sensory ecology of ocean navigation. *The Journal of Experimental Biology, 211*, 1719–1728. https://doi.org/10.1242/jeb.015792

Love, M. (2021). Tabu, MPA and community: nomenclature and the political ecology of marine conservation in Vanuatu. *Journal of Political Ecology, 28*, 91–110. https://doi.org/10.2458/jpe.2301

Martínez-Loustalot, P., Audley, K., Cheeseman, T., De Weerdt, J., Frisch-Jordán, A., Guzón, et al. (2022). Towards the definition of the humpback whale population units along the Mexican and Central American coasts in the Pacific Ocean. *Marine Mammal Science*. https://doi.org/10.1111/mms.12980

Mate, B. R., Lagerquist, B. A., & Calambokidis, J. (1999). Movements of North Pacific Blue Whales During the Feeding Season off Southern California and Southern Fall Migration. *Marine Mammal Science, 15*(4), 1246–1257. https://doi.org/10.1111/j.1748-7692.1999.tb00888.x

Mate, B.R., Palacios, D.M., Baker, C.S., Lagerquist, B.A., Irvine, L.M., Follettet, T.M. et al. (2020). *Humpback whale tagging in support of marine mammal monitoring across multiple navy training areas in the Pacific Ocean: Preliminary summary of field tagging effort in Washington in September–October 2019*. Prepared for Commander, U.S. Pacific Fleet. Submitted to Naval Facilities Engineering Command, Southwest, under Cooperative Ecosystem Studies Unit, Department of the Navy Cooperative Agreement No. N62473-19-2-0002. Oregon State University, Newport, Oregon, 9 January 2020. 23 pp.

McCrea-Strub, A., Zeller, D., Rashid Sumaila, U., Nelson, J., Balmford, A., Pauly, D., et al. (2010). Understanding the cost of establishing marine protected areas. *Marine Policy, 35*(1), 1–9. https://doi.org/10.1016/j.marpol.2010.07.001

Mead, H. (2003). *Tikanga Maori. Living by Maori values*. Hirini Moko Mead Copyright. 382 pp. Huia Publishers. ISBN: 9781775502227.
Miller, K., Thompson, K., Johnston, P., & Santillo, D. (2018). An overview of seabed mining including the current state of development, environmental impacts, and knowledge gaps. *Frontiers in Marine Science, 4*, 418. https://doi.org/10.3389/fmars.2017.00418
Mizroch, S., Perryman, W., & Waite, J. (2009). Distribution and movements of fin whales in the North Pacific ocean. *Mammal Review, 39*, 193–227. https://doi.org/10.1111/j.1365-2907.2009.00147
Morato, T., Hoyle, S. D., Allain, V., & Nicol, S. J. (2010). Seamounts are hotspots of pelagic biodiversity in the open ocean. *Proceedings of the National Academy of Sciences USA, 107*, 9707–9711. https://doi.org/10.1073/pnas.0910290107
Nordhoff, C. (1930). Note of the off-shore fishing of the Society Islands. *Journal of Polynesian Society, 39*, 137–173.
Nortarbartolo di Sciara, G., Agardy, T, Hyrenbach, D., Scovazzi, T. van Klaveren, P. (2008). The Pelagos sanctuary for Mediterranean Marine mammals. Aquatic Conservation Marine and Freshwater Ecosystems 18(4):367 – 391. https://doi.org/10.1002/aqc.855.
O'Connor, S., Campbell, R., Cortez, H., & Knowles, T. (2009). *Whale watching worldwide : Tourism numbers, expenditures and economic benefits. A special report from the International Fund for Animal Welfare*. International Fund for Animal Welfare (IFAW).
O'Leary, B. C., Brown, R. L., Johnson, D. E., Von Nordheim, H., Ardron, J., Packeiser, T., et al. (2012). The first network of marine protected areas (MPAs) in the high seas: the process, the challenges and where next. *Marine Policy, 36*, 598–605. https://doi.org/10.1016/j.marpol.2011.11.003
Palacios, D. M., Mate, B. R.,Baker, C. S., Hayslip, C.E., Follett, T. M., Steel, D., et al. (2019a). *Tracking North Pacific Humpback Whales to unravel their basin-wide movements*. Final Technical Report. Prepared for Pacific Life Foundation. Marine Mammal Institute, Oregon State University. Newport, Oregon, USA. 30 June 2019. 58 pp. https://doi.org/10.5399/osu/1117. https://ir.library.oregonstate.edu/concern/technical_reports/z890s0924
Palacios, D., Bailey, H., Becker, E. A., Bograd, S. J., DeAngelis, M. L., Forney, K. A., et al. (2019b, 7). Ecological correlates of blue whale movement behavior and its predictability in the California Current Ecosystem during the summer-fall feeding season. *Movement Ecology*, 26. https://doi.org/10.1186/s40462-019-0164-6
Palacios, D., Felix, F., Montecinos, Y., Najera, E., Kelez, S., Samaniego, J., et al. (2023). *Blue corridor of the Eastern Pacific: Opportunities and actions for the protection of Migratory Whales* (69 pp). A Technical Report. World Wide Fund for Nature (WWF).
Pallin, L., Baker, S., Steel, D., Kellar, N., Robbins, J., Jonston, D., et al. (2018). Climate-related perturbations to the North Pacific ecosystem, known as marine heatwaves, also appear to have impacted birth rates of humpback whale populations. *Royal Society Open Science*. https://doi.org/10.1098/rsos.180017
Pelayo-González, L., Herra-Miranda, D., Pacheco-Polanco, J. D., Guzmán, H. M., Goodman, S., & Oviedo, L. (2022). Decreases in encounter rate of endangered Northeast Pacific humpback whales in Southern Costa Rica: Possible changes in migration pattern due to warming events. *Frontiers in Marine Science, 9*. https://doi.org/10.3389/fmars.2022.927276
Pitman, R. L., & Stinchcomb, C. (2002). Rough-toothed Dolphins (*Steno bredanensis*) as Predators of Mahimahi (*Coryphaena hippurus*). *Pacific Science, 56*(4), 447–450. https://doi.org/10.1353/psc.2002.0043
Pompa, S., Ehrlich, P. R., & Ceballos, G. (2011). Global distribution and conservation of marine mammals. *Proceedings of the National Academy of Sciences USA*. https://doi.org/10.1073/pnas.1101525108
Rasmussen, K., Palacios, D. M., Calambokidis, J., Saborío, M. T., Dala Rosa, L., Secchi, E. R., et al. (2007). Southern hemisphere humpback whales wintering off Central America: Insights from water temperature into the longest mammalian migration. *Biology Letters, 3*, 302–305. https://doi.org/10.1098/rsbl.2007.0067

Reed, J., Harcourt, R., New, L., & Bilgmann, K. (2020). Extreme effects of extreme disturbances: a simulation approach to assess population specific responses. *Frontiers in Marine Science, 7*, 519845. https://doi.org/10.3389/fmars.2020.519845

Riisager-Simonsen, C., Rendon, O., Galatius, A., Tange Olsen, M., & Beaumont, N. (2020). Using ecosystem-services assessments to determine trade-offs in ecosystem-based management of marine mammals. *Conservation Biology, 34*, 1152–1164. https://doi.org/10.1111/cobi.13512

Risch, D., Norris, T., Curnock, M., & Friedlaender, A. (2019). Common and Antarctic Minke Whales: conservation status and future research directions. *Frontiers in Marine Science Section Marine Megafauna, 6*. https://doi.org/10.3389/fmars.2019.00247

Ritter, F., & Panigada, S. (2019). Chapter 28-Collisions of Vessels with cetacean-the underestimated threat. In *World seas: An environmental evaluation (second edition). Volume III: Ecological issues and environmental impacts* (pp. 531–547). https://doi.org/10.1016/B978-0-12-805052-1.00026-7

Rogers, A. (2018). The biology of seamounts : 25 years on. *Advances in Marine Biology, 79*, 137–224. https://doi.org/10.1016/bs.amb.2018.06.001

Rogers, A., Baco, A., Griffiths, H., Hart, T., & Hall-Spencer, J. M. (2007). Corals on seamounts. In T. J. Pitcher, T. Morato, P. J. B. Hartet, et al. (Eds.), *Seamounts: Ecology, fisheries and conservation* (pp. 141–169). Blackwell.

Roman, J., Estes, J. A., Morissette, L., Smith, C., Costa, D., McCarthy, J., et al. (2014). Whales as marine ecosystem engineers. *Frontiers in Ecology and the Environment, 12*, 377–385. https://doi.org/10.1890/130220

Ruddle, K., Hviding, E., & Johannes, R. (1992). Marine resources management in the context of customary tenure. *Marine Resource Economics, 7*, 249–273. https://doi.org/10.1086/mre.7.4.42629038

Savoca, M. S., Czapanskiy, M. F., Kahane-Rapport, S. R., Gough, W. T., Fahlbusch, J. A., Bierlich, K. C., et al. (2021). Baleen whale prey consumption based on high-resolution foraging measurements. *Nature, 599*, 85–90. https://doi.org/10.1038/s41586-021-03991-5

Sears, R., & Perrin, W. F. (2009). Blue whale (*Balaenoptera musculus*). In W. F. Perrin, B. Würsig, & J. G. M. Thewissen (Eds.), *Encyclopedia of marine mammals* (pp. 120–124). Academic Press. ISBN: 9780080919935.

Simmonds, M., Nunny, L., Sangster, G., & Luksenburg, J., (2020). *The real and imminent extinction risk to whales, dolphins and porpoises: an open letter from (over 250) cetacean scientists, statement of concern*. 3/9/2020. https://www.researchgate.net/publication/344136973

Smith, C. R., (2006). Bigger is better: The role of whales as detritus in marine ecosystems. In: *Whales, whaling and ocean ecosystems*, ed. JA Estes, P DeMaster, DF Doak, TM Williams, RL Brownell Jr, pp. 286–300. University of California Press.

Smith, C. R., & Baco, A. R. (2003). Ecology of whale falls at the deep-sea floor. *Oceanography and Marine Biology: An Annual Review, 41*, 311–354. ISBN 9780429217715.

Smith, C. R., Gaines, S., Watling, L., Friedlander, A., Morgan, C., Thurnherr, A. et al. (2008). Rationale and recommendations for the establishment of preservation reference areas for nodule mining in the Clarion-Clipperton Zone. *Fourteenth session. Kingston, 26 May–6 June 2008, sessions/2008/documents: Legal and Technical Commission, International Seabed Authority.Technical document no. ISBA/14/LTC/2.* Available online at: https://www.isa.org.jm/(click/contractors/exploration-areas/ClarionClippertonFractureZone)

Smith, C. R., Bernardino, A. F., Baco, A., Hannides, A. K., & Altamira, I. (2014). The seven-year enrichment: Macrofaunal succession in deep-sea sediments around a 30-tonne whalefall in the Northeast Pacific. *Marine Ecology Progress Series*. https://doi.org/10.3354/meps10955

Smith, C. R., Glover, A., Treude, T., Higgs, N., & Amon, D. (2015). Whale-fall ecosystems: recent insights into ecology, paleoecology and evolution. *Annual Review of Marine Science, 2015*(7), 571–596. https://doi.org/10.1146/annurev-marine-010213-135144

Southall, B. L., Benoit-Bird, K. J., Moline, M. A., & Moretti, D. (2019). Quantifying deep-sea predator-prey dynamics: Implications of biological heterogeneity for beaked whale conservation. *Journal of Applied Ecology, 56*(5), 1040–1049. https://doi.org/10.1111/1365-2664.13334

Stafford, K., Fox, C., & Clark, D. (1998). Long-range acoustic detection and localization of blue whale calls in the Northeast Pacific Ocean. *Journal of the Acoustical Society of America, 104*(6), 3616–3625. https://doi.org/10.1121/1.423944

Stafford, K. M., Nieukirk, S. L., & Fox, C. G. (1999). An acoustic link between blue whales in the eastern tropical pacific and the northeast pacific. *Marine Mammal Science., 15*, 1258–1268. https://doi.org/10.1111/j.1748-7692.1999.tb00889.x

Stafford, K. M., Moore, S. E., & Fox, C. G. (2005). Diel variation in blue whale calls recorded in the eastern tropical Pacific. *Animal Behaviour, 69*, 951–958. https://doi.org/10.1016/j.anbehav.2004.06.025

Stafford, K. M., Nieukirk, S. L., & Fox, C. G. (2006). An acoustic link between blue whales in the eastern tropical pacific and the northeast pacific. *Marine Mammal Science, 15*(4), 1258–1268. https://doi.org/10.1111/j.1748-7692.1999.tb00889.x

Stewart, J., Joyce, T., Durban, J., Calambokidi, J., Fauquier, D., Fearnbach, H., et al. (2023). Boom-bust cycles in gray whales associated with dynamic and changing Arctic conditions. *Science, 382*(6667), 207–211. https://doi.org/10.1126/science.adi1847

Tawake, A., & Tuivanuavou, S. (2004). Section 2: Community involvement: Community-based co-management of marine resources - case study 9 - Community involvement in implementation of ocean policies: the Fiji Locally Managed Marine Areas network. In *SPC Traditional Marine Resource Management and Knowledge Information Bulletin #17 – December 2004* (Noumea: SPC), 26–27.

Tawake, A., Radikedike, P., Aalbersberg, B., Vuki, V., & Salafsky, N. (2001). Harvesting clams and data involving local communities in monitoring can lead to conservation success in all sorts of unanticipated ways: a case in Fiji. *Conservation in Practice, 2*, 32–35. https://doi.org/10.1111/j.1526-4629.2001.tb00020.x

Thaman, B., Thaman, R., Balawa, A., & Veitayaki, J. (2017). The recovery of a tropical marine mollusk fishery: A transdisciplinary community-based approach in Navakavu, Fiji. *Journal of Ethnobiology, 37*, 494. https://doi.org/10.2993/0278-0771-37.3.494

Thompson, K. F., Miller, K. A., Wacker, J., Derville, S., Laing, C., Santillo, D., et al. (2023). Urgent assessment needed to evaluate potential impacts on cetaceans from seabed mining. *Frontiers in Marine Science Section Marine Magafauna, 10*. https://doi.org/10.3389/fmars.2023.1095930

Tilot, V., Cazalla, E., Casado, J. M., Catala, T., & Bernal C. (2016). Assessment of whale watching activities in the Gibraltar Strait and drafting of new guidelines for whale watching. Accobams IUCN final report. Funded by MAVA. 68pp.

Tilot, V. (2004). Management plan for the marine mammals sanctuary in the Mediterranean Sea. *Pelagos*. Adopted by the Governments of France, Italy and Monaco in October 2004, Signatories to the Agreement on the Creation of a Sanctuary for Marine Mammals, 107pp.

Tilot, V. (2006a). *Biodiversity and distribution of the megafauna Vol 1 The polymetallic nodule ecosystem of the Eastern Equatorial Pacific Ocean*. Intergovernmental Oceanographic Commission. Technical Series 69, Project Unesco COI/Min, Vlanderen. Available online at: http://unesdoc.unesco.org/images/0014/001495/149556f.pdf

Tilot, V. (2006b). *Biodiversity and distribution of the megafauna Vol 2 annotated photographic atlas of the echinoderms of the Clarion-Clipperton Fracture Zone*. Intergovernmental Oceanographic Commission. Technical Series 69, Project Unesco/COI/Min Vlanderen. Available online at: http://unesdoc.unesco.org/images/0014/001495/149556f.pdf

Tilot, V. (2010). *Biodiversity and distribution of faunal assemblages Vol 3 Options for the management and conservation of the nodule ecosystem in the Clarion Clipperton Fracture Zone (NE Pacific Ocean): Scientific, legal and institutional aspects*. Technical Series 69, Project Unesco COI/Min Vlanderen. Available online at: http://unesdoc.unesco.org/images/0014/001495/149556f.pdf

Tilot, V. (2013). Tools for marine spatial management in deep seas (A guide for Research, Conservation and Development Practitioners). In *Proceedings of the International Marine Protected Areas Congress (IMPAC3), IUCN/Agence des Aires Marines Protégées, Marseille, France.*

Tilot, V. (2014). An innovative Multilayer Rapid Ecological Assessment adapted to monitoring and management of Deep Sea Ecosystems. In *Proceedings deep sea mining summit 2014, London, UK*.

Tilot V. (2019). Assessment of deep-sea faunal communities-indicators of environmental impact. Chapter 4. In *Environmental issues of deep-sea mining. Impacts, consequences and policy perspectives* (577pp). Perspectives in deep-sea mining https://www.springer.com/us/book/9783030126957, https://doi.org/10.1007/978-3-030-12696-4

Tilot,V., Cazalla, E., Casado, J. M., Catala, T., & Bernal, C. (2016). *Assessment of whale watching activities in the Gibraltar Strait and drafting of new guidelines for whale watching*. Accobams IUCN final report. Funded by MAVA. 68pp.

Tilot, V., Ormond, R., Moreno-Navas, J., & Catala, T. (2018). The benthic megafaunal assemblages of the CCZ (Eastern Pacific) and an approach to their management in the face of threatened anthropogenic impacts. *Frontiers in Marine Science, 5*, 7. https://doi.org/10.3389/fmars.2018.00007

Tilot, V., Willaert, K., Guilloux, B., Chen, W., Mulalap, C. Y., Gaulme, F., et al. (2021a). Traditional dimensions of seabed resource management in the context of Deep Sea Mining in the Pacific: Learning from the socio-ecological interconnectivity between island communities and the ocean realm. *Frontiers in Marine Science, Special Issue: Integrated Marine Biosphere Research: Ocean Sustainability, Under Global Change, for the Benefit of Society, 8*, 637938. https://doi.org/10.3389/fmars.2021.637938

Tilot, V., Willaert, K., Guilloux, B., Mulalap, C. Y., Gaulme, F., Bambridge, T., et al. (2021b). Traditional and cultural/societal dimensions of seabed resource management and present applicable legal frameworks in the Pacific. In R. Sharma (Ed.), *Perspectives in deep-sea mining*. Springer. ISBN: 10303087981X.

Tilot, V., Guilloux, B., Willaert, K., Mulalap, C. Y., Gaulme, F., Bambridge, T., et al. (2021c). The concept of Oceanian Sovereignity in the context of Deep sea mining in the Pacific. *Frontiers in Marine Science, Special Issue: Integrated Marine Biosphere Research: Ocean Sustainability, Under Global Change, for the Benefit of Society*. https://doi.org/10.3389/fmars.2021.756072

Tilot, V., Fourchault, L., Jeudy de Grissac, A., Mallefet, J., & Moreno Navas, J. (2023). Multilevel assessment and management of cumulative impacts on pelagic ecosystems in the NE Pacific. In R. Sharma (Ed.), *Deep-sea mining and the water column*. Springer. In press.

Towers, J. R., McMillan, C. J., Malleson, M., Hildering, J., Ford, J. K. B., & Ellis, G. M. (2013). Seasonal movements and ecological markers as evidence for migration of common minke whales photo-identified in the eastern North Pacific. *Journal of Cetacean Research and Management, 13*(3), 221–229. https://doi.org/10.47536/jcrm.v13i3.539

Tulloch, V. J. D., Richardson, A. J., Matear, R., & Brown, C. (2019). Future recovery of baleen whales is imperiled by climate change. *Global Change Biology, 25*(4), 1263–1281. https://doi.org/10.1111/gcb.14573

UNEP-WCMC and IUCN. (2017). Protected Planet: The World Database on Protected Areas (WDPA) (online), September, 017, Cambridge, UK: UNEP-WCMC. Available at www.protectedplanet.net.

Van Weelden, C., Towers, J., & Bosker, T. (2021). Impacts of climate change on cetacean distribution, habitat and migration. *Climate Change Ecology, 1*, 100009. https://doi.org/10.1016/j.ecochg.2021.100009

Veitayaki, J., Aalbersberg, W., Tawake, A., Rupeni, E., & Tabunakawai, K. (2003). *Mainstreaming resource conservation: The Fiji locally managed marine area network and its influence on national policy development*. Australian National University. Available online at: https://openresearchrepository.anu.edu.au/bitstream/1885/40990/3/rmap_wp42.pdf

Veitayaki, et al. (2011). On cultural factors and marine managed areas in Fiji. *Terra Australis, 35*. https://doi.org/10.13140/2.1.2018.1122

Vierros, M., Harrison, A. L., Sloat, M., Ortuno Crespo, G., Moore, J., Dunn, D., et al. (2020). Considering indigenous peoples and local communities in governance of the global ocean commons. *Marine Policy, 119*, 104039. https://doi.org/10.1016/j.marpol.2020.104039

Wang, Z.-T., Supin, A. Y., Akamatsu, T., Duan, P.-X., Yang, Y.-N., Wang, K.-X., et al. (2021). Auditory evoked potential in stranded melon-headed whales (Peponocephala electra): With sever hearing loss and possibly caused by anthropogenic noise pollution. *Ecotoxicology and Environmental Safety, 228*. https://doi.org/10.1016/j.ecoenv.2021.113047

Wedding, L. M., Friedlander, A. M., Kittinger, J. N., Watling, L., Gaines, S. D., et al. (2017). From principles to practice: A spatial approach to systematic conservation planning in the deep sea. *Proceedings of the Royal Society B, 280*, 20131684. https://doi.org/10.1098/rspb.2013.1684

Wedding, L. M., Reiter, S. M., Smith, C. R., Gjerde, K. M., Kittinger, J. N., Friedlander, A. M., et al. (2015). Managing mining of the deep seabed. *Science, 349*(6244), 144–145. https://doi.org/10.1126/science.aac6647

Weaver, P. P. E., Aguzzi, J., Boschen-Rose, R. E., Colaço, A., de Stigter, H., Gollner, S., et al. (2022). Assessing plume impacts caused by polymetallic nodule mining vehicles. *Marine Policy, 139* (2022) 105011. https://doi.org/10.1016/j.marpol.2022.105011.

Whitehead, H., Coakes, A., Jaquet, N., & Lusseau, S. (2008). Movements of sperm whales in the tropical Pacific. *Marine Ecology Progress Series, 361*, 291–300. https://doi.org/10.3354/meps07412

Willaert, K. (2020). Crafting the perfect deep sea mining legislation: A patchwork of national laws. *Marine Policy, 119*, 104055. https://doi.org/10.1016/j.marpol.2020.104055

Wiley, D. N., Ware, C., Bocconcelli, A., Cholewiak, D. M., Friedlander, A., & Thompson, M. (2011). Underwater components of humpback whale bubble-net feeding. *Behaviour, 148*(5-6), 575–602. https://doi.org/10.1163/000579511X570893

Würsig, B., & Kovacs, K. M. (2018). *Encyclopedia of marine mammals book* (3rd ed., B. Würsig, J. G. M. Thewissen, & K.M. Kovacs (Eds.)). 1157 pp. https://doi.org/10.1016/C2015-0-00820-6.

Dr. Virginie Tilot holds a PhD in biological oceanography, an ecological engineer degree (conservation and fisheries) and an MSc. She is honorary Attaché of the National Museum of Natural History, Paris (France) and is a corresponding member of the Académie des Sciences d'Outre-Mer of France and a titulary member of the Académie Royales des Sciences d'Outre-Mer of Belgium. Presently, she is a research scholar and senior international expert on marine environment, fisheries, deep-sea ecology, coral reefs, biological oceanography, conservation and sustainable development for diverse national and international organizations, institutions and universities, in particular for EU and agencies of the United Nations (UNEP, UNDP, FAO, Unesco/IOC, etc.). She has more than 35 years of experience and has produced a total of 115 publications. She has participated in 37 regional projects in the world, mainly in the Indian and Pacific oceans and the red sea.

Concerning the domain of deep-sea mining, Dr. Tilot has issued recently four peer-reviewed publications, several communications in international meetings, a series of three volumes in the Unesco/IOC Technical Series on *the Biodiversity and Distribution of faunal assemblages associated to polymetallic nodule ecosystem and options for the conservation and the management facing deep sea mining* as well as two chapters in two open access books edited by Springer on *Environmental Issues of Deep-Sea Mining. Impacts, Consequences and Policy perspectives* and on *Perspectives in Deep-Sea Mining*. This is the third commitment with two chapters for the Springer book on *Deep-Sea Mining and Water Column*.

Lea Fourchault holds an Erasmus Mundus MSc in tropical ecosystems and biodiversity. She is a young researcher interested in socio-ecological interactions, now working at the Royal Belgian Institute of Natural Sciences. Her current focus is on the resolution of taxonomic and phylogenetic uncertainties in under-researched vertebrates, using molecular techniques. Key research outcomes include a better understanding of the interrelation between humans, biodiversity and ecosystem health.

She gained experience in marine spatial planning and spatial prioritization techniques under the mentorship of Professor A. J. Richardson (Queensland University) and has recently published a paper on generating affordable protection to high seas biodiversity. Economic and political feasibility is a key part of her research, wherefore she developed a cross-sectoral prioritization approach that reduces the projected opportunity cost of MPA implementation to the fishing, shipping and deep-sea mining sectors simultaneously.

Dr. Alain Jeudy de Grissac holds a PhD in marine geology, sedimentology and oceanology from the University of Marseilles (France). With more than 40 years of experience with different international organisations or institutions, he is presently a corresponding member of the Académie des Sciences d'Outremer (France) and still active as an independent expert specialized in projects related to marine and coastal conservation, natural and cultural heritage, local community development and planning for sustainable development. His main focus is on marine and coastal protected areas, from the identification, creation, management and evaluation of individual sites or networks of sites to the development of national policies or action plans, improvement of national legislation and training of national administrations. He participated in the creation of more than 70 marine protected areas in the world.

Part V
Legal, Policy and Economic Issues of Deep-Sea Mining

Chapter 15
The "Water Column" and Seabed Mining in the Area: Selected Environmental Legal Considerations

Philomène Verlaan

Abstract The definition of the Area in Part I of the 1982 Law of the Sea Convention (the Convention) and further elaborations on the Area elsewhere in the Convention do not specifically name or include the water column overlying the Area, or include its specifically named and arguably likely equivalent in Part XI (The Area), i.e., "superjacent waters." Waters overlying the Area are part of the marine environment under the Convention. They are addressed not only in Part XI, but also in other Parts of the Convention, including, e.g., Parts VII (High Seas), XII (Protection and Preservation of the Marine Environment), and XIII (Marine Scientific Research). These waters are also subject to these Parts. They involve multiple actors. Given this complex legal context for these waters in the Convention, this chapter reflects on selected legal considerations under the Convention regarding the regulation of "activities in the Area" in and under these waters. Particular attention is given to the marine environmental protection requirements set out in the Convention for conducting activities associated with seabed mining in the Area.

Keywords Law of the Sea · Legal considerations · Marine environmental protection · Seabed mining

P. Verlaan (✉)
Department of Oceanography, University of Hawaii, Honolulu, HI, USA
e-mail: verlaan@hawaii.edu

R. Sharma (ed.), *Deep-Sea Mining and the Water Column*,
https://doi.org/10.1007/978-3-031-59060-3_15

1 Introduction

In "establishing … 'a legal order for the seas and oceans,'"[1] the 1982 United Nations Convention on the Law of the Sea (hereinafter LOSC or the Convention)[2] sets up a set of "sub-regimes,"[3] each with a named Part or Section, including the one of particular interest here: Part XI on "the Area."[4] Absent from this set of formally labeled "sub-regimes," and undefined in the Convention, is the water column, whose relationship with seabed mining (SBM)[5] is the topic of the present book. This chapter identifies and reflects on selected legal considerations arising under the Convention that apply to SBM in the Area insofar as it interacts with the overlying water column/superjacent waters. Particular attention is given to marine environmental protection requirements set out in the Convention for and/or relevant to SBM in the Area. This chapter does not address SBM or the water column/superjacent waters overlying the seabed within national jurisdiction.

2 Use of Terms in the Convention: "Water Column" and "Superjacent Waters"

The Convention refers to the "water column" only in Article 257.[6] Here it is undefined and specifically limited to its relationship with the Exclusive Economic Zone (EEZ): "beyond the limits of the [EEZ]" for purposes of Article 257, which does not refer to the Area.[7]

---

[1] LOSC Preamble, 4th *chapeau.*

[2] United Nations Convention on the Law of the Sea (Montego Bay, December 10, 1982, in force November 16, 1994) 1833 *UNTS* 396. All references to the Convention include and are compatible with the 1994 Part XI Implementing Agreement.

[3] LOSC "sub-regimes" include, e.g., Part II: Territorial Sea and Contiguous Zone; Part III: Straits Used for International Navigation; Part V: Exclusive Economic Zone (EEZ); Part VI: Continental Shelf; Part VII: High Seas; Part VIII: Islands; Part XII: Protection and Preservation of the Marine Environment (MEPP); Part XIII: Marine Scientific Research (MSR); Part XIV: Development and Transfer of Marine Technology; Part XV: Dispute Settlement.

[4] The Area is defined as: "the seabed and ocean floor and subsoil thereof, beyond the limits of national jurisdiction"; LOSC Part I (Introduction), Article 1 *Use of terms and scope*, 1. (1).

[5] In this chapter, SBM is used as a shorthand "term of art" to encompass all "activities in the Area" as defined in ibid., Article 1 *Use of terms and scope*, 1. (3): i.e., "all activities of exploration for, and exploitation of, the resources of the Area." See also International Tribunal for the Law of the Sea (ITLOS) Sea-Bed Disputes Chamber (SDC) Advisory Opinion, *Responsibilities and Obligations of States Sponsoring Persons and Entities with respect to Activities in the Area*, Case No. 17 (1 February 2011), ITLOS Reports 2011, p. 10 (SDC Advisory Opinion), at para 82.

[6] LOSC Part XIII (MSR), Article 257: *Marine scientific research in the water column beyond the exclusive economic zone.*

[7] Ibid., Article 256: *Marine scientific research in the Area.* It does not refer to the water column.

In two Articles specific to the Area,[8] i.e., one regarding the EEZ[9] and one regarding the continental shelf,[10] the Convention associates a "legal status" with a part of ocean space referred to as "superjacent waters." However, the Convention provides neither an eponymous specific "sub-regime" (or Part or Section), nor a definition for "superjacent waters." Nevertheless, these two Articles appear to be similar in their intent to distinguish the "seabed and subsoil [and ocean floor]"[11] from both "superjacent waters" and "the water column."[12] Under the Convention, it would seem to be appropriate to consider "superjacent waters" and "the water column" as being both physically (this would also be scientifically accurate) and legally different from the "seabed."[13]

It is unclear whether or not the Convention imputes or intends to impute a legally material distinction between "superjacent waters" and "the water column." The present author is unaware of any evidence, including any judicial interpretation, to this end. Other than raising the point for the sake of completeness, addressing it further here is unnecessary. Thus, for the purposes of this chapter, and for ease of

---

[8] LOSC Part XI (The Area), Article 135: *Legal status of the superjacent waters and air space.* "Neither this Part nor any rights granted or exercised pursuant thereto shall affect the legal status of the waters superjacent to the Area or that of the air space above those waters." Ibid., Article 155 (2): *The Review Conference.* "The Review Conference ... shall also ensure the maintenance of the principles laid down in this Part with regard to the legal status of the waters superjacent to the Area and that of the air space above those waters ..."

[9] LOSC Part V (EEZ), Article 56 (1): *Rights, jurisdiction and duties of the coastal State in the exclusive economic zone.* "In the [EEZ], the coastal State has: (a) sovereign rights for the purpose of exploring and exploiting, conserving and managing the natural resources, whether living or non-living, of the waters superjacent to the seabed ..."

[10] LOSC Part VI (Continental Shelf), Article 78 (1): *Legal status of the superjacent waters and air space and the rights and freedoms of other States.* "The rights of the coastal State over the continental shelf do not affect the legal status of the superjacent waters or of the air space above those waters."

[11] See *supra* note 4. Only the Preamble, 6th *chapeau* and Part VI (Continental Shelf), Article 76(1) *Definition of the continental shelf*, refer to the ocean floor.

[12] The Convention also distinguishes "the seabed and subsoil" from "waters" in, e.g., Part II (The Territorial Sea and Contiguous Zone); Part III (Straits Used for International Navigation); and Part IV (Archipelagic States).

[13] The present author has addressed issues arising from the difference between scientific and legal definitions in the international law, the law of the sea, and the Convention elsewhere (see, e.g., P. Verlaan, "New seafloor mapping technology and Article 76 of the 1982 United Nations Convention on the Law of the Sea," *Marine Policy* 21(5):425–434 (1997); P. Verlaan, "The Interface of Science and Law: A Challenge to the Privileging of 'Marine Biodiversity' over 'Marine Environment'" (invited paper), in R.A. Barnes and R. Long (eds), *Frontiers in International Environmental Law: Oceans and Climate Challenges: Essays in Honour of David Freestone*, pp. 409–429, Brill, Leiden, 2021 (https://brill.com/view/title/38675?language=en), most generously made available in open-access form by the World Maritime University (WMU) and the Sasakawa Global Ocean Institute at the WMU; P. A. P. Verlaan, "'The Interface of Science and Law: Protecting, Preserving and Conserving Biodiversity with the Law of the Sea Convention" (invited paper), *The Korean Journal of International and Comparative Law*, 10(2):222–239 (2022), doi: https://doi.org/10.1163/22134484-12340170

reading, it is assumed that no such distinction exists; when appropriate they are herein referred to collectively with the term of art "overlying waters.".[14]

3 Legal Regime and Selected Legal Considerations Applicable to "Overlying Waters" of the Area Under the Convention

3.1 Legal Regime

Given that neither term is defined, and given that the waters delineated by either term do not fall under national jurisdiction as defined in the Convention, the "sub-regime" of the High Seas is applicable to both superjacent waters and the water column, and also includes the term of art "overlying waters" proposed in Sect. 2 above to describe the waters overlying the seabed and subsoil of the Area.[15]

3.2 Selected Legal Considerations

3.2.1 Legal considerations deriving from the High Seas "sub-regime" applicable to these "overlying waters" include:

(a) SBM in the Area, insofar as it involves overlying waters, must take into account (at least)[16] two LOSC "sub-regimes": Parts VII[17] and XI. Part XI does not refer to the High Seas.

---

[14] This term is not found in the Convention. The present author employs it to describe the element common to both these terms, neither of which have, to the best of the author's knowledge, been judicially interpreted in this context, so as not to risk confusion as to any possible legal implications by using one or the other of them here, unless the context requires such a use. The author considers this approach to be consistent with the Vienna Convention on the Law of Treaties (Vienna, adopted May 23, 1969, in force January 27, 1980) 1155 *UNTS* 331, Section 3. Interpretation of treaties, Article 31: *General rule of interpretation.* "1. A treaty shall be interpreted in good faith in accordance with the ordinary meaning to be given to the terms of the treaty in their context and in the light of its object and purpose."

[15] LOSC Part VII (High Seas), Article 86: *Application of the provisions of this Part.* "The provisions of this Part apply to all parts of the sea that are not included in the [EEZ], in the territorial sea or in the internal waters of a State, or in the archipelagic waters of an archipelagic State." The Convention does not define the High Seas as such.

[16] See further *infra* on other LOSC "sub-regimes" applicable to the Area and its overlying waters.

[17] Ibid., Article 87 (2): *Freedom of the high seas.* "These freedoms shall be exercised by all States with due regard for the interests of other States in their exercise of the freedom of the high seas, *and also with due regard for the rights under this Convention with respect to activities in the Area.*" [Emphasis supplied.]

(b) "[t]he high seas are open to *all* States ..."[18] (emphasis supplied).
(c) in and on the High Seas, *all* States have the right to "[exercise]" ... "the freedom of the high seas ... under the conditions laid down by this Convention and by other rules of international law."[19] The Convention sets out a non-exhaustive list of six High Seas freedoms, none of which are specifically made subject to Part XI (the Area),[20] which must nevertheless be taken into account, as per item (d) immediately below.
(d) *all* States have the right to exercise all High Seas freedoms in and on waters overlying the Area "with due regard for the interests of other States in their exercise of the freedom of the high seas, and also with due regard for the rights under this Convention with respect to activities in the Area."[21]

3.2.2 Legal considerations deriving from the interaction between the High Seas "sub-regime" applicable to these "overlying waters" and Part XI (the Area) with regard to activities in the Area and the mandate of the International Seabed Authority (the ISA or the Authority) include:

(a) LOSC Part XI (the Area) Article 147 (1) and (3) introduces a further nuance to the Part VII (High Seas) "due regard" requirement for the Area.[22] Article 147 recognizes and distinguishes between two types of

[18] Ibid., Article 87 (1): *Freedom of the high seas.* "All States" includes non-parties to the LOSC. These freedoms may only be exercised by States.

[19] Ibid. These conditions include, e.g., LOSC Part XI (The Area) and Part XII (MEPP), as these are not specifically excluded. See, e.g., B. Oxman, "The High Seas and the International Seabed Area," *Michigan Journal of International Law* 10:526–542 (1989). Available at: https://repository.law.umich.edu/mjil/vol10/iss2/8 (accessed May 7, 2023).

[20] Ibid. Three of the six listed freedoms are made subject to specifically named other LOSC Parts (laying of submarine cables and pipelines: Part VI; construction of artificial islands and other installations: Part VI; scientific research: Parts VI and XIII; the word "marine" is not used here) and one (fishing) is made subject to Section 2 (Conservation and Management of the Living Resources of the High Seas) of the present Part VII.

[21] Ibid., Article 87 (2): *Freedom of the high seas.* The meaning of "due regard" in the context of the Area is discussed in the scholarly literature. See, e.g., Oxman, *supra* note 19; T. Treves, "'Due regard' obligations under the 1982 UN Convention on the Law of the Sea: the laying of cables and activities in the Area," *The International Journal of Marine and Coastal Law* 34(2):167–194 (2019), doi: https://doi.org/10.1163/15718085-23421085 (accessed May 7, 2023); D. Kroon, "'Due regard"'in the high seas: the tension between submarine cables and deep seabed mining. *Australian International Law Journal*, vol. 24, annual 2018, pp. 35+ (2018), Gale Academic OneFile, link.gale.com/apps/doc/A629439118/AONE?u=anon~3fc0bb87&sid=googleScholar&xid=0a23320d (accessed May 7, 2023).

[22] LOSC Part XI (The Area), Article 147, *Accommodation of activities in the Area and in the marine environment.* LOSC Article 147 (1): "Activities in the Area shall be carried out with reasonable regard for other activities in the marine environment." LOSC Article 147 (3): "Other activities in the marine environment shall be conducted with reasonable regard for activities in the Area." The present author does not address the legal implications, if any, regarding the use of "reasonable regard" in LOSC Part XI (The Area), and "due regard" in LOSC Part VII (High Seas), and regarding the use of "carried out" and "conducted" in, respectively, LOSC Article 147 (1) and (2) with

activities under Part XI (the Area) in terms of their location: "activities in the Area" (defined in LOSC Article 1)[23] and "other activities in the marine environment." The latter activities are not defined in the Convention. The Seabed Disputes Chamber (SDC) of the International Tribunal for the Law of the Sea (ITLOS) addresses the former in some detail,[24] but does not specifically define the latter. The SDC offers a rationale and examples[25] of what are *not* "activities in the Area." It does not exclude the interpretation that those activities are, or could be, part of "other activities in the marine environment."

(b) The SDC limits the ISA's mandate to "adopt and uniformly apply rules, regulations and procedures [hereinafter RRP] regarding those "other activities" if their actual or potential harmful effects on the marine environment "do not arise from such "activities [in the Area]."[26] Nevertheless, the ISA "shall [mandatory] adopt and uniformly apply [RRP] ... on ...: prevention of interference [by SBM] with other [legitimate] activities in the marine environment."[27] It must do so because the Convention permits non-contractor actors to operate in and on the Area's "overlying waters," and, under appropriate circumstances, in the Area itself. It also permits contractors engaged in "activities in the Area" within the meaning of LOSC Part XI (the Area) to do so.

(c) LOSC Part XI and the ISA's mandate cover the regulation of activities in the Area as defined in the Convention. Other activities are "permitted," in the sense that they fall under provisions or freedoms elsewhere in the Convention, are not included in the definition of "activities in the Area," and other than the "reasonable regard" requirement,[28] are covered by the regime of the Area.

---

regard to the Area. This is because any that may exist are unlikely to be relevant for the purposes of the present paper. Oxman, *supra* note 19, does not refer to any potentially legally material difference in this context.

[23] See *supra* note 5 for LOSC definition of "activities in the Area."

[24] See SDC Advisory Opinion, *supra* note 5, at paras 82–97.

[25] Not included as "activities in the Area" according to the SDC are, for example, "transporting, processing and marketing of minerals recovered from the Area" (LOSC Annex IV, Statute of the Enterprise, Article 1, *Purposes* (1)), at para 84.

[26] See SDC Advisory Opinion, *supra* note 5, at paras 85–88.

[27] LOSC Annex III, Basic Conditions of Prospecting, Exploration and Exploitation (hereinafter, Basic Conditions), Article 17 *Rules, regulations and procedures* [RRP] *of the Authority.* Article 17 (1)(b)(ix). The Convention frequently employs, without definition, the concept of "interference," including for activities in the Area, e.g., in Part XI, Article 145 (a) *Protection of the marine environment* (see further below in Section 4 on MEPP), LOSC Article 147(2)(b) (on "installations used for carrying out activities in the Area"), LOSC Annex III (Basic Conditions), Article 16, *Exclusive right to explore and exploit,* and ibid., Article 17 (2)(d)(ix). The present author interprets this requirement as at least providing further detail regarding the Authority's duty to develop and adopt RRP to address such interference.

[28] LOSC Article 147 (3), *supra* note 22.

4 Protection and Preservation of the Marine Environment and "Overlying Waters" of the Area Under the Convention

Under the Convention, these "Overlying Waters" are part of the marine environment. Hence, in addition to Parts VII (High Seas) and XI (the Area), a third LOSC sub-regime governs the overlying waters of the Area and thus also governs activities carried out there that are related to exploration for, and/or exploitation of resources of the Area (including seabed mining): Part XII (MEPP). Selected legal considerations deriving from the applicability of Part XII to these "overlying waters" are discussed in this Section.

The MEPP regime is perhaps the only sub-regime in the Convention[29] that not only considers, but also fully treats the world ocean, from internal waters seaward, as a unified whole for MEPP purposes, in accordance with the ocean's fundamentally interconnected nature, as recognized in the Preamble.[30] The Convention requires States "to protect and preserve the marine environment" without exceptions or qualifications,[31] including within and/or for geopolitical jurisdictions and other "sub-regimes" established in the oceans elsewhere in the Convention.[32] In addition to being assigned their own lengthy chapter (Part XII), strong MEPP provisions are found throughout the Convention, beginning with its very first Article,[33] including for the Area[34] and for SBM, as it is an "activity in the Area."[35]

---

[29] It is, to the best of the author's knowledge, also the only international multi-lateral treaty that does so.

[30] The third *chapeau* of the Preamble to the Convention states: "*Conscious* that the problems of ocean space are *closely interrelated,*" followed by an ambitiously inclusive objective: "and need to be considered *as a whole, ...*" [Emphasis supplied.] See also Verlaan (2022), *supra* note 13.

[31] LOSC Part XII (MEPP), Article 192.

[32] See, e.g., *supra* note 3 for examples of such 'sub-regimes'. See also: "the obligations in Part XII apply to all States with respect to the marine environment in all maritime areas, *both inside the national jurisdiction of States and beyond it.*" [Emphasis supplied.] Permanent Court of Arbitration (PCA) Case No. 2013–19, The South China Sea Arbitration Award, *The Republic of the Philippines v. The People's Republic of China* (July 12, 2016) [SCS Award], at para 940; see further paras 939–946 and references therein.

[33] LOSC Part I (Introduction), Article 1 *Use of terms and scope*, 1. (4) defines pollution of the marine environment. See also, e.g., SCS Award, *supra* note 31, at para 943.

[34] LOSC Part XI (The Area), Article 142 *Rights and legitimate interests of coastal States*; ibid., Article 145 *Protection of the marine environment*; LOSC Part XII (MEPP), Article 209 *Pollution from activities in the Area*; ibid., Article 215 *Enforcement with respect to pollution from activities in the Area*; LOSC Annex III (Basic Conditions) Article 17 *Rules, regulations and procedures of the Authority* Article 1(b)(xii); Article 2(f). With regard to MEPP in the Area, see also P. Verlaan, "Future of Deep-Seabed Mineral Resources: Environmental Issues" in A. Ascencio-Herrera and M.H. Nordquist (eds) *The United Nations Convention on the Law of the Sea, Part XI Regime and the International Seabed Authority: A Twenty-Five Year Journey,* pp. 198–216. Brill/Nijhoff, Leiden, 2021, doi: https://doi.org/10.1163/9789004507388_014

[35] Ibid. See also *supra* note 5: "activities in the Area" include exploration for and exploitation of resources in the Area. See also SDC Advisory Opinion (*supra* note 5), at para 96: "... transportation within that part of the high seas, when directly connected with extraction and lifting, should be included in activities in the Area. In the case of polymetallic nodules, this applies, for instance, to transportation between the ship or installation where the lifting process ends and another ship or

The Convention does not qualify or make exceptions to the Part XII (MEPP) provisions for the Area, its "overlying waters,"[36] or "activities in the Area" (e.g., SBM) in any of its designated "sub-regimes" or elsewhere.[37] The MEPP provisions for "overlying waters" of the Area regarding "effective protection for the marine environment from harmful effects *which may arise from such activities*" [emphasis supplied], i.e., "activities in the Area," may under appropriate circumstances extend as far as "the coastline,"[38] i.e., well beyond the Area itself, given that "overlying waters" are part of the marine environment, as explained above.[39] Note that although the ISA's MEPP jurisdiction is *spatially* much wider than the Area itself, its *subject matter* jurisdiction for MEPP is specifically limited to addressing "harmful effects potentially or actually arising from activities in the Area."[40] Implementation of the MEPP provisions for the Area, and for its "overlying waters," requires, inter alia, adoption by the ISA of, e.g., "appropriate" [RRP] for 'prevention of damage to the flora and fauna of the marine environment',[41] as well as "(a) the prevention, reduction and control of pollution and other hazards to the marine environment, including the coastline, and of interference with the ecological balance of the marine environment, …"[42] but always linked to effects from activities in the Area.[43]

The Convention's Part XII (MEPP) provisions apply to *all* actors operating in both domains, the "overlying waters" and the underlying seabed/subsoil, of the world ocean, which includes the Area. Furthermore, contractors within the meaning of LOSC Part XI (The Area) are additionally subject to the MEPP provisions of LOSC Part XI (The Area), Article 145 and the RRP adopted by the ISA to

installation where the evacuation of water and the preliminary separation and disposal of material to be discarded take place."

[36] "Overlying waters" as High Seas are subject to, e.g., the provisions of Part XII (MEPP): "Freedom of the high seas is exercised *under the conditions laid down by this Convention* and by other rules of international law." [Emphasis supplied.] LOSC Part VII (High Seas), Article 87(1). No exceptions are made for "overlying waters" of the Area.

[37] No exceptions are made for "activities in the Area," either as such or insofar as they affect/are conducted in or on the "overlying waters," either in Part XII (MEPP) or in LOSC Part XI (the Area), Article 145: "Necessary measures shall [must] be taken *in accordance with this Convention* with respect to activities in the Area to ensure effective protection for the marine environment from harmful effects which may arise from such activities." [Emphasis supplied.] The emphasized phrase eliminates any uncertainty regarding applicability of Part XII (MEPP) to "activities in the Area" …

[38] LOSC Part XI (the Area), Articles 142(3) and 145(a).

[39] See *supra* Sect. 3.2 in this chapter.

[40] See SDC Advisory Opinion, *supra* notes 5 and 34, and related discussion in text and notes.

[41] LOSC Part XI (the Area), Article 145 (b); "prevention" is the *only* authorized RRP measure here, "reduction and control" are not included, and therefore are not options.

[42] Ibid., Article 145 (a).

[43] In this context the inclusion of "climate change" as a relevant criterion in the current version of the Draft Exploitation Regulations on MEPP is legally problematic, given the requirement to demonstrate the harmful effects on the marine environment *attributable to SBM* from climate change. This problem with the Draft Exploitation Regulations is not contradicted by the SDC Advisory Opinion (*supra* note 5); indeed, it is arguably reinforced.

implement those provisions. However, non-contractor actors, i.e., States,[44] in both domains that are exercising High Seas freedoms set out in and pursuant to LOSC Part VII (High Seas) and other parts of the Convention applicable to these freedoms, are not necessarily obliged to comply with MEPP provisions *adopted by the ISA* for contractors under LOSC Part XI (the Area), Article 145. These freedoms include for both domains (in alphabetical order): cable-laying, construction of artificial islands and other installations, fishing, navigation, and scientific research.

This environmentally and legally unsatisfactory situation could be addressed by developing a regulatory regime based on the legal premise that *all* actors at sea are obliged under the Convention (as well as under international law[45]) to be under the jurisdiction and[46]/or[47] control[48] of a State.[49] Furthermore, for purposes of LOSC Part XI (the Area), if the actor is a contractor,[50] that actor must be "effectively controlled" by [States Parties] or their "nationals."[51] *All* Parties[52] to the Convention are *also* members of the ISA.

With regard to the MEPP provisions of the Convention, at least, these apply equally to States, States-Parties and, for purposes of LOSC Part XI (the Area), States-Parties sponsoring or operating as LOSC Part XI contractors,[53] as well as to other parties to the Convention,[54] especially where implementation of LOSC Part XI (the Area) Article 145 and the ISA's RRP to that end are concerned. However, the specific rights and duties of these actors may differ depending on the type of party concerned and the circumstances involved.

The "due/reasonable regard" provisions of the Convention[55] and the rules on treaty interpretation[56] may not suffice to address potential questions of overlapping/

---

[44] See *supra* note 18.

[45] Only the requirements under the Convention are discussed here.

[46] Jurisdiction and control: LOSC Part VII (High Seas), Article 94 *Duties of the flag State.* Article 94 (1) and (6);

[47] Jurisdiction or control: LOSC Part XII (MEPP), Article 194 *Measures to prevent, reduce and control pollution of the marine environment.* Article 194 (2); Article 196 *Use of technologies or introduction of alien or new species.* Article 196 (1); Article 206 *Assessment of potential effects of activities.*

[48] The distinction between the two is not discussed here as the point made is relevant to both.

[49] Though not necessarily of a State that is a party to the Convention.

[50] Ibid.

[51] As defined in LOSC Part XI (the Area), Article 153 *System of exploration and exploitation,* Article 153(2)(b).

[52] Parties to the Convention include international organizations that meet the requirements of LOSC Part XVII (Final Provisions) Article 305 *Signature* and Annex IX (Participation by International Organizations). The European Union is a party to the Convention pursuant to these provisions.

[53] See *supra* note 51.

[54] See *supra* note 52.

[55] See discussion *supra* in Sect. 3.2 and accompanying notes.

[56] See *supra* note 14.

conflicting rights and duties predictably and consistently.[57] Recourse to judicial review is likely to be necessary.[58]

5 Conclusion

The Area's "overlying waters" are subject to the High Seas provisions of the Convention, to which the legal consequences (including rights and duties) thereof, as set out in, in particular, LOSC Part VII (High Seas), apply, both for SBM and other "activities in the Area." All States, including non-Parties to the Convention, have the right to exercise High Seas freedoms in the Area's "overlying waters," as well as in the Area itself, subject to "due/reasonable regard" conditions and other requirements set out in the Convention.

As well as contractors engaged in "activities in the Area" within the meaning of LOSC Part XI (the Area), the Convention permits non-contractor actors to operate in and on the Area's "overlying waters," and in the Area itself, under certain conditions, as set out in, e.g., LOSC Part XI (the Area) and other "sub-regimes" in the Convention.

The Convention's MEPP provisions apply to all actors in both domains—"overlying waters" and the seabed/subsoil—and to the various "sub-regimes" set up in the Convention. The potential and/or actual presence of multiple actors with different interests, rights, and duties in the two domains contributes to the regulatory complexity of implementing the Convention's MEPP provisions. This complexity faces States-Parties to the Convention as members of the ISA, States-Parties when acting as non-ISA members, and as States in their own right (the latter includes States who are not Parties to the Convention).

The Convention offers several avenues to address potential or actual operational uncertainties for different users that result from the lack of clarity regarding the MEPP rights and obligations on this multi-dimensional legal interface between the Area and its "overlying waters." These include, in particular, the extensive range of dispute settlement options set out in Part XV (Settlement of Disputes). Part XI (the Area) also sets out dispute settlement options for SBM,[59] including establishment of the SDC[60] specifically for matters related to SBM, of which a particular advantage is that its jurisdiction also covers disputes involving certain defined natural or juridical persons[61] under defined circumstances. The Assembly and the Council of the

---

[57] The current [as of 30 June 2023] version of the Draft Exploitation Regulations also provides that *in conjunction with member States*, the Authority shall take measures to ensure that *other activities in the marine environment* shall be conducted with reasonable regard for the activities of Contractors in the Area (DR 31 (2); emphasis supplied). Unfortunately, this does not solve the problem of *non-Contractor* actors.

[58] See discussion *infra* on dispute resolution (Conclusion) in the present paper.

[59] See, e.g., LOSC Part XI (the Area), Section 5, *Settlement of disputes and advisory opinions*.

[60] LOSC Article 186.

[61] LOSC Article 187 (c), (d) and (e). These "persons" include "private" contractors as defined in LOSC Article 153 (2) (b).

ISA may also request an advisory opinion from the SDC "on legal questions arising within the scope of their activities."[62]

Thus, the undoubtedly complicated legal situation presented by "overlying waters" in the context of SBM is provided with ample means under the Convention to achieve "the peaceful uses of the seas and oceans, the equitable and efficient utilization of their resources, the conservation of their living resources, and the study, protection and preservation of the marine environment."[63]

Acknowledgments The excellent comments from three reviewers are greatly appreciated.

Dr. Philomène Verlaan is an oceanographer, specialized in the biogeochemistry and ecology of deep-sea ferro-manganese nodules and crusts (Ph.D., Imperial College London) with extensive sea-going experience (23—so far—oceanographic research cruises and nine submersible dives). She is also an attorney-at-law specialized in international law of the sea (J.D., Florida State University; Member of the Florida Bar). She assists international public and private organizations in negotiating the complex interface between marine science and the law of the sea to achieve environmentally and commercially responsible uses of marine resources compatible with the Law of the Sea Convention, the world's "Constitution for the Oceans." Author of over 50 (so far) refereed publications, she is a Visiting Colleague at the Department of Oceanography, University of Hawai'i, Senior Technical Adviser, Advisory Committee on Protection of the Sea, Coordinator of the International Marine Minerals Society's Code of Environmental Management of Marine Mining, on the editorial board for Marine Georesources and Geotechnology and the International Journal of Marine and Coastal Law, Member, World Commission on Environmental Law, and Fellow of the Institute of Marine Engineering, Science, and Technology.

[62] LOSC Article 191.

[63] LOSC, Preamble, 4th chapeau.

Chapter 16
Economic Policy Considerations for Deep-Sea Mining

Kaja von der Leyen, Mads Fjeld Wold, and Taryn Ann Galloway

Abstract Bypassing the normative question of whether exploration for, and mining of, deep-sea minerals is desirable, this chapter dwells on the following question: How should governments manage these resources to maximize the economic benefits to society? We use theoretical and empirical contributions in economic science to discuss matters of taxation and possible public interventions related to deep-sea mineral activities. We delve particularly on considerations related to the high uncertainty about profitability and the related lack of data about the deep-sea areas concerned as well as the possible emergence of a resource rent.

Keywords Economic science · Deep-sea minerals · Resource rent taxation · Market failure · Norwegian case

1 Introduction

Although the topic is controversial, there is considerable interest in establishing large-scale deep-sea mining (DSM) of minerals from the seabed. It is therefore timely to consider relevant policy issues since the economic consequences of DSM-related activity not only depend on the value of the resources on the seabed but also on how governments chose to manage the economic activities that these resources bring about. By looking into this, we do not suggest that deep-sea minerals should be mined. Rather, we acknowledge that individuals that are interested in the

K. von der Leyen (✉) · M. F. Wold
NTNU Social Research, Center for Economic Research, Trondheim, Norway
e-mail: kaja.leyen@samforsk.no

T. A. Galloway
Asplan Viak AS, Sandvika, Norway

R. Sharma (ed.), *Deep-Sea Mining and the Water Column*,
https://doi.org/10.1007/978-3-031-59060-3_16

potential of DSM-related activities to impact society in a beneficial way should also be informed about the consequences of accompanying choices made by governments.

While no commercial-scale DSM exists to date, there is a wealth of relevant research to draw on when considering the possible economic and social impacts of DSM-related activities. We believe that economic science is particularly well positioned to contribute. Much theoretical and empirical work has been done in microeconomics and public economics on taxation, the scope of government, as well as the management of natural resources. The goal of this chapter is hence to provide a summary of the main lessons from this literature as well as to relate its findings to DSM-related activity.

We will draw on the Norwegian state's approach to managing natural resources, in particular petroleum resources, to provide the reader with specific examples. The reason for choosing the Norwegian case is twofold. The country has abundant natural resources and hence extensive experience with the management of them. Moreover, deep-sea mineral exploration and mining in Norway is of particular interest at the moment as the government has just proposed opening an area the size of 281,200 km^2 of the Norwegian continental shelf to commercial exploration for minerals (Meld. St. nr. 25 (2022–2023)).[1] The issue is yet to be debated in parliament.

This chapter is organized as follows: Part two deals with the income side of the public budget by looking at optimal taxation of economic activity, zooming in on resources rent taxation. We then move on to examine the effects and side-effects of public expenditure and regulations. Part three discusses the possible justifications for government interventions in private markets. Part four relates this discussion to characteristics of deep-sea mineral activities, whereas part five discusses the pros and cons of specific types of government interventions. Part six concludes the chapter with our recommendations for public management of DSM-related activity.

2 Optimal Tax Policy

On average, tax revenue as a percentage of GDP in the OECD area was 34.1% in 2021 (OECD, 2022). Hence, it goes without saying that taxation has important economic impacts. Taxation not only enables public spending but also has other impacts mainly because it affects relative prices, which in turn affects the incentives that underlie individual behavior. In turn, this has effects on macroeconomic variables such as aggregate savings, investments, and, ultimately, the production. Yet, it is not only the size of tax system that matters. Economists Akgun et al. (2017) for instance used cross-country data and showed that the tax mix matters both for inequality and growth, which is explained in this part of chapter.

[1] While this document also contains the government's proposed strategy for managing potential DSMs, it was released too late for us to use much to exemplify solutions to regulatory challenges linked to DSM activity in this study.

2.1 Lessons from Economic Studies of Taxation

The tax system has several economic benefits, and the tax mix will determine which of them is the most palpable. By working in a counter-cyclical way, taxes contribute to stabilizing the business cycle so that large downturns and periods of overheating with high inflation can be avoided. This is possible because tax revenues automatically go up in periods of strong income growth, thereby reducing aggregate demand so long as they are not spent over the public budget and vice versa during slumps. Furthermore, taxes can reduce inequality by redistributing income from high-income earners to low-income earners. Many taxes are progressive—the tax rate increases with the value of the tax base—and often the tax base itself (for example, property, wealth, or consumption items) is primarily owned by individuals with higher income. Finally, an important task of the tax system is to enable public spending.

A tax that is *immobile* is—all else equal—preferable to a tax base that can easily be moved between tax jurisdictions to the one with the lowest tax rates. Such tax avoidance behavior causes a drop in public revenues and hence the room for public spending either through a drop in the tax base or by bringing about a lower tax rate than would otherwise be the case.

Most taxes affect the choices individuals make—for better and for worse. In some instances, they can lead to more efficient outcomes. Environmental taxes are an example of taxes that can yield more efficient outcomes since they force individuals to consider the environmental impact of their behavior. However, most taxes create what economists refer to as a deadweight loss to society because with taxes individuals' behavior depends on market prices and hence is sub-optimal. To achieve more efficient use of society's resources, it is paramount that decisions about consumption, production, investment, and work effort reflect the true incomes and costs thus generated. Ideally, they are reflected in market prices.

There are however taxes that are so called neutral, meaning that they do not affect individual behavior to a great extent. *Neutral* taxes provide the state budget with revenues without causing a deadweight loss. One example of this is resource rent taxation, which we will shortly return to. Such revenues can make it possible to reduce other non-neutral taxes that cause a greater efficiency loss to society, thereby leading to more efficient use of resources and thus in finite increase societal well-being.

This goes to show that one cannot see the various tax bases in isolation. This is all the more important when deciding how a new type of economic activity should be subject to taxation. An important principle for an efficient overall tax system is that of *sector-neutrality*. If taxation varies between sectors, it will have negative societal effects because investment- and production decisions not only depend on the value-creation happening in each sector of the economy. Furthermore, for the same reason, *geographical neutrality* is preferable, since if taxation varies between geographical areas within a country, it will also distort investment and production decisions away from maximizing value creation.

Finally, given multiple tax bases, a goal of the tax system is to *avoid double-taxation* meaning that taxes are paid twice on the same source of income. That can occur when, for example, income is taxed at both the corporate level and personal level. Intuitively, it makes sense that one should not pay more than once for the same income, but another argument for not having double-taxation is to avoid that effective tax rates become too high, pushing up the deadweight loss associated with taxation.

2.2 The Nature of Resource Rents

Resource rent is a common theoretical concept in economics. Economic rent is the term used to denote the profit a firm is left with after all production factors, including physical capital and labor, have received their market remuneration. When applied to an activity where natural capital, i.e., a natural resource, is involved, economists usually refer to economic rent as resource rent.

We highlighted that resource rent is a theoretical concept because in practice it can be difficult to determine what share of the returns from an economic activity involving a natural resource should be attributed to the resource itself. A necessary, but not sufficient, condition for there to be a resource rent is scarcity. If the supply of, for instance, oil is somehow restricted and demand is greater, then supply prices will increase, and this could generate an extra surplus of funds for oil producers.

There are, however, several other conditions that must be fulfilled for there to be resource rents in an industry. Costs play an important role in this regard. Take the fisheries industry as an example; for a long time, most capture fisheries around the world were not profitable, even though a central, and arguably scarce, input—the fish—was provided for free. In later years, the industry in several countries has become more professional, and the resource stock better managed such that costs have been reduced. This has led to higher returns for fishermen. Costs, therefore, play an important role in the potential for resource rents in a resource industry.

Further, scarcity could come about due to government intervention and not because the natural resource itself is limited in supply. Production in natural resource industries can be capped to reduce emissions, prevent overuse of the resource in question, or to protect vulnerable nature, among other reasons. In aquaculture, for instance, it is common for regulators to put a cap on total allowed biomass in production facilities in an attempt to reduce salmon lice and minimize the environmental impact on the surrounding fjords. This can give rise to what some economists call *regulation rent*. The idea is that regulation limits output thereby creating a scarcity. This often leads to higher prices for a given level of demand.

2.3 Existing Resource Rent Taxation in Norway

Taxing natural capital is often preferred by governments over taxing other types of capital. One reason is that the tax base (the natural capital) in many cases is less mobile than other types of capital. Consequently, a high tax on natural resources in one jurisdiction is less likely to push activity to another jurisdiction, where the tax base is lower.

Another, more ideological, reason for why resource rent taxation is popular is that governments often regard natural resources as belonging to the people. The income from taxing the use of natural resources can in this regard be seen as a renumeration to the people, from the companies that make use of their natural resources.

The most well-known example of resource rent taxation in Norway is that of the petroleum industry. Norway's history as an oil-producing nation started in 1969 when the oil field *ekofisk* was declared commercially viable. Today, it is by far the largest industry in Norway. The industry is taxed at 78%—22 percent in ordinary taxes and a 56% resource rent tax. Figure 16.1 shows the net government cash flow from petroleum activities between 1971 and 2021.

In Norway, there is a broad consensus that petroleum tax system should be neutral—meaning that a company looking to invest in an oil-related project should not refrain from investment due to resource rent taxation. To ensure that this is the case, there are several mechanisms in the tax system, such as the ability for companies to carry forward costs that can later be used to reduce their tax base. The regulators' goal with the system is to capture as much as the resource rent as possible without stifling investments.

The immobility of the petroleum resource is a crucial feature of this the tax system. International firms cannot move one of their inputs, the oil resource, to another

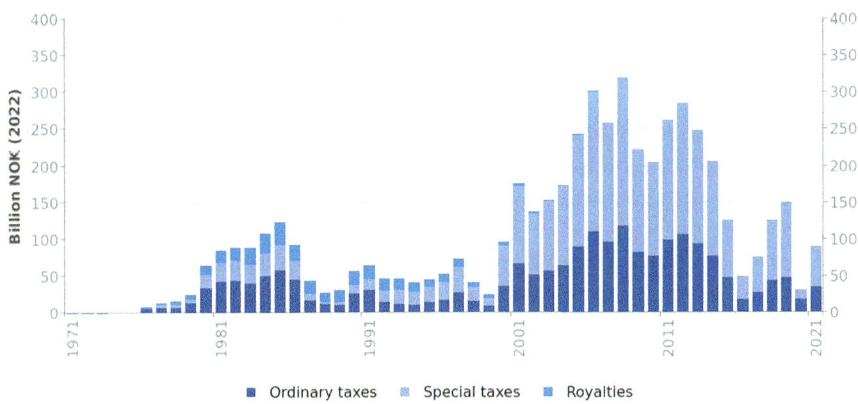

Source: www.norskpetroleum.no

Fig. 16.1 Net government cash flow from petroleum activities in Norway. (Source: www.norskpetroleum.no)

tax jurisdiction to avoid the high tax rate. If the firms are able to generate a profit that is similar to, or higher, than what is normal in other industries, they will operate in Norway and generate rents for both themselves and the government. The high tax rate clearly has not deterred companies from investing and operating in Norway, so it appears that the petroleum tax system has worked.

2.4 Implications for Taxation of DSM-Related Activities

Commercial DSM-related activities can generate tax revenues from a variety of tax bases. The composition and amount of these revenues will differ from country to county depending on the tax mix as well as tax rates. Furthermore, the revenues can accrue to both central and local governments depending on the institutional structure of the given country.

Companies that establish or expand production and hence earnings generate corporate income taxation revenues. The owners of these companies, when they choose to take out dividends, may pay capital taxation. The accompanying increase in employment generates increases in revenues from personal taxation and, for some countries, contributions to social insurance. To the extent that these corporate and individual revenues increase savings, they are also in some countries subject to wealth taxation. These effects will occur both in the DSM sector and in the sectors that supply its production inputs, yet the way these taxes are designed will have important consequences.

Resource rent taxation is a common way for governments to raise additional revenue from economic activity that makes use of natural resources such as oil, hydropower, or deep-sea minerals. The tax is usually levied on top of other taxes, and governments tend to motivate this special treatment by classifying the resource as belonging to the people. As discussed above, taxing natural resources is theoretically appealing for many reasons, one being their supposed neutrality, i.e., they don't distort investment decisions. Also as discussed above, it is not given that natural resources are associated with significant resource rents.

Metals found in deep-sea minerals are used for a wide range of applications, such as batteries and wind turbines, that are important for lower carbon emissions. Many of these metals can be found on the EU-commission's list of critical raw materials, which identify materials that are both important to the EU economy *and* are associated with considerable risk on the supply side. The high demand for metals such as nickel and copper, for which deep-sea minerals are a source, combined with the limited global supply from existing deposits, will—other things equal—put a strong upward pressure on prices for these metals. Higher prices will normally motivate firms to increase supply, and the metals that can be retrieved from nodule deposits is estimated to be significantly higher than the current global resource base on land (Hein et al., 2020). At the same time, increasing production might be a challenge;

there is much uncertainty surrounding the potential supply of metals from the seabed globally, and the viability of current and proposed production technologies is largely unknown.

If supply of certain minerals is unable to respond to the high prices, it could lead to high resource rent generation in the DSM industry. This of course depends on firms being able to generate profits. A firm's ability to generate profits will largely depend on its production costs, which in deep-sea mining are currently highly uncertain and probably strongly related to the type of deposit (nodules or massive sulfides). In sum, there is a potential for resource rent generation from DSM and the subsequent processing of deep-sea minerals, but large uncertainties on the supply side will affect how large it will be.

Regardless of whether a significant resource rent emerges, it is clear from the discussion on optimal taxation above that taxation of DSM-related activities should align with the tax system of the country that has ownership of the minerals. There may however also be good reasons to deviate from optimal taxation principles such as that of neutrality. The tax system, as well as other types of public interventions, can be used as a tool by governments to address market failures, which we will now address.

3 When Should Governments Seek to Influence Commercial Activity?

As a starting point for the discussion that will follow, we clarify our understanding of free markets. They are markets in which prices are free to adjust without legal restriction, not markets without any form of government intervention. As a minimum, the state should enforce property rights as well as laws against fraud. The government interventions that we focus on in this chapter therefore either directly or indirectly affect the price of a given good or service. The latter refer to regulations that are tantamount to price regulations, such as the need for a license to perform certain economic activities.

According to Adam Smith's invisible hand, selfishness leads markets to produce whatever people want; to get rich, you have to sell what the public is eager to buy. When markets work as they should, prices give companies and consumers correct signals about the opportunity costs of resources and hence ensures that—although each individual acts self-interestedly—the social optimum is obtained. Yet, markets do not always work as they should. In the following, we give an overview of these so-called market failures.

3.1 Externalities[2]

Externalities occur when the (market) interaction of certain agents impact others not directly involved in the interaction. Externalities undermine the social benefits of individual selfishness. If selfish consumers do not have to pay producers for benefits, they will not pay; and if selfish producers are not paid, they will not produce. A valuable product thus fails to appear. The problem, as David Friedman aptly explains, "is not that one person pays for what someone else gets but that nobody pays and nobody gets, even though the good is worth more than it would cost to produce" (Friedman, 1996, p. 278). Admittedly, the real world is rarely so stark. Externalities usually lead to underproduction rather than the nonexistence of the good.

Research and development (R&D) is a standard example of a positive externality, air pollution of a negative externality. Ultimately, however, the distinction is semantic. It is equivalent to say "clean air has positive externalities and so clean air is underproduced" or "dirty air has negative externalities and so dirty air is overproduced."

While the concept of externalities is not very controversial in economics, its application is. Defenders of free markets usually argue that externalities are manageably small; critics of free markets see externalities as widespread, even ubiquitous.

Instead of arguing about whether something is an externality, it can therefore be more productive to ask about transactions costs. If transactions costs are reasonably low, then the affected parties negotiate tolerably efficient solutions without government intervention. In the book "The Problem of Social Cost," Coase (1960) bypasses the earlier view that it is literally impossible to charge for some benefits. Instead, he observes that every exchange has some transactions costs, which vary from negligible—such as putting coins into a vending machine—to enormous—such as negotiating a contract with six billion signatories to improve air quality.

It is worth noting that while externalities are often blamed for market failure, they are also a source of government failure. Many economists who study politics decry the large negative externalities of voter ignorance. An economic illiterate who votes for protectionism hurts not just himself but also his fellow citizens (Caplan, 2003; Downs, 1957). Other economists believe that externalities in the budget process lead to wasteful spending. A congressman who lobbies for federal funds for his district improves his chances of reelection but hurts the financial health of the rest of the nation.

[2] Much of this text is taken from the following blogpost, which we find very useful to understand this concept (downloaded on July 4th): https://www.econlib.org/library/Enc/Externalities.html

3.2 Incomplete Information

Incomplete information means that the free-competition assumption of complete information of all aspects that will impact the supply and the demand of a good or a service is violated. It is common to distinguish between two forms of incomplete information: unavailable information and asymmetrical information. The former relates to a situation where a consumer does not know of the existence of a good or does not have adequate price information in his or hers decision-making process, whereas the latter refers to a situation where different agents are privy to different types of information.

Both lead to sub-optimal market outcomes. An obvious shortcoming of asymmetric information is that certain agents can exploit their informational upper hand to their advantage. A less obvious—but nonetheless important—consequence is that rational agents will be aware of the informational disadvantage and hence refuse to pay more than a given price for goods and services, which in turn will repel high-quality goods and services from markets (Akerlof, 1970).

Market-based solutions can in some cases resolve problems related to information. However, they mostly rely on repeated interactions between agents (for instance reputation effects), which is not always possible (for instance in the case of medical treatments) or efficient (for instance in the case of house refurbishments).

3.3 Public Goods

Public goods (also referred to as a social good or collective good) is a good that is both non-excludable and non-rivalrous. The former implies that individuals cannot be barred from accessing or using them and the latter means that the use of the good by one person neither prevents its access nor reduces its availability to others.

Because of these characteristics, it is hard to get paid to produce collective goods, which in turn leads to under-production—or no production—of such goods in private markets. This is both because it is difficult to divide the total production into sellable units and because consumers all have an incentive to act as so-called free-riders and enjoy the good without paying for it. It is therefore mainly the public sectors that supplies public goods.

Certain forms of knowledge can be considered public goods (i.e., information that is easily accessible in books or on the internet).

3.4 Incomplete Competition

Free markets encourage competition, which—according to basic economic theory—leads to lower prices for consumers, higher quality goods and services, greater variation in consumer choice, and more innovation. Yet, competition between producers is not always possible.

One reason for imperfect competition is high infrastructure costs that imply the presence of a so-called natural monopoly, where the largest (and often the first) producer has a large advantage over potential competitors. Other reasons for imperfect competition are strategic behavior among firms and patents following technological innovation. Regardless of the origin, in markets with imperfect competition, firms can hold a position of market power, which in economic thinking implies that the companies are price-makers, not price-takers.

It can be preferable to have public production that follows other goals that profit maximization in markets with incomplete competition. Yet, this is not always the case. Sometimes, the possibility of market power is a necessary condition for investing in innovative activities, particularly if these are high-cost investments.

4 Characteristics of DSM-Related Activities and Potential Markets

Economic activity in connection to deep-sea minerals may be subject to several market failures that warrant monitoring and possible follow-up.

4.1 High Upfront Investment Costs

A possible outcome of high investment costs is the prevention of investment in what could be profitable exploration projects for firms and contributors to growth for the national economy.

Simulations of the costs of exploring seafloor massive sulfide deposits in Bang and Trellevik (2022) suggest that they will be substantial and strongly affect the overall profitability of such undertakings. According to Sharma (2024), different studies have found that a capital expenditure for mining polymetallic nodules at 1.5 Mt./year (including the cost of mining system, ore transfer and processing plant), based on estimates of 2008 corrected for inflation as of 2022, works out to $ 2.58 billion, whereas a capital expenditure for mining at 3.0 Mt./year (including pre−/feasibility study, mining system, ore transfer, and processing plant), based on estimates of 2018 corrected for inflation for 2022, works out to $ 4.72 billion.[3]In comparison, total investments for Norwegian oil and gas companies were $36 billion in the peak year of 2014 (Norsk Petroleum, 2023).

Importantly for economic policy, the simulations in Bang and Trellevik (2022) also point out that the cost of exploration is incurred long before there is potential to

[3]It is important to consider that the figures are based on estimated input prices in 2008 and 2018. While inflation-corrected, it is not sure that the prices of these specific goods have followed inflation since then.

gain income from the extracted minerals. All else equal, this increases the chances of a market in a state of natural monopoly, where the first mover(s) gain(s) lasting market power, i.e., the ability to set prices above the welfare maximizing level that would have had in a competitive market. Furthermore, the estimations presented in Sharma (2024) indicate that the pro unit cost is lower when doubling the production volume, which indicates economies of scale. Economies of scale are a characteristic of natural monopolies, yet there are also many competitive industries with economies of scale.

In perfect markets that satisfy the assumptions of complete markets, perfect rationality of agents, and full information, high upfront investments costs would not be an issue. In reality, we know that markets are not perfect, and we now turn to the challenges caused by imperfect capital markets.

4.2 Imperfect Capital Markets

Imperfect capital markets are fundamentally a problem of information. It is often spoken of as a problem of adverse selection. Lenders have limited information about the borrower, whether they have the capacity to pay back their debt and/or whether they are willing to pay. However, in the case of DSM-related activities, it seems more appropriate to talk about incomplete information; it is hard for anyone to predict at the time of signing a financial contract how profitable an investment in exploration and possibly future mining will be.

While imperfect capital markets affect all economic activities, the problems they cause become larger, and there is more uncertainty concerning the costs and benefits. The uncertainty surrounding costs and benefits of exploration after, and later mining of, deep-sea minerals are very high. The one project that came close to commercial-scale mining (Solwara 1 by Nautilus Mining Ltd) filed for bankruptcy in 2019 (Trellevik, 2023).

This uncertainty has its roots in a wide array of factors. Ellefmo (2022) lists several technical variables that will influence the economic value of mineral deposits, the mining method being a key factor. Trellevik (2023) also finds that the techno-operational concepts for exploration add other factors that will matter for future profitability of DSM. Which technological options will be chosen is currently unknown, but this uncertainty might be resolved as technological progress continues.

Furthermore, simulations of net presented values of seafloor massive sulfide industry on the Norwegian continental shelf in Bang and Trellevik (2022) show large variation depending on the underlying assumptions. These assumptions concern the resource scenario and whether investment in exploitation technology[4] follows a "wait and see" policy requiring a substantial amount of confirmed resources

[4] In the model, the companies finance these investments themselves. Yet, the model result is also relevant for cases where investments are (partly) financed by external capital. In such a case, it would be the investors that could either adopt a "wait and see" policy or an "anticipatory" policy.

before investing or an "anticipatory" policy showcasing a more risk-taking behavior. There is currently little data available to assess the realism of these assumptions. Hence, Trellevik (2023, p. 7) notes that "it is not given that deep-sea SMS [seafloor massive sulfide] exploration and extraction on the NCS [Norwegian continental shelf] will be profitable. NPV [net present value] varies from a negative 980 million USD low to a profit of 2.53 billion USD through the simulation horizon."

On top of the incertitude caused by insufficient geological knowledge about the seabed comes the risk linked to future developments in attitudes (political and that of the public opinion) toward DSM.

Capital market imperfections limit the number of financial contracts that can be signed and hence they lead to aggregate underinvestment. On top of this inefficiency, they can also lead to an unwanted selection of the projects that are invested in. Lenders charge higher interest rates for riskier projects. With imperfect information, as the interest rate rises, the possibility of selecting riskier borrowers increases as they are less likely to pay it back.

4.3 *Underproduction of Knowledge Spillovers*

The seabed is relatively unchartered territory. Therefore, general insight into the geological and biological conditions of the seabed could be very valuable for society. The hearing document and ministerial press release in connection with the Norwegian Ministry of Petroleum and Energy's impact assessment for exploring and extracting deep-sea minerals on the Norwegian continental shelf highlight the need for more knowledge and data for responsible resource management in the case of deep-sea minerals.

There is a potential for knowledge spillover effects when private companies explore to find deep-sea minerals, and the information they gather can be applied for other purposes, which again the Norwegian Ministry of Petroleum and Energy notes in the same document by assuming that opening up for mineral exploration will significantly increase data retrieval and knowledge generation.

Such knowledge is however comparable to a public good, meaning that private exploration for deep-sea minerals will tend to underproduce it. Private companies only have incentives to collect information with a commercial interest. In a free market without any form of government intervention, there will therefore also be an underproduction of such knowledge spillover.

4.4 *Information Asymmetry Between Regulators and Firms May Occur*

Moreover, if companies gather information of general interest when exploring the seabed, undesirable information asymmetry between regulators and firms may occur. The firms will have control over which information they choose to gather and share with the regulators. This asymmetry can be used to their advantage when applying for mining licenses, for example, by underestimating the environmental damages of mining. If the regulator has limited knowledge about the biological conditions of the area, it will be difficult to ensure that the firm is truthful in its assessment. As such, information asymmetry could have a negative impact on the environment and the society as a whole.

5 Public Interventions with Examples from Norwegian State Resource Management

Market failure may warrant government intervention related to commercial exploration and exploitation of the seabed. Governments throughout the world use various policy instruments in an attempt to, among other things, correct for market failure. In this section, we discuss various instruments that may be useful, and we discuss some interesting previous examples of how government intervention has influenced or shaped industrial development.

The Norwegian experience with oil and gas is often cited as an example of successful national development and management of valuable natural resources. Petroleum resources were first discovered in Norway the late 1960s. Norway was, as such, a latecomer to the oil industry. Despite this fact, Norwegian actors were able to gain a strong foothold in this capital intensive and highly competitive industry. The Norwegian petroleum industry was even at the forefront in the development of offshore technology from the 1970s to roughly the mid-1990s (Ryggvik, 2015) and had a dominant position in the international market. Even though Norway no longer has the same dominant position today, it remains internationally competitive, with Norwegian suppliers still acquiring contracts within highly valuable parts of the value chain for offshore oil production throughout the world. The Norwegian oil company Equinor (formerly Statoil, see Sect. 5.4) is now also a major international oil companies active in projects around the world.

5.1 Command and Control Instruments

Command and control instruments involve a government issuing a rule and controlling adherence by monitoring the affected individuals or companies (Thornley, 2012). For instance, the before-mentioned underproduction of knowledge spillovers can be addressed by requiring commercial actors to share and make data public by law.

In its proposal to open part of the Norwegian continental shelf to commercial exploration for minerals, the Norwegian government states that it wants to ensure that relevant data (including environmental data) collected by actors holding an exploration license are shared with the authorities in accordance with the Norwegian Deep-Sea Mineral Act. This law stipulates that holders of licenses must send Norwegian authorities (or someone appointed by them) the following: samples of minerals extracted from the seabed, investigation reports, copies of data, tests and results including their interpretation, information on the minerals extracted including their market value, if sold.[5]

A challenge with command and control instruments is that the actors themselves do not necessarily have an intrinsic motivation to adhere to the rules and regulations. Hence, monitoring costs to ensure compliance can weigh heavily on public budgets.

5.2 *Public Subsidies*

The World Trade Organization (WTO, 1995) defines subsidies as "a financial contribution by a government or any public body [...] where

(i) government practice involves a direct transfer of funds (e.g. grants, loans, and equity infusion), potential direct transfers of funds or liabilities (e.g. loan guarantees);
(ii) government revenue that is otherwise due is foregone or not collected (e.g. fiscal incentives such as tax credits);
(iii) a government provides goods or services other than general infrastructure, or purchases goods;
(iv) a government makes payments to a funding mechanism or entrusts or directs a private body to carry out one or more of the type of functions illustrated in (i) to (iii) above which would normally be vested in the government and the practice, in no real sense, differs from practices normally followed by governments".[6]

Some forms of subsidies, as well as taxes, can be a response or attempt to correct for market failure; see the discussion of possible sources of market failure above. For instance, Davis and Lund (2018) discuss basic considerations and principles for authorities wishing to capture natural resource rents through taxation in the presence of high upfront investment costs, with Norwegian petroleum taxation as an example. They show that there needs to be generous tax deductions for upfront investment costs in order to avoid disincentives to investment. The taxes paid later when oil production starts up more than make up for the upfront deductions so that net taxes over the full course of investment and production for a particular oil field are positive.

---

[5] Information (in Norwegian) in § 9–11 here (downloaded 30th of June 2023): https://lovdata.no/dokument/NL/lov/2019-03-22-7

[6] The WTO Agreement on Subsidies and Countervailing Measures also mentions income or price support as a form of subsidy and underlines that a financial contribution must confer a benefit to be considered a subsidy. See https://www.wto.org/english/docs_e/legal_e/24-scm.pdf

However, if a subsidy is *not* a response to market failure, it will create a distortion in the workings of the economy such that society's resources are not utilized as effectively as possible. As such, society loses out on value that could have been created. The presence of a market failure can however be debatable, and both the interpretation of a policy instrument as a subsidy and legitimacy of the use of subsidies can therefore also be contentious.[7]

In Norway, there seems to be broad agreement that taxation and policies directed at the highly lucrative petroleum industry should, in total, be neutral or non-distortionary. However, the complete system of taxes, tax breaks, and similar policy instruments is complicated, so there is often room for disagreement about the details about what is a subsidy and what may create distortions.

Over time, various details of the tax system for the Norwegian petroleum sector have, for example, been subject to revision, due in part to aims of lessening distortions. Just recently, for example, the government passed a bill that made adjustments to the rules for handling depreciation of assets and tax breaks for investments and interest as the previous rules appeared to create unfortunate incentives and could, thus, be distortionary (Prop. 88 LS, 2021–2022).

Because they are not likely to generate distortions in the same way, it is useful to distinguish between subsidies for R&D and subsidies for exploration and exploitation.

5.2.1 Subsidies for Research and Development (R&D)

In economics, productivity is a measure of efficiency that relates output—the amount of goods and services produced—to the use of inputs in production. R&D activities are often able to contribute to the improvement of existing production processes or to the creation of entirely new ways of doing things. As such, R&D is generally considered an important driver of economic growth and development and can contribute to improvement in standards of living and human welfare.

Companies and other actors that engage in R&D are often able to move the global frontier of technological possibilities thereby increasing productivity, but productivity can also accrue when existing technologies and effective new processes are adopted by many different actors and diffuse through the economy. In fact, catching up to the current technological frontier through diffusion will often be the main channel by which countries and sectors can improve productivity. This is because, with the exception of very large (and affluent) countries, most countries will not be able to maintain the necessary level of expertise in order to move the technological frontier in all areas. As such, there is likely to always be room for

[7] Aarsnes and Lindgen (2012) analyze, for example, a number of policy instruments that they claim are subsidies for the petroleum sector, but Lund (2012) disagrees on the interpretation of these as true subsidies. He claims that one must see the entire system of taxes, tax breaks, etc., as a whole in order to access the potential for distortions.

improving economic productivity through the adoption of cutting-edge technology and innovation developed elsewhere.

The vital role played by R&D in productivity growth and economic development makes it an area in which many countries, governments, and companies may wish to channel resources, including through various forms of R&D subsidies and support. When discussing possible financial contributions/subsidies to R&D activities from government sources, it is often useful to distinguish between two broad categories of R&D activities: basic/pure research and applied research. Basic research is undertaken primarily for the sake of gaining further knowledge or insight without a clear view of possible (commercial) applications and uses. It is freely available, at least for those with the necessary knowledge and background to understand it. In economics, it is therefore generally considered a non-rivalrous (public) good. As such, there is good reason to believe that it may be underfunded and underproduced without government intervention and subsidies. Recent studies suggest, in fact, that public R&D policies and support are effective and also stimulate private R&D spending and activities (Becker, 2015).

Not all applied research is directly related or motivated by commercial or industrial uses and purposes, but some is. When that is the case, the line between subsidies for R&D and subsidies that more generally support businesses and industrial activity can be blurred. For that reason, there is a greater chance that some types of subsidies for applied research for commercial or industrial use can turn out to be distortionary.

Some scholars working in the field of research on research, development, and innovation emphasize the intrinsic risk and uncertainty involved in R&D and innovation processes. According to these, R&D and innovation are part of an "evolutionary process" that takes place in a broader "system of innovation" by means of creative disruption (Mazzucato, 2015). This approach differs somewhat from other economic theories and thoughts on innovation by not making strict assumptions about the form of uncertainty involved. It also emphasizes that it is solely the amount of R&D and innovation activities that take place that matter, but also the ways and means by which knowledge diffuses through various innovation networks in the economy. This way of thinking has given rise to policies that are called "mission-oriented" and which have the ultimate aim of inducing more profound innovation and change in the economy.[8]

5.2.2 Subsidies for Exploration and Production

Once subsidies cross the line from subsidies for R&D to subsidies that support specific businesses or industrial activities, they become harder to justify.

[8]The EU has shown considerable interest in this approach, which is highlighted here: https://research-and-innovation.ec.europa.eu/funding/funding-opportunities/funding-programmes-and-open-calls/horizon-europe/eu-missions-horizon-europe/mission-oriented-policy-studies-and-reports_en

An important topic in international economics is protectionism—economic policy imposed by regulators to support the domestic industry. Proponents of protectionism state that nascent industries should be shielded (for instance through subsidies) from foreign, more established, competitors until they have developed and matured sufficiently to meet this competition (Meltiz, 2005). This is known as the infant industry argument, an economic rationale for protectionism first introduced by Alexander Hamilton in 1791. Both the USA and Britain adopted numerous protectionist policies in the nineteenth and first half of the twentieth century based on this rationale (Chang, 2005).

The infant industry argument rests on the concept of economies of scale; larger firms are often able to produce goods and services at a lower cost than small ones. The cost advantage of larger firms has implications for countries that wish to develop domestic industries in which their trading partners are well established. Consider, for example, a scenario where one country has a highly developed DSM industry sector that extracts and refines minerals at a large scale. If we assume that there are some cost advantages of being large, then this industry would be able to sell refined minerals at a lower cost than the firms in a different country, where DSM activity is just starting up. To make up for the cost-disadvantage of their nascent DSM industry, the government in the latter country could provide subsidies enabling the domestic DSM industry to claim equal or lower prices compared to the established producers until it had grown sufficiently large. This last point is an important element of the nascent industry theory: protective measures should be removed as soon as the domestic industry has reached a scale that allows it to compete with foreign firms.

Protectionist policy recommendations, however, are often controversial and highly debated both within the field of economics and by policy makers. For example, a government subsidizing the domestic industry might face similar restrictions from its trading partners as retaliation, and this could have a negative impact on the domestic economy—in some cases even on the nascent industry that the government initially sought to protect. Further, the policies could be exploited by agents in ways that could harm the economy through, for example, inefficiencies or loss of government revenue.

It might be tempting to interpret the Norwegian experience with oil and gas as a typical application of the infant industry argument and approach. Some features of the Norwegian experience suggest interventionism and protectionism in an early phase (roughly the 1970s) and liberalization in a later phase. Ryggvik (2015) argues against such a strict interpretation. He claims that early oil policy in Norway was rather open and that strict regulation and active involvement by the state continued even in the 1990s, albeit in a different form. Discussions below on Norwegian licensing procedures and state-ownership of key companies show that the details of the story were, indeed, more complicated and not as straightforward as the typical infant industry argument would suggest.

5.3 Licensing Procedures

Licensing practices and procedures for activities involving the use of natural resources can also be harnessed to promote certain overall aims in industrial development and resource management. In Norway, licenses for activities that make use of natural resources are not usually sold or auctioned off, although there are examples of such practices, too. Licensing procedures can vary depending on the resource. Licenses for oil production have typically been awarded after an elaborate application procedure and competition based on a number of different criteria that the government sets. The licenses are themselves awarded by the government based on an overall, often somewhat subjective, assessment of the competing companies based on these criteria (Lægreid et al., 1989). A similar procedure has just been applied to the licensing round for floating wind turbines off the island of Utsira on the western coast.

The allocation of licenses for oil production in Norway allowed the government to grant certain advantages to Statoil or other companies with a predominance of Norwegian shareholders and interests and contribute thereby to successful development of the Norwegian oil industry, see also the discussion in Sect. 5.4. When developing the procedures for petroleum licensing, Norway was, however, aware of the need to not go too far with such differential treatment, as it was necessary for Norway to also attract foreign companies with the technical expertise and know-how that the fledgling Norwegian companies needed to learn from (Lægreid et al., 1989). The criteria for allocating licenses and the somewhat subjective application of these criteria also allowed the government to further other aims. In the fourth petroleum licensing round (in 1978), for example, weight was placed on the proportion of Norwegian suppliers in the value chain for development and production of oil fields. This helped Norwegian companies gain a foothold in various parts of the overall supply chain, the more lucrative elements of which had previously been dominated by American suppliers (Ryggvik, 2015).

5.4 State Ownership

One way for the government to intervene is by acting itself as an agent in the market. State ownership is a more direct market intervention than providing subsidies or regulating the product market. The goal is nevertheless the same: to steer the development of an industry in the direction that the government wants with the aim of achieving various societal goals. Greenhouse gas emission reduction, for example, is high on most governments' agenda. A government can instruct companies it owns to reduce their greenhouse gas emissions by, for instance, adopting new technologies that they would not otherwise have adopted if they were privately owned. Often, governments will point to market failures that prevent the best socioeconomic results being realized as reason for taking direct ownership in companies.

The history of the Norwegian oil company Statoil, now known as Equinor, is one interesting example of how state ownership can be used to foster the creation and development of a specific industry with certain (national) goals and interests in mind. It also illustrates the many twists, turns, pitfalls and controversies potentially involved in extensive state ownership.

Despite strong traditions of state-ownership and active government involvement in industry in Norway, the creation of a state-owned oil company was not entirely uncontroversial, and other strategies were also debated in the first years after oil was discovered off the coast. Early on, the Norwegian government had also acquired the majority of stocks in an existing private company, Norsk Hydro, which had aims of establishing itself in the oil industry. In addition, the government sent signals and made efforts to channel and concentrate private investments (from Norwegian companies and citizens) into just one or two ventures. This led to the creation of yet another private oil company, Saga Petroleum (Lægreid et al., 1989). Saga coexisted and also functioned as operator alongside both Statoil and Hydro until it was bought up by the other two in the late 1980s. Norsk Hydro's petroleum-related activities were eventually incorporated into Statoil in 2007.

Statoil was created as a limited liability company with all stocks initially owned by the Norwegian government. In the mid-1970s, there was political controversy surrounding how quickly the company should develop and expand. Two factors seem to have played a decisive role in Statoil being able to establish itself as competitive oil company: the Norwegian system for granting licenses for oil extraction, see Sect. 5.3 above, and a special arrangement with the international oil company Mobil that allowed Statoil to gain the necessary knowledge and expertise to establish itself as an operator in the strategically important Statfjord oil field (Ryggvik, 2015). The government has throughout time granted Statoil a large proportion of the petroleum production licenses, and Statoil has functioned as both operator and (passive) investor in various oil fields. With time, Statoil was also instrumental in allowing Norwegian suppliers to capture key contracts that helped them gain the necessary experience and know-how to grow and become (internationally) competitive (Ryggvik, 2015).

Debates about state ownership and involvement in Statoil/Equinor remain to this day. The company was partly privatized in 2001, and the Norwegian government currently owns 67 percent of the shares in the company. In response to concerns that Statoil has become too large, the government created and moved a substantial share of its passive petroleum-related investments from Statoil into a new institution called the State's Direct Financial Interest (SDFI). The SDFI is currently managed by another entirely state-owned company Petoro. Petoro is not a fully-fledged oil company in the same way as Statoil/Equinor, as it does not take on responsibilities as operator of oil fields.

5.5 Management of Public Incomes from Natural Resources

History shows that economic activity that make use of natural resources can generate large public incomes that need to be properly managed.

Historically, natural resources prices have fluctuated more than those of manufactured goods (Jacks et al., 2011). Crude oil prices, for example, rose from $58 to $145 per barrel between January 2007 and May of 2008, before plummeting to $40 per barrel in January of 2009.This revenue volatility can adversely affect public budgets when they are dominated by tax income from natural resource industries. This has negative consequences for the economy because of the economic uncertainty it fosters, and because it could force the government into debt in periods of low commodity prices when tax revenues are low. Furthermore, a resource-rich country might expand their debt burden as the inflow of revenues from resource exploitation increases, but this leaves them in a vulnerable position should commodity prices suddenly drop, as they often do.

Avoiding imbalances in the economy was a central tenet of Norwegian petroleum politics in the nascent years of the industry. The authorities wanted to uncouple the activity on the Norwegian continental shelf from the rest of the mainland economy as much as possible, so in June of 1990, they established what is now known as the Government Pension Fund Global. Public revenue from oil and gas production is transferred to the fund and so does not directly impact the budget each year. The revenues come from the extra petroleum tax, see Sect. 2.3, as well as dividends pay out by Equinor and Petoro/SDFI, see Sect. 5.4. The government adheres to a policy rule that it can only use a small part (up to 3% per year) of the fund to cover annual budget costs. The idea is to gradually phase petroleum revenues into the economy and avoid the imbalances that are often associated with government budgets dominated by natural resource revenues. The majority of the revenue is invested in stocks and real estate abroad. The value of the fund was $1.25 trillion in 2022, and it invested in more than 9000 companies across 70 countries (NBIM, 2019). Since 2009, most of the fund's growth has come from investment returns, and not directly from petroleum revenue, although the direct contributions from petroleum activity continue to be substantial.

6 Recommendations for Public Management of DSM-Related Economic Activity

Governments have a wide array of policy instruments at their disposal to intervene in and possibly shape markets. Economists justify such interventions based on the existence of so-called market failures that prevent the free market allocation of goods and services from reaching the socially optimal use of resources.

At the current early stage of DSM development, public support and initiatives to encourage R&D can foster technological progress and move the industry more quickly along a path to commercial feasibility. Public investment support might also be necessary to make up for the lack of risk capital, given the high degree of uncertainty surrounding the economic prospects of DSM that worsen capital market imperfections. Government's large capital reserve allows it to spread the risk across many projects, which is not always possible for private firms.

This type of intervention would simultaneously legitimize that the government requires firms to gather and share information beyond their immediate commercial interest. As previously noted, due to the public good character of such knowledge, economic theory suggests that knowledge spillover may not happen in and of itself.

Another, more market-based option to ensure knowledge spillovers when private companies explore the seabed for minerals would be to offer them compensation for information concerning the geological and biological conditions of the area in question. This solution has two additional advantages: It reduces the income-risk of engaging in exploration activities and it creates an incentive for sharing information, which is in line with the hypothesis of profit-maximization behavior and avoids concerns related to failing compliance with public regulation.

If DSM activity reaches the stage of economic viability, a resource rent can emerge. It is too soon to predict its potential size. While high and probably increasing demand for minerals in the future should—all else equal—lead to higher resource rents, potentially high production costs and prospects for international competition between suppliers in a global mineral market can be a limited factor.

Yet, also in the case of limited natural resource exploitation, a resource rent tax can be both legitimate and efficient. In the case of a resource rent of a certain size, resource rent taxation is recommended since it is an efficient way of raising public revenue. However, even if the resource rent and hence the resulting income is low in comparison to the administrative burden of the tax, it can be argued that it is good to establish resource rent taxation early on since the resource rent can increase over time, and it can be difficult to introduce taxation ex-post.

History tells us that it is important to specify the tax system that will be applicable, and possible government market interventions, as early as possible for a new industry. The reasons for this are twofold: first, it increases the predictability for market actors, which is particularly important in the case of commercial DSM-related activities given the already high level of inherent uncertainty. Second, it can be difficult to change the regulatory framework after the activity has become profitable.

The recent Norwegian case of resource rent taxation in the aquaculture illustrates this. In line with its taxation of the hydropower and petroleum industries, the Norwegian government initiated a regulatory process in the fall of 2022 with the goal of introducing a resource rent for Norwegian fish farmers. In contrast to the two other industries, however, this proposal came long after the aquaculture industry had matured and become profitable. The industry, various political parties, and many local communities with strong interest in aquaculture, strongly lobbied against the proposal, which after several amendments—and a considerable reduction in the tax rate—was accepted by a narrow majority in the Norwegian parliament. Several of the political parties in the minority have signaled that they will work to repeal or amend the tax in the future.

This chapter has treated all deep-sea minerals equal, which of course is a simplification. Market failures—and hence the need for government regulation—will be different between polymetallic nodules and seafloor massive sulfides. Given the current state of knowledge, commercial actors and their financers are probably in a better position to assess the resource potential of—and subsequently mine—polymetallic nodules. This would explain why polymetallic nodules so far have attracted the most attention in terms of the number of contracts signed. However, there is still considerable uncertainty surrounding the economic viability of exploiting these mineral resources, due in large part to the need for developing production infrastructure (Sharma, 2024). Moreover, the problem caused by lack of knowledge spillovers from the commercial sector is—all else equal—larger the more data we lack on the seabed in question.

The benefits of interventions must also be weighed against both the administrative burden they cause, the alternative use of public funds to support deep-sea mineral activities, "lock-in-effects" over time, as well as regulatory failures that arise due to, for instance, political cycles and the difficulties for regulatory bodies to "pick the winners," meaning the best projects.

References

Aarsnes, F., & Lindgren, P. (2012). *Fossil fuels – At what cost? Government support for upstream oil and gas activities in Norway* (Report for the Global Subsidies Initiative). Pöyry Management Consulting.

Akerlof, G. A. (1970). The Market for "Lemons": Quality Uncertainty and the Market Mechanism. *The Quarterly Journal of Economics, 84(3), 488–500.* https://doi.org/10.2307/1879431

Akgun, O., Cournède, B., & Fournier, J. (2017). *The effects of the tax mix on inequality and growth* (OECD economics department working papers, no. 1447). OECD Publishing. https://doi.org/10.1787/c57eaa14-en

Bang, R. N., & Trellevik, L.-K. L. (2022). Perspectives on exploration and extraction of seafloor massive sulfide deposits in Norwegian waters. *Mineral Economics*. https://doi.org/10.1007/s13563-022-00346-y

Becker, B. (2015). Public R&D policies and private R&D investment: A survey of the empirical evidence. *Journal of Economic Surveys, 29*(5), 917–942.

Caplan, B. (2003). The logic of collective belief. *Rationality and Society, 15*(2), 218–242.

Chang, H. (2005). Kicking away the ladder: Infant industry promotion in historical perspective. *Oxford Development Studies, 31*(1), 21–32. https://doi.org/10.1080/1360081032000047168

Coase, R. H. (1960). The problem of social cost. *The Journal of Law and Economics, III*, 1–44.

Davis, G., & Lund, D. (2018). Taxation and investment decisions in petroleum. *Energy Journal, 39*(6), 189–208.

Downs, A. (1957). *An economic theory of democracy*. Harper.

Ellefmo, S. L. (2022). Chapter 16: Conceptual 3D modeling and direct block scheduling of a massive seafloor sulfide occurrence. In R. Sharma (Ed.), *Perspectives on deep-sea mining*. Springer. https://doi.org/10.1007/978-3-030-87982-2

Friedman, D. (1996). *Hidden order: The economics of everyday life*. HarperBusiness.

Hein, J. R., Koschinsky, A., & Kuhn, T. (2020). Deep-ocean polymetallic nodules as a resource for critical materials. *Nature Reviews – Earth and Environment, 1*, 158–169.

Jacks, D. S., O'Rourke, K. H., & Williamson, J. G. (2011). Commodity price volatility and world market integration since 1700. *The Review of Economics and Statistics, 93*(3), 800–813.

Lægreid, P., Olsen, J., & Stenberg, A.-M. (1989). Næringsorganisering og partsrepresentasjon. In J. Olsen (Ed.), *Petroleum og politikk: det representative demokratiets møte med oljealderen* (pp. 143–166). Otta.

Lund, D. (2012). Is the petroleum activity subsidized? Samfunnsøkonomen. ISSN 1890-5250. *126(4), 22–31.*

Mazzucato, M. (2015). *The entrepreneurial state: Debunking public vs. private sector myths*. Public Affairs.
Meld. St. nr. 25. (2022–2023). *Mineralverksemd på norsk kontinentalsokkel—opning av areal og strategi for forvaltning av ressursane*. Ministry of Petroleum and Energy.
Meltiz, J. M. (2005). When and how should infant industries be protected? *Journal of International Economics, 66*(1), 177–196. https://doi.org/10.1016/j.jinteco.2004.07.001
Norges Bank. (2019). *About the fund*. NBIM. https://www.nbim.no/en/the-fund/about-the-fund/. Downloaded 27th of February 2019.
Norsk Petroleum. (2023). *Investments and operating costs*. Norwegian Petroleum. https://www.norskpetroleum.no/en/economy/investments-operating-costs/. Downloaded 11th of May 2023.
OECD. (2022). *Revenue statistics 2022: The impact of COVID-19 on OECD tax revenues*. OECD Publishing. https://doi.org/10.1787/8a691b03-en
Prop. 88 LS. (2021–2022). *Omlegging av særskatten for petroleumsvirksomhet tik en kontantstrømskatt (endringer i petroleumsskatteloven mv.)*. Ministry of Finance.
Ryggvik, H. (2015). A short history of the Norwegian oil industry: From protected National Champion ot internationally competitive multinationals. *Business History Review, 89*(1), 3–41.
Sharma, R. (2024). Economic potential of polymetallic nodules mining. In R. Sharma (Ed.), *Deep-sea mining and the water column*. Springer. pp. 589–598 (this volume).
Thornley, P. (2012). Bioenergy policy development. In A. Sayigh (Ed.), *Comprehensive renewable energy*. Elsevier. https://doi.org/10.1016/B978-0-08-087872-0.00527-8
Trellevik, L.-K. L. (2023). Exploring exploration—How to look for deep-sea minerals. *Mineral Economics*. https://doi.org/10.1007/s13563-023-00379-x
WTO. (1995). *Agreement on subsidies and countervailing measures, part 1, Article 1*. WTO.

Kaja von der Leyen is a researcher working at NTNU Social Research in Trondheim. She holds a Ph.D. in economics from the University of Göttingen, Germany. She has previously worked for the Norwegian Ministry of Finance and the Organisation for Economic Co-operation and Development. Her main research interests all lie in the fields of public economics.

Mads Fjeld Wold is a researcher working at NTNU Social Research in Trondheim. He holds an M.Sc in economics with a specialization in Energy, Natural Resources, and the Environment from the Norwegian School of Economics. Wold's main research interests are in the fields of environmental and resource economics, with a special interest in the policy implications of resource management. He has also worked on several projects on local government finances and efficiency analyses of local government service production.

Taryn Ann Galloway is an economist currently employed by the Norwegian engineering consulting firm AsplanViak. She holds a Ph.D. in economics from the University of Oslo and has previously worked in the Norwegian Ministry of Labor and in the research development at Statistics Norway. Her current professional interests and activities include labor market research and policy analysis, analysis of regional economic development, and methodological developments in cost-benefit analyses.

Chapter 17
A Comparative Economic Scenario of Nodule Mining in Pacific and Indian Oceans, Associated Challenges and Their Prospects

T. R. P. Singh

Abstract This chapter presents a comparative economic scenario of nodule mining and metallurgical processing projects in the Pacific and Indian Oceans. The baseline case assumes a nodule abundance of an initial 20 years of mining operation with the best values for both cases. The capital cost, operating cost and annual revenue figures are drawn from relatively recent publications in the public domain for a 3 million tonnes per annum of dry nodules mining and metallurgical plant for a project in the Pacific ocean. Based on corresponding data of nodule abundance and grades for the initial 20 years of mine-plant operation, the capital cost (capex), operating cost (opex) and the annual revenue have been estimated for a project in the Indian Ocean for the same capacity. Using these data, simplified computations of internal rate of return (IRR) on constant value basis have been made for both cases. It is concluded that for identical mine capacity, the capital and operating cost for mining and transport operations of a project in the Indian ocean are higher by 55% compared to a project in the Pacific Ocean. The implications of high capital cost for mining operations in the case of the Indian Ocean on the royalty payments have been highlighted. Finally, some technical challenges and possible prospects of nodule mining in medium term have also been discussed.

Keywords Nodule mining · Economic analysis · Capex · Opex · IRR · Pacific Ocean · Indian Ocean

T. R. P. Singh (✉)
Independent Consultant, New Delhi, India
e-mail: trp_singh@yohoo.com

R. Sharma (ed.), *Deep-Sea Mining and the Water Column*,
https://doi.org/10.1007/978-3-031-59060-3_17

1 Introduction

Manganese nodules represent a vast source of mineral resources in the Clarion-Clipperton Zone (CCZ) in the Pacific Ocean as well as in the Indian Ocean. These nodules, also called polymetallic nodules, typically contain 27% Mn, 1.2% Cu, 1.3% Ni and 0.25% Co in the CCZ. Thus, these nodule resources may be best described as low-grade manganese ores containing Cu, Ni and Co as additional metals. The polymetallic nature of these vast resources with valuable metals like Ni and Co has attracted the attention of most countries for decades. Consequently, a large number of countries or their sponsored private entities have obtained exclusive exploration rights from the International Seabed Authority (ISA) and are designated as contractors. In view of highly rich resources particularly in terms of nodule abundance (expressed as kg/m^2), the number of such contractors is very large (18) in the CCZ, while there is only one contractor in the Indian Ocean represented by the Government of India.

A number of studies on the economic evaluation of deep seabed mining of nodules are available in the public domain. However, all these studies are limited to CCZ only and are generally based on the global average of nodule abundance. In the ultimate scenario of nodule mining, it is logical to assume that an investor/contractor would identify the part of the respective contract area that will have the highest nodule abundance and grades for an initial 20 years of operation of the mine and process plant (the term mine or mining is used to include transportation to land).

Typically, in all the studies, the capacity of the nodule mining and metallurgical plant has been assumed as 3 million tonnes per annum (mtpa) of dry nodules. The difference in the nodule abundance values between a typical contract area in CCZ and in the Indian contract area will influence the area requirement for 20 years of mining operation. It will also impact the capital cost (capex) and operating cost (opex) of mining operations in the two areas under identical technological and operational conditions. The capex and opex for the metallurgical plant may not change, but the revenue for the Indian case may be affected due to marginally lower grades of nodules in the Indian contract area. A comparative economic evaluation of nodule mining will be discussed against such a background for a typical contract area in CCZ and in the Indian contract area. In addition, the current evaluation of nodule mining will also be reviewed from the contractor's perspective.

2 Quantity of In situ Nodules Required for 20 Years of Operation

For a mining capacity of 3 mtpa of dry nodules and assuming 60% net collection efficiency with 20% moisture content in wet nodules, the quantity of in situ wet nodules required for 20 years of mining operation Q20 works out to:

$$Q20 = 3 \times 20 / 0.8 \times 0.6 = 125 \, \text{million tonnes of wet nodules}$$

2.1 Mean Nodule Abundance for Initial 20 Years of Mining Operation

From the study reported by Singh and Sudhakar (2015), it may be inferred from their graphic presentation with some extrapolation that for the above quantity of wet nodules, the best areas containing the above quantity of nodules in the CCZ contract areas are likely to be more than 20 kg/m^2 (say 22 kg/m^2). The corresponding value of nodule abundance for 20 years of operation in the Indian contract area may be less than 7 kg/m^2. Such a high value of nodule abundance for the initial 20 years of operation in CCZ is also supported by high abundance values in the contract area of Nauru Ocean Resources Inc (NORI) in the range of 17–18.6 kg/m^2 (TMC, 2022) as an average value for the contract area and also 20.4 kg/m^2 abundance value for the area identified for test mining.

2.2 Areas Required for Initial 20 Years of Operation

For contract areas in CCZ, the size of the area required for an initial 20 years of operation is given by

$$Q_{20} = \mathrm{Ab} \times \mathrm{A} \times 10^3,$$

where

Q_{20} = Quantity of wet nodules in tonnes = 125×10^6t
Ab = Abundance in kg/m^2 = 22 kg/m^2
A = Area required in km^2
10^3 = km^2 to m^2 and kg to tonnes factor

Thus, the area A works out to

$$A = \frac{125 \times 10^6}{22 \times 10^3} = 5682 \text{ km}^2$$

Or say, 6000 km^2 for CCZ

Similarly, area A for the Indian contract area works out to

$$A = \frac{125 \times 10^6}{7 \times 10^3} = 17{,}857 \text{ km}^2$$

Or say, 18,000 km^2 for Indian contract Area.

Thus, while the area required for 20 years of operation in CCZ is only 6000 km^2 (less than 8% of the contract area), the same required in the Indian contract area is 18,000 km^2 (24% of the contract area).

3 Estimates for Capex, Opex, Annual Revenue, and IRR in CCZ

It is assumed that with high abundance value of 22 kg/m^2, a single mining system will be able to produce an annual capacity target of 3 mtpa of dry nodules during the initial 20 years. (MIT (2020) assumed 3.86 mtpa mining capacity with 2 independent mining systems corresponding to a mean abundance value of 10.9 kg/m^2 over a lifetime of exploitation activity). The estimates for capex and opex for mining and for the metallurgical plant have been drawn from Van Nijen (2018) as well as from MIT (2018). Both the sources give similar values from which the following figures have been adopted.

3.1 *Capex*

Mining and transport:	1400 million USD
Metallurgical plant:	2600 million USD
Total:	4000 million USD

3.2 *Opex*

Mining and transport:	300 million USD
Metallurgical plant:	700 million USD
(All estimates at 2018 level)	

3.3 *Estimates for Annual Revenue for CCZ*

Based on the mean grades of nodules (on in situ basis) as well as metal prices for Cu, Ni, Co and Mn (at 2018 level) given by MIT (2018) and the modified metal recovery factors used by MIT (2020), the revenue for the 3 mtpa plant is estimated as under:

Metal	Composition	Recovery	Quantity (t) of metal recovered
Mn	27%	90%	7,29,000
Ni	1.3%	90%	35,100
Co	0.25%	80%	6000
Cu	1.20%	90%	32,400

Revenue per year

Metal	Quantity (t)	Price (USD/t)	Revenue (million USD)
Mn	7,29,000	1600	1166
Ni	35,100	16000	560
Co	6,000	50000	300
Cu	32,400	7000	227
		Total	**2253**

Say, 2250 million USD.

3.4 *Computation of IRR for CCZ*

The capex, opex and annual revenue for 3 mtpa capacity project for a Contract area in CCZ are summarized below:

Capex:	4000 Million USD
Opex:	1000 Million USD
Revenue:	2250 Million USD/Year

With the above data, a simplified computation for internal rate of return (IRR) with commonly used assumptions has been made considering 20 years of operation. The results are presented in Table 17.1.

4 Estimation of Capex, Opex, Annual Revenue and IRR for Indian Contact Area

4.1 *Capex for Mine and Process Plant*

For a low value of abundance (7 kg/m^2) in the initial 20 years of operation, it is assumed that three independent mining systems and associated facilities will be needed, each system with a capacity of 1 mtpa of dry nodules considering the proportion of abundance values in the Indian contract area and in the CCZ contact area.

Since the capex data in the public domain is available only for 3 mtpa case for CCZ, the capex for 1 mtpa mining system is computed indirectly using the following commonly used formula:

$$\text{Cost of Plant A} = \frac{(\text{capacity of Plant A})^{0.6}}{\text{capacity of Plant B}}$$

Table 17.1 Computation of IRR for CCZ

Cash flow statement

Years	1	2	3	4	5	6	7	8	9	10	11	12	13	14	15	16	17	18	19	20	21	22	23	24
In flow																								
Sales revenue					1 125	1 800	2 250	2 250	2 250	2 250	2 250	2 250	2 250	2 250	2 250	2 250	2 250	2 250	2 250	2 250	2 250	2 250	2 250	2 250
Salvage value																								200
Out flow																								
MI + MM	802	1 604	1 203	401																				
Cost of production (Incl. int of S.T)					502	802	1 002	1 002	1 002	1 002	1 002	1 002	1 002	1 002	1 002	1 002	1 002	1 002	1 002	1 002	1 002	1 002	1 002	1 002
Gross margin	−802	−1 604	−1 203	−401	623	998	1 248	1 248	1 248	1 248	1 248	1 248	1 248	1 248	1 248	1 248	1 248	1 248	1 248	1 248	1 248	1 248	1 248	1 448
Depreciation					190	190	190	190	190	190	190	190	190	190	190	190	190	190	190	190	190	190	190	190
Amortisation					2	2	2	2	2	2	2	2	2	2										
Net margin	−802	−1604	−1203	−401	431	806	1056	1056	1056	1056	1056	1056	1056	1056	1058	1058	1058	1058	1058	1058	1058	1058	1058	1258
interest on LT loan						80	160	164	154	144	134	123	112	101	89	76	63	0						
Net profit (PBT)	−802	−1604	−1203	−401	431	726	896	892	902	912	922	933	944	955	969	982	995	1058	1058	1058	1058	1058	1058	1258
Tax											304	319	331	342	351	361	383	385	386	387	387	388	388	451
Profit after tax (PAT)					431	726	896	892	902	912	618	613	612	614	618	621	612	673	672	671	671	670	670	807
L.T debt repayment								201	201	201	201	201	201	201	201	201	201							
PAT + Depreciation					621	916	1086	1082	1092	1102	808	803	802	804	808	811	802	863	862	861	861	860	860	997

Depriciation for tax					823	617	463	348	261	196	147	110	83	62	47	27	20	15	11	9	7	5	4	4
Net margin for tax	−802	−1604	−1203	−401	−201	379	783	398	985	1050	1099	1136	1163	1184	1201	1221	1228	1233	1237	1239	1242	1243	1244	1444
Interest on LT loan					80	160	164	154	144	134	123	112	101	89	76	63								
Net profit (PBT)					−281	218	619	744	841	916	976	1023	1062	1095	1125	1158	1228	1233	1237	1239	1242	1243	1244	1444
Tax liability						68	193	232	262	286	304	319	331	342	351	361	383	385	386	387	387	388	388	451
Profit after tax					−281	150	426	512	579	630	671	704	731	753	774	797	845	848	851	853	854	855	856	994
Gross margin after tax	−802	−1604	−1203	−401	623	930	1055	1016	986	962	944	929	917	906	897	887	865	863	862	861	861	860	860	797
Inputs	**USD (In Million)**																							
Capex	**$ 4000**																							
Opex	**$ 1000**																							
Working capital @3% of OPEX	**$ 30**																							
Sales revenue	**$ 2250**																							
Interest rate on ST loan	**10.00%**																							
Interest rate on LT loan	**8.00%**																							

(continued)

Table 17.1 (continued)

Cash flow statement																								
Financial indices																								
Internal rate of return (%) pre tax	19.58																							
Internal rate of return (%) post tax	16.38																							
Pay back period (Years) pre tax	3.91																							
Return on investment (%)	44.14																							
Sensitivity analysis (pre tax IRR)																								
IRR with 15% increase in sales revenue	23.37%																							
IRR with 15% decrease in sales revenue	15.13%																							
Sensitivity analysis (post tax IRR)																								
IRR with 15% increase in sales revenue	19.74%																							
IRR with 15% decrease in sales revenue	12.67%																							

where

Capacity of Plant A = 1 mtpa
Capacity of Plant B = 3 mtpa
Capex of Plant B = 1400 million USD (as used for CCZ)

Thus, the capex of Plant A (mining system of 1 mtpa capacity) = $(1/3)^{0.6} \times 1400$ Million USD.

The total capex for 3 mining systems (each with 1 mtpa capacity) amounting to 3 mtpa capacity will be:

$$3 \times (1/3)^{0.6} \times 1400 \,\text{million}\, USD = 2172 \,\text{Million}\, USD.$$

Thus, the capex for a total capacity of 3 mtpa mining system for the Indian Contact area is 2172 Million USD against the capex of 1400 million USD for the same capacity in CCZ, implying a rise of 55%. Since the capacity of onland process plant in the Indian case will remain unchanged at 3 mtpa, the capex for the process plant will therefore remain unchanged at 2600 Million USD as in the CCZ case.

Thus, the total capex for the mine and process plant for Indian contract area works out to 4772 million USD.

4.2 *Opex for Mine and Process Plant for Indian contract Area*

As discussed earlier, there will be three independent mining systems, each with a capacity of 1 mtpa against one mining system of 3 mtpa for the CCZ case. Consequently, the ratio of opex for the mining system in CCZ and the Indian contract area is assumed to be the same as the ratio of capex in the two cases. The opex for the mining system for the Indian case is taken as 55% higher than the CCZ case. The opex for the mining system for the Indian contract area therefore works out to 465 million USD (300x1.55 = 465). The capacity of the process plant being the same as the CCZ case, the opex for the process plant remains unchanged at 700 million USD for the Indian case also.

The total opex for the mine and process plant for the Indian contract area thus works out to 1165 million USD (700 + 465).

4.3 *Estimate for Annual Revenue for Indian contract Area*

The metal grade of nodules in the Indian contract area (Sudhakar, 1993) is at least 10% lower than the grades in the contract area in CCZ (MIT, 2018). Accordingly, the annual revenue for the Indian contract area is taken as 90% of the corresponding value for CCZ. This works out to 2025 million USD (2250 × 0.9)

4.4 Computation of IRR for 3 mtpa Mine Plant for Indian contract Area

The capex, opex and annual revenue for 3 mtpa capacity of the mine plant for the Indian contract area are summarized below:

Capex:	4772 Million USD
Opex:	1165 Million USD
Revenue:	2025 Million USD

With the above data, a simplified computation for IRR with commonly used assumptions has been made considering 20 years of commercial operation. The results are presented in Table 17.2.

5 Discussion on the Results

The following points emerge from the results presented above:

(a) The capex as well as opex for the mining system in the Indian contract area are significantly higher (55%) than those of a contract area in CCZ, under identical technological and operational efficiency conditions.
(b) It is also interesting to note that while the ratio of capex for the mine and process plant in CCZ is about 1:1.9, the same ratio for the Indian case is 1:1.2. This is the effect of high mining costs in the Indian contract area.
(c) The high capex and opex together with lower revenue (on account of lower metal grades by 10%) significantly bring down the IRR for the Indian Case. While the IRR (post tax) for the CCZ case is 16.48%, it drops to 9.81% for the Indian case. It is important to clarify that while all the comparative computations are made on the nodule abundance value representing the best part of a contract area for the initial 20 years in both cases, the results will directionally be similar even if mean values of nodule abundance over a lifetime of mining activities in the contract area for both cases are used with the number of independent mining systems in the two cases undergoing proportionate changes.

It may be noted that in the ultimate financial model, these are two additional variables that must be included in the capex, opex and revenue structure. First, any payment to resource owner in terms of royalty or other payments towards environmental protection funds that are currently under discussion in the ISA must enter into the cost structures. The second relates to actual/mineable grades of nodules. In the current models, the metal grades are on the in situ basis. However, in spite of the best system design, it is still likely that there may be some dilution due to associated sediments/non-nodule particles being collected and transported along with nodules. Further, during the transport of nodules to the surface in a rigid riser system, there

Table 17.2 Computation of IRR for Indian contract Area

Cash flow statement																								
Years	1	2	3	4	5	6	7	8	9	10	11	12	13	14	15	16	17	18	19	20	21	22	23	24
In flow																								
Sales revenue					1 013	1 620	2 025	2 025	2 025	2 025	2 025	2 025	2 025	2 025	2 025	2 025	2 025	2 025	2 025	2 025	2 025	2 025	2 025	2 025
Salvage																								200
Out flow																								
MI + MM	957	1 914	1 435	478																				
Cost of production (Incl. int of S.T)					585	934	1 167	1 167	1 167	1 167	1 167	1 167	1 167	1 167	1 167	1 167	1 167	1 167	1 167	1 167	1 167	1 167	1 167	1 167
Gross margin	−957	−1 914	−1 435	−478	428	686	858	858	858	858	858	858	858	858	858	858	858	858	858	858	858	858	858	1 058
Depreciation					190	190	190	190	190	190	190	190	190	190	190	190	190	190	190	190	190	190	190	190
Amortisation					2	2	2	2	2	2	2	2	2	2										
Net margin	−957	−1914	−1435	−478	236	494	666	666	666	666	666	666	666	666	668	668	668	668	668	668	668	668	668	868
Interest on LT loan						96	191	195	184	172	160	147	134	120	106	91	75	0						
Net profit (PBT)	−957	−1914	−1435	−478	236	398	474	471	482	494	506	519	532	546	562	577	592	668	668	668	668	668	668	868
Tax											175	191	204	215	225	236	261	263	264	265	266	266	266	329
Profit after tax (PAT)					236	398	474	471	482	494	331	328	328	331	337	341	331	405	404	403	402	402	401	539
L.T debt repayment								239	239	239	239	239	239	239	239	239	239							
PAT + depreciation					426	588	664	661	672	684	521	518	518	521	527	531	521	595	594	593	592	592	591	729

(continued)

Table 17.2 (continued)

Cash flow statement																								
Depriciation for tax					823	617	463	348	261	196	147	110	83	62	47	27	20	15	11	9	7	5	4	4
Net margin for tax	−957	−1914	−1435	−478	−397	67	393	508	595	660	709	745	773	793	811	831	838	843	846	849	851	853	854	1054
Interest on LT loan					96	191	195	184	172	160	147	134	120	106	91	75								
Net profit (PBT)					−492	−125	197	324	423	500	562	611	653	688	720	756	838	843	846	849	851	853	854	1054
Tax liability						−39	62	101	132	156	175	191	204	215	225	236	261	263	264	265	266	266	266	329
Profit after tax					−492	−86	136	223	291	344	386	421	449	473	495	520	576	580	582	584	586	587	588	725
Gross margin after tax	−957	−1914	−1435	−478	428	725	796	757	726	702	683	667	654	643	633	622	596	595	594	593	592	592	591	529
Inputs	**USD (In Million)**																							
Capex	**$ 4772**																							
Opex	**$ 1165**																							
Working capital @3% of OPEX	**$ 35**																							
Sales revenue	**$ 2025**																							
Interest rate on ST loan	**10.00%**																							
Interest rate on LT loan	**8.00%**																							
Financial indices																								

Internal rate of return (%) pre tax	12.16																							
Internal rate of return (%) post tax	9.81																							
Pay back period (years) pre tax	6.28																							
Return on investment (%)	24.32																							
Sensitivity analysis (pre tax IRR)																								
IRR with 15% increase in sales revenue	16.02%																							
IRR with 15% decrease in sales revenue	7.34%																							
Sensitivity analysis (post tax IRR)																								
IRR with 15% increase in sales revenue	13.14%																							
IRR with 15% decrease in sales revenue	5.65%																							

may be attrition generating finer particles leading to a possible loss of metal values through the process of dewatering. The impact on the net mineable grades on account of possible losses and dilution will need to be determined through actual experimentations during mining trials.

The above two factors will directionally be negative for IRR for both the cases.

5.1 Implications of High Capex for Mining System in Indian contract Area

The significantly higher mining cost on account of a very low nodule abundance value in the Indian contract area makes the nodule mining project highly uncompetitive with respect to the projects in CCZ. Consequently, any payment structure for royalty, for instance, cannot logically be uniform for both CCZ as well as the Indian contract area. The cost of mining must enter into any equation while formulating the royalty structure. Such a view was also shared in the ISA information webinar for council members and other States of the ISA (ISA, 2020), which stated thus (MRG Consulting, 2020):

> *Royalty levied on production either value or volume. Value most common and calculated based on the sale value often with some simplified productions costs deductible (2 to 3% upto 10 to 12%)*

In view of the substantially higher cost of mining in the Indian contract area, there is a logical case for formulating any payment structure for royalty or other charges taking into consideration such a reality besides incentivizing such a project for a contract area in a developing country like India.

5.2 Basis and Level of Currently Available Economic Analyses in Public Domain

The economic analyses of a potential nodule mining project available currently in the public domain have drawn data largely from historical information, relevant land-based plants, and bench-scale process tests by some investors, supplemented by theoretical modelling. From investors' perspective, while these estimates are directionally meaningful, they are preliminary and do not lend themselves to be used as a guidance on the economic viability of the project nor are so claimed by the authors for such analyses.

As an illustration, it is pertinent to mention that even for establishing technical feasibility, there is a complexity in relation to Mn, the most dominant metal in the nodules. First, whether the Mn/Mn products assumed for revenue estimates have global markets for their consumption.

Second, if the products so assumed are produced through specific processes for which capex and opex have been estimated. It is assumed (Van Nijen, 2018; MIT, 2018) that the Cuprion process is used to recover Cu, Ni and Co, leaving the manganese in $MnCO_3$ form, which may be converted to electrolytic manganese metal (EMM). For a 3 mtpa plant, the quantity of EMM so produced is unlikely to be absorbed by the market (Van Nijen, 2018). Additionally, the estimates for capex and opex for converting $MnCO_3$ to EMM are likely to be tentative without any experimental backup data generated from nodule processing. MIT (2018) also considers the Cuprion process and uses Mn Metal for revenue estimation. However, MIT (2020), under four metal system considers Co, Cu, Ni and a basket of Mn metal and metal alloys without identifying the processes assumed. While the revenue estimates for Mn in any product mix are considered fair (ignoring the market consumption), the corresponding estimates for capex and opex are challenging without identifying and relating to specific processes producing multiple products. Thus, these estimates are preliminary, and the problem lies in the availability of relevant process data.

5.3 *Prospects of Nodule Mining Project*

Before discussing the prospects of nodule mining project (including metallurgical plant), it is in order to look at the current status of development. While there has been extensive work undertaken by the contractors in the field of exploration and environmental studies in the last 20–30 years, not a single contractor has completed integrated mining tests as well as pilot scale metallurgical process tests for producing ultimately saleable products. Only after experiments in the mining and metallurgy at a reasonable scale are done along with preliminary engineering, the resultant data will allow a pre-feasibility study (PFS) of a project that may also demonstrate if the mineral resources can be converted to mineral reserves and to what extent.

It appears that such studies are not even planned for the next 5 years by most contractors. There could be a number of factors that are holding most contractors from undertaking such assessments, as described here. These activities are highly demanding in terms of cost and efforts. With the current global financial conditions under stress, the contractors may not be willing to prioritize this project. It may also be that the contractors do not recognize urgency in the near term. Perceived environmental risks may be yet another reason for not moving with speed. In addition, one of the problems in establishing technical feasibility is to be able to produce high-value Mn products having bulk application and enough market to absorb such products. It appears that the only Mn product that may have bulk application is ferromanganese.

However, it will need to be proved through experiments that the Mn-bearing slag (after recovery of Cu, Ni and Co) produced from low-grade Mn nodules will be similar to land-based high-grade manganese ore, which will be good for the production of ferromanganese. From investors' perspective, there are only two conditions which will attract investors to invest in a project, assuming that there are no serious external risks. Either the project should prima facia appear financially attractive with a potential of good returns compared to other competing projects/options, or the minerals/metals in the project are so critical that the security of supply may become a compelling reason for an individual investor even if the project is unlikely to give competing returns. In the given circumstances, when the available data do not qualify to be good for PFS, any judgement on the financial viability is beset with obvious challenges. Regarding the criticality of minerals/metals, it is true that with growth in many high-end applications, including energy transition and electrical vehicles (EVs), the demand for Ni and Co is likely to grow by an order of magnitude.

It is, therefore, of interest to review the supply-side scenario. The current global consumption of Ni is about 2.5 mtpa. The reserves and resources of nickel are estimated at 90 mt and 300 mt respectively (Nickel Institute, 2023). Additionally, nickel is recyclable endlessly. For cobalt, the current global consumption is about 135 ktpa. The reserves and resources of cobalt are estimated at 8.3 mt and 25 mt, respectively (Investing News Network, 2023; USGS, 2022). Co-recycling may also add to the supply chain. Additionally, to meet the growing demand of Ni and Co for EVs, alternate battery cathode chemistry mixes are being developed and tested. Thus, in addition to nickel, manganese and cobalt (NMC), other alternatives like nickel, cobalt, aluminium oxide (NCA) and lithium, iron, phosphate (LFP) are competing chemistry mix for batteries. NMC batteries require more cobalt but less nickel compared to NCA, and LFP batteries do not need Ni and Co. While each mix may have limitations, they represent potential alternatives in addition to some non-lithium solutions.

Thus, given the supply situation, recycling potential and development of alternative technologies for reducing the consumption of Ni and Co, it would emerge that globally the availability of Ni and Co seems adequate for the current as well as future generations. However, given the highly skewed distribution of these metals geographically as well as concentration of mineral value chains in a few hands have the potential to disrupt the global supply chain in a politically charged international environment. It may be for this reason that every country is developing its own list of critical minerals. This may prompt an individual country/contractor to consider nodule project as a potential source of supply of these metals, if all other options fail.

Finally, the availability of renewable energy and carbon neutrality of the process will be other potential determinants for the future of a nodule mining project.

Thus, given the current status of development in the field of deep seabed mining and processing of nodules and the challenges ahead, it seems unlikely that there will be a project that will meet all the legal, technical and financial criteria and consequently be supported by financial institutions for funding on medium-term period. Exceptions as isolated cases with specific objectives may not be ruled out. Once an integrated mining trial is conducted, a process test on a pilot scale for producing final saleable products is undertaken along with preliminary engineering and a comprehensive market study, the resultant data will allow a PFS which will provide a direction on the financial viability of nodule mining project. The global metal industry will be keenly watching the results of such studies.

Acknowledgements The author thanks Dr. Rahul Sharma for his suggestion to write this chapter. The author also thanks Dr. Magnus Ericsson of MRG Consulting for authorization to quote from their presentation material. The author thankfully acknowledges the support provided by Dr. PK Guru, visiting faculty, Utkal University, for IRR computations. This chapter reflects the views of the author only and not of any organization, even if the author has any indirect association with an organization.

References

Investing News Network. (2023). https://investingnews.com

ISA. (2020). *Information webinar for council members and other states of the ISA, June 12, 2020.*

MIT. (2018). *Understanding the economics of Seabed mining for polymetallicnodules.* Report submitted by MIT to InternationalSeabed Authority.

MIT. (2020). *Financial model updates – Polymetallic nodules.* Report submitted toInternational Seabed Authority.

MRG Consulting. (2020). *Tax regimes in countries mining cobalt, manganese and nickel.*

Nickel Institute. (2023). *The story of nickel.* www.nickelinstitute.org

Singh, T. R. P., & Sudhakar, M. (2015). Estimating polential of additional mine sites for polymetallicnodules in Pacific and Indian Oceans. *International Journal of Earth Sciences and Engineering, 8*(4), 1938–1941.

Sudhakar, M. (1993). *Geostatistical analysis of polymetallic nodules from the Indian Ocean and a comparative studies from the Pacific Ocean deposits.* Unpublished PhD thesis, Indian School of Mines Dhanbad, P287.

TMC. (2022). *The metal company scoping document for NORI – D Social Impact Assessment (SIA).* https://metals.co/nori

USGS. (2022). *Mineral Commodity summaries 2022 – Cobalt.* https://pubs.usgs.gov

Van Nijen, K. (2018). *Marine Policy.* https://doi.org/10.1016/J.marpol.2018.02.027

Dr. T. R. P. Singh obtained his Master's in applied geology from IIT Kharagpur; Diploma in mining and exploration; MSc in ore evaluation from Delft, The Netherlands; and PhD in mining geostatistics from Osmania University. After serving NMDC, Dr. Singh joined Engineers India Limited (EIL) in 1981, where he headed the Mining and Metallurgy Division. While at EIL, Dr. Singh worked as a Project Manager for the Seabed Mining Programs of the Dept. of Ocean Development, Govt. of India and also served as a member of the UN Group of Technical Experts on Deep Seabed Mining during the PREPCOM period. Dr. Singh continues to be a member of the standing committee of the Ministry of Earth Sciences, Govt. of India.

Chapter 18
Economic Potential of Polymetallic Nodules Mining

Rahul Sharma

Abstract Deep-sea minerals such as polymetallic nodules, ferromanganese crusts and hydrothermal sulphides are emerging as potential sources of metals such as Cu, Ni, Co, Mn, Fe and rare earth elements that could contribute towards transitioning to green energy alternatives in the future. Although many of these mineral deposits lie in the international waters in the oceans of the world, some of these are also found within the exclusive economic zones of several countries. Mapping of these resources has been underway for several decades using sophisticated sounding, sampling and imaging techniques, leading to resource evaluation of these minerals that are found in millions of tonnes, and estimates show that these contain strategic metals worth billions of dollars. This paper assesses the economic potential of mining the polymetallic nodules from deep-sea regions, as an alternative source of metals for meeting the demands in the future.

Keywords Deep-sea minerals · Resource potential · Metal production · Estimated expenditures

1 Introduction

Deep-sea minerals such as polymetallic nodules (Kuhn et al., 2017), ferromanganese crusts (Halbach et al., 2017) and hydrothermal sulphides (Cherkashov, 2017) have emerged as alternative sources of critical metals (such as Cu, Ni, Co, Mn, Fe and rare earth elements) that could contribute towards meeting the world's growing demands and transitioning to green energy alternatives (Hein et al., 2020; Cronan, 2022). This becomes more critical in view of depleting land resources and increasing production rates of different metals ranging from 1.1% for Cd to 8.3% for Co (Sharma & Smith, 2019). While many of these metals are required for various

R. Sharma (✉)
Consultant, Deep-sea Mining and Former Chief Scientist, National Institute of Oceanography, Dona Paula, Goa, India
e-mail: rsharmagoa@gmail.com

R. Sharma (ed.), *Deep-Sea Mining and the Water Column*,
https://doi.org/10.1007/978-3-031-59060-3_18

Table 18.1 World reserves, production rates and availability of selected metals

Metal	World reserves on land in 2018[a]	Production rate in 2016[a]	Increase in production rate/year (%)[b]	No. of years for which land reserves will last[c]
Mn	680 mill t (Mn content in ore)	15,700,000 t	4.3	23
Cu	790 mill t	20,100,000 t	3.1	24
Ni	74 mill t	2,090,000 t	3.7	21
Co	7,100,000 t	111,000 t	8.3	21
Pb	88 mill t	4,710,000 t	2.6	14
Zn	230 mill t	12,600,000 t	2.9	13

[a]www.usgs.gov
[b]Sharma and Smith (2019)
[c]Sharma et al. (2019)

applications, the available land reserves of some of these metals will last for just over two decades (e.g. Mn, Cu, Ni and Co), whereas a few others for just about a decade (e.g. Pb and Zn) (Table 18.1), unless new technology for processing the lowgrade minerals is developed, recycling of metals with improved technology is possible and new resources are found or known resources can be converted to reserves. Estimates show that by 2050, the demand for cobalt will be 423%, nickel 136% and lithium 280% of the currently known reserves (www.mineralsindepth.org), which could be met from the metals available from deep-sea minerals, estimated to be in millions of tonnes, which will be worth billions of dollars (Van Nijen et al., 2018; Hein et al., 2020).

Different types of deep-sea minerals have been identified in the Exclusive Economic Zones (EEZs) of some countries and in the International waters in all the oceans of the world. Some of the countries that have deep-sea minerals within the EEZs include Japan, Cook Islands, Kiribati (Hein et al., 2005, 2015), Korea and France (Fouquet & Lacroix, 2014). Cook Islands has issued licenses for the exploration of polymetallic nodules covering an area of 254,654 km^2 within their EEZ (https://www.sbma.gov.ck), whereas an area of 650,000 km^2 is covered by exploration contracts within EEZs of different coastal states in the Pacific Ocean (SPC, 2013; Petersen et al., 2018).

Similarly, New Zealand government is issuing prospecting licenses to investigate hydrothermal sulphide deposits rich in silver and gold in their EEZ (Boschen-Rose et al., 2022), and Norway is in the process of inviting applications for exploration and exploitation of seafloor deposits consisting of 38 million tonnes of copper and 45 million tonnes of zinc in polymetallic sulphides, as well as 24 million tonnes of magnesium, 3.1 million tonnes of cobalt and 1.7 million tonnes of cerium (https://www.reuters.com, http/:oceanminingintel.com). Besides these, the International Seabed Authority (ISA) has signed 19 contracts for polymetallic nodules (each measuring 75,000 km^2), 7 for hydrothermal sulphides (each measuring 2500 km^2) and 5 for ferromanganese crusts (each measuring 1000 km^2) in the international waters of Pacific, Indian and Atlantic Oceans, totaling to 1,447,500 sq.km. (Sharma & Smith, 2019).

It is estimated that the Clarion-Clipperton Zone (CCZ) in the Pacific Ocean alone has nodule deposits that contain key metals such as Mn, Ni, Co and Y that are higher by 1.15–3.4 times than the entire global resources on land (Hein et al., 2020). The total resource available in a typical contract area of 75,000 sq. km. for polymetallic nodules with a cutoff abundance of 5 kg/sq. m; would be 375 Mt. (wet) or 281.25 Mt. (dry) with a total metal equal to 67.081 Mt. (@ Mn = 22%, Ni = 1.0%, Cu = 0.78%, Co = 0.1%). From this 281.25 Mt. (dry nodules), for mining at the rate of 1.5 million tonnes/year (ISA, 2008) or three million tonnes/year (UNOET, 1987), only 10.6 or 21.2% (i.e. 30–60 million tonnes) will be actually mined, respectively, over 20 years period lifetime of a mine site, with a large balance (78.8–89.4%) that could be mined for more than a century (Sharma, 2017). Cobalt-rich crusts found on one of the seamounts in the North-Western Pacific Ocean (Takuyo Seamount No. 5) contain enough cobalt and nickel to meet Japan's demand for 88 years and 12 years, respectively (www.jogmec.go.jp). Similarly, deep-sea sediments found near Minamitorishima island in the western North Pacific Ocean have an estimated resource of 1.2 Mt. of rare earth oxides that account for 62, 47, 32 and 56 years of annual global demand of Y, Eu, Tb and Dy (Takaya et al., 2018).

As polymetallic nodules have attracted maximum attention in terms of the number of contracts signed, as well as being relatively easy to mine due to their occurrence as loosely lying objects on the surface of the seafloor, this chapter deals with the resource potential and production estimates of metals from these deposits.

2 Estimating the Resource Potential of Polymetallic Nodules

A general criteria for identifying polymetallic nodules as a resource is as follows (UNOET, 1987):

- Cutoff grade: 1.8% Ni + Cu
- Cutoff abundance: 5 kg/sqm
- Topography: acceptable (< 3° slope)
- Life of a mine site: 20 years
- Mining rate: 1–3 mi.t/year

Estimates have shown that for nodule mining to be economically viable, the average abundance is expected to be between 8 and 12 Kg/m^2 (Singh & Sudhakar, 2015) and extraction of metals such as manganese besides copper, nickel and cobalt is suggested (Sen, 2017). Mining of polymetallic nodules is relatively easier than crusts and sulphides as they are loosely strewn on the surface of the seafloor and can be scooped up. However, the economic viability of a deep-sea mining venture would depend on the development of infrastructure and its performance as it will be required to operate for 250–300 days per year, at depths of more than 5000 m underwater, under extreme conditions such as 500 bars of pressure, 2 °C temperature, no natural light; in inclement weather conditions that can include strong winds, rainfall, cyclones and/or hurricanes on the surface; cross-cutting currents in the water column, in which the riser pipes and communication channels will operate; as well as undulating topography and heterogenous mineral distribution (Sharma, 2019); combined with large distances of operating area (~1000 km) from the shore.

The calculation of the total turnover from a nodule mine site will be as follows:

$$TO = Q((Ni\%^{*} NiP) + (Cu\%^{*} CuP) + (Co\%^{*} CoP) + (Mn\%^{*} MnP),$$

in which

TO = Turnover at the mouth of the mine (in situ)
Q = Quantity mined (tonnes)
NiP, CuP, CoP = Average price of Ni, Cu and Co (USD/tonne) at the time of mining
MnP = Average price of Mn (USD/tonne) from metal trading websites

As described earlier, in a typical nodule 'contract' area of 75,000 sq. km with a cutoff abundance of 5 kg/m^2, the total available resource would be 375 Mt. (wet) or 281.25 Mt. (dry), containing a total metal resource of 67.53 Mt. (@Mn = 22%, Ni = 1%, Cu = 0.78%, Co = 0.23%). For a mining rate of 1.5 Mt./year and metal prices as of January 2023, the gross in-place value of these four metals would be $ 24.67 billion, whereas the same would be $ 49.35 billion for a mining rate of 3.0 Mt./year (Table 18.2). However, nickel, copper and cobalt are expected to be extracted as metals, whereas manganese (taken as silicomanganese here) will be extracted as an alloy, and hence, the value for the finally produced alloy will have to be considered as per the process adopted for extraction.

3 Estimated CAPEX and OPEX for Mining of Polymetallic Nodules

Projections from various contractors (ISA, 2008) had shown that for mining at 1.5 Mt./year and a single mining system, the Capital Expenditure (CAPEX) for the mining system, ore transfer and processing plant was expected to be $ 2.58 billion and the Operating Expenditure (OPEX) worked out to $ 679.64 million per year (Table 18.3). Alternatively, a study for nodule mining in CCZ (Van Nijen et al., 2018), based on the mining rate of three million tonnes per year using two vessels with independent mining systems and nodule collectors, estimates production of 37,050 tonnes of nickel, 32,400 tonnes of copper and 6375 tonnes of cobalt annually. For this, whereas the estimated CAPEX for pre-feasibility and feasibility phases is about 9%, construction of the commercial scale mining system is about 14.4%, surface vessels about 17% and the processing plant about 59.60% of the total capital cost of $ 4051 million, the offshore OPEX is estimated at $

Table 18.2 Resource potential and metal production estimates

Nodule/Metal	Mean concentration[a]	Resource potential t[b]	Metal production per year t		Price of metal ($/t)[c]	Gross in-place value of metal $/year		Gross in-place value of metal $/20 years	
			@1.5 Mt./year	@3 Mt./year		@ 1.5 Mt./year	@ 3 Mt./year	@1.5 Mt./year	@ 3 Mt./year
Wet nodules	–	375,000,000	–	–	–	–	–	–	–
Dry nodules	75% of wet nodules[d]	281,250,000	–	–	–	–	–	–	–
Manganese(SiMn)	22% of dry nodules	61,875,000	330,000	660,000	1517[c]	500,610,000	1001,220,000	10,012,200,000	20,024,400,000
Nickel	1.0% of dry nodules	2,812,500	15,000	30,000	30500[c]	457,500,000	915,000,000	9,150,000,000	18,300,000,000
Copper	0.78% of dry nodules	2,193,750	11,700	23,400	8400[c]	98,280,000	196,560,000	1,965,600,000	3,931,200,000
Cobalt	0.23% of dry nodules	646,875	3450	6900	51455[c]	177,519,750	355,039,500	3,550,395,000	7,100,790,000
Total (metals)	24.01%	67,528,125	360,150	720,300	–	1,233,909,750	2,467,819,500	24,678,195,000	49,356,390,000

Modified after Sharma (2022)

[a]Source: Morgan (2000) for Clarion-Clipperton Zone in Pacific Ocean

[b]@5 kg/sqm for 75,000 sq. km (75×10^9 sqm)

[c]Three-month average metal prices as of January 2023 (Source: www.lme.com for Ni, Cu, Co, and www.metalshub.com for Mn)

[d]Mero (1977)

Table 18.3 Estimated CAPEX/OPEX for nodule mining (@1.5 mi. t./year)

Item	Capital expenditures	% of Total	Operating expenditures	% of Total
Mining system	$ 550 mi.[a] ($ 372–562 mi.)	29	$ 100 mi/year[a] ($ 69–96 mi.)	20
Ore transfer	$ 600 mi.[a] ($ 495–600 mi.)	31.5	$ 150 mi/year[a] ($ 93–132 mi/year)	30
Processing plant	$ 750 mi.[a]	39.5	$250 mi/year[a]	50
Total	$ 1.90 billion[a] $ 2.58 billion[b]	100	$ 500 mil/year[a] $ 679.64 mi/year[b]	100

Figures in brackets show the estimated range for different systems
[a]Figures taken from ISA report (ISA, 2008) and rounded off to nearest fifty of the highest value
[b]Recalculated for inflation @ 35.9% for 2023 (www.usinflationcalculator.com)

Table 18.4 Estimated CAPEX/OPEX for nodule mining (@3.0 mi. t./year)

Item	Capital expenditures	% of Total	Operating expenditures	% of Total
Pre/feasibility	$ 365 mi.	9	$ 325 mi./year	32
Mining system	$ 583 mi.	14.4		
Ore transfer	$ 689 mi.	17		
Processing plant	$ 2414 mi.	59.6	$ 688.7mi/year	68
Total	$ 4051 million $ 4.051billion $ 4.72 billion	100	$ 1013.7 million $ 1.0137 billion $ 1.18 billion	100

Figures from Van Nijen et al. (2018)
#Recalculated for inflation @ 16.5% for 2023 (www.usinflationcalculator.com)

1013 million, which translates to $4.72 billion and $ 1.18 billion, respectively, considering the inflation rate (Table 18.4).

It is important to consider that the figures proposed for CAPEX and OPEX for the above are based on the prices of various items at the time of making the estimates. Whereas the estimate for 1.5 MT/year was based on the figures given by different contractors during a workshop conducted by ISA in 2008, the estimates for 3 MT/year are from the mining company GSR (Global Seabed Resources) in 2018. However, in both cases, the figures arrived at in the respective years have been recalculated based on inflation rates as of the year 2023.

A comparison of the components-wise distribution of different CAPEX and OPEX (Table 18.5) between 2008 and 2018 shows that:

1. The estimated cost for offshore components (mining and transport) was high in 2008 as estimates would have been made for the development of new systems, whereas the same is lower in 2018, probably due to the availability of offshore systems that are not being used currently due to the recent recession in shipping industry.
2. The estimated cost for onshore components (processing) has increased, probably due to an increase in the cost of land and for setting up new processing plants in recent years.

Table 18.5 Comparison on estimated investments

	CAPEX		OPEX	
Component	2008	2018	2008	2018
Mining	29%	23.4%	20%	32%
Ore transfer	31.5%	17%	30%	(included in mining)
Processing	39.5%	59.6%	50%	68%

3. The major expenditure in both estimates is for the processing plant (CAPEX—40–60%, OPEX—50–68%), which is an onshore component.

However, the CAPEX and OPEX may not only be functions of the mining rate but will also depend on the number of vessels and collectors and other infrastructure used in the mining operation. In an assessment, Shuboyo et al. (2013), while considering the mining rate of 2.14 million tonnes of wet nodules per year, and assuming 20 h of operation for 250 operating days, estimated a production of 488 tonnes per hour and 8650 tonnes per day, evaluated the cost and time required for nodule transport from offshore locations to different ports by different sizes of supply vessels and suggested 'having the onshore location as close to the offshore production site as possible' to reduce costs and make transfer of nodules less complex. However, all logistics requirements will vary with the location and distance of each mining area with respect to the respective onshore processing sites. Many more such studies are required for deciding the feasibility of the location of onshore operations so as to arrive at more realistic economic estimates. Moreover, a detailed exercise is required based on all such factors, including the global demand and supply, in order to decide the optimum mining rate, as well as the timing of the commencement of mining. Due provision will also be required for incorporating additional costs in the form of royalties, insurances, bonds or other expenses that may be levied on the contractors from time to time in order to get the mining license or to ensure the safety of the environment (Lodge et al., 2019).

4 Mining Estimates Based on Different Mining Rates

If polymetallic nodules are mined at mining rates ranging from 1 to 3 MT/y, ores between 3000 and 10,000 tonnes could be recovered per day from a mine site ranging between 4267 km^2 and 12,800 km^2 in area (Table 18.6), which is 5.68–17.06% of the 'Contract' area (75,000 sq.km) for a cutoff nodule abundance of 5 Kg/m^2 and minimum 25% efficiency of the mining system. Moreover, the touchdown (contact) area that would be actually scraped would vary between 200 and 600 km^2 per year for mining rates between 1.0 and 3.0 million tonnes per year, which is just 0.66–2.0 Km^2 per day and works out to 5–16% of the Contract area (Sharma, 2017). However, the large volume of sediment that will be disturbed on the seafloor (60,000–180,000 m^3 per day) is of a serious concern not only in terms of the environmental impact that it would cause but also in terms of additional costs that may

Table 18.6 Estimates for mining of polymetallic nodules at different mining rates

Estimates for operation of 300 days year^{-1}	Mining rate					Remark/implication of mining
	1.0 Mt year^{-1}	1.5 Mt year^{-1}	2.0 Mt year^{-1}	2.5 Mt year^{-1}	3.0 Mt year^{-1}	
Ore production/day	3333.3 t day^{-1}	5000 t day^{-1}	6666.6 t day^{-1}	8333.25 t day^{-1}	10,000 t day^{-1}	Proportionate storage and transport facility required
Area (size) of mine site[a]	4267 Km^2	6400 Km^2	8533 Km^2	10,667 Km^2	12,800 Km^2	Negligible (5.68–17.06%) of the contract area
Area of contact per year[a]	200 Km^2	300 Km^2	400 Km^2	500 Km^2	600 Km^2	i.e. 0.66–2 km^2 day^{-1}
Volume of sediment disturbed at seafloor	60,000 m^3 day^{-1}	90,000 m^3 day^{-1}	120,000 m^3 day^{-1}	150,000 m^3 day^{-1}	180,000 m^3 day^{-1}	Major source of environmental impact
Wt. of disturbed sediment (wet) (@1.15 g cm^{-3} density)	69,000 t day^{-1}	103,500 t day^{-1}	138,000 t day^{-1}	172,500 t day^{-1}	207,000 t day^{-1}	In slurry form that can travel with bottom currents to adjacent areas
Wt. of disturbed sediment (dry) (@ 80% water content)	13,800 t day^{-1}	20,700 t day^{-1}	27,600 t day^{-1}	34,500 t day^{-1}	41,400 t day^{-1}	Dominant (50–60%) fine clays, may remain suspended for longer periods

From Sharma (2017)

[a]For cutoff abundance of 5 kg m^{-2} (minimum required for nodule mining, UNOET, 1987), Contract area of 75,000 km^2, and mining for 20 years for 300 days of operation per year

have to be incurred to mitigate these by incorporating necessary modifications in the mining system (Sharma, 2024). Mining in areas of higher nodule abundances and with systems having higher efficiency could lead to a smaller area being covered (Hong et al., 2019; Kirchain et al., 2020), hence reducing the environmental impact as well as the overall cost of the operations, proportionately.

5 Conclusions

1. Deep-sea minerals can serve as alternative resource for critical metals in order to meet the increase in metal demands due to industrial growth and rising living standards, as well as the need to transition the energy, transport and industrial infrastructure from fossil fuel to green and renewable energy options.
2. There is a projected demand for millions of tonnes of metals (including Mn, Ni, Cu and Co) to meet these requirements.
3. Mean production rate of different metals has been rising in the last 20 years (e.g. Co = 8.3% per year, Cu = 3.1%, Ni = 3.7%, Mn = 4.3%, Pb = 2.6%, Zn = 2.9%).
4. Estimates have shown that the available reserves of some of the metals on land such as Mn, Cu, Ni and Co would last for just over two decades, whereas Pb and Zn would last over one decade.
5. Polymetallic nodules mined at different mining rates (1–3 MT/year) would yield ores between 3000 and 10,000 tonnes per day and the area (size) of the mine site would range between 4267 sq.km and 12,800 sq.km.
6. The area of contact that would be actually scraped would vary between 200–600 km^2 per year for mining rates between 1.0 and 3.0 million tonnes per year, which is 0.66–2.0 Km^2 per day and works out to 5–16% of the Contract area (75,000 sq.km).
7. In a typical nodule contract area of 75,000 sq. km with a cutoff abundance of 5 kg/m^2, the total available resource would be 375 MMT (wet) or/281.25 MMT (dry), containing a total metal resource of at least 67.53 MMT (@Mn = 22%, Ni = 1%, Cu = 0.78%, Co = 0.23%).
8. At the metal prices of 2023, the gross in-place value of these four metals would be $ 24.67 billion at the mining rate of 1.5 Mt./year, or twice as much ($ 49.35 billion) for 3 Mt./year.
9. The CAPEX for mining at 1.5 Mt./year (including the cost of the mining system, ore transfer and processing plant), based on estimates of 2008 corrected for inflation as of 2022, works out to $ 2.58 billion, whereas the CAPEX for mining at 3.0 Mt./year (including pre/feasibility study, mining system, ore transfer and processing plant), based on estimates of 2018 corrected for inflation for 2022, works out to $ 4.72 billion, which is 1.83 times of the cost for 1.5 Mt./year.
10. Similarly, the OPEX for mining at 1.5 Mt./year, based on 2008 estimates corrected for inflation as of 2022, works out to $ 680 million per year, whereas the OPEX for mining at 3.0 Mt./year, based on estimates of 2018 corrected for

inflation of 2022, works out to $ 1.18 billion per year, which is 1.74 times of the cost for 1.5 Mt./year,

11. The estimates show that the CAPEX for the onshore component (processing plant) is 40–60% and the OPEX of the same is 50–60%, whereas CAPEX for offshore component (mining system and ore transfer) is 41–60% and the OPEX for the same is 32–50%. It is important to note that the overall cost of the onshore component is either equal to or higher than the offshore component in these models.
12. While the actual area to be mined (<20%) and the actual area of contact (<2 km/day) is quite small in comparison to the total contract area (75,000 sq. km), it is important to address the issue of resuspension of the associated sediment. For this necessary modifications will have to be incorporated into the design and operational practices of deep-sea mining systems so as to minimize the environmental impact on the deep-sea ecosystems. This will add to the overall cost of a deep-sea mining project.

References

Boschen-Rose, R. E., Clark, M. R., Rowden, A. A., & Gardner, J. P. A. (2022). Integrated environmental management of the ecological impacts from seafloor massive sulfide mining – Perspectives from the Kermadec Volcanic Arc, New Zealand. In R. Sharma (Ed.), *Perspectives on deep-sea mining – Sustainability* (pp. 373–422). Springer International Publishers AG.

Cherkashov, G. (2017). Seafloor massive sulfides: Distribution and prospecting. In R. Sharma (Ed.), *Deep-sea mining; Resource potential, technical and environmental considerations* (pp. 143–164). Springer International Publishing AG.

Cronan, D. S. (2022). Deep-sea mining: Historical perspectives. In R. Sharma (Ed.), *Perspectives on deep-sea mining – Sustainability, technology, environmental policy and management.* Springer International Publishers AG. (this book).

Fouquet, Y., & Lacroix, D. (2014). *Deep-sea marine mineral resources.* Springer.

Halbach, P. E., Andreas, J., & Cherkashov, G. (2017). Marine ferromanganese crust deposits: Description and formation, occurrence and distribution, estimated world-wide resources in. In R. Sharma (Ed.), *Deep-sea mining; resource potential, technical and environmental considerations* (pp. 65–142). Springer International Publishing AG.

Hein, J. R., McIntyre, B. R., & Piper, D. Z. (2005). *Marine mineral resources of Pacific Islands – A review of the exclusive economic zones of islands of US affiliation, excluding the state of Hawaii.* US Geological Survey. http://pubs.usgs.gov/circ/2005/1286/

Hein, J. R., et al. (2015). Critical metals in manganese nodules from The Cook Islands EEZ, abundances and distributions. *Ore Geology Review, 68*, 97–116.

Hein, J. R., Koschinsky, A., & Kuhn, T. (2020). Deep-ocean polymetallic nodules as a resource for critical materials. *Nature Reviews – Earth and Environment, 1*, 158–169.

Hong, S., Hyung-Woo, K., YeuTaekyung, C. J.-S., Hee, L. T., Sung, B. D., & Jong-Gap, L. (2019). Technologies for safe and sustainable mining of deep-seabed minerals. In R. Sharma (Ed.), *Environmental issues of deep-sea mining – Impacts, consequences and policy perspectives* (pp. 95–143). Springer International Publishers AG.

ISA. (2008, February). *Executive summary of the International Seabed Authority's workshop on polymetallic nodule mining technology: Current status and challenges ahead* (p. 20). International Seabed Authority.

Kirchain, R., Roth, R., Peacock, T., Field, F., & Royo, C. M. (2020, October). *Financial model updates – Polymetallic nodules.* MIT presentation at International Seabed Authority Webinar.

Kuhn, T., Wegorzewski, A., Ruhlman, C., & Vink, A. (2017). Composition, formation and occurrence of polymetallic nodules. In R. Sharma (Ed.), *Deep-sea mining; resource potential, technical and environmental considerations* (pp. 23–64). Springer International Publishing AG.

Lodge, M., Segerson, K., & Squires, D. (2019). Environmental policy for deep-sea mining. In R. Sharma (Ed.), *Environmental issues of Deep-sea mining – Impacts, consequences and policy perspectives* (pp. 347–380). Springer International Publishers AG.

Mero, J. L. (1977). Economic aspects of nodule mining. In G. P. Glasby (Ed.), *Marine manganese deposits* (pp. 327–355). Elsevier.

Morgan, C. L. (2000). Resource estimates of the clarion-Clipperton manganese nodule deposits. In D. S. Cronan (Ed.), *A handbook of marine mineral deposits* (pp. 145–170). CRC Press.

Petersen, S., Kratschell, A., Augustin, N., Jamieson, J., Hein, J. R., & Hannington, M. D. (2018). News from the seabed – Geological characteristics and resource potential of deep-sea mineral resources. *Marine Policy, 70*, 175–187.

Sen, P. K. (2017). Sustainable processing of deep-sea polymetallic nodules. In R. Sharma (Ed.), *Deep-sea mining; Resource potential, technical and environmental considerations* (pp. 395–422). Springer International Publishing AG.

Sharma, R. (2017). Deep-sea mining: Current status and future considerations. In R. Sharma (Ed.), *Deep-sea mining: Resource potential, technical and environmental considerations* (pp. 3–22). Springer International Publishing AG.

Sharma, R. (2019). Environmental factors for design and operation of deep-sea mining system – Based on case studies. In R. Sharma (Ed.), *Environmental issues of Deep-sea mining – Impacts, consequences and policy perspectives* (pp. 315–344). Springer International Publishers AG.

Sharma, R. (2022). Approach towards deep-sea mining – Current status and future prospects. In R. Sharma (Ed.), *Perspectives on Deep-sea mining – Sustainability* (pp. 13–51). Springer International Publishers AG.

Sharma, R. (2024). Deep-sea mining and the water column – An introduction. In R. Sharma (Ed.), *Deep-sea mining and the water column – Advances, monitoring and related issues* (pp. 1–38). Springer International Publishers International Publishers AG. https://doi.org/10.1007/978-3-031-59060-3_18

Sharma, R., & Smith, S. (2019). Deep-sea mining and environment – An introduction. In R. Sharma (Ed.), *Environmental issues of Deep-sea mining – Impacts, consequences and policy perspectives* (pp. 3–22). Springer International Publishers AG.

Sharma, R., Mustafina, F., & Cherkashov, C. (2019). Review of mining rates, environmental impacts, metal values and investments for polymetallic nodules mining. In R. Sharma (Ed.), *Environmental issues of Deep-sea mining – Impacts, consequences and policy perspectives* (pp. 519–546). Springer International Publishers AG.

Shuboyo, P., van Hassel, E., & Vanelslander, T. (2013). Logistical assessment of Deep-Sea polymetallic nodules transport from an offshore to an onshore location using a multiobjective optimization approach. *Sustainability, 15*(14), 11317. https://doi.org/10.3390/su151411317

Singh, T. R. P., & Sudhakar, M. (2015). Estimating potential of additional mine-sites for polymetallic nodules in Pacific and Indian Oceans. *International Journal of Earth Sciences and Engineering, 8*, 1938–1941.

SPC. (2013). Deep-sea minerals and the green economy. In E. Baker & Y. Beaudoin (Eds.), *Secretariat of the Pacific community* (Vol. 2).

Takaya Yutaro, Yasukawa Kazutaka, Kawasaki Takehiro, Fujinaga Koichiro, Ohta Junichiro, Usui Yoichi, Nakamura Kentaro, Kimura Jun-Ichi, Chang Qing, Hamada Morihisa, DodbibaGjergj, Nozaki Tatsuo, Iijima Koichi, Morisawa Tomohiro, Kuwahara Takuma, Ishida Yasuyuki, Ichimura Takao, Kitazume Masaki, Fujita Toyohisa, Kato Yasuhiro (2018). The tremendous potential of deep-sea mud as a source of rare-earth elements. *Scientific Reports, 8,* 5763. https://doi.org/10.1038/s41598-018-23948-5

UNOET, 1987. Delineation of mine sites and potential in different sea areas. : UN Ocean Economics and Technology Branch and Graham & Trotman Limited. 27 pp 27.

Van Nijen, K., Passel, S. V., & Squires, D. (2018). A stochastic techno-economic assessment of seabed mining of polymetallic nodules in the Clarion Clipperton Fracture Zone. *Marine Policy, 95*, 132–141.
www.lme.com – Average metal prices for Ni, Cu, Co. Accessed 1 Jan 2023.
www.metalshub.com – Average metal price of Mn. Accessed 1 Jan 2023.
www.mineralsindepth.org – Minerals in Depth – The Story of Seabed Minerals. Accessed 1 May 2022.
www.oceanminingintel.com/news/industry/norwegian-government-to-facilitate-new-ocean-industry-seabed-mineral-activities. Accessed 28 June 2023.
www.reuters.com/markets/commodities/norway-finds-substantial-mineral-resources-its-seabed. Accessed 27 Jan 2023.
www.sbma.gov.ck/exploration. Accessed 3 Aug 2022.
www.usinflationcalculator.com – Recalculated cumulative inflation rate for 2021 Accessed 6 May 2021.

Dr. Rahul Sharma retired as Chief Scientist from the CSIR-National Institute of Oceanography in Goa, India with a career spanning more than 40 years in the field of deep-sea mineral exploration and exploitation. He has led a multi-disciplinary group on 'environmental studies for marine mining'. He has a master's degree in geology and a doctorate in marine science. His professional interests include application of exploration and environmental data to deep-sea mining.

He has edited three special issues of journals, published 38 scientific papers, authored 30 articles and 41 technical reports, and presented more than 60 papers at national and international conferences. He has also edited three books, viz. *Deep-Sea Mining* (2017), *Environmental Issues of Deep-Sea Mining* (2019) and *Perspectives on Deep-Sea Mining* (2022), all published by Springer International Publishers, containing 18–23 chapters contributed by experts from around the world.

His international assignments include: visiting scientist to Japan, visiting professor to Saudi Arabia, member of the UNIDO mission 'to assess the status of Deep-sea mining technologies' in Europe, the USA and Japan and invited speaker and consultant for the International Seabed Authority, Jamaica. He has contributed to the 'World Ocean Assessment Report I' of the United Nations and has also been invited to contribute a chapter on 'Potential Impacts of Deep-Sea Mining on Marine ecosystem' for the Oxford Encyclopedia for environmental science. In addition to his research career, he has been involved with several activities relating to science communication and outreach, as well as training programmes for international participants, professionals and students.

Currently, he is a freelance consultant, advising international agencies on various aspects of deep-sea mining.

Index

R. Sharma (ed.), *Deep-Sea Mining and the Water Column*,
https://doi.org/10.1007/978-3-031-59060-3

The manufacturer's authorised representative in the EU is Springer Nature Customer Service Centre GmbH, Europaplatz 3, 69115 Heidelberg, Germany. If you have any concerns regarding our products, please contact ProductSafety@springernature.com

Printed and bound by CPI Group (UK) Ltd, Croydon, CR0 4YY
30/09/2026
02225563-0017